W9-DHX-417

# Handbook of Laboratory Animal Science *Second Edition*

## Volume I

### Essential Principles and Practices

Edited by

## Jann Hau and Gerald L. Van Hoosier, Jr.

## CRC PRESS

Boca Raton   London   New York   Washington, D.C.

Senior Editor: John Sulzycki
Project Editor: Susan Fox
Project Coordinator: Pat Roberson
Cover Designer: Dawn Boyd
Marketing Manager: Nadja English

### Library of Congress Cataloging-in-Publication Data

Handbook of laboratory animal science / edited by Jann Hau, Gerald L. Van Hoosier,
Jr.--2nd ed.
        p.  cm.
      Includes bibliographical references and index.
      Contents: v. 1. Essential principles and practices.
      ISBN 0-8493-1086-5 (alk. paper)
        1. Laboratory animals. 2. Animal experimentation. 3. Animal models in research. I.
Hau, Jann. II. Van Hoosier, G. L.

QL55 .H36 2002
599′.07′24--dc21                                                                 2002031315

This book contains information obtained from authentic and highly regarded sources. Reprinted material is quoted with permission, and sources are indicated. A wide variety of references are listed. Reasonable efforts have been made to publish reliable data and information, but the author and the publisher cannot assume responsibility for the validity of all materials or for the consequences of their use.

Neither this book nor any part may be reproduced or transmitted in any form or by any means, electronic or mechanical, including photocopying, microfilming, and recording, or by any information storage or retrieval system, without prior permission in writing from the publisher.

All rights reserved. Authorization to photocopy items for internal or personal use, or the personal or internal use of specific clients, may be granted by CRC Press LLC, provided that $1.50 per page photocopied is paid directly to Copyright clearance Center, 222 Rosewood Drive, Danvers, MA 01923 USA. The fee code for users of the Transactional Reporting Service is ISBN 0-8493-1086-5/03/$0.00+$1.50. The fee is subject to change without notice. For organizations that have been granted a photocopy license by the CCC, a separate system of payment has been arranged.

The consent of CRC Press LLC does not extend to copying for general distribution, for promotion, for creating new works, or for resale. Specific permission must be obtained in writing from CRC Press LLC for such copying.

Direct all inquiries to CRC Press LLC, 2000 N.W. Corporate Blvd., Boca Raton, Florida 33431.

**Trademark Notice:** Product or corporate names may be trademarks or registered trademarks, and are used only for identification and explanation, without intent to infringe.

**Visit the CRC Press Web site at www.crcpress.com**

© 2003 by CRC Press LLC

No claim to original U.S. Government works
International Standard Book Number 0-8493-1086-5
Library of Congress Card Number 2002031315
Printed in the United States of America  1  2  3  4  5  6  7  8  9  0
Printed on acid-free paper

# The Editors

**Jann Hau** is Professor in Comparative Medicine at University of Uppsala in Sweden. Dr. Hau graduated in experimental biology from University of Odense in Denmark after medical and biology studies in 1977, and specialized in laboratory animal science. Following research fellowships at University of Odense, he did his doctorate (Dr. Med) at this university. In 1983, he joined the Department of Pathology at The Royal Veterinary and Agricultural University (RVAU) in Copenhagen as Associate Professor and Head of the Laboratory Animal Science Unit. He was later Head of the Department of Pathology and Dean of the Faculty of Animal Husbandry and Veterinary Science at the RVAU.

In 1991, he moved to the Royal Veterinary College (RVC) in London as Professor in the London University Chair in Laboratory Animal Science and Welfare. At the RVC, he was responsible for the undergraduate and postgraduate teaching in laboratory animal science and welfare, which included a specialist Master of Science course in Laboratory Animal Science that attracted a number of postgraduate students from many parts of the world.

In 1996, Dr. Hau was appointed Professor in Comparative Medicine in Uppsala and Head of a new Department of Comparative Medicine. Following amalgamations of departments at the medical faculty, Comparative Medicine is presently integrated with the Department of Physiology of which Dr. Hau is presently Head. In Uppsala, he has established a number of courses for undergraduate students and postgraduate students, including specialist education programs.

Dr. Hau has organized several international meetings and courses on laboratory animal science. He is the editor-in-chief of the *Scandinavian Journal of Laboratory Animal Science* and editor of the laboratory animals' section of the UFAW journal *Animal Welfare*. He is a member of a number of laboratory animal science organizations and former president of the Scandinavian Society of Laboratory Animal Science and the Federation of European Laboratory Animal Science Associations.

Dr. Hau has supervised many postgraduate master's students and Ph.D. students and published several hundred scientific papers and chapters in books. Together with Dr. P. Svendsen, he wrote the first Danish textbook on laboratory animals and animal experiments published in 1981, 1985, and 1989, and they co-edited the first edition of the *CRC Handbook of Laboratory Animal Science* published in 1994.

Dr. Hau's current research interests include development of refined laboratory animal models for studies of biological mechanisms in reproductive biology and infections as well as development of methods to assess stress and welfare in animals. His research activities also include projects focused on ways to replace, reduce. and refine the use of animals in antibody production.

**Gerald Van Hoosier** is Professor of Comparative Medicine in the School of Medicine at the University of Washington in Seattle, Washington. Dr. Van Hoosier graduated from the College of Veterinary Medicine at Texas A&M University at College Station, Texas, in 1957 and subsequently obtained postdoctoral training in virology and epidemiology at Berkeley, California, and in pathology at Baylor College of Medicine in Houston, Texas. From 1957–1962, he served as a Commissioned Officer in the U.S. Public Health Service assigned to the biologics program at the National Institutes of Health in Bethesda, Maryland, where he focused on the development and safety evaluation of poliomyelitis and measles vaccine. Following five years in the Public Health Service, Dr. Van Hoosier joined the faculty of the Division of Experimental Biology at Baylor College of Medicine in Houston, Texas, and did research on the role of viruses in the etiology of cancer. In 1969, he moved to Pullman, Washington, where he was a faculty member in the Department of Veterinary Pathology in the School of Veterinary Medicine and Director of Laboratory Animal Resources at Washington State University. He introduced a course on laboratory animals into the third year of the veterinary school curriculum, taught a graduate course on the pathology of laboratory animals, and began the development of a series of audio tutorials in collaboration with the American College of Laboratory Animal Medicine. In 1975, Dr. Van Hoosier was invited to develop an experimental animal program at the University of Washington. He obtained a training grant for veterinarians from the National Institutes of Health and established the Department of Comparative Medicine, which offers an M.S. degree. He served as the department chairman and Attending Veterinarian until 1995.

After becoming a diplomate of the American College of Laboratory Animal Medicine in 1968, he served as President in 1977–1978. Other professional activities have included serving as Chairman of the Board of Trustees of the American Association for Accreditation of Laboratory Animal Care in 1981–1982, President of the American Association of Laboratory Animal Science in 1992, and a member of the Governing Board of the International Council for Laboratory Animal Science from 1995–1999. In addition to approximately 100 scientific papers, Dr. Van Hoosier was a co-editor of *Laboratory Hamsters*, one of a series of texts by the American College of Laboratory Animal Medicine and served as editor of *Laboratory Animal Science* from 1995–1999. He is currently a member of the Editorial Council of the *Baltic Journal of Laboratory Animal Science* and *Animales de Experimentacion*.

He is the recipient of the Griffin Award from the American Association of Laboratory Animal Science and a Distinguished Alumni Award from the College of Veterinary Medicine at Texas A&M University.

# Contributors

**Vera Baumans**
Utrecht University
Utrecht, The Netherlands

**Kathryn Bayne**
AAALAC International
Rockville, Maryland

**André Chwalibog**
The Royal Veterinary
  and Agricultural University
Frederiksberg, Denmark

**Paul deGreeve**
Ministerie van VWS
Veterinair Account
Gravenhage, The Netherlands

**Gilles Demers**
Canadian Council on Animal Care
Ottawa, Ontario, Canada

**Kimberly Drnec**
Johns Hopkins University
Baltimore, Maryland

**Nicole Duffee**
Washington University School of Medicine
St. Louis, Missouri

**Heather Elliott**
GlaxoSmithKline
Hertfordshire, United Kingdom

**Ricardo E. Feinstein**
The National Veterinary Institute
Uppsala, Sweden

**Michael F.W. Festing**
MRC Toxicology Unit
University of Leicester
Leicester, United Kingdom

**Alan M. Goldberg**
Johns Hopkins University
Baltimore, Maryland

**Franziska B. Grieder**
National Center for Research Resources
National Institutes of Health
Bethesda, Maryland

**Axel Kornerup Hansen**
The Royal Veterinary and Agricultural
  University
Frederiksberg, Denmark

**Jann Hau**
University of Uppsala
Uppsala, Sweden

**Patricia Hedenqvist**
Karolinska Institutet
Stockholm, Sweden

**Ludo J. Hellebrekers**
Department of Equine Sciences
  and Department Clinical Sciences
  of Companion Animals
Utrecht, The Netherlands

**Coenraad Hendriksen**
National Institute of Public Health and
  Environment
Bilthoven, The Netherlands

**Jack R. Hessler**
New York University School of Medicine
Laytonsville, Maryland

**Urban Höglund**
Uppsala University
Uppsala, Sweden

**Nancy A. Johnston**
University of Washington
Seattle, Washington

**Sue Knoblaugh**
University of Washington
Seattle, Washington

**Warren Ladiges**
University of Washington
Seattle, Washington

**Ronald E. Larson**
Covance
Madison, Wisconsin

**John G. Miller**
AAALAC International
Rockville, Maryland

**David B. Morton**
University of Birmingham
Edgbaston, Birmingham, United Kingdom

**Timo Nevalainen**
University of Kuopio
Kuopio, Finland

**I. Anna S. Olsson**
Institute for Molecular and Cell Biology
Porto, Portugal

**Cynthia A. Pekow**
Veterans Affairs Puget Sound Health Care
  System
Seattle, Washington

**Richard M. Preece**
AstraZeneca
Macclesfield, Cheshire, United Kingdom

**Kathleen Pritchett**
The Jackson Laboratory
Bar Harbor, Maine

**Anne Renström**
Karolinska Institutet
Stockholm, Sweden

**Merel Ritskes-Hoitinga**
University of Southern Denmark
Odense, Denmark

**Paul Robinson**
Lexicon Editorial
Surrey, England

**Charsa Rubin**
University of Washington
Seattle, Washington

**Peter Sandøe**
The Royal Veterinary
  and Agricultural University
Frederiksberg, Denmark

**Alison C. Smith**
Medical University of South Carolina
Charleston, South Carolina

**Daniel A. Steinbrüchel**
Rigshospitalet
University of Copenhagen
Copenhagen, Denmark

**John D. Strandberg**
National Center for Research Resources
National Institutes of Health
Bethesda, Maryland

**M. Michael Swindle**
Medical University of South Carolina
Charleston, South Carolina

**Harry H. van Herck**
Utrecht University
Utrecht, The Netherlands

**Gerald L. Van Hoosier, Jr.**
University of Washington
Seattle, Washington

**Patri Vergara**
Universitat Autonoma de Barcelona
Bellaterra (Barcelona), Spain

**Kimberly S. Waggie**
Zymogenetics
Seattle, Washington

**Harry Bryan Waynforth**
GlaxoSmithKline
Hertfordshire, United Kingdom

**Benjamin J. Weigler**
Fred Hutchinson Cancer Research Center
Seattle, Washington

# Preface

Most of our knowledge in physiology, microbiology, immunology, pharmacology, and pathology has been derived from studies of animals — from studies of genetics in fruit flies to studies of life-threatening infections in nonhuman primates. Biomedical research involving animals remains essential for the advancement of the medical, veterinary, agricultural, and biological sciences. All drugs prescribed for use in humans and animals have been developed and tested in laboratory animals. And, new surgical techniques and materials are developed and tested in animals before they are accepted for humans or domestic animals.

In 1959, W.M.S. Russell and R.L. Burch published their famous book on humane experimental technique. The scientist should aim at replacing experiments on live animals with alternative methods whenever possible, reduce the number of animals needed to obtain valid results within each experiment, and refine techniques to reduce the discomfort to the animals used. The three Rs, replacement, reduction and refinement, have since become the cornerstones of laboratory animal science, and the concept has been integrated in numerous laws and guidelines regulating the use of animals in research.

Efficient and humane use of animals in animal experiments requires skillful and conscientious staff, including specialist veterinarians. In many parts of the world, the authorities require that all staff working with laboratory animals must have competence obtained through formal teaching and training programs. Many universities have established mandatory courses for scientists who wish to use animals in their research, and some universities have specialist educations, often masters' courses, for staff training for laboratory animal specialist competence. In the United States and Japan as well as in Europe, laboratory animal medicine is now a recognized veterinary specialty.

This handbook is a thoroughly revised second edition of the handbook first published in 1994 and edited by Per Svendsen and Jann Hau. Per Svendsen has since retired from the University of Odense in Denmark and spends a good deal of his time flying his airplanes and traveling to far away places. Thus, he did not wish to use the considerable time necessary for co-editing a revised version of the handbook.

Jann Hau and his old friend and colleague Gerald Van Hoosier in Seattle thus decided to join forces and prepare a new and completely revised and updated handbook in laboratory animal science. Most of the chapters in the first edition of the book were written by Scandinavian and European experts. With the complete revision of the book, the editors broadened the faculty of authors and included many eminent experts, in particular from North America. The editors wish to thank all of the authors for their valuable contributions and also many anonymous colleagues for their valuable assistance as reviewers of the texts. The individual chapters focus on an important subdiscipline of laboratory animal science, and the chapters can be read and used as stand-alone texts without the necessity of consulting other chapters for information. This approach has resulted in a slight overlap of the contents of certain chapters, but the editors feel that this was a small price to pay in order to make the book as user friendly as possible.

It is our hope that this handbook will be useful all over the world as a course book on laboratory animal science courses for postgraduate and undergraduate students and as a useful handbook for scientists using animals in their research, university veterinarians, and other specialists in laboratory animal science.

**Jann Hau**

**Gerald Van Hoosier**

# Table of Contents

# CHAPTER 1

# The Contribution of Laboratory Animals to Medical Progress — Past, Present, and Future

Franziska B. Grieder and John D. Strandberg

## CONTENTS

## INTRODUCTION

Animal-based research has significantly contributed to the advancement of scientific knowledge in general and to biomedical progress specifically. Studies on animals have provided basic information on animal biology and physiology; this information, in turn, has important relevance to human biology. Animal models of human disorders have added invaluable information about many aspects of the pathophysiology, treatment, and diagnosis of human diseases. Moreover, experimental results from animal studies have served as the basis for many key clinical trials and will lead to future therapeutic interventions. Established animal models, as well as others that will be developed in the future, help address increasingly challenging new questions in biomedical science. This is a time of change in animal-based research as the scientific community proceeds from the deciphering of the genetic code of a range of animal species to the determination of the functions of the genes that are identified with increasing frequency.

The total yearly value of laboratory animal sales for rodents is in the multimillion-dollar range and may increase by 10 to 20 percent annually according to experts.[1] Given the magnitude of biomedical research, with its economic and political impacts, a clear understanding of the central function that laboratory animals play in biomedical research is of fundamental importance. This chapter focuses first on past and present contributions of laboratory animals in biomedicine. Then, we discuss the essential future roles of laboratory animals in contributing to potential advancements of biomedical research. Finally, we discuss the necessity of expanding issues related to laboratory animal research, including funding and training of young investigators.

## PAST ROLE OF LABORATORY ANIMALS

### Overview

Any attempt to assemble a comprehensive or detailed chronological overview of all the contributions to biomedical research that are based on experiments involving laboratory animals would result in either an endless list or a vast volume of medical accomplishments, and would be neither complete nor useful. However, the Foundation for Biomedical Research has compiled a list of Nobel Prize winners from the year 1901 to the present; this includes 67 examples, ranging from determinations of basic biologic mechanisms to studies that have led to the cure and prevention of important infectious diseases. Therefore, this section will start with a brief overview of past contributions of laboratory animals to medical progress and will highlight examples that have significantly influenced human health and reduced suffering.

The use of animals in biomedical research and, in particular, the use of the mouse in such studies dates as far back as the 1600s. During the subsequent centuries, as scientific methods and research practices evolved, laboratory animals have continuously contributed to virtually all aspects of biomedical progress. This fact can be illustrated with history-making contributions from diverse areas of science, as represented in the following list:

Achieved or improved diagnosis of infectious diseases (e.g., rabies, yellow fever)
Understanding of susceptibility and resistance to microbial agents leading to antimicrobial agents
Knowledge of immune biology and deficiencies (e.g., histocompatibility, severe combined immunodeficiency)
Understanding of transplantation immunology and development of related technologies
Development of vaccines (e.g., smallpox, polio)
Development of advanced technologies in heart surgery and other cardiovascular or stroke-related inventions (e.g., open heart surgery)
Development of cancer treatments
Identification of metabolic dysfunctions
Characterization of neurological defects
Achievements in space medicine

### Examples from the Past

A few specific examples will underscore that many past accomplishments based on scientific observations or experiments performed on laboratory animals formed the foundation for medical progress, thus leading to some of today's sophisticated health-science technologies.

In 1877, the German scientist Robert Koch built upon observations that were already over a decade old, namely that anthrax could be transmitted from animal to animal. Koch's animal experiments proved that the *Bacillus anthracis* bacterium caused a particular disease, and that the isolated and purified bacterium from an initial host could cause the same disease in a new, second host. Hence, Koch's postulates were born.[2] Koch's observations gained general acceptance and helped lay the foundation for the more-intensive use of laboratory animals, especially for investigations of infectious diseases. Even today, over a century later, Koch's postulates form a cornerstone for infectious-disease biology and research. These simple principles are applied in their original form or in an updated, more molecular version to emerging diseases, such as the Four-Corner-disease in 1993, which is caused by the hantavirus *sin nombre*, and *Kaposi sarcoma*, which was etiologically linked to human herpes virus type 8 in 1995.

Without animal research, most of the effective vaccines against infectious microbes or their toxins would not have been developed. Two major milestones were the independent developments of the first vaccines against smallpox and rabies, centuries-old human diseases that resulted in either severe, potentially fatal illness or in 100% mortality, respectively.

In 1798, English physician Edward Jenner conducted experimental vaccination for the prevention of smallpox by inoculating the closely related vaccinia, or cowpox virus.[3] Persons inoculated with cowpox virus showed complete resistance to a challenge with the deadly smallpox virus. A century and a half later, Jenner's smallpox vaccine formed the basis for the World Health Organization's 1958 program of global eradication of smallpox. The freeze-dried vaccine, the simple application with the bifurcated needle, and the concept of mass vaccination, combined with surveillance and containment, resulted in the eradication of smallpox in 1979. The subject remains timely, however, given the potential use of this virus in biowarfare and bioterrorism.

In Paris, Louis Pasteur adapted the wild-type, or street rabies, virus to laboratory animals, resulting in a change in viral properties that today would be called attenuated virus strains. Pasteur and his colleagues subsequently developed concepts and experimental approaches that led to the first protective vaccination against rabies. Furthermore, in 1885, Pasteur successfully used the first post-exposure treatment, or passive immunization, against rabies.[4] Other excellent examples include the development of vaccines against diphtheria and poliomyelitis, further illustrating the critical role that animal experiments played in the history of immune prophylaxis.[5]

Much of the knowledge about the structure of immunoglobulins or antibodies — molecules of central importance in immunology and host defense — was derived from extensive investigations into neoplastic plasma cells derived from mouse myelomas. Data on the Y-shaped protein structure and the potential for large-scale production of homologous antibodies were put forth by Kohler and Milstein in 1976.[6] The hybridoma technique that these two scientists developed has provided a method of antibody production that is now widely utilized. The basic principle of hybridoma technology relies on the fusion of immortal myeloma cells with antibody-producing spleen cells harvested from a previously immunized mouse against the antigen of interest. Successful hybridoma clones will produce one type of target-specific, or monoclonal, antibody in unlimited quantities. Such monoclonal antibodies have been used for a wide range of applications, including diagnosis of pathogens, identification of physiological cell components, treatments of diseases, and purification of biological materials, to name just a few. Moreover, the production of large quantities of identical immunoglobulins allowed for a better understanding of their molecular structure critical to our understanding of immune-response induction, including the development of a secondary immune response (e.g., antibody class switch). More recently, the traditional production method of monoclonal antibody as mouse ascites is being replaced by alternative *in vitro* techniques that abolish the need for using live mice.

Another application of our expanded understanding of the immune system is in the field of tissue transplantation, which was significantly advanced by experimental findings in mice. At the Jackson Laboratory, scientists performed pioneering work by restoring to health a mouse with a blood disorder after performing a bone marrow transplant.[7] Furthermore, working at the same institution, George Snell discovered genetic factors recognized by the immune system that determine the possibilities of transplanting tissue from one individual to another. This pioneering work on the concept of the H antigen and the major histocompatibility complex later resulted in the shared 1980 Nobel Prize in Physiology and Medicine.

## Specific Examples from the Past — Nonhuman Primates

As noted previously, the eradication of poliomyelitis from the human population was dependent upon the development of an effective vaccine. Critical to this was the use of nonhuman primates, specifically, rhesus macaques, to study the pathogenesis of the disease and to test the efficacy of candidate vaccines. They have more recently continued to be of use, although in greatly reduced numbers, in ongoing testing of the candidate AIDS vaccines currently produced.

Several types of viral hepatitis are of major importance as human health problems worldwide. Again, nonhuman primates have been and continue to be essential in development of control methods for these conditions. Chimpanzees, susceptible to hepatitis B virus, were of critical importance in development of an effective vaccine against this agent. This vaccine is widely used and has helped to bring this disease under control. As close human relatives, chimpanzees are also susceptible to other human hepatitis viruses. One of the most important of these, hepatitis C virus, is a major cause of acute disease and, more importantly, of chronic liver failure. It is present on all continents and is responsible for extensive suffering and economic damage. Efforts are now underway to determine the pathogenesis of this disease in chimpanzees and to develop control methods, including an effective vaccine.

As close human relatives, nonhuman primates have been important in neurobiological studies in a wide range of fields, including perception, behavior, and basic neurologic studies. Many of these experiments have been of relatively long duration and require training of animals in controlled experiments, which can help to pinpoint the effects of specific treatments on various parts of the complex primate brain. Nonhuman primates are of special value in determining the addictive potential of specific compounds and in dissecting the mechanisms of such addictions.

## Specific Examples from the Past — Nontraditional Species

Several unlikely species have been and continue to be of significance in biological and medical studies. The armadillo, native to the southern United States, is one of few species susceptible to the mycobacterium that causes human leprosy. Thus, the animals can serve as sources of this agent and also can be used in the attempts to identify the pathogenetic mechanisms that it employs and to assess potential therapies.

The chinchilla, a South American rodent raised for its pelt and used in the fur industry, has a large and accessible acoustic system, which has been exploited to study basic mechanisms of hearing and to assess the effects of factors that are toxic or otherwise detrimental to hearing.

Ferrets, now common as pets, are also of importance in the study of influenza, which continues to be a major threat to the human population. The development of efficacious vaccines is dependent upon a susceptible host; the ferret acts as such a host.

## Public Concerns

Over the past decade, the use of cats and dogs in biomedical research has significantly diminished for a variety of reasons. First and foremost is the changing focus of much research to basic biologic questions that are best investigated in the rodent model. However, there is also continuing and increasing public concern about the use of dogs and cats in research. The general public — primarily urban — views these animals as members of the family; thus, there is opposition to their use by certain vocal groups. It is of interest to note that such opposition to the use of these animals in research is not new; in fact, Congressional hearings in 1900 on the subject of vivisection were a forum in which founders of the Johns Hopkins University School of Medicine were active participants.

Public concern has also focused on the use of nonhuman primates for research studies (see "Specific Examples from the Past – Nonhuman Primates"). Cultural differences contribute to these concerns for certain species, as reflected by the intensity and vocalization of objections to animal-based research in Europe and the United States as compared to other regions.

It is worthy of note that dogs and cats (and their owners) have benefited greatly from research done to improve human health. Numerous therapeutic technologies and drugs used to treat conditions of all types were developed for human medicine, but are of critical importance in the veterinary arena, which itself does not have the financial resources to support such research.

## PRESENT ROLE OF LABORATORY ANIMALS

As discussed previously, rodents, i.e., mice and rats, account for the overwhelming majority of all laboratory animals. This is reflected in the present scientific development of the exponentially growing number of genetically engineered mouse strains produced. Therefore, after outlining the contributions of a wide range of laboratory animals to Physiology and Medicine Nobel Prizes, we discuss the present role of laboratory animals in medical science, with the focus on mice. However, a final section will introduce nontraditional species in the laboratory animal sciences.

### Contributions to Nobel Prizes

Looking back at the Nobel Prizes for Physiology and Medicine between 1996 and 2001, five of the six selected Nobel awards were given to achievements that, at least in part, were based on experiments utilizing laboratory animals. The 2001 Nobel Prize in Physiology and Medicine was awarded to Leland Hartwell, Timothy Hunt, and Paul Nurse for their work on "key regulators of the cell cycle." Using sea urchins and yeast cells in independent experiments, their research led to the identification of key regulatory genes and their products, which are responsible for cell growth and division. Over 100 cell-division control proteins, later named cyclin-dependent kinases, or CDKs, were identified, as were molecules called cyclins that bind to CDKs to control their phosphorylation and activity. The cyclins are conserved in eukaryotic organisms, including yeast, plants, animals, and humans. This discovery may lead to the development of future therapeutic agents that target the cell cycle machinery and kill cancer cells.

The 2000 Physiology and Medicine Nobel Prize awarded to Arvid Carlsson, Paul Greenberg, and Eric Kandel recognized pioneering discoveries on slow synaptic transmissions between neurons. Eric Kandel received his prize for his contributions to the understanding of the neuronal pathways of learning and memory, which was deduced from research conducted in the sea slug *Aplysia*.

Nitric oxide formed the basis of the 1998 Nobel Prize in Physiology and Medicine, which was awarded to Robert Furchgott, Louis Ignarro, and Ferid Murad. In a serendipitous experiment conducted on rabbit aorta preparations, Robert Furchgott discovered an endothelium-derived factor that was later recognized as nitric oxide (NO). NO acts as a signaling molecule in the cardiovascular system, causing local vasodilation, thereby displaying protective properties for the vascular system against atherosclerosis via various effects on leukocyte vascular permeability. However, as a universal signaling molecule, NO is also involved in critical roles in inflammation, apoptosis, and neurotransmission.

Stanley Prusiner received the 1997 Physiology and Medicine Nobel Prize for his discovery of prions, tiny protein molecules that are etiologically linked to a variety of slow-acting, inevitably fatal diseases in humans and animals. "Prion" is an acronym for "proteinaceous infectious particles," and forms an entirely new genre of disease-causing agents. Prions cause neurodegenerative disorders in sheep and goats (scrapie), cattle (mad cow disease), elk and deer (chronic wasting disease), and also humans (Creutzfeldt-Jakob disease and kuru). Moreover, findings gained through the study of these diseases may contribute to the understanding of more common dementia disorders, such as Alzheimer's disease.

Discoveries elucidating interactions between viruses and the immune system resulted in the 1996 Nobel Prize awarded to Peter Doherty and Rolf Zinkernagel. Using both inbred and outbred strains of mice, Doherty, a veterinarian, and Zinkernagel pioneered work that explored cell-mediated lysis of virus-infected cells. Results from their experiments uncovered the central function of the major histocompatibility antigens in signaling self to the immune system. Discoveries associated with this work are significant for the understanding of autoimmune diseases and immunological surveillance mechanisms crucial in transplantation medicine.

These are just five recent examples in which biomedical research, utilizing a wide range of laboratory animals, contributed to significant progress in medical knowledge. Such research contributions leading to Nobel Prizes, the highest accolade in science, are testimony to the important role of laboratory animals in medical progress, as well as to the wide acceptance of animal-based research among scientists. This latter fact was further demonstrated in a petition to the then Surgeon General C. Everett Koop that was signed by 30 Nobel Prizewinners in 1989.[8] The Nobelists strongly endorsed the use of animals in medical research, a recommendation prompted by the increasingly visible actions of animal-rights activists.

## Rodent Embryonic Stem Cells and the Advances of Molecular Biology

One of the key advances in animal-based research in the last two decades was the development of technologies that allowed researchers to grow embryonic stem (ES) cells in tissue culture. In mammals, the fertilized oocytes and the 4- or 8-cell-stage blastomere are totipotent, i.e., when transferred by injection into early embryos, they can develop into the entire complex animal. Bradley[9] first demonstrated in the 1980s that ES cells could give rise to all somatic cell types of the embryo. This finding formed the basis for targeted gene technology, which resulted in the creation of "knockout" and other transgenic animals.[10]

Such transgenic animals can be made by randomly inserting a new gene, called a transgene, into the genome of the recipient animal embryo. Alternative technologies include direct mutagenesis of the transgenic animal's genome using targeted approaches, which results in a gene knockout animal, or the use of spontaneous mutations in animals that results in disease models for specific conditions. Mice occupy the key position in this expansion of transgenic animal development, but efforts continue to employ these and other technologies in rats and other species.

## Genetically Altered Mice

The development of genetically engineered mice is one of the most significant achievements that is moving biomedical research forward into a new area. This accomplishment, combined with the publication of the first drafts of the human and murine genome maps (see "Murine Genome Map Completed"), represents a pivotal step in understanding the genetic components of human diseases. Such understanding of the contributions of genes and their products to the pathogenesis of disease will further help identifying illnesses earlier, leading to advances in prevention, prognostication, and therapeutic development. The potential of altering the genetic makeup of laboratory mice allows researchers to create animal models that are not just "workable" approximations, but are, in fact, close replicas of the human disease under study. Genetic alterations in animal models allow the study of single or multiple gene effects of human diseases. However, since many complex diseases are multifactorial, not every single pathologic condition or genetic defect may be replicated in a genetically altered laboratory animal model.

A surprisingly large number of the estimated 30,000–40,000 human genes have murine equivalents. Scientists who study how these genes function in animal models will discover the roles that homologue animal genes play. Such new knowledge deduced from experiments using laboratory mice and their genetically engineered siblings, as well as other species, will allow scientists to study a wide variety of illnesses. It is expected that thousands of new mouse strains with altered genetic backgrounds will be generated and used within the next decade.

Among the many examples of genetically engineered mouse strains that have resulted in advances in biomedical sciences are the following models:

The *Cystic fibrosis knockout mouse* provides an outstanding animal model for cystic fibrosis (CF). Human CF is caused by mutations in the gene encoding the cystic fibrosis transmembrane regulator (CFTR). The CF knockout mouse, CFTR-/-, was created by expressing CFTR under an intestinal-specific promoter in a mouse that carries the mutated CF gene, thereby becoming a double transgenic animal. This site-specific expression of CFTR corrected the single-knockout animal's early death because they now show functional ileal goblet and crypt cells. The double CFTR-/- mouse maintains the pulmonary CF phenotype, thus providing an animal model for CF.[11] Recent investigations demonstrated reversion of the lethal CF phenotype in this double CFTR-/- mouse model by *in utero* gene therapy using an adenovirus containing the CFTR gene.[12]

The *atherosclerotic lesion mouse* contributes to progress in the fight against atherosclerosis and related illnesses. Apo E3 transgenic mice express the human dysfunctional apo E variant of the apolipoprotein. The Apo E3 transgenic mouse develops hyperlipidemia and atherosclerosis on a high fat/high cholesterol diet. Plasma cholesterol levels in the apo E3 mice increase in 4 months, and they develop atherosclerotic lesions in their aortas. Because these lesions resemble those in human patients with atherosclerosis, the apo E3 mouse is a valuable animal model for studies on the genetics, development, and therapy of atherosclerosis.[13]

The *Amyotrophic Lateral Sclerosis* (ALS) *mouse* serves as an animal model for Lou Gehrig's disease, which is caused by a mutation in the enzyme superoxide dismutase-1, or SOD-1. Transgenic mice expressing mutant forms of the SOD-1 gene, detectable in familial ALS human patients, display clinicopathological features of the disease. The SOD-1 mouse model has helped to elucidate the pathogenesis of ALS and establish new therapeutic agents.[14] Similar studies to those that resulted in the creation of the SOD-1 mouse led to a model for another neurodegenerative disorder, Huntington's disease.[15]

*Murine models with Germ-line p53 mutations* contribute to a better understanding of tumorigenesis. A variety of different lines of mice carrying the p53 alteration on different background strains is available. Specifically, one mouse strain carrying a mutant p53 transgene (135Valp53) develops pulmonary tumors more frequently after exposure to carcinogens as compared to controls. Therefore, the mutant p53 transgene may have a negative effect, and the mouse model offers a potentially useful tool for studies on chemoprevention and chemotherapy.[16]

## Murine Genome Map Completed

In April 2001, the private enterprise Celera Genomics announced that they completed sequencing the genome of three different mouse strains, covering 15.9 billion base pairs for approximately 99% coverage of the full murine genome.[17] Among the three sequenced mouse strains, Celera identified about 2.5 million single-nucleotide polymorphisms, or SNPs. It is believed that this genetic diversity will help in the characterization of mouse models of human disease. Comparative genomics will enable researchers to align the assembled and annotated human genome[18] and the assembled mouse genome[19] to identify pathologic mechanisms of diseases and their therapeutic treatment via drug development.

The public-private Mouse Sequencing Consortium (MSC), an international effort composed of six NIH institutes, three private companies, and the Wellcome Trust, announced in May 2001 that it also has successfully completed the genetic map of the mouse. The data, which were generated using the same sequencing techniques — called shotgun approach — that produces random bits of sequences, are now being assembled into the finished sequence of the mouse genome. The annotated data in this murine genetic map will be made freely available for the unrestricted use of researchers worldwide.[20,21]

## Nontraditional Species — Zebrafish, Xenopus, Drosophila

In recent years, increased attention has been devoted to studies using fish. Fish are used for several reasons, including the fact that many species are egg layers and produce large numbers of eggs on a frequent basis. These eggs are transparent and thus, early developmental events can be closely monitored. The effects of a variety of chemical and other stimuli can be detected early, and the pathogenesis of the resulting abnormalities can be dissected in controlled environments. Furthermore, a single pair of fish can produce an extraordinary number of offspring. The major types of fish used in such studies have been derived from species formerly based in the aquarium trade and include zebrafish (*Danio rerio*), swordtails and platys (*Xiphophorus sp.*), and the Japanese medaka. Large numbers of inbred stocks of animals have been developed and are used in genetic studies, including determination of those changes that correspond to genetic abnormalities in human and other mammalian populations. They are also used in large studies of chemical mutagenesis and carcinogenesis. Advances in the genomics of zebrafish are contributing significantly to these efforts.

Fish-based studies pose unique problems. Water quality is of paramount importance, and intercurrent diseases that affect the various fish species are often not well defined. However, advances continue to be made in these areas, especially in the large, centralized laboratories that serve as centralized repositories.

African clawed toads (*Xenopus sp.*) are other aquatic animals that are of considerable importance in studies of embryogenesis, early development, and of factors that can affect these events. Embryos of these species are accessible for surgical and chemical manipulations, and the effects of manipulating or removing specific cells in the early embryo can be determined. In *Xenopus*, early developmental events occur more slowly than in other species, and thus, they can be more easily dissected out. As frequently pointed out by the scientific community using *Xenopus*, much of the understanding of early embryonic development has come from studies done using these animals.

Work with the fruit fly, *Drosophila melanogaster*, has been carried out over many decades and has resulted in major advances in animal genetics. More recent efforts have focused on delineation of the *Drosophila* genome, which will then be followed by studies on expression of products of specific genes. As with many other nonmammalian species, the costs of maintaining individual animals are greatly reduced and large numbers of individuals can be used in studies.

## FUTURE DIRECTIONS FOR LABORATORY ANIMALS IN BIOMEDICAL RESEARCH

### Genetically Engineering Technologies (see also "Genetically Altered Mice")

As we have seen, the genomic revolution has opened new and seemingly unlimited possibilities for biomedical research. Reading and understanding the genomic blueprints associated with specific pheno-types, including pathologic or advantageous characteristics, provide the possibility of intervening at the gene level in certain disease conditions. This contrasts with the traditional approach of treating signs of illness and alleviating disease manifestations. Undoubtedly, the sequencing and manipulation of the genome will play a significant role in biomedical research of the future. Using targeted mutagenesis, specific mutations can be introduced and foreign genes can be expressed, resulting in a desired, potentially pathologic phenotype in a wide range of different species. Alternatively, in-depth knowledge of the genome will open unprecedented possibilities for treatment modalities for genetic disorders. These advances in the genetic area of biomedical research will require the expansion and development of repositories and stock centers to supply genetically engineered laboratory animal models available for the scientific research community. Thus, future research will make use of mice, rats, and other rodent models, hence requiring the continued support of programs that improve, oversee, and expand laboratory animal research programs (see "Support of Biomedical Research Directed Towards labaratory Animals, Funding Opportunites"). Maintaining increasingly sensitive and specialized animal strains, which require sophisticated and expen-sive infrastructures, will form an important part of future biomedical research investigations.

### *Targeted Transgenic and Knockout Technologies*

At the present time, the mouse is the only species in which genetic manipulations at the level of gene knockouts have been successfully completed resulting in live animals. This now-widespread technology has, once again, given the murine species a significant advantage in its leading role as laboratory animal of choice. However, other species have been successfully utilized for targeted transgenic manipulations, resulting in the expression of foreign genes in "cloned" offspring. Among the higher vertebrate species, rats, rabbits, pigs, goats, and rhesus monkeys have been successfully used for such research work, although with widely varying in efficiency and cost. Zebrafish and the nematode *C. elegans* need to be added to this list of model species used for transgene expression. These technologies will be expanded in the future to include other species and also to provide potentially new avenues for biomedical research directions.

### *Stock Centers*

The recent explosion of research tools and technologies available to biomedical scientists has created unprecedented opportunities. The capacity to modify the genetic makeup of biologic models and explore gene function provide powerful tools to determine gene function and its modulation by a host of factors. The resulting proliferation of genetically modified research animals, including zebrafish, mice, and other species, has significantly increased the need for centralized repositories for genetically altered animals, with special emphasis on genetic monitoring, phenotyping, and the control of intercurrent infections. Recently established repositories for transgenic and knockout mice and for zebrafish have only begun to address this rapidly increasing need. There are other species, such as *Drosophila*, that can be maintained only as reproductive colonies. Cryopreservation technologies could significantly reduce the cost and labor of maintaining animals of this type. It is essential that groups work together to develop consistent approaches to housing, health monitoring, and characterization of mutant animals. Harmoni-zation of guidelines for laboratory animal care is thus necessary. Moreover, research has become globalized and requires validation of biologic models used in research. Minimizing or eliminating unwanted duplications of effort will lead to cost-effective approaches and will increase the need to share research resources internationally.

### Nontraditional Species and Approaches

The future of biomedical research will involve existing traditional laboratory animal models, traditional models that will be improved or modified, and nontraditional model species and approaches. One such nontraditional and nonanimal model organism is the small, flowering plant in the mustard family called *Arabidopsis thaliana*. *Arabidopsis* is not of agricultural significance, but it has been used and is being further developed for both basic genetic and molecular research in biology. Research utilizing *Arabidopsis* offers several advantages, including its small genome size of 125 megabases with only five chromosomes, the extensive genetic and physical maps of the genome, and the large number of available mutant lines. Furthermore, *Arabidopsis* has a short life cycle with a prolific seed production. Since 1965, when the first international *Arabidopsis* conference was convened, the number of laboratories conducting research on *Arabidopsis* and the amount of research money spent on *Arabidopsis* research have significantly increased. Stock centers, databases, and publications related to *Arabidopsis* have moved this plant to a position where it may become a potential future model organism for studies in cellular and molecular biology that will cover questions well beyond the world of plants.

Nontraditional future approaches supplementing existing laboratory animal models will include the cell culture systems. The recent advances in embryonic stem cell culture technology exemplify one small step towards the limitless expansion that single- or multiple-cell systems could play in advancing biomedical research. The power of computers and their potential to model physiological and pathological situations and conditions, as well as their power to store and analyze data, will allow future biomedical research to move into areas of modeling that will provide opportunities beyond today's imagination.

## Support of Biomedical Research Directed Towards Laboratory Animals, Funding Opportunities

Continued strong support from both federal and private sources is necessary to ensure the steady growth and progress of our ability to decipher the secrets of biology. Funding from many agencies targets research aimed at basic biomedical questions, specific disease modalities, or their model systems. Research applications with broader focus, including shared research tools or applications, specific laboratory animals, animal colonies or their infrastructure support, are supported through the National Center for Research Resources (NCRR), a component of the National Institutes of Health (NIH).

The Division of Comparative Medicine at NCRR, as its name suggests, has the goal and responsibility of developing and supporting an array of mammalian and nonmammalian models to both underpin and facilitate biomedical research. In addition to traditional research grants, the division awards competitive grants aimed at the development of specialized research tools, centralized animal colonies, research training for veterinarians, and career-development programs for individuals across the educational spectrum. The variety of NCRR-supported biorepositories is quite broad, ranging from micro-organisms through invertebrates, to higher vertebrates, including nonhuman primates. One of the major challenges presented to these resources is to remain responsive in a timely manner to the ever-changing needs of investigators and to not become a rate-limiting barrier for research.

Animal-based research is increasingly dependent on technologies that are continuously being developed, e.g., MRI, ultrasound, and arrays. Such tools and equipment are important in basic science as well as clinical settings, and lead to improved diagnostic and therapeutic outcomes.

One of the means to increase and improve in the present and future is bioinformatics. The power of computer-based research technologies, data analysis, and data transmission is not only significant in its growth, but also in its future potential. However, this comes with significant costs and needs for coordination and standardization of approaches used to preserve and manipulate a wide variety of data.

## SUMMARY – IMPLICATIONS FOR LABORATORY ANIMAL MEDICINE (TRAINING, RESEARCH, CAREER DEVELOPMENT)

With the advances of modern technologies in medical research, new ways of thinking in terms of combating diseases have emerged. Today, many investigations can be conducted *in vitro,* or outside the living animal body, especially since the advancement of technologies in cell culture and molecular biology techniques allows performing assays in test tubes, thereby mimicking metabolic reactions in an isolated setting. Many biomedical subspecialists, including anatomists, biochemists, geneticists, immunologists, microbiologists, pharmacologists, and physiologists, perform *in vitro* experiments to understand how cells and their subcellular components function, how cells and their membranes and receptors react to other cells or chemicals, or what external factors affect their metabolism and growth rate. Scientists need to understand how these pieces fit together into the big picture; that is, how the entire animal body functions as a whole. For these reasons, laboratory animals, as they have in the past, will continue to be an essential component of biomedical research and to contribute vitally to continued exponential growth in biomedical progress.

To ensure this continued and successful progress in animal-based biomedical research, training of broadly based scientists is growing increasingly important. There are numerous demands for individuals with research backgrounds in such areas as organismic biology, including laboratory animal medicine and comparative pathology. Individuals trained in emerging technologies, including bioinformatics and biotechnology transfer, will also become increasingly important. At this time, there is a significant need for such individuals, and this need will only increase. Challenges in this area include the duration and cost of extended research training. The impediments that these factors impose to interested and qualified candidates must be lowered through mechanisms of debt forgiveness, integration of research training with professional educations, and provision of quality mentoring of candidates by established investigators early in their educational careers.

The rapidly proliferating volumes of biomedical data place increasing demands on databases and associated technologies. Harmonized approaches will be necessary to reduce costs and to increase exchange of information among widely divergent groups. Much of these data is in a format that can be readily exchanged electronically; this provides advantages in terms of ease of sharing, but possible hazards related to data alteration or corruption. Individuals trained in biological sciences must increasingly become familiar with the bioinformatics tools and their application for data management, ranging from data visualization to modeling and validation of experiments.

There is a need for increased synergy on national and international levels. This can be done through sharing biomaterials internationally and by scientific collaborations over state-of-the-art networks in cyberspace, supplemented with access to Web-based databases that provide carefully recorded information about genetic models. As research becomes more complex, research teams become a necessity to provide the wide-ranging expertise required in this setting. The coming decade has great promise for biomedical research of all types. It also will pose significant challenges and demands on the biomedical research community, upon which individuals with expertise in veterinary and comparative biomedical science can have a major impact.

## REFERENCES

1. Malakoff, D., The rise of the mouse, biomedicine's model mammal, *Science,* 288, 248–253, 2000.
2. Koch, R., Die Aetiologie der Milzbrand-Krankheit, begrundet auf die Entwicklungsgeschichte des *Bacillus anthracis, Beitr Biol Pflanz,* 2, 277–310, 1877.
3. Jenner, E., An inquiry into the causes and effects of the variolae vaccine, London, Samson, Low, 1798.
4. Meslin, F., Fishbein, D., and Matter, H., Rationale and prospects for rabies elimination in developing countries, *Current Topics in Microbiology and Immunology,* 187, 1–26, 1994.
5. Hendriksen, C., A short history of the use of animals in vaccine development and quality control, *Dev Biol Stand,* 86, 3–10, 1996.
6. Kohler, G. and Milstein, C., Derivation of specific antibody-producing tissue culture and tumor lines by cell fusion, *Eur J Immunol,* 6, 511–519, 1976.
7. Russell, E., Smith, L., and Lawson, F., Implantation of normal blood-forming tissue in radiated genetically anemic hosts, *Science,* 124, 1076–1077, 1956.
8. Culliton, B., Nobelists back animal research, *Science,* 244, 524, 1989.
9. Bradley, A., Evans, M., Kaufman, M., and Robertson, E., Formation of germ-line chimaeras from embryo-derived teratocarcinoma cell lines, *Nature,* 309, 255–256, 1984.
10. Thomas, K. and Capecchi, M., Site-directed mutagenesis by gene targeting in mouse embryo-derived stem cells, *Cell,* 6, 503–512, 1987.

11. Zhou, L., Dey, C., Wert, S., DuVall, M., Frizzell, R., and Whitsett, J., Correction of lethal intestinal defect in a mouse model of cystic fibrosis by human CFTR, *Science*, 266, 1705–1708, 1994.

12. Larson, J., Delcarpio, J., Farberman, M., Morrow, S., and Cohen, J., CFTR modulates lung-secretory cell proliferation and differentiation, *Am J Physiol Lung Cell Mol Physiol,* 279, L333–41, 2000.

13. Leppanen, P., Luoma, J., Hofker, M., Havekes, L., and Yla-Herttuala, S., Characterization of athero-sclerotic lesions in apo E3-leiden transgenic mice, *Atherosclerosis*, 136, 147–152, 1998.

14. Shibata, N., Transgenic mouse model for familial amyotrophic lateral sclerosis with superoxide dismutase-1 mutation, *Neuropathology*, 21, 82–92, 2001.

15. Levine, M., Klapstein, G., Koppel, A., Cepeda, C., Vargas, M., Jokel, E., Carpenter, E., Zanjani, H., Hurst, R., Efstratiadis, A., Zeitlin, S., and Chesselet, M., Enhanced sensitivity to N-methyl-D-aspartate receptor activation in transgenic and knocking mouse models of Huntington's disease, *J Neurosci Res* 58, 515–32, 1999.

16. Zhang, Z., Liu, Q., Lantry, L., Wang, Y., Kelloff, G., Anderson, M., Wiseman, R., Lubet, R., and You, M., A germ-line *p53* mutation accelerates pulmonary tumorigenesis: p53-independent efficacy of chemopreventive agents green tea or dexamethason/*myo*-inositol and chemotherapeutic agents taxol or adriamycin, *Cancer Res,* 60, 901–907, 2000.

17. Washington Fax, Celera completes 99% of the mouse genome (Genomics comply looks ahead to next stage – annotation), *Washington Fax*, May 4, 2001.

18. Venter, J.C., Adams, M.D., Myers, E.W., Li, P.W., Mural, R.J., et al. The sequence of the human genome, *Science*, 291, 1304–1351, 2001.

19. Nadeau, J.H., Balling, R., Barsh, G., Beier, D., and Brown, S.D.M., Functional annotation of mouse genome sequence, *Science*, 291, 1251–1255, 2001.

20. Washington Fax, Public – private mouse-sequencing consortium completes mouse genome (Complete sequence will be released to researchers worldwide), *Washington Fax*, May 10, 2001.

21. Rogers, J.R. and Bradley, A., The mouse genome sequence: status and prospects, *Genomics*, 77, 117–118, 2001.

# CHAPTER 2

# Animal Research Ethics

Anna S. Olsson, Paul Robinson, Kathleen Pritchett, and Peter Sandøe

## CONTENTS

## INTRODUCTION

Contemporary research in the life sciences, and particularly biomedicine, involves experimentation on live animals.* This research is considered an important tool in the progress of science. Much of it is directed towards the discovery of new ways to prevent, alleviate, or cure human diseases. However, the animals on which experiments are performed are often housed in such a way that they have limited freedom, or are subjected to distressing or severely painful interventions, or are killed. The overwhelming majority of these animals are mammals with highly developed nervous systems. They cannot, of course,

---

* The relationship between human beings and other animals is a central theme in this chapter. In referring to non-human animals as "animals," we do not mean to deny that human beings are animals. Following many pieces of legislation, we use the term "laboratory animal" to refer to vertebrates only.

0-8493-1086-5/03/$0.00+$1.50
© 2003 by CRC Press LLC

consent to their own participation in research. Nor, generally, will they benefit from such participation. And they appear to be capable of experiencing not only pain, but other forms of suffering as well.

These familiar facts present both the scientific community and society in general with a question: with the ultimate aim of alleviating or preventing human suffering, scientists carry out experiments causing pain and distress to animals, but are we, as human beings, morally justified in acting in this way?

This question has many dimensions, as we hope to show in this chapter. Some advocates of animal rights insist that we are not entitled to harm animals, even when our purpose is noble. Those who think that the human misery caused by the more serious kinds of ill health is a more urgent concern, on the other hand, respond that it would be unwise or unethical to abolish any animal-based experimental program that may lead to the effective treatment of a human disease. Most people appear to accept neither of these positions. Preferring to take a middle course, they regard the benefits gained through research as too important not to be pursued, but also believe we have a duty not to cause animals to suffer unduly.

In this chapter, we offer a survey of the ethical issues that animal research raises. We have not tried to provide a review paper, with comprehensive coverage of approaches and views. Nor have we attempted at any point to present a fully argued case for a particular conclusion about the ethics of animal experimentation — although it will be obvious at times where our sympathies lie. Our aim has instead been to set out prominent ideas in this area and to indicate how these ideas have been developed by specialists in animal ethics.

## HUMAN BENEFITS OF ANIMAL EXPERIMENTATION

Broadly speaking, modern animal experimentation began in seventeenth-century England and France. It has been central to our understanding of animal and human physiology ever since. A famous early example is William Harvey's investigation of the role of the heart in blood circulation. Observing the hearts of live animals with opened thoraxes, Harvey was able to see that the blood circulates in the body as a result of contractions of the heart.

> In the first place, then, when the chest of a living animal is laid open and the capsule that immediately surrounds the heart is slit up or removed, the organ is seen now to move, now to be at rest; there is a time when it moves, and a time when it is motionless...We are, therefore, authorized to conclude that the heart, at the moment of its action, is at once constricted on all sides, rendered thicker in its parietes and smaller in its ventricles, and so made apt to project or expel its charge of blood.

**William Harvey (1628)** *On the Motion of the Heart and Blood in Animals*

It is difficult to imagine how discoveries such as this could have been made in Harvey's time without vivisection in its true sense — that is, without the cutting open of live animals.

More recently, experiments on animals have played a central role in the development of vaccines and therapeutic treatments for a number of infectious diseases, including anthrax, smallpox, rabies, yellow fever, typhus, and polio.[1] They have been equally important in the study of noninfectious diseases, playing a part in the development of insulin in the treatment of diabetes, techniques of blood dialysis for patients with kidney failure, transplantation techniques, and advances in various types of surgery.[2] And there is little doubt that, if it continues to be pursued, animal research will make important contributions to the development of new medical treatments at the initial stages of research, in the subsequent development of treatments, and in the safety testing of pharmaceutical products.

Live animal experimentation is also conducted outside the field of biomedicine. Animals are used in fundamental research in the life sciences, not only in studies pursued within a broadly biomedical perspective, but also in basic work in biology and psychology. They are also used to test new products and substances for toxicity and other possible negative effects on human health prior to marketing — although alternatives to animal use in toxicology studies are constantly being developed, and where animals are still used, refinement of the techniques has reduced the number of animals required and reduced suffering.[3]

We can summarize the current situation, then, by saying that most experimental animals are used for three main purposes: to develop pharmaceutical and other medical products; to advance fundamental research in the life sciences; and to test the safety of potentially toxic products and substances. These, at least, are the uses of animals on which we shall concentrate in this chapter. We readily acknowledge that animals are used in biomedicine in other ways as well: for example, in the breeding of more animals for research, the education and training of scientists and veterinary personnel, and the diagnosis of disease and the production of biological matter, such as cells and antibodies.

The proportion of experimental animals used for each of the purposes just distinguished varies from country to country. It depends on the presence and activity of biomedical research and the pharmaceutical industry in that country and, less directly, on the way "animal experimentation" is defined in the relevant legislation. According to the latest available statistics from the European Union, in 1996, most experimental animals (44%) were used in the research, development, and quality control of products for human and veterinary medicine. Fewer animals (25%*) were involved in fundamental biological research, and fewer still (9%) in toxicology and other safety testing.[4]

In connection with any research project falling into one of the three categories identified, we can ask: must this project involve animal experimentation? We can also ask: must this project be carried out at all? In some cases, the answer to one of these questions will be negative. It might be the case, for example, that the scientists are accustomed to proceeding in the way they currently do and are unaware of alternative methods. Similarly, regulatory bodies might continue to require animal tests that have always been applied even though there are now alternatives to using live animals as a means for research. Again, a comprehensive review of the literature may reveal that the results of an animal experiment involve duplication or are unnecessary for some other reason. To give more specific examples, it might be possible to exploit cell lines, or some other replacement method, or to use human volunteers.

Inevitably, however, there will be many cases in which it is essential to study an intact, living organism and in which the procedure or substance is too risky or too invasive to permit human volunteers to be used. In these cases especially, we need to consider carefully what costs we are imposing on the relevant animals.

## THE COSTS TO ANIMALS

Animal experiments can be costly in economic terms, and from an ethical point of view, it will sometimes be appropriate to ask whether the resources a study consumes could have been used more effectively or for some other purpose altogether. The cost we want to focus on here, however, is one carried not by those who fund or perform animal research, but by the experimental subjects — the animals. It is a cost measured not in monetary terms, but in suffering. Animals may suffer because the relevant experimental interventions provoke one or another of a wide range of unpleasant states. Pain, for instance, may be the result of surgical interventions, noxious stimuli, the application of irritating or corrosive substances, certain progressive diseases, genetic disorders, or infectious diseases. The systemic administration of test substances may provoke nausea and general discomfort. Fear is common in experimental situations because the animals are exposed to procedures to which they are averse and from which they cannot escape. And even when they are not specifically painful, most medical conditions occurring through provocation or spontaneously in laboratory animals are likely to be accompanied by some general discomfort. Finally, experimental animals are often housed in restrictive conditions in which they experience frustration when they cannot carry out species-specific behaviors.

Some countries have made data available that show the impact of various kinds of experimentation on animals. Examples are given in Table 2.1. This kind of information has a significant bearing on the ethics of animal research. Data from Canada and Switzerland, for instance, show that toxicology studies and studies performed in the research, development, and quality control of pharmaceutical and other medical products are those that most frequently expose animals to severe discomfort.

In general, statistical information of this kind suggests that the vast majority of animals suffer little or no distress in experimental settings. Some observers have used this finding to defend the position that animal research is unproblematic because relatively few animals suffer greatly. However, this argument seems weak. First, the plight of the 5 to 10% of animals described in the data as being exposed to severe pain or distress certainly seems, on the face of things, to be an ethical problem. Generally speaking, morally objectionable conduct does not become unobjectionable when it is directed at only a few victims. Second, in connection with the remaining cases, in which the animal distress is minor, we need to ask:

---

* This figure excludes projects on the genetic modification of animals and immunological studies.

**Table 2.1   Estimated Degree of Discomfort in Laboratory Animals**

|  | Switzerland[a] 2000 | | Canada[b] 1999 | |  | U.S.[c] 2000 | |
|---|---|---|---|---|---|---|---|
| Category | N animals | % | N animals | % | Category | N animals[d] | % |
| No/Minor | 296,888 | 70.2 | 1,128,866 | 64.6 | No pain/distress | 897,226 | 63.3 |
| Moderate | 103,653 | 24.5 | 558,912 | 32.0 | Pain/distress Drugs for relief | 415,215 | 28.6 |
| Severe | 31,668 | 5.3 | 58,828 | 3.4 | Pain/distress No drugs for relief | 104,202 | 7.3 |

[a] Swiss Federal Veterinary Office
[b] Canadian Council on Animal Care
[c] United States Department of Agriculture
[d] Includes vertebrates except rats, mice, birds, and animals used for food or fiber research

in what way are these experiments unproblematic? The suggestion that this kind of suffering simply does not matter morally looks dubious. Surely it does matter — it is just that it carries less weight than more serious suffering. The alternative suggestion that minor suffering is justifiable is more plausible, but it is implicit in it that even minor suffering needs to be justified and hence, matters. In this sense, such suffering is not unproblematic.

The moral acceptability of animal experimentation is bound to be less questionable where animal suffering is minimized. To reduce such suffering, however, we will need to be quite clear about its nature — about the way or ways in which animals suffer. Our intuitive perceptions of what causes distress to animals are not always reliable. Procedures that appear unpleasant to human beings may cause an animal little distress. Thus, for example, no increase in levels of the stress-related hormone corticosterone was observed in mice kept in a room where other mice were being killed.[5] Conversely, procedures that strike human beings as pleasant may be demonstrably stressful for an animal. Thus, the standard husbandry practice of cage-cleaning results in an immediate, although not necessarily prolonged, increase in aggression among male mice.[6] Again, photographs of animals with electrodes attached to their heads may appear shocking when shown in animal protection campaigns, but cranial implants often look more inhumane than they, in fact, are. If the implants are positioned correctly and kept clean and free of infection, the most that the animal experiences in day-to-day life is some itching at the site of the implant. Brain tissue is not itself innervated, so the implant does not cause pain beneath the skull.

If we are to avoid these anthropomorphic assessments of animal distress and suffering, we will need to employ technically defined measures of the impact of experimental procedures on the subjective experience — measures defined in research on laboratory animal welfare. Over the last three decades, the study of animal welfare has become an established scientific discipline. Methods of assessing animal welfare have emerged from studies of the ways in which animals react, behave, and function in different experimental and everyday husbandry situations. Within the field, there is some variation in the approach to animal welfare, in that some animal welfare scientists emphasize health and biological functioning (e.g., Reference 7), whereas others hold that animal welfare is primarily a matter of the feelings of the animals (e.g., Reference 8). These approaches may diverge less than they appear to, because in practice, the factors they appeal to generally correspond. Nevertheless, in defining animal welfare, it may be helpful to ask why we are interested in animal welfare and suffering in the first place. The following remarks set out an answer to this question:

Animal welfare involves the subjective feelings of animals. The growing concern for animals in laboratories, farms, and zoos is not just concern about their physical health, important though that is. Nor is it just to ensure that animals function properly, like well-maintained machines, desirable though that may be. Rather, it is a concern that some of the ways in which humans treat other animals cause mental suffering and that these animals may experience "pain," "boredom," "frustration," "hunger," and other unpleasant states perhaps not totally unlike those we experience.

**Marian Dawkins (1990) From an Animal's Point of View: Motivation, Fitness, and Animal Welfare,** *Behavioral and Brain Sciences*

This reasoning looks cogent. It suggests that any acceptable definition of animal welfare will take subjective feelings into account. If this is right, it follows that when animals are used as research models, it is the prospect of their experiencing pain or other kinds of unpleasant mental states that is the main cause for concern.

It should be acknowledged, however, that the occurrence of unpleasant mental states does not, by itself, imply that there is suffering. Such states are an unavoidable part of normal animal life and often serve as signals or behavioral prompts that help the animal to satisfy its biological needs. Sometimes, negative experiences are compensated for by corresponding positive experiences — few would argue that a hungry animal that finds food is suffering, even though the experience of being hungry is not pleasant. Unpleasant states, therefore, represent a welfare problem only when they are not compensated by corresponding positive feelings, or persist for an extended period of time, or occur frequently. Thus, a captive hungry animal that is not fed raises a welfare problem. So, too, does an animal that is strongly motivated to build a nest, or explore, but is kept in an environment where it has no opportunity to exercise these kinds of behavior.

Let us now turn to the measurement of animal welfare. Obviously, it is impossible to measure the intensity or duration of negative mental states directly. We therefore have to rely on physiological, behavioral, pathological, and other indirect parameters of feeling. Parameters of this kind include: changes in normal behavior; the occurrence of abnormal behaviors, such as stereotypies; altered activity of the hypothalamo-pituitary-adrenal axis or the sympatho-adrenal system; other hormonal changes; and modifications in body temperature, immunocompetence, plasma ion levels and body weight.[9-13] These parameters are connected with the activation of various physiological and behavioral systems. It is worth pointing out that, since these systems affect not just welfare levels, but also the way an animal will react in an experimental situation, an animal's welfare status might well affect its suitability as a research model.[14-16]

Much animal suffering can be avoided through the proper use of anaesthetics and analgesics, and by careful handling procedures and improved housing systems. However, the fact that a great deal of pain can be controlled neither removes nor diminishes the remaining discomfort. An animal that has been operated on or is developing a medical condition, such as ascites or tumors, will inevitably experience a degree of discomfort. Equally, it is almost always necessary to deprive animals of some freedom and control over their environments to experiment upon them, and experimental interventions will, in many cases, involve a certain amount of distress even when they are carried out with care. Welfare problems caused by restrictive housing are often overlooked, but they are clearly important, because they will affect not just the experimental animals, but control groups and breeding stock as well. Behavioral restrictions imposed by standard housing might be stressful either because animals experience frustration when species-specific behaviors cannot be carried out or because animals have no opportunity to control their environments.[16,17] In most cases, the experimental protocol does not require restrictive housing.

In most countries, direct cruelty and the infliction of unnecessary suffering are illegal. In practice, the focus on *unnecessary* suffering will obviously permit some animal suffering — namely, that which secures a sufficiently important benefit for human beings. But in this formula, what is to count as a sufficiently important benefit? Is it enough that we will understand the function of all the genes in the mouse genome, or find an effective vaccine for HIV, or develop new agricultural pesticides and test their toxicity? And what should we say about the common situation in which the benefits of a line of research are, as the research gets underway, uncertain? Can a potentially large benefit that cannot be guaranteed be sufficiently important to license animal suffering?*

## THE ETHICAL DILEMMA

Behind questions such as these lies an "ethical dilemma." Ordinarily, when we describe someone as facing a dilemma, we mean that the person is in a no-win situation: whatever he or she does, the result will be unsatisfactory in some way. An ethical dilemma is a special case of this. It again involves a no-win situ-

---

* Questions like these are examined at greater length at the end of the section entitled "Claim (1): The Need for Animal Research."

ation, but this time, whatever the person does, the result will be *morally* unsatisfactory in some way. The ethical dilemma that animal experimentation presents is summed up by the following four claims:

1. Live animal research is the only effective way of bringing certain important benefits to mankind, particularly in the prevention and therapeutic treatment of serious human diseases.
2. It is morally imperative to find new ways to prevent or treat serious human diseases.
3. In the course of live animal research, individual animals will inevitably be caused suffering or distress, and the interventions will not benefit the animals concerned.
4. It is morally imperative to preserve the welfare of animals, and in particular, one should not cause an animal to suffer if that suffering is not compensated for by a corresponding benefit of some kind.

It is easily seen that these claims are in conflict with one another. To ease the tension between them, we need to show that at least one of the claims is false, or at least an overstatement. Unsurprisingly, people who oppose animal research are normally skeptical about one or other of the first two claims. Some argue that there are ways of obtaining the necessary research results without animal experimentation and that the first claim is therefore incorrect. Cosmetics are an obvious example of a product for which this might be claimed. Much less commonly, it is suggested that we do not need to refine new medical responses to serious human diseases and hence that the second claim is misguided. This attitude is sometimes found among those with deeply held religious or ideological convictions.

By contrast, people who argue for continued animal experimentation reject the third and fourth claims. They maintain, in other words, either that animals do not suffer at all because they are not conscious in the way required for feeling, or that animals, unlike humans, do not matter from a moral point of view. Either way, the conclusion is that there is no dilemma because what we do to animals does not matter from a moral point of view.

As already mentioned, serious skepticism about the suggestion that it is morally imperative, or at least highly desirable, to discover ways of preventing or treating life-threatening human diseases is rare. We can, therefore, assume for the purposes of this chapter that the second claim is overall correct. The first, third, and fourth claims are a great deal more contentious, however, and we therefore need to examine them with some care.

## Claim 1: The Need for Animal Research

It is often said that live animal research *must* be performed if we are to enjoy human benefits of the kind it brings. How true is this? Obviously, we cannot here scrutinize specific projects involving animal experimentation. Let us, instead, first note that it is extremely implausible to suggest that live animal research is totally unnecessary — that all the human benefits it delivers could be secured using methods that do not involve animals. With this noted, we can now look at reservations about the need for animal research that cannot be dismissed out of hand.

One reservation is so obvious it hardly merits mention. Some experiments are badly designed, or are carried out in unsuitable conditions, or unnecessarily repeat previous research. It cannot be claimed that these experiments are essential in bringing important benefits to mankind because they do not produce new benefits at all. Ill-conceived experiments of this sort may indeed be morally wrong, regardless of any animal suffering involved, because they waste material resources. Causing animal distress under such conditions can never be justified.

Doubts about the need for animal research can arise even when an animal experiment is well designed and capable of delivering valuable results. Suppose an alternative research method — a method involving, say, cell lines, bacteria, or human volunteers — will deliver the sought-after results equally readily. Here, it cannot be said that animal experimentation is required. Again, it might be possible to secure certain health benefits by encouraging people to change their lifestyles or avoid risky behaviors. Where ill health can be avoided in this way, it would be misleading at best to insist that we need to develop drug treatments using animal research.

In certain cases, doubts of the kind described in the last paragraph are fuelled by the fact that economic profit is a dominant motive within the pharmaceutical industry. Is it not conceivable that the R&D departments of pharmaceutical companies are guided as much by potential monetary gain as the aim to relieve human suffering caused by disease? And similar suspicions arise when we turn to toxicology and safety testing. The immediate goal of such testing, whether or not it involves animals, is to protect human health and the environment by preventing hazardous products from being marketed and thus allowed to enter the biosystem. However, some products that undergo toxicological analysis and safety testing are of questionable human importance. How important is it to provide a new garden herbicide? Do we need a new kitchen disinfectant or a shampoo with a different formula? Where these products do not offer substantial human benefits, any connected animal research can hardly be described as essential in bringing important benefits to mankind.

Finally, we have already mentioned that some animal experimentation is undertaken in fundamental research in the life sciences. This research produces information that may come to be useful in the understanding of disease, but it is mainly pursued with the aim of advancing general knowledge. Some might deny that such experimentation plays a vital role in the delivery of substantial new human benefits. Against this, it should be pointed out that efforts to combat human ailments always depend to some degree on knowledge gained through more fundamental research.

Perhaps this problem — the problem of predicting benefits — is quite general. In many experiments, fundamental and applied, it can be hard to know at the outset whether the hours spent in the laboratory will result in human benefit. It might be said that in applied studies, at least, it is often possible to guarantee benefits at a late stage in the research process: for instance, when studying the dose effect of a substance that has already been proven efficient. However, it is an unavoidable fact that the later stages of research cannot be reached without going through previous ones, and while they go through previous stages, researchers will often have to follow leads that will turn out to be fruitless. This complicates the assessment of experimental necessity. It means that we will not always be able to look merely at the human benefits of single experiments: at times, the net gains of research projects, or even of entire theoretical approaches, will be what we should try to assess. Often, scientific quality will be the only criterion to which we can appeal if we wish to know whether the relevant experiment will secure important benefits for mankind.

## Claim 3: The Animal's Experience of Pain

Let us now turn to the claim that individual animals are inevitably caused suffering and distress in the course of live animal research. There is a long tradition of animal experimentation in the natural sciences. For centuries, such experimentation must, at its worst, have caused terrible pain and suffering to animals because anaesthetics and analgesics were virtually unknown. Had they been available, however, these palliatives might still not have been used, since for a long time, it was believed that animals were automata and incapable of feeling pain in the way human beings do.* In the following passage, a seventeenth-century eyewitness describes the undeniably grim implications of this view for experimental animals:

> They administered beatings to dogs with perfect indifference, and made fun of those who pitied the creatures as if they had felt pain. They said that the animals were clocks; that the cries they emitted when struck, were only the noise of a little spring which had been touched, but that the whole body was without feeling. They nailed poor animals up on boards by their four paws to vivisect them and see the circulation of the blood which was a great subject of conversation.

**Fontaine (1968/1738)** *Mémoires pour Servir à L'Historire de Port-Royal*

How did scientists come to think in this way? Why did they adopt the view that sophisticated animals were mere mechanisms, rather like clocks, capable of producing visible behaviors, such as crying, but incapable of feeling? The view has its origins in the work of the seventeenth-century French mathematician and philosopher René Descartes. Descartes is associated with what is referred to today as mind-body dualism. He believed that men and women consist of a material body and an immaterial soul. He could not accept, however, that animals have a soul, so in effect, he was materialist about nonhuman animals. The upshot of this can easily be guessed. According to Descartes, feelings are properties of the soul. Since animals do not possess souls, they cannot have any feelings: animals are, in effect, machines.

* Although physicians from the ancient Egyptians onwards were aware of the sedative qualities of both alcohol and opium, genuine anaesthetics suitable for human use were not available until the 1840s. It was not until the late 1800s that anaesthetics were routinely used with animals.

Generations of natural scientists inherited this belief. Their modern heirs no longer believe, of course, that human beings have immaterial souls. More to the point, most would deny that possession of such a soul is a prerequisite of feeling. However, the view that animals are devoid of feeling has persisted in parts of the scientific world into the twentieth century.[18]

Descartes' view that animals cannot suffer prevents an ethical dilemma from arising even in the most invasive animal experiments. It does so by implying that our third claim is false. However, is it at all plausible today to deny that animals experience feeling? Descartes' own case for such a denial was, primarily, that in the absence of language animals cannot communicate feelings. However, we do not rely exclusively on linguistic behaviors in diagnosing other people's feelings: nonlinguistic behaviors and facial expressions often communicate as much as speech. Thus, seeing a sprinter limping off the track with distorted facial features, we rarely feel the need to await a verbal pain report. We know enough already to be sure that he or she is feeling pain. There seems to be no reason why we should not draw a similar conclusion about, say, a dog that holds an injured paw close to its body, whimpers, and turns to bite anyone who attempts to touch the leg.

Today's scientists largely agree that all vertebrates, and some invertebrates (such as octopuses), have the capacity for pain, but a further complication arises at this point. The complication arises because it can sensibly be asked whether the pain that animals have is actually felt as an unpleasant mental state — and it can therefore be asked whether animal pain involves suffering. This is a contentious issue, and one to which both philosophers of mind and scientists might be expected to contribute. In defense of the view that animal pain does not involve feeling, it has been claimed that to feel pain, it is necessary to have cognitive capacities, and hence, a developed prefrontal cortex of a kind that most animals other than anthropoid apes do not possess.[19] On the other hand, it is clear that animal pain, like human pain, is causally connected with aversive behavior, and some observers believe that this causal connection between peripheral nociceptive nerve signalling and centrally controlled aversive behavior could not occur in the absence of unpleasant feelings, and thus in the absence of suffering.

As we have said, this issue is contentious. Nevertheless, most of us, including most scientists, are convinced that animals can suffer. In view of the seriousness of the issue, agnostics should probably also adopt a reasonable measure of precaution and give animals the benefit of the doubt. They should act as though animals are capable of suffering and assume that a procedure that is painful to humans is also painful for animals. The adoption of this working assumption does not necessarily force us to accept the third claim. For clearly, even if animals have the capacity to suffer, it does not follow that in the course of live animal research, they will inevitably suffer. And, as will be obvious to most scientists reading this, animal suffering and distress during experimentation can be reduced or eliminated in several ways. Refinements in experimental methods, such as more rapid and exact sampling techniques, or the introduction of noninvasive sampling methods, as well as extended use of anaesthetics and analgesics, reduce the distress an experiment causes. Likewise, improvements to the conditions under which experimental animals are housed using so-called environmental enrichment mitigate animal stress. And through improved animal models and the correct use of statistics, the number of animals needed to obtain valid results can be reduced, thus reducing the total amount of any inevitable suffering.

## Claim 4: The Moral Status of Animals

Suppose none of the aforementioned pain-reduction strategies were available. Would this show that the relevant research should cease, or be radically limited, on moral grounds? If we were discussing the pain of human volunteers, the reply would almost certainly be "yes."* But where animals are concerned, matters are less straightforward. This is because, traditionally, animals have been thought to be less important than human beings, morally speaking. It is this claim that we need to examine now.

The attitude that animals matter less than human beings is widespread in western society. It is often linked with the Judeo-Christian tradition upon which much of our culture is based, for according to the Bible, man occupies a special position in the world: he was created in the image of God and given dominion over other living creatures (Genesis 1:26–28). However, other reasons can be given for the view that human beings have a different moral status from animals. A common argument runs as follows: only human beings are known to possess language and to be able to reason in abstract terms. Because

---

* Although complications turning on consent would arise. Of course, animals cannot give consent. Interestingly this seems to have implications for the regulation of animal research: see the section entitled "Ethics Committees".

animals can neither reason nor communicate reasons, they cannot act morally, and therefore we have no moral obligations towards them.

But should the capacity to act morally determine whether an individual should be given moral consideration? The notion that it should is not obviously correct. As early as the eighteenth century, the English philosopher Jeremy Bentham asked why we deny animals moral rights that we ascribe to ourselves. He wrote:

> Is it the faculty of reason, or, perhaps, the faculty of discourse? But a full-grown horse or dog is beyond comparison a more rational, as well as a more conversible animal, than an infant of a day, or a week, or even a month, old...The question is not, Can they *reason*?, nor, Can they *talk*? but Can they *suffer*?

**Bentham (1789)** *The Principles of Morals and Legislation*

In this well-known passage, Bentham does two things. First, he offers a philosophical argument designed to embarrass those who suggest that human beings are morally superior to animals because they possess intelligence and language. This argument is simple and proceeds in the following way: certain human beings — Bentham speaks of infants, but we could also mention the mentally impaired — have lower levels of intelligence and linguistic ability than some higher animals. Therefore, intelligence and linguistic ability cannot be the criteria of human moral superiority. This argument is basically sound. It obliges us either to offer an alternative rationale for the view that human beings matter more than animals or to drop that view. Bentham himself takes the latter course, for the second thing he does is to suggest that it is the capacity to suffer that confers moral status. This suggestion brings infants back into the moral realm. It also brings in any animals that are able to suffer. Bentham would have regarded both of these implications as welcome.

The contemporary Australian philosopher Peter Singer is Bentham's modern heir. Having explored the options at length, he claims that it is impossible to identify a difference between human beings and animals that separates them morally. And he concludes that when we imagine that animals have no moral standing or a lower moral status than human beings, we are laboring under a moral prejudice similar to that found among racists or sexists:

> I am urging that we extend to other species the basic principle of equality that most of us recognize should be extended to all members of our own species...The racist violates the principle of equality by giving greater weight to the interests of members of his own race when there is a clash between their interests and the interests of those of another race. Similarly, the speciesist allows the interests of his own species to override the greater interests of members of other species. The pattern is the same in each case.

**Peter Singer (1989) All Animals are Equal, in** *Animal Rights and Human Obligations*

Again, the problem for those who prioritize human interests is to explain what they take to be the moral difference between animals and human beings. Singer's point is that just pointing to a difference in species does not seem to be sufficient.

Most of us assume, most of the time and more or less consciously, that human beings deserve special moral consideration — consideration that is not due to animals. In this section, we have seen, however, how difficult it is to provide a compelling rationale for this assumption. The discussion has been more exploratory than conclusive. To make further progress with the issues, we need to look at morality in general terms. We must enquire into its basis and purpose. In other words, we must examine ethical theories.

## IS ETHICAL THEORY NECESSARY?

At this point, it may be wondered whether animal researchers really need to be familiar with ethical theory. What possible objection could there be to the scientist who simply proceeds in an intuitively humane manner? Surely the ethical theorizing can be left to philosophers and theologians. This attitude is understandable. However, there are, in fact, several ways in which scientists can benefit from explicit appreciation of ethical theory. Here, we shall briefly sketch three such benefits.

In today's society, there are many different views about what we are entitled to do to animals in the name of scientific progress. Animals and animal materials continue to be used in laboratories, yet this usage is repeatedly challenged. Gruesome images of cats, dogs, and monkeys in experimental conditions have been put before the general public by animal rights organizations. They often evoke strong feelings in observers, but there is absolutely no doubt that people also want access to effective medical treatments and safe chemical products. Indeed, they may even be willing to support the research such access entails through taxes and fund-raising campaigns. Likewise, when asked if scientists should be allowed to continue to experiment on animals, 64% of the participants in a British survey opposed the use of living animals in research.[20] But when the question was prefaced with the statement, "Some scientists are developing and testing new drugs to reduce pain, or are developing new treatments for life-threatening diseases, such as leukaemia and AIDS. By conducting experiments on live animals, scientists believe they can make more rapid progress than would otherwise have been possible," disapproval dropped to 41%.

The first problem, then, with being led by one's feelings, rather than approaching matters through ethical theory, is simply that people's feelings about animal research are often unstable or ambivalent. Such feelings cannot be relied upon as a rational guide. This immediately leads to a second problem. This ambivalence encourages double standards, and these standards are both morally objectionable and logically indefensible.

However, the third problem is perhaps the most serious. It is clear that, at present, we are engaged in the West in an increasingly serious debate about the rights and wrongs of animal use. However, it seems unlikely that scientists and others taking part in this debate will be able to communicate effectively while they merely press their intuitively held beliefs. These beliefs are normally sincere, and often strongly held, but they can be extremely difficult to understand and highly resistant to change. The ideal of meaningful and transparent discussion that leads to mutual understanding of the arguments is attainable, however. For people's gut feelings about matters such as animal research are often based on underlying ethical theories, and these theories are much more susceptible to rational assessment than the individual beliefs to which they give rise. The suggestion we wish to make here, then, is that if laypeople and scientists are willing to think a little about fundamental ethical theory, they will have a much greater prospect of communicating with one another effectively, articulating their convictions in a coherent manner, and perhaps even reaching a compromise upon which all can agree.

Moral philosophers distinguish a number of types of ethical theory, and, in principle, any of these might underlie a person's views about animal experimentation. Here, we will discuss three prominent theoretical positions: contractarianism, utilitarianism, and rights views. These have been selected because they have direct and obvious implications for the ongoing debate over animal use.

## CONTRACTARIANISM

Why should we act morally? This is a central question in moral philosophy, and one to which the contractarian gives a straightforward answer: one should act morally because it is in one's self-interest. The outlook underlying contractarianism is egoism. According to the egoist, when one is obliged to show consideration for other people, this is really for one's own sake. In general, by respecting the rules of morality, one contributes to the maintenance of a society that is essential to one's own welfare. The moral rules are thus those that best serve the self-interest of all members of the society. Contractarian morality is confined to those individuals who can "contract in" to the moral community, so it is important to define who these members are:

On the contract view of morality, morality is a sort of agreement among rational, independent, self-interested persons, persons who have something to gain from entering into such an agreement…

A major feature of this view of morality is that it explains why we have it and who is party to it. We have it for reasons of long-term self-interest, and parties to it include all and only those who have *both* of the following characteristics: 1. They stand to gain by subscribing to it, at least in the long run, compared with not doing so. 2. They are *capable* of entering into (and keeping) an agreement... Given these requirements, it will be clear why animals do not have rights. For there are evident shortcomings on both scores. On the one hand, humans have nothing generally to gain by voluntarily refraining from (for instance) killing animals or "treating them as mere means." And on the other, animals cannot generally make agreements with us anyway, even if we wanted to have them do so....

**Narveson (1983) Animal Rights Revisited, in *Ethics and Animals***

On this view, there is clearly a morally relevant difference between my relationship to other human beings and my relation to animals. I am dependent on the respect and cooperation of other people. If I treat my fellow humans badly, they will respond by treating me badly. By contrast, the animal community will not strike back if, let us say, I use some of its members in painful experiments. From an egoistic point of view, I need only treat the animals well enough for them to be fit for my own purposes. And in any case, as Narveson points out, nonhuman animals cannot enter into a contract, or agreement, governing future conduct, so they cannot join the moral community.

For the contractarian, since neither animal suffering nor the killing of animals is an ethical problem *per se*, animal experimentation is in itself ethically acceptable. It may even be ethically desirable, since, as long as the experiments are effective, it is certainly in the interest of the moral community to run animal experiments to find treatments for diseases that cause human suffering. The lack of standing of animals in the moral community does not necessarily mean that the way animals are treated is irrelevant from the contractarian point of view: if people *like* animals, for example, and dislike the practice of their being used in this or that way, animal use can become an ethical issue, because it is in a person's interests to get what he or she likes. Nevertheless, the contractarian view of animals is highly anthropocentric, since any rights to protection animals have will always be dependent on human concern. Inevitably, we tend to like some types of animals more than others. We are more troubled by the suffering of our favorite sorts of animals. Hence, levels of protection will differ across different species of animal. For example, because most people like cats and dogs more than rats and mice, causing distress to cats and dogs is likely to turn out to be a more serious ethical problem than causing the same amount of distress to rats and mice. Likewise, nonhuman primates will probably receive more protection than other animals, because (perhaps because they are perceived as closer to humans) their plight is of considerable concern to people.

Since it is egoistic human concern that determines how animals should be treated on the contractarian approach, this approach requires an open dialogue between those who use animals and those who are concerned about their welfare. Both activists in animal protection organizations and the general public as taxpayers and consumers of animal-tested products should be permitted free access to information about the ways in which animals are used in research and other activities.

The contractarian view agrees with certain attitudes towards animal treatment that are prevalent in our society. Thus, it serves to explain why legislation, allegedly for the protection of animals, usually protects the animals that matter most to humans, such as cats and dogs. Contractarianism can, however, seem inadequate. Can it really be correct to hold that causing animals to suffer, even for a trivial reason, or for no particular reason, is morally unproblematic as long as no human being is bothered by the relevant conduct? Many would want to insist that it is immoral as such to cause another to suffer for little or no reason, whether one's victim is a human being or an animal. An ethical theory that captures this insistence is utilitarianism.

## UTILITARIANISM

According to the utilitarian, the interests of every individual affected by an action count morally and deserve equal consideration. In utilitarian writings, the notion of an interest is usually defined in terms of "the capacity for suffering or enjoyment or happiness".[21] Thus, individuals have an interest in acts that will enhance their enjoyment or reduce their suffering. From this it follows that all sentient beings, human and nonhuman, have interests. And, since for the utilitarian, all interests count morally and deserve equal consideration, this implies that the impact of one's actions on all sentient creatures, including animals, is a matter of moral concern.

Many philosophers have proposed the principle of equal consideration of interests, in some form or other, as a basic moral principle; but…not many of them have recognized that this principle applies to members of other species as well as to our own…. If a being suffers, there can be no moral justification for refusing to take that suffering into consideration. No matter what the nature of the being, the principle of equality requires that its suffering be counted equally with the like suffering — in so far as rough comparisons can be made — of any other being.

**Peter Singer (1989) All Animals are Equal, in *Animal Rights and Human Obligations***

For the utilitarian, then, ethical decisions require us to strike the most favorable balance of benefits and costs for all the sentient individuals affected by what we do. However, doing the right thing, according to the utilitarian, is not only a matter of doing what is optimal. It is also essential to do something rather than nothing: if something can be done to increase well-being, we have a duty to do it. This utilitarian duty to act always to bring about improvements has important consequences for society. In contemporary Western society, we have a general tendency to give ourselves priority over animals. A thoroughgoing utilitarian will regard this tendency as essentially wrong. However, the anthropocentric outlook is obviously well established, and in view of this, it may well be that, for the time being at least, any attempt to ensure that sentient animals are accorded the same status as human beings is bound to fail. It may be that the best thing a utilitarian can do is to secure higher levels of animal welfare within the current system. To give a specific illustration, in the case of laboratory animals, a utilitarian realist might be willing to apply the so-called "principle of the three Rs" — that is, endorse actions and policies leading to the *replacement* of existing live-animal experiments with alternatives, or *reductions* in the number of animals used, or *refined* methods that cause animals less suffering.[22] It can be seen, then, that less-invasive sampling techniques, improved housing systems, and more precise models requiring fewer animals to be used are likely to be viewed as morally attractive developments within the realist utilitarian perspective.

In the ethical conflicts prompted by animal research, human interest in obtaining some benefit stands against the animal's interest in avoiding suffering. Sometimes, however, the utilitarian will want to weigh not just animal interests against human interests, but also the interests of different animals against each other. Animal experiments can benefit animals as well as humans: many of the insights underlying veterinary medicine have been derived from experiments on animals. When a pet cat receives a vaccination against FIV (feline immunodeficiency virus), it benefits from immunology research done on other cats, even though the primary purpose of this research was to develop treatments for HIV. It can be seen, then, that in deciding whether an animal experiment is ethically justifiable, it is sometimes necessary to take into account both the animals whose interests are sacrificed in the experiment and the animals that may benefit from the results.

Animal-based research is just one of the many ways in which we make use of animals. The overwhelming majority of domestic animals are kept for food production. Most are kept under restrictive conditions in which basic behavioral or physiological needs are thwarted. Laying hens, for example, are commonly kept in battery cages where they cannot perform strongly motivated nesting behaviors before egg laying and where the restriction of their movement results in bone brittleness and a high incidence of broken bones. Similarly, breeding sows are often confined to crates in a way that limits most movements other than simply lying down and standing up. It seems beyond doubt that food production under the conditions currently prevalent in commercial farming causes considerable animal distress. Naturally, this cost must be weighed against the benefit, to human beings, of access to cheap meat and eggs. However, given that the average citizen in the developed world consumes far more protein than is physiologically necessary, and often more animal fat than is healthy, low-cost meat cannot be considered a vital human interest.

We have gone into this matter in some detail because the welfare implications of present-day commercial farming have a significant bearing on the utilitarian response to animal experimentation. This bearing is easily seen: the abandonment of intensive animal husbandry practices will probably promote animal well-being (without jeopardizing vital human interests) much more effectively than the abolition of animal-based research.

In this section, we have described a pragmatic utilitarian approach. We have suggested that realistically the utilitarian should perhaps accept that animal interests are best sacrificed where that leads to the satisfaction of vital human interests — as happens in much biomedical research. But for all that has been said, a more radical utilitarianism might be worth exploring. Animal experimentation sometimes means sacrificing vital animal interests in continued life and in the avoidance of abject suffering. Insisting that human and animal interests deserve equal consideration, Singer concludes that the sacrifice of such vital animal interests is acceptable only where the benefits are extraordinarily important:

> ...If a single experiment could cure a major disease, that experiment would be justifiable. But in actual life the benefits are always much, much more remote, and more often than not they are nonexistent...an experiment cannot be justifiable unless the experiment is so important that the use of a retarded human being would also be justifiable.

**Singer (1975)** *Animal Liberation*

It is evident, then, that within the utilitarian approach to animal experimentation a wide range of views are represented. Some utilitarian observers accept most animal experiments as long as we do our utmost to prevent and alleviate animal suffering. Others, like Singer, setting the demand for human benefit higher, would prefer to see nearly all such experiments abolished. What all utilitarians agree on, however, is the methodological precept that ethical decisions in animal research require us to balance the harm we do to laboratory animals against the benefits we derive for humans and other animals. Interestingly, some moral philosophers have attacked this very precept — the notion that we can work out what is ethical by trading off one set of interests against another. The allegation is that such trade-offs violate the rights of the individuals whose interests are in the moral balance. To get clearer about this, we need to turn to rights theory.

## RIGHTS VIEW

There is an obvious sense in which, in focusing on overall improvements in welfare, the utilitarian treats sentient beings as mere instruments. The utilitarian believes that it is ethically justifiable to sacrifice the welfare of one individual when this sacrifice is outweighed by connected gains in welfare. Rights theorists object to this, holding that it is always unacceptable to treat a sentient being merely as a means to obtain a goal. Historically, rights theory is associated with the eighteenth-century German philosopher, Imman-uel Kant. In Kant's view, human beings have "an intrinsic worth, i.e., dignity" and should therefore be treated "always as an end and never as a means only." Clearly, this view is at variance with the utilitarian's willingness to sacrifice one individual's welfare when this leads overall to welfare gains. Kant himself confined the right to be treated as an end to human beings, but later rights theorists, such as the American philosopher Tom Regan,[23] have argued that the principle of dignity should be extended to animals:

> ...Attempts to limit its scope to humans only can be shown to be rationally defective. Animals, it is true, lack many of the abilities humans possess. They can't read, do higher mathematics, build a bookcase, or make *baba ghanoush*. Neither can many human beings, however, and yet we don't (and shouldn't) say that they (these humans) therefore have less inherent value, less of a right to be treated with respect, than do others. It is the *similarities* between those human beings who most clearly, most noncontroversially, have such value (the people reading this, for example), not our differences that matter most. And the really crucial, the basic similarity, is simply this: we are each of us the experiencing subject of a life, a conscious creature having an individual welfare that has importance to us whatever our usefulness to others. We want and prefer things, believe and feel things, recall and expect things. And all these dimensions of our life, including our pleasure and pain, our enjoyment and suffering, our satisfaction and frustration, our continued existence or our untimely death — all make a difference to the quality of our life as lived, as experienced, by us as individuals. As the same is true of those animals that concern us (the ones that are eaten and trapped, for example), they, too, must be viewed as the experiencing subjects of a life, with inherent value of their own.

**Regan (1989) The Case for Animal Rights, in *Animal Rights and Human Obligations***

What implications does the rights view have for animal experimentation? The answer to this question will depend on whether we are prepared to go along with Regan and ascribe rights to animals. If we refuse to take this step, rights theory will have little to tell us about animal research. However, if we allow that animals possess intrinsic dignity and have rights, various things will follow. To begin with, the balancing of human benefits against animal suffering that has been central in our discussion so far becomes, to some extent, a background issue. No benefit can justify disrespect for the rights of an individual — human or animal — so where an experiment violates an animal's rights, there is no reason to look for its expected benefits. To find out whether an experiment is morally justified, we need only ask whether it is respectful and preserves the animal's dignity. The implications of this way of looking at matters are radical, as Regan[23] explains:

> …Having set out the broad outlines of the rights view, I can now say why its implications for farming and science, among other fields, are both clear and uncompromising. In the case of the use of animals in science, the rights view is categorically abolitionist. Lab animals are not our tasters; we are not their kings. Because these animals are treated routinely, systematically, as if their value were reducible to their usefulness to others, their rights are routinely, systematically, violated. This is just as true when they are used in trivial, duplicative, unnecessary, or unwise research as it is when they are used in studies that hold out real promise of human benefits…The best we can do when it comes to using animals in science is — not to use them. That is where our duty lies, according to the rights view.

**Regan (1989) The Case for Animal Rights, in *Animal Rights and Human Obligations***

This view is radical enough to merit repeating. It does not matter that an experiment will cause only minor harm to the animals it involves. It does not matter that this experiment is of extraordinary importance to humanity at large. Animal experiments are simply unacceptable because they treat animals as means to an end.

Categorical abolitionism of this sort probably goes further in its attempt to limit the utilitarian trade-offs than most of us would consider necessary. After all, weighing costs against benefits is part of our daily life. Every day, we balance outcomes and seek what is best overall in private decisions that involve friends, family, and ourselves. We expect others — for example, employers and govern-ment bodies — to do the same. In all this, we accept that we are not treated, and do not treat others, purely as ends. On the other hand, most people would presumably allow that certain rights are sacrosanct, and that there are limits to the extent to which an individual can be sacrificed for an overall benefit. Only (what we might call) a *moderate* rights view is likely to command widespread acceptance.

How would such a moderate view apply to animal research? The detail would depend on what rights we take to be fundamental. The right to life — or more accurately, the right not to be killed — is often regarded as basic. Curiously, however, this does not appear to be a basic right that people would ascribe to animals: after all, most of us happily eat animals that have been killed just for this purpose. Something like a right to protection from suffering, or significant suffering, seems to be much more promising. We might agree that all animals should be protected from suffering if this involves intense or prolonged pain or distress that the animal cannot control. We might conclude that such suffering in experimental conditions is always unjustified. This would be consistent with the toleration of some balancing of animal and human interests.

In essence, this is the view on which animal research legislation is based in some countries, such as the U.K. and Denmark. Through this combination of utilitarianism and the moderate rights view, animal research that promises to deliver human benefits is allowed as long as the animals are guaranteed protection from serious pain and distress.

## TWO CONCERNS ABOUT THINGS OTHER THAN SUFFERING

We have seen that the notion of individual suffering is central to utilitarianism, and that a moderate rights theory with application to animal research might also need to refer to animal pain and distress.

However, some aspects of animal experimentation appear to raise ethical issues even though they need not involve suffering of any kind. In this section, we shall discuss two practices of this kind: genetic modification and, in particular, transgenesis and euthanasia.*

Turning to the first of these, genetic interventions, and particularly the production of transgenic animals, concern many observers. Often the animals themselves do not suffer as a result of having their genome modified. Even so, people feel uneasy at the thought of introducing spinach genes into the pig genome. In this unease, there seems to be a sense that the "naturalness" of an animal is important: a pig should be a pig and nothing else. In trying to capture this attitude, the philosopher Bernard Rollin[24] refers to the "telos" of an animal. Using this notion, we might say that the creation of transgenic animals is morally unacceptable because animals have the right to have their telos respected. Against this, it has been suggested that the idea that there is a genuine ethical issue here is an illusion: human beings altered the genetic makeup of various animals through breeding long before anybody knew how to manipulate genes directly, and the only significant difference between traditional breeding techniques and direct manipulation of the genome is that the latter is faster. This is a fascinating, if rather intractable, issue. We cannot devote any more space to it in the present chapter, but interested readers might like to consult Rollin,[24] Sandøe et al.,[25,26] and Appleby.[27]

We now comment briefly on animal euthanasia. At the end of an experiment, animals are often killed in a manner that, as far as possible, prevents their feeling any fear or pain. Properly carried out, euthanasia involves no suffering, but some people feel that the killing of sentient creatures in this way nevertheless raises ethical issues. Is this correct? Should we worry about euthanasia?

To answer this question satisfactorily, we would need to stand back from the animal question and ask in general terms why killing is unethical. This question is too complex to deal with here, but let us just sketch one relevant line of thought. It has been suggested that killing an individual is wrong because it prevents him or her from realizing future desires. This account of the wrongness of killing can be used to defend the killing of animals because it is plausible that animals lack a concept of the future and therefore possess only immediate desires. On the other hand, if we can justify the killing of animals on the basis of their limited cognitive capacities, it would appear that we can also justify the killing of human beings, such as infants or mentally deficient adults, with similar cognitive limitations. One way to block this unwelcome implication is to say that killing a person tends to cause more harm than killing an animal because the person will be mourned by relatives and friends, and because knowledge about the killing will provoke fear among other people. For more extensive discussion of the ethics of killing animals, we refer readers to References 28 through 31.

## SOCIETY AND THE ETHICAL DILEMMA

About any given animal research project — whether it involves experimental techniques that are standard practice or puts animals to novel use — the scientist can always ask: would it be morally acceptable to use animals in this way? Ultimately, in individual cases, researchers and research teams will make up their own minds about this question, and people will undoubtedly come to different conclusions. This need not be a problem in a pluralistic, democratic society. Nevertheless, society needs legislation and professional codes of practice that the majority of citizens can agree on, and it is therefore necessary to find a minimal consensus among the different views. In a democracy, compromise is normally the best way of achieving consensus. (If this seems too obvious to be worth mentioning, it should be borne in mind that under tyranny, consensus is generally reached without compromise.) In this chapter, we have tried to set out the issues that inform this compromise.

Broadly speaking, the compromise towards which much of what we have said points is one in which animal experimentation is held to be acceptable where, and only where, it is the case both that substantial human benefits are at stake and that animal suffering is minimized. From the perspective of this compromise position, research projects causing great suffering in which pain relief cannot be offered are unacceptable even if they are crucial for the advancement of knowledge. Equally, as was urged in the previous edition of this book, "If doing without a harmful animal experiment involves only a slight risk or loss to humans, we should do without it."[32] "Great suffering" and "slight risk" are relative terms;

---

* Notice that, because they do not involve animal suffering, these issues take us outside the ethical dilemma represented in claims 1 through 4.

their definition is itself an ethical decision, and one that must be made before the acceptability of any research project can be determined.

The three Rs proposed by Russell and Burch[22] — the replacement of existing experiments with animal-free alternatives, or reductions in the number of animals used, or refined methods that cause animals less suffering — will help animal researchers to ensure that their work is acceptable. In actual fact, this threefold rubric often figures in codes of practice governing animal research, and in most Western countries, at least, scientists are required by law to apply something like it. However, if Russell and Burch's recommendations are to be effective, scientists will have to take them seriously. Scientists will need to ensure that they are up-to-date with developments in experimental methodology, bearing in mind, particularly, the design of alternative methods. Again, those responsible for the housing and daily care of laboratory animals will ideally be equipped with a thorough understanding of the behavioral and physiological needs of the relevant animal species and know how to implement various forms of environmental enrichment. In view of this, it is not stretching matters unduly to say that the study of laboratory animal science is a moral duty for researchers and others involved in animal experiments.

Perhaps the main thing is to keep the channels of communication open. In the twenty-first century, transparency and accountability are watchwords. They are expected, and indeed demanded, in most areas of collective human endeavour. Thus, faced with questions about their work, the worst thing animal researchers can do is try to shut the enquirer out. A society in which animal experimentation enjoys a secure, unchallenged role is likely to be one in which there is open dialogue between the scientists and lay observers — somewhat paradoxically, it is likely to be one in which scientists welcome challenges from all sides.

## ETHICS COMMITTEES

Finally, we need to say something about regulation. Animal research is not like sex. Many people think that certain sexual practices are immoral or distasteful, but it is generally acknowledged that what consenting adults do in their bedrooms is up to them. Perhaps, in good part, because animals are not "consenting adults" and have limited ability to defend their own interests, we do not take the same attitude to animal experiments. The recognition that animal treatment in the laboratory raises ethical questions leads quickly to a demand for regulation. This demand is surely reasonable. First, it is in the interest of society to ensure that ethical norms that prevail among us are adhered to. Second, if a research proposal raises moral issues, any decision about its acceptability must be made by a third party — an individual or body of individuals that is not involved in the relevant project and does not stand to gain from its completion. The researcher has a vested interest in his proposed investigation, since his career may depend on the results. Certainly, it is up to him or her to present the ethical and scientific case for the line of experimentation being proposed, but the decision about acceptability must be made by people who are independent and who can represent society as a whole.

For obvious reasons, verdicts on acceptability will ideally be based on extensive knowledge of the scientific area in question, the relevant animal welfare parameters, and ethics. In the majority of cases, a single individual is highly unlikely to possess such knowledge, and so the best way to ensure that decisions are properly informed is by forming an ethics committee. Such committees are indeed now mandatory in many countries. They usually employ researchers, animal specialists, people with training in law and ethics, animal advocates, and representatives of the general public. And they are not, in general, enemies of the researcher. They can, for example, be quite helpful in making suggestions about how to minimize animal distress.

Naturally, each member of the ethics committee will look at the ethical dilemmas research proposals raise in his or her own personal way. Committee members will agree, however, both that it is important to minimize any harm to animals and that animal experiments need to be justified by (primarily) human benefits. Once these broad principles are agreed upon, the committee needs to find a common language in which to describe animal costs and human benefits. However, if there are significant negative effects on animal welfare, the committee should — perhaps in dialogue with the researchers — find out whether it is possible to derive the same benefits in a way that will have a less harmful impact on the animals. And, at the end of the day, the committee must decide whether or not the experiment is acceptable. For interesting attempts to develop instruments for decision-making in ethics committees, see References 3, 33, and 34.

When they play a proactive role in the development of animal research projects, competent ethics committees function as a crucial interface between the research community and society in general. Researchers, whose future may depend on the ethical approval of the committee, are forced to present their projects in an accessible way and to think of alternatives. Because of this, the committee does not remove ethical responsibility from scientists — rather, it helps them to ensure that their activities are transparent and challenges them to proceed in an ethically sound way.

## CONCLUSION

We have suggested that the ethical dilemma at the heart of animal research can be captured in four straightforward propositions. How did these help? The idea was that because these propositions are in conflict, we can make progress in thinking about the ethics of animal experimentation by asking which of them is incorrect, or at least an overstatement. We looked at three ethical theories — contractarianism, utilitarianism, and the rights view — and noticed that these generate different conclusions about the rights and wrongs of animal experimentation. Potentially, the contractarian viewpoint seems the most liberal. At the other end of the scale, the rights view places severe restrictions on animal use, restrictions essentially the same as those we would expect to govern the use of human "guinea pigs." Midway between these two approaches lies utilitarianism. Within utilitarianism, animal suffering is treated as no less important, morally, than human suffering. However, looking at the overall balance of suffering and benefit, the utilitarian concludes that research projects in which animal suffering is minimized and the human dividends are substantial are best permitted.

In the course of the discussion, we have noted several ways in which, in practice, the researcher can keep animal suffering to a minimum: by devising experiments that use no animals at all, by using fewer animals, and by refining experimental techniques so that the pain or distress they cause is lessened. We have also observed that, while there is absolutely no doubt that animal research has delivered significant human benefits, in some research, the anticipated benefits do not justify continuance — for example, because they are too unimportant, or because there is a good chance that they will not be secured.

These days, more and more people take an interest in what is happening in the laboratory, and a substantial number of people have grave concerns about animal welfare. As a result, scientists who do animal research often need to explain and justify their work to others — to friends and colleagues or, more formally, to funding or ethics committees. In our view, researchers are better equipped to account for their methods when they understand both the ethical dilemma those methods pose and the ethical theories that lie behind this dilemma. In this chapter, we have tried to set out a theoretical framework that, we hope, will help interested readers to acquire this understanding.

## ACKNOWLEDGMENTS

The authors are grateful to Merel Ritskes Hoitinga and Gerald Van Hoosier for comments and suggestions that improved this chapter.

## REFERENCES

1. Kiple, K.F. and Ornelas, K.C., Experimental animals in medical research: a history, in, Frankel, P.E. and Paul, J., Eds., *Why Animal Experimentation Matters: The Use of Animals in Medical Research*, New Brunswick and London: Transaction Publishers, 2001.
2. Research Defence Society, 2002, www.rds-online.org.uk.
3. Smith, J.A. and Boyd, K.M., *Lives in the Balance: The Ethics of Using Animals in Biomedical Research*, Oxford, 1991.
4. European Union, Second report on the statistics on the number of animals used for experimental and other scientific purposes in the member states of the European Union, COM, (99)191, 1999.
5. Tuli, J.S., Smith, J.A., and Morton, D.B., Corticosterone, adrenal, and spleen weight in mice after tail bleeding, and its effect on nearby animals, *Laboratory Animals*, 29, 90–95, 1993.
6. Gray, S. and Hurst, J.L., The effects of cage cleaning on aggression within groups of male laboratory mice, *Animal Behaviour*, 49, 821–826, 1995.

7. Broom, D.M., Animal welfare defined in terms of attempts to cope with the environment, *Acta Agricultura Scandinavicae*, Sect. A, *Animal Science*, Suppl. 27, 22–28, 1996.
8. Duncan, I.J.H., Animal welfare defined in terms of feelings, *Acta Agricultura Scandinavicae*, Sect. A, *Animal Science*, Suppl. 27, 29–35, 1996.
9. Manser, C.E., The assessment of stress in laboratory animals, *RSPCA*, 1992.
10. Broom, D.M. and Johnson, K.G., *Stress and Animal Welfare*, London: Chapman and Hall, 1993.
11. Mench, J.A. and Mason, G.J., Behaviour, in Appleby, M.C. and Hughes, B.O., Eds., *Animal Welfare*, Oxon: CAB International, 1997.
12. Terlouw, E.M.C., Schouten, W.G.P., and Ladewig, J., Physiology, in Appleby, M.C. and Hughes, B.O., Eds., *Animal Welfare*, Oxon: CAB International, 1997.
13. Clark, J.D., Rager, D.R., and Calpin, J.P., Animal well-being, *Laboratory Animal Science*, 47, 564–597, 1997.
14. Spinelli, J.S. and Markowitz, H., Prevention of cage-associated distress, *Laboratory Animals*, 14, 19–24, 1985.
15. Würbel, H., Ideal homes? Housing effects on rodent brain and behaviour, *Trends in Neuroscience*, 24, 207–211, 2001.
16. Sherwin, C.M., Comfortable quarters for mice in research institutions, in Reinhardt, V. and Reinhardt, A., Eds., *Comfortable Quarters for Laboratory Animals*, Washington: Animal Welfare Institute, 2002.
17. Olsson, I.A.S. and Dahlborn, K., Improving housing conditions for laboratory mice: a review of "environmental enrichment," *Laboratory Animals*, 36, 243-270, 2002.
18. Rollin, B., *The Unheeded Cry: Animal Consciousness, Animal Pain, and Science*, Oxford and New York: Oxford University Press, 1989.
19. Bermond, B., A neuropsychological and evolutionary approach to animal consciousness and animal suffering, *Animal Welfare,* 10, Supplementum S31-S40, 2001.
20. Aldhous, P., Coughlan, A., and Copley, J., Let the people speak, *New Scientist*, 22 (May), 26–31, 1999.
21. Singer, P., All animals are equal, in Regan, T. and Singer, P., Eds., *Animal Rights and Human Obligations*, Englewood Cliffs, NJ: Prentice Hall, 1989.
22. Russell, W.M.S. and Burch, R.L., *The Principles of Humane Experimental Technique*, London: Methuen, 1959.
23. Regan, T., The case for animal rights, in Regan, T. and Singer, P., Eds., *Animal Rights and Human Obligations*, Englewood Cliffs, NJ: Prentice Hall, 1989.
24. Rollin, B., Animal welfare, science, and value, *Journal of Agricultural and Environmental Ethics*, 6 (Suppl 2), 44–50, 1993.
25. Sandøe, P., Forsman, B., and Kornerup, Hansen A., Transgenic animals: the need for ethical dialogue, *Scandinavian Journal of Laboratory Animal Science*, 1, 279–285, 1996.
26. Sandøe, P., Nielsen, B.J., Christensen, L.G., and Sørensen, P., Staying good while playing God – the ethics of breeding farm animals, *Animal Welfare*, 8, 313–328, 1999.
27. Appleby, M.C., Tower of Babel: variation in ethical approaches, concepts of welfare, and attitudes to genetic manipulation, *Animal Welfare,* 8, 381–390, 1999.
28. Johnson, E., Life, death, and animals, in Regan, T. and Singer, P., Eds., *Animal Rights and Human Obligations*, Englewood Cliffs, NJ: Prentice Hall, 1989.
29. Cigman, R., Why death does not harm animals, in Regan, T. and Singer, P., Eds., *Animal Rights and Human Obligations*, Englewood Cliffs, NJ: Prentice Hall, 1989.
30. Regan, T., Why death does harm animals, in Regan, T. and Singer, P., Eds., Animal *Rights and Human Obligations*, Englewood Cliffs: Prentice Hall, 1989.
31. Rodd, R., Biology, *Ethics and Animals*, Oxford: Clarendon, 1990.
32. Sandøe, P., Animal research and ethics, in Svendsen, P. and Hau, J., Eds., *Handbook of Laboratory Animal Science, Volume 1, Selection and Handling of Animals in Biomedical Research*, Boca Raton, FL: CRC Press, 1994.
33. Stafleu, F.R., Tramper, R., Vorstenbosch, J., and Joles, J.A., The ethical acceptability of animal experiments: a proposal for a system to support decision-making, *Laboratory Animals*, 33, 295–303, 1999.
34. Van Hoosier, G.L., Principles and paradigms used in human medical ethics can be used as models for the assessment of animal research, *Comparative Medicine*, 50, 103–105, 2000.

# CHAPTER 3

# An Overview of Global Legislation, Regulation, and Policies on the Use of Animals for Scientific Research, Testing, or Education

Kathryn Bayne and Paul deGreeve

## CONTENTS

0-8493-1086-5/03/$0.00+$1.50
© 2003 by CRC Press LLC

## INTRODUCTION

With the steady increase in international collaborations in animal research, testing, and teaching, there is a concomitant increase in a need to be familiar with the standards of oversight in different countries. This task has been made difficult by the lack of a single, concise compilation of the laws, rules, regulations, policies, and guidelines in use around the world, and by their constant evolution toward a more restrictive stance. The goal of this chapter is to provide an introduction to the governance of animal use in a variety of countries, with more detail provided for certain parts of the world than others, due to the greater complexity of standards to be met in those countries.

## EUROPE

The first legislation concerning animal experimentation in Europe was enacted in the United Kingdom in 1876 in the form of the Cruelty to Animals Act. This statute emerged as a result of a long debate between scientists and animal-rights activists. The U.K. was thus the first and, for many years, the only country with legislation protecting animals used for scientific purposes. However, concern for laboratory animals has since grown. In the 1980s, two important documents controlling the use of animals in experiments were issued in Europe. In 1985, after several years' discussion, the 26 countries of the Council of Europe (a political intergovernmental organization set up in 1949) in Strasbourg reached agreement on the Convention for the Protection of Vertebrate Animals used for Experimental and Other Scientific Purposes (ETS 123).[1] A convention, however, is not a binding document and has no legislative force. Rather, a convention should be considered as soft law. A convention becomes effective when it is signed and ratified by a member state, i.e., when the parliament or government of a country has approved the instrument. Then, the member state becomes a party to that convention and is legally bound under international law to implement its provisions.

Convention ETS 123 contains the provision that parties should hold multilateral consultations to examine the progress of its implementation and the need for revision or extension of any of its provisions on the basis of new facts or developments. During the last decade, three multilateral consultations have been held. At these multilateral consultations, parties adopted the following resolutions:

Resolution on the interpretation of certain provisions and terms of the convention (1992): This resolution gives some precision on the scope of the convention with regards to animals carrying harmful genetic modifications, and a common interpretation of certain terms used in the statistical tables. Furthermore, it provides sample statistical tables used to facilitate communication of statistical information.

Resolution on education and training of persons working with animals (1993): This resolution presents guidelines for topics included in educational and training programs for four categories of persons working with laboratory animals (from animal caretaker to animal science-specialist).

Resolution on the acquisition and transport of laboratory animals (1997): This resolution states criteria for granting exemption for the acquisition of animals in supplying or breeding establishments that are not registered. It also contains principles of best practices in the packing and transport of laboratory animals, with particular emphasis on the necessary competence and responsibilities of the persons involved to ensure a proper transport with the shortest delays.

Resolution on the accommodation and care of laboratory animals (1997): This resolution contains guidelines, complementary to those of Appendix A of the convention, for the housing and care of laboratory animals, as defined in the light of scientific evidence and practical experience (e.g., environmental enrichment and group housing).

In 1986, the Directive for the Protection of Vertebrate Animals used for Experimental and Other Scientific Purposes (86/609/EEC)[2] was adopted by the Council of Ministers of the European Economic Community. This document was based upon the convention, although its text is more concise and its requirements more stringent. The aim of this directive is to ensure that where animals are used for experimental or other scientific purposes, the provisions laid down by law, regulation, or administrative provisions in the member states for their protection are approximated so as to avoid affecting the establishment and functioning of the common market, in particular, by distortions of competitions or barriers to trade. All EEC member states are compelled to implement the provisions of the EEC directive through their national legislation. These provisions must be seen, however, as minimum requirements.

All laboratory-animal protection legislation is based on the premise that, under certain conditions, it is morally acceptable to use animals for experimental and other scientific purposes. Most laws, however, contain provisions ensuring that the number of animals used is kept to a minimum. In addition, most regulatory systems have the following general objectives:

- To define legitimate purposes for which laboratory animals may be used
- To ensure competence of all laboratory personnel and researchers
- To limit animal use where alternatives are practicably available
- To prevent unnecessary pain or distress to animals
- To provide for the inspection of facilities and procedures
- To ensure public accountability

In some regulatory systems, the principles of the 3Rs of Russell and Burch[3] have been implemented explicitly.

## The EC Directive

### Scope

The provisions of the Directive apply to vertebrate animals used in experiments that are likely to cause pain, suffering, distress, or lasting harm. In several member states, the use of an invertebrate animal has been considered, too. For example, in the U.K., the use of *Octopus vulgaris* as an experimental animal is covered by the Animals (Scientific Procedures) Act. In the Netherlands, if, on the basis of scientific evidence, it is likely that an invertebrate species can experience discomfort, this species could be designated by an order to be covered by the Experiments on Animals Act. However, until now, no invertebrates have been so designated. The directive also covers the development of genetically modified animals likely to suffer pain and distress.

The killing or marking of an animal using the least-painful method is *not* considered an experiment. However, in France, Sweden, and the Netherlands, the definition of *procedure* is broader and includes the killing of an animal without any previous intervention.

The Directive is restricted in its application to experiments undertaken for the development and safety testing of drugs and other products, together with the protection of the natural environment. However, in 1987, the European Parliament passed a resolution stating that the provisions of Directive 86/609/EEC should also apply, through national legislation, to all animal experiments, including those undertaken for basic research and educational purposes.

In 2000, the European Commission became a party to the convention ETS 123. Therefore, a second amendment to Directive 86/609/EEC had to be prepared to bring the Directive in line with the convention. To that end, in 2001, the European Commission invited the National Competent Authorities to consider revising the scope of the Directive to include, for example, the use of animals for training purposes, forensic inquiries, and routine production; animals to be killed for tissue and organs; and the commercial breeding of animals destined for experiments (housing and care). Others issues to be dealt with were specific provisions concerning the use of nonhuman primates, dogs, and cats; specific conditions relating to the use of transgenics and cloning; authorization and inspection; statistical requirements; ethical review of protocols; and applications of the 3Rs. At this time, legislation on animal experimentation is under revision in several member states.

Most national laws are put into operation by a government-controlled "authority," as described by the directive. The control mechanism may rest with the authority itself, through a central licensing system, such as that operating in, for example, the United Kingdom, or may be (partially) deployed at an institutional level, such as through institutional (ethics) committees.

## Accommodation and Animal Care

Article 5 of the Directive contains provisions to ensure that animals are treated humanely not only prior to, but also during and after, any experimental procedure. Detailed guidelines for the implementations of these provisions are set out in Annex II of the directive. These guidelines are based principally upon common laboratory practices. They can be amended as new scientific or other evidence emerges with improved methods for the housing and care of animals. In 1997, the parties to the Convention ETS 123 started a process of revising the guidelines for the housing and care of laboratory animals. It can be concluded that the revised parts will contribute to the well-being of these laboratory animals.

## Competence

Article 5 states that a competent person must oversee the well-being and state of health of the animals. Article 19 stipulates that a veterinarian or other competent person should be charged with advisory duties in relation to the well-being of animals. The provisions on competence warrant special attention. Laws and regulations are poor tools if not based upon an understanding of what constitutes humane and responsible animal use. Therefore, additional education and training provide the opportunity for gaining this understanding and also for an evaluation of the ethical considerations.

Article 7 of the Directive states that animal experiments should be performed solely by a person considered to be competent or under the direct responsibility of such a person. This provision is amplified by Article 14, which states that persons undertaking experiments, taking part in procedures, or caring for experimental animals (including supervision) should have appropriate education and training. It is essential that the people involved in the design and conduct of experiments should have received an education in a scientific discipline relevant to the experimental work. They also need to be capable of handling and taking care of laboratory animals. Each member state must specify how the provision of competence is to be implemented within national legislation. The Federation of European Laboratory Animal Science Associations (FELASA)[4] has prepared a proposal concerning educational and training requirements for scientists. Some countries have already introduced strict regulations regarding competence based upon these FELASA guidelines.

## Alternatives to Animal Experiments

Article 7 of the Directive not only deals with competence, but also with alternatives to animal use. Performance of an experiment is not permissible if the result can be reasonably and practically obtained without the use of animals. If there is no alternative to animal use, then animals with the lowest degree of neurophysiological sensitivity (or the least capacity for suffering), compatible with the scientific objective, must be selected. Animals taken from the wild may not be used unless other animals would not fulfill the aims of the experiment. Furthermore, all experiments must be designed to minimize distress and suffering.

The directive clearly states that the EC and member states should encourage research into the development and validation of alternatives. In 1991, the European Centre for the Validation of Alternative Methods (ECVAM) was established. In several member states, a national institute has been established for the development and application of alternative methods. These institutes are working closely together, with ECVAM as the coordinating body.

## Anesthesia

All experiments must be carried out under general or local anesthesia (Article 8), unless anesthesia is judged to be more traumatic than the experiment itself, or is incompatible with the aims of the experiment. If anesthesia is not possible, then pain, distress, or suffering must be limited, and analgesics or other appropriate methods should be used. No animal should be subjected to severe pain, distress, or suffering.

## Euthanasia

At the end of the experiment, a veterinarian or other competent person must decide whether the animal should be kept alive or be humanely killed. No animal is to be kept alive if it is likely to remain in permanent pain or distress, or if its well-being is otherwise jeopardized (Article 9). It is not permissible to use animals more than once in experiments entailing severe pain, distress, or equivalent suffering.

## Registration

There is an obligation to notify the (governmental) authority in advance about the proposed use of animals in experiments and who will be conducting them. If an animal is expected to experience severe pain that is likely to be prolonged, the experiment must specifically be declared and justified to, or specifically authorized by, the authority. Such an experiment is permitted only if it is of sufficient importance in meeting the essential needs of man or animal.

## Statistics

The authority must collect information on the total number of animals used and statistics detailing the number used for specific purposes. As far as possible, this must be made available to the public in the form of published statistics. In addition, the information has to be sent to the European Commission.

In 1997, after several years of negotiation, the EC and the member states reached agreement on the (scientific) purposes for which data on animal use should be collected. Specific EU tables have been designed and distributed for the collection of such data. From 1999 on, data on animals used for scientific purposes is collected in all member states. Every two years, the EC sends a report on animals used for scientific purposes in Europe to the European Parliament.

## Supply of Animals

Only establishments approved by the authority in each member state are allowed to breed or supply animals for research. Such establishments must keep records of the number and species of animals sold or supplied, and the names and addresses of the recipients. Dogs, cats, and nonhuman primates must be supplied with an individual identification mark.

## Animal Facilities

Establishments for animal use must be registered with or approved by the authority. Each user establishment must have sufficient numbers of trained staff and provision for adequate veterinary support. Only animals bred within the animal facility or from authorized breeding or supplying establishments may be used. The use of stray animals is not allowed. Records must be kept of all animals used.

## *Inspectorate*

In most countries, the authority for supervising compliance with the regulations is a governmental inspectorate. The inspectors are mainly veterinarians or biologists with experience in research and training in laboratory-animal science. For purposes of public accountability, some member states report to Parliament each year on the animals used for the life sciences.

## *Ethics Committees*

There is no specific provision in the Directive requiring the establishment of such committees. However, during the last decade, as a result of animal welfare concerns of the general public, the ethical review of proposed animal experiments has been considered as the next step in improving animal welfare. An assessment of likely benefit and potential for animal suffering is an essential part of the review process. In Belgium, Sweden, and the Netherlands, the ethical review process has a legal basis. In Denmark, the inspectorate has the right to reject an application for a project if the suggested activity is judged not to be of "essential value or benefits." In France, the United Kingdom, and Finland, institutional committees are specifically dedicated to review the ethical aspects of animal experimentation prior to the commencement of the animal experiment.

In Europe, several scientific organizations have prepared guidelines for the ethical use of animals in research. The European Science Foundation (ESF), an association of 67 member organizations devoted to scientific research in 24 European countries, has stipulated that member organizations will adopt guidelines for the ethical use of animals in research. According to a survey done by the ESF, there is already an ethical review committee in most member states. The competence of the committee members, the committee composition, and the committee members' responsibilities vary considerably. It can be expected that the ethical review process will be adopted by the European Commission and, therefore, will be mandatory for all member states in the near future.

Because the provisions of the directive are considered minimum standards, member states are free to regulate specific issues more strictly. As a consequence, considerable differences among legislative procedures can occur. Furthermore, in some member states (e.g., Finland, Germany, and Portugal), the Minister of Agriculture is responsible for the protection of laboratory animals. In other member states, the responsible minister can be the Home Office (U.K.), the Minister of Justice (Denmark), the Minister of Science and Education (France), or the Minister of Public Health (The Netherlands).

## NORTH AMERICA

## Canada

There is no federal legislation pertaining to the use of animals in research, testing, or education in Canada. However, some provinces have established provincial laws, and there is a voluntary assessment program for the care and use of experimental animals. In addition, although there is no federal requirement to participate in the Canadian Council on Animal Care (CCAC) assessment program, some funding agencies require grantee institutions to comply with CCAC guidelines, and contractors performing work for the federal government are required to adhere to CCAC guidelines, as specified in the Public Works and Government Services Canada Standard Acquisition Clauses and Conditions Manual, Section 5, Subsection A, Clause A9015C: Experimental Animals. The CCAC, founded in 1968, places responsibility for humane animal care and use with the animal-care committee at each institution, which is granted specific authority and provided with terms of reference under which it operates (e.g., membership, authority, responsibilities, and functioning). The CCAC's mission is "to act in the interests of the people of Canada to ensure, through programs of education, assessment, and persuasion, that the use of animals, where necessary, for research, teaching, and testing employs optimal physical and psychological care according to acceptable scientific standards, and to promote an increased level of knowledge, awareness, and sensitivity to relevant ethical principles." Thus, the CCAC has two principal functions: the development of guidelines and policies to govern experimental animal care and use; and to monitor compliance with those guidelines and policies. The CCAC is an independent organization and receives funding from the Medical Research Council (MRC) and the Natural Sciences and Engineering Research Council (NSERC).

The CCAC establishes guidelines for its certified institutions to follow, currently contained in the two-volume *Guide to the Care and Use of Experimental Animals*. Adjunct guidelines address topics such as animal-use protocol review, transgenic animals, selecting appropriate endpoints, and developing an animal user-training program. The CCAC also has established several policies, the earliest of which addresses the ethics of animal research (1989). Other subjects covered by policies include review of scientific merit, social and behavioral requirements of experimental animals, acceptable immunological procedures, and categories of invasiveness.

On-site assessments using panels of experts from the animal care and use community and a representative nominated by the Canadian Federation of Humane Societies are conducted triennially. An institution is deemed to be in compliance if the CCAC report prepared by the assessment panel and approved by the assessment committee — a standing committee composed of at least four council members — contains only Regular, Minor, and/or Commendatory recommendations, and the institution submits an implementation report for any regular recommendations that are judged to be satisfactory. Institutions that have been found to be in Compliance or Conditional Compliance, will receive a CCAC Certificate of Good Animal Practice™.[7] If the CCAC report contains major or serious recommendations whose corrections do not require verification by an on-site reassessment, but rather can be verified through documentation, and the institution provides to the CCAC an implementation report that is judged to be satisfactory, then compliance is maintained. Increasingly problematic status-determinations are conditional compliance, probation, and noncompliance. Recently, the MRC and NSERC revised the policy on compliance and noncompliance such that all funding agencies and government ministries and departments are notified of an institution's noncompliance with CCAC guidelines (2000).[5] Sustained nonconformance with CCAC guidelines and policies can ultimately result in withdrawal of all animal-based research funding to the institution.

## The United States

In the United States, oversight of animal care and use for research, testing, and teaching is achieved by numerous laws, regulations, policies, and guidelines from two principal governmental organizations — the United States Department of Agriculture (USDA) and the U.S. Public Health Service (PHS)*. Other guidance may be derived from scientific panels and endorsed by the government as required standards. Federal laws are annually compiled and categorized into their respective subjects (e.g., agriculture) and published as the United States Code (USC). The USC includes a discussion of the intent of Congress for establishing the law and any interpretations from the courts. Regulations are promulgated to enforce the corresponding law. Proposed regulations are published in the Federal Register for public comment. After the responsible agency reviews and addresses the public comments, the regulations are again published in the Federal Register in final format and then incorporated into the Code of Federal Regulations (e.g., 9 CFR).[6,7] In general, laws address two specific areas: animal welfare and procurement, and animal importation and shipment.

### U.S. Department of Agriculture

Federal laws for the humane treatment of animals have been in place since 1873, when Congress passed a law governing the treatment of livestock during shipment for export. The law was called the "28-Hour Law," after the maximum length of time animals could be transported before receiving food, water, and rest.[8] This law was later repealed, and a new "28-Hour Law" was passed in 1906, which is still in effect today. However, the first federal law to protect non-farm animals was not passed until 1966 and was called the Laboratory Animal Welfare Act, administered by the Animal and Plant Health Inspection

---

* U.S. Congress, Office of Technology Assessment, 1986, Federal regulation of animal use, pp. 275-301, in *Alternatives to Animal Use in Research, Testing, and Education*, U.S. Government Printing Office: Washington, D.C.

U.S. Congress, Office of Technology Assessment, 1986, State regulation of animal use, pp. 305-331, in *Alternatives to Animal Use in Research, Testing, and Education*, U.S. Government Printing Office: Washington, D.C.

Service (APHIS), USDA. At the time, this law was primarily directed at dog and cat dealers, as it required that individuals or corporations that bought or sold dogs or cats for laboratory activities be licensed and adhere to certain minimum standards for the care of animals, and that users of cats or dogs for research register with the USDA and also meet minimum standards for animal care. For animal users, the law applied only to animals held prior to or after the laboratory activity. Interestingly, the New York Anticruelty Bill of 1866 addressed the use of animals in research and predated federal interest in this subject.[9]

The Laboratory Animal Welfare Act of 1966 was amended in 1970, 1976, 1985, and 1990 to broaden coverage of the law. Public Law 91–579, the Animal Welfare Act of 1970, increased the species of animals covered under the law to include all warm-blooded animals and increased the scope of applicability of the law to include the time animals were held in the facility. Specifically exempted were horses not used in research and agricultural animals used in food and fiber research, retail pet stores, state and county fairs, rodeos, purebred cat and dog shows, and agricultural exhibitions. Public Law 94–279, the Animal Welfare Act Amendments of 1976, included common commercial carriers, such as airlines, under the law, which subsequently led to the development of standards for shipping containers and conditions of shipment. Public Law 99–198, the Improved Standards for Laboratory Animals Act, added several new provisions to the law, including: minimization of animal pain and distress; consideration of alternatives to painful procedures; consultation with a doctor of veterinary medicine for any practice that could cause pain to animals; limitation on conducting more than one major survival surgery on an animal; establishment of an Institutional Animal Care and Use Committee (IACUC) to provide oversight of the animal care and use program and facilities; provision of specific training to personnel; provision of exercise to dogs; and a stipulation to promote the psychological well-being of nonhuman primates. The most recent amendment to the Animal Welfare Act, Public Law 101–624, Food, Agriculture, Conservation, and Trade Act of 1990, Section 2503, Protection of Pets, established a holding period for dogs and cats at shelters and other holding facilities prior to sale to dealers. The law also requires dealers to provide written certification to the recipient regarding the background of each animal.

Of increasing debate has been the exclusion of rats and mice from the Animal Welfare Act. The 1970 amendment to the Animal Welfare Act stated that an animal was defined as: "Any live or dead dog, cat, monkey (nonhuman primate animal), guinea pig, hamster, rabbit, or other such warm-blooded animal as the Secretary may determine is used, or is intended for use, for research, testing, experimentation, or exhibition purposes, or as a pet." In this way, the Secretary of the Department of Agriculture was given the authority to determine which animals would be covered by the Act. In 1977, the USDA promulgated regulations that specifically excluded rats, mice, and birds from the definition of "animal." To date, rats (*Rattus rattus*), mice (*Mus sp.*), and birds remain excluded from coverage of the Act.

Since the 1966 Act, Congress has vested the USDA with both promulgation and enforcement authority. The USDA is required to conduct unannounced annual inspections of research facilities, with follow-up inspections until any cited deficiency has been corrected. Exempt from this provision are federal research facilities. Research, intermediate handlers, and common carriers are required to register with the USDA, while animal dealers and exhibitors must be licensed. Research facilities and U.S. government agencies are required to purchase animals only from licensed sources, unless the source is exempted from obtaining a license. Failure to comply with regulatory requirements, despite formal notification of the violation and an opportunity to effect a correction, can result in fines levied on the facility, suspension of authority to operate, and even permanent revocation of the facility's license to operate. Penalties can be imposed on the facility such that some portion of the fine is mandated to be spent on the operation of the facility (e.g., physical plant repairs) to improve the program of animal care and use. Thus, the enforcement arm of the USDA's oversight responsibility is strong, and has been used over the years to improve animal welfare at dealers, exhibits, and research facilities.

### Public Health Service Policy

The other federal agency charged with oversight of research-animal care and use is the Public Health Service (PHS). The PHS Policy on Humane Care and Use of Laboratory Animals was implemented in 1973, and was revised in 1979 and 1986. Today, the PHS authority is derived from Public Law 99–158, the Health Research Extension Act of 1985, Section 495, Animals in Research. Under this Act, institutions conducting animal research using PHS funding, such as through the National Institutes of Health, must comply with the PHS Policy. The Policy requires submission by the funding recipient (referred to as an "awardee institution") of an Animal Welfare Assurance Statement, which must be approved by the PHS's Office of Laboratory Animal Welfare (OLAW), National Institutes of Health, which commits to following the U.S. Government Principles for the Utilization and Care of Vertebrate Animals Used in Testing, Research, and Training (IRAC 1985) (see Table 3.1) and the *Guide for the Care and Use of Laboratory Animals* (NRC 1996). Based on the *Guide*, the PHS Policy covers all vertebrate animals used in research, testing, or teaching. In addition to stating a commitment to animal welfare, the Assurance Statement must designate clear lines of authority and responsibility for institutional oversight of the work, inclusive of a designated "Institutional Official," who is ultimately responsible for the animal care and use program; must identify a qualified veterinarian who is involved in the program; provide a description of the occupational health and safety program for relevant personnel in the program; provide a description of mandated training; and provide a description of the facility. In the Assurance Statement, the institution must indicate whether the animal care and use program is reviewed by a third party, such as the Association for Assessment and Accreditation of Laboratory Animal Care International, or that the program and facilities are reviewed by internal systems of the institution. Institutions in this latter category must provide a copy of their most recent semiannual report with the Assurance Statement. The Assurance

**Table 3.1  U.S. Government Principles for the Utilization and Care of Vertebrate Animals used in Testing, Research, and Training (IRAC 1985)[34]**

**The development of knowledge necessary for the improvement of the health and well-being of humans, as well as other animals, requires *in vivo* experimentation using a wide variety of animal species. Whenever U.S. government agencies develop requirements for testing, research, or training procedures involving the use of vertebrate animals, the following principles shall be considered; and whenever these agencies actually perform or sponsor such procedures, the responsible Institutional Official shall ensure that these principles are adhered to:**

| | |
|---|---|
| I. | The transportation, care, and use of animals should be in accordance with the Animal Welfare Act (7 U.S.C. 2131 et seq.) and other applicable federal laws, guidelines, and policies. |
| II. | Procedures involving animals should be designed and performed with due consideration of their relevance to human or animal health, the advancement of knowledge, or the good of society. |
| III. | The animals selected for a procedure should be of an appropriate species and quality and the minimum number required to obtain valid results. Methods such as mathematical models, computer simulation, and *in vitro* biological systems should be considered. |
| IV. | Proper use of animals, including the avoidance or minimization of discomfort, distress, and pain when consistent with sound scientific practices, is imperative. Unless the contrary is established, investigators should consider that procedures that cause pain or distress in human beings may cause pain or distress in other animals. |
| V. | Procedures with animals that may cause more than momentary or slight pain or distress should be performed with appropriate sedation, analgesia, or anesthesia. Surgical or other painful procedures should not be performed on unanesthetized animals paralyzed by chemical agents. |
| VI. | Animals that would otherwise suffer severe or chronic pain or distress that cannot be relieved should be painlessly killed at the end of the procedure or, if appropriate, during the procedure. |
| VII. | The living conditions of animals should be appropriate for their species and contribute to their health and comfort. Normally, the housing, feeding, and care of all animals used for biomedical purposes must be directed by a veterinarian or other scientist trained and experienced in the proper care, handling, and use of the species maintained or studied. In any case, veterinary care shall be provided as indicated. |
| VIII. | Investigators and other personnel shall be appropriately qualified and experienced for conducting procedures on living animals. Adequate arrangements shall be made for their in-service training, including the proper and humane care and use of laboratory animals. |
| IX. | Where exceptions are required in relation to the provisions of these principles, the decisions should not rest with the investigators directly concerned, but should be made, with due regard to Principle II, by an appropriate review group, such as an institutional-animal care and use committee. Such exceptions should not be made solely for the purposes of teaching or demonstration. |

Statement is renegotiated with OLAW every five years. OLAW can approve, disapprove, restrict, or withdraw approval of the Assurance Statement.

PHS-funding agencies, such as the NIH, may not make an award for an activity involving live vertebrate animals unless the prospective awardee institution and all other institutions participating in the animal activity have an approved Assurance Statement with OLAW and provide verification that the IACUC has reviewed and approved those sections of the grant application that involve the use of animals. Applications from organizations with approved Assurance Statements must address five specific points pertaining to the use of animals:

- A detailed description of the proposed work, including species, strain, sex, age, and number of animals to be used in the proposed work
- A justification of the use of animals, species, and number of animals
- Information on the veterinary care for the animals
- A description of the procedures for ensuring that discomfort, distress, pain, and injury will be minimized
- A description of the method of euthanasia and the reason for the selection of that method, including a justification for any method that does not conform with the American Veterinary Medical Association's (AVMA) Report of the Panel on Euthanasia

Awardee institutions that do not comply with the standards of the *Guide*, the USDA Animal Welfare Regulations, and other standards referenced in the PHS Policy (e.g., the AVMA's Report of the Panel on Euthanasia), may have their Assurance Statements restricted, which, in turn, can limit access to PHS funding for research. Sustained noncompliance with the PHS Policy can result in withdrawing the approval of the Assurance Statement and cessation of all PHS funding for animal-based activities.

The awardee institution must also submit an annual report. Institutions must report any change in category status from that noted in the Assurance Statement. Institutions indicate the dates of their IACUC's semiannual program reviews and facility inspections, and provide copies of any "minority views" filed by IACUC members with the annual report. The role of the IACUC in providing local oversight of animal care and use is a key element of the PHS Policy. Although the required composition of the IACUC for the PHS differs slightly from USDA requirements, due to the Memorandum of Understanding concerning laboratory-animal welfare among APHIS/USDA, the Food and Drug Administration (FDA), and the NIH that sets forth procedures for cooperation among the three agencies in their oversight of animal care and use programs, the general functions and responsibilities of the IACUC are similar (see Table 3.2).

**Table 3.2   Composition and Functions of the Institutional Animal Care and Use Committee**

| USDA (minimum of three members) | PHS (minimum of five members) |
|---|---|
| Chairman | Doctor of Veterinary Medicine |
| Doctor of Veterinary Medicine | Practicing animal-research scientist |
| Not affiliated with the institution | Nonscientist |
|  | Not affiliated with the institution |

1.   Review at least once every six months the institution's program for humane care and use of animals.
2.   Inspect at least once every six months all of the institution's animal facilities (including satellite facilities and animal-study areas).
3.   Prepare reports of the IACUC evaluations conducted, and submit the reports to the Institutional Official.
4.   Review and investigate concerns involving the care and use of animals at the institution.
5.   Make recommendations to the Institutional Official regarding any aspect of the institution's animal program, facilities, or personnel training.
6.   Review and approve, require modifications in (to secure approval), or withhold approval of those components of activities related to the care and use of animals.
7.   Review and approve, require modifications in (to secure approval), or withhold approval of proposed significant changes in ongoing activities regarding the care and use of animals.
8.   Be authorized to suspend an activity involving animals.

OLAW conducts site visits of awardee institutions, both "for cause" and "not for cause." In addition, an ongoing significant mission of OLAW is the educational outreach it performs in collaboration with awardee institutions. Jointly sponsored workshops focus on information of value to Institutional Officials and IACUCs to provide appropriate oversight of animal care and use. OLAW also provides guidance through articles in journals, commentary on other articles, NIH Guide Notices, and a list serve.

## Other Laws, Regulations, and Policies

In 1978, the FDA promulgated regulations for the conduct of animal-based research of new or existing pharmaceutical agents, food additives, or other chemicals. These regulations, known as the Good Laboratory Practice (GLP) regulations, specify appropriate diagnosis, treatment, and control of disease in animals used in this work. The Environmental Protection Agency (EPA) has issued companion regulations for conducting research pertaining to health effects, environmental effects, and chemical fate testing in a separate set of GLP regulations. Both FDA and EPA GLP regulations rely heavily on adequate and detailed record keeping. Records must include standard operating procedures, animal identification, food and water analysis, documentation that any pesticides or chemicals used near the animals do not interfere with the study, and documentation of any disease and treatment animals experience. On-site inspections are conducted to ensure compliance with the GLP standards.

The Department of Defense (DoD) developed a "Policy on Experimental Animals" in 1961 to ensure that all research at DoD facilities involving animals was conducted in accordance with certain principles of animal care.[9] Later versions of this policy included overseas sites. Subsequently, a joint regulation, entitled "The Use of Animals in DoD Programs," from the Army, Navy, Air Force, Defense Nuclear Agency, and Uniformed Services University, required all DoD facilities to "seek accreditation by AAALAC" and to establish local institutional-animal care and use committees.

State laws to protect animals have a long history, with the first anticruelty law passed in 1641 in the Massachusetts Bay Colony to prevent riding or driving farm animals beyond established limits.[10] All 50 states and the District of Columbia have enacted anticruelty laws. The overarching goals of these laws are to protect animals from cruel treatment, require that animals have access to suitable food and water, and require that animals have shelter from extreme weather. Some state laws define "animal," and some do not. The state laws encompass a diversity of approaches to providing protection to animals. Some states have additional provisions for animals used in research, and many states prohibit the sale of pound animals into the research stream. In general, criminal penalties are imposed for offenses. On occasion, state anticruelty laws have been used against research facilities. In recent years, state and federal laws have been used by private citizens or citizen groups claiming "standing" to sue on behalf of animals. The issue of "standing" has undergone a long litigation process, and a chronology of court decisions on this issue has been compiled by the National Association for Biomedical Research.[11]

Because animal research can involve a variety of different species, several other federal acts, laws, and treaties have bearing on animal use. These include the U.S. Endangered Species Act, which restricts the research conducted on these animals to those studies that would directly benefit the species under investigation; the Marine Mammal Protection Act, which provides authority for scientific research on marine mammals by special permit; the Convention on International Trade in Endangered Species of Wild Fauna and Flora (CITES), which requires signator countries to obtain a permit for the import or export of certain species; the Lacey Act, which governs import, export, and interstate commerce of foreign wildlife; and the Migratory Bird Treaty Act, which makes it unlawful to take or possess any protected bird except by permit.

## AAALAC International

The Association for Assessment and Accreditation of Laboratory Animal Care (AAALAC) International is a nonprofit organization incorporated in 1965 to provide a voluntary, confidential, peer review of animal care and use programs. The Association is comprised of a Board of Trustees of more than 60 scientific organizations, patient advocacy groups, animal welfare organizations, and research lobby groups; a Council on Accreditation and a group of ad hoc Consultants made up of veterinarians, animal researchers, research administrators, and facility managers who are experts in the field of laboratory animal science and medicine; and an office staff. The Board of Trustees sets the vision and general direction of the Association; the Council on Accreditation is responsible for the conduct of site visits and for determining the accreditation status of institutions; ad hoc Consultants assist in the conduct of on-site evaluations of animal care and use programs; and the office staff (in the United States and Europe) serve as a point of coordination of these activities and as an information resource to the laboratory animal-using community.

AAALAC does not establish policies with which institutions must conform. Rather, AAALAC International relies on the *Guide*; laws, regulations, and policies; and numerous scientifically based standards, referred to as "reference resources," and which address specific subject areas (e.g., recombinant DNA, surgery, euthanasia) for evaluation of animal care and use programs around the world. AAALAC International has developed a limited number of "position statements" pertaining to adequate veterinary care, occupational health and safety, multiple major survival surgical procedures, and *Cercopithecine herpesvirus*-1, to name a few, to provide clarification to the *Guide* and other reference resources. The accreditation process includes an extensive internal review conducted by the institution, which is summarized in a Program Description. On-site visits are announced and conducted every three years. There is also an annual report requirement. The standards and process for accreditation are described in the Rules of Accreditation. Nonconformance with AAALAC International standards results in formal notification that full accreditation has not been granted and a provision of a timeline for correcting identified deficiencies is provided. Revocation of accreditation can ultimately result from sustained nonconformance.

AAALAC International also offers a Program Status Evaluation (PSE) service. The PSE is designed to assist institutions in determining if their animal care and use programs meet AAALAC International standards for accreditation. The PSE service also familiarizes institutions around the world with the AAALAC accreditation process. The evaluation is more consultative in tone than the accreditation site visit, and has occasionally been referred to as a "pre-AAALAC" visit.

### Institute for Laboratory Animal Research

The Institute for Laboratory Animal Research (ILAR) is a component of the National Research Council, National Academy of Sciences. ILAR was founded in 1952. It publishes consensus reports on subjects of importance to the animal care and use community through the use of expert committees and publishes the *ILAR Journal* on a quarterly basis. ILAR manages the Animal Models and Genetic Stocks Information Program, which includes the maintenance of an international database on laboratory registration codes. The committee reports generated by ILAR have shaped substantially the standards of laboratory-animal care and use. Principal among these reports is the *Guide*,[12] which is currently in its seventh edition and has been translated into six languages in addition to English. The *Guide* has proven to have international application because of its emphasis on "performance standards" that "define an outcome and provide criteria for assessing that outcome," but which are less prescriptive than an "engineering" standard (NRC 1996). The *Guide* provides direction on appropriate institutional policies and responsibilities; the animal environment, housing, and management; veterinary care; and the physical plant. ILAR has also published in-depth reports on rodent management, dog management, the recognition and alleviation of pain and distress, the psychological well-being of nonhuman primates, monoclonal antibody production, cost-containment methods for animal research facilities, and occupational health and safety.

### Cornerstones of an Animal Care and Use Program

An institution that uses animals for research, education, or testing purposes must determine which of the many federal and state regulations, policies, and guidelines it must follow, in addition to determining if it will participate in a voluntary accreditation program. Despite the variety of standards an institution may be obligated to follow, there are some consistent elements that distinguish successful animal care and use programs.

### Institutional Animal Care and Use Committee

The committee that is designated to review proposed uses of animals has been variously called the Animal Care Committee (Canada), Ethics Committee (Europe), and the Institutional Animal Care and

Use Committee (United States). This group of individuals, representing institutional and public interests, has the responsibility for oversight and evaluation of the entire animal care and use program and facilities. Because committee members act on behalf of the institution, their role is pivotal to engendering a humane and progressive animal care and use program. The successful program is overseen by a committee that is engaged and knowledgeable and that receives strong administrative support. Because the committee is responsible for investigating reports of concern regarding animal welfare, the committee's functions must be well-known throughout the institution and there must be ready (and confidential) access to the committee.

## Training

The importance of adequate training for all those involved in the animal care and use program is underscored by the emphasis it receives in the Animal Welfare Regulations, PHS Policy, and the *Guide*. The Animal Welfare Regulations and PHS Policy require institutions to ensure that people caring for or using animals are qualified to do so. The Animal Welfare Regulations stipulate several key topics be included in the institution's training program. They are:

- Humane methods of animal maintenance and experimentation, including the basic needs of each species of animal, proper handling and care for the various species of animals used by the institution, proper pre-procedural and post-procedural care of animals, and aseptic surgical methods and procedures.
- The concept, availability, and use of research or testing methods that limit the use of animals or minimize animal distress.
- Proper use of anesthetics, analgesics, and tranquilizers for any species of animal at the institution.
- Methods to report any deficiencies in animal care and treatment.
- Use of the services at the National Agricultural Library, such as appropriate methods of animal care and use, alternatives to the use of live animals in research, prevention of unintended and unnecessary duplication of research involving animals, information regarding the intent, and requirements of the Animal Welfare Act.

The *Guide* urges adequate training be provided to members serving on the IACUC so that they can appropriately discharge their responsibilities. In addition to the IACUC members, the *Guide* recommends that the professional and technical personnel caring for animals be trained, as well as investigators, research technicians, trainees (including students), and visiting scientists. The *Guide* also endorses training in occupational health and safety, in procedures that are specific to an employee's job, and in procedures specific to research (e.g., anesthesia, surgery, euthanasia, recognition of the signs of pain or distress, etc.).

## Occupational Health and Safety

Although not mandated by the Animal Welfare Regulations, the *Guide*, and thus the PHS Policy, require that an occupational health and safety program be part of the larger animal care and use program. The details of the occupational health and safety program will vary among institutions, but will be predicated on the experimental and nonexperimental hazards identified at each institution and an assessment of the risks posed to personnel (by either job classification or the health of the individual) by these hazards. The *Guide* emphasizes that health professionals (doctors or nurses, as appropriate) should be involved in the design and implementation of the program. Participation by individuals involved in the animal care and use program should be based on "the hazards posed by the animals and materials used; on the exposure, intensity, duration, and frequency; on the susceptibility of the personnel; and on the history of occupational illness and injury in the particular workplace" (NRC 1996). Several federal standards and regulations have been published that must be incorporated into the occupational health and safety program, depending on the species and hazardous agents in use (e.g., the Biosafety in Microbiological and Biomedical Laboratories (CDC/NIH 1999),[13] Occupational Safety and Health Administration's Bloodborne Pathogen standards (2001),[14] and recombinant DNA guidelines (NIH 1994).[15]

## Adequate Veterinary Care

The Animal Welfare Regulations and the PHS Policy stipulate that the veterinarian must have the authority to oversee several key components of the animal care and use program, including animal procurement and transportation; quarantine, stabilization, and separation of animals; surveillance, diagnosis, treatment, and control of disease; surgery; the selection of analgesic and anesthetic agents; method of euthanasia; animal husbandry and nutrition; sanitation practices; zoonosis control; and hazard containment. The veterinarian must be qualified through either experience or training in laboratory-animal medicine or in the species used. The veterinarian brings a specific perspective to the deliberations of the IACUC, and is a voting member of the IACUC. The Animal Welfare Regulations describe the program of adequate veterinary care as including:

- The availability of appropriate facilities, personnel, equipment, and services.
- The use of appropriate methods to prevent, control, diagnose, and treat diseases and injuries, inclusive of the availability of emergency, weekend, and holiday care.
- Daily observation of all animals to assess their health and well-being.
- Guidance to researchers regarding handling, immobilization, anesthesia, analgesia, tranquilization, and euthanasia.
- Nutrition.
- Pest and parasite control.
- Adequate pre-procedural and post-procedural care in accordance with current professional standards (see also APHIS Technical Note, March 1999, Animal Care Policy #3, and APHIS Form 7002).[16,17]

The *Report of the American College of Laboratory Animal Medicine on Adequate Veterinary Care in Research, Testing, and Teaching* (1996)[18] describes a program of adequate veterinary care as including: disease detection and surveillance, prevention, diagnosis, treatment, and resolution; providing guidance on anesthetics, analgesics, tranquilizer drugs, and methods of euthanasia; the review and approval of all preoperative, surgical, and postoperative procedures; the promotion and monitoring of an animal's well-being before, during, and after its use; and involvement in the review and approval of all animal care and use at the institution. This report is used by AAALAC International as a reference standard in its assessments of animal care and use programs.

## Resources

The infrastructure of the animal care and use program, including facilities, equipment, number and qualifications of personnel, and the genetic and health status of the animals used, has a significant influence on the quality of the program. The Animal Welfare Regulations specify the size of the primary enclosures in which animals are to be kept. The *Guide* describes several environmental variables, such as temperature, ventilation, illumination, sanitation standard, and cage size, as well as the components of the physical plant that can facilitate the research, testing, or teaching goals of the institution, and states that "a well-planned, well-designed, well-constructed, and properly maintained facility is an important element of good animal care and use" (NRC 1996). AAALAC International has identified that physical plant deficiencies (specifically the operation of the heating, ventilation, and air conditioning systems) rank as the third most-common concern (following the functioning of the IACUC and the scope of the occupational health and safety program) requiring correction before a full accreditation status can be granted. In this manner, the impact of the facility on the safety and well-being of the animals is underscored.

## ASIA

## Japan

In Japan, the Prime Minister's Office is responsible for animal-protection laws and regulations. Although animal protection in Japan is based predominantly on ethical principles (derived in large part from their

Buddhist traditions), the Law for the Protection and Management of Animals (1973, translated into English in 1982)[19] and the Standards Relating to the Care and Management of Experimental Animals (1980)[20] are the framework for animal care and use for all universities and public and private research institutions.[21] The law is designed to prevent cruelty to animals and foster "a feeling of love" and "respect for life." When an animal is used for education or research, methods should be employed that minimize pain, and euthanasia must be by a method that causes a minimum of pain. The standards apply to mammals and birds, and address transportation, quarantine, animal health and safety, minimization of pain during experimental procedures, occupational health, emergency planning, waste disposal, breeding, and protection of the environment. There is no law that requires a formal review and approval of proposed animal experimentation. However, animal experimentation committees, first recommended by the Ministry of Education, Science, and Culture for all universities, provide guidance to researchers. Creation of an animal-experimentation committee is not a legal requirement, although many institutions have voluntarily complied with the Ministry recommendations. The committee's composition includes personnel with veterinary, research, regulatory, and ethical expertise. The Japanese Association for Laboratory Animal Science has provided further guidance in its *Guidelines on Animal Experimentation*, which institutions voluntarily use as a reference document.

## Korea

The Korean Animal Protection Law of 1991 permits the use of animals for teaching, research, "or other scientific study." The minimization of pain is inherent in Article 10, "Experiments with Animals." The law encourages the use of methods, whenever possible, that cause no pain, and the inspection of animals after an experimental procedure with timely euthanasia, should it be determined that the animals will suffer chronic pain or be "inviable." In addition, several institutions, such as the Korea Food and Drug Administration, have their own standards for animal use. The Korean Association of Laboratory Animal Science has published guidance on animal experimentation and provides a technician certification program. The Korean Academy of Medical Sciences published the *Guide for Animal Experimentation* in 2000, which is applied to animal research published in any Korean journal of medical science. Several institutions have determined to adhere to an international standard of quality animal care and use by becoming accredited by AAALAC International.

## People's Republic of China

The Regulations for Administration of Laboratory Animals were approved by the State Council in 1988[22] and issued by the State Science and Technology Commission. The Ministry of Health subsequently published *Implementing Detailed Rules of Medical Laboratory Animal Administration*. In general, these regulations are designed to ensure high-quality animals for research. Standards are provided regarding construction of the animal housing areas; separation of animals by source, species, strain, experiment, and pathogen status; quality of food, water, and bedding provided to the animals; quarantine procedures; preventive medicine; and animal transportation. However, personnel qualifications and occupational health and safety are also addressed in the regulations. The Beijing Municipality has additional regulations regarding the administration of laboratory animals (1997).[23] These regulations are specific to "artificially raised and bred animals with control of microbes and parasites carried by them and definite genetic background and clear sources that are used for scientific researches, teaching, production, examinations and other scientific experiments." The Beijing regulations require a license be obtained from the Beijing Municipal Science and Technology Commission for the use of laboratory animals in breeding, research, or testing. The municipal regulations require personnel training, and technical staff must complete a technical competence assessment. Proper care, handling, and treatment of the animals are emphasized throughout the regulations.

## Taiwan, ROC

In Taiwan, the Animal Protection Law (1998)[24] has provisions that address animals used for commercial purposes (e.g., meat, milk, fur, etc.), science (teaching and research), and animals kept as pets. Chapter

II, Article 12 of the Animal Protection Law precludes the killing of animals, with certain exceptions — including killing for scientific purposes. Chapter III, Articles 15–18 specify the conditions for the "scientific utilization of animals." Included in this chapter is the mandate that the minimum number of animals necessary will be used in ways that cause the minimum amount of pain or injury. Article 16 requires that the institution using animals form an "animal-experimentation management unit" to oversee the scientific utilization of the laboratory animals. In addition, the institution must establish an ethics committee, which must include a veterinarian and one representative of a private "animal protection group." Under this law, the institution is entitled to employ an "animal protection inspector" or use voluntary "animal protectors" to assist with the supervision of animal use, including inspection of locations where animals are housed and used.

## AUSTRALIA AND NEW ZEALAND

### Australia

The Australian Code of Practice for the Care and Use of Animals for Scientific Purposes (1997)[25] provides Commonwealth standards for animals used in research, testing, and teaching. It is codified by the Animal Research Act. The Code of Practice, now in its sixth edition, is a report of a panel of experts convened by the National Health and Medical Research Council (NHMRC). The NHMRC is a statutory authority within the Commonwealth Minister for Health and Family Services and advises the government. The NHMRC established an Animal Welfare Committee in 1985 to provide advice to the Council on issues related to the conduct and ethics of animal experimentation. The Animal Welfare Committee participates in the revision of the Code of Practice and has developed numerous policies, guidelines, and publications on animal welfare topics. Subject areas addressed include: independent members on Animal Ethics Committees (AECs); guidelines on monoclonal antibody production; policy on the care of dogs used in research; guidelines on the use of animals for training surgeons and demonstrating new surgical equipment and techniques; policy on the use of nonhuman primates in research; and methods of minimizing pain and distress in research animals.

The Code of Practice covers all live nonhuman vertebrates and has six stated purposes:

- Emphasize the responsibilities of investigators, teachers, and institutions using animals.
- Ensure that the welfare of animals is always considered.
- Ensure that the use of animals is justified.
- Avoid pain or distress for each animal used in scientific and teaching activities.
- Minimize the number of animals used in projects.
- Promote the development and use of techniques that replace animal use in scientific and teaching activities.

The Code of Practice endorses the principles of replacement of animals with other methods; reduction of the number of animals used; and refinement of techniques used to reduce the impact on animals. It also requires the implementation of an AEC at the institution, which ensures adherence to the Code of Practice and the principles of replacement, reduction, and refinement. The scope of responsibilities for the AEC (referred to as its Terms of Reference) include oversight of all aspects of the animal care and use program; review and recommend approval, modification, or rejection of animal use proposals; withdraw approval of a project or authorize treatment or euthanasia of animals; maintain a register of all approved projects; review and provide input to institutional plans and policies that may impact animal welfare; and make recommendations to the institution to ensure conformance with the Code of Practice. The Code of Practice describes general principles for the care and use of animals, responsibilities of institutions and their AECs, responsibilities of investigators and teachers, acquisition and care of animals in breeding and holding areas, wildlife studies, care and use of livestock for scientific and teaching activities, and the use of animals in teaching.

The Animal Research Act also requires oversight of animal research by an AEC, and stipulates that one member of the committee be an individual who is not involved in animal research, is not an animal supplier, and is not affiliated with the institution other than by service on the committee.[26] The Animal

Research Review Panel was also established by the Act. This 12-member panel is comprised of representatives from the scientific community, animal welfare groups, and the government. The panel has broad responsibilities, including serving as the conduit for community input to policy development; advising on the resolution of animal care and use-based complaints; and oversight of institutional self-regulation, through on-site inspections and audits of institutions' conformance with the Code of Practice, in collaboration with government veterinary inspectors.[26]

In addition to Commonwealth law, state and territory legislation have been enacted that regulate the use of animals for scientific purposes. Related Commonwealth legislation includes the Australian Wildlife Protection (Regulation of Exports and Imports) Act (1982); Export Control Act 1982, including Export Control (Animals) Order 1987; and the Quarantine Act (1908). All states and territories also have passed legislation pertaining to occupational health and safety. The laws require the employer to make the work environment as safe as possible, provide training to workers, and ensure supervision of employees and the workplace. Similarly, the employee is required to follow safety precautions, use protective equipment, and cooperate with the employer on occupational health and safety issues.[27]

## New Zealand

Animal use in New Zealand was initially regulated by the Animals Protection Act (1960) and, in particular, by its 1983 amendment, which stipulated conformance with a code of ethical practices when using animals for research or testing. The act covered all vertebrate animals kept in captivity or that are dependent upon humans for care. The 1983 amendment also established a National Animal Ethics Advisory Committee (NAEAC).[28] Specific requirements regarding animal use were contained in the Animals Protection (Codes of Ethical Conduct) Regulations (1987). A code of ethical conduct was considered specific to the institution, but had to be approved by the Minister of Agriculture upon the advice of the NAEAC. The 1987 regulations required that codes of ethical conduct address several topics, including: alternatives to the use of animals, a justification for the choice of species, minimization of the number of animals consistent with obtaining sound data, assurance of the general health and welfare of the animals, minimization of pain and distress, and scientific merit of the project.[29]

The Animal Welfare Act of 1999[30] replaced the Animals Protection Act to meet changing societal views toward the use of animals. It also expanded the types of animals covered under the law to most of those that can feel pain. The Animal Welfare Act requires that an institution hold a Code of Ethical Conduct (CEC) before research, testing, or teaching with animals can be done. The CEC typically describes administrative procedures for the committee, as well as general policies and procedures for animal care, and must be renewed every five years. The 1999 act also calls for an independent, periodic review of the program to ensure it conforms with the CEC, the act, and other relevant regulations. An AEC implements the code. The committee is appointed by the code holder, who may be the chief executive of the institution, or his or her nominee. The AEC must be comprised of at least four members, including a senior member of the organization; an outside veterinarian; a person representing animal welfare groups who is not affiliated with the institution and is not involved with animal research, testing, or teaching; and a person to represent the public. The AEC is responsible for overseeing research conducted at the institution, including reviewing proposed projects, monitoring the project conduct, reviewing project renewals, monitoring management practices and facilities to ensure conformance with the CEC, suspending or revoking project approval, and recommending to the code holder changes to the CEC. Each proposed project must include a harm/benefit analysis and must address reduction, replacement, and refinement of animal use. If the animals are euthanized at the end of a manipulation, Part 6 does not require that the AEC consider the ethical question of killing the animals. Rather, the harm/benefit analysis that must be considered with each proposal is limited to the pain or distress that the animal may experience. Part 6 of the Act provides for circumstances where pain, distress, and "compromised care" of the animals may be allowed such that the researcher cannot be prosecuted for not conforming with Parts 1 or 2 of the Act.

The Act also mandates codes of welfare, which describe appropriate care for different species of animals, different uses of animals, and different management situations. The National Animal Welfare Advisory Committee, a statutory advisory committee to the Minister of Agriculture and Forestry, recommends the content of the codes of welfare. Of relevance to the use of animals in research, testing, and teaching is the Code of Recommendations and Minimum Standards for the Care and Use of Animals for Scientific Purposes.[31] Codes address five basic needs of animals: freedom from thirst, hunger, and malnutrition; the provision of appropriate comfort and shelter; the prevention, or rapid diagnosis and treatment, of injury, disease, or infestation with parasites; freedom from distress; and the ability to display normal patterns of behavior. The New Zealand Code is based on the Australian Code of Practice and describes the general principles for the care and use of animals; details the responsibilities of investigators and the institution; and states the terms of reference, membership, and operation of the AEC.

## ELSEWHERE AROUND THE WORLD

### Costa Rica

Costa Rica passed an Animal Welfare Law in 1994, the Ley de Bienestar de los Animales (Ley No. 7451). The implementing regulations were published as a presidential decree, Decreto No. 26668-MICIT. Briefly, the law requires an ethics committee to review the proposed animal use in both private and public institutions. The law also requires consideration of the value of the experiment vis-à-vis the promotion of human or animal health, selection of the most appropriate species and number of animals, minimization of pain, and humane euthanasia. Consideration must also be given to the use of analgesics and anesthetics, using current veterinary professional judgment and the provision of adequate veterinary care. Individuals using animals are expected to be qualified to do so. Alternatives to animal use are encouraged in the law. Experiments with animals must be registered with the Minister of Science and Technology.

### India

The Animal Welfare Board of India was set up in accordance with Section 4 of the Prevention of Cruelty to Animals Act (1960) (No. 59 of 1960).[32] The Ministry of Food and Agriculture constituted the Animal Welfare Board of India in 1962. Since 1998, oversight of the board is the purview of the Ministry of Social Justice and Empowerment. The functions of the board include advising the government on promulgating rules with a view to preventing unnecessary pain or suffering of captive animals and on potential amendments to the law. Chapter 4 of the Act addresses experimentation on animals. Included in the Act is the authority for the government to appoint a Committee for the Purpose of Control and Supervision of Experiments on Animals. The Committee must ensure that animals are not subjected to unnecessary pain or suffering before, during, or after the performance of experiments on them. To achieve this, the Committee may, subsequent to notification in the *Gazette of India,* develop rules regarding the conduct of experiments. In general, the rules for animal experimentation pertain to appropriate qualifications of individuals conducting the experiment, minimization of animal pain by the use of anesthetics, euthanasia, consideration of alternatives to animal experimentation, ensuring that pre- and post-procedural care be provided to the animals, and ensuring that suitable records are maintained. The Committee can authorize inspection of the location of the experiment and can suspend animal work by an individual or an institution. The Indian National Science Academy is responsible for the development of guidelines for the operation of Institutional Animal Ethics Committees (IAEC). For example, protocols must be provided to the IAEC 30 days in advance of the committee meeting. The IAEC's principal responsibility is the review and authorization of proposed animal experimentation. Each IAEC includes a member of the Committee for the Purpose of Control and Supervision of Experiments on Animals. Most experimentation is conducted on small laboratory animals (e.g., mice, rats, guinea pigs, rabbits); permission must be obtained from a subcommittee of the Committee for the Purpose of Control and Supervision of Experiments on Animals to conduct research on larger animals.

### Mexico

Technical specifications for the production, care, and use of laboratory animals (Norma Oficial Mexicana para la Producción, Cuidado y Uso de los Animales de Laboratorio) are contained in the federal Mexican law, NOM-062-ZOO-1999. The law covers dogs, cats, pigs, and nonhuman primates. In addition, Mexico City enacted a law in 1981 for the prevention of cruelty to animals (Ley del Distristito Federal para la Prevención de la Crueldad a los Animales).

## Russia

Russian regulations (No.1045–73), Sanitary Regulations for the Organization, Equipment, and Maintenance of Animal Facilities for Experimental Biology (Vivaria),[33] describe the location and design of animal facilities, sanitation requirements of animal facilities, housing and husbandry requirements, acquisition and quarantine of animals, standards for personal hygiene, and standards for the humane treatment of animals. Included in the latter section is the requirement to minimize pain animals experience through the use of anesthetics and analgesics.

## ACKNOWLEDGMENTS

The authors acknowledge the efforts of Ms. Darlene Brown for her research in international laws and regulations, Dr. Liliana Pazos for the information she provided regarding the laws and regulations in Costa Rica, and Rafael Hernandez Gonzalez for information regarding Mexican laws pertaining to laboratory animal welfare. The information provided by our European colleagues on their national system is highly appreciated.

## REFERENCES

1. Council of Europe, Convention for the protection of vertebrate animals used for experimental and other scientific purposes (ETS 123), Strasbourg, Council of Europe, 1986.
2. European Commission, Directive for the protection of vertebrate animals used for experimental and other scientific purposes (86/609/EEC), *Off. J. Eur. Comm.,* L 358, 1, 1986.
3. Russell, W.M.S. and Burch, R.L., *The Principles of Humane Experimental Technique*, Methuen & Co. Ltd., London, 1959.
4. FELASA guidelines: www.felasa.org
5. CCAC guidelines: www.ccac.ca
6. Johnson, D.K., Martin, M.L., Bayne, K.A.L., and Wolfle, T.L., Laws, regulations, and policies, in *Nonhuman Primates in Biomedical Research: Biology and Management,* Bennett, B.T., Abee, C.R., and Henrickson, R., Eds., Academic Press, Inc., New York, 1995, 15.
7. Johnson, D.K., and Morin, M.L., U.S. laws, regulations, and policies important to managers of nonhuman primate colonies, *J. Med. Primatol.,* 12, 223, 1983.
8. McPherson, C.W., Laws, regulations, and policies affecting the use of laboratory animals, in *Laboratory Animal Medicine*, Fox, J.G., Cohen, B.J., and Loew, F.M., Eds., Academic Press, New York, 1984, 19.
9. Rozmiarek, H., Origins of the IACUC, in *The IACUC Handbook,* Silverman, J., Suckow, M.A., and Murthy, S., Eds., CRC Press, LLC, New York, 2000, 1.
10. Office of Laboratory Animal Welfare, National Institutes of Health, *Public Health Service Policy on Humane Care and Use of Laboratory Animals*, Bethesda, MD, 1986, reprinted 2000.
11. National Association for Biomedical Research, *Animal Legal Defense Fund (ALDF), et al.v. Glickman, et al. and NABR,*U.S. District Court for the District of Columbia, Civil Action No. 96–408 (CRR), U.S. Court of Appeals No. 97–5009 consolidated with 97–5031 and 97–5074, summary as of June 8, 1999, Washington, D.C., 1999.
12. National Research Council, *Guide for the Care and Use of Laboratory Animals*, National Academy Press, Washington, D.C., 1996.
13. Centers for Disease Control and Prevention/National Institutes of Health, *Biosafety in Microbiological and Biomedical Laboratories*, 4th Edition, DHHS Pub. No. (CDC) 93–8395,U.S. Government Printing OfficE, Washington, D.C., 1999.

14. Occupational Safety and Health Administration, Department of Labor, Occupational exposure to bloodborne pathogens, needlestick, and other sharps, *29 CFR,* 66, 5317, 2001.

15. National Institutes of Health, *NIH Guidelines for Research Involving Recombinant DNA Molecules,* Bethesda, MD, 1994.

16. Animal Plant Health Inspection Service, Ensuring adequate veterinary care: roles and responsibilities of facility owners and attending veterinarians, *Tech Note,* March 1999.

17. Animal-Plant Health Inspection Service, Form 7002, Program of Veterinary Care for Research Facilities or Exhibitors/Dealers, June 1992.

18. American College of Laboratory Animal Medicine, Report of the American College of Laboratory Animal Medicine on Adequate Veterinary Care in Research, Testing, and Teaching, 1996.

19. Law for the Protection and Management of Animals (in English), *Exp. Anim.,* 31, 221, 1982.

20. Standards Relating to the Care and Management of Experimental Animals (in English). *Exp. Anim.,* 31, 228, 1982.

21. Nomura, T., Laboratory animal care policies and regulations, Japan, *ILAR Journal,* 37, 60, 1995.

22. State Science and Technology Commission, Regulations of People's Republic of China for Administration of Laboratory Animals, 1988.

23. Beijing Municipal Science and Technology Commission, Regulations of Beijing Municipality for Administration of Laboratory Animals, People's Republic of China, 1997.

24. Animal Protection Law, Taiwan, November 4, 1998.

25. National Health and Medical Research Council, Australian Code of Practice for the Care and Use of Animals for Scientific Purposes, Publication Number 1960, Australian Government Publishing Service, Canberra, ACT, 1997.

26. Larkin, R.A. and Brooks, R.M., Laboratory Animal Welfare around the Globe: Australia, *Lab. Animal,* 25, 24, 1996.

27. James, T., Occupational health and safety in the animal house and associated laboratory facilities, *ANZCCART News Insert,* 113, 1, 1998.

28. National Animal Ethics Advisory Committee, Guidelines for Institutional AnimaL Ethics Committees, Ministry of Agriculture and Fisheries: Wellington, New Zealand, 1988.

29. Reid, C.S.W., Laboratory Animal Care Policies and Regulations: New Zealand, *ILAR Journal,* 37, 62, 1995.

30. Animal Welfare Act Public Act 1999, Number 142, New Zealand, 1999.

31. Animal Welfare Advisory Committee, Code of Recommendations and Minimum Standards for the Care and Use of Animals for Scientific Purposes, Ministry of Agriculture and Forestry: Wellington, New Zealand, 1995.

32. The Prevention of Cruelty to Animals Act, 1960 (No. 59 of 1960), amended by Central Act 26 of 1962, Republic of India, 1962.

33. Sanitary regulations for the organization, equipment, and maintenance of animal facilities for experimental biology (vivaria), No. 1045–73, Russia, 1973.

34. Interagency Research Animal Committee, U.S. Government Principles for the Utilization and Care of Vertebrate Animals Used in Testing, Research, and Training, 1985.

# CHAPTER 4

# Assessment of Animal Care and Use Programs and Facilities

John G. Miller, Harry H. van Herck, and Ronald E. Larson

## CONTENTS

## INTRODUCTION

"Good Animal Care and Good Science Go Hand in Hand." This caption on a poster developed by the United States National Institutes of Health's (NIH's) Office of Animal Care and Use, was intended to remind NIH employees and visitors that sound science relies on high-quality animal care and use practices. More than 50 years before this poster was produced, Dr. Milton J. Rosenau, the director of NIH's predecessor organization (the Hygienic Laboratory), sent much the same message to his employees when he wrote in 1904, "Animals are to be used in the proper work of the lab, but anything [that] inflicts pain upon them will not, under any circumstances, be allowed."[1]

This recognition that consistent, high-quality animal care and use programs are vital in producing research results of corresponding high quality, has led to the development of several programs that we will refer to in this chapter as "assessment systems." The three such systems to be presented here are: accreditation by the Association for Assessment and Accreditation of Laboratory Animal Care

0-8493-1086-5/03/$0.00+$1.50
© 2003 by CRC Press LLC

International (AAALAC International), Good Laboratory Practice (GLP) regulations, and certification or registration by the International Organization for Standardization (ISO). In addition to describing these three systems, this chapter will provide information on the historical bases for their development and how the differences in origins influence their procedures and ultimate aims. We will also attempt to show that, despite these differences, all three can be integrated at the institutional level.

Why is consistent, high-quality animal care important as it relates to biomedical and behavioral research? There are at least three different motivators that underlie the need for quality: legal necessity, an ethical imperative, and the validity of scientific results.

For most scientists and institutional staff who support the activities of research involving animals, the baseline for an animal care and use program is established in laws or other governmental requirements. Minimum requirements for animal welfare in science are laid down in the United States by the Animal Welfare Act[2] and the Public Health Service Act[3], and in countries of the European Union (EU) by Directive EEC 86–609.[4] Other countries have animal welfare laws of varying stringency, ranging from general anticruelty statutes to near-prohibition of using animals in research. Additional legal requirements are imposed if the experimental data generated through animal research (as well as *in vitro* methods) are to be submitted to a governmental authority, such as the United States Food and Drug Administration (FDA), for pre-marketing approval.

However, meeting the minimum legal requirements for animal research is increasingly viewed by both the public and the scientific community as merely a foundation on which to build programs that optimize animal welfare. To those who hold this view, using animals in research brings with it an ethical imperative to go beyond the strict letter of the law to strive for the highest standards of animal care and use. Indeed, scientists themselves recognize that using animals to further scientific knowledge is a privilege that they are ethically obliged as individuals to not only acknowledge, but also to constantly factor into their research decisions. Validation that institutions and individuals are striving to achieve high standards in this specific area is one function of external assessment groups, such as AAALAC International.

Scientific inquiry is guided by the scientific method, one principle of which is that a hypothesis must withstand the scrutiny of other scientists. This scrutiny involves not only the validity of the hypothesis itself, but also the methods originally used to test the hypothesis. When scientific methods include animal experiments, a number of variables are introduced that can significantly affect the reproducibility of data, and hence, the ability to validate a study. These variables include the genetic "purity" of the animal strain; the animals' health and immune status, including latent infections; potential animal stressors, including those involved with routine maintenance as well as experimental manipulations; and the animals' macro- and micro-environments. Minimizing these and other animal-related variables is a goal that is advanced by applying consistent high standards of procurement, husbandry, veterinary care, and experimental use. Systems such as GLP and ISO provide methods and measures to ensure the requisite consistency and, along with participation in AAALAC International's accreditation program, demonstrate that high standards in these broader areas are met and maintained.

## BACKGROUND

The three systems to be discussed here each occupy an individual niche in promoting and assuring quality animal care and use practices. To understand both the functions of each system and the nature of their niches, it is useful to examine how each was conceived and developed.

### History of ISO

In the book, *Friendship among Equals*, by Jack Latimer,[5] one of the founders of ISO, Willy Kuert, writes,

> ISO was born from the union of two organizations. One was the ISA (International Federation of the National Standardizing Associations), established in New York in 1926, and administered from Switzerland. The other was the UNSCC (United Nations Standards Coordinating Committee), established in 1944, and administered in London.

The conference of national standardizing organizations [that] established ISO took place in London from 14 to 26 October, 1946.

The result of this founding meeting of delegates from 25 countries was the creation of a new international organization, "the object of which would be to facilitate the international coordination and unification of industrial standards"; in other words, to ease the movement of goods (or services) across international boundaries. The new organization, ISO, which began functioning officially on February 23, 1947, has evolved to become a worldwide federation of national standards bodies from 140 countries.[6] Literally hundreds of ISO standards exist that define the requirements for everything from foods to film speed. In fact, many of us will recognize the film-speed measurement called ISO 100, ISO 200, and so on.

The mission of ISO is "to promote the development of standardization and related activities in the world with a view toward facilitating the international exchange of goods and services, and to developing cooperation in the spheres of intellectual, scientific, technological, and economic activity. ISO's work results in international agreements, which are published as International Standards."[6] Further, ISO defines its standards as documented agreements containing technical specifications or other precise criteria to be used consistently as rules, guidelines, or definitions of characteristics, to ensure that materials, products, processes, and services are fit for their purpose.[7] ISO standards cover a huge area of subjects. The subject of most interest and applicability for our purposes is that of "quality requirements in business-to-business dealings," which is covered in the ISO 9000–2000 standard.

## History of AAALAC International

The origin of AAALAC International can also be traced to the WWII period. The postwar boom in science, including the dramatic growth in public funding of research in the United States by the NIH, brought with it a commensurate burst of animal experimentation. So it was that in the late 1940s, five veterinarians involved in managing laboratory-animal facilities in major institutions in Chicago, Illinois, first met to discuss and share information about the care of laboratory animals. Before the end of the decade, a movement emerged from this group to form a national organization to address the issues facing the growing field of laboratory-animal science.[8]

The first meeting of AAALAC International's progenitor, the Animal Care Panel (ACP), took place in 1950. At that meeting, Carl Schlotthauer of the Mayo Foundation emphasized the need to "establish some uniformity in animal handling."[8] Although ACP members initially recognized the need for standards, certification, and accreditation, the organization struggled with how to implement such programs during the period between 1950 and 1960. As early as 1951, the ACP had a Committee on Animal Care Standards and a Committee on Regulations for the Care of the Dog, but it was not until 1958 that the committee finally decided that the ACP should undertake accreditation functions for laboratory-animal maintenance and care.

Finally, in the fall of 1964, a subcommittee of the ACP issued an in-depth report, *Accreditation of Laboratory Animal Facilities*, recommending establishment of an autonomous entity (not part of the ACP) to conduct an accreditation program. The American Association for Accreditation of Laboratory Animal Care was established as a nonprofit corporation in April 1965.

## History of GLP Regulations

Good Laboratory Animal Practice (GLP) regulations differ from ISO and AAALAC International in several basic ways. First, they are legal requirements, as opposed to the voluntary nature of ISO and AAALAC International. Second, they have their roots not in the hopes of scientists to better their professional specialties, but rather in governmental actions to prevent harm to the general public.

The United States FDA was created with the enactment of the Food and Drug Act of 1906. Its principal mission, then and now, is to protect consumers from harmful foods and drugs, which prior to that time were frequently misbranded and often impure. FDA's powers were expanded in 1938 in response to a 1937 incident in which an incorrectly labeled elixir sulfanilamide containing diethylene glycol killed 100 people in two months.[9] And in 1962, another tragedy: the widespread and severe birth defects caused by the European approved drug thalidomide led to additional FDA powers to require manufacturers to prove effectiveness and provide post-approval reports.

However, it was actual criminal activities that brought about the GLP regulations. Audits authorized by the 1962 law showed that 618 of 867 studies were invalid because of numerous discrepancies between actual study procedures and data. As a result, the FDA prosecuted four individuals. These toxicology-testing laboratory managers were found guilty and sentenced to prison. The subsequent FDA decision to regulate laboratory testing resulted in proposed GLP regulations in 1976. These were finalized in 1978[10] and took effect in 1979. As knowledge and new technologies have grown, GLPs have been modified several times (in 1987 and 1999).

During this period, concerns about the safety of food and drugs were not confined to the United States. At the international level, the Organization for Economic Cooperation and Development (OECD) was working at the same time to develop guidance in this area. The OECD is comprised of 30 countries from Asia, Europe, North America, and the Pacific Rim whose goals are to "build strong economies in its member countries, improve efficiency, hone market systems, expand free trade, and contribute to development in industrialized as well as developing countries."[11] In addition to the 30 members, OECD has active relationships with 70 additional countries and many nongovernmental organizations (NGOs), giving it a truly global influence. An integral part of its 1981 "Council Decision on the Mutual Acceptance of Data in the Assessment of Chemicals" was the establishment of the OECD "Principles of GLP," which were subsequently revised in 1997.[11] To further ensure harmonization of the procedures related to GLP compliance-monitoring in preclinical safety studies, the OECD developed guidance documents that strive to ensure that studies are carried out according to the "Principles of GLP." These include a 1989 council decision, "Recommendation on Compliance with Good Laboratory Practice," which requires the establishment of national compliance-monitoring programs based on laboratory inspections and study audits, and recommends the use of the "Guide for Compliance-Monitoring Procedures for Good Laboratory Practice" and "Guidance for the Conduct of Laboratory Inspections and Study Audits." Both these guidance documents were revised in 1995.[11]

## SYSTEM DESCRIPTIONS

### ISO

In scientific research in general, results should be accurate, reproducible, and obtained in a clear and transparent way. This demands consistency and high quality during all phases of an animal experiment, such as those shown in Table 4.1. With its emphasis on high-quality product development and consumer satisfaction, the ISO process can be a powerful management tool for an organization.

The ISO standard most commonly applied to organizations with animal care and use programs is ISO 9000, the standard for quality systems. The former ISO 9000 family of standards was upgraded in 2000 and is now identified as ISO 9000–2000. According to the official ISO Web site, ISO 9000–2000 contains "standards and guidelines related to management systems, and related supporting standards on terminology and specific tools, such as auditing (the process of checking that the management system conforms to the standard)."[12]

For organizations with animal care and use programs, ISO 9001–2000 is most applicable, because it "specifies quality system requirements for use where a supplier's capability to design and supply conforming products needs to be demonstrated."[13] It is also a standard against which requirements of a system can be certified by an external organization. In the animal research context, researchers are the "consumers" of the products — which can range from providing and maintaining appropriate animals, to performing a complete study (e.g., contract research organizations).

ISO 9001–2000 has eight parts, with five of the eight identifying the requirements of the standard. These five parts are titled Quality Management System, Management Responsibility, Resource Management, Product Realization, and Measurement, Analysis, and Improvement. The core philosophy of the standard is defined by taking a slightly closer look at each of the five parts.

The quality-management system part of ISO 9001–2000 includes, in addition to general requirements, that a company document what the system is and how it works. In addition, the quality system must be defined in writing in a quality manual. Finally, there must be a control system for documents and records.

The management-responsibility part clearly and specifically assigns the responsibility for creation, implementation, documentation, and improvement of the quality system squarely to an organization's

**Table 4.1 Phases of an Animal Experiment to which ISO is Applicable**

Preliminaries
  Check financial and scientific solidity
  Design outline of study protocol
  Check legal and ethical aspects
Plan, define, and verify requirements
  "Test substance"
  Animals (e.g., housing, feed, destination at end of study)
  Facilities (e.g., equipment, operating rooms, biohazards)
  Staff (e.g., number, techniques, trained)
  Paperwork (e.g., study protocol, instruction, registration)
Perform study
  Animals: order, receive, verify that specifications are met, house
  Perform techniques needed (e.g., administration of test substance)
  Obtain data and samples (e.g., collection of blood, feces, urine, body weight, observations)
Analyze samples and data
Write report, paper
Archive

senior management. This is one of the most significant parts of ISO 9001–2000. Management's commitment to quality, listening to the customer, and planning for quality are included. In addition, a regular, documented review of the quality program, with an eye toward continuous improvement, is required. While all of senior management is identified as responsible, a single person (e.g., a director of animal laboratories or a study director) is identified as the management representative appointed by top management to ensure that the quality-management system is established, implemented, maintained, and improved.

Resource management emphasizes that human resources must be competent and available in sufficient numbers to assure quality work. This part also requires that the facilities, equipment, supporting services, and training programs are sufficient to assure product quality (e.g., healthy animals, humanely cared for and appropriately used animals). The need for an appropriate work environment is also required.

Product realization focuses on the quality of the product and the need to listen carefully to the customer to assure that requirements are well understood. How design decisions are made, reviewed, validated, and controlled is emphasized. The purchase of anything used to produce data, healthy animals, or good animal care is included in this part. Vendor audits must be conducted to help assure that purchased materials meet specifications. The last requirement of this part is for a company to take total responsibility for the quality of its goods and services.

The fifth part is termed Measurement, Analysis, and Improvement. Decisions about processes and goods and services must be made by reviewing adequate data obtained from rigorous measurements and audits. The data is then analyzed to assure continual improvement in all areas. This part also contains the need for a strong corrective and preventative action program. An analogy to fixing a roof may help in understanding this. If a leaking roof is fixed, it is a repair. If the leaking roof is replaced, it is corrective action. If a roof is inspected at appropriate intervals and replaced before it leaks, that is preventative action.

Registering with ISO 9001–2000 is a voluntary process. An organization begins by creating a quality system, a quality manual, and all the processes and systems that assure the quality of its goods and services. One important distinction is that management responsibility in ISO 9001–2000 extends to support groups. In other words, everything purchased is subject to the standard, which in the case of animal care and use programs, would include animals, feed, bedding, caging, etc. An ISO inspection is a rigorous process, entailing such elements as a visit to the facility's boiler room to determine if the building maintenance staff has adequate training and processes and is following specific written procedures.

## AAALAC International

As noted in the Background section of this chapter, AAALAC International is a nonprofit corporation. Governance of the corporation is through a board of trustees comprised of representatives from scientific, professional, and nonprofit organizations involved with or otherwise interested in the humane care and use of laboratory animals, including the American Association for Laboratory Animal Science, Federation of European Laboratory Animal Science Associations, International Council on Laboratory Animal Science, International Association for Gnotobiology, Federation of American Societies for Experimental Biology, et al. A complete list of these organizations can be found on the AAALAC International Web site.[15]

Through this structure, the direction and policies followed by AAALAC International are established and driven by the communities to which AAALAC International programs are directed. The founders of AAALAC International recognized that the involvement of end-users of an accreditation program would be beneficial in several ways. First, involving laboratory-animal specialists and other scientists in the process would ensure that AAALAC International programs would be based on scientific principles, with the standards and procedures employed more likely to be based on empirical scientific data and professional judgment. Second, those who would be subject to the accreditation process would be far more likely to be receptive to standards and procedures they had a role in developing.

For practically the entire 37 years of its existence, AAALAC International has relied on the *Guide for the Care and Use of Laboratory Animals (Guide)*[16] as the principal standard against which animal programs are evaluated. This continues to be true in the United States and in countries that do not have at least equivalent standards. For countries with existing laws, regulations, or other standards dealing with research-animal care and use, the *Guide* serves as an adjunct in areas not covered by national standards. In addition to the *Guide,* AAALAC International employs a number of resources (referred to as "Reference Resources") in specific areas that provide more detailed information than the *Guide,* e.g., for agricultural animals, occupational health and safety, euthanasia, and numerous other topics.[17] It is important to note that national requirements always serve as the baseline for AAALAC International accreditation, and organizations must be in full compliance with their own national laws, regulations, and policies before they can hope to achieve AAALAC International accreditation.

Similar to the ISO registration process, the AAALAC International accreditation process begins with a form of internal self-assessment. For accreditation, this involves a description of an organization's animal care and use program through completion of a standard outline provided by AAALAC International, and which closely follows the sections of the *Guide* (see Table 4.2). In 2002, the *Guide* was available in eight languages, and those seeking accreditation outside the United States may prepare the Program Description (PD) in their native language.

Upon acceptance of the Program Description (and its translation, if necessary) an on-site evaluation by AAALAC International representatives is scheduled. The site-visit team is comprised of at least one member of AAALAC International's Council on Accreditation, along with one or more additional individuals from a group of AAALAC International ad hoc consultants and specialists. The Council on Accreditation is comprised of 32 individuals with knowledge and experience in research-animal care, use, or oversight and are elected by their peers. All are recognized experts active in their fields of endeavor. Ad hoc consultants and specialists are also elected by the Council on Accreditation, and are chosen based on their knowledge, experience, and area of expertise to allow AAALAC International to tailor site-visit teams to meet any special circumstances at the organization seeking accreditation.

The site visit is the first of several levels of peer review in the AAALAC International process. The on-site evaluation begins with a review of the organization's Program Description with appropriate staff, and additional verification that actual practices correspond to their descriptions. Review of records, interviews with personnel, and evaluations of the animal facility, support areas, and research laboratories are also conducted to assess the degree to which AAALAC International standards are met. The site visit concludes with an exit briefing, at which the site visitors provide a summary of their observations and preliminary findings. This also provides the organization with an opportunity to correct misperceptions and ensure that there are no errors-in-fact in the site visitors' preliminary findings. Following the site visit, the team prepares a draft report, which is distributed to several other council members for review. This second level of peer review precedes a meeting of the full council (which meets three times a year), at which the council member who led the site visit presents his or her report, along with a

**Table 4.2 AAALAC International Program Description Outline**

A. Institutional policies and responsibilities
    1. Monitoring the care and use of animals
    2. Veterinary care
    3. Personnel qualifications and training
    4. Occupational health and safety of personnel
B. Animal environment, housing, and management
    1. Physical environment
    2. Behavioral management
    3. Husbandry
    4. Population management
C. Veterinary medical care
    1. Animal procurement and transportation
    2. Preventive medicine
    3. Surgery
    4. Pain, distress, analgesia, and anesthesia
    5. Euthanasia
    6. Drug storage and control
D. Physical plant

recommendation for accreditation status. The reviewer's comments are discussed, and the council votes on final accreditation status, completing the third level of peer review.

Organizations are informed of their accreditation status through a letter report that includes commendations, findings, and recommendations. For new applicants, possible outcomes include full accreditation, provisional status, or withhold accreditation. Fully accredited organizations are revisited every three years, and may receive continued, full, deferred continued, or probationary accreditation, or, under extreme circumstances, may receive an Intent to Revoke notice. Any report of status other than full accreditation will include a description of findings that must be corrected before full accreditation can be restored (known as "mandatory items") and any additional suggestions for improvement. Full accreditation reports may also include suggestions for improvement. To maintain accreditation, annual reports are required, and organizations must complete updated Program Descriptions and be revisited every three years.

## GLP

As might be expected from regulatory agencies such as FDA, OECD, and the U.S. Environmental Protection Agency (which has its own set of GLP regulations similar to the FDA's), the GLP requirements are extensive and detailed.[10] Compliance with GLPs is accomplished and verified through both internal and external procedures and inspections. Internally, Quality Assurance (QA) units are required to perform a variety of specified duties. Externally, an authorized regulatory agency official can inspect a laboratory's facility, records, and specimens at any time and without prior notification.

The subparts of GLP regulations categorize requirements under General Provisions, Organizations and Personnel, Facilities, Equipment, Testing Facilities Operation, Test and Control Articles, Protocol for and Conduct of a Nonclinical Laboratory Study, Records, and Reports, and Disqualification of Testing Facilities. Subparts and special issues are described in more detail in the "Guidance Documents" (see, e.g., www.barqa.com). At organizations that conduct animal studies in support of applications for research or marketing permits from regulatory agencies, GLP-compliance activities will be required in most of these areas.

A frequently heard maxim that may have its roots in GLP compliance is, "If you didn't document it, you didn't do it." Documentation is, in fact, a hallmark of GLP compliance. There are anecdotal reports of pharmaceutical companies delivering entire truckloads of documents to the FDA when requesting marketing approval for a new drug. A key element of the documentation process is the Standard Operating Procedure (SOP), and the GLP regulations require SOPs in 12 specific areas (see Table 4.3). At least seven of these deal directly with animal-related aspects of the study. GLP SOPs must state exactly how the procedure is to be done each and every time. Regulations also require that historical records of SOPs, including all revisions be maintained.

**Table 4.3 Areas Requiring SOPS under GLPS**

| |
| --- |
| Animal room preparation |
| Animal care |
| Receipt, identification, storage, handling, mixing, and method of sampling of the test and control articles |
| Test system observations |
| Laboratory tests |
| Handling of animals found moribund or dead during study |
| Necropsy of animals or postmortem examination of animals |
| Collection and identification of specimens |
| Histopathology |
| Data handling, storage, and retrieval |
| Maintenance and calibration of equipment |
| Transfer, proper placement, and identification of animals |

Specific requirements in GLPs under the heading of animal care are all standard practices in well-run animal care and use programs. They include provisions for quarantine of newly arrived animals, health-status determinations, identification, separation of species, equipment cleaning and sanitization, feed and water analysis, and bedding changing. Although these are standard practices for many, strict adherence to SOPs and documentation of all key actions differentiate GLP studies from non-GLP studies.

Aside from the areas previously identified, other provisions of GLP regulations are also applicable in an animal care and use program. These include the appropriate training of personnel (although the scope and intensity are not specified); adequate numbers of personnel; use of personal protective equipment; adequate and appropriate storage areas; clean, well-maintained laboratory space; and inspected, well-maintained equipment. Maintenance of records, both of animal health and research data, are also critical components of GLP studies that should be part of any well-run animal care and use program. Given the basic GLP requirements in this area, one cannot assume *a priori* that full GLP compliance guarantees a high-quality animal care and use program.

The rigid nature and complexity of GLP requirements make them inherently difficult to comply with, especially for the duration of long-term studies that may last one or two years. The regulatory agencies at least tacitly acknowledge this through their provisions in GLPs for internal QA units. A QA unit may be "any person or organization element, except the study director, designated by testing-facility management," and is responsible for monitoring each study to assure that facilities, personnel, equipment, methods, practices, records, and controls conform with GLPs.[10] The regulations further require that the QA unit be entirely separate from and independent of the personnel directing or conducting the study. QA units maintain schedules and protocols, and are required to conduct periodic inspections of studies, with reports provided to study directors and facility management. This ongoing internal oversight mechanism plays a significant role in assuring the facility, study sponsors, and regulatory agencies that the strict requirements of GLPs are met.

## DISCUSSION

Attaining and maintaining high standards should be the goal of all organizations that involve animals in research, testing, and teaching. The three systems described can all be applied to help achieve this goal, either individually or in combination. Many pharmaceutical companies, contract research organizations (especially those conducting toxicology testing), and centralized laboratory institutes[18,19] integrate all three into their animal care and use programs.

The fact that three distinct assessment systems can be employed simultaneously is not surprising when one examines their origins, procedures, and ultimate aims. Although aspects of the systems differ to greater and lesser degrees, they should be viewed as complimentary rather than exclusionary.[20,21]

Beginning with an examination of their histories, it is clear that the GLP system came into existence in response to serious adverse events — illness, deformities, and even death — related to the development and sale of products consumed by humans. Thus, GLPs are heavily oriented toward avoiding harm to humans. Given the complex nature of drug development, it is logical that this assessment system involves strict adherence to rather inflexible requirements at multiple points in that process. The principal focus of GLPs is on high-quality data.

Review of the origin of ISO reveals that organizational founders were primarily concerned with the need for the international harmonization of standards. Standard development continues to be its main orientation, and the list of ISO standards is both extensive and diverse. Although the vast majority of ISO standards are in areas unrelated to animals, the 9000–2000 series dealing with quality management can clearly be applied to the operation of animal care and use programs, especially for multinational corporations wishing to demonstrate consistency among their operational locations. The principal focus of ISO 9000–2000 is on developing a high-quality product and ensuring consumer satisfaction with that product.

From its inception, AAALAC International has been primarily concerned with how animals are cared for when they are involved in research, testing, and teaching. AAALAC International was founded on the belief that sound science requires that animals used in its pursuit should be maintained in a uniform manner under high standards. It is not surprising then that the AAALAC International accreditation process continues to emphasize animal care and use, with the goal of minimizing the "animal variable" in science. Animals themselves, and the uses to which they are put, vary in a myriad of ways. This requires that AAALAC International standards and procedures be sufficiently flexible to accommodate these diversities. The principal focus of AAALAC International is on animal welfare.

The differing focuses of GLPs, ISO, and AAALAC International quite naturally lead to variations in how the three systems accomplish their goals. Minimizing (hopefully to zero) the possibility that a new drug or device will cause unintended harm to human patients leaves the FDA with few options other than requiring absolute adherence to means-related standards. Almost no deviation from highly prescriptive "engineering" standards is allowed, with verification of adherence to GLP requirements accomplished by both strict internal controls and unannounced inspections by representatives of national authorities. These national authorities have ultimate approval power and are the final arbiters of whether or not an organization has complied with the requirements.

With its focus principally on products and consumer satisfaction, ISO is primarily concerned with processes and procedures. Customer feedback is critical to enhancement, on an ongoing basis, of the processes that will lead to a consistent, high-quality product. When ISO management systems are applied to animal care and use, the products are both "hard," i.e., healthy animals, and the service provided to maximize accurate, reproducible scientific results. As the body that both develops standards and evaluates their implementation, ISO utilizes professional external evaluators to assess conformance to its standards.

Accomplishing the goal of promoting animal welfare internationally requires a level of flexibility generally greater than that associated with ISO and largely forbidden under GLPs. Accommodating different species and circumstances of their use is manifested in both the standards employed by AAALAC International and the procedures followed to assess organizations' adherence to them. The *Guide* and Reference Resources that form the "AAALAC International Standards" are almost exclusively performance- or outcome-based, providing (sometimes wide) latitude in the ways that organizations may choose to meet the performance standards.[22,23] Under these circumstances, the only practical method for determining whether or not an organization has met the desired outcome is through the process of peer review by individuals with proven abilities to apply professional judgment in a consistent manner.[24]

## SUMMARY

Consistent, high-quality animal care and use programs are critical to the research effort at several levels. In this chapter, we have presented three systems that affect the quality of such programs: the ISO process, AAALAC International accreditation, and GLP regulations.

Most developed countries have some form of national legislation dealing with the humane treatment of animals used for scientific purposes, thereby establishing a required minimum level of what might be termed "quality." Those planning to submit animal-related data to a national authority for pre-marketing approval must meet additional GLP regulatory requirements of the United States' FDA or the OECD.

A second driving force for quality in this area is an ethical imperative to care for and use animals in such a way that pain, distress, and discomfort are minimized to the greatest extent practicable. This includes recognizing that using animals in the advancement of science to improve human health is a privilege to be aware of at all times.

Third, high-quality animal care and use practices are important elements in producing statistically valid and reproducible scientific data. Through the application of uniform high standards in the areas of animal procurement, husbandry, veterinary care, and experimental use, animal-related variables may be minimized.

GLP regulations have the longest history of the three systems, with roots in the United States Food and Drug Act of 1906. The actual regulations were finalized in 1978, after being developed over several decades in response to harmful incidents involving human illnesses and deaths. The ISO process originated in 1946 and evolved from the recognition that uniformity of industrial standards on an international basis would facilitate global cooperation and trade. AAALAC International traces its origins to the late 1940s, when rapid expansion of research involving animals led laboratory-animal scientists to recognize the need for standardized practices of animal care.

The recently consolidated ISO 9000–2000 is the ISO standard most applicable to animal care and use programs, as it deals broadly with management systems. Recognizing that a high-quality animal care and use program involves a successful management system, ISO 9000–2000 can be readily applied to such programs. Key elements of ISO 9000–2000 are the product produced and consumer satisfaction with the product. For animal care and use programs, the product may be healthy animals or the service of providing their care. Implementation of ISO 9000–2000 involves putting in place procedures that meet ISO's detailed standards, conducting an extensive self-assessment and, if desired, certification by an external organization.

AAALAC International is a nonprofit corporation governed by a board of trustees that represents a broad cross-section of scientific, professional, and other nonprofit organizations interested in animal welfare in science. Using science-based standards, including the *Guide for the Care and Use of Laboratory Animals* and additional international Reference Resources, AAALAC International conducts onsite evaluations of animal care and use programs. A pre-site visit self-assessment by the applicant organization serves as the starting point for discussions during the on-site part of the accreditation process. The site-visit team includes at least two expert peer-reviewers, including one from AAALAC International's Council on Accreditation, to which the site visit report is presented for a final determination of accreditation status. Several categories of accreditation are possible for new applicants and pursuant to re-visits, which take place every three years.

Regulatory compliance for organizations conducting animal studies in support of products destined for consumers involves meeting stringent national or international GLP regulations. Hallmarks of GLP regulations are internal controls and documentation. QA units within the organization — required by GLPs — not only oversee schedules and protocols, but conduct periodic inspections to ensure adherence to study elements and other GLP requirements. The aspects of GLPs that relate specifically to animal care and use are basic, and deal principally with procurement, husbandry, and veterinary care. What sets GLP regulations apart from the other assessment systems described in this chapter is their strong emphasis on SOPs and documentation. Verification of adherence to GLPs is through internal oversight mechanisms and unannounced audit inspections by officials from regulatory authorities.

Although they represent three distinct assessment systems, ISO registration, AAALAC International accreditation, and compliance with GLP regulations should be considered complimentary and not in an "either/or" manner. Integration of all three systems into institutional animal care and use programs, when appropriate (for instance, when GLP compliance is legally required), serves to enhance the quality of the overall program. Each system focuses on a different aspect of the use of animals in the advancement of science, and each employs a unique approach to verifying that institutions are meeting its requirements. A key component of many animal care and use programs is the committee charged with oversight responsibilities, known as the Ethics Committee, Animal Care and Use Committee, Ethical Review Processes, or other names. In addition to serving as another form of internal quality monitoring, these bodies provide a natural locus for integration of the three systems into an overall institutional program.

The histories of the three systems presented are instructive in understanding their differences and how they may fit together as a whole. Questionable and even deadly practices in industry led to GLP regulations designed to reduce risks to humans. These regulations took the form of strict compliance with highly prescriptive record keeping and other documentation requirements. This emphasis on consistent, well-documented internal procedures and controls is highly congruent with the ISO 9000–2000 system, which grew from its founders' desire to facilitate uniformity through adherence to international standards. Animal care and use programs operating under ISO 9000–2000 rely on frequent feedback from "customers," i.e., scientists using animals, to continuously improve the quality of service provided and management of the overall program. Rooted in the desire to improve laboratory animal care, AAALAC International focuses specifically on institutional mechanisms and practices that directly affect animal welfare. Due to great variations in species, and even individual animals, evaluating animal welfare is best done through the application of outcome- or performance-based standards, rather than through a prescriptive, means-based approach. To accomplish this, AAALAC International involves expert peer-reviewers in on-site evaluations of animal care and use programs, facilities, and the animals themselves.

Organizations that consistently apply well-documented, high-quality management practices (such as those promoted by GLP and ISO) that are also responsive to the needs of animals will find achieving AAALAC International accreditation a natural next step. Thus, far from being exclusionary, the three assessment systems presented here can work in concert to improve the quality of all aspects of animal care and use programs.

## REFERENCES

1. Gordon, H., The history of the Public Health Service Policy on Humane Care and Use of Laboratory Animals, in *50 Years of Laboratory Animal Science*, McPherson, C.W. and Mattingly, S.F., Eds., AALAS, Memphis, Chap. 21, 1999.
2. Animal Welfare Act, 7 U.S.C. 2131 et seq., Public Law 89–544, August 24, 1966.
3. Health Research Extension Act, Public Law 99–158, November 20, 1985.
4. Council Directive on the Approximation of Laws, Regulations, and Administrative Provisions of the Member States Regarding the Protection of Animals used for Experimental and Other Scientific Purposes, Directive 86/609/EEC, European Union, 1986.
5. Latimer, J., *Friendship among Equals*, 1997, http://www.ISO.CH/iso/en/aboutiso/introduction/how-started/fifty/friendship.html
6. What is ISO?, http://www.ISO.CH/iso/en/aboutiso/introduction/whatisISO.htm
7. What are standards?, http://www.ISO.CH/iso/en/aboutiso/introduction/index.html
8. Miller, J. and Clark, J.D., The history of the Association for Assessment and Accreditation of Laboratory Animal Care International, in *50 Years of Laboratory Animal Science*, McPherson, C.W. and Mattingly, S.F., Eds., AALAS, Memphis, Chap. 6, 1999.
9. Cook, J.D., Good Laboratory Practice (GLP) versus CLIA, Guest Essay, http://www.westgard.com/guest16.htm
10. U.S. Food and Drug Administration, Non-Clinical Laboratory Studies, Good Laboratory Practice Regulations, *U.S. Federal Register*. Vol. 43, No. 247, pp. 59,986–60,020, December 22, 1978.
11. Organization for Economic Cooperation and Development, OECD Principles of Good Laboratory Practice, http://www.oecd.org/ehs/glp.htm
12. ISO 9000 and ISO 14000 in plain language, http://www.ISO.CH/iso/en/iso9000–14000/tour/plain.html
13. Selection and Use of the ISO 9000–2000 Family of Standards, http://www.ISO.CH/iso/en/iso9000–14000/iso9000/selection_use.html
14. Implementing Your ISO 9001–2000 Quality Management System, http://www.ISO.CH/iso/en/iso9000/selection_use/iimplementing.html
15. About AAALAC, AAALAC's Member Organizations, http://www.aaalac.org/memorgs.htm
16. National Research Council, Institute of Laboratory Animal Resources, *Guide for the Care and Use of Laboratory Animals*, National Academy Press, Washington, D.C., 1996.
17. The Accreditation Program, AAALAC's Reference Resources, http://www.aaalac.org/resources.htm
18. Van Velden-Russcher, J.A. and van Herck, H., Quality management in the CLAI, *Der Tierschutzbeauftragte,* 1(01), 48, 2001.
19. van Herck, H., Implementation of a quality management system in the Central Laboratory Animal Institute (GDL) of the Utrecht Universiteit: Why and how? in *Science and Responsibility* (abstracts of the 29th annual symposium of the Scandinavian Society for Laboratory Animal Science), 33, 1999.
20. Jansen, C.C., A quality system for laboratory animal facilities, *Scand. J. Lab. Anim. Sci.*, 26(1), 17, 1999.

21. Ritskes-Hoitinga, M., Kalisie-Korhonen, E., and Smith, A., Evaluation of quality systems for animal units: Report of the Scand-LAS working group, *Scand. J. Lab. Anim. Sci.*, 26(3), 117, 1999.
22. Bayne, K.A. and Martin, D.P., AAALAC International: Using performance standards to evaluate an animal care and use program, *Lab. Animal*, 27(4), 32, 1998.
23. Bayne, K.A. and Miller, J.G., Assessing animal care and use programs internationally, *Lab. Animal*, 29(6), 27, 2000.
24. Miller, J., International harmonization of animal care and use: The proof is in the practice, *Lab. Animal*, 27(5), 28, 1998.

# CHAPTER 5

# Education and Training

**Nicole Duffee, Timo Nevalainen, and Jann Hau**

## CONTENTS

## INTRODUCTION

Individuals who work with laboratory animals must have appropriate skills and qualifications for performing experimental procedures on animals. This requirement is derived from national laws and regulations relating to the protection of research animals, although the laws of individual geographical regions may define this requirement differently. Despite differences in the text of the laws, the intent and rationale of each law are the same. The proper training of individuals to handle and restrain animals and to perform experimental procedures is essential for fulfilling legal requirements that laboratory animals be treated humanely when used for research, testing, and education. Insufficient training is known to result in substantial harm to animals and may cause occupational injuries from animal scratches and bites.

0-8493-1086-5/03/$0.00+$1.50
© 2003 by CRC Press LLC

Competence in animal methodologies is also recognized for benefitting in the outcome of research and testing. Changes in the physiological status of research animals may have an impact on research data as sources of nonexperimental variation. Inadequate handling and poor methodology, which may distress and even injure an animal, may alter the immune system due to chronic stress and may activate inflammation due to injury or infection (see also Chapter 18).

Although the intent of each nation's laws may be similar, the approach toward training differs among countries or regions. Differences in the specifics of laws and regulations have led to the development of different systems for complying with these requirements. The first two sections of this chapter describe the system of training in Europe and the United States. The third section addresses training resources that are useful for complementing a training program. Many countries require competence of staff in contact with laboratory animals, but the present chapter is focused on conditions in Europe and the U.S.

## EUROPE

### Legal Requirement for Training

The European Union (EU) and Council of Europe (CoE) are federations including most, but not all, European countries. Both have statutes to govern and regulate the use of vertebrate animals for scientific purposes[1,2] (see also Chapter 3). In addition to these statutes each nation has its own legislation. National legislation should be harmonized with the European regulations, and may exceed the requirements of the European regulations, which serve as the minimum standard.

The Council of Europe Convention 123, Article 26, states that: "Persons who carry out, take part in, or supervise procedures on animals, or take care of animals used in procedures, shall have had appropriate education and training."

The CoE Convention ETS 123 contains the provision that parties should hold multilateral consultations to examine the progress of its implementation and the need for revision or extension of any of its provisions on the basis of new facts or developments. During the last decades, three multilateral consultations have been held. At these multilateral consultations, the parties in 1993 adopted the following resolution on education and training of persons working with animals: "This resolution presents guidelines for topics to be included in educational and training programs for four categories of persons working with laboratory animals (from animal caretakers to specialists in animal science." This requirement is reiterated in the European Directive.

Article 5 of the European Directive states that a competent person must oversee the well-being and the state of health of the animals. Article 19 stipulates that a veterinarian or other competent person should be charged with advisory duties in relation to the well-being of the animals. The provisions on competence warrant special attention. Laws and regulations are poor tools when not based upon an understanding of what constitutes humane and responsible animal care and use. Therefore, well-directed education and training provide the means for gaining this understanding and for evaluating the ethical ramifications.

Article 7 of the Directive states that only a person considered to be competent, or under the direct responsibility of such a person, should perform animal experiments. This provision is amplified by Article 14, which states that persons conducting, collaborating in, or supervising experiments or the care of laboratory animals should have appropriate education and training. It is essential that the people involved in the design and conduct of experiments should have received an education in a scientific discipline relevant to the experimental work. They also need to be capable of handling and taking care of laboratory animals. Each member state must specify how the provision of competence is to be implemented within national legislation.

### FELASA Guidelines for Teaching and Training

The Federation of European Laboratory Animal Science Associations (FELASA) has prepared proposals concerning educational and training requirements for staff and personnel working with laboratory animals. Some countries have already introduced strict regulations regarding competence based upon these FELASA guidelines.

The categories are:

Category A[3] - persons taking care of animals (animal technicians)
Category B[4] - persons carrying out animal experiments (research technicians)
Category C[3] - persons responsible for directing animal experiments (scientists)
Category D[5] - specialists in laboratory animal science or laboratory higher-management (specialists)

These recommendations, published as FELASA Working Party Reports in the journal *Laboratory Animals* (U.K.), are in the form of syllabi for the training of each category of personnel. The recommendations for categories A and D are in-depth, career-type educations, while categories B and C are relatively short courses. Because of the multilingual nature of the continent and differences in job titles, recommendations for training objectives are handled through descriptions of duties and responsibilities instead of a defined nomenclature for each category. CoE has adopted the competence categories; hence they can be regarded as the basic competence classification in Europe.

FELASA Category A training is organized to address four levels of staff needs and experience: Level 1 for basic laboratory animal care, Level 2 for those with at least two years of work experience, Level 3 for those with an additional three years work experience (five years total), and Level 4 for those in higher management or specialization.

FELASA Category B guidelines contain a set of topics and subtopics to be taught during 40 hours. Practical exercises are emphasized by means of a recommendation that half of the instruction be devoted to hands-on exercises or demonstrations. Category B guidelines have been published as the last of the four categories, and hence the very first courses have recently been established.

Category C training has a prerequisite of a full university degree in a biomedical discipline, such as animal biology, medicine, or veterinary medicine. The Category C curriculum is double the volume of Category B, i.e., 80 hours, or an equivalent. The multilateral consultations of CoE have adopted this curriculum.

Category D training has a prerequisite of a degree in biomedical or veterinary sciences, demonstrated competence at Category C level, and appropriate experience in the field of laboratory animal science. Since laboratory animal science combines knowledge of the scientific method and animal welfare principles, as well as specific research on laboratory animals, the Category D curriculum includes completion of a scientific project to be published in a peer-reviewed journal. In all, the curriculum is expected to take two years.

## Common Approaches to Training

In order to avoid confusion or inappropriate reference to FELASA guidelines, the FELASA Board established a working group, the purpose of which was to make recommendations for the delivery of education and training in accordance with FELASA. This document, which will soon be published, should assure the quality of education and training in laboratory animal science and promote further harmonization within Europe. This scheme is tailored to accredit courses, not institutes or individuals.

At the same time, the European Science Foundation (ESF), the umbrella organization of the European research coordinating and funding organizations, issued a statement that reads: "Investigators and other personnel involved in the design and performance of animal-based experiments should be adequately educated and trained. ESF member organizations should encourage the development and organization of accredited courses on laboratory animal science, including information on animal alternatives, welfare, and ethics." The FELASA accreditation system on laboratory animal science training will be the only one in existence.

A specialty board for laboratory animal veterinarians has been established in Europe. Following an application period for de facto specialists, European board examination is based on work experience and passing an examination on topics of FELASA D-category, supplemented with specific, veterinary-oriented topics.

## UNITED STATES

### Legal Requirement for Training

The regulatory environment for the care and use of laboratory animals, and therefore the basis for the training of personnel to work with these animals, is anchored on a series of federal mandates: two laws, one regulation, and a policy. One federal law is the Animal Welfare Act, which is supported by the Code of Federal Regulations, Title 9, Subchapter A — "Animal Welfare," and enforced by the U.S. Department of Agriculture (USDA).[6,7] The second federal law, the Health Research Extension Act of 1985, "Animals in Research," is implemented by the Public Health Service Policy on Humane Care and Use of Laboratory Animals, which is enforced by the Office of Laboratory Animal Welfare (OLAW) of the National Institutes of Health.[8,9] Most research institutions in the United States are covered by one of these laws and the corresponding regulations or policy, and many institutions are indeed covered by both sets of mandates.

In addition, many institutions are voluntarily accredited by the Association for Assessment and Accreditation of Laboratory Animal Care, International (AAALAC), including many of those that are not covered by either federal law. (AAALAC accreditation is also available to institutions in other countries.) An additional mandate that addresses personnel training and qualifications is the *Guide for the Care and Use of Laboratory Animals* from the National Research Council.[10] Although a guideline, both the Public Health Service Policy and AAALAC accreditation require compliance with the *Guide's* standards. For the protection of farm animals used for nonagricultural purposes, the United States Department of Agriculture/Animal and Plant Health Inspection Service adopted this guideline, as well as the *Guide for the Care and Use of Agricultural Animals in Agricultural Research and Teaching*, published by the Federated Animal Science Societies.[11,12]

Altogether, there is a great deal of overlap in institutional coverage by these federal mandates. There is also similarity in the requirements for staff training and qualifications, although the precise language varies somewhat among these mandates. The Animal Welfare Regulations specify the training topics in greater detail than does the Public Health Service Policy on Humane Care and Use of Laboratory Animals. Because the policy requires institutions to comply with the Animal Welfare Act and other federal statutes and regulations relating to animals, it refers specification of training requirements to these other mandates, such as the Animal Welfare Act regulations and the *Guide for the Care and Use of Laboratory Animals*.

From the Code of Federal Regulations, Title 9, Subchapter A — Animal Welfare, Sec. 2.32, Personnel qualifications:[7]

(a) It shall be the responsibility of the research facility to ensure that all scientists, research technicians, animal technicians, and other personnel involved in animal care, treatment, and use are qualified to perform their duties. This responsibility shall be fulfilled, in part, through the provision of training and instruction to those personnel.

(b) Training and instruction shall be made available, and the qualifications of personnel reviewed, with sufficient frequency to fulfill the research facility's responsibilities under this section and Sec. 2.31.

(c) Training and instruction of personnel must include guidance in at least the following areas:

*(1) Humane methods of animal maintenance and experimentation, including:*
(i)  The basic needs of each species of animal
(ii)  Proper handling and care for the various species of animals used by the facility
(iii) Proper pre-procedural and post-procedural care of animals
(iv) Aseptic surgical methods and procedures

*(2) The concept, availability, and use of research or testing methods that limit the use of animals or minimize animal distress.*

*(3) Proper use of anesthetics, analgesics, and tranquilizers for any species of animals used by the facility.*

*(4) Methods whereby deficiencies in animal care and treatment are reported, including deficiencies in animal care and treatment reported by any employee of the facility. No facility employee, committee member, or laboratory personnel shall be discriminated against or be subject to any reprisal for reporting violations of any regulation or standards under the Act.*

*(5) Utilization of services (e.g., National Agricultural Library, National Library of Medicine) available to provide information:*
(i) On appropriate methods of animal care and use
(ii) On alternatives to the use of live animals in research
(iii) That could prevent unintended and unnecessary duplication of research involving animals
(iv) Regarding the intent and requirements of the Act

In addition to the training of research staff, there is also an obligation to provide training to the members of the Institutional Animal Care and Use Committee (IACUC). Neither USDA or PHS have issued guidelines, formally or informally, on IACUC training. Nevertheless, these regulatory agencies and AAALAC commonly expect that IACUC members receive training specific to their role in the animal-research program.

## Common Approaches to Training

Collectively, the United States mandates on laboratory animal welfare apply a performance standard to the assessment of compliance with all programmatic requirements, including the training of personnel. The opposite of performance standards is engineering standards, in which regulatory requirements define the procedural details in the conduct of a process: for example, in training, what is taught, how, when, and by whom. A performance standard instead focuses on the outcome of a process. In such a staff-training program, the outcome is measured in how well-qualified personnel are to carry out the animal-related procedures. The institution can determine how that program is constituted in terms of staff resources and training objectives, activities, and frequency.

From the *Guide for the Care and Use of Laboratory Animals* (National Research Council), Institutional Policies and Responsibilities — Personnel Qualifications and Training:[10]

AWRs [Animal Welfare Act Regulations] and PHS [Public Health Policy] require institutions to ensure that people caring for or using animals are qualified to do so. The number and qualifications of personnel required to conduct and support an animal care and use program depend on several factors, including the type and size of institution, the administrative structure for providing adequate animal care, the characteristics of the physical plant, the number and species of animals maintained, and the nature of the research, testing, and educational activities.

Personnel caring for animals should be appropriately trained (see Appendix A, "Technical and Professional Education"), and the institution should provide for formal or on-the-job training to facilitate effective implementation of the program and humane care and use of animals. According to the programmatic scope, personnel will be required to have expertise in other disciplines, such as animal husbandry, administration, laboratory animal medicine and pathology, occupational health and safety, behavioral management, genetic management, and various other aspects of research support.

Not only can training services be customized for the type of procedures and species used, but training can also be adapted to the level of staff experience, competence, and the degree of staff turnover. Training requirements can be satisfied by formal or on-the-job training. This provides flexibility for each institution to meet its needs and to utilize personnel and other resources as best fits the institution. In an informal statement, the Office for Laboratory Animal Welfare (OLAW) addressed the question of how much flexibility an institution may have in the development of a training program to satisfy federal requirements:[13]

Each assured institution is responsible for training its staff to meet the performance requirements cited in paragraph IV.C.1.a-g. of the PHS policy, and guidelines have been developed to assist institutions to meet these objectives. OLAW recognizes research programs vary from one institution to another, and are relative to the size and nature of the institution, staffing, numbers of species and individual animals maintained, and the kinds of research conducted. Therefore, the scope and depth of instructional programs and the frequency at which they are offered will also vary. At a minimum, however, the policy requires institutions to ensure that individuals who use or provide care for animals are trained and qualified in the appropriate, species-specific housing methods, husbandry procedures, and handling techniques. The institution must ensure that research staff members performing experimental manipulation, including anesthesia and surgery, are qualified through training or experience to accomplish such procedures humanely and in a scientifically acceptable fashion. They must also provide training or instruction in research and testing methods that minimize the number of animals required to obtain valid results and minimize animal distress. Institutions must also ensure that professional staff whose work involves hazardous biological, chemical, or physical agents have training or experience to assess potential dangers and select and oversee the implementation of appropriate safeguards.

As an example of the degree of flexibility in institutional training programs, if influxes of new staff are frequent, a program may emphasize entry-level training. In an institution with a low turnover of staff, there is an opportunity to present advanced topics for continuing-education purposes. Furthermore, depending on the amount of service needed, training can be assigned as a principal job responsibility or as a part-time duty along with other research or veterinary responsibilities. When used to evaluate a training program, performance standards direct attention toward assessing staff expertise in animal care and use procedures as a means of determining the effectiveness of a training program. An effective training program is expected to result in staff competence and therefore in the humane and appropriate treatment of animals. A lack of competence in how animals are handled or used points to a need for improvement in an institution's training program. Typically, governmental and accrediting inspectors will observe and query staff working with animals as a means of evaluating the effectiveness of a training program.

## Types of Personnel and Related Training Requirements

Individuals who are about to work with animals in the research program should be assessed for training needs based on the nature of their contact with the animals. General topic areas for training in an institutional program are presented in Table 5.1 for staff in the following categories: animal care technician, researcher, and staff involved in maintenance, transportation, and administration.[14] Animal facility managers and directors are excluded from this table because it is assumed that these individuals will receive training appropriate for their positions from sources outside of the institutional training program. For example, attending veterinarians are typically trained through programs for professional degrees and specialty board certification in laboratory animal medicine.

To compare the training needs of those served by the institutional training program, husbandry staff and research staff have similar requirements for training in animal behavior, care, and handling. Both husbandry and research staff require an orientation to the animal welfare laws, regulations, policies, and guidelines. However, husbandry staff require more detailed training on animal housing requirements and housekeeping practices for the animal facility environment.

Researchers, in turn, need to be qualified in the particular procedures that they will carry out on animals in their experimental studies. A "procedure" is any activity performed on the animal, such as behavioral observations, venipuncture, or surgery. The care that must be provided to the animals during experimental manipulations includes preparing the animal to humanely undergo the procedure, supporting and monitoring the animal's physiological function during the procedure, providing adequate analgesia to minimize pain, and providing additional supportive monitoring and care to aid the animal in recovering from the procedure.

Researchers should also have instruction on animal alternatives and on the conduct of an alternatives search, i.e., research and testing methods that minimize the number of animals required to obtain valid results, that minimize animal pain and distress, and that replace animals with alternative models. It is recognized that many institutions assign their staff to positions that combine husbandry and research functions, in which case, the training requirements would reflect the range of their duties.

For researchers, the animal-use protocol serves as a vehicle to formally identify research personnel who will have contact with animals and to describe their individual qualifications for performing their animal-related duties. The animal-use protocol is then used by a training coordinator or the attending veterinarian to assess the training needs of the investigator's staff in the proposed research activity. Assessment of training needs should take place in collaboration with the principal investigator or staff.

Husbandry and research staff should receive training in the hazards involved in animal research and related safety practices and equipment. These hazards are classified as biological, chemical, radiological, or physical agents. A federal law called the Occupational Safety and Health Act of 1970 requires that a safe working environment be provided to employees.[15] That training on the nature of the occupational hazards and related safety practices is integral to compliance with that law. This standard applies to all staff with regard to hazards associated with the animal facility. For specific recommendations on hazard assessment and safety practices in an animal facility refer to the "Occupational Health and Safety in Animal Care and Use Programs" from the National Research Council.[16]

Staff who only work in the vicinity of animals or who work with biological samples obtained from animals also require training as relates to the institutional occupational health and safety program. This includes individuals who perform tasks involving maintenance, transportation, or administration. Such staff may be exposed to allergens, animal wastes and biological samples, or physical and radiological hazards used in research. The level of risk should be assessed for each type of staff to determine whether safety training is necessary and what topics should be included.

Staff having indirect contact with animals should also be offered training on how the animals used in the research program are appropriately handled and treated in compliance with the relevant laws. Workers with duties peripheral to an animal facility may fear the nature of the work performed on the animals, or they may have qualms that the care of the animals may be inappropriate. These individuals can benefit from learning about the general nature of the regulatory mandates and the institution's commitment to these standards. This training may also be helpful to the organization to avert the development of animal rights activism in persons who have no preconceived bias against animal research, but may be receptive to animal rights notions in an institutional environment where information is not forthcoming about the animal care and use program.

## Professional Qualifications

Certification in competence areas by professional organizations in the United States are encouraged by regulatory authorities (USDA and OLAW) and by accrediting bodies (AAALAC). The *Guide for the Care and Use of Laboratory Animals* identifies technician certification as an option for the provision of staff training (Institutional Responsibilities, Personnel Qualifications, and Training, page 13):[10]

> There are a number of options for the training of technicians. Nondegree training, with certification programs for laboratory animal technicians and technologists, can be obtained from the American Association for Laboratory Animal Science (AALAS).

Professional societies have developed three certification systems to certify the knowledge and competence of animal research staff in key job descriptions:

Laboratory animal technician certification (three levels) by the American Association for Laboratory Animal Science (AALAS):

Assistant Laboratory Animal Technician (ALAT)
Laboratory Animal Technician (LAT)
Laboratory Animal Technologist (LATG)

Laboratory animal facility manager certification by the American Association for Laboratory Animal Science (AALAS), Laboratory Animal Management Association (LAMA), and the Institute for Certified Professional Managers (ICPM):

**Table 5.1    General Training Objectives Recommended for Each Staff**

Adapted from *Education and Training in the Care and Use of Laboratory Animals: A Guide for Developing Institutional Programs, a report of the Institute of Laboratory Animal Resources Committee on educational programs in laboratory animal science (U.S. National Research Council).*[14]

### Animal Care Technicians

| Regulatory Matters | Animal-Related Procedures | Occupational Safety and Health |
|---|---|---|
| Be aware of animal welfare laws, regulations, policies, and guidelines. Know that all animals are covered by a protocol. Know and fill out cage card information. Know how to report perceived deficiencies of animal care and use. | Know healthy behavior and appearance of animals, and recognize abnormalities. Handle and restrain animals humanely and safely. Identify, sex, and mark individual animals. Perform sanitation procedures for caging and facilities. Monitor and record room conditions, e.g., temperature and humidity. Perform animal use procedures humanely, e.g., blood collection and injections, if these are appropriate to the job functions. | Identify workplace hazards. Know precautions to take for each hazard. Use protective equipment properly. Know how and where to get medical aid. |

### Researchers

| Regulatory Matters | Animal-Related Procedures | Occupational Safety and Health |
|---|---|---|
| Know the principles of laws, regulations, policies, and guidelines that apply to animal research. Know the animal issues that are covered by a protocol, i.e., pain and distress; justification of animal use, species, and animal numbers; consideration of alternatives; etc. Understand responsibilities of a principal investigator for overseeing animal welfare compliance and occupational safety by staff. Understand the authority and function of the IACUC. Know how to report perceived deficiencies of animal care and use. | Understand species anatomy and physiology. Perform humane procedures for animal use methodologies, i.e., restraint, anesthesia, asepsis, surgery, euthanasia, etc. Know methods to alleviate pain and distress related to the procedures conducted on animals. Know how to provide support for the animal before, during, and after a procedure, such as, but not limited to, anesthetic monitoring and postsurgical care. Assure that staff are qualified to perform animal use methodologies (principal investigators). | Identify hazards and use precautions related to research projects. Oversee training on proper precautions for staff (principal investigators). Ensure all staff are enrolled in an occupational-health program (principal investigators). |

### Institutional Staff Having Indirect Contact with Animals
### (Technical, Administrative, Transportation, and Maintenance)

| Regulatory Matters | Occupational Safety and Health |
|---|---|
| Know animal welfare laws, etc., ensure the humane care of animals in research. Know how to report perceived deficiencies of animal care and use. Be aware of animal-related and other hazards in facility areas. | Know safety procedures and practices. Know how and where to get medical aid. |

Certified Manager of Animal Resources (CMAR)

Surgical Technician Certification by the Academy of Surgical Research (ASR)

Surgical Research Specialist (SRS)

In each of these certification programs, educational and work-experience criteria are specified for eligibility, and candidates must pass an examination to attain certification. Certified laboratory animal technicians may maintain an additional credential as a registered animal technician (at all three certification levels) by continuing their education in a voluntary program known as the AALAS Technician Certification Registry. To maintain their credentials, certified managers of animal resources and certified surgical research specialists must periodically recertify by complying with continuing-education requirements. AALAS provides educational resources to support its certification programs and to offer continuing-education opportunities.

Many institutions encourage their staff to attain certification in the previously mentioned specialty areas to support compliance with the training requirements of the federal laws and policies. Oftentimes, institutions offer financial incentives, such as pay raises, bonuses, and payment of certification fees. Some institutions may even require certifications as a job requirement or a promotion criterion.

Although the training for professional qualifications (i.e., as veterinarians or veterinary technicians) is beyond the scope of an institutional training program, the institution can enhance its compliance with U.S. national animal-welfare mandates by hiring technical staff with degrees and licensing as veterinary technicians. For example, veterinary technicians who have a two-year degree in veterinary technology and are state-licensed are well qualified to fill positions as research technicians, in addition to assuming duties in veterinary technical and animal care positions.

Institutions should provide opportunities and support for continuing education to staff who are professionally licensed, certified, or registered, so that they may maintain their qualifications. In informal statements on the attributes of a training program, the NIH Office for Laboratory Animal Welfare (OLAW) refers to AALAS technician certification and state professional licensing as indications of staff qualification:[13]

> [OLAW] strongly recommends that institutions offer their staff access to training leading to certification in animal technology, such as that available from AALAS or a formally designated academic program. Institutions should also know and ensure compliance with any initial and continuing-education [s]tate requirements for the licensing of veterinary or animal health technicians.

Likewise, the *Guide for the Care and Use of Laboratory Animals* stresses the importance of continuing education (page 13) to maintain staff qualifications:[10]

> Personnel using or caring for animals should also participate regularly in continuing-education activities relevant to their responsibilities. They are encouraged to be involved in local and national meetings of AALAS and other relevant professional organizations.

## Program Trainers

The legal responsibility for meeting training requirements related to animals used in research technically lies with the institution, and so this responsibility falls to the Institutional Animal Care and Use Committee (IACUC). Since the IACUC has oversight responsibilities over all aspects of the animal care and use program, it takes on the role of assuring that staff training meets the standards imposed by federal mandates. A common approach is to designate a staff member as a training coordinator to assure that all staff who work with animals are provided with appropriate training services. Oftentimes under the direction of a training coordinator, additional individuals on staff at U.S. animal facilities are involved in providing training services on species and methods for which they have particular expertise. When expertise is not available within the animal facility, training may be sought from outside sources, such as researchers in academic departments, i.e., individuals with no reporting relationship to the animal facility, and experts from other institutions. In areas where multiple research institutions are in close proximity, training programs should capitalize on the expertise available at other institutions. Whether training is provided by in-house staff or outside experts, it is a best practice that the institutional training coordinator oversee all training activities.

The training of the IACUC to perform its program oversight may be beyond the scope of many program trainers. The IACUC's training should encompass the review of animal protocols and activities, review of program policies, and inspection of facilities. Often, the IACUC chair, the IACUC administrator, or the attending veterinarian have the task of orienting and training new IACUC members to their role. It is also desirable if an experienced IACUC member can be assigned as a mentor to the new member; this is especially important for the nonscientist or nonaffiliated member, since such persons require an additional orientation to the science and terminology of animal research.

## Verification of Training — Animal Use Competence

In response to the U.S. federal requirement that animal research staff be qualified to work with laboratory animals, many institutions verify the competence of staff in animal-use activities however the skills are obtained, e.g., via the institutional training program or training by colleagues. Verification of competence is relatively straightforward with animal care staff, due to the lines of authority within the animal facility. A greater challenge is to verify competence for research staff who have no reporting relationship with the animal facility unit. Implementing a system for research staff necessitates an institution-wide policy and administrative support. Methods to assess competence in animal use among research staff generally follow two basic approaches. Some institutions utilize a "certification" process whereby designated trainers visit the lab and observe the conduct of animal-use procedures. Individual researchers, or in some cases entire labs, receive a recognition, or "certification," of their competence for specific procedures. This recognition may encompass the authorization to train others in the same procedure. Institutions using this method typically maintain a documentation of "certified" individuals or labs. A second approach couples competence verification with a generalized assessment of compliance with animal welfare mandates. Designated compliance staff visit labs in a rotating schedule to assess compliance and to offer training as needed. Such visits should have the objectives of verifying that animal-use procedures are conducted in accordance with relevant mandates, verifying that drugs and medical materials are current with respect to product expiration dates, providing information on federal mandates and institutional policies, and distributing related literature, including guidelines and news bulletins. Whatever the approach used by an institution, the balance between training and compliance should emphasize foremost the training service. Both training and compliance staff should have a cooperative and helpful attitude toward the research staff, commensurate with principles of customer service. Training and compliance staff should remain mindful that their primary goal is to provide a support service to animal research through their role of assuring compliance with federal and institutional mandates.

## Recordkeeping

In the United States, there is no specific statement in federal laws, regulations, or policies to maintain documentation of training. That is, documentation of training is not included among the types of facility records that must be maintained by an institution and which are inspectable by federal authorities, according to federal laws or regulations. Nevertheless, federal and accrediting authorities consistently expect to have access to training records during an inspection of a research institution. The USDA has affirmed this expectation in published articles on the subject of training for compliance with the Animal Welfare Act.[17,18] Because training programs are federally mandated, a system of documentation is the only practical way for an institution to prove, and for inspectors to verify, compliance with the training requirements. Record keeping of training activities is therefore a practical necessity for compliance with federal animal welfare mandates.

There is no mandate for the system of training documentation, i.e., as an engineering standard. That is, there is no specification for the format to be used for the records, who should maintain the records, or where or how the records should be stored. The training records may be electronic (e.g., residing in a database, spreadsheet, etc.) or paper-based. If electronic, they may be stored locally on a computer, or on an intranet, or on a host server inside or outside of the institution. Records may be accessed or filed primarily by individuals, departments, or by training activities. The training records may be generated and archived by a training coordinator, the IACUC, or another administrative unit. The records may be stored centrally or segregated by administrative units.

As in the operation of a training program, federal and accrediting authorities apply a performance standard for the documentation of training activities. The expectation is that the documentation system reflects a complete profile of the training program, that the records accurately reflect the training activities, that the records are comprehensible, and that all documentation are readily accessible on demand. To be adequate, the documentation should demonstrate that the institution's training program meets the objectives of government mandates (see previous sections).

Many institutions use database software systems as a means to track and document staff training activities. Commercial database systems for animal facility management often incorporate a module for training documentation. Such systems may allow some software customization to enhance the system's accommodation of the institution's training program features. In addition, some database systems offer connectivity of the database with e-mail to facilitate the communication with staff on training matters. For institutions that choose a stand-alone training records database, a system customized for training in the laboratory animal field is available for free from the Laboratory Animal Welfare Training Exchange (LAWTE). The program, based on Microsoft Access 1997, can be downloaded from the LAWTE Web site (http://www.lawte.org) from the "Exchange" page. This program contains two tracking systems to document staff training and compliance with institutional programs of occupational health and safety.

## Training Resources

Investigators should look to the veterinary or training staff of the animal facility for information about and training on animal care and use procedures. The veterinary and training staff members offer expertise in basic methodologies, and they can recommend other resources that may best address a specialized training need.

A growing number of resources are available for enhancing a training program. Instructional media (videotapes, CD-ROM, and Web-based) offer content that can be used to either augment in-class teaching or be used for self-directed learning. Models and devices that simulate anatomical structures are valuable aids for teaching skills for performing animal procedures.

The following organizations offer information on training materials that are available from a variety of sources.

## ORGANIZATIONS

### American Association for Laboratory Animal Science (AALAS)

As a membership organization of laboratory animal professionals, AALAS (http://www.aalas.org) offers an exchange of information and expertise in the care and use of laboratory animals:

- AALAS provides training materials for animal researchers and animal care technicians, including an online learning system, the AALAS Learning Library (http://www.learninglibrary.org).
- The AALAS National Meeting includes a Learning Resource Center that has a large assortment of instructional media for viewing by meeting attendees.
- Educational sessions at the AALAS National Meeting offer many training opportunities in the laboratory animal field, including Train the Trainer workshops, which focus on basic and advanced methods of teaching.
- The COMPMED and TECHLINK listservs hosted by AALAS provide a forum for discussion of laboratory animal issues and methodologies.
- The IACUC Web site (http://www.iacuc.org) includes links to training programs on laboratory animal welfare at U.S. research institutions and links to online information on training media for the laboratory animal training field.

## International Network for Humane Education (InterNICHE)

InterNICHE (http://www.interniche.org) is a network of students, teachers, and animal campaigners that focuses on animal use and alternatives within biological science, medical, and veterinary medical education. The InterNICHE Alternatives Loan System maintains a library of multimedia for lending to teachers and students in any country. These materials include CD-ROMs, videos, models, and mannequins that can be used for the teaching of anatomy, physiology, and surgery.

## Laboratory Animal Welfare Training Exchange (LAWTE)

LAWTE is an organization for trainers and training coordinators in the laboratory animal field. Members use a listserv for discussions about training issues and methods. The Web site (http://www.lawte.org) offers information on training media and methodologies. A conference is held every two years in the U.S.

## Norwegian Inventory of Alternatives (Norina)

The Norwegian Reference Centre for Laboratory Animal Science and Alternatives maintains the Norina database (http://oslovet.veths.no/NORINA/), which is an English-language archive of information on training materials for use in biological science education. This database offers an overview of training media (computer programs, CD-ROMs, interactive videos, films, and traditional teaching aids) that can serve as alternatives or supplements to the use of animals in student teaching at all levels of education.

## REFERENCES

1. European Convention for the Protection of Vertebrate Animals Used for Experimental and Other Scientific Purposes, Council of Europe, European Treaty Series (ETS) No. 123, March 18, 1986.
2. European Comission, Directive for the Protection of Vertebrate Animals Used for Experimental and Other Scientific Purposes (86/609/EEC), *Off. J. Eur. Comm.,* L 358, 1, 1986.
3. Federation of European Laboratory Animal Science Associations, Working Group on Education, FELASA Recommendations for the Education and Training of Persons Working with Laboratory Animals: Category A and C, *Lab. Anim.,* 29, 121, 1995.
4. Federation of European Laboratory Animal Science Associations, Working Group on Education of Persons Carrying Out Animal Experiments, FELASA recommendations for the education and training of persons carrying out animal experiments (Category B), *Lab. Anim.,* 34, 229, 2000.
5. Federation of European Laboratory Animal Science Associations, Working Group on Education of Specialists: FELASA recommendations for the education of specialists in laboratory animal science (Category D), *Lab. Anim.,* 33, 1–15, 1999.
6. Animal Welfare Act as Amended, USC, Title 7, Sections 2131 to 2156, U.S. Government Printing Office, Washington, D.C.
7. Code of Federal Regulations, Title 9, Chapter 1 (Animals and Animal Products), Subchapter A (Animal Welfare), Washington, D.C.: Office of the Federal Register, 1985.
8. Health Research Extension Act of 1985, U.S. Public Law 99–158, Section 495, "Animals in Research," U.S. Government Printing Office, Washington, D.C.
9. Public Health Service, Public Health Service Policy on Humane Care and Use of Laboratory Animals, Washington, D.C.: U.S. Department of Health and Human Services, 28 [PL 99–158, Health Research Extension Act, 1985], 1996.
10. *Guide for the Care and Use of Laboratory Animals,* Institute of Laboratory Animal Research, National Research Council, National Academy Press, Washington, D.C., 1996.
11. *Guide for the Care and Use of Agricultural Animals in Agricultural Research and Training,* Federation of Animal Science Societies, Savoy, IL, January 1999.
12. Department of Agriculture, Animal and Plant Health Inspection Service, Animal Welfare: Farm Animals Used for Nonagricultural Purposes, *Federal Register,* Vol. 65, No. 23, February 3, 2000.

13. Potkay, S., Garnett, N.L., Miller, J.G., Pond, C.L., and Doyle, D.J., Frequently asked questions about the public health service policy on humane care and use of laboratory animals, *Lab. Animal*, 24, 24, 1995.

14. *Education and Training in the Care and Use of Laboratory Animals: A Guide for Developing Institutional Programs*, Committee on Educational Programs in Laboratory Animal Science, Institute of Laboratory Animal Research, Commission on Life Sciences, National Research Council, National Academy Press, Washington, D.C., 1991.

15. Occupational Safety and Health Act of 1970, as amended, USC, Title 29, Chapter 15, Section 651. U.S. Government Printing Office, Washington, D.C.

16. *Occupational Health and Safety in Animal Care and Use Programs*, Committee on Occupational Safety and Health in Animal Research Facilities, Institute of Laboratory Animal Research, Commission on Life Sciences, National Research Council, National Academy Press, Washington, D.C., 1997.

17. Slauter, J.E., Evaluation of a training program: A USDA perspective, *Lab. Animal*, 29, 25, 2000.

18. Slauter, J.E., When the USDA veterinary medical officer looks at your training program, Animal Welfare Information Center Bulletin, 10, No. 1–2, Summer 1999.

# CHAPTER 6

# Laboratory Animal Science and Service Organizations

Patri Vergara and Gilles Demers

## CONTENTS

0-8493-1086-5/03/$0.00+$1.50
© 2003 by CRC Press LLC

## INTRODUCTION

In parallel with the development of biomedical research, laboratory animal science and service organizations have been in constant evolution and development over the last 40 years. Demands for a higher quality of animals, together with a greater concern for animal welfare, are the driving force behind the development of organizations that provide support for people working in the field of laboratory animal science.

The first laboratory animal organizations were created in the 1950s and 1960s in North America, Japan, and Europe: AALAS (formerly ACP) in 1950, JALAS in 1952, LASA in 1963, ICLAS in 1967, and CCAC in 1968. Since then, increasing levels of biomedical research in other countries, mainly from Asia and Central and South America, has created an explosion of new laboratory animal science and service organizations around the world.

The role of the International Council for Laboratory Animal Science (ICLAS) as an international umbrella organization is important in this worldwide development. In several parts of the world, regional organizations are created to maintain links between national scientific organizations and to lead the field in providing policies and guidelines related to laboratory animal care and use. FELASA in Europe has played an important role in this respect.

Several countries now have more than one laboratory animal science organization serving different goals, viz. continuing education, training, production of guidelines, scientific communication, accreditation, and certification programs.

In order to give the reader a useful and practical overview of the principal laboratory animal science organizations around the world, the authors have classified them according to their primary aims and scope, i.e., international organizations, laboratory animal science associations, professional organizations, animal care and welfare organizations, and miscellaneous associations.

However, this chapter is limited to providing basic information about the main laboratory animal science organizations. It is neither all inclusive nor exhaustive, as the continuous growth of laboratory animal science and the ongoing creation of new organizations make this impossible. The reader should visit organization Web sites for further information. When a Web site was not available at the time of publication, a mailing or e-mail address is given.

# INTERNATIONAL ORGANIZATIONS

## International Council for Laboratory Animal Science (ICLAS)

### Background

Established in 1956 under the auspices of the Council for International Organizations of Medical Sciences (CIOMS), the International Union of Biological Sciences (IUBS), and the United Nations Educational, Scientific, and Cultural Organization (UNESCO), ICLAS is a nongovernmental organization for international cooperation in laboratory animal science.

### Aims

As stated in its constitution, the aims of ICLAS are:

- To promote and coordinate the development of laboratory animal science throughout the world
- To promote international collaboration in laboratory animal science
- To promote quality definition and monitoring of laboratory animals
- To collect and disseminate information on laboratory animal science
- To promote the humane use of animals in research through recognition of ethical principles and scientific responsibilities

### Membership

ICLAS has national, scientific, union, and associate members, which number about 32, 29, 6, and 37, respectively. National members represent national perspectives; scientific members represent national and regional laboratory animal science associations; union members represent international nongovernmental unions; and associate members represent commercial, academic, and scientific organizations that support the aims of ICLAS.

### Strategic Plan

The ICLAS governing board has developed a strategic plan to guide the organization through the next several years. The strategic plan includes the mission statement of ICLAS: "The International Council for Laboratory Animal Science advances human and animal health by promoting the ethical care and use of animals in research worldwide." ICLAS strives to act as a worldwide resource for laboratory animal science knowledge; to be the acknowledged advocate for the advancement of laboratory animal science in developing countries and regions; and to serve as a premier source of laboratory animal science guidelines and standards, and as a general laboratory animal welfare information center.

### Meetings

An international scientific meeting is held in association with the general assembly every four years. It is organized by a national or scientific member and is often held in association with regional or local organizations. Other regional meetings and courses are organized by individuals in the various regions of the world under the auspices of six ICLAS regional committees for the following regions: Europe, Asia, Africa (French and English regions), Oceania, and the Americas. This has allowed ICLAS to focus on each region and to apply local intellectual resources to issues within each of the regions.

### Publications

- ICLAS News
- ICLAS FYI Bulletin
- Web site: www.iclas.org

## LABORATORY ANIMAL SCIENCE ASSOCIATIONS

### Membership

In terms of membership, laboratory animal science associations consist of professionals whose work is related to laboratory animals, i.e., scientists, veterinarians, animal technicians, educators, etc. Although there are some exceptions, members are normally from the geographical area where the association is based. Some associations, such as AALAS or LASA, also have an international membership.

### Aims

The aims shared by most laboratory animal science associations are:

- To promote a better and more rational use of laboratory animals, following the ethical principles of the 3Rs
- To provide training and education in laboratory animal science and welfare
- To advance the knowledge, skills, and status of those who care for and use laboratory animals
- To promote informed public discussion regarding the use of animals in research
- To inform scientists about the appropriate use and care of animals in research

## LABORATORY ANIMAL SCIENCE ORGANIZATIONS BY REGIONS OF THE WORLD

### Europe

#### *Federation of European Laboratory Animal Science Associations: FELASA*

##### Background

FELASA is composed of independent European national and regional laboratory animal associations, and was established by them in 1978. It is currently composed of 12 European associations and represents laboratory animal scientists and technologists in at least 20 European countries. FELASA is managed solely by representatives of its constituent associations.

##### Aims

- To represent common interests of constituent associations in the furtherance of all aspects of laboratory animal science by coordinating the development of education, animal welfare, health monitoring, and other aspects of laboratory animal science in Europe by such means as meetings, study groups, and publications
- To act as a focus for the exchange of information on laboratory animal science among European states
- To establish and maintain appropriate links with international and national bodies, as well as with other organizations concerned with laboratory animal science
- To promote the recognition and consultation of FELASA as the specialist federation in laboratory animal science and welfare throughout Europe

##### Programs

Within Europe, FELASA sees as its role not only to respond rapidly to European Union and Council of Europe developments, but also to guide the thinking of these bodies by offering them timely and authoritative advice. Within the purpose of establishing a common European standard for laboratory animals, FELASA supports education and training programs inside Europe.

FELASA has published recommendations on education and training for persons responsible for the welfare of laboratory animals, for persons responsible for the design and conduct of studies involving animals, and for specialists in laboratory animal science. It has published several recommendations for the health monitoring of breeding and experimental units, for the detection, relief, and control of any pain or suffering in them, and a guidance paper for the accreditation of laboratory animal diagnostic laboratories. It has also recently approved guidance for accreditation of education and training programs.

### Meetings

A FELASA international scientific meeting is organized every three years by each of the constituent organizations in turn. Proposals for more frequent meetings are currently under discussion.

### Publications

FELASA's recommendations and guidelines are published on FELASA's Web site, www.felasa.org, and in *Laboratory Animals* (www.lal.org.uk), the official journal of FELASA, LASA, GV-SOLAS, NVP, SGV, and SECAL. *Laboratory Animals* is a peer-reviewed journal, published quarterly, and is a leading publication in the field of laboratory animal science.

## FELASA Constituent Associations

### Association Française des Sciences et Techniques de l'Animal de Laboratoire: AFSTAL (French Association of Laboratory Animal Sciences and Techniques)

#### Background

It was known until 1999 as the Société Française d'Expérimentacion Animale (SFEA).

#### Membership

AFSTAL has over 500 members, including scientists, veterinarians, laboratory and animal technicians, specialists in laboratory animal science, breeders, and suppliers.

#### Meetings

Two annual meetings are held, including a scientific meeting and a one-day technical meeting.

#### Publications

*Journal Sciences et Tecnhniques de l'Animal de Laboratoire, STAL.*

#### Address

28, Rue Saint Dominique, 75007, Paris. E-mail: afstal@free.fr

### Associazione Italiana per Scienze degli Animale da Laboratorio: AISAL (Italian Association for Laboratory Animal Sciences)

#### Aims

AISAL has mainly focused on organizing scientific meetings and on educational activities for various professionals.

#### Membership

AISAL has over 150 members, including scientists, graduate students, and technicians who perform scientific work or teach, as well as some institutions related to AISAL's interests.

#### Meetings

An annual symposium is organized in autumn in conjunction with the general assembly of AISAL.

#### Publications

These include teaching texts, three books in Italian, and national guidelines on Educational Programs in Laboratory Animal Science and Health and Safety in Animal Experimentation. The *AISAL News* is a quarterly booklet including information and news on AISAL programs and activities, as well as on national and international events and publications.

### Address

Website: www.aisal.org

## Baltic Laboratory Animal Science Association: Balt-LASA

### Background

Balt-LASA serves scientists and specialists in three Baltic countries — Estonia, Latvia, and Lithuania — whose professional activities are connected with the breeding, research, and use of laboratory animals.

### Membership

There are 75 members, mostly from the Baltic countries, but also from other Nordic countries, as well as from Russia and Eastern Europe.

### Programs

Balt-LASA is mainly focused on promoting European standards in the care and use of laboratory animals in Eastern Europe, and has participated actively, both with ICLAS and FELASA, to promote education and training in Central and Eastern Europe.

### Publications

*Baltic Journal of Laboratory Animal Science* (vip.latnet.lv/journalLAS/) is the official periodical (quarterly) publication of Balt-LASA. It publishes both reviews and original articles, and is the only English-Russian–language journal in Eastern Europe devoted to laboratory animal science. Balt-LASA holds an annual conference and organizes seminars and courses on laboratory animal science.

### Address

Balt-LASA, 53 Krustpils St, Riga, LV-1057, Latvia. E-mail: gjakobsonegrindeks.lv

## Belgian Council for Laboratory Animal Science: BCLAS

### Membership

There are over 100 regular members, as well as sponsoring and honorary members.

### Programs

The main activities of BCLAS are to organize symposia on problems related to laboratory animals, and courses and working groups focused on education, health, pathology, feeding, and ethical problems.

### Address

John Verstegen, Department of Small Animal Clinical Studies, Veterinary College of the University of Liège, Blvd Colonster 20 Blg 44, B-4000 Liège/Belgium. E-mail: j.verstegen@ulg.ac.be

## Czech Laboratory Animal Science Association: CLASA

### Membership

There are 35 members.

### Programs

Training courses are the main focus of CLASA activities. CLASA holds a scientific conference every two years and special workshops annually.

### Publications

*CLASA Newsletter* is published in Czech.

### Address

Lukas Jebavy, President, Institute of Pharmacology and Biochemistry, Konarovice u Kolina 279, CZ-28125 Konarovice, Czech Republic. E-mail: jebavy@biotest.cz. Michael Boubelik, boubelik@biomed.cas.cz.

## Gesellschaft für Versuchstierkunde: GV-SOLAS (German Society for Laboratory Animal Science)

### Background

GV-SOLAS is a society for the enhancement of laboratory animal science.

### Aims

GV-SOLAS sees itself as a source of information on laboratory animals and animal welfare for government officials and the general public in Germany.

### Membership

Ordinary members (currently about 600) are scientists with an interest in laboratory animal science. Nonscientists may be accepted as extraordinary members and companies and organizations as sponsors.

### Meetings

The General Assembly of GV-SOLAS is held yearly in conjunction with its scientific meeting.

### Publications

GV-SOLAS has no journal of its own, but shares *Laboratory Animals* with FELASA and other European Associations. GV-SOLAS also publishes a series of reports as the result of the activities of working parties.

### Awards

Awards are given for studies in the furtherance of laboratory animal science.

### Address

Website: www.mh-hannover.de/einrichtungen/tierlabor/gv-solas.

## Hellenic Society of Biomedical and Laboratory Animal Science: HSBLAS

### Membership

There are approximately 80 members.

### Programs

HSBLAS organizes training sessions consisting of lectures and videos on the basic principles of laboratory animal science. HSBLAS also organizes round tables in conjunction with the Annual Medical Congress of Greece and participates in the organization of courses for scientists in conjunction with other groups.

### Address

15b Agiou Thoma Street, GR 115 27 Athens, Greece. E-mail: Ismene Dontas, ismene@prodeco.gr; Nickolas Kostomitsopoulos, nkostom@hol.gr.

## Laboratory Animal Science Association: LASA

### Background

LASA was founded in the U.K. in 1963 by representatives from industry, universities, government ministries, and research councils.

### Aims

The goals of LASA are to provide information and a forum for ideas on the science of animals used in research.

### Membership

There are over 400 members who are experienced in various aspects of laboratory animal science, provision, care, and use.

### Meetings

The main scientific meetings of LASA consist of a one-day meeting in the spring and a two-day residential meeting in autumn or winter.

### Publications

The *LASA Newsletter*, which is sent free to members, contains notices of forthcoming meetings, reports on conferences, new items on animal science, and correspondence from members. Members also receive *Laboratory Animals*. LASA creates special study groups to prepare guidelines for publication in *Laboratory Animals*.

### Address

Website: www.lasa.co.uk.

## Nederlandse Vereniging voor Proefdierkunde: NVP (Dutch Association for Laboratory Animal Science)

### Aims

NVP is a scientific association that promotes a responsible attitude towards the use of laboratory animals for experimental purposes. A unique feature of NVP is the open atmosphere between NVP and the main animal welfare organizations in the Netherlands, in which both parties are willing to listen to each other's arguments and accept differences in points of view.

### Membership

Among the 200 NVP members are the Dutch animal welfare officers, members of IACUCs, representatives of the Inspectorate of the National Competent Authority, scientists of various disciplines, and technicians.

### Meetings

National biannual symposia are held, of which abstracts are published in *Laboratory Animals*. NVP has two committees working on a permanent basis on the organization of the NVP symposia and the publication of the *NVP Newsletter*, which is aimed at a public audience. There are also temporary working groups on specific issues.

### Address

Website: www.proefdierkunde.nl.

### Scandinavian Society for Laboratory Animal Science: Scand-LAS

#### Membership

There are approximately 430 members, mainly from the Nordic and Baltic countries. Membership is open to everyone working within the field of laboratory animal science.

#### Programs

Standing working groups on education, health monitoring, and pain, stress, and discomfort, as well as policy making.

#### Meetings

Every spring, in conjunction with the general assembly. Following a fixed order, Denmark, Norway, Finland, and Sweden are the host countries for these meetings. Scand-LAS has working groups on education, pain and stress, and information.

#### Publications

Scand-LAS publishes a quarterly journal, *Scandinavian Journal of Laboratory Animal Science*, that is free to members of Scand-LAS. The Scand-LAS Web site provides other publications and guidelines, including Tech Tips.

#### Address

Website: www.scandlas.org.

### Sociedad Española para las Ciencias del Animal de Laboratorio: SECAL (Spanish Society for Laboratory Animal Science)

#### Membership

The membership includes scientists and animal technicians working within the field of laboratory animal science. The current membership is over 200, primarily from Spain, but also from Latin American countries.

#### Programs

SECAL has focused strongly on training and education, organizing and supporting specific courses, and also supporting applications to Spanish or European agencies for grants to promote education and training in Spain and in Latin American countries.

#### Meetings

SECAL organizes a symposium every two years in conjunction with its general assembly.

#### Publications

SECAL has adopted *Laboratory Animals* as its official journal, but also publishes *Animales de Laboratorio*, a newsletter with information about relevant national issues, international events, publications, etc. and review articles in Spanish about special topics.

#### Address

Website: www.secal.es.

## *Schweizerische Gesellschaft für Versuchtierkunde: SGV (Swiss Laboratory Animal Science Association)*

### *Membership*

The membership includes professionals interested and engaged in laboratory animal science. Organizations or persons intending to support the activities of SGV are welcome as institutional members. SGV has about 130 members.

### *Meetings*

Every year, a scientific meeting is organized, usually as a symposium at the annual meeting of the Swiss Union of Societies for Experimental Biology, of which SGV is a member. A two-day training course or workshop on a selected topic is held in the autumn.

### *Publications*

An informal newsletter is published twice a year. SGV has adopted *Laboratory Animals* as its official journal.

### *Awards*

An award is given every year for an outstanding contribution in the field of laboratory animals.

### *Address*

Website: www.sgv.org.

## Other European Associations Not Affiliated With FELASA

### *Croatian Society for Laboratory Animal Science: CSLAS*

### *Address*

Jadranka Bubic, PLIVA, Istrazivacki Institut, Prilaz Baruna Filipovica 25, 10000 Zagreb, Croatia.

### *Finland Laboratory Animal Science: FinLAS*

### *Background*

FinLAS was founded in 1986.

### *Membership*

There are approximately 50 members, who have academic backgrounds as biologists, physicians, or veterinarians.

### *Meetings*

FinLAS organizes two annual meetings with one-day seminars. Every fourth year, FinLAS is responsible for organizing the Scand-LAS meeting. FinLAS has several working groups; the most important one is on education.

### *Address*

Website: www.uku.fi/jarjestot/FinLAS.

### Hungarian Laboratory Animal Scientists' Society: HLS

#### Address

University of Veterinary Science Budapest, Deptartment Animal Breeding, Nutrition and Laboratory Animal Science, P.O. Box 2, H-1400 Budapest, Hungary. E-mail: iszekell@hs.univet.hu

### Sociedad Española de Experimentación Animal: SEEA (Spanish Society of Animal Experimentation)

#### Address

J.R. Morandeira, Unidad Mixta de Investigación, Domingo Miral, s/n, 50009 Zaragoza, Spain. E-mail: jmoran@posta.unizar.es.

## Americas

### North America

### American Association for Laboratory Animal Science: AALAS

#### Background

The American Association for Laboratory Animal Science strives to be the premier forum for the exchange of information and expertise in the production, care, and use of laboratory animals.

#### Aims

AALAS advances responsible laboratory animal care and use to benefit people and animals.

#### Membership

Established in 1950, AALAS now comprises more than 10,000 clinical veterinarians, technicians, technologists, biomedical scientists, educators, and businesspeople representing all aspects of the laboratory animal research field. The three levels of membership — bronze, silver, and gold — are available to individuals, businesses, and institutions.

#### Programs

AALAS offers certification at the technician and management levels, a certification registry, continuing-education opportunities, and a public outreach program. The AALAS Technician Certification Program is a nationally recognized, professional, and authoritative endorsement on three levels of technical knowledge: Assistant Laboratory Animal Technician (ALAT), Laboratory Animal Technician (LAT), and Laboratory Animal Technologist (LATG). The AALAS Management Certification program recognizes individuals committed to excellence in animal resource management with the Certified Manager Animal Resources (CMAR) designation. The AALAS Certification Registry indicates that participants maintain a current level of knowledge.

#### Meetings

Continuing-education programs include the Institute for Laboratory Animal Management (ILAM), an annual educational program concentrating on industry-management concepts; the AALAS national meeting, the largest annual gathering of professionals concerned with the production, care, and responsible use of laboratory animals; and the Management and Technology Conference, an annual gathering offering presentations on technologies that impact the laboratory animal facility manager. The AALAS Campus (www.aalascampus.org) is an online learning forum offering over 650 courses in personal and career development, as well as laboratory animal science courses developed by leading educators, industry experts, and clinicians.

### Awards

AALAS presents several professional and technical awards for excellence in the field of laboratory animal science each year at the national meeting.

### Publications

AALAS offers numerous publications to its members. *Comparative Medicine*, a bimonthly journal listed in *Index Medicus*, is a leading publication in the field of comparative and experimental medicine. The bimonthly journal *Contemporary Topics in Laboratory Animal Science* features articles on clinical, technical, managerial, and philosophical subjects, as well as association news. *Tech Talk* is a quarterly newsletter offering current information and technology. The *AALAS Reference Directory* includes a Products and Services Guide section, technical information, and certification/education information and forms. AALAS also publishes training manuals for each certification level and other materials for professional development and education, with Spanish versions and CD-ROMs currently in development.

### Public Outreach

AALAS seeks to inform the public about the importance of animal research to animals and people alike. Representatives from AALAS frequent many local, state, and national teachers-association meetings to provide classroom resources, and a variety of free and inexpensive educational materials are offered to members and the public.

### Address

Website: www.aalas.org. Related sites maintained by AALAS are www.IACUC.org, www.kids4research.org, and http://foundation.aalas.org. Additionally, AALAS manages three laboratory animal science-related listservs: CompMed, TechLink, and IACUC-FORUM.

## Canadian Association for Laboratory Animal Science - Association canadienne pour la science des animaux de laboratoire: CALAS/ACSAL

### Background

The Canadian Association for Laboratory Animal Science (CALAS/ACSAL) was founded in 1962. The association is composed of a multidisciplinary group of people, institutions, and regional chapters concerned with the care and use of laboratory animals in research, teaching, and testing.

### Aims

CALAS/ACSAL is dedicated to the elimination of inhumane and unnecessary use of animals in research and to the improvement of their standard of care. The aims can be summarized as follows:

- To advance the knowledge, skills, and status of those who care for and use laboratory animals
- To improve the standards of animal care and research
- To provide a forum for the exchange and dissemination of knowledge of sound animal care and research

### Membership

CALAS/ACSAL membership is at several levels, as follows: Institutional membership is intended for institutions that are concerned with the care and use of laboratory animals and wish to participate in efforts to improve the scientific quality of animal research and the educational standards of animal-care personnel. The institutions obtaining such membership are issued two membership cards for named representatives and receive a certificate of recognition. National membership carries full voting rights and the entitlement to be elected to the board of directors. Registered Technicians are national members who have successfully met the requirements of the registry board. Student membership is offered at a reduced rate for students.

### Programs

To advance the knowledge and skills of those who care for and use laboratory animals, the CALAS/ACSAL registry board, through its examination process, enables technicians working in a variety of laboratory animal settings to become registered at one or more of the following levels of registration: Associate Registered Laboratory Animal Technician, Registered Laboratory Animal Technician, Registered Laboratory Animal Technician (Research), Registered Master Laboratory Animal Technician, and Registered Master Laboratory Animal Technician (Research).

### Meetings

Each year, CALAS/ACSAL hosts a national symposium in a different part of Canada to provide a forum for the exchange and dissemination of knowledge in the areas of sound animal care and research. Scientific and technical papers and workshops are based on a main theme and complemented by exhibits mounted by commercial firms to display their products. The annual general meeting of members is held at the national symposium.

### Publications

Four times a year, every member of CALAS/ACSAL receives a *Member's Magazine*, which is intended to keep members informed of CALAS/ACSAL events and to provide information on pertinent topics. The magazine also includes scientific and technical articles presented at annual CALAS/ACSAL symposia, as well as articles on continuing education.

### Address

Website: www.calas-acsal.org.

## Central America and Mexico

### Asociación Centroamericana del Caribe y Mexicana de la Ciencia de Animales de Laboratorio: ACCMAL (Central American, Caribbean, and Mexican Association of Laboratory Animal Science)

### Membership

ACCMAL members (over 200) come from Costa Rica, Mexico, Honduras, El Salvador, Nicaragua, Panama, Cuba, and Colombia.

### Programs

ACCMAL is highly active, organizing training programs directed at people working on laboratory animal science.

### Awards

In collaboration with Purina, ACCMAL provides awards to young scientists and experts on laboratory animal science from the Americas.

### Address

Liliana Pazos, Universidad de Costa Rica, Laboratorio de Ensayos Biologicos, San Jose, Costa Rica. E-mail: lpazos@cariari.ucr.ac.cr.

### Associación Mexicana para las Ciencias del Animal de Laboratorio: AMCAL (Mexican Association for Laboratory Animal Science)

### Membership

There are 50 members.

## Programs

One of the main aims of AMCAL is the organization of education programs. AMCAL courses deal with genetics, animal diseases, introduction to animal experimentation, and topics for technicians.

## Publications

AMCAL publishes in several journals, such as *Animales de Experimentación* and the *Mexican Association of Veterinarian Specialists in Small Species (AMMVEPE)* journal. AMCAL will soon publish its own bulletin.

## Meetings

AMCAL organizes a scientific meeting every two years.

## Address

Website: amcal2000.tripod.com.

# South America

# Federation of South American Societies and Associations of Laboratory Animal Science Specialists: FESSACAL

## Aims

The aims of FESSACAL are to coordinate efforts and provide a common representation to governments and the international scientific community and to organize joint meetings and disseminate scientific information among members.

## Membership

The membership includes over 570 professionals from several South American laboratory animal associations.

## Address

Website: www.cemib.unicamp.br.

## FESSACAL Constituent Associations

## Asociación Argentina de Especialistas en Animales de Laboratorio: AADEAL (Argentinean Association of Researchers Who Use Laboratory Animals)

## Aims

The aims are to promote laboratory animal science in Argentina and to unify concepts and procedures in experimental animal breeding, care, and management, as well as to organize education and training courses.

## Membership

There are currently approximately 80 biomedical professionals.

## Programs

AADEAL is working on a project for a national law for the care and use of experimental animals to be presented to the Argentinean government.

## Publications

A quarterly bulletin is published for members.

### Address

Adela Rosenkranz, Parana 557, Piso 3 A, 1017 Buenos Aires, Argentina. E-mail: adelar@decanato.de.fcen.uba.ar

## Asociación Uruaguaya de Ciencias del Animal de Laboratorio: AUCAL (Uruguayan Association for Laboratory Animal Science)

Information on AUCAL can be found on the FESSACAL Web site.

## Asociación de Veterinarios Especialistas en Animales de Laboratorio: AVECAL (Chilean Association of Veterinary Specialists in Laboratory Animal Science)

Information on AVECAL can be found on the FESSACAL Web site.

## Colégio Brasileiro de Experimentação Animal: COBEA (Brazilian Laboratory Animal Science Association)

### Aims

The aims are to promote training courses, meetings, congresses, workshops, and symposia in order to spread scientific information on laboratory animal science.

### Membership

There are approximately 260 members who are researchers, animal technicians, students, people working in life sciences, and commercial organizations associated with the breeding, care, study, or manufacturing of goods for laboratory animals.

### Programs

COBEA has committee-organized programs on education and training, ethics and legislation, and organization and management of laboratory animal facilities, as well as a scientific committee and an ad hoc committee.

### Meetings

There is a congress every two years.

### Publications

A newsletter is published twice a year.

### Address

Website: www.meusite.com.br/cobea.

## Sociedad Venezolana de Ciencias del Animal de Laboratorio: SOVECAL (Venezuelan Society for Laboratory Animal Science)

Information on SOVECAL can be found on the FESSACAL Web site.

## Asia

## Bangladesh Society for Laboratory Animal Science: BSLAS

### Meetings

A scientific meeting is held every year.

*Address*

Dr. M.A. Awal, Bangladesh Agricultural University, Mymensingh, Bangladesh.

## Chinese Association for Laboratory Animal Science: CALAS

*Meetings*

A CALAS youth scientists symposium is organized every year.

*Address*

Dr. Yinong Liu, Functional Genomics Center, Institute of Genetics Chinese Academy of Sciences, Datun Road A5, Chaoyang District, Beijing 100101, P.R. China.

## Chinese Society for Laboratory Animal Sciences: CSLAS

*Address*

Dr. Chung-Nan Weng, Pig Research Institute Taiwan, P.O. Box 23, Chunan, Miaoli 350, Taiwan 115 R.O.C.

## Japanese Association for Laboratory Animal Science: JALAS

*Background*

JALAS was organized in 1952 to promote laboratory animal science in Japan.

*Membership*

At present, there are about 1,600 individual members and about 140 corporate members whose work is related to laboratory animals, i.e., scientists, veterinarians, animal technicians, educators, etc.

*Aims*

- To inform scientists about the appropriate use and care of animals in research
- To provide training and education in laboratory animal science and welfare
- To advance the knowledge, skills, and status of those who care for and use laboratory animals

*Meetings*

A scientific meeting is held every year. Workshops are also held on the basic topics for this field.

*Publications*

The official journal of JALAS, *Experimental Animals*, is published quarterly in English.

## International Information Exchange

JALAS sends experts in the field of laboratory animal science to Asian countries to promote mutual information exchange.

*Address*

Website: www.jalas.or.jp.

## Korean Association for Laboratory Animal Science: KALAS

*Meetings*

An annual scientific meeting is held.

*Address*

Dr. Byung-Hwa HYUN, Genetic Resources Center, P.O. Box 115, Yusong, Taejon, 305–600, Korea.

### Laboratory Animal Science Association of India: LASAI

*Background*

LASAI was founded in 1984.

*Publications*

*Laboratory Animal Science of India* is published quarterly.

*Address*

Dr. K.R. Bhardwaj, Central Drug Research Institute, Lucknow, India.

### Malaysian Association for Laboratory Animal Science: MALAS

*Address*

Dr. G. Baskaran, University of Malaya, Institute of Medical Research, Division of Laboratory Animal Resources, Kuala Lumpur, Malaysia.

### Philippine Association for Laboratory Animal Science: PALAS

*Meetings*

A scientific meeting is organized every year.

*Publications*

PALAS Code of Practice for the Care and Use of Laboratory Animals in the Philippines (1993). This Code of Practice (COP) was prepared by PALAS to set locally applicable standards on the proper care and use of laboratory animals and to upgrade the status of laboratory animal science in the Philippines.

*Address*

Dr. Angel Antonio B. Mateo, President, c/o Medical Affair Division, United Laboratories, Mandaluyong City, Philippines.

## Australia and New Zealand

### Australian and New Zealand Society for Laboratory Animal Science: ANZLAS

*Background*

ANZLAS was originally known as ASLAS (Australian Society for Laboratory Animal Science). The society changed its name in 1995 to reflect the active participation of its New Zealand members.

*Membership*

The membership includes veterinarians, scientists, and laboratory animal technicians in senior management positions, who hold an interest or work in the field of laboratory animal science. Membership categories exist for institutions and for students.

*Programs*

The main achievements of ANZLAS include the promotion of education and training; the establishment of laboratory animal microbiological and genetic monitoring services; the establishment of a program for the routine monitoring of constituents of diet for all rodents, rabbits, and guinea pigs; the organization of post-graduate courses; proposing the establishment of ANZCCART; the development of legislation related to animal-based research in several Australian states and territories; and the establishment of ethics committees and advisory work to institutions and professionals.

## Meetings

ANZLAS holds an annual conference and jointly sponsors specific workshops with other organizations, such as ANZCCART and the Australian Animal Technicians Association (AATA).

## Publications

A quarterly newsletter is circulated to members.

## Address

E-mail: biochem.otago.ac.nz/anzslas/.

# Africa

## Kenyan Laboratory Animal Technician Association: KLATA

### Address

Aggrey Otieno, P.O. Box 20750, Nairobi, Kenya.

## South African Association for Laboratory Animal Science: SAALAS

### Background

SAALAS serves all South Africans involved in the breeding, husbandry, welfare, and use of laboratory animals.

### Aims

SAALAS and Technikon Pretoria constitute part of SAQA's Standards Generating Body (SGB) for veterinary and paraveterinary professions. The goal of SAALAS in this partnership is to attain internationally acceptable standards in laboratory animal science.

### Membership

The membership includes persons concerned with the production, care, or study of laboratory animals and who hold appropriate qualifications in biomedical science or technology or, who by their experience and attainments, are eligible for membership.

### Programs

SAALAS, together with Technikon Pretoria, coordinate a laboratory animal technology course. SAALAS played a decisive role in the establishment of the national code for the handling and use of animals in research, education, diagnosis, and testing of drugs and related substances in South Africa.

### Publications

A quarterly newsletter and annual SAALAS bulletin are published.

### Address

Dr. F. Potgieter, President of SAALAS, Laboratory Animal Centre, P.O. Box 339, Bloemfontein 9300 South Africa. E-mail: gnpefjpmed.uovs.ac.za.

# PROFESSIONAL ORGANIZATIONS

This section lists the organizations formed by professional groups, i.e., veterinarians, technicians who work in the field of laboratory animal science, etc. The main objectives of these organizations are to promote excellence and recognition of these professional groups. Some of these organizations have a well-established accreditation system.

## American College of Laboratory Animal Medicine: ACLAM

### Background

ACLAM is an organization of board-certified veterinary medical specialists who are experts in the humane, proper, and safe care and use of laboratory animals. ACLAM establishes standards of education, training, experience, and expertise necessary to become qualified as a specialist, and recognizes that achievement through board examination.

### Aims

The aims are to encourage education, training, and research in laboratory animal medicine to establish standards of training and experience for qualification of specialists in this field and to recognize qualified specialists by certification.

### Membership

Open to all veterinarians who are graduates of a college or school of veterinary medicine accredited or approved by the AVMA, or who posses an Educational Commission for Foreign Veterinary Graduate (ECFVG) certificate, or who are qualified to practice veterinary medicine in a state, province, or possession of the United States, Canada, or other country; have satisfactory moral character and impeccable professional behavior; and have been certified as Diplomates in accordance with ACLAM bylaws.

### Programs

ACLAM has an active and extensive program of continuing education for its members and other interested scientists. ACLAM conducts a continuing-education forum and various symposia and seminars on different topics to keep Diplomates informed of new discoveries and innovations in laboratory animal medicine. In addition, ACLAM has published a series of auto-tutorial materials, CDs, books, and other educational materials directed at veterinarians, biomedical scientists, students, and technicians.

### Address

Website: www.aclam.org.

## American Society of Laboratory Animal Practitioners: ASLAP

### Background

ASLAP was founded in 1966.

### Aims

- To provide a mechanism for the exchange of scientific and technical information among veterinarians engaged in laboratory animal practice
- To encourage the development and dissemination of knowledge in areas related to laboratory animal practice
- To act as a spokesperson for laboratory animal practitioners within the American Veterinary Medical Association (AVMA) and to work with other organizations involved in the care and use of laboratory animals in representing common interests and concerns to the scientific community and the public at large
- To actively encourage its members to provide training for veterinarians in the field

## Membership

ASLAP membership is open to any veterinarian who is interested or engaged in laboratory animal practice and who maintain membership with a national veterinary medical association. Membership is also open to veterinary students.

## Meetings

Business meetings are held at the annual AVMA and AALAS meetings.

## Address

Website: www.aslap.org.

# Canadian Association for Laboratory Animal Medicine: CALAM - L'Association canadienne pour la médecine des animaux de laboratoire: ACMAL

## Background

CALAM/ACMAL is a national organization of veterinarians with an interest in laboratory animal medicine.

## Aims

The aims are to advise interested parties on all matters pertaining to laboratory animal medicine, to further the education of its members, and to promote ethics and professionalism in the field. CALAM/ACMAL is committed to the provision of appropriate veterinary care for all animals used in research, teaching, or testing.

## Programs

CALAM has compiled a comprehensive collection of standard operating procedures (SOPs) and a veterinary information kit.

## Meetings

A CALAM/ACMAL seminar is held annually in conjunction with the CALAS/ACSAL symposium.

## Publications

CALAM publishes a newsletter, *Interface*, four times a year within the CALAS/ACSAL newsletter.

## Address

Website: www.calas-acsal.org/calam.

# European College of Laboratory Animal Medicine: ECLAM

## Background

ECLAM is the veterinary college within Europe dedicated to the specialty of laboratory animal medicine and, like other European veterinary specialty colleges, is overseen by the European Board of Veterinary Specialization.

## Aims

The primary objectives of ECLAM are to promote high standards for laboratory animal medicine by providing a structured framework to achieve certification of professional competence and by stressing the need for scientific inquiry and exchange via progressive continuing-education programs.

*Membership*

In accordance with the policies of the European Board of Veterinary Specialisation, only veterinarians may become diplomates of European veterinary specialty colleges. The ECLAM constitution defines a diplomate as a veterinarian who satisfies the ECLAM requirements with regards to training, experience, and competence in laboratory animal medicine, as required by the constitution. The recent establishment of this college (2000) also required the extraordinary recognition of charter and de facto-recognized specialists that constitute the base of ECLAM members at present.

*Address*

Website: www.eclam.org.

### European Society of Laboratory Animal Veterinarians: ESLAV

*Aims*

ESLAV objectives are to give veterinarians working in the field of laboratory animal medicine a forum to discuss their issues, to support the creation of ECLAM, and to promote and disseminate expert veterinary knowledge within the field of laboratory animal science.

*Membership*

Members are veterinarians dedicated to laboratory animal medicine in Europe.

*Meetings*

ESLAV holds an annual scientific meeting in conjunction with other laboratory animal science organizations in Europe.

*Publications*

ESLAV publishes its news in *LAVA Briefing*, the publication of the Laboratory Animal Veterinary Association.

*Address*

Website: www.eslav.org.

### Institute of Animal Technology: IAT

*Background*

IAT was created in the U.K. to attain proper recognition of the essential contributions to science of those employed to care for laboratory animals.

*Aims*

To advance and promote excellence in the technology and practice of laboratory animal care and welfare. In education, IAT makes provision for animal technicians, technologists, and others professionally engaged in the field of animal science to receive appropriate training and qualifications, thus ensuring that they may contribute to advancing standards of laboratory animal welfare.

*Membership*

IAT started in the U.K., but now welcomes applications from elsewhere. IAT welcomes application for membership from animal technicians, technologists, animal scientists, veterinarians, and others engaged in the field of animal technology and its supporting industries. Applicants must be proposed by an existing member of the institute and be employed in the field of animal technology, or one of the mentioned professions or supporting industries.

## Meetings

IAT holds an annual congress and develops a varied program of meetings throughout the year in conjunction with other scientific organizations. IAT is also involved, together with similar organizations inside Europe, in the European Federation of Animal Technologists (EFAT, efat@iat.org.uk), and as such, participates actively in the European forums of discussion, i.e., Council of Europe.

## Publications

IAT publishes *Animal Technology* that is circulated worldwide. A monthly bulletin is also published as a mechanism to keep members informed of the activities of the institute and its branches. IAT has also published educational aids, books, videos, etc.

## Address

Website: www.iat.org.uk.

## Japanese Association for Laboratory Animal Medicine: JALAM

### Background

JALAM is an academic organization for laboratory animal medicine in Japan.

### Aims

The association promotes the advancement of research and education for laboratory animal medicine and the health of laboratory animals.

### Membership

Members are registered veterinarians and specialists in laboratory animal medicine, in particular, scientists who are teaching laboratory animal medicine in veterinary schools.

### Meetings

JALAM holds a meeting once a year, and several education and training programs are held several times a year.

### Publications

The *JALAM Newsletter* is published biannually in Japanese, but the contents in English are available on their Web site.

### Address

Website: hayato.med.osak-u.ac.jp/index/societies/jalam.html.

## Japanese College of Laboratory Animal Medicine: JCLAM

### Aims

One of the tasks of JALAM is to establish a college for the accreditation of diplomates in laboratory animal medicine.

### Membership

The college is formed by charter diplomates selected to establish the college and the certification program. Development of the college can be followed through the JALAM Web site.

## Animal Care and Welfare Organizations

### *Association for Assessment and Accreditation of Laboratory Animal Care International (AAALAC International)*

#### *Background and Aims*

AAALAC International is a private, nonprofit organization that promotes the humane treatment of animals in science through voluntary accreditation and evaluation programs. More than 630 companies, universities, hospitals, government agencies, and other research institutions in 17 countries have achieved AAALAC accreditation, demonstrating their commitment to responsible animal care and use. Institutions volunteer to participate in AAALAC's programs, in addition to complying with the local and national laws that regulate animal research.

#### *Membership*

Institutions that use animals in research, teaching, or testing (or supply animals for these purposes) are eligible for AAALAC accreditation. Accredited institutions include universities, pharmaceutical and biotechnology companies, contract laboratories, breeders, hospitals, government agencies, and agricultural research programs.

#### *Programs*

AAALAC accreditation is a voluntary, confidential, peer-review accreditation program for institutions working with animals in science. Those that meet or exceed AAALAC standards are awarded accreditation.

Program Status Evaluation (PSE) is a voluntary, confidential, peer-review service that helps assess the quality of all aspects of an animal research program, including animal husbandry, veterinary care, institutional policies, and the facilities where animals are housed and used. The review helps institutions understand AAALAC standards and is often used to prepare institutions for accreditation.

#### *Publications*

*AAALAC Connection* is published three to four times annually for institutions working with animals in science. *AAALAC E-Brief* is published bimonthly for AAALAC-accredited institutions.

#### *Address*

Website: www.aaalac.org.

### *Australian and New Zealand Council for the Care of Animals in Research and Teaching: ANZCCART*

#### *Aims*

The primary role of ANZCCART is of an advisory nature. The terms of reference of ANZCCART are to establish a national forum for effective communication between groups with concerns for the care and use of animals in research and teaching, and to provide information and advice on optimal standards for the care and use of such animals. Priority is given to the establishment and development of an associated information resource unit.

#### *Membership*

Membership of ANZCCART is based on the following criteria:

- Organizations involved in the administration or funding of substantial amounts of animal-based research or teaching
- Commonwealth and state bodies involved in the regulation of animal-based research and teaching

- Professional groups whose membership consists predominantly of persons involved in animal-based research or teaching
- Organizations with an established commitment to furthering the welfare of animals used in research and teaching

## Programs

Through all its activities, ANZCCART seeks to promote alternatives to the use of animals in research and teaching, using the principles of replacement, reduction, and refinement. ANZCCART provides expert information to the scientific and lay community, as well as to government. It is represented in relevant state, territory, and national government committees. It liaises with an international network of similar organizations.

## Meetings

ANZCCART holds an annual conference, as well as workshops, and seminars.

## Publications

ANZCCART publishes a quarterly newsletter, *ANZCCART News*. ANZCCART also publishes the proceedings of its annual meeting and monographs relating to the use of animals in research and teaching.

## Address

Website: www.adelaide.edu.au/ANZCCART/.

## Canadian Council on Animal Care: CCAC - Conseil canadien de protection des animaux: CCPA

### Background

Experimental animal use of vertebrates and cephalopods in Canada is subject to the requirements of the Canadian Council on Animal Care (CCAC), a national, peer-review organization founded in Ottawa in 1968. Every year in Canada, approximately two million animals are used for research, teaching, and testing purposes. The CCAC policy statement on the Ethics of Animal Investigation (October 1989) includes the following basic principles:

The use of animals in research, teaching, and testing is acceptable only if it promises to contribute to understanding of fundamental biological principles or to the development of knowledge that can reasonably be expected to benefit humans or animals.

Animals should be used only if the researcher's best efforts to find an alternative have failed. A continuing sharing of knowledge, review of the literature, and adherence to the Russell-Burch "3R" tenet of replacement, reduction, and refinement are also requisites. Those using animals should employ the most humane methods on the smallest number of appropriate animals required to obtain valid information.

### Aims

The aims of the CCAC are to ensure that these basic principles are followed by:

- Publishing guidelines on the care and use of animals in science
- Assessing scientific institutions using animals
- Providing educational materials and activities to those who use and care for animals, as well as information to the public

## Membership

The council comprises 22 member organizations, whose representatives include scientists, educators, veterinarians, and delegates from industry and the animal welfare movement. Scientific institutions funded by Canada's federal granting councils and other funding agencies must be full participants in the CCAC program: these include universities, hospitals, and colleges. Private-sector institutions and governmental units choose to be participants in the CCAC program. All institutions assessed by the CCAC and which achieve a CCAC status of compliance or conditional compliance receive a CCAC Certificate of GAP — Good Animal Practice.®

## Programs

There are three main programs:

- Assessment Program: The CCAC conducts assessments of institutions using animals at least every three years. Assessments are based on CCAC guidelines and policies, and are conducted by panels composed of scientists, veterinarians, and community representatives.
- Guidelines Development Program: New CCAC guidelines are drafted, and older ones are reviewed, by groups of experts in the relevant field. They are then submitted to an expert peer review, followed by a wider review, in which they are provided to all CCAC constituents before finalized.
- Education and Training Program: The CCAC provides educational materials, particularly through its Web site, and organizes workshops across Canada to provide information on all aspects of animal care and use, and to address questions from its constituents and from the public.

## Publications

The CCAC publishes guidelines and policies on the care and use of animals in science in both English and French. Most of the guidelines are contained in the two volumes of the CCAC *Guide to the Care and Use of Experimental Animals*. These are gradually being updated in separate guidelines on various topics, and emerging issues, such as appropriate endpoints and transgenic animals, are also the subject of separate guidelines. CCAC guidelines are gradually being translated into Spanish; volume 1 of the CCAC *Guide to the Care and Use of Experimental Animals* is available in Spanish, as are the guidelines on appropriate endpoints.

The CCAC also publishes and distributes widely a newsletter, *Resource*, available in English and French at no cost to any interested persons.

Finally, the CCAC regularly produces educational documents on various aspects of animal care and use, which are generally available on the CCAC Web site, as are all guidelines and policies.

## Address

Website: www.ccac.ca.

## The Universities Federation for Animal Welfare: UFAW

## Background

UFAW was founded in 1926 as the University of London Animal Welfare Society (ULAWS) by Major Charles Hume, based on his belief that "animal problems must be tackled on a scientific basis, with a maximum of sympathy but a minimum of sentimentality." Since then, UFAW has played a considerable role in improving conditions for animals. This success is not only the result of UFAW's approach to animal welfare problems, but also because it is independent and not beholden to universities, government departments, or commercial enterprises. This independence is possible due to the generous support it has received over the years in the form of donations and legacies from concerned individuals. About 50 percent of UFAW's annual income is from legacies.

## Aims

UFAW's expertise is increasingly called upon to tackle animal welfare problems in the emerging field of biotechnology and wildlife management, as well as in the more traditional areas of farm, zoo, laboratory, and the home.

## Meetings

Over the years, several meetings were organized, including in 1999, the First Annual Vacation Scholarship meeting, held at Edinburgh, and in 2000, the UFAW Conference on Cognition and Welfare.

## Publications

In 1992, UFAW established the quarterly scientific and technical journal, *Animal Welfare*, which now has a worldwide distribution and a high citation index. UFAW has published several publications, including the fourth edition of *Management and Welfare of Farm Animals* and the seventh edition of the *UFAW Handbook on the Care and Management of Laboratory Animals*.

## Address

Website: www.ufaw.org.uk.

# MISCELLANEOUS ORGANIZATIONS

## Institute of Laboratory Animal Research: ILAR

### Background

Founded in 1952, the Institute for Laboratory Animal Research (ILAR) prepares authoritative reports on subjects of importance to the animal care and use community; serves as a clearinghouse for information about animal resources; and develops and makes available scientific and technical information on laboratory animals and other biological research resources to the scientific community, institutional animal care and use committees (IACUCs), the federal government, science educators, students, and the public.

### Publications

*Guide for the Care and Use of Laboratory Animals (1996).* The *Guide*, ILAR's most widely distributed work, is accepted by the scientific community as the main resource on animal care and use. The *Guide* is recognized as the standard reference by the Association for the Assessment and Accreditation of Laboratory Animal Care International, the Office of Laboratory Animal Welfare (NIH), and other private organizations and federal agencies. Its guidelines are based on established scientific principles, expert opinion, and experience with methods and practices consistent with high-quality, humane animal care. The *Guide* has been translated into several languages.

*ILAR Journal* is the quarterly, peer-reviewed publication of ILAR. The *ILAR Journal* provides thoughtful and timely information for all those who use, care for, and oversee the use of laboratory animals. The audience of *ILAR Journal* includes investigators in biomedical and related research, institutional officials for research, veterinarians, and members of animal care and use committees.

All other publications and ILAR reports are listed on the ILAR Web site at: http://dels.nas.edu/ilar/ and may be purchased through the National Academy Press.

### Address

ILAR, 2101 Constitution Avenue, NW, Washington, D.C. 20418. E-mail: ILAR@nas.edu.

### Laboratory Animal Management Association: LAMA

#### Aims

The Laboratory Animal Management Association's mission is to advance the laboratory animal profession through education, knowledge exchange, and professional development.

#### Membership

There are individual and institutional memberships.

#### Programs

LAMA is an active participant and co-sponsor, with AALAS, of the Certified Manager, Animal Resources program.

#### Meetings

Two meetings a year are organized: one in spring or summer, a two-day event, seminar, workshop, or a tour of local facilities; the other, in fall, is in conjunction with the AALAS meeting.

#### Publications

The *LAMA Review* is a quarterly publication dedicated to providing the highest quality management information. The *LAMA Lines* is published six times a year and includes noteworthy and educational items about happenings within the organization.

#### Address

LAMA, P.O. Box 877, Killingworth, CT 06419.

### Laboratory Animal Welfare Training Exchange: LAWTE

#### Background

The Laboratory Animal Welfare Training Exchange (LAWTE) is a nonprofit educational association of persons professionally concerned with laboratory animal welfare-training issues.

#### Aims

LAWTE aims to promote an information exchange among laboratory animal welfare trainers on training programs, systems, materials and services for the purpose of promoting the highest standards of laboratory animal care and use.

#### Membership

LAWTE is an organization of trainers, training coordinators, and IACUC administrators. By sharing ideas on methods and materials for training, members can learn together how best to meet the training and qualification requirements of national regulations and guidelines.

#### Meetings

Since 1994, conferences have been held every two years for trainers to exchange information on their training programs in the U.S. and abroad. They have been attended by institutional representatives from the U.S. and around the world.

#### Programs

The purpose of the listserv is to provide an electronic forum for discussion by LAWTE members on the issues and methods of training of research personel in the proper and humane use of laboratory animals.

## *Publications*

Programs and proceeding of meetings are maintained on the LAWTE Web site.

## *Address*

Website: www.lawte.org.

# ACKNOWLEDGMENTS

Special thanks to all of our colleagues from the various organizations who provided us with most of the information in this chapter.

# CHAPTER 7

# Laboratory Animal Allergy

**Richard M. Preece and Anne Renström**

## CONTENTS

0-8493-1086-5/03/$0.00+$1.50
© 2003 by CRC Press LLC

# INTRODUCTION

Laboratory animal allergy (LAA) is a common health problem in biomedical research and the prevalence of allergy has been found to be as high as 56% of animal-exposed workers.[1] Sensitization to laboratory animal allergens can give rise to both severe, acute (anaphylactic) reactions[2] and disabling chronic illnesses (dermatitis and asthma). Both acute and chronic reactions to allergen exposure can have a significant adverse impact on affected workers and their employers. Several authors have reviewed this topic.[3-5]

Allergy to many different animals has been described.[6] The allergens are present in hair, skin, feces, urine, and other material from the animals.[3] The most important laboratory animal allergens have been identified and characterized.[7] These include the major rat and mouse allergens, which are urinary proteins.

The immunological mechanisms that give rise to symptoms of allergy have been characterized.[5] Allergen exposure can lead to the production of specific immunoglobulin E (IgE) against allergens. The interaction between an allergen and specific IgE initiates a cascade of events that leads to the symptoms of allergy. This cascade will be repeated on each subsequent exposure to the allergen. The allergy may present as disorders of the nose and eyes (rhinoconjunctivitis), skin (urticaria and contact dermatitis), and chest (asthma) or, more rarely, as acute anaphylactic reactions.

There is conflicting evidence for personal risk factors for development of LAA, especially tobacco smoking and a family history of allergic reactions.[8] The majority of studies have indicated that workers who have a tendency to produce IgE against allergens and to develop allergic symptoms (atopics)[9] are more likely to develop LAA.[10] Also, atopics appear to develop LAA after a shorter time of exposure.[11] However, even atopy is not a sufficiently good predictor of LAA to be used in pre-placement selection.[6,12-14]

A number of occupational risk factors for exposure to allergen have been identified, which include high-risk tasks (such as handling animals or cleaning cages)[15,16] and some work practices (e.g., choice of bedding).[17] Studies have demonstrated a positive relationship between exposure to animal allergens and the prevalence of sensitization.[18-20] The combination of preexisting allergy or atopy and environmental risk factors (e.g., exposure level) may further increase the risk for LAA.[18,21,22] The exposure-response relationship is not linear, and further work is needed before the role of different features of exposure, such as peak and troughs of exposure, in the development of allergy is fully understood.

The prevention of allergy is a significant challenge while uncertainty remains about which are the critical characteristics of exposure to control. Allergy-prevention strategies have generally focused on measures to control total exposure to allergens, through a combination of engineering, procedural, and personal controls. The cost of measures to reduce allergen exposure, such as robots and sophisticated ventilation, may be prohibitive, while their ability to reduce the incidence of allergy is uncertain. No studies of the cost effectiveness of allergen-control programs have been reported.

Although a high incidence and prevalence of LAA has been reported, where comprehensive measures have been introduced to reduce personal exposure to allergen, this has led to a decrease in the incidence of allergy to low levels.[12,23] These reports suggest that for the majority of workers exposed to animals, the development of allergy can be prevented by effective control of allergen exposure.

# EPIDEMIOLOGY AND PATHOGENESIS

## Prevalence: How Many People have LAA?

Many cross-sectional studies have investigated the proportion of cases of work-related allergy in laboratory animal-exposed populations. Pooled data from the studies of 4,988 persons at risk has indicated a prevalence of LAA of about 20%, although the spread of reported prevalence was broad.[3] Many factors contribute to the wide variation in prevalence, including differences in investigators' methods and definitions of allergy, the tendency to retain unaffected workers in the exposed population, employment policies for workers already allergic to laboratory animals, and the nature and degree of allergen exposure.[10]

## Incidence: The Risk of Getting LAA

There have been few studies of the number of newly developing cases of LAA in exposed populations, but in those published, incidences of LAA among subjects in the first years of animal work vary between 5 to 40%, when no special prevention strategies have been employed.[10, 14,24] Where comprehensive programs have been introduced, a reducing incidence of allergy has been observed.[12,23] Nose, eye, and skin symptoms have a higher incidence than chest symptoms.[25] In a recent study of 373 apprentices, the incidence of probable occupational asthma was 2.7% after 8 to 44 months of follow-up.[26] This incidence is similar to that reported elsewhere.[25,27,28]

Most workers who develop LAA do so within the first three years of exposure (see Figure 7.1).[3,29] The mean exposure before symptoms develop is shortest for nasal symptoms and longest for chest symptoms.[19] However, there is great variation in the length of the period before allergy is clinically present. While some workers develop almost immediate symptoms, others may have contact with the animals for 15 to 20 years before they react.[3] However, the time to development of symptoms appears to depend upon atopic status, as was demonstrated in a retrospective study, in which atopics developed LAA after a median time of 2.2 years and nonatopics after 8.2 years.[11]

## Exposure-Response Relationship

Exposure as a risk factor for LAA has been studied, but only recently using actual aeroallergen measurements in cohorts of exposed subjects. Exposure proxies, such as job title, work years, or hours of animal work per week, have been used in many studies, with partially contradictory results. For instance, some studies have found an inverse relationship between exposure intensity or length of employment and LAA.[27,30] This has usually been attributed to healthy worker selection: those in the highest exposure groups who develop LAA have left the workforce. This may, however, suggest that length of exposure (i.e., cumulative exposure) is less important than intensity.

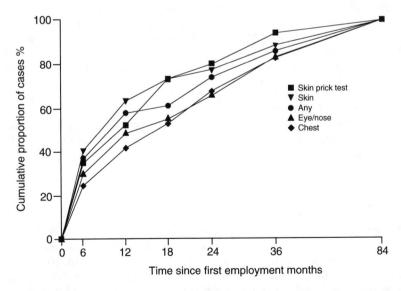

**Figure 7.1**    Latency of new, work-related symptoms or development of sensitization to rat urinary allergens. (Cullinan, P., et al., *Eur. Respir. J.*, 1999; 13:1139–1143. Reproduced with permission.) [25]

As methods to measure aeroallergen exposure have been developed, studies are beginning to explain the relationship of measured allergen exposure with development of allergy, symptoms of allergy, and biological evidence of sensitization.[18,20,25] The prevalence of LAA has been shown to be associated with length and intensity of exposure to aeroallergens.[18,20] Similar findings were reported in the one study that has examined the relationship between exposure and incidence of LAA.[25]

Although an exposure-response relationship has now been demonstrated for animal allergens, the nature of this relationship is not fully understood. The total dose of allergens accumulated over a period of animal exposure is a function of exposure intensity and exposure duration. In the case of LAA, it has been suggested that intensity and duration are not equivalent and peaks of high-intensity exposure may be more important.[25,31] Both studies of the exposure-prevalence relationship and the study of incidence are consistent with a key role for peaks of exposure. However, there appears to be an exposure-response plateau at high exposure, for which there may be other explanations than healthy worker selection. Some individuals may develop immunological tolerance during high or constant exposure, as has recently been observed in the development of allergy to cats;[32] this hypothesis must be studied longitudinally regarding LAA, which has not yet been done.

Apart from inhalation, it is not known what other mechanisms of exposure (such as mucocutaneous contact) have in the development of LAA. Skin contact is certainly relevant in the development of skin symptoms, such as contact urticaria,[33] and may have a role in respiratory tract symptoms. Respiratory sensitization following skin exposure to chemicals has been described,[34] and percutaneous exposure to animal allergen has been associated with acute (anaphylactic) respiratory symptoms.[2,35,36]

These exposure-response observations have important implications for strategies to control allergen risks. Studies suggest that reducing total exposure to aeroallergen may prevent allergy.[18,20,25] However, reductions in aeroallergen exposure alone are likely to be inadequate. Measures should also be taken to reduce the number and intensity of peaks of exposure and the possibility of exposure by and through skin contact and by ingestion.

## Animal Allergens

Several sources of allergens have been found in rodents, such as urine, fur, saliva, skin, and serum.[3] However, about 90% of those who are allergic to rats or mice react to the closely related allergens Rat n 1.01 (alpha$_{2u}$-globulin) and Rat n 1.02 (prealbumin) from rat[37] and Mus m 1 (or the MUP complex) from mouse.[38] Sensitization to one of the species is often associated with sensitization to the other.[22] Thus, developing allergy against, for instance, rat increases the risk of soon becoming allergic also to mouse. The proteins are mainly excreted in urine, but minor amounts are also excreted in saliva and by perianal and other glands.[39] The excretion is sex-, age-, and diet-dependent, and post-pubertal male urinary levels may be more than a hundredfold higher than in mature females or pre-pubertal animals.[38,40] These allergens are transport pheromones, which are important for sexual communication and will influence other males, pregnant and non-pregnant females and pre-pubertal animals in a range of ways.[41]

The allergens belong to a rapidly growing group of molecules called lipocalins. Many other major fur-animal allergens, such as Equ c 1 from horse and Can f 1 from dog, also belong to the lipocalin superfamily.[42] The molecules are structurally similar and have the capacity to bind or transport hydrophobic molecules within a barrel-shaped pocket. Other allergens from rodents are, for instance, rat albumin (68 kD), to which about 30% of rat-allergic subjects react.[43,44] Allergens of 10–40 kD in urine, saliva, fur, and dander from guinea pigs and rabbits also cause LAA symptoms.[45–47]

## Pathogenesis

Subjects who are hypersensitive to animals usually have an immediate type, IgE-mediated allergic reaction. This reaction is preceded by a sensitization process. Upon exposure to allergens, local cells belonging to the immune system (antigen-presenting cells) internalize the allergens. The molecules are processed, and pieces of the antigens are presented to T-helper lymphocytes. The production of signalling molecules called cytokines, especially Interleukin-4 (IL-4), will induce B-lymphocytes to proliferate and produce immunoglobulin E (IgE) antibodies with specificity against these allergen epitopes. IgE antibodies are released and bound to IgE receptors on the surfaces of certain white blood cells, notably mast cells. Next, mast-cell contact with the relevant allergens initiates a series of events. The cross-linking of IgE receptors on the mast-cell surface through binding of allergen causes degranulation: the numerous granules within the mast cell rapidly eject their contents. The mast-cell granules contain several pre-formed molecules, such as histamine, tryptase, and cytokines, and new synthesis of, e.g., leukotrienes and prostaglandins is stimulated. Histamine and other released factors have immediate effects on local blood vessels, mucous membranes, and airway smooth-muscle tissue, causing leakage and swelling, bronchoconstriction, etc. In highly sensitive subjects, the allergic reaction can cause rapid systemic effects (anaphylactic shock). The released molecules also mobilize other types of white blood cells. A late-phase response may follow, peaking at 4 to 8 hours after exposure, and involve inflammatory cells, such as eosinophils, attracted to the target tissues. Immune cells are activated, which potentiates and prolongs the allergic inflammation.

## CLINICAL MANIFESTATIONS

### Symptoms

Symptoms during work with laboratory animals are mainly, but not exclusively, mediated by mechanisms initiated by the interaction of allergen with specific IgE. Symptoms of LAA arise from the release of biochemical mediators that lead to an inflammatory response. The nature of the symptoms vary from person to person, but form three principal patterns that affect the nose and eyes, chest, and skin.

The most commonly reported symptoms affect the nose and eyes (rhinitis and conjunctivitis). The symptoms include sneezing, nasal congestion and discharge, redness of the conjunctiva, and itching, watery eyes. If the lower airways are affected, then the presenting symptoms are those of asthma, with cough, wheezing, production of sputum, and shortness of breath.

Skin reactions include urticaria, with itching and circumscribed red lesions, and contact dermatitis.[33] Typically localized lesions appear quickly on the skin exposed to allergen, such as the face, neck, and arms, but skin reactions can be more generalized.[48]

Anaphylactic reactions are a rare manifestation of LAA that have been reported in association with both rat and mouse bites[2,35] and a puncture wound from a needle used on a rabbit.[36] These reactions can lead to generalized itching and urticaria, swelling of the face, lips, and tongue (angioedema), obstruction of the airways, and shock (low blood pressure). The symptoms may vary from mild skin rashes to life-threatening cardiorespiratory reactions.

Pooled data from 13 studies revealed a consistent picture of symptom distribution.[3] Of 10 persons with symptoms of LAA, about eight will have rhinoconjunctivitis (range 53 to 100%), about four will have skin reactions (13 to 70%), and about three to four will have asthma (13 to 71%). Subsequent studies of symptom incidence suggests this 2:1:1 ratio of symptoms remains typical.[25,29] There is inevitably overlap between symptoms; most subjects have more than one affected target organ, and asthma rarely occurs in the absence of the prior development of rhinoconjunctivitis (see Figure 7.2).[29,49]

### Diagnosis

Strong evidence for LAA is provided by a clear history of symptoms that are associated with work. Where there are immediate symptoms on contact with the relevant animal, then the diagnosis is usually self-evident. However, care must be taken to make sure that it is LAA and not symptoms arising from contact with another allergen found in the animal facility, such as latex from gloves or wood dust from bedding materials.

Workers who have delayed reactions to allergen may only experience symptoms after leaving the animal facility. They may not attribute these symptoms to allergen exposure. When symptoms are those of LAA, a careful history will normally reveal a temporal association with workplace exposure. Symptoms are typically worse at the end of the working week and after periods of intense animal handling, and usually improve during periods away from work, such as weekends and vacations. It is common for persons who care for animals at work to keep pets at home; this should be taken into account when considering the pattern of symptoms during "exposure-free" periods. Health workers will easily recognize the symptoms of allergy, such as the typical rashes, although there may be few findings of allergy on examination in the absence of obvious symptoms.

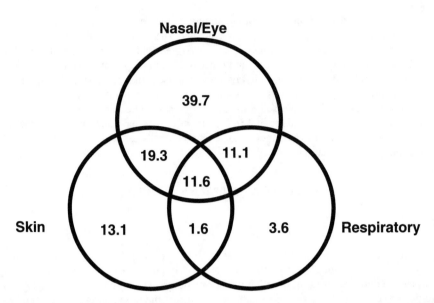

**Figure 7.2.**    Comparison of symptoms in LAA subjects (%). (Aoyama, K., et al., *Br. J. Ind. Med.*, 1992; 49:41–47. Reproduced with permission.)[29]

Immunological tests are valuable, as they provide additional confirmatory evidence of the diagnosis. The most widely used tests are skin-prick tests and immunoassay tests for specific IgE in the serum. In skin-prick testing, an extract of allergen is placed on the skin and the underlying skin is then punctured. The skin reaction is then compared with a positive (histamine) and negative control. More than 60 percent of cases of LAA (and almost all asthmatic individuals) will be positive by immunoassay. In cases of asthma with no specific IgE to laboratory animals, the symptoms may be due to reactions to other agents that are present in the working environment, e.g., dust, ammonia, formaldehyde, or disinfectants.[50]

When occupational asthma is suspected, this is usually confirmed with pulmonary function tests. Animal workers with suspected asthma are invited to record their peak expiratory flow rate every two hours throughout the day. They should continue recording this for at least four weeks and include a period away from work. A difference between the mean peak flow while at work and away from work of greater than 15% is indicative of occupational asthma. An alternative technique is cross-shift spirometry, in which pulmonary function is measured before and after work. A deterioration during the working day is suggestive of an occupational cause. However, this technique is often not practicable and the results may be confounded by the normal diurnal increase in pulmonary function from morning to afternoon. The results may also be confounded in asthmatics with a late-phase bronchoconstriction, which may appear several hours after getting off work. Rarely, pulmonary-challenge testing may be necessary to confirm the diagnosis, but this is a risky technique that is only ever done in designated specialist centers.

## Prognosis

The symptoms of rhinoconjunctivitis and urticaria are a nuisance. Unless they are effectively managed, they may make it difficult for the affected persons to continue to work with the animals to which they are allergic. If the affected persons cease to be significantly exposed to the allergen, then symptoms will not reappear. However, subjects who have developed sensitivity to one type of fur animal are at increased risk to develop allergies to others. In a study of 100 subjects diagnosed with occupational asthma and followed up after a mean 5.8 years after ceasing exposure, significantly more subjects had developed symptoms against other animals.[51]

Continued exposure to allergen may lead to the insidious development of asthma, which is of greater concern. The reduction in pulmonary function may be persistent, even after exposure ends,[52] with substantial impact on the LAA sufferer's quality of life.[53] Continued exposure may lead to permanent loss of function.

## Case Management

The workplace management of LAA should be focused on the reduction of exposure to a clinically insignificant level. The medical management of LAA is generally focused on the relief of symptoms, and is not curative. Successful treatment of symptoms should not be allowed to confound efforts to reduce allergen exposure. The exposure can only be regarded as sufficiently low when the LAA sufferer is free of symptoms in the absence of treatment. The degree of exposure precipitating symptoms in allergic workers varies.[54,55]

Employers and employees should be aware that a risk of anaphylaxis exists and this may be life-threatening. Arrangements for the immediate treatment of anaphylactic reactions should be in place in all animal facilities.

Any person who develops LAA should be counseled by a knowledgeable physician. The worker's job should be analyzed and control measures reviewed and, if necessary, enhanced, so that exposure is reduced (to the benefit of all exposed workers). The capacity of the individual to reduce the risk, through good personal hygiene, careful use of personal protective equipment, and by performing tasks in less exposure-generating ways should be reemphasized. It may be possible to relocate the worker to a lower-risk area, such as work with isolators or in an area with lower aeroallergen levels,[56] or to transfer the worker to an area where a different species is present. However, the risk for a rodent-sensitive worker to develop allergies against other rodents is considerable,[22] and if symptoms of LAA persist, then the fitness of the worker to continue any animal work should be reviewed carefully. Redeployment into a different role where key skills can be utilized, such as quality assurance or training, may be necessary.

The necessary reduction of exposure to a clinically insignificant level can usually be achieved by changes in working practices or redeployment away from the relevant animal. Only in the most extreme cases will the affected persons have to leave their employment.

## CONTROL STRATEGY

Allergy is costly to both employers and employees, and controlling the risks animal allergens pose is desirable from both a moral and economic perspective. In most jurisdictions, it is a legal duty. Unfortunately, as neither the vulnerable people can be identified with certainty in advance, nor the hazard (allergen exposure) completely eliminated under normal circumstances, there is inevitably residual risk to be managed.

Ideally, the risk posed by allergens should be considered in a comprehensive management system that addresses all the safety, health, and environmental risks in the animal facility. The components of this management system should include setting policy, establishing an appropriate organization, making and implementing plans, and measuring and reviewing performance. An effective system will be dependent on the leadership and commitment of senior management and the cooperative involvement of people at all levels of the organization.

A control strategy should be introduced prior to the construction of the animal facility so that it can be implemented in the design and influence key purchasing decisions. In many cases, this is not possible, as the facility is already established, however, the strategy should still influence future refurbishment decisions. The control strategy should be dynamic and responsive to changes in technology and understanding. Even when the initial design has been strongly influenced by the allergen-control strategy, as this evolves, there may be implications for the facility.

The principal mechanism by which allergen initiates LAA is assumed to be inhalation. Control measures should be mainly, but not exclusively, aimed at the control of aeroallergen in the worker's breathing zone. Ambient levels of allergen may not be representative of personal exposure. In a microscopy room, for example, ambient levels of allergen could be low, but, because the microscopist works with a source of allergen close to the breathing zone, personal exposure may be significantly greater. Controls should reduce both the intensity and the duration of exposure. Several studies have described the intensity of exposure associated with different tasks.[15, 57–59] Directly handling animals (especially during close-up, detailed work) and cleaning and changing dirty cages are associated with exposure to high concentrations, whereas work on animal tissues postmortem is associated with lower exposures (see Table 7.1).

**Table 7.1.  The Likelihood of Exposure in the Absence of Specific Control Measures**[15,58,59]

| Exposure | Task |
| --- | --- |
| Low | Postmortem and surgery |
|  | Slide preparation |
|  | Laboratory work (low number of animals) |
|  | Automated cage cleaning |
| Medium | Cleaning |
|  | Indirect contact in animal room |
|  | Feeding |
|  | Taking specimens |
| High | Injections |
|  | Handling animals |
|  | Changing and cleaning cages |
|  | Changing filters |
|  | Washing cages |

### Preventing Asthma in Animal Handlers (NIOSH 1998)[78]

#### Animal handlers should:

Perform animal manipulations within ventilated hoods or safety cabinets when possible.
Avoid wearing street clothes while working with animals.
Leave work clothes at the workplace to avoid potential exposure problems for family members.
Keep cages and animal areas clean.
Reduce skin contact with animal products, such as dander, serum, and urine, by using gloves, lab coats, and approved particulate respirators with face shields.

#### Employers of animal handlers should:

Modify ventilation and filtration systems:
  Increase the ventilation rate and humidity in the animal-housing areas.
  Ventilate animal-housing and -handling areas separately from the rest of the facility.
  Direct airflow away from workers and toward the backs of the animal cages.
  Install ventilated animal cage racks or filter-top animal cages.
Decrease animal density (number of animals per cubic meter of room volume).
Keep cages and animal areas clean.
Use absorbent pads for bedding. If these are not available, use corncob bedding instead of sawdust bedding.
Use an animal species or sex that is known to be less allergenic than others.
Provide protective equipment for animal handlers: gloves, lab coats, and approved particulate respirators with face shields.
Provide training to educate workers about animal allergies and steps for risk reduction.
Provide health monitoring and appropriate counseling and medical follow-up for workers who have become sensitized or have developed allergy symptoms.

#### Principal Elements of an Occupational Health and Safety Program:[6]

Administrative procedures
Facility design and operations
Exposure/control methods
Education and training
Occupational health services
Equipment performance testing
Information management networks
Emergency procedures
Program evaluation and audit

A simple hierarchy for risk management can be applied. Controls based on engineering solutions are preferable to those based on procedures or people, because they are less reliant on human factors. An effective strategy will be based on all three.[6] It should take account of controls during normal operations, controls for operations when conditions are not normal (e.g., spillage or breakdown), and controls for exposures that may occur beyond the controlled areas (fugitive exposures).[6]

# ENGINEERING CONTROLS

Animal facilities should be designed to incorporate engineering controls to the extent feasible. The most likely limitations to the introduction of engineering controls are the constraints imposed by the existing facility and the need for significant capital investment. In existing facilities, the costs of retrofitting may be prohibitive, not least because to do so may mean business operations have to stop temporarily.

One of the problems associated with evaluating engineering controls is that there is relatively little evidence that specific building (ventilation or architect design) systems that may contribute to reduced total exposure to allergen actually help to prevent allergy. For instance, there is a general assumption that ventilation design contributes to reductions in particle counts and thereby leads to less allergy, but there is, until now, little published evidence regarding the biological significance of the various forms of technology, other than the indirect evidence provided that prevalence of symptoms is lower among low-exposed, than among medium- or high-exposed workers.[18]

## Separation

Concerning allergen spread within facilities, the first consideration should be separation of the potential population at risk from the hazard. The hazard is not just the animals, but, most importantly, the allergen they produce. At the facility level, this can be interpreted as construction of the facility away from nonanimal workers. Within the facility, this can be achieved by clear segregation of work with animals from other work, such as administration and rest facilities. Boundaries can be established and, where necessary, access controls introduced to prevent exposure of people who are not directly involved in animal work or support of the animal areas. Even within the areas that animal work is directly carried out, it may be possible to widely separate the majority of workers from the areas where the potential exposure is highest (e.g., the animal-holding rooms). Separation can be facilitated by a two-corridor system where this is feasible, one corridor used for "clean," the other for "dirty"' activities.

Separation should also be considered in the specific context of the allergens. Where allergen exposure is foreseeable despite the absence of animals, this should be controlled. Key areas include places where cage waste is handled, in the laundry, and at the exhaust points from ventilation systems. Ventilation inlets, especially those to clean and nonanimal areas, should not be placed in proximity to or downwind from outlets from contaminated areas.

## General Ventilation

Facility ventilation, including the control of temperature and humidity, contributes to the general control of allergen. General ventilation has an important influence on the microenvironment in the animal cage, and it is this factor more than allergen control that has usually been more influential in the development of ventilation systems. Task-specific local exhaust ventilation, rather than general ventilation, is the principal control method, being more effective, less costly, and probably easier to implement.

Studies have demonstrated that an increase in air-change frequency can reduce allergen levels.[60] However, many different approaches to the general ventilation of animal facilities have been shown to be effective, not all of which are dependent on expensive high-frequency air changes.[61] One-way airflow systems with sliding perforated screens, behind which are the cage racks and exhaust vents, have been shown to effectively draw allergens behind the screens, leaving minimal allergen levels in the room.[62]

Pressure gradients are an important adjunct to the control of allergen spread; these are a common feature of animal facilities. However, there are potential conflicts between the gradients required to protect the animals' health and those required to protect human health. In general, it is desirable to use a gradient that minimizes spread of allergen (and pathogens) from cage-cleaning areas and into "clean areas," such as offices and restrooms. Negative pressure "sinks" adjacent to animal-holding rooms can also be used.

Increasing the relative humidity has been shown to reduce the levels of airborne allergen.[63,64] Presumably, in conditions of higher humidity, particles weigh more, are more adhesive, and will settle more readily. High humidity is, however, more uncomfortable for workers, increases growth of molds and mites, and may have an adverse effect on animal health.

Exhaust air from animal areas will be contaminated with allergens (and possibly pathogenic organisms). In some circumstances, it will be necessary or desirable to filter exhaust air. Exhaust air should not be recirculated without filtration. Exhaust air and filters are important sources of fugitive exposure. Controls should be in place to prevent exposure to exhaust air and to minimize the risk to people involved in the maintenance of ventilation systems and changing of filters.

## Task Ventilation

Task ventilation, or local exhaust ventilation, is one of the most important control measures. These systems remove allergen at the source and can be designed (usually at relatively low cost) to accommodate the tasks with potential for the highest exposure. They contribute to reductions in both the spread of allergens and other contaminants.

Task ventilation includes biosafety cabinets, fume cupboards, and ventilated workstations that use downdraft or backdraft systems.[65] Often, these systems have the advantages of being mobile and suitable for installation in established facilities. However, it may be difficult to demonstrate the effectiveness of these systems, especially novel designs, such as downdraft benches that are reliant upon undisturbed laminar flow. It is easier to demonstrate that novel ventilation systems function effectively when not in use than when used by operators under normal work conditions. For instance, covering too great a proportion of the ventilated surface of a downdraft table is likely to reduce the effectiveness of the exhaust system. If the effectiveness of the ventilation system cannot be confirmed under operational conditions, then it should not be relied upon as a primary control measure.

## Automation

New technology is enabling the automation of many tasks. This benefit of automation is especially interesting where the tasks are labor-intensive and pose significant risk. These risks may be high allergen levels or other factors, such as exposure to potentially harmful pathogens or test substances or ergonomical risks. For instance, when cleaning cages and bottles, the risks will be due to both allergens and ergonomics. Automated cage cleaning and waste handling systems have now been introduced in some animal facilities.[61] Automated cage cleaning systems have been shown to greatly reduce ambient levels of allergen and personal exposure of operators under normal operating conditions.[59]

## Cage Systems

The introduction of filters to conventional open-top cages is associated with reductions in allergen concentrations of greater than 75% (see Figure 7.3).[17,55,58,66] Individually ventilated cage systems are now widely available,[67] and these have been shown to effectively reduce background aeroallergen levels in a number of studies.[55,68,69] The most impressive reductions in aeroallergen levels in undisturbed animal rooms — almost 100% — arise when the system is operated with the cages under negative pressure.[55,69,70]

## PROCEDURAL CONTROLS

The objective of engineering controls is to minimize the influence of human factors. As the risk of allergy cannot be eliminated by removing the allergen hazard completely at the source or by engineering solutions, additional controls are essential. The emphasis of procedures is to control choice so that work is carried out in a way that minimizes the levels and spread of environmental allergen.

### Reduction of the Number of Exposed Persons

A first strategy of prevention is to minimize the number of animal-exposed personnel. One way is to reduce use of animals, increasing the use of alternative methods, such as using cell lines in the production of antibodies, or toxicity tests that do not involve animals. Another way is concentrating animal work so that fewer individuals need be exposed. An example: through the installation of a labor-saving device,

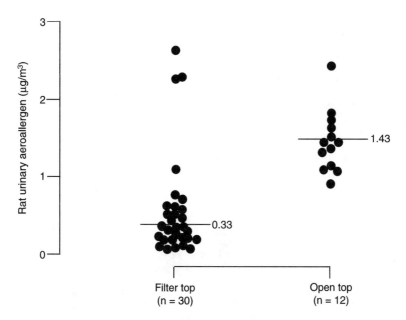

**Figure 7.3.** Comparison of cage design. Rat urinary aeroallergen concentrations measured when 30 rats were housed on woodchip bedding in filter-top and open-top cages, excluding those measurements made on cleaning-out days for the open-top cages. The geometric mean (GM) is indicated. (Gordon, S. et al., *Br. J. Ind. Med*, 49:416–422, 1992. Reproduced with permission.)[17]

such as a cage wash robot (and at the same time reducing ergonomics problems and allergen exposure), technicians have more time that can be used to increase their competence through more involvement in research. These workers, who have excellent animal-handling competence, could perform work procedures that would otherwise expose fresh and fumbling postgraduate students.

## Animals, Stock Density, and Bedding

Mature male animals have been shown to generate higher concentrations of allergen in urine[40,44] and in animal rooms.[66] A recent study suggests that working with male animals may increase the risk for LAA.[22] If it is feasible, considering the scientific question at hand, substitution with younger or female animals is likely to reduce aeroallergen levels and possibly LAA.

Several studies have shown an association between stock density and allergen levels (see Figure 7.4).[17,57,60,63] The usefulness of this information is slight, as density is much more likely to be dictated by business factors. The advent of individually ventilated cage systems now means it is feasible to increase stock density while maintaining control of ambient levels of aeroallergen.[69] However, the allergen and potentially pathogen-releasing cage-changing task still requires practical solutions to minimize contamination.

Bedding has an influence on allergen concentrations, although other factors, such as toxicological implications for the animals, will also influence the choice. Absorbent pads are associated with lower allergen levels than wood chips or sawdust (see Figure 7.5).[17] Wood chips give lower aeroallergen concentrations than sawdust.[71] Crushed corncob was found to give lower levels than wood shavings.[66] The impact of animal cage enhancements on allergen exposure has not been reported, however, preliminary findings suggest that enrichment measures may increase allergen levels during cage changing.[72]

## Housekeeping

Animal facilities should be designed so that they can be effectively and safely cleaned. Examples of this are the use of closed vacuum cleaners that deliver dust into a closed-pipe conveyor with deposition into a sealed container, the use of moist mopping, or damp sweeping. Dry-cleaning procedures, such as

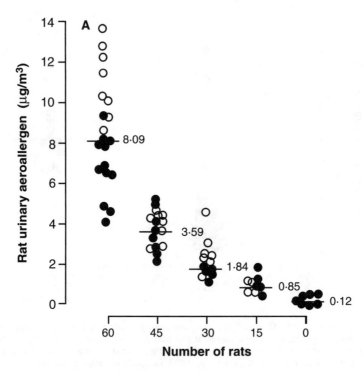

**Figure 7.4.** Effect of reducing stock density. Rat urinary aeroallergen concentrations measured when the stock density was reduced from 3 (1 to no rats m3 (60 to no rats). The measurements made on cleaning-out days are shown as open circles. The geometric mean (GM) is indicated. (Gordon, S. et al., *Br. J. Ind. Med.*, 49:416–422, 1992. Reproduced with permission.)[17]

brushing and the use of portable vacuum cleaners, should be avoided, as they can create high amounts of airborne particles contaminated with allergen. The use of power-washing systems, using high-pressure water, can generate contaminated aerosols and should be avoided where possible. Where power washing is used, appropriate personal protective equipment should be worn.

The contaminated outputs from animal work must be controlled. There should be procedures to control exposure to allergen in the handling of contaminated documents and in the disposal and handling of animal carcasses and tissues, animal waste and bedding, and contaminated personal protective equipment. Measures should be implemented to minimize the need for documents in animal holding and handling areas, such as using computers with washable-plastic-covered keyboards for documentation, connected to printers outside of the animal department. Clean documents can be created in a scanning (but not photocopying) process. If documents need to be retained (and archived), then measures should be taken to minimize spread of allergen from the facility. Contaminated records and archives are an important potential source of allergen exposure because they provide a mechanism by which workers with serious LAA can be inadvertently exposed

Procedures should be implemented to reduce exposure to allergen during the laundry of reusable protective clothing. The risk to laundry workers, who may not be aware of animal allergen risks, should be considered. Soluble laundry bags that can be sealed in the animal facility and dissolve during the laundering process are available.

Cage cleaning has the potential to expose workers to high aeroallergen levels.[15] Care must be taken to reduce allergen spread from this source at all points of handling. Dedicated equipment that minimizes contact with soiled bedding should be provided in the cage-washing area. Several commercially available systems have been described. Common to all of these is that cages are emptied into closed transport systems that deliver bedding to different types of sealed containers. Vacuum cleaning systems that are designed not to generate local contaminated exhaust have been used, and result in low levels of allergen during floor cleaning.[56] However, these measures will not necessarily contribute to reductions in personal exposure if they do not enable efficient completion of the task.[69]

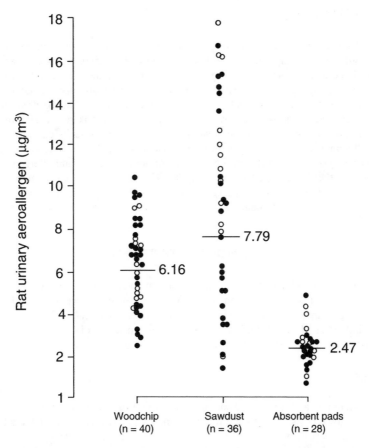

**Figure 7.5.** Comparison of litter type. Rat urinary aeroallergen concentrations measured when 60 rats were housed on woodchip and sawdust in contact litter, and absorbent pad, non-contact litter. The measurements made on cleaning-out days are shown as open circles. The geometric mean (GM) is indicated. (Gordon, S. et al., *Br. J. Ind. Med.*, 49:416–422, 1992. Reproduced with permission.)[17]

## Movements Within the Facility

Procedures should be designed to prevent the spread of allergens into the environment and to adjacent areas, such as corridors, offices, and rest areas. Transport of soiled cages from the animal room to the cleaning area should be done using a closed or covered system, preferably only in designated "dirty corridors." Animals should be transported in suitable transport cabinets equipped with filters that prevent the spread of allergens along corridors and in elevators. Single cages with animals should be transported using a filter top. Single cages should likewise be covered with a filter top when standing freely in laboratory or procedure rooms.

## Work Permits and Visitors

A mechanism should exist whereby unusual exposures are adequately controlled. These can involve regular animal workers carrying out irregular tasks, such as during equipment breakdown, or involve workers not normally exposed to animals, e.g., ventilation maintenance technicians. These tasks should be individually assessed and may need to be controlled under a work-permit procedure.

## Environmental Monitoring

Hygiene studies have improved understanding of the exposure characteristics of different workplaces and activities. Hygiene monitoring can give confidence that controls are reducing allergen concentrations and that changes in controls do not have a negative effect. In some circumstances, it may be necessary to demonstrate that a change in working practice for a reason other than improved allergen control does not have a significant adverse effect on levels. Unfortunately, the analysis of allergen samples is expensive and time-consuming. Analytical services are not widely available. It is relatively simple to measure concentrations of particles, and this has the advantage that it can be done in real time, giving a clearer indication of the fluctuating personal exposure (to particles) during the working day (see Figure 7.6).[73] There is, however, no correlation between the concentrations of particles and allergen. For example, it has been pointed out that handling clean and dirty bedding may both generate high concentration of particles, but in the first case, no animal allergen is present at all. At present, there are no internationally recognized standards for analytical methods or acceptable aeroallergen exposure levels.

## PERSONAL CONTROLS

In addition to engineering and procedural controls, for some tasks, it may be necessary to implement controls that are targeted at the workers themselves. Although some of these, such as choice of personal protective equipment, may be incorporated into procedures, their effectiveness is wholly within the influence of the individual. The focus of the personal control measures is on influencing behavior, and includes the correct use of personal protective equipment and training.

### Personal Hygiene

Eating, drinking, smoking, and the application of cosmetics should not be permitted in animal facilities. Changing routines should be established that minimize the spread of allergen around and from the animal areas and the introduction of pathogens into the facility. Workers heavily exposed to allergen should shower before leaving the facility to remove allergen adsorbed to exposed skin and hair. If gloves are used, care must be taken to ensure that allergens do not contaminate the skin by entry via the open-arm end of the glove or through puncture holes. Workers should wash their hands regularly and talways after

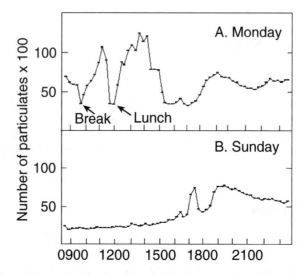

**Figure 7.6.** Variations in particles during the day in a typical mouse room. (Kacergis, J.B. et al., *Am. Ind. Hyg. Assoc. J.*, 57:634–640, 1996. Reproduced with permission.)[73]

handling animals without gloves. However, care must be taken to prevent small skin fissures from developing through frequent washing, since these will increase the risk of systemic allergen exposure. Barrier creams and hand-friendly washing substances are widely available.

## Protective Clothing

Most modern laboratory animal facilities require personnel to use special clothing while working in the facility. The main objective of this is to prevent microbiological contamination of the animals within the facility. The type of clothing will vary according to the degree of protection required and may consist of full surgical clothing with gown, cap, mask, and shoe covers to conventional changes of laboratory coat. The design of protective clothing is important. Coat or coverall arms should be designed such that allergens are not trapped inside the arm surface. The sleeves should have an elastic hem or other suitable mechanism. Alternatively, coverall arms should be rolled back so as to allow personnel to wash both hands and forearm. Several types of gloves are available on the market. Non-latex gloves are preferable to latex gloves, due to the risk of developing latex allergy.[74]

## Respiratory Protective Equipment

Half-face particle filter respirators are a significant control measure used in successful programms.[12,23] Such respirators of type P2 have been shown to reduce inhaled allergen by 90%.[59] This should be sharply contrasted with the disposable surgical masks that are occasionally used. These are primarily designed to protect surgical fields from exhaled droplets. They provide a poor seal around the edge and, although they may be comfortable to wear, provide limited, if any, protection.

Several types of ventilated masks and helmets have been developed in which filtered air is delivered to the operator. These can be highly effective in reducing exposure to airborne allergen. Air-stream respirator helmets have been shown to relieve symptoms of LAA in sensitized workers.[75] Supplied air systems have an important place in both the prevention of allergy and the control of symptoms. They should be available for workers at high risk and for workers carrying out high-risk tasks, such as cleaning up spillages of heavily contaminated bedding or maintaining cage emptying systems. Workers with LAA can continue to work in animal areas, as long as they remain symptom-free, if they use these devices so their exposure to allergen is negligible. Any person using a ventilated device should be trained in its use. Equipment should be cleaned and stored and their filters changed without spread of allergen or cross-contamination.

Although these air-supplied devices can reduce personal exposure to aeroallergen to negligible levels, they do have major drawbacks: they do not easily facilitate detailed work up close to the animal when the risk of exposure is high; they may be uncomfortable if used for long periods, leading to headaches and neck discomfort; and they may not be suitable for some tasks, such as socialization.

## Training and Education

The success of any risk-control program is dependent on the support of those at risk. Effective training and education, at the start of working with animals, and regularly repeated, is an important feature of successful allergy prevention programms.[12,23] Workers at risk of LAA must understand the nature of the risk if they are to be able to appreciate the actions they can take to protect themselves. These actions include wearing protective equipment when handling the animals, reducing allergen spread, taking care with personal hygiene, and promptly reporting symptoms of allergy. Improved use of protective equipment has been reported following the implementation of education programs.[23]

## Pre-Placement Assessment: Assessing Risk

The prevention of allergy should begin before exposure to allergen occurs. Assessment of potential animal workers prior to commencing work with animals is good practice, and also a legal duty in some jurisdictions. Partly, the purpose of this assessment is to consider individual vulnerability to allergy. It is also to consider the worker's capability to do the proposed work and determine the need for interventions and adjustments (such as immunization against infectious diseases and provision of lifting aids). If individuals who will definitely develop allergy could be identified reliably prior to exposure then they could be excluded from the workplace. The identification of these vulnerable people is highly desirable, but it it feasible?

Atopy is a genetic predisposition to develop specific IgE and allergic reactions (e.g., skin rashes, rhinoconjunctivitis, asthma).[9] The majority of studies have indicated that workers who have a personal history of atopy are more likely to develop LAA, although this has not been a universal finding.[10] Some investigators have identified an association between family history of atopy and the development of LAA, but others have found no association.[10] It is likely that any association between family history and the development of LAA is weak.

Some studies have examined the association between biological indicators of atopy (skin testing and immunoassay) and allergy. Most of these studies have been a cross-sectional design and have examined the association of atopy in established cases of LAA. This limits their usefulness in establishing the predictive value of these indicators in workers without LAA. There is a clear association between skin-test reactivity to animal allergens and allergic symptoms. Pooled data from seven studies has shown 51 to 69% of people allergic to rats or mice have positive skin tests to allergens from these animals.[4]

An association between total IgE and the later development of allergy has been reported.[14] Only one study has examined the predictive value of radioallergosorbent tests (RAST) for IgE specific to the animal allergen. A combination of a positive RAST and positive skin test was 87.4% predictive of the development of LAA.[12]

Even using the best predictive tests (personal history and biological indicators of atopy) to exclude vulnerable workers, more people who would never develop symptoms would be excluded than people who would become allergic.[12,13] While it is possible to identify asymptomatic workers at the start of employment who are at increased risk of developing LAA, it is not practical nor ethical to implement effective screening criteria and exclude them from work. If the implementation of a comprehensive allergen-control program has reduced the incidence of allergy to low levels, the value of these tests as predictors of LAA is similarly reduced.

Pre-placement assessment is still worthwhile. It is the first opportunity to assess the vulnerability of the candidate and counsel him or her on the measures he or she should take to minimize the risk of developing allergy. It is an opportunity to establish baseline data and carry out baseline investigations against which future assessments can be compared. Serum banking is, however, not recommended.[6] Some candidates will have a history of exposure to laboratory animals (either from their studies, work, or from pets) and may report preexisting allergies. Many of these will be able to begin their intended occupation (with appropriate adjustments and restrictions), but some will not. If a candidate reports anaphylactic reactions or occupational asthma, then the risk of continued exposure to the relevant allergen is likely to be unacceptable.

Counseling at health assessment should be an integral part of any allergy-prevention program. In addition to helping people exposed to allergens understand the potential health effects and the need for early reporting, it is an opportunity to explain the steps they can take to protect their own health and to reinforce the importance of engineering, procedural, and behavioral controls. It provides an opportunity to explain individual risk in the context of the proposed work (and exposure pattern), the local experience of allergy incidence, and the individual's tendency to develop allergy. This information allows the candidates to make their own informed decision about the risks of the proposed employment.

## Health Surveillance

Regular health surveillance of workers significantly exposed is worthwhile. It provides an opportunity to raise awareness of the potential effects of allergen on health, investigate symptoms of allergy that the worker reports, and reinforce the need to report relevant symptoms if they do develop. Annual surveillance of exposed workers is typical. The majority of workers who develop allergic symptoms do so within two years of first exposure.[3] During this period, more frequent surveillance may be warranted.

The basis of surveillance should be a questionnaire.[76] Some centers may perform lung function tests, skin testing, and immunoassay, but there is no evidence that they have a quantitative value as routine screening tests. Individuals who have developed signs of LAA or asthma who do not cease working with animals should be carefully monitored at regular intervals while major efforts to reduce exposure are made and they are equipped with effective personal protection. Additional tests may have value in helping to reinforce the educational messages.

## INTEGRATED HEALTH, SAFETY, AND WELFARE

It is imperative that efforts to reduce staff exposure to animal allergens with the goal of reducing incidence of LAA do not give rise to new health problems, either among staff or animals. If allergen levels are reduced, but with increased ergonomic strain and compromised animal welfare (potentially affecting research), nothing has been gained. Few studies have as yet focused on these balances. Recently, cage systems for containing mice were compared, evaluating allergen levels, ergonomics features, biting frequency, weight gain, and cage climate. It was shown that although allergen levels in undisturbed mouse rooms were low using IVCs, they were less ergonomically suitable, compared to regular open shelving or ventilated cabinets.[70] Also, some evidence was found that cage climate and animal welfare may vary between IVC systems.[77] We conclude that all planned installations should be evaluated beforehand regarding both animal and worker health, and alternatives weighed. Investments are usually costly, and worker relocation or compensation costs or loss of research animals through contamination or stress cannot be ignored.

## REFERENCES

1. Bryant, D., Boscato, L.M., Mboloi, P.N., and Stuart, M.C., Allergy to laboratory animals among animal handlers, *Med. J. Aust.*, 163, 415, 1995.
2. Teasdale, E.L., Davies, E.G., and Slovak, R., Anaphylaxis after bites by rodents, *Br. Med. J.*, 286, 1480, 1993.
3. Hunskaar, S. and Fosse, R.T., Allergy to laboratory mice and rats: A review of the pathophysiology, epidemiology, and clinical aspects, *Lab. Anim.*, 34, 358, 1990.
4. Hunskaar, S. and Fosse, R.T., Allergy to laboratory mice and rats: A review of its prevention, management, and treatment, *Lab. Anim.*, 27, 206, 1993.
5. Bush, R.K., Mechanism and epidemiology of laboratory animal allergy, *ILAR J*, 42, 4, 2001.
6. National Research Council (NRC), *Occupational Health and Safety in the Care and Use of Research Animals*, Washington, D.C.: National Academy Press, 1997.
7. Wood, R.A., Laboratory animal allergens, *ILAR J*, 42, 12, 2001.
8. Seward, J. P., Medical surveillance of allergy in laboratory animal handlers, *ILAR J*, 42, 47, 2001.
9. Johansson, S.G.O., Hourihane, J. O'B., Bousquet, J., Brujnzeel-Koomen, C., Dreborg, S. et al., position paper, A revised nomenclature for allergy, an EAACI position statement from the EAACI nomenclature task force, *Allergy*, 56, 813, 2001.
10. Seward, J. P., Occupational allergy to animals, *Occup. Med.*, 14, 247, 1999.
11. Kruize, H., Post, W., Heederik, D., Martens, B., Hollander, A., and van der Beek, E., Respiratory allergy in laboratory animal workers: A retrospective cohort study using preemployment screening data, *Occup. Environ. Med.*, 54, 830, 1997.
12. Botham, P.A., Lamb, C.T., Teasdale, E.L., Bonner, S.M., and Tomenson, J.A., Allergy to laboratory animals: A follow-up study of its incidence and of the influence of atopy and preexisting sensitization on its development, *Occup. Environ. Med.*, 52, 129, 1995.
13. Newill, C.A., Evans, R. III., and Khoury, M., Preemployment screening for allergy to laboratory animals: Epidemiologic evaluation of its potential usefulness, *J. Occup. Med.*, 28, 1158, 1986.
14. Renström, A., Malmberg, P., Larsson, K., Sunblad, B-M., and Larsson, P.H., Prospective study of laboratory-animal allergy: Factors predisposing to sensitization and development of allergic symptoms, *Allergy*, 49, 548, 1994.
15. Niewenhuijsen, M.J., Gordon, S., Harris, J. M., Tee, R.D., Venables, K.M., and Newman Taylor, A.J., Variation in rat urinary aeroallergen levels explained by differences in site, task, and exposure group, *Ann. Occup. Hyg.*, 39, 819, 1995.
16. Hollander, A., Van Run, P., Spithoven, J., Heederik, D., and Doekes, G., Exposure of laboratory animal workers to airborne rat and mouse urinary allergens, *Clin. Exp. Allergy*, 27, 617, 1997.

17. Gordon, S., Tee, R.D., Lowson, D., Wallace, J., and Newman Taylor, A.J., Reduction of airborne allergenic urinary proteins from laboratory rats, *Br. J. Ind. Med.*, 49, 416, 1992.
18. Hollander, A., Heederik, D., and Doekes, G., Respiratory allergy to rats: Exposure-response relationships in laboratory animal workers, *Am. J. Respir. Crit. Care Med.*, 155, 562, 1997.
19. Cullinan, P., Lowson, D., Nieuwenhuijsen, M.J., Gordon, S., Tee, R. D., Venables, K.M., McDonald, J. C., and Newman-Taylor, A.J., Work-related symptoms, sensitization, and estimated exposure in workers not previously exposed to laboratory rats, *Occup. Environ. Med.*, 51, 589, 1994.
20. Heederik, D., Venables, K.M., Malmberg, P., Hollander, A., Karlsson, A-S., Renström, A., Doekes, G., Nieuwenhuijsen, M., and Gordon, S., Exposure-response relationship for work-related sensitization in workers exposed to rat urinary allergens: Results from a pooled study, *J. Allergy Clin. Immunol.*, 103, 678, 1999.
21. Hollander. A., Gordon, S., Renstrom, A., Thissen, J., Doekes, G., Larsson, P.H., Malmberg, P., Venables, K.M., and Heederik, D., Comparison of methods to assess airborne rat and mouse allergen levels, I. Analysis of air samples, *Allergy*, 54, 142, 1999.
22. Renström, A., Karlsson, A-S., Malmberg, P., Larsson, P.H., and van Hage-Hamsten, M., Working with male rodents may increase risk for laboratory animal allergy, *Allergy*, 56, 964, 2001.
23. Fisher, R., Saunders, W.B., Murray, S.J., and Stave, G.M., Prevention of laboratory animal allergy, *J. Occup. Environ. Med.*, 40, 609, 1998.
24. Gautrin, D., Ghezzo, H., Infante-Rivard, C., and Malo, J. L., Incidence and determinants of IgE-mediated sensitization in apprentices, A prospective study, *Am. J. Respir. Crit. Care Med.*, 163, 1222, 2000.
25. Cullinan, P., Cook, A., Gordon, S., Nieuwenhuijsen M.J., Tee, R.D., Venables, K.M., McDonald, J. C., and Newman Taylor, A.J., Allergen exposure, atopy, and smoking as determinants of allergy to rats in a cohort of laboratory employees, *Eur. Respir. J.*, 13, 1139, 1999.
26. Gautrin, D., Infante-Rivard, C., Ghezzo, H., and Malo, J-L., Incidence and host determinants of probable occupational asthma in apprentices exposed to laboratory animals, *Am. J. Respir. Crit. Care Med.*, 163, 899, 2001.
27. Kibby, T., Powell, G., and Cromer, J., Allergy to laboratory animals, *J. Occup. Med.*, 31, 842, 1989.
28. Fuortes, L.J., Weih, L., Pomrehn, P., Thorne, P.S., Jones, M., Burmeister, L., and Merchant, J. A., Prospective epidemiologic evaluation of laboratory animal allergy among university employees, *Am. J. Ind. Med.*, 32, 665, 1997.
29. Aoyama, K., Ueda, A., Manda, F., Matsushita, T., Ueda, T., and Yamauchi, C., Allergy to laboratory animals: An epidemiological study, *Br. J. Ind. Med.*, 49, 41, 1992.
30. Venables, K.M., Upton, J. L., Hawkins, E.R., Tee, R.D., Longbottom, J.L., and Newman-Taylor, A.J., Smoking, atopy, and laboratory animal allergy, *Br. J. Indust. Med.*, 45, 667, 1988.
31. Venables, K.M., Epidemiology and the prevention of occupational asthma, *Br. J. Ind. Med.*, 44, 73, 1987.
32. Platts-Mills, T., Vaughan, J., Squillace, S., Woodfolk, J., and Sporik, R., Sensitization, asthma, and a modified Th2 response in children exposed to cat allergen: A population-based cross-sectional study, *Lancet*, 357, 752, 2001.
33. Agrup, G., and Sjöstedt, L., Contact urticaria in laboratory technicians working with animals, *Acta. Derm. Venerol.*, 65, 111, 1985.
34. Kimber, I., The role of the skin in the development of chemical respiratory sensitivity, *Tox. Letters*, 86, 89, 1996.
35. Hesford, J.D., Platts-Mills, T.A.E., and Edlich, R.F., Anaphylaxis after laboratory rat bite: An occupational hazard, *J. Emerg. Med.*, 13, 765, 1995.
36. Watt, A.D., and McSharry, P., Laboratory animal allergy: Anaphylaxis from a needle injury, *Occup. Environ. Med.*, 53, 573, 1996.
37. Bayard, C., Holmquist, L., and Vesterberg, O., Purification and identification of allergenic $alpha_{2u}$-globulin species of rat urine, *Biochim. Biophys. Acta.*, 1290, 129, 1996.
38. Lorusso, J.R., Moffat, S., and Ohman, Jr., J.L., Immunologic and biochemical properties of the major mouse urinary allergen (Mus m 1), *J. Allergy. Clin. Immunol.*, 78, 928, 1986.
39. Mancini, M.A., Majumdar, D., Chatterjee, B., and Roy, A.K., $Alpha_{2u}$-globulin in modified sebaceous glands with pheromonal functions: Localization of the protein and its mRNA in preputial, meibomian, and perianal glands, *J. Histochem. Cytochem.*, 37, 149, 1989.

40. Vandoren, G., Mertens, B., Heyns, W., van Baelen, H., Rombauts, W., and Verhoven, G., Different forms of $\alpha_{2u}$-globulin in male and female rat urine, *Eur. J. Biochem.*, 134, 175, 1983.

41. Keverne, E.B., Vomeronasal/accessory olfactory system and pheromonal recognition, *Chem. Senses*, 23, 491, 1998.

42. Virtanen, T., Zeiler, T., Mäntyjärvi, R., Important animal allergens are lipocalin proteins: Why are they allergenic?, *Int. Arch. Allergy Immunol.*, 120, 247, 1999.

43. Wahn, U., Peters, Jr., T., and Siraganian, R.P., Studies of the allergenic significance and structure of rat serum albumin, *J. Immunol.*, 125, 2544, 1980.

44. Gordon, S., Tee, R.D., and Newman Taylor, A.J., Analysis of rat urine proteins and allergens by sodium dodecyl sulfate-polyacrylamide gel electrophoresis and immunoblotting, *J. Allergy Clin. Immunol.*, 92, 298, 1993.

45. Price, J.A., and Longbottom, J.L., ELISA method for measurement of airborne levels of major laboratory animal allergens, *Clin. Allergy*, 18, 95, 1988.

46. Walls, A., Newman Taylor, A., and Longbottom, J., Allergy to guinea pigs. II. Identification of specific allergens in guinea pig dust by crossed radio-immunoelectrophoresis and investigation of possible origin, *Clin. Allergy*, 15, 535, 1985.

47. Swanson, M.C., Agarwal, M.K., Yunginger, J.W., and Reed, C.E., Guinea-pig-derived allergens. Clinicoimmunologic studies, characterization, airborne quantitation, and size distribution, *Am. Rev. Respir. Dis.*, 129, 844, 1984.

48. Rudzki, E., Rebandel, P., and Rogozinski, T., Contact urticaria from rat tail, guinea pig, streptomycin, and vinyl pyridine, *Contact Dermatitis*, 7, 86, 1981.

49. Fuortes, L.J., Weih, L., Jones, M.L., Burmeister, L.F., Thorne, P.S., Pollen, S., and Merchant, J.A., Epidemiologic assessment of laboratory animal allergy among university employees, *Am. J. Ind. Med.*, 29, 67, 1996.

50. Das, R., Tager, I.B., Gamsky, T., Schenker, M.B., Roycce, S., and Balmes, J.R., Atopy and airways reactivity in animal-health technicians. A pilot study, *J. Occup. Med.*, 34, 53, 1992.

51. Perfetti, L., Hébert, J., Lapalme, Y., Ghezzo, H., Gautrin, D., and Malo, J.L., Changes in IgE-mediated allergy to ubiquitous inhalants after removal from or diminution of exposure to the agent causing occupational asthma, *Clin. Exp. Allergy*, 28, 66, 1998.

52. Venables, K.M., Occupational asthma, *Lancet*, 349, 1465, 1997.

53. Malo, J-L., Boulet, L-P., Dewitte, J-D., Cartier, A., L'Archevêque, J., Côté, J., Bédard, G., Boucher, S., Champagne, F., Tessier, G., Constanopoulos, A-P., Juniper, E.F., and Guyalt, G.H., Quality of life of subjects with occupational asthma, *J. Allergy Clin. Immunol.*, 91, 1121, 1993.

54. Anon allergy to laboratory animals, *Vet. Record*, 107, 122, 1980.

55. Reeb-Whitaker, C.K., Harrison, D.J., Jones, R.B., Kacergis, J.B., Myers, D.D., and Paigen, B., Control strategies for aeroallergens in an animal facility, *J. Allergy Clin. Immunol.*, 103, 139, 1999.

56. Thulin, H., Björkdahl, M., Karlsson, A-S., and Renström, A., Reduction of exposure to laboratory animal allergens in a research laboratory, *Ann. Occ. Hyg.*, 2002 (in press).

57. Eggleston, P.A., Newill, C.A., Ansari, A.A., Pustelnik, A., Lou, S-R., Evans, III., R., Marsh, D.G., Longbottom, J. L., and Corn, M., Task-related variation in airborne concentrations of laboratory animal allergens: Studies with Rat n I, *J. Allergy Clin. Immunol.*, 84, 347, 1989.

58. Hollander, A., Heederik, D., Doekes, G., and Kromhout, H., Determinants of airborne rat and mouse urinary allergen exposure, *Scand. J. Work. Environ. Health*, 24, 228, 1998.

59. Renström, A., and Kallinn, C., Characterisation of the work environment in an animal laboratory facility, Stockhom: NIWL (unpublished report), 2000.

60. Swanson, M.C., Campbell, A.R., O'Hollaren, M.T., and Reed, C.E., Role of ventilation, air filtration, and allergen production rate in determining concentrations of rat allergens in the air of animal quarters, *Am. Rev. Respir. Dis.*, 141, 1578, 1990.

61. Harrison, D.J., Controlling exposure to laboratory animal allergens, *ILAR J*, 42, 17, 2001.

62. Lindqvist, C., Persson, L., Iwarsson, K., Lustig, G., Renström, A., and Larsson, P.H., A sliding curtain system for improving air distribution of animal rooms in relation to working environment and cage climate, *Scand. J. Lab. Anim. Sci.*, 23, 135, 1996.

63. Edwards, R.G., Beeson, M.F., and Dewdney, J. M., Laboratory animal allergy: The measurement of airborne urinary allergens and effects of different environmental conditions, *Lab. Anim.*, 17, 235, 1983.

64. Jones, R.B., Kacergis, J.B., MacDonald, M.R., McKnight, F.T., Turner, W.A., Ohman, J.L., and Paigen, B., The effect of relative humidity on mouse allergen levels in an environmentally controlled mouse room, *Am. Ind. Hyg. Assn. J.*, 56, 398, 1995.

65. Skoke, H.H., Ventilated tables' impact on workstation and building design, *Lab. Anim.*, 24, 22, 1995.

66. Sakaguchi, M., Inouye, S., Miyazawa, H., Kimura, M., and Yamazaki, S., Evaluation of countermeasures for reduction of mouse airborne allergens, *Lab. Anim. Sci.*, 40, 613, 1990.

67. Lipman, N.S., Isolator rodent caging system (state of the art): A critical view, *Contemporary Top. Lab. Anim. Sci.*, 38, 9, 1999.

68. Clough, G., Wallace, J., Gamble, M.R., Merryweather, E.R., and Bailey, E., A positive, individually ventilated caging system: A local barrier system to protect both animals and personnel, *Lab. Anim.* 29, 139, 1995.

69. Gordon, S., Wallace, J., Cook, A., Tee, R.D., and Newman Taylor, A.J., Reduction of exposure to laboratory animal allergens in the workplace, *Clin. Exp. Allergy*, 27, 744, 1997.

70. Renström, A., Höglund, U., and Björing, G., Evaluation of individually ventilated cage systems for laboratory rodents. Occupational health aspects, *Lab. Anim.*, 35, 42, 2001.

71. Platts-Mills, T.A.E., Heyman, P., Longbottom, J.L., and Wilkins, S.R., Airborne allergens associated with asthma: Particle size carrying dust mite and rat allergens measured with a cascade impactor, *J. Allergy Clin. Immunol.*, 77, 850, 1986.

72. Renström, A., unpublished data.

73. Kacergis, J.B., Jones, R.B., Reeb, C.K., Turner, W.A., Ohman, J.L., Ardman, M.R., and Paigen, B., Air quality in an animal facility: Particulates, ammonia, and volatile organic compounds, *Am. Ind. Hyg. Assoc. J.*, 57, 634, 1996.

74. Brehler, R. and Kutting, B., Natural rubber latex allergy: A problem of interdisciplinary concern in medicine, *Arch. Intern. Med.*, 161, 1057, 2001.

75. Slovak, A.J.M., Orr, R.G., and Teasdale, E.L., Efficacy of the helmet respiratory in occupational asthma due to laboratory animal allergy (LAA), *Am. Ind. Hyg. Assoc. J.*, 46, 411, 1985.

76. Bush, R.K., Wood, R.A., and Eggleston, P.A., Laboratory animal allergy, *J. Allergy Clin. Immunol*, 102, 99, 1998.

77. Höglund, U. and Renström, A., Evaluation of individually ventilated cage systems for laboratory rodents. Cage environment and animal health aspects, *Lab. Anim.*, 35, 51, 2001.

78. National Institute for Occupational Safety and Health, NIOSH Alert: Preventing asthma in animal handlers, Publication No. 97–116, 1998.

# CHAPTER 8

# Laboratory Animal Facilities and Equipment for Conventional, Barrier, and Containment Housing Systems

Jack R. Hessler and Urban Höglund

## CONTENTS

0-8493-1086-5/03/$0.00+$1.50
© 2003 by CRC Press LLC

## INTRODUCTION

The objective of this chapter is to provide an overview of facilities and equipment required for housing laboratory animals. Other recent publications on the subject that provide additional details include a chapter in the 2nd edition of *Laboratory Animal Medicine* by Hessler and Leary,[1] an entire issue of *Lab Animal*[2] that is dedicated to animal facility design and planning, and a 1991 book by Ruys.[63] A review of the progress made in research animal facilities and equipment during the latter half of the 20th century is summarized by Hessler in a chapter of a book published to celebrate the 50th anniversary of the American Association of Laboratory Animal Science (AALAS).[3]

A major objective of laboratory animal science is to control the laboratory animal's environment. Environmental variables can alter the animal's biology, resulting in background "noise" that can mask the biological response to experimental variables, thus confounding the interpretation of the experimental data. Of course, animal comfort and well-being is paramount not only for moral reasons but also scientific reasons. Distressed animals make poor research subjects, but the fact is that many biological responses to environmental factors are not manifested in terms of stress, distress, or any overt pathologies. For this reason, the degree of control required for the research animal's environment is primarily science driven, going well beyond that required to assure the animal's well-being. It is the responsibility of laboratory animal specialists not only to provide for the comfort and well-being of the animals, but also to assist the scientist with controlling animal-related variables that may confound the science. Figure 8.1 illustrates the many environmental factors that must be considered, including genetic, microbial, chemical, and physical. Control of genetic variables is primarily a matter of biology, but control of other variables is dependent to a significant degree on the design and management of the research animal facility and equipment. Environmental standards and design concepts for animal facilities are constantly evolving toward higher levels of performance with regards to controlling the animal's environment and operational efficiency.[3] Properly designed and equipped facilities greatly facilitate effective management and consistent day-to-day animal care, which is required to optimally support animal research and testing. In spite of the many choices and wide berth for creativity in designing research animal facilities, the general research animal facility and caging standards have evolved to become well defined.[4–7]

There are a variety of ways to categorize facilities for housing laboratory animals. The three most common are addressed in this chapter — conventional, barrier, and containment facilities. For the sake of clarity, because these are not necessarily universally defined terms, they are defined here as follows:

*Barrier (keep out)* — Animal housing systems designed and managed to protect the animals from undesirable microbes.

*Containment (keep in)* — Animal housing systems designed and managed to contain experimental or naturally occurring hazards, e.g., biological, chemical, and radiation, in order to protect workers, other animals, and the general environment.

*Conventional* — Standard housing systems for laboratory animals that do not offer the added level of control provided by barrier and containment systems.

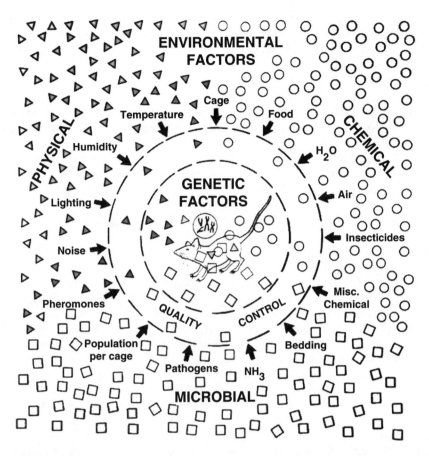

**Figure 8.1** Conceptual depiction of the laboratory animal and how its biology, while based on genetics, is readily influenced by environmental factors. Controlling these environmental factors, any of which may confound animal research data, is a major component of laboratory animal science. (From Baker, H.J., Lindsey, J.R., and Weisbroth, S.H., Housing to control research variables, in *The Laboratory Rat: Vol. I, Biology and Diseases*, Baker, H.J., Lindsey, J.R., Weisbroth, S.H., Eds., Academic Press, New York, 1979, pp. 169–192. With permission.)

These categories do not pertain to any particular species; however, the descriptions in this chapter pertain primarily to standard laboratory animals.

## FACILITIES

### General Considerations

#### *Location*

Animal facilities should be located in a secure area isolated from the rest of the research facility for many reasons, including public health, public relations, security, human comfort, animal health, and animal husbandry considerations. Because animal allergens pose a potentially serious health risk, exposure to animals, animal dander, equipment soiled by animals, and animal waste products must be limited to personnel whose job requires exposure; and for those individuals, steps should be taken to limit exposure.[8] Careful planning must reconcile the necessity for isolating the animal facility with the desirability of locating animal facilities as near as possible to the research laboratories. In addition, access and egress patterns for research staff, supplies, animals, and trash need to be carefully planned to facilitate efficiency, reduce contamination between animal rooms, and prevent unnecessary exposure of personnel to animals and animal waste products.

## *Arrangement*

A single-story centralized facility with direct access to ground-level transportation is the most efficient facility to operate. Alternative arrangements include a central facility with dedicated elevators to gain access to ground level transportation; a central facility on multiple floors arranged around dedicated elevators; multiple autonomous units that contain all the necessary animal care and use support services; and satellite facilities that rely to varying degrees on a primary facility for some support services. If properly planned and managed, almost any arrangement can be made to work, but the more an arrangement varies from the single-floor facility with direct access to ground-level transportation, the less the operational efficiency and the greater the operational cost for the overall animal care and use program.

## *Circulation*

Vertical circulation in facilities without direct access to ground-level transportation and multilevel animal facilities should include a minimum of two dedicated freight elevators, one for transporting "clean" items and one for "soiled" items, and more importantly, one for backup, while the other is being serviced. The focus of traffic flow in an animal facility revolves around the cage sanitation facility and the flow of cages between it and the animal rooms. The horizontal circulation pattern to be used is one of the early decisions to be made in the facility planning process. There are two basic horizontal circulation patterns, single corridor and dual corridor. Dual corridors are also known as "clean–dirty" corridors. The objective of the dual corridor circulation pattern is to decrease the potential for cross contamination between animal rooms. Theoretically, dual corridors are superior to single corridors in terms of reducing cross contamination; however, as compared to single corridors, they come at a high cost in terms of the ratio of animal housing space to circulation space. Figure 8.2 illustrates this point. Whether or not dual corridors are cost-effective is a complex issue, and the answer will vary according to the relative weight assigned to each of the many pros and cons.[9] Clearly each has advantages, disadvantages, and limitations (Table 8.1). Few would disagree that a dual-corridor plan is the best choice if cost and space are not an issue. However, many single-corridor barrier and containment facilities appear to function effectively in terms of providing adequate contamination control.

## Function Areas

Animal facility space may be divided into two major functional types: animal housing space and space that supports animal housing and use. The ratio of support space to animal housing space varies considerably from facility to facility depending on the programmatic requirements, but it typically ranges between 30:70 and 70:30. In general, the smaller the facility, the higher the percent of space devoted to support.

## *Support Areas*

### *Administrative, Training, and Personnel Health and Hygiene*

Managing an animal facility is a complex business that requires the coordinated effort of a variety of staff, many of whom require office space, including professional, management, supervisory, training, and clerical staff. It is highly desirable to provide the administrative and training space in a consolidated suite adjacent to the animal facility but outside of the security perimeter. The suite should include space for office equipment and storage of office supplies and files and amenities such as an office kitchen unit. This is also a highly desirable location to place the training space. Given the importance of training, for

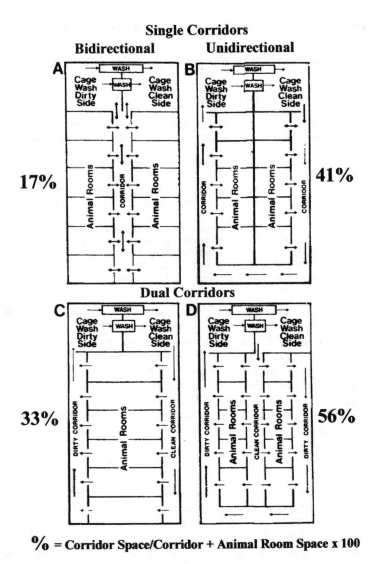

$$\% = \text{Corridor Space/Corridor} + \text{Animal Room Space} \times 100$$

**Figure 8.2**   Schematics of four basic types of circulation patterns. The arrows indicate the direction of cage traffic between the animal rooms and the cage sanitation area. "A" illustrates a single-corridor bidirectional pattern. "B" illustrates a single-corridor unidirectional pattern. "C" illustrates a dual-corridor pattern with relatively large animal rooms. "D" illustrates a dual-corridor pattern with relatively small animal rooms. All four are drawn within the same footprint to illustrate the relative "cost" of the different circulation patterns and small versus large animal room sizes in terms of the ratio of corridor space to corridor plus animal room space. The percentages only serve to illustrate the significance of choosing a combination of circulation patterns and animal room sizes, and do not necessarily apply to a particular plan.

animal care and animal use staff, such space should have high priority. It should include space for conferencing and training, training equipment, and storage of library and training materials. Space for animal procedure training may best be located in the animal facility. Animal technician supervisor offices may be in the administrative suite or scattered throughout animal housing areas, depending on the size of the facility. Office space for veterinary technicians may be located inside the surgery suite or in the diagnostic laboratory space, which may also be included inside the administrative suite.

A safe, efficient, and healthy working environment must be provided for personnel working in the facility.[8] The primary safety issues are animal allergens, infectious agents, chemical hazards, and physical hazards. To protect personnel and animals and to reduce the potential for transporting hazardous agents

**Table 8.1   Advantages and Disadvantages of Dual Corridors as Compared with Single Corridors**

**Advantages of Dual Corridors**

The separation of clean cages and supplies from soiled cages and trash eliminates the potential for cross contamination in the corridors.

They facilitate the flow of supplies and cages through the facility.

They allow for reduced congestion in the corridors.

**Disadvantages of Dual Corridors**

The higher ratio of circulation space to animal room and support space is costly.

Labor costs are higher when managed to maintain a strict unidirectional flow of personnel between clean and soiled sides.

The additional door in the animal room limits room layout options and decreases space utilization efficiency.

**Comment**

The smaller the animal rooms, the higher the space cost of dual corridors, and vice versa.

The potential for airborne cross contamination from the corridor to the room is similar for dual- and single-corridor configurations.

Contamination control is the primary issue. Most would agree that a dual corridor system is the best model for contamination control; however, effective contamination control can be provided in a single-corridor system by using an appropriate combination of management procedures and equipment options.

Which is the most cost effective? The answer to that question depends on the individual situation and how much weight is put on the various advantages and disadvantages.

---

between home and the animal facility, animal care technicians are required to wear work uniforms, and other personnel working the facility are typically required to at least wear protective outer garments prior to entering animal rooms. All uniforms and protective outerwear are provided and laundered by the facility. In addition, eating and drinking are not permitted in animal housing areas or in most support areas. To accommodate these requirements, support facilities should include lavatory, shower, locker rooms, and a break area. A laundry room for laundering uniforms and surgical linens is useful, even if a commercial laundry service is to be used. In addition, amenities and aesthetic considerations throughout the facility and especially in the break area that make for a quality work environment and enhance the recruitment and retention of staff should be provided.

## Animal Care

*Cage Servicing and Sanitation* — This is one of the most important spaces in the animal facility. Mobile animal cages are typically transported between animal rooms and the cage sanitation area from one to three times per week. This makes it the busiest and one of the most important areas in the facility. Where this area is located relative to the animal rooms and how it is designed and equipped has a major impact on how effectively and efficiently adequate animal care can be provided. Typically, the main portion of the cage sanitation area is divided into two sides — soiled side and clean side — separated by pass-through cage sanitation equipment and a wall (Figure 8.3). Single-room cage sanitation areas are not recommended. In addition, it is desirable, especially if dust-generating automatic bedding dispensers are used, to divide the clean side into two areas, separating the area in front of the discharge side of the sanitation and bedding dispensing equipment from the clean cage storage area. In some large rodent barrier facilities, there may be a bulk autoclave between the discharge area and the clean cage storage area. The type of cage sanitation equipment (described later) and the amount of space required in the cage sanitation area depends on the species housed, cage types, cage rack capacity of the facility, and cage sanitation program.[10] Space for bulk storage of cage sanitation chemicals is also required. Often, this space is provided adjacent to the cage sanitation area, but another good location is near the facility loading dock from where it can readily be piped to the cage sanitation equipment.

*Feed and Bedding Storage* — Feed and bedding is typically delivered to the facility on pallets and then taken out of the storage space one bag at a time; therefore, the ideal location is not at the dock, but rather as close as possible to where it will be used, which in the case of bedding, is the clean side of the cage sanitation area (Figure 8.3). Usually, this proves to be the best location for feed storage. Of course, all circulation space and doors in the path between the dock and the storage areas must be wide

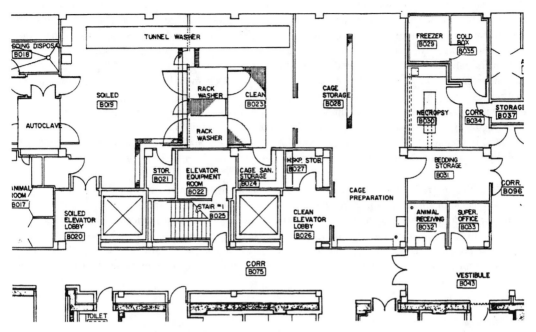

**Figure 8.3**    Schematic drawing of a cage sanitation area with two cage and rack washers and one tunnel washer. The area consists of the soiled side (B019), sanitation chemical storage (B021), clean side (B023), clean cage storage and cage preparation (B028), bedding storage (B031), and a small cage parts storage room (B024). It is located adjacent to a pair of elevators that connects the cage sanitation area to a corridor just outside a rodent barrier facility on the floor above. One elevator is used for transporting soiled cages (elevator lobby B020) and one for transporting clean cages (elevator lobby B026). The bedding storage has two doors: one connects to the cage sanitation area for convenient access to the automatic bedding dispenser in the clean side at the end of the tunnel washer (Figure 8.23), and the other connects to a corridor that leads to the nearby receiving dock. The floors on the soiled and clean sides slope toward grate-covered drain troughs the width of the rooms. The grate-covered drain pit on the discharge end of the rack washers spans the width of the two washers and extends out 2.4 m (8 ft) from the washers. The floor in the cage storage area gently slopes to a grate-covered drain trough in the center of the room. Immediately to the left is a containment facility (Figure 8.4).

enough to accommodate the size pallets to be used. The maximum recommended storage temperature for natural ingredient feed is 21°C (70°F).[4] Purified and chemically defined diets, though dry, are often less stable, and thus, their shelf life may be significantly less than that of natural ingredient diets unless stored at 4°C (39°F).[11] Refrigerated storage space is also required for fresh meats, fruits, and vegetables. The need for food preparation capabilities ranges from none to complex, depending on the research being supported. Safety testing laboratories that administer test compounds in feed require highly specialized preparation areas that allow for safe mixing of potentially hazardous compounds with animal feed.

*Housekeeping and Supply Storage* — This space is required to support sanitation of animal rooms, corridors, and other support areas. These include storage rooms for sanitation supplies and equipment, including floor scrubbers, and janitorial/mop closets strategically located in corridors and self-contained areas such as the surgery suite, biocontainment, and rodent barriers.

*Receiving and Shipping* — For most facilities, a dedicated, strategically located, and well-designed receiving and shipping area is essential for handling the large volume of supplies, e.g., bedding, feed, sanitation chemicals and supplies, disposables, and animals routinely received into an animal facility. A nearly equal volume of materials, mostly in the form of trash, exits the facility. Ideally, a separate dock or, at least, an isolated portion of the receiving and shipping area should be provided for trash disposal.

Animal shipments out of the facility are a given for animal production facilities, but it is becoming increasingly common in research facilities because of sharing transgenic animals between research institutions. At a minimum, the receiving and shipping area should include a dock, an enclosed receiving room immediately adjacent to the dock, and a room for short-term housing of animals in shipping containers until they can be delivered to an animal room or be picked up for shipment. In order to accommodate a wide variety of delivery vehicle sizes, the dock should be equipped with a scissor lift that ranges from ground level to the height of large trucks. An overhang extending at least 2 m out from dock bumper is required to protect animals and supplies from rain. Consideration should be given to fully enclosing docks that are exposed to a high volume of public traffic or that are located in cold climates. In addition to a standard hinged door for personnel entrance, automatic rollup doors equipped with flying insect air shields should be provided.

*Waste Storage/Removal/Disposal* — A large amount of waste material, including soiled bedding, general trash, and animal carcasses, is generated in animal facilities. It needs to be removed from the facility without being transported through common corridors or elevators outside the facility. Often, the point of exit is a dock inside the animal facility. Soiled bedding typically makes up the bulk of the waste. The most common method is to dump it into a trash container, preferably inside a high-efficiency particulate air (HEPA) filtered bedding disposal cabinet (Figure 8.4), and then manually (transport it to

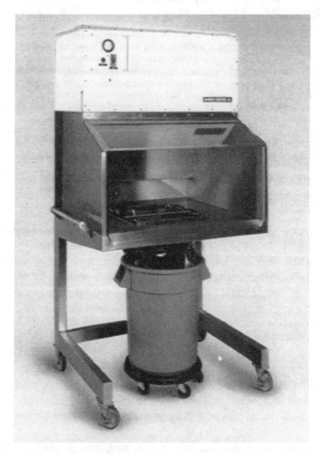

**Figure 8.4**    Shown is a soiled bedding dumping station that uses mass air displacement to contain bedding dust, drawing it away from the operator standing in front of the cabinet, while dumping soiled bedding from a cage inside the cabinet. The air draws the dust into the back of the cabinet, where it is filtered out from the air that is first passed through a coarse filter and then a HEPA filter before being returned to the room. (Courtesy of Allentown Caging Systems Co., Inc.)

**Figure 8.5** From the view of the soiled side of a cage sanitation facility, shown is a bedding dump and disposal station that disposes of bedding directly to the sanitary sewer system. To the left of the picture is the load end of a tunnel washer. To the near side of the disposal unit is a wall-mounted stainless steel sink typical of a type often recommended for use in animal rooms. The wall and floor finish is ceramic tile with epoxy grout.

a trash container outside the facility. Other methods used for disposing soiled bedding include dumping it inside the soiled side of cage sanitation either directly into the sanitary sewage system (Figure 8.5), local codes permitting, or into a vacuum system that transports it directly to a disposal container outside the facility. Similar vacuum systems may be used for transporting clean bedding into the facility. Such vacuum systems require dedicated space for the vacuum equipment, preferably outside the facility in order to contain noise and dust potentially generated by the equipment. At one time, incinerators inside the animal facility or on-site were commonly used to dispose of soiled bedding, animal carcasses, and certain hazardous wastes, but environmental protection codes often preclude using incinerators. Today, most hazardous waste and animal carcasses are packed inside the animal facility into special containers and incinerated off site, typically by commercial disposal companies. Space needs to be provided for safely packing the containers and, as was noted previously, refrigerated space is required for storing the containers, preferably near the dock, until they can be picked up for final disposal elsewhere. An alternative means for disposing of animal carcasses includes chemical digestion in specially designed equipment that prepares the carcasses for disposal through the sanitation sewerage system.

## Animal Use

*Surgery* — In most localities, dedicated space is required for conducting major survival surgical procedures on nonrodent mammalian species.[4,12] Typically, this should be a surgical suite. The design of the surgery suite will depend on the species, number, and complexity of procedures likely to be performed. In addition to operating rooms, the surgery suite should include rooms, or areas within in a room or rooms, for preparation and storage of sterile supplies, surgeon preparation, animal preparation, immediate postsurgical recovery, and equipment and supply storage. Ideally, these are separate rooms, but at a minimum, it is essential to limit activities in the surgery room to those required to conduct the surgical procedure, and to separate "clean" and "dirty" activities. Depending on the size of the surgery suite, office space for veterinary technicians and veterinarians may be required. Ideally, the surgery suite should be located near where the nonrodent mammalian species are likely to be housed and arranged to preclude unnecessary traffic through it. See Hessler[13] for a detailed sample program description of a surgery facility. Standards for conducting survival surgical procedures on rodents may be less stringent in some localities in that a dedicated space is not required. Even then, aseptic procedures are required. If it is anticipated that a large number of rodent surgical procedures will be conducted, a surgical room located near the rodent housing areas should be provided. It need not necessarily be part of the surgery suite or even be dedicated to this purpose, but it should be designed so that it can readily be sanitized prior to use as a surgery room. See Cunliff-Beamer[14] and Brown[15] for a description of surgical facilities and management procedures for rodents.

*Diagnostic Laboratories and Necropsy* — Diagnostic laboratory facilities are an essential component of an adequate veterinary care program. The size and complexity of the laboratory space may vary from a simple wet laboratory used to process samples for delivery to a comprehensive diagnostic laboratory, or may be adequate to support a comprehensive diagnostic laboratory, or anything in between. It is efficient and convenient for diagnostic laboratory space to be immediately adjacent to or a part of the Administrative and Training suite. A necropsy laboratory is required in most facilities by veterinary and investigative staffs. Ideally, this should be located in a relatively isolated area adjacent to refrigerated space used for storing animal carcasses.

*Imaging and Special Research Support Facilities* — An adequate veterinary care program may require imaging equipment such as x-ray and ultrasound. In addition, MRI, CT scanners, PET scanners, and rodent whole-body irradiators are also often used as animal research tools. Diagnostic imaging for larger animals is typically located inside the surgery suite. Imaging of transgenic mice with MRI, CT scanners, and micro-PET scanners, etc., is rapidly becoming an essential research tool. If such equipment is to be used with animals housed in a barrier facility, consideration should be given to including space for the equipment inside the barrier because many rodent barriers are managed such that animals are not returned once removed. An even more useful arrangement would be to locate the imaging suite so that it can be directly accessible from inside as well as outside the barrier. Properly managed, this arrangement would increase access to the equipment without compromising the barrier. In addition to imaging equipment, a whole-body irradiator for small rodents is a frequently required research tool that should be located either inside the rodent barrier or even better, in a location directly accessible from inside and outside the barrier.

*Animal Procedure Laboratories* — Research animal facilities have increasingly become more active extensions of the research laboratory with most if not all survival animal procedures being conducted inside the animal facility. The primary drivers for this change are human health issues raised by taking the animals out of the facility, and animal health issues raised by returning animals to the animal facility once removed. Other concerns relate to security and public relation issues, and the biological effects that the transport and change of environment has on the animal, potentially confounding the research data derived from the animals. For these reasons and others, an increasing percentage of animal facility space is being devoted to animal procedure space. For most nonrodent species, shared animal procedure laboratories are useful. A procedure room for every four to eight animal rooms works well. These rooms are equipped with procedures lights and examinations and other amenities that facilitate performing animal procedures. Shared procedure rooms for use with "clean" mice and rats are not advised because of the increased potential for spreading infectious agents from room to room. For these animals, most procedures are performed in the animal room or in dedicated animal procedure space adjacent to each room housing rodents. In rooms housing rodents in microisolation cages, animal procedures are performed in mobile clean benches or biosafety cabinets, referred to here as "animal transfer cabinets" that are primarily used for changing cages. Even though most routine procedures on rodents can be performed in the animal transfer cabinets in the animal room, rodent facilities require a considerable amount of procedure space for performing more complex procedures than can practically be performed in the animal room, including ones that involve extensive equipment. Examples of animal procedure space that may be required in a rodent barrier facility include surgery laboratories, laboratories for diagnostic and experimental imaging, a laboratory for whole body irradiation, and a transgenic and knockout (TG/KO) animal procedure laboratory. See Hessler[13] for detailed sample program descriptions and layouts of various types of animal procedure space.

## Animal Housing Areas

### General Animal Housing Concepts

*Types and Sizes of Animal Rooms* — Basically, animal rooms can be divided into two types: rooms for housing animals using dry bedding cage systems, generally for housing small animals from rodents to rabbits, and rooms for housing animals using hose-down caging systems, generally for housing nonhuman primates, canines, and small agriculture animals in cages or floor pens. One design approach is to design all animal rooms to accommodate either type of housing system and another is to design rooms for either one system or the other. The hose-down caging system requires floors sloping to floor drains, preferably in troughs, and the presence of a hose, preferably on a hose reel. The dry bedding housing system does not necessarily require floor drains or sloped floors. The obvious advantage of designing for both types of housing systems is maximum flexibility. The disadvantage is that a room designed for both types of housing systems is not optimal for either. If it is known with a reasonable degree of certainty that only rodents will be housed in the facility, e.g., a rodent barrier facility, then a reasonable choice is to design it with no floor drains. The size and shape of the animal room can vary depending on many factors, including the species to be housed, the types of housing systems to be used, and the arrangement of the cages and racks in the room. There is no one best or ideal size, but it is important to decide on the cage type to be used as well as the placement in the room prior to deciding on sizes and shapes of the animal rooms. For example, double-sided rodent racks are typically arranged library style with multiple racks parked parallel, with the end of each rack against a common wall or two rows on opposite walls with an aisle between them. Single-sided cage racks are typically parked with the back of the rack. A combination of both types combines the advantages of both. A room with both types may have single-sided racks lined against both side walls with double racks placed end to end in the center of the room, forming two aisles between facing cage racks.

*Animal Cubicles* — This is an animal room concept that provides maximum flexibility for animal isolation within minimal space by dividing animal rooms into multiple small spaces, typically each large enough to hold one rack and occasionally two cage racks (Figures 8.6 and 8.7). Cubicles help solve the problem of what to do when a facility has plenty of animal housing space but too few spaces to provide the necessary separation of species, source, microbiological status, project, and experimental hazards. They were first described in 1961[16] and have been used extensively since then, especially for specialized housing areas, where isolation of small groups of animals and containment of hazardous or potentially hazardous agents are priorities, e.g., quarantine, biocontainment, and chemical and radioisotope containment areas. Animal cubicles typically have three solid sides, with the fourth side comprised of full panel glass doors, either vertical stacking doors or a pair of conventional hinged doors. The most common cubicle size is approximately 1.2 m deep by 1.8 m wide (4 ft × 6 ft), although, larger cubicles, e.g., 2.1 m × 2.1 m (7 ft × 7 ft), that can hold two racks and in which a person could perform simple tasks with the doors closed, are useful. The size of the room depends on the size and number of cubicles. It is recommended that the aisle between facing cubicles be maintained at a minimal width of 1.5 m (5 ft). Typically, animal cubicles are used to house smaller animals in cages on mobile cage racks using dry bedding aging systems, but the concept can also be applied to housing large animals requiring hose-down caging systems.[17,18]

Extensive experience over many years suggests that cubicles effectively prevent airborne infectious agents from spreading between cubicles in the same room. The reason is probably related to the brief window of opportunity for cross-contamination when a cubicle door is open and substantial dilution of the contaminant with large volumes of air ventilating the aisle and cubicles. The usefulness of animal cubicles has decreased with the advent of microisolation cages for rodents; however, cubicles continue to be useful for conventional housing of rodents and other species. Animal cubicles can be built in place or commercially prefabricated. Prefabricated cubicles typically come complete with lighting and internal ventilation, with and without HEPA filtration and the ability to switch between positive and negative relative air pressures. The many options regarding architectural and engineering features for animal cubicles and animal cubicle rooms along with pros and cons have been described in detail.[17,18,19,20]

### Conventional Animal Housing

In this context, "conventional" is a generic term with no specific definition that refers to almost any type of laboratory animal housing facility, area within a facility, or animal room that is not specially designated

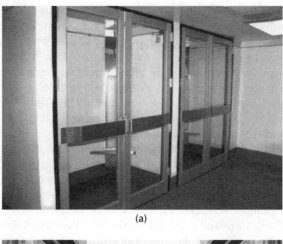

(a)

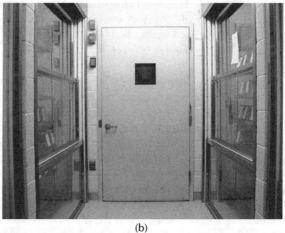

(b)

**Figure 8.6**   Figure 8.6a shows two animal cubicles side by side. Immediately across the 1.52 m (5 ft) aisle
are two identical cubicles (not shown). Note the lights in each back corner of the cubicles, the
low air returns, and the pair of hinged doors with bumper guards. Figure 8.6b shows two facing
animal cubicles with vertical-stacking three-panel sash doors.

otherwise, e.g., barrier and containment. All animal rooms, whether conventional, barrier, or containment,
should be designed for ease of cleaning and have minimal built-ins. Usually, a sink is all that is required.

### Barrier Animal Housing

In the jargon of laboratory animal science, a "barrier facility" has come to be known as an animal
housing system designed and managed to protect animals from undesirable microbes. In other words,
"barrier" equates to "keep out." Until recently, the primary use of barriers was for the production of
laboratory rodents; however, the need to maintain a similar level of barrier housing in the research
environment extended the need for barrier housing to the research facility. The need has been expanded
with the extensive use of immune-compromised animals and transgenic and knockout (TG/KO) mice.

The "barrier" may be at the cage level, the room level, at the level of an area within a facility, or
the entire facility. For example, it is common to create a barrier in a conventional animal room using
various types of cages and equipment, including microisolation caging systems, high-efficiency particulate air (HEPA) filtered mass air displacement racks, and flexible film isolators of the type used
for maintaining germ-free animals. All of these approaches work reasonably well but are much more
labor intensive to manage than a barrier designed as an area within a large animal facility or as an
entire facility. The primary difference is that with the room level barrier system, the cages and supplies

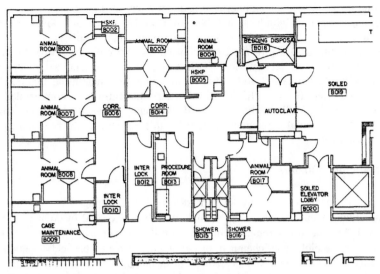

**Figure 8.7** Schematic drawing of a containment area. This containment area was designed to provide maximum flexibility. It consists of six animal rooms. One is a standard animal room (B004) and five (B001, B003, B007, B008, and B017) are animal cubicle rooms, each of which is divided into four animal cubicles, an area for changing cages and conducting animal procedure in a biosafety cabinet, and an area for a sink and storage of feed containers and sanitation equipment and supplies. There are three entry and exit vestibules with interlocking doors. One vestibule (B010) enters into corridor B006, which along with animal rooms B001, B007, and B008, can be isolated from the rest of the containment area to serve as an Animal Biosafety Level 2 (ABSL-2) facility in which research staff enter and exit through vestibule B010. The second vestibule (B012) enters into corridor B014 that also may be entered from two private shower and locker rooms (B015 and B016). This is an ABSL-3 area. It includes three animal rooms (B003, B004, and B017), a laboratory (B013), a housekeeping closet (B005), and a bedding disposal room (B018). The bedding disposal room is for disposing of bedding soiled with hazardous chemicals or radioisotopes. The third vestibule is between the autoclave and the bedding disposal room. It enters the soiled side of the cage sanitation area so that cages contaminated with hazardous chemicals or radioisotopes can be taken directly to the cage and rack washer to be decontaminated without having to be being transported through corridors. The bulk autoclave is large enough to hold two racks. Cages contaminated with biohazards are autoclaved out of the containment area. The door between the ABSL-2 and ABSL-3 areas allows for the entire area to be operated as an ABSL-3 facility.

are wrapped and autoclaved elsewhere in the facility before being transported to the animal rooms. In a barrier area of a larger facility, the cages and supplies are autoclaved into the barrier area; once inside the barrier, they are handled in a conventional manner, thus eliminating the need for wrapping and unwrapping. Some larger barrier facilities are designed with cage sanitation equipment inside the barrier, which offers the option of not autoclaving cages and relying on the level of sanitation provided by the cage washing equipment. Large rodent barriers may include a cage sanitation facility inside the barrier. This offers the option of relying on the level of sanitation provided by the cage washing equipment operating at a minimum temperature of 82.2°C (180°F) and not routinely autoclaving cages unless there is a disease outbreak. The use of irradiated feed and bedding also eliminates the need for autoclaving them.

Barrier facilities are designed and managed at various levels of microbiological control, which translates to the degree of control over how supplies and personnel enter the facility. The highest level barrier facilities may have one or more double-door pass-through autoclaves, preferably pit mounted, floor loading bulk autoclaves; one or more ventilated entry and exit vestibules with interlocking doors, where packaged sterile supplies and animals in filtered containers are passed into the barrier after having the exterior surface of the package chemically sanitized, or soiled equipment and trash are passed out of the barrier. Sometimes a pass-through dip tank filled with high-level disinfectants may be used to pass sterile items packaged in watertight containers into the barrier. Personnel may be required to shower and change clothing prior to entering the barrier, but more typically, at least in research barrier facilities, personnel enter through a vestibule with interlocking doors where they put on sterile outergarments over street clothes or uniforms along with head and shoe covers, a face mask, and gloves. Air showers using mass quantities of HEPA filter air may be added to a personnel entry vestibule.

Depending on the intended use of the barrier, space may be required inside the barrier for wet laboratories, animal procedure laboratories, TG/KO laboratories, specialized imaging equipment, irradiation equipment, etc. A research rodent barrier may require a quarantine area inside the barrier. This is especially important for a TG/KO facility, because quarantine is recommended for all foster mothers coming out of the TG/KO laboratory until the young are weaned and the mothers' health statuses are determined. Animal cubicles are useful for this purpose, even if the animals are housed in microisolation cages.

### Containment Animal Housing

"Containment" refers to animal housing systems designed and managed to prevent the escape of experimental hazardous agents to which the animals have been exposed in order to protect workers, other animals, and the general environment. In other words, "containment" equates to "keep in." The hazardous agents may be biological, chemical, or radiological. Like a "barrier," "containment" can be achieved at the cage level, the room level, an area within an animal facility, or it can be the entire facility, all of which when used together can be considered to provide increasing levels of containment. Figure 8.7 is a schematic of a flexible containment facility designed for containment of all three classes of hazardous agents. The design features of a containment facility are similar to a barrier facility. At all levels of containment, the primary objective is to contain the hazardous agent as close to the source as possible, ideally, at the cage level, e.g., a microisolation cage. Animal cubicles are particularly well suited for use in containment facilities. Of course, the more levels of containment, the higher the safety level. For example, when using a conventional room, housing experimentally infected mice in a microisolation cage, the cage provides the first level of containment and the room door the second level. If housed in a microisolation cage inside an animal cubicle in an animal cubicle room in a barrier area located inside a larger animal facility, there could be at least five levels of containment — the microisolation cage, the cubicle doors, the cubicle room door, and the two doors of the entry vestibule (Figure 8.7). It is important to provide appropriate laboratory and animal procedure space inside containment facilities to avoid having to remove live animals from the facility.

*Biohazard Containment* — Microbiological agents are classified into four biosafety levels (BSL) according to the degree of risk to humans (classified by the CDC-NIH in the publication "Biosafety in Microbiological and Biomedical Laboratories"[21]). They are BSL one to four, with one being agents considered to have very low or no pathogenicity for humans and four being the highest risk level. The same publication describes combinations of laboratory practices and techniques, safety equipment, and facilities required for working with agents and animals in each classification level. When animals are infected with microbial agents, the corresponding facilities and management practices are referred to as animal biosafety levels (ABSL) one to four. Animal studies with BSL-2 agents are relatively common and recently have become more so with the use of viral vectors for gene therapy studies, most of which are classified as BSL-2 agents, even if they are referred to as being "replication deficient."[22,23] Animal studies with BSL-3 agents are less common than ABSL-2 studies, however, even research facilities that will never need to support an ABSL-3 study could benefit from having an ABSL-3 facility. Studies with BSL-4 agents are rare and are limited to approximately 20 ABSL-4 facilities in the entire world. Studies with BSL-2 agents can be conducted in conventional animal rooms using appropriate equipment and ABSL-2 practices; however, they are more efficiently and consistently conducted at a higher level of safety in an ABSL-3 facility. The primary reasons is that contaminated cages, supplies, and wastes are autoclaved directly out of the facility, eliminating the time-consuming and potentially hazardous practice of having to bag them before transporting them out of the facility to a remote autoclave. In addition, an ABSL-3 facility is highly desirable for quarantine of rodents infected with adventitious agents, or that are of unknown health status. These agents are not hazardous to humans but have the potential to be devastating for many if not most of the rodent studies in the facility.

ABSL-2 is the highest level of biocontainment that can practically be achieved in a conventional room with appropriate equipment and management practices. An ABSL-3 facility has all the design features of a high-level barrier facility as described above. In fact, infectious containment facilities are often managed as both a barrier and containment facility, in that cages and supplies are autoclaved in and soiled cages and wastes are autoclaved out. ABSL-3 facilities should have ventilated entry and exit vestibules with interlocking doors, an autoclave in the facility, and a hand washing sink in each animal room. In addition, a number of design features are required to facilitate keeping agents in, such as an effective sealed envelope around each room and around the entire facility except for the doors (gasketed doors are not required), and air balancing that directs the movement of air from the least contaminated areas to the most contaminated areas. HEPA filtering of exhaust air is not required but is recommended, not only because it increases the degree of safety, but also because it helps to allay public concerns about the existence of the facility in their neighborhood. Exhaust air filters should be the bag-in, bag-out type to facilitate safe replacement of contaminated filters. More details regarding animal biosafety facilities and practices can be found in the literature.[21,24–31]

## Chemical and Radioisotope Containment

As with biohazard containment, appropriate equipment and management practices are critical, but the physical characteristics of the facility influence the level of safety that can be attained and the consistency at which it can be maintained. Work with chemicals and radioisotopes in animals may often be carried out safely in conventional animal rooms, however, there are some exceptions. For example, HEPA filtering of exhaust air may be required for working with concentrated levels of especially potent carcinogens, or special shielding may be required for working with certain radioisotopes. It is desirable and sometimes essential to isolate such studies to help prevent cross-contamination. When small numbers of animals are required, it is inefficient to use an entire animal room for a single study. Animal cubicles, semiridged isolators, microisolation cages, etc., can provide the isolation necessary to prevent cross-contamination, while housing multiple studies involving small numbers of animals within a relatively small area as compared with using conventional animal rooms for each study. An area planned for supporting chemical and radioisotope studies may utilize one or more rooms with cubicles or equipped with other containment devices and one or more procedure rooms equipped with radioisotope and chemical fume hoods.

Decontamination of cages can usually be accomplished safely with the use of conventional mechanical cage washers, especially cage and rack washers, taking advantage of the dilution factor that occurs due to the large volume of water used by the washers. Ideally, the chemical and radioisotope containment area should be near the dirty side of the cage sanitation area to minimize the need to transport contaminated cages through corridors (Figure 8.7). Also recommended is a separate room, where contaminated bedding can be removed from cages or pans inside a laminar air flow cabinet in which the aerosolized contaminant is drawn away from the operator into a HEPA filter (Figure 8.4).

## Quarantine

Most laboratory animals used today are purpose bred using disease control measures equal to or superior to that in the research facility; therefore, most research facilities do not require special quarantine for the vast majority of the animals received into the facility. However, there typically are exceptions. One common exception is the result of the increased use of transgenic and knockout (TG/KO) animals and the sharing of these unique rodents (usually mice) between research institutions. Most facilities today require a high-level rodent quarantine facility for holding animals until they can be documented to be "clean" or the genetic line can be rederived by C-section or embryo transfer. Ideally, rodents of unknown health status, or worse, animals known to be infected with agents hazardous to other rodents in the facility, are best maintained in an ABSL-3 facility. At a minimum, the rodent quarantine facility should be well isolated from other rodent housing areas. Animal cubicles are a good option for quarantine areas, even when used in conjunction with microisolation cages.

## Housing for Nonhuman Primates

Housing for nonhuman primates is somewhere between conventional housing utilizing a hose-down caging system and biohazard housing, because of their potential for carrying zoonotic diseases and the high level of noise that at least certain species can generate. For these reasons, the ideal arrangement is to house them in an isolated area under ABSL-2 standards. At a minimum, rooms housing nonhuman primates must be arranged and located to avoid the necessity of transporting animals or cages and equipment soiled by the animals through corridors or on elevators outside the animal facility. The objective is to avoid exposing individuals who do not have an occupational requirement to be exposed to nonhuman primate associated diseases. Special features for a nonhuman primate housing area or room may include additional security, and an entry vestibule to animal rooms, typically made of chain-link fencing, that prevents animals that get out of their primary enclosure from escaping when the room door is opened. Lights and any other fixtures in the animal room must be mounted such that animals free in the room cannot damage them and so that they do not impede capturing the animals.

## Housing for Canines and Small Agriculture Mammals

Housing for canines and small agricultural mammals (e.g., swine, ovine) is typically designed for a hose-down housing system. It should be isolated from other animal housing and human occupancy areas because of their relatively "dirty" microbial status as compared with rodents, and the fact that some species, e.g., swine, generate high noise levels. The zoonotic disease concern, while present, is not as great as with nonhuman primates with the exception of sheep, which because of concerns for Q fever, especially in connection with pregnant sheep, should be maintained under ABSL-2 standards. Animal procedure space should be provided in this area. Because these animals are commonly used as surgical research models, they should be housed near the surgical suite. Rooms may be provided for postoperative recovery and intensive care of surgical patients. Generally, these species are housed in mobile double-tiered cages, mobile single-tier pens, or fixed-floor pens (Figures 8.8 through 8.14). Even when dry bedding systems are used to house these species, routine cage and room sanitation still requires floors sloped to floor drain troughs.

# Architectural Features

The primary focus for the following architectural features, especially the interior surface features, is to create a durable, easy to maintain, sanitizable surface, capable of withstanding scrubbing, chemical cleaning and disinfecting agents, and impact from high-pressure water. All surface junctions and penetrations should be sealed to facilitate air balancing and vermin control. Animal facility interior surfaces are exposed to much abuse in the normal conduct of animal care and use. Selections made with an eye toward "saving" money on architectural features rarely prove to be wise and could easily cost many times more in long-term maintenance costs than the initial cost "savings." It is also true that "expensive" does not necessarily guarantee a satisfactory performance.

## *Interior Surfaces*

### *Floors*

Floors should be monolithic, slip resistant even when wet, yet relatively smooth and easy to sanitize. Commonly used flooring materials include troweled on or broadcast polymer (typically epoxy but methyl-methacrylates are also used) composites ranging in thickness from 1/8 to 1/4 in. Many floor coverings work well in a rodent room, including vinyl with sealed seams if the cage racks are not to heavy, but few materials work consistently well in hard use, high-moisture areas such as cage sanitation. Ceramic tile with epoxy grout top dressing (Figures 8.5 and 8.23) has proven to be a relatively maintenance-free floor for cage sanitation areas, where seamless composite polymer floors too often fail. Grouted tile

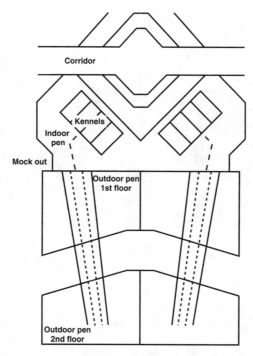

**Figure 8.8**    Schematic drawing of dog housing facility showing areas for feed preparation and storage, kennels where dogs may be housed individually when fed, and where they can be group housed in indoor and outdoor pens.

**Figure 8.9**    View from above outdoor pens, which corresponds to Figure 8.8.

floors are not suitable for corridors, because the joints cause excessive noise when cage racks roll across it. There should be a minimum 10 cm (4 in) high $^1/_2$ in. radial coved base to form a watertight seal at the floor-to-wall junction and facilitate sanitation.

## Walls

The most commonly used wall material is masonry blocks coated with block filler to eliminate pits and sealed with epoxy paint. This wall performs well in most areas of the facility, with the exception of high-moisture areas, such as animal rooms in which hose-down caging systems are used, and cage sanitation areas, where coatings tend to peel from the block. Structural glazed facing blocks, or ceramic tiles over a water-resistant foundation, in which the grout is top dressed with epoxy, makes a maintenance-

**Figure 8.10**   Shown are dogs that are being held individually when fed, which corresponds to the indoor pens
noted in Figure 8.8. This strategy ensures that dogs housed in groups for most of the day may
be fed without disturbances from fellow dogs. It also allows caretakers to identify if some dogs
have a decreased appetite.

**Figure 8.11**   Indoor pen for dogs, which corresponds to the indoor pens noted on Figure 8.8. Wood shavings
are spread on the floor to facilitate cleaning. Note the door on the right side of the picture connecting
the indoor pen with the outdoor pen.

free wall that performs exceptionally well in these high-moisture areas as does masonry block covered
with mineral fiber composite panels. Gypsum board on studs has rarely proved suitable for any area of
an animal facility. However, newer sheet materials made of a variety of mineral fiber composite panels
mounted directly on metal studs or in combination with fiberglass-reinforced gypsum board on stud
walls is a viable alternative in many areas of the facility. Such walls are especially useful in earth-
quake-prone locations.

Protective guardrails or wall curbs are required in corridors and may also be cost-effective in animal
rooms and other areas, where wall damage from caging and other equipment is likely. Guardrails should
be sturdy, sanitizable, and constructed to avoid providing harborage for cockroaches and other pests.
Extruded solid aluminum rails fastened to the wall with I-beam standoffs have proved very useful in
animal facilities (Figures 8.15 and 8.16). Guardrail height should be carefully matched to the equipment
used in the facility. A double row of guardrails may be provided; however, if there is to be only one
row, its height should be determined by a careful examination of the rolling equipment to routinely be
used in the facility.

**Figure 8.12**  Cabinets are provided in the indoor kennels for dogs, noted in Figure 8.8, where they may sleep (bottom) or sit (top).

**Figure 8.13**  View of an outdoor dog pen that corresponds to the same in Figure 8.8. Note the large space provided for a few dogs.

## Ceilings

Gypsum board ceilings sealed with epoxy paint are adequate for relatively dry areas of the facility, including rodent rooms, but are generally not suitable for high-moisture areas like cage sanitation. A drop ceiling with lay-in panels is generally not recommended for animal housing rooms, because they impede sanitation and vermin control. However, in recent years, composite panels made of lightweight water-impervious materials and sealed to fiberglass "T" bars with gaskets and clamps have proved to be a satisfactory, virtually maintenance-free choice for ceilings. These are particularly cost-effective for

**Figure 8.14**   Given a large area, dogs and pigs may be held together. Outdoor pens such as the one shown here stimulate the animals to exercise and to engage in social interactions.

**Figure 8.15**   A corridor in an animal facility showing extruded solid aluminum bumper guardrails at two levels to protect the walls and the door. If only one guardrail is to be used, the lower one may provide the most protection from the widest variety of mobile equipment. Note that the rail extends further out from the wall in one location to protect wall-mounted equipment that protrudes further from the wall than the typical rail mounting of 3 in from the wall.

use in high-moisture areas such as cage sanitation and animal rooms with hose-down type animal housing systems. In all cases, the ceiling to wall junction should be sealed. The minimal recommended ceiling height is 2.7 m (9 ft) and may need to be higher in rodent and nonhuman primate rooms, depending on the height of rodent racks or nonhuman primate cages to be used.

**Figure 8.16**  An individually ventilated microisolation cage rack with the cage exhaust filter and fan unit connected directly to the room exhaust at the ceiling, which minimizes odors in the room. The supply and exhaust fan and filter units are situated on the top of the rack. Note the extruded aluminum rail protecting the wall at the left side of the photo.

## Doors

The minimum door size should be 107 cm (42 in) wide by 2.1 m (7 ft) high; however, nonhuman primate cages and ventilated high-density rodent cage racks may require wider and higher openings. Doors measuring 122 cm (48 in) wide by 2.4 m (8 ft) high frequently prove useful for animal rooms. If 8-ft high doors are provided for animal rooms, it is important to make certain that all doors in the facility through which the higher cage racks will be transported are also at least 8-ft high. This includes all corridor doors, doors in and out of the cage sanitation area, the rack washer doors, and dock doors. Stainless steel or fiberglass-reinforced polyester doorframes are the most cost-effective choice. They should have hospital stops to facilitate cleaning. Jamb guards may be mounted on the corridor side. There must be no doorsill, as this seriously impedes the movement of cage racks through the door.

Like the frames, stainless steel or fiberglass-reinforced polyester doors prove more cost-effective than less durable materials, including painted hollow metal doors. The doors should be sealed and have flush finished tops and bottoms. If the doors are not SS or fiberglass, they should be outfitted with stainless steel kick plates on both sides and edge guards on the strike side. Automatic drop bottoms should be surface-mounted on the animal room side of the door, leaving no gaps larger than $^1/_4$ in. A view panel is highly desirable, if not essential, for security and personnel safety. Size and shape of the view panel is a matter of choice, but it should provide a clear view of the room from the corridor. Light control through the view panel may be desirable and can best be provided with carefully selected red laminated glass, e.g., 1/8 in clear annealed glass with an inner layer of Opti-Color™ film #5557 (Monsanto Chemical Co., St. Louis, MO). Other options include a variety of solid blackout view panel coverings attached with magnets or hinges and latches, most of which are inconvenient and high maintenance. Hospital, lever-type door openers are a good choice. Push and pull plates should be mounted on both sides of the door. Strike plates should have a cup design. If fire codes permit, it may be preferable to eliminate the latch. If access to the animal room is controlled via a security system, magnetic locks are generally found to require less maintenance than electric strikes. Assuming that doors swing into the room, a crash rail extending the width of the door should be mounted just below the door handle on the corridor side and protrude away from the door enough to protect the door handle. A heavy-duty surface-mounted, self-closing door closer with variable delays and hold opens is essential. Hinges should be stainless steel, heavy-duty, standard, or continuous. Swing-clear hinges can be used to optimize door width. Door seals of various types may be required to control air movement around the door to facilitate balancing the ventilation system.

Automatic sliding or hinged doors should be provided in doorways with a high traffic of rolling stock, such as cage sanitation, the loading dock, and selected corridor doors. Depending on the situation, they may be opened with sensors that detect movement or with wall-mounted push plates or ceiling-mounted pull chords.

## Vermin Control

Careful planning and construction will go a long way toward facilitating the control of vermin and insects without the use of organic insecticides and baits, especially wild or escaped rodents and cockroaches. Organic insecticides and baits should not be used in research animal facilities, because they have the potential to change biological baselines and alter the animal's response to experimental variables. The basic control approach is to seal vermin and insects out of the facility and eliminate hiding and nesting places within the facility. All cracks, joints, utility penetrations, lights, wall switches, communication, and power outlets must be sealed. Animal rooms should have a minimal amount of "built-ins" consisting of little more than a paper towel dispenser, utility hangers, and possibly a sink. These should be sealed to the wall or mounted away from the wall to eliminate hiding places and allow cleaning between the wall and the mounted item. Animal rooms should not have casework. Casework for animal procedure rooms and other laboratory spaces in the animal facility should be of an open design type to reduce hiding places under and in back of and to facilitate cleaning. Boxed-in casework should be avoided. The control of cockroaches and vermin starts during construction by keeping the construction site free of garbage on which they feed. This requires having a zero tolerance for eating or drinking in the facility during construction. In addition, all hollow dead spaces in the facility, including inside concrete blocks and studded walls, should be treated with amorphous silica to preclude the harborage of cockroaches. There should be high-pressure sodium (not mercury vapor) lamps, or dichrome yellow (not incandescent flood) lamps located at exterior doors or vents to reduce the influx of vermin and insects into the facility. Air curtains with a velocity of 490 m (1600 ft) per minute can help reduce the influx of flying insects at frequently used exterior entrances that may be open for extended periods of time, such as loading dock doors.

## Noise Control

Noise is another potential variable in the animal's environment that can also be stressful for the staff. The primary noise producers are the cage sanitation area, canine-housing rooms, and sometimes, depending on the species, nonhuman primate rooms. Design features such as strategically locating these areas to buffer them, and architectural measures that reduce sound transmission should be carefully considered, including double-entry doors, soundproof walls, locating corridors and support areas around the noise-generating areas, and locating the noise-generating areas next to outside walls or mechanical spaces. Conventional acoustical materials impede sanitation and vermin control and should be avoided; however, sound attenuating panels that can easily be removed, washed, and sanitized in mechanical cage washers are commercially available and should be considered for use in especially noisy areas of the facility.[33] All in-room activities, including cage changing, must be conducted in a manner that generates as little noise as possible. Background noise, e.g., soft music, can help to buffer unavoidable noise inherent in routine care and use procedures.

Other common sources of avoidable excessive noise include improperly sized ventilation ducts and outlets, improper air balancing that results in whistling around the room door, and improperly sealed room penetrations that also result in whistling. Vacuum equipment and the conduit used to transport bedding generate a large amount of noise and should be isolated or insulated or both to assure adequate sound attenuation. Fire alarms selected for animal housing areas should disturb the animals as little as possible. Most rodent species cannot hear frequencies below 1000 kHz, although guinea pigs are capable of hearing down to 200 kHz. Fire alarms that operate between 400 kHz and 500 kHz should be used in facilities that house rodents.

## Engineering Features

### Heating, Ventilation, and Air Conditioning (HVAC)

The function of the HVAC system is to control the laboratory animal's macroenvironment (the room) and microenvironment (the cage) and to maintain a healthy work environment for personnel. The HVAC system must supply clean air to the animal rooms with a consistent temperature and humidity all year, while effectively removing heat, particulate, and gaseous contaminants generated in animal rooms. Many of the animal facility planning related references cited earlier in this chapter cover HVAC systems to some degree.[1–6] The American Society of Heating, Refrigerating, and Air Conditioning Engineers, Inc. recognized the unique design requirements of HVAC systems for research animal facilities and included a separate section, "Laboratory Animal Rooms," in the ASHRAE Handbook 1999 HVAC Applications.[32]

### Air Quality

The quality of air delivered to the facility is determined to a large extent by the source of the air and the degree of filtration. The source of the supply air must be selected to avoid contamination with exhaust air from other buildings or the same building, especially the animal faculty, incinerator smokestacks, vehicle exhaust fumes, etc. The quality of filters used for filtering incoming air varies from 85 to 99.97% high-efficiency particulate air (HEPA) filters depending on the type of facility or area of the facility. For example, the air being delivered to rodent barrier facilities and surgery rooms may be HEPA filtered, while the air to other areas of the facility may be filtered with 85 or 95% efficient filters. The need for HEPA air even in rodent barrier facilities is not well documented, and its cost-effectiveness is questionable. Task-directed HEPA filtering, e.g., using HEPA filters on ventilated racks and in cage change cabinets, may be more cost effective than HEPA filtering all the air coming into the facility.

Another use of task-directed HEPA filter air is in "mass air displacement (MAD) clean rooms," similar to but typically at a lower quality to that used in electronic fabrication plants. In MAD rooms, air is recirculated within the room through HEPA filters at volumes sufficient to change the air 150 (most common) to 600 times per hour, depending on the type of system and clean room class desired.[34] Fresh air exchanges are superimposed over the recirculated air at a rate similar to that in a conventional room. MAD rooms effectively control the animal's airborne microbial environment, thereby reducing cross-contamination. MAD rooms may be "hard wall" or "soft wall" units, the size of rooms large enough to house multiple cage racks or soft wall units just large enough to house a single cage rack. Recently, there has been an increased interest in using multiple soft wall units in large open warehouse type spaces to gain maximum flexibility at a minimal cost.

### Ventilation

Animal facilities must have dedicated supply and exhaust air handling units. The supply air must be 100% outside makeup air. The ventilation rate that has proven effective for most animal rooms expressed in terms of fresh air changes per hour (cph) is around 15 cph. However, this varies between 10 and 20 cph depending on the heat load as well as microbial, particulate, and gaseous contaminants generated in the room, which is dependent on the species and density of animals to be housed in the room. Control of the heat load in the room is the most critical concern, because high temperatures are stressful for all animals and may be lethal for laboratory species, especially rodents, not adapted to high temperatures, even at temperatures not normally dangerous for most species. The minimal lethal temperature for laboratory rodents is time and relative humidity dependent but may start at temperatures as low as 29.4°C (85°F). It is important to note that the temperature in the animal's microenvironment inside of microisolation caging can be several degrees higher than the macroenvironment. The prominent gaseous contaminant is ammonia, which is generated by urease positive bacteria from the feces splitting each urea molecule from urine to form two ammonia molecules. Ammonia production depends on many factors, including the species and density of animals, the sanitation level, and the relative humidity in the room and cage. As a general rule, a ventilation rate that adequately controls the heat load when air is delivered to the room at 12.8°C (55°F) is adequate to control the gaseous and particulate contaminants. Heat loads for various species of animals are listed in the ASHRAE Handbook.[32]

The ventilation rates noted above are not necessarily to be considered absolutes. Variable air volume (VAV) ventilation systems in which ventilation volume is based on actual heat load may achieve the objective while conserving energy. This would be consistent with the idea of using performance standards as noted in the Guide,[4] as opposed to inflexible engineering standards. The same applies to other rooms in the facility, e.g., the cage sanitation area where loads range from very high when the sanitation equipment is being used to very low when it is not. If VAV is to be used, consideration must be given to how varying the volume may alter the ventilation efficiency in terms of distribution of fresh air in the room and removing particulates, including allergens and infectious agents.

Room ventilation patterns, with regard to the location and type of supply diffusers and location of return and exhaust grills, significantly affect the room ventilation efficiency; however, the most efficient pattern has yet to be definitively defined. The dogma for many years has been to supply high, typically from the ceiling down the center of the room, and exhaust low near the floor, preferably in all four corners. This dogma has been called into question by the result of some computational fluid dynamic studies (CFD) but is supported by other studies. CFD is the use of highly complex mathematical models to predict air circulation patterns in a space.[35,36,37] It appears to be a power design tool for determining the optimal animal room ventilation pattern given the room configuration, the species and number of animals to be housed, and the type of caging. One published study suggests that high returns, preferably in each corner or above each cage rack, are the most effective,[35] and another suggests that low returns, one in each corner, are the most effective.[36] The problem is complex, and these CFD studies used different assumptions for key features, thus, additional study will be required to clarify this important issue. The Hughes'[35] study suggests that an even more efficient configuration is to supply and exhaust room air from a soffit mounted in the center of the ceiling extending the full length of the long axis of the room. In this CFD model, supply air is directed from radial diffusers in the bottom of the soffit toward the floor. Exhaust inlets located along both sides of the soffit capture the air as it curls from the floor, up the wall parallel with the soffit, across the ceiling, and into the soffit, where it is removed from the room. A full-scale test model of an animal room fitted with this type of soffit is reported to have performed even better than predicted by the CFD model.[35] Given the uncertainty, the high returns in each corner or the soffit configurations are tempting options in that they are less costly to construct than low returns and do not take up floor space; however, the best current answer is to do CFD studies specifically for the animal rooms in the facility being planned.

There is fixed equipment commonly used in research animal facilities that have special ventilation requirements. Fume hoods and certain types of biosafety cabinets require independent direct exhaust systems. Autoclaves require canopy exhaust hoods immediately above the autoclave doors with sufficient airflows to capture the heat, moisture, and odors that emanate from the autoclave when opened. This is especially important if the organic materials are to be autoclaved because they generate high odor levels. The cage sanitation area has unique ventilation requirements, because high heat and moisture loads are generated in the room by cage washwater temperatures that are 82°C (180°F) or higher. Tunnel washers and often the cage and rack washers are connected directly to the exhaust system, and in addition, cage and rack washers must have exhaust canopies above the doors to capture the heat and moisture that emanates from the machines when the doors are opened (Figure 8.20). The high moisture levels in the cage sanitation area dictate having a dedicated independent exhaust system for this area, including the exhaust fan and the ducts. The canopies and all ducts must be nonferrous and acid resistant, and the ducts must be watertight and slopped and fitted to drain of the large amount of condensate released from the water-saturated hot air coming from the washers. The overall ventilation requirements for the cage

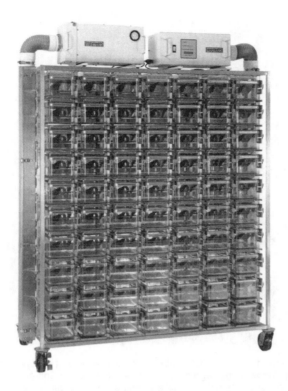

**Figure 8.17** An individually ventilated microisolation cage rack with two filter and blower units on top of the rack. One supplies HEPA-filtered air to the cages, the other captures the air coming from the cages and HEPA filters it before blowing it into the room.

sanitation area must take into consideration the enormous heat load in the room that may include a significant mass of stainless steel coming out of the washers at temperatures of 82°C (180°F) or higher.

Ventilated rodent cage racks are an example of mobile equipment that may be connected directly to the ventilation system. Ventilated racks may be used as freestanding equipment with blower and filter units that supply HEPA-filtered room air to the cages. They may also be equipped with blower and filter units that capture air coming from the cages and HEPA-filter it before blowing it back into the room (Figure 8.17). The blower and filter units can be mounted on top of the cage racks but, ideally, are mounted on wall shelves and connected to the racks with flexible ducting (Figure 8.18). HEPA filtering

**Figure 8.18** Upper portion of an individually ventilated microisolation cage rack showing the cage supply and exhaust filter and blower units sitting on a wall-mounted shelf. This arrangement facilitates rack changes by eliminating the need to transfer the filter and blower units from rack to rack.

**Figure 8.19**   An individually ventilated microisolation cage rack with two filter and blower units mounted or a
                  mobile rack alongside the cage rack. The cage exhaust filter and fan unit is connected directly to
                  the room exhaust at the ceiling, similar to that shown in Figure 8.16.

the exhaust air from the cages removes particulate contaminants but does not remove gaseous contaminants and heat. This is best accomplished by coupling the rack exhaust directly to the room exhaust
(Figures 8.16 and 819). There are many strategies for integrating supply and exhaust air of ventilated
racks with the ventilation system.[1,38,39] Regardless of which strategy is selected, it is important to decide
early in the planning process because the design of the room ventilation system must be matched with
the equipment to gain maximum benefit. Not only does the decision affect the physical couplings; it
also impacts on the cubic feet of air per minute (cfm) of supply air that will be required in the room.

*Air Balancing* — Appropriate relative air pressures throughout the facility must be maintained to control
airborne contaminants.[9,30,31] This involves balancing supply and exhaust to maintain predetermined relative
air pressures between adjoining spaces, typically between the room and corridor. Table 8.2 summarizes
various balancing options, depending facility type, and corridor plan. Maintaining proper balance requires
proper sealing of the room envelope and maintenance of the appropriate volumetric offset between supply
and exhaust air to achieve adequate differential pressures, typically between 0.08 and 0.2 cm (0.03 and
0.075 in) of water. Proper air balance is important in controlling contaminants, but it has limitations.[9]
Most significant is to realize is that the relative air pressure in the spaces on either side of an opened door
is essentially zero, allowing airborne contaminants to move freely between the spaces.

Relative air pressures in animal rooms of a single-corridor facility are dependent on how the facility
is to be managed: conventional, containment, or barrier. In a single-corridor conventional facility, animal
rooms are typically balanced negative to the corridor, except for rooms that are designated as "barrier"
or "clean" rooms, which are then balanced positive to the corridor. For this reason, the ability to
automatically reverse room air pressure relative to the corridor without having to rebalance the entire
system is a highly desirable feature in a single-corridor conventional animal facility. In a single-corridor
containment facility, where the objective is to contain airborne contaminants, the relative air pressure in
the animal rooms will be balanced negative to the corridor. The opposite does not necessarily hold for
a single-corridor barrier facility, where the choice depends more on management philosophy. One
philosophy calls for balancing animal rooms positive to the corridor in an effort to keep airborne
contaminants out; the other calls for balancing animal rooms negative to the corridor with the objective

**Table 8.2    Relative Air Pressure Between the Corridor and the Animal Rooms**

| Managed as a: | Single Corridor | Dual Corridor | |
| --- | --- | --- | --- |
| | | Clean | Soiled |
| Conventional facility | + or − | + | − |
| Barrier facility | + or − | + | − |
| Containment facility | - | + | + or − |

+ Corridor positive to animal room
− Corridor negative to animal room

#### + or − Single-Corridor Conventional

In a conventional facility, the air pressure in the corridor is generally maintained positive to the animal rooms. The exceptions are facilities with mixed "conventional" and "barrier" rooms, where the air pressure in the "barrier rooms" is maintained positive to the corridor and in the "conventional rooms" is maintained negative to the corridor.

#### + or − Single-Corridor Barrier

Both options are used. Following is a rationale for each:
Corridor negative to animal rooms — To keep airborne contaminants out of the animal room
Corridor positive to animal rooms — To contain inadvertent contaminants
Infectious agents of concern are not ordinarily present in a barrier facility, so the rationale "to keep airborne contaminants out of the animal room," does not ordinarily apply as it does in a mixed facility. However, it must be assumed that a "break" will occur in a barrier room at some time. When this happens, the management objective is to contain the infectious agent, like in a biocontainment facility, until it can be detected and eliminated from the room and the facility. Keeping air pressure in the corridor positive to the animal room has the added benefit of reducing animal allergens and odors in the corridors and throughout the facility.

#### + or − Double-Corridor Containment

Both options are used, with negative being more common, but positive may be preferred in some situations.

being to contain a disease break until it can be detected and eliminated. Both management philosophies have merit, and neither is clearly right or wrong. However, one advantage to the latter is that it maintains corridors relatively free of animal allergens, which are well documented as a serious and common occupational hazard.[8,40] In dual-corridor facilities, regardless of facility type, relative air pressures are typically balanced with the clean corridor positive to animal rooms and animal rooms positive to the soiled corridor; however, in some instances, both corridors may be balanced positive to the animal rooms.

## Temperature and Relative Humidity (RH) Control

Each animal room should have individual temperature control to allow for environmental temperature requirements for different species and differences in heat loads between rooms because of species differences and animal density. The standard design temperature range for animal rooms is between 18 to 29°C (65 to 85°F). This is not to be confused with temperature variations around a set point. The temperature control system should be capable of maintaining temperature ±1°C (±2°F) around any set point selected from the designed temperature range.[4] Designing for a narrower temperature range may be acceptable for facilities intended for a single purpose, e.g., rodent production. Room temperatures as low as 18°C (65°F) are desirable for some commonly used species, e.g., rabbits, but occasions for room temperatures over 26.6°C (80°F) are rare, and usually involve the maintenance of relatively exotic species.

Relative humidity (RH) in animal rooms should be maintained between 30 and 70%[4] with no specific set point generally required within this range. A well-designed HVAC system that supplies 12.8°C (55°F) air nearly saturated with water vapor can maintain this range of RH in multiple rooms without HR control in each animal room. Zonal control may be desirable in some situations, e.g., separating rooms where dry bedding systems from those where hose-down housing systems will be used. Steam free of boiler chemicals should be used for humidification in order to avoid the potentially confounding effects of chemical additives often used in boilers.

## Miscellaneous HVAC Issues

*Redundancy* — Mechanical systems require routine preventative maintenance, often requiring shut-down of the system. In addition, mechanical systems are prone to fail. To account for such down time and still assure consistent control of the research animal's environment, the HVAC system should be designed with redundant critical components such as air handlers, pumps, chillers, and heat sources. There are many options for supplying redundancy, including parallel or N +1 air handling systems; dual chillers; boilers and pumps installed as parallel or N+1; cross connecting with other lower priority sources to access available chilled water or steam; and having spare parts available for quick replacement. *Energy Conservation* — Because of the high fresh air exchange rates required by animal faculties, an energy recovery system often proves to be cost-effective, depending on local climatic conditions. Recover systems should be limited to types that preclude contaminating incoming air with outgoing air.

## Power and Lighting

### Power and Emergency Power

The demand for electrical outlets in animal rooms has increased with the increased use of ventilated racks, data processing equipment, scales, research equipment, HEPA-filtered mass air displacement cabinets, and powered sanitation equipment in the animal room. The outlets should have water-resistant covers. The location of the outlets needs to be carefully planned, especially if ventilated cages are to be used. Ground fault interrupters (GFI) should be used for every circuit in areas of the facility where water will be routinely used, which is most of the facility.

Emergency power should be adequate to maintain all essential services in the event of a main power failure. At a minimum, emergency power should include HVAC at 100% capacity, including chillers; any animal housing equipment that relies on power to maintain airflow, e.g., ventilated racks; all environmental control and monitoring systems, at least one light fixture per animal room and other safety lighting as required by code, the security system, the surgery room, and freezers.

### Lighting

Photoperiods are a critical component of maintaining the animal's environment. Therefore, automatic control of lighting in windowless animal rooms is the norm. Most research facilities will require independent lighting control for each animal room. Conventional fluorescent lighting is standard for animal rooms. The ceiling fixtures may be recessed or surface mounted. The light fixtures should be water resistant and arranged to provide uniform lighting throughout the room. A digital light control system located in a secure location remote from the animal rooms is best. A dark room light independently controlled with a timer switch at the room may be used to facilitate activities that must be conducted during the dark cycle.

High light levels cause retinal damage in albino rodents. In recognition of this phototoxic effect, the ILAR Guide[4] recommends that light levels in rooms housing albino animals be 325 lux (30 fc), 1 m (3.3 ft) above the floor. This level is generally sufficient for animal care, and task lighting can be provided for performing procedures that require higher levels of light. A bilevel low/high (325 lux/800 lux) lighting system may be considered with the intent of using the high level to facilitate working in the animal room; however, it must be noted that even brief periods of high light levels may result in retinal damage for albino animals. Retinal damage does not occur in animals with normally pigmented eyes at typical indoor lighting levels. Therefore, it is acceptable and even desirable to provide light levels of 800 to 1100 lux (75 to 100 fc) in animal rooms designed to house only dogs, nonhuman primates, or other animals that normally have pigmented eyes.

## Plumbing and Drainage

There are two primary questions relative to plumbing and animal rooms: (1) does every animal room require a sink? (2) Does every animal room require a floor drain?

### Sinks

Sinks are desirable in most animal rooms, but are required[21,31] only in ABSL-2 and ABSL-3 animal rooms. Besides being useful for hand washing and miscellaneous uses that come up when working with animals in the room, the most beneficial use of sinks in an animal room is for dumping mop buckets. It is considered good sanitation practice for each animal room to be equipped with dedicated equipment for routine cleaning of the floor. For this reason, a stainless steel mop sink is desirable. The sink should be mounted on the wall to avoid impeding floor sanitation (Figure 8.5). Hands off controls are desirable. A coldwater hose bib mounted on the wall under or near the mop sink at a height suitable for filling mop buckets is also useful. If there is uncertainty regarding the installation of sinks in the animal rooms, an option is to fit each room with plumbing for hooking up mobile sinks. Alternatively, several rooms in a suite of animal rooms could share a single sink so long as there is no concern of carrying infectious agents between rooms. If there is uncertainty regarding the need for sinks in the animal rooms, an option is to fit each room with plumbing that could accommodate portable sinks.

### Floor Drains

Rooms to be used for housing animals using dry bedding cage systems do not require drains. Whether or not to include one is a matter of choice. There are advantages and disadvantages to both, with flexibility being the primary advantage to including them. Other than that, the disadvantages of having them may easily outweigh any other advantages of having them. Disadvantages include installation cost, confounding pest control, especially the control of cockroaches, the potential for sewage backing up into the room, slopping floors that can cause problems when trying to park racks with wheels, underutilized traps drying and thus allowing sewer gas to escape (capping the drain with airtight seals can alleviate this problem, but then the drains are not convenient to use), and taking up space, especially trough drains.

In animal rooms designed for hose-down caging systems, the location of the floor drain is critical to efficient cleaning. Ideally, the drain should be at the low point of an open floor trough located against the sidewalls of the room so that the cages or pens back up to the drain trough but do not cover it. If floor pens are used, the trough should be uncovered and outside the pens, leaving a minimum 46 cm (18 in) access aisle between the pens and the wall. The room floor should be sloped at a minimum of 1.5 cm per m (3/16 in per ft) from a crown in the center of the room to the floor trough on each side of the room or from one side of the room to a trough on the opposite side. The bottom of the trough should slope a minimum of 2 cm per m (1/4 in to the ft) toward a minimum 10 cm (4 in) diameter drain, or 15 cm (6 in) diameter if the drain will service a large number of animals. The drain should have rim and trap flush fittings. In addition, there should be a water source at the highpoint of the trough controlled with the same ball-type valve that controls the flow of water to the flush drain fittings.

## Special Plumbing Considerations

### Animal Drinking Water

Treatment options could include none, acidification, chlorination, and reverse osmosis. Options for providing drinking water to the animals include water bottles or automatic watering. Special plumbing considerations and equipment are required for both.

## Hoses

Hose reels in areas where hoses are to be used routinely in addition to an independent pressurized recirculating warm water plumbing system supplying water to all hoses are highly desirable features.

## Safety

Safety eyewash and shower stations are required any place caustic chemicals may be used, including both sides of the cage sanitation, near animal water bottle filling equipment, near the reverse osmosis water production unit, and in most laboratories, especially those containing chemical fume hoods.

## Bulk Detergent Delivery

Detergents, acid, and neutralizing agents may be piped to cage sanitation areas from vats or barrels located at or near the receiving dock.

## Miscellaneous Features

### Communications

Essential communications design features include telephone lines strategically located throughout the facility to include most rooms (but not animal rooms), computer network lines in most rooms (including all animal rooms), and video cable lines in selected rooms (surgical and training rooms).

### Environmental Monitoring

The importance of environmental control was emphasized at the beginning of this chapter. Monitoring, alarming when out of range parameters are detected, especially temperature, and documenting that the systems are working properly is equally important. At a minimum, environmental monitoring includes monitoring and documenting animal room temperature, relative humidity, and lighting (sensing light or absence of light without regard to intensity). Relative air pressures monitoring in critical areas such as biocontainment is also important. Environmental monitoring, alarming, and documenting may be accomplished through the environmental control system or with a totally independent system with redundant probes and sensors and central processing unit. Both are acceptable; however, the redundancy offered by having separate control and monitoring systems is highly desirable.

### Security and Access Control

Access control into the animal facility managed by the institutional security department is essential for all research animal facilities. Preferably, it should be a digitally controlled system with some type of physical identification feature such as a swipe or transponder proximity card or best of all a biometric reader. Access control within the animal facility to animal rooms and areas within the facility, e.g., biocontainment, may enhance security, but its primary purpose is to manage traffic in the facility. Only people who have a reason to enter a particular area or animal room should have access to that area or room. A key system to accomplish this in a facility with more than a few users is unmanageable. A central processor control system maintained by the animal facility staff provides the most manageable system for most facilities.

# EQUIPMENT

## Cage Sanitation and Sterilization

The cage washing area is one of the most important areas of a laboratory animal facility but, unfortunately, an area that is often not planned carefully enough. Insufficient space and inadequate equipment are the most significant planning errors. The area needs to be large enough to hold the required space demanding equipment as well as the dirty and clean cages in a configuration that facilitates work efficiency (Figure 8.3). It should have one or more pieces of cage sanitation equipment and may have an autoclave, bottle filling equipment and equipment for sanitizing automatic watering devices. The type of equipment required depends on the size of the facility in terms of cage rack capacity and types of cages, which, of course, is dependent on the species to be housed. See earlier sections in this chapter for architectural and HVAC considerations.

Cage cleaning includes several steps to ensure that cages are freed from urine salts, feces, and vegetative microorganisms. This can be accomplished by washing the cages by hand, but not efficiently and not without the use of chemical disinfectants that are best avoided if possible. Mechanical cage washers sanitizing with high temperature water accomplish the job more effectively, efficiently, and more safely for personnel and the environment. Commercial mechanical cage washers, while in many respects similar to restaurant dishwashers and hospital cart washers, have design features specifically suited for washing cages. Cage washing cycles typically start with a prewash rinse to get rid of loosely attached items such as bedding material and feces. Depending on the washer, the rinse may be performed with cold tap water or with warm water recycled from the final rinse water. The second cycle is intended to wash cages free from fatty products using an alkaline (basic) detergent that is automatically dispensed into the wash water. The desirable water temperature in this cycle is 60 to 70°C (140 to 158°F). The last cycle is always the final rinse with a water temperature of at least 82°C (180°F) to sanitize the equipment being washed and render it free of viruses and vegetative bacteria. A fourth cycle involving an acid rinse may be interspersed between the detergent wash cycle and the final rinse cycle to neutralize the alkaline detergent that may result in hydrolyses of polycarbonate during autoclaving, especially if strong caustic sodium or potassium hydroxide alkaline detergents are not adequately rinsed from the cages. Hydrolysis of polycarbonate material during autoclaving is less likely to occur following washing with milder sodium bicarbonate alkaline detergents, thus eliminating the need for an acid rinse. Acid treatment of cages as an acid rinse cycle during mechanical washing or as a prewash treatment applied by hand can effectively reduce the buildup of urine salts on the cages.

There are two basic types of mechanical washers, "batch washers" and "continuous belt washers." Batch washers cover all the cycles within a single chamber. Continuous belt washers, also known as "tunnel washers," transport materials to be washed on a belt through a tunnel sectioned into various rinse and wash cycles. The type of washer selected depends on the facility size and species to be housed. It is common for facilities to have both. Larger facilities will require two or more of one or the other or both types of washers.

### Batch Washers

Batch washers come in single-door and double-door pass-through models. Pass-through models are much preferred, because they provide for the separation of clean and soiled sides of the cage sanitation area. They come in two basic sizes and types. The most common is the floor-loading models, often referred to as "cage and rack washers" because entire racks of cages can be rolled into the washer. Typically, they have washing chambers sized wide enough for one cage rack, deep enough to hold one or two cages racks, and tall enough to hold the tallest mobile cage racks in the facility (nominal maximum rodent rack size = 0.6 m wide, 1.8 m long, 2 m high; dog and primate cages may be larger, e.g., 1 m wide). Cage and rack washers are best suited for washing large cages used for housing rabbits, dogs, nonhuman primates, etc. They are typically pit mounted to make the washer floor level with the room floor (Figure 8.20), and while it is possible to ramp up to the washer floor, this presents a potentially dangerous ergonomic problem (Figure 8.21). With appropriate wash racks, they can also be used effectively for washing rodent shoebox-type cages (Figure 8.21). A second type of batch washer, often called

**Figure 8.20**   A view of a pair of batch-type cage and rack washers from the clean side of the cage sanitation area that are pit mounted to make the washer floor level with the room floor. Note the exhaust fume hoods at the ceiling and the grate-covered pit in front of the washers.

**Figure 8.21**   A batch-type cage and rack washer that is not pit mounted, thus requiring a ramp to load rolling equipment into the washer. Pit mounting is much preferred in that it avoids the potentially dangerous ergonomic problems associated with the ramp. The wash rack alongside the washer is useful for holding small cages and other small equipment in the cage and rack washer.

a "cabinet washer," may have most of the features of the cage and rack washer except that it has a much smaller chamber, e.g., 122 cm (48 in) wide × 79 cm (31 in) high × 86 cm (34 in), with the bottom of the wash chamber being approximately 91 cm (36 in) off the floor. This type of washer is suitable for very small facilities that use only cages that can be safely lifted up into the chamber. Some larger facilities may have one for washing animal watering bottles, feed pans, and small cage parts. An optional feature on cage and rack washers is to have saddle tanks that store the water from either or both the alkaline detergent wash and acid rinse cycles so that it can be reused to conserve water and chemicals.

## Continuous Belt Washers

Continuous belt washers, also referred to as tunnel washers, while not as versatile as cage and rack washers, are more efficient at washing and sanitizing solid-bottom shoebox-type cages (Figure 8.22). Therefore, they are commonly used in facilities that have a large number of shoebox-type rodent cages. This type of washer is also well suited for sanitizing cage pans, water bottles, and other small equipment. Cages and equipment to be cleaned and sanitized are placed on a conveyor belt that moves through a tunnel divided into sections, e.g., prerinse, detergent wash, rinse, and final rinse. Water is .usually

**Figure 8.22**    A continuous belt tunnel washer designed for use with robots that load and unload cages from the belt. It has two features that are required to accommodate robots: (1) it is an indexing tunnel washer that stops periodically for loading and unloading the belt, and (2) it has exceptionally long load and unload extensions that can accommodate several rows of cages as opposed to the more typical extensions that may hold only one row of cages. The dividing wall for separating the clean and soiled sides of the cage sanitation area is not yet in place.

**Figure 8.23**    In the center of this figure is a bedding dispenser at the discharge end of a tunnel washer. As cages exit the tunnel washer (top left) open side down, they are automatically turned over, as they tumble onto the lower bedding dispenser belt so the open side of the cage is up. Upon exiting the bedding dispenser, the cages collect on the roller conveyor to the right of the dispenser. Above and to the right of the dispenser is a vacuum cleaner that collects bedding dust generated by the dispenser. Note the ceramic tile floor and wall coverings with epoxy grout and a corner of a grate-covered floor pit at the bottom left of the figure in front of the rack washer.

recirculated: the recirculating rinse water is used for prerinse, which is discarded, and the final rinse water flows into the recirculating rinse water to freshen it. Often, a dryer and bedding dispenser section are added to the tunnel washer (Figure 8.23). Tunnel washers come in a variety of sizes; however, a typical tunnel washer may have a conveyor about waist high, a tunnel opening of 61 cm (24 in) to 107 cm (42 in) wide by 61 cm (24 in) high. The washer section is typically a minimum of 4.6 m (15 in) and may be longer. The width and length of the washer determines the washing capacity. The longer the washing tunnel, the faster the conveyor belt can run and still provide an effective exposure time in each section. A dryer, automatic bedding dispenser, and load and unload sections are optional features. With a washer, dryer, bedding dispenser, and load and unload extensions, a typical tunnel washer assembly may be 10 m (33 ft) long or longer.

### *Robotic Cage Washing*

The rapidly increasing use of mice in molecular biology research and the resulting large increase in the mouse cage census of many facilities has led to the initiation of robotics into the care of laboratory animals.[41,42,43] To date, the use of robots has been limited to having a robot on each end of a tunnel washer and dryer (Figure 8.24). On the soiled side, a robot picks soiled cages off of a cart, typically one to four cages at a time, dumps the bedding into a disposal unit that automatically transports and deposits the bedding in a disposal container outside the facility, and then places the cages on the tunnel washer belt. At the clean end of the tunnel washer, a second robot picks the cages off of the washer belt, places them under an automatic bedding dispenser, and then stacks them filled with bedding on a cart. This requires using an indexing tunnel washer and dryer that stops while the robot is loading and unloading the belt on long extensions on both ends that hold, for example, 16 to 20 cages at a time. Because the belt does not run continuously, indexing tunnel washers may have automatic sliding doors that separate the various sections of the washer.

### *Sterilization Equipment*

The usual way to sterilize cages, bottles, lids, and other material necessary for laboratory animal care that can withstand high temperatures is to use an autoclave. An autoclave is a pressure vessel that uses a combination of above atmospheric pressure and live steam to generate temperatures above boiling. A complete sterilization cycle for nonliquid materials typically includes several cycles of high vacuum followed by live steam injection to rid the chamber of air, the high temperature and pressure sterilization cycle, typically 120°C (250°F) for 20 min or 135°C (275°F) for 15 min, followed by time for steam withdrawal, pressure equilibration, and cooling. High vacuum autoclaves are far more effective and more practical than gravity autoclaves in the laboratory animal environment. In addition, "clean steam" generated from domestic water (not necessarily as high a quality as pharmaceutical grade "clean steam") should be used in the autoclave to avoid contaminating the cages, etc., with chemical additives (typically filming amines used to coat the boiler and steam pipes to reduce corrosion) normally put in steam boilers. Amines from live steam damage polycarbonate and add an avoidable variable to the animal's environment.

**Figure 8.24**    A robot stacking clean bedded cages four at a time after having picked the cages off of the indexing tunnel washer belt in back of the robot and placing them under a bedding dispenser that is pneumatically fed from a larger bedding storage hopper located near the dock where bedding is received. The cages are being stacked on special carts that rest on rails, all designed to precisely align the stacks of cages. On the soiled side, an identical combination of robot, carts, and rails are used to load soiled cages onto the tunnel washer belt after the robot dumps the soiled bedding into a hopper, from where it is pneumatically transported to a disposal container outside the facility.

For effective sterilization, the material being autoclaved needs to be not wrapped, wrapped with materials that are not airtight, or the wrapped with the wrapping partially open to allow the vacuum cycle to extract the air and to prevent the creation of air pockets that impede sterilization. Autoclave chamber sizes used in animal facilities vary considerably from ones that will hold only a few cages to ones that will hold three, four, or even more racks of cages. When planned for use in an infectious containment facility or a rodent barrier facility where routine husbandry relies on the sterilization of cages, it is essential to carefully calculate output capacity, which is dependent on the size of the autoclave chamber and the autoclave cycle times. Pit-mounted floor loading bulk autoclaves large enough to hold one or more cage racks are commonly used in biocontainment and barrier facilities (Figure 8.25). For productivity calculations, a cycle time of 1 h 30 min should be used to determine the number of cycles possible in a typical workday. Such heavy use of autoclaves increases maintenance downtime. If money and space is available, two autoclaves are preferred.

Autoclaves used for infectious and containment purposes should have two doors so that materials can be passed through. The autoclave may be located in the cage sanitation area, either between the soiled and clean sides, or between the clean side and the "sterilized" equipment storage area. The ideal location is as an integral part of the infectious containment (Figure 8.7) or rodent barrier area of the facility where cages, etc., can be autoclaved into or out of the area, thus eliminating the need to wrap the cages, etc., for transport between the area of the facility where used and the autoclave.

Materials that may not withstand the high temperatures of autoclaving may be sterilized with gases such as para-formaldehyde, ethylene oxide, and chlorine dioxide. Irradiation is commonly used for sterilizing feed and bedding.

## Animal Watering

Potable water from municipal water treatment plants is commonly used for laboratory animals on the assumption that if it is good enough for humans it is good enough for laboratory animals; however, this is not necessarily so. It is possible that potable water may contain chemical contaminants at a level too low to be considered toxic, but that may be at high enough levels to confound study results when given to laboratory animals. The use of reverse osmosis (RO) to consistently provide high-quality drinking water for the animals greatly reduces the potential for such and therefore is rapidly becoming the standard. In addition, because water quality and content varies from community to community, the use of RO water provides a practical and economical means of standardizing drinking water for research animals throughout the world. It is common to acidify animal drinking water, especially water delivered to the animals in bottles, to a pH of 2.5 to 3 in order to limit the growth of microorganisms in the water during the time between bottle changes, particularly if the bottles are to be changed less frequently than every 3 to 4 days. Hyperchlorination up to 10 ppm also effectively limits microorganism growth, but it tends to dissipate from water in a bottle within a few days.

There are two commonly used methods to provide water to laboratory animals: in glass or polycarbonate bottles with holes in the bottle or equipped with stainless steel sipper tubes or through automatic watering devices. Both systems have pros and cons. Water bottles are labor intensive, needing to be exchanged with freshly sanitized bottles at least weekly and, if not acidified, at least two times a week. Automatic watering devices have a relatively high front-end cost and also present a quality control challenge because of biofilms that form on the inside of the distribution system and bacterial growth that can and does build up in the water of the low-pressure low-flow distribution system, especially in dead end segments and in the rack water manifolds. The quality of automatic watering valves has greatly improved in recent years, reducing the problem of cage flooding from leaking watering valves, but this problem has not been totally eliminated. Of course, water bottles can also leak their contents into the cage, and while it will not flood the cage with water, it will soak the bedding material and deprive the animals of water. Choosing a watering strategy is a critical decision that needs to be made early in the facility planning process.[44,45]

(a)

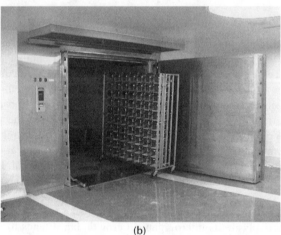

(b)

**Figure 8.25**   Figures 8.25a and 8.25b each picture a double door pass through bulk autoclave. The autoclave
pictured in Figure 8.25a has sliding doors and is not pit mounted, thus requiring a lift in order to
load and unload carts and cage racks. The autoclave in Figure 8.25b has hinged doors and is pit
mounted, which greatly facilitates loading and unloading. A steel plate bridges the gap between
the floor of the autoclave and the facility floor. This autoclave has the capacity to hold three of
the type of mouse rack shown in the figure. Note the exhaust hood in the ceiling above the
autoclave door in Figure 8.25b.

## Caging

### *Rodents*

Small rodents, e.g., rats, mice, hamsters, gerbils, and guinea pigs, are usually housed in solid bottom
shoebox-type cages with several centimeters of various types of bedding materials covering the cage
bottom. Open stainless steel wire-bottom cages without bedding material are considered less desirable
in terms of animal welfare but are occasionally used when required to achieve the scientific objectives.

**Figure 8.26** Rats group housed in an old rabbit cage. The complexity of the environment has been increased with hay and a branch on which the rats can climb to a shelf.

Large guinea pigs are sometimes housed in rabbit cages, especially ones with perforated plastic floors. Rabbit cages are particularly adaptable to creating a highly enriched environment for housing rats (Figure 8.26). Rodent shoebox-type cages are made of various types of plastic materials. The material most frequently used at this time is polycarbonate [Lexan® (GE) or Makrolon® (Bayer)]. This material is transparent, rigid, durable, and sanitizable. Its "softening temperature" is 152 to 157°C (305 to 315°F), and it withstands autoclaving at 120°C (250°F) but eventually becomes brittle after multiple exposures to autoclave temperatures. A newer copolycarbonate [Apec® (Bayer)] better withstands higher temperatures, thus it is often referred to as "high temperature polycarbonate," is reported to hold up better under repeated autoclaving. Recently, polysulfone [Radal® (Amoco)] was introduced as a cage material. It is a durable plastic that is reported to be able to handle thousands of repeated autoclaving while maintaining impact strength and transparency. Polysulfone, due to its brownish color, has about 35% less light penetration than polycarbonate. This might not be a problem in many laboratory facilities; it may, on the contrary, be beneficial, because light intensity may be high to make caretaking easier without illuminating the animals too much. The sole drawback with polysulfone is the initial cost, which is high in comparison to the other materials but may be cost effective in the long run because of the high durability. Other types of plastic are also used for rodent caging. Polypropylene is a light, flexible, material with high chemical inertia and thermal resistance up to 120°C. It may be translucent or opaque, depending on whether it is a copolymer or a reinforced copolymer. Polystyrene is rigid, with low impact and heat resistance. It is usually used to form disposable cages that are suitable for toxic or radioactive applications when decontamination of the cage is impractical or too dangerous for personnel.

Much emphasis has recently been given to environmental enrichment for all species, including rodents. The value of enrichment for the well-being of rodents is controversial and will probably be a source of debate for years to come. Experiments with pen housing and different ways to increase the complexity of the primary environment for the animals are currently being performed (Figure 8.26). The challenge is to balance seemingly conflicting requirements to provide for animal well-being through cage enrichment with features required to provide for routine animal care and cage sanitation, to assure animal health, and to successfully achieve the research goals.

Solid bottom shoebox-type cages may be covered with wire bar lids that leave the cage interior open to the room environment, or may be covered with filter tops that provide a barrier between the cage microenvironment and the room macroenvironment that prevents the spread of airborne infections between cages. The most common type of filtered top is a rigid inverted shoebox with an air filter insert that covers the top of the cage and overlaps the sides to form the microbiologic equivalent of a petri dish. These are known as microisolation cages. Microisolation cages have proven to be effective at protecting rodents from microbial contamination when used in combination with a HEPA-filtered mass air displacement cabinet used any time the microisolation cage is opened for cage changes or performing

**Figure 8.27**    A rack with static microisolator cages adjacent to a Type II A biosafety cabinet with a microisolator cage inside the cabinet. The combination of the microisolation cage and cabinet makes up what is known as a "microisolation cage system." The microisolation cages are only opened inside the cabinet, including for cage changes or research procedures.

procedures on the animals (Figure 8.27);[46–50] however, in the static air microenvironment of the cage, animals are subjected to elevated levels of ammonia, carbon dioxide, moisture, and heat.[51,52,53] The longer the time between cage changes, the higher the ammonia levels. There are many variables that affect the ammonia levels, e.g., days between cage changes, relative humidity levels, and the type of bedding, but in general, two changes per week is considered the minimum number of changes for a microisolator cage housing the maximum capacity of mice. Ventilated microisolation caging was designed to better control the microenvironment.[54,55] Directly ventilating each microisolation cage with HEPA-filtered air significantly slows the buildup of ammonia, etc., in the cage, thus decreasing the cage change frequency to once a week or even once every two or three weeks.[56–60] Ventilated cage racks are fitted with HEPA filter blower units that deliver HEPA-filtered air to each cage through a system of manifolds. Some ventilated racks also capture the air coming from each cage and HEPA filters it before dumping it into the room (Figures 8.17 and 8.18) or, even better, directs the air from the cages directly into the room exhaust (Figures 8.16 and 8.19). Some ventilated microisolation cage racks that control the cage supply and exhaust air can selectively maintain the air pressure in the cage, either positive or negative relative to the room air pressure. When exhausted directly into the room exhaust ducts, HEPA filtration is not necessary, but furnace or higher grade filters are desirable to reduce the amount of dust dumped into the exhaust ducts and especially onto the heat recovery system. A significant "fringe benefit" of ventilated cage racks that HEPA filter air coming from the cages or directly exhaust the cage air from the room is the reduction of animal allergens in the workers' environment because allergies to animal allergens are the most significant occupational hazard for personnel working with animals.[40,61] Another advantage is that it allows for high-density housing. Mobile double-sided ventilated mouse cage racks commonly hold up to 140 cages (7 cages wide × 10 rows high × 2 sides) each with the capacity for up to five mice. Static racks may even be stacked higher, further increasing housing density but also increasing the ergonomic problems and possibly injuries associated with reaching the upper cages.

### *Rabbits*

The traditional way to keep rabbits is to house them one to a cage to facilitate handling, preclude fighting, and minimize the risk for spread of infections. During the last 10 years, there has been movement away from fabricating rabbit cages with stainless steel toward warmer plastics, such as NORYL® (GE), which is opaque, lightweight, nontoxic, and withstands the sanitizing temperatures of cage washers. In-cage enrichment strategies have included a shelf in the cage, making it possible for the rabbits to utilize more of the cage volume than just the floor and to find a hiding or "burrowing" place underneath. the shelf. Single housing is usually the sole alternative for housing male rabbits that have a strong tendency to fight when housed together. Female rabbits housed together may also fight, but with careful management

**Figure 8.28**    Rabbits group housed on the room floor. Certain types of studies that involve holding rabbits for long periods of time, such as polyclonal antibody production protocols, allow for group housing of rabbits. Experience shows that most female rabbits readily adapt to being housed in groups, where they benefit from social interactions and exercise.

and observation to reduce fighting and the resulting fight wounds, it is possible to successfully house them in groups of two or more in cages or on the floor (Figure 8.28). Some rabbit cage racks are designed with pairs of cages separated with a removable divider. This allows doubling the floor area for housing two adult females together or for using it as a kindling cage and for housing a female and her nursing young. Pair housing females is especially beneficial when rabbits are kept for longer periods, as those that are used for antibody production.

## Dogs

Some types of studies with dogs require individual housing for at least part if not all of the day. When studies permit, dogs are ideally suited to group housing. Because dogs are large animals with large needs for exercise, a dog facility is space demanding (Figures 8.8 through 8.14). The facility is ideally divided into three parts: indoor pens with resting places, indoor cages where the dogs can be housed individually when fed, and outdoor pens for exercise. U.S. Animal Welfare Act Regulations specify that dogs housed in cages providing less than two times the required floor space must be provided the opportunity to exercise outside of the cage on a routine basis as established by the attending veterinarian.

## Cats

Cats, especially females, are also well suited to group housing (Figure 8.29). A regular animal room furnished with shelves, boxes, tunnels, branches, or other items where the animals may hide or climb is suitable for group housing of cats and is commonly used. For some types of studies, it is necessary to individually house cats, for example, if they have been subjected to some kind of surgical procedure, or if it is necessary to have male cats. In these cases, stainless steel cages with wood or plastic resting boards are commonly used.

## Nonhuman Primates

Group or pair housing is most desirable for most species of nonhuman primates; however, single housing in stainless steel caging has been the more traditional way to keep large nonhuman primates such as macaques. The main advantage for single caging of the larger species is safety, both for personnel and the animals. These animals are strong, have sharp teeth, and may have infections that are deadly for humans, e.g., herpes B virus in macaques. Experimental procedures often require frequent handling of the animals. Individually housed animals can be more safely restrained and captured. In addition, the single animal per cage housing regime avoids the risk of severe harm that these large animals inflict on

**Figure 8.29**    Group housing of cats. Most female cats readily adapt to group housing. The room is a regular laboratory animal facility animal room furnished with shelves and tunnels.

each other when fighting. In spite of these serious considerations, pair or group housing is often possible when carefully arranged and is preferred, because these are social, highly developed animals with needs for exercise and environmental complexity. For these reasons, individual caging is best limited to situations requiring individual housing based on scientific requirements and individual animal behavior.

The most optimal housing may be to have two primates in a large cage with branches, toys, and bedding material into which food may be placed to stimulate foraging behavior. A transport cage may be attached to such cages, and the animals may be urged to go into the transport cage with some training. With two primates in one cage, it would be simple enough to get the primate wanted for a particular experiment without too much distress of nonhuman primate and personnel. It usually is possible to find two animals that like to be housed together. Larger groups of nonhuman primates may be formed even from animals that have no or little experience of group housing. Care must be taken to find a good structure of such a group, which needs a strong and reliable leader.

### Pigs, Sheep, and Goats

A facility for housing pigs, sheep, and goats can be similar to a dog facility. These animals are ideal to house in groups (Figure 8.14). They need a large space for exercise and need outdoor and indoor pens. The main difference is that pigs do not need to be housed individually when fed. While species are typically separated for housing, pigs and dogs have proven to be compatible when sharing an outside exercise pen (Figure 8.14).

## Equipment and Caging for Biological Control (Barrier and Containment)

### Microisolation Caging System (Microisolation Cages and Cage Change Cabinets)

Microisolation cages that were described in the rodent housing section of this chapter provide a highly effective barrier and containment caging system. Static microisolation caging may be more suited for containment than positive pressure ventilated microisolation cages. If ventilated cages are to be used for containment, they should be tightly sealed.

To be fully effective at controlling cross contamination in barrier or containment situations requires using the "microisolation caging system." The system includes using microisolation cages, static or ventilated, in combination with HEPA-filtered air "clean bench" or a Class II Type A biocontainment cabinet that protects the interior of the cage from external microbial contamination when the cage is opened, either for performing procedures with the animals or transferring the animals to a clean cage. These are commonly referred to as cage change cabinets or animal transfer cabinets. The Class II Type A biocontainment cabinet has the advantage of protecting the operator and the operator's environment, as well as the product in the cabinet, which in this case, is the animals in the open cage, and must be used if biocontainment is the objective (Figure 8.27). In addition to the microisolation cage and the cage change cabinet, the microisolation caging "system" includes operational procedures such as opening one cage at a time in the cabinet except when transferring animals to a clean cage, which requires having the soiled and clean cages open in the cabinet; and between working with cages in the cabinet, using a fast-acting high-level disinfectant on the cabinet work surface, gloved hands, and any instrumentation that touches the animals.

## Isolators

Isolators provide the most effective protection against the spread of infectious agents, whether used as a barrier to protect an animal in the isolator, even to the point of maintaining "germ-free" or gnotobiotic animals, or for containment to protect the macroenvironment (room) from biohazards in the isolator. They are of two basic types: flexible film isolators and rigid wall isolators (Figures 8.30 and 8.31). The original germ-free isolators were rigid stainless steel cylinders, but most rigid isolators in use today are made with clear polycarbonate. Currently, most isolators are made from transparent poly-vinyl-chloride (PVC) flexible plastic sheeting supported by a metal or plastic frame and positive pressure air, and rigid isolators from stainless steel or polycarbonate. HEPA-filtered air is blown into the isolator, and when used for containment, the air coming from the isolator is also HEPA filtered. Isolators come in many sizes. In effect, the isolator is a room within a room that can be provided with sterile air, water, diet, cages, and bottles, etc., minimizing the risk of contaminating animals with infectious agents from outside the isolator. Personnel work inside the isolators through portholes fitted with sleeves and gloves made of latex or other similar material. There is a well-established system involving equipment and procedures designed for moving sterile materials into the isolator and soiled materials and trash out of the isolator.

## Autoclaves

Autoclaves were covered previously in this chapter (Figure 8.25). They are an essential component of an effective barrier or containment facility, required for sterilizing cages, cage parts, bedding, feed, and other supplies. In the case of barriers, cages and supplies are autoclaved into the barrier area. In the case of containment, they are autoclaved out of and in many cases are also autoclaved into the containment area. If the autoclave is not an integral part of the barrier or biocontainment area, items to be autoclaved must be wrapped or otherwise protected or sealed to safely transport them between the autoclave and the barrier or containment area.

## Biological Safety Cabinets

Biological safety cabinets (BSC) provide primary containment for working with infectious agents, or as noted above, are also useful for working with microisolation cages. The CDC-NIH publication "Biosafety in Microbiological and Biomedical Laboratories"[21] includes a detailed description of the various types of BSCs along with installation requirements.

There are three classes of biosafety cabinets: Class I, II, and III. Class I and II BSCs have inward airflow at velocities of 75 to 100 linear ft per min through an open front. Exhaust air from the cabinet passes through HEPA filters before being discharged into the room or into the laboratory exhaust system.

**Figure 8.30**    Flexible film isolator. The isolator in the photo has a large square entrance port, which is easier to load than the more common circular ports.

**Figure 8.31**    A rigid stainless steel isolator that was specially adapted to perform hysterectomies for rederivation of animals. The isolator is divided into two parts, one in which the hysterectomy is performed and one where the pups are resuscitated. The two parts are separated with a dip tank containing a disinfectant.

Class II BSCs have the additional benefit of protecting objects in the cabinet from extraneous microbial contamination. Class I and II BSCs are suitable for working with up to BSL-3 infectious agents. Class III BSC cabinets provide the highest degree of personnel and environmental protection from infectious aerosols, as well as protection from extraneous microbiological contaminants for the materials in the cabinet. They essentially are a totally enclosed, gas tight, ventilated cabinet. All operations in the cabinet are performed through attached rubber sleeves with surgical-type gloves. Supply air is HEPA filtered, and exhaust air is filtered through two HEPA filters. Class III BSCs are suitable for working with infectious agents classified at the highest biosafety level, BSL-4.

Class II BSCs, which are the type most commonly used in research animal facilities, come in two types, A and B. Type A is suitable for containing particulate hazardous agents only and may be exhausted into the room through HEPA filters or to the outside via a thimble connection to the building exhaust ductwork. Class II Type A BSCs, as noted previously in this chapter, are often used as rodent cage change cabinets. Class II Type B BSCs are suitable for containing infectious agents, volatile chemicals, and radionuclides. They have a face velocity of 100 linear ft per min and are hard-ducted to the exhaust system. Class II Type B BSCs are further subdivided into types B1, B2, and B3 depending on multiple features, including the degree of air recirculated within the cabinet versus that discarded, i.e., B1, 70% recirculation; B2, 30% recirculation; and B3, 0% recirculation, 100% exhausted. Of course, the higher the percent exhausted, the greater the control of volatile hazards.

## CONCLUSION

The emphasis in this chapter is on research animal facilities required to support contemporary biomedical research, but it must be noted that sound management of the facilities is at least equally important. Control of some of the environmental factors noted in Figure 8.1 requires properly designed facilities, but all require sound management. The better the facility is designed to facilitate sound management, the lower the cost of animal care and the more likely environmental variables will be adequately controlled. It is theoretically possible for good management to overcome design features contrary to efficient management, but even the best management cannot overcome human nature, which dictates that if a routine task is difficult to do, it will guarantee that it will not be routinely done. A properly controlled environment for research animals means more reliable and reproducible research, thus reducing the number of animals required to achieve the research goals. Good facility design and sound management facilitates high-quality contemporary biomedical research, sound research economics and, most importantly, humane care and use of laboratory animals.

## REFERENCES

1. Hessler, J.R. and Leary, S.L., Design and management of animal facilities, in *Laboratory Animal Medicine*, 2nd ed., Fox, J. and Lowe, F., Eds., Academic Press, New York, 2002, chap. 21.
2. Shalev, M., Ed., *Facility Design and Planning*, Lab Animal, New York, NY, Fall 2001.
3. Hessler, J.R., The history of environmental improvements in laboratory animal science: caging systems, equipment, and facility design, in *Fifty Years of Laboratory Animal Science*, McPherson, C. and Mattingly, S., Eds., American Association for Laboratory Animal Science, Memphis, TN, 1999, chap. 15.
4. Institute of Laboratory Animal Resources, Commission on Life Sciences, National Research Council, National Academy of Sciences, *Guide for the Care and Use of Laboratory Animals*, National Academy Press, Washington, DC, 1996.
5. Council of Europe, European Convention for the Protection of Vertebrate Animals Used for Experimental and Other Scientific Purposes, Council of Europe (Convention ETS 123), 1985 (adopted May, 1999).
6. Canadian Council on Animal Care, *Guide to the Care and Use of Experimental Animals*, Volume 1 (2nd ed.), Canadian Council on Animal Care, Ottawa, Ontario, Canada, 1993 (adopted May, 1999).
7. Federation of Animal Science Societies, *Guide for the Care and Use of Agricultural Animals in Agricultural Research and Teaching*, 1st revised ed. (adopted January, 1999) FASS, Savoy, IL, 1999.
8. National Research Council, National Academy of Sciences, *Occupational Health and Safety in the Care and Use of Research Animals* (adopted September, 1997) National Academy Press, Washington, DC, 1997.

9. Hessler J.R., Single- versus dual-corridor systems: advantages, disadvantages, limitations, and alternatives for effective contamination control, in *Handbook of Facilities Planning, Vol. 2, Laboratory Animal Facilities*, Ruys, T., Ed., Van Nostrand Reinhold, New York, NY, 1991, pp. 59–67.

10. Leary, S.L., Majoros, J.A., and Tomson, J.S., Making cagewash facility design a priority, *Lab Animal*, 27, 28–31, 1998.

11. Fullerton, F.R., Greenman, D.I., and Kendall, D.C., Effects of storage conditions on nutritional qualities of semipurified (AIJN-76) and natural ingredient (NIH-07) diets, *J. Nutrition*, 12, 567–573, 1982.

12. American Veterinary Medical Association, Guidelines for animal surgery in research and teaching, *Am. J. Vet. Res.*, 54, 1544–1559, 1993.

13. Hessler, J.R., Facilities to support research, in *Handbook of Facilities Planning, Vol. 2, Laboratory Animal Facilities*, Ruys T., Ed., Van Nostrand Reinhold, New York, NY, 1991, pp. 35–54.

14. Cunliffe-Beamer, T.L., Applying principles of aseptic surgery to rodents. *AWIC Newsletter*, 4, 2, 3–6, 1993.

15. Brown, M.J., Aseptic surgery for rodents, in *Rodents and Rabbits: Current Research Issues*, Niemi, S.M., Venable, J.S., and Guttman, H.N., Eds., Scientist Center for Animal Welfare, Bethesda, MD, 1994, pp. 67–72.

16. Dolowy, W.C., Medical research laboratory of the University of Illinois, *Proc. Anim. Care Panel*, 11, 267–290, 1961.

17. Hessler, J.R., Animal cubicles, in *Handbook of Facilities Planning, Vol. 2, Laboratory Animal Facilities*, Ruys, T., Ed., Van Nostrand Reinhold, New York, NY, 1991, pp. 135–154.

18. Hessler, J.R., Animal cubicles: questions, answers, options, opinions, *Lab Animals*, 22, 21–32, 1993.

19. Ruys, T., Isolation cubicles: space and cost analysis, *Lab Animal*, 17, 25–23, 1988.

20. Curry, G., Hughes, H.C., Loseby, D., and Reynolds, S., Advances in cubicle design using computational fluid dynamics as a design tool, *Laboratory Animals*, 32, 117–127, 1998.

21. Center for Disease Control and Prevention, Biosafety in Microbiological and Biomedical Laboratories, NIH, DHHS, U.S. Government Printing Office Pub. No. (CDC) 93–8395, Washington, DC, 1999.

22. Evans, M.E. and Lesnaw, J.A., Infection control in gene therapy, *Infection Control and Hospital Epidemiology*, 20, 568–576, 1999.

23. Webber, D.J. and William, A.R., Gene therapy: a new challenge for infection control, *Infection Control and Hospital Epidemiology*, 20, 530–532, 1999.

24. Barkley, W.E., Abilities and limitations of architectural and engineering features in controlling biohazards in animal facilities, in *Symposium on Laboratory Animal Housing*, Institute for Laboratory Animal Research, National Academy of Science, National Research Council, Washington, DC, 1979, pp. 158–163.

25. Barkley, W.E. and Richardson, J.H., Control of biohazards associated with the use of experimental animals, in *Laboratory Animal Medicine*, Fox, J.G., Cohen, B.J., and Loew, F.M., Eds., Academic Press, Orlando, FL, 1984, pp. 595–602.

26. Richmond, J.Y., Hazard reduction in animal research facilities, *Lab Animals*, 20, 23–29, 1991.

27. Hessler, J.R., Methods of biocontainment, in *Current Issues and New Frontiers in Animal Research*, Bayne, K.A.L., Greene, M., and Prentice, E.D., Eds., Scientist Center for Animal Welfare, Greenbelt, MD, 1995, pp. 61–68.

28. White, W.J., Special containment devices for research animals, in *Proceedings of the Fourth National Symposium on Biosafety: Working Safely with Research Animals*, Richmond, J.Y., Ed., Center for Disease Control, Atlanta, GA, 1996, pp. 109–112.

29. Richmond, J.Y., Animal biosafety levels 1–4: an overview, in *Proceedings of the Fourth National Symposium on Biosafety: Working Safely with Research Animals*, Richmond, J.Y., Ed., Center for Disease Control, Atlanta, GA, 1996, pp. 5–8.

30. Hessler, J.R., Broderson, R., and King, C., Animal research facilities and equipment, in *Anthology of Biosafety 1. Perspectives on Laboratory Design*, Richmond, J.Y., Ed., American Biological Safety Association, Mundelein, IL, 1999, chap. 13.

31. Hessler, J.R., Broderson, R., and King, C., Rodent quarantine: facility design and equipment for small animal containment facilities, *Lab. Animal*, 28, 34–40, 1999.

32. ASHRAE (American Society of Heating, Refrigerating, and Air Conditioning Engineers, Inc.), Laboratory animal facilities, in *ASHRAE Handbook: Heating, Ventilation, and Air Conditioning Applications*, ASHRAE, Atlanta, GA, 1999, pp. 13.13–13.19.

33. Carlton, D.L., Affordable noise control in a laboratory animal facility, *Lab. Animal*, 31, 47–48, 2002.

34. Hessler J.R. and Moreland, A.F., Design and management of animal facilities, in *Laboratory Animal Medicine*, Fox, J.G., Cohen, B.J., and Loew, F.M., Eds., Academic Press, Orlando, FL, 1984, chap. 17.

35. Hughes, C.H., Reynolds, S., and Rodrigues, M., Designing animal rooms to optimize airflow using computational fluid dynamics, *Pharmaceutical Engineering*, March/April, 44–65, 1996.

36. Memarzadeh, F., Ventilation Design Handbook on Animal Research Facilities Using Static Microisolators, Animal Facility Ventilation Handbook Volumes I and II, NIH, Division of Engineering Services, Bethesda, MD, 1998.

37. Jackson, C.W., Rehg, Rock, Henning, and Reynolds, Computational fluid dynamics optimizes ventilation in animal rooms, *Lab. Animal*, 50–53, Fall, 2002.

38. Lipman, N.S., Strategies for architectural integration of ventilated caging systems, *Contemp. Topics Lab. Anim. Sci.*, 32, 7–10, 1993.

39. Bilecke, B., Integrating ventilated caging equipment with facility HVAC and monitoring systems, *Lab. Animal*, 42–47, Fall, 2001.

40. Reeb-Whitaker, C.K. and Harrison, D.J., Practical management strategies for laboratory animal allergy, *Lab. Animal*, 28, 25–30, 1999.

41. Ruggiero, R.F., Considerations for an automated cage-processing system, *Lab. Animal*, 28–31, Fall 2001.

42. Corey, M., Davey, R., and Faith, R., Case study: automated cagewash design at Baylor College of Medicine, *Lab. Animal*, 32–35, Fall 2001.

43. Roe, P., Cage processing and waste management: a cost-analysis and decision-making exercise, *Lab Animal*, 31, 43–46, 2002.

44. Lempken, B., Drinking water, in *Handbook of Facilities Planning, Vol. 2, Laboratory Animal Facilities*, Ruys, T., Ed., Van Nostrand Reinhold, New York, NY, 174–180, 1991.

45. Novak, G., Selecting an appropriate watering system for your facility, *Lab. Animal*, 28, 43–46, 1999.

46. Lipman, N.S., Newcomer, C.E., and Fox, J.G., Rederivation of MHV and MEV antibody positive mice by cross-fostering and use of the microisolator caging system, *Lab. Anim. Sci.*, 37, 195–199, 1987.

47. Dillehay, D.L., Lehner, N.D., and Huerkamp, M.J., The effectiveness of a microisolator cage system and sentinel mice for controlling and detecting MHV and Sendai virus infections, *Lab. Anim. Sci.*, 40, 367–370, 1990.

48. Borello, D'Amore, Panzini, Mauro, and Nello, Individually ventilated cages — microbiological contaminant testing, *Scand. J. Lab. Anim. Sci.*, 27, 142–152, 2000.

49. Whary, Cline, King, Corcoran, Xu, and Fox, Containment of *Helicobacter hepaticus* by use of husbandry, *Comparative Medicine*, 50, 78–81, 2000.

50. Otto, G. and Tolwani, R.J., Use of microisolator caging in a risk-based mouse import and quarantine program: a retrospective study, *Contemp. Topics in Lab. Anim. Sci.*, 41, 20–27, 2002.

51. Corning, B.F. and Lipman, N.S., A comparison of rodent caging systems based on microenvironmental parameters, *Lab. Anim. Sci.*, 41, 498–503, 1991.

52. Lipman, N.S., Microenvironmental conditions in isolator cages: an important research variable, *Lab. Anim. Sci.*, 21, 23–27, 1992.

53. Hasenau, J.J., Baggs, R.B., and Kraus, A.L., Micro-environments in micro-isolation cages using BALB/c and CD-1 mice, *Contemp. Top. Lab. Anim. Sci.*, 32, 11–16, 1993.

54. Lipman, N.S., Isolator rodent caging systems (state of the art): a critical view, *Contemp. Topics in Lab. Anim. Sci.*, 38, 1–17, 1999.

55. Novak, G.R. and Sharpless, L.C., Selecting an individually ventilated caging system, *Lab. Animal*, 36–41, Fall 2001.

56. Keller, L.S.F., White, Snider, and Lang, An evaluation of intra-cage ventilation in three animal caging systems, *Lab. Anim. Sci.*, 39, 237–241, 1989.

57. Huerkamp, M.J., Dillehay, D.L., and Lehner, N.D.M., Comparative effects of forced air, individual cage ventilation or an absorbent bedding additive on mouse cage microenvironment, *Contemp. Topics in Lab. Anim. Sci.*, 33, 58–61, 1994.

58. Hasegawa, M.Y., Kurabayashi, Ishii, Yoshida, Eubayashi, Sato, and Kurosawa, Intra-cage air change rate on forced-air-ventilated micro-isolation system-environment within cages: carbon dioxide and oxygen concentrations. *Exp. Anim.*, 46, 251–257, 1997.

59. Reeb-Whittaker, C.K., Paigen, Beamer, Bronson, Churchill, Schweitzer, and Myers, The impact of reduced frequency of cage changes on the health of mice housed in ventilated cages, *Lab. Anim.*, 35, 58–73, 2001.

60. Baumans, V., Schlingmann, Vonck, and Van Lith, Individually ventilated cages: beneficial for mice and men?, *Contemp. Topics in Lab. Anim. Sci.*, 41, 13–19, 2002.

61. Renström, A.G., Björing, G., and Höglund, U., Evaluation of individually ventilated cage systems for laboratory rodents: occupational health aspects, *Lab. Anim.*, 35, 42–50, 2001.

62. Baker, H.J., Lindsey, J.R., and Weisbroth, S.H., Housing to control research variables, in *The Laboratory Rat: Vol. I, Biology and Diseases*, Baker, H.J., Lindsey, J.R., and Weisbroth, S.H., Eds., Academic Press, New York, NY, 169–192, 1979.

63. Ruys, T., *Handbook of Facility Planning: Vol. 2, Laboratory Animal Facilities*, Van Nostrand Reinhold, New York, NY, 1991.

# CHAPTER 9

# Laboratory Animal Genetics and Genetic Quality Control

Michael F.W. Festing

## CONTENTS

0-8493-1086-5/03/$0.00+$1.50
© 2003 by CRC Press LLC

## THE DEVELOPMENT OF LABORATORY ANIMAL GENETICS

The remarkable advances in the science of genetics over the past few decades are having a strong impact on laboratory animal science. These advances are due to the development of a range of molecular techniques making it possible to map, clone, and sequence many genes, but progress has been facilitated by the classical period of mouse genetics which started soon after the rediscovery of Mendel's work in 1900 and continued until about 1980. This laid a firm foundation on which the new molecular methods could be based. A large proportion of "Laboratory Animal Genetics" is in fact "Mouse Genetics."[1] This should not imply that the genetics of other species are unimportant, but simply that for many technical reasons, including small body weight, high reproductive performance, small space requirements, and the availability of a wide range of strains and mutants, the mouse has been used more extensively than any other mammalian species.

Following the rediscovery of Mendel's paper in 1900 setting out the laws of inheritance of discrete characters in garden peas, the validity of these laws was soon confirmed in the mouse by Cuenot[2] and others using coat color in pet mice as the unit characters. Since that time, visible mouse mutants have often been preserved, even if the mutation appeared to be bizarre and have no obvious biomedical significance.

The development of inbred strains of mice by C.C. Little and rats by Wilhelmina Dunning both in 1909 and guinea pigs by G.M. Rommel, later taken over by Sewall Wright, starting in 1906, was a major advance, because for the first time pure-breeding lines became available.

Several of these strains, and ones developed by other investigators in this early period, are still available and are widely used in biomedical research. Inbreeding, usually due to many generations of brother and sister mating, has the effect of fixing the genotype within a strain, while maximizing the differences between strains. Some of these mouse strains were selected for a high incidence of various types of cancer, such as mammary tumors in C3H, and DBA, leukemia in strain AKR, and lung tumors in strain A. These strains were widely used in cancer research.

As early as 1903, it was found that tumors could often be transplanted within the strain in which they originated, without being rejected, but they were usually rejected when transplanted into a different strain. These early studies had been done using Japanese waltzing mice, which had apparently become inbred accidentally by fancy mouse breeders. Subsequent studies by Little and Tyzzer[3] showed that tumor rejection was dependent on a number of genes with a dominant mode of inheritance.

George Snell, at the Jackson Laboratory in Maine, continued these studies and identified some of the gene loci responsible for this rejection by backcrossing the ability to resist tumor grafts from one strain into a strain where the grafts would normally be accepted (called the inbred partner). His so-called "congenic-resistant" stains were usually found to differ from the inbred partner at a single genetic locus. In most cases, this was what is now known as the major histocompatibility complex (MHC), also designated the *H2* locus in mice. Not only did Snell identify this important locus, but his work also promoted the use of backcrossing to an inbred strain as a method of fixing the genotype in order to provide stable material for further study. Snell and his generation of transplantation immunologists developed several hundred of these congenic-resistant strains, which are still widely used in research involving the immune system. Congenic strains developed by backcrossing mutants and, more recently, transgenes to an inbred genetic background are now widely used.

Another significant advance was the development of sets of "Recombinant Inbred (RI) Strains" by Donald Bailey in 1971.[4] He crossed two standard inbred strains and then brother × sister mated the offspring so as to produce a whole set of new inbred strains in which the genes from the parental strains had recombined. His first set consisted of seven strains derived from a cross between inbred strains BALB/c and C57BL/6By. Larger sets of strains were later developed by B.A. Taylor,[5] and were used extensively first to determine whether a given phenotypic (i.e., observed) difference between strains could be attributed to a single genetic locus, and if so, whether it was linked to any known genetic markers. These sets of RI strains have been used to identify and map gene loci, which were polymorphic between the two parental inbred strains. Their use is discussed in more detail below.

RI strains are quite good for resolving the genetics of characters controlled by one or two loci, but they have some limitations for studying many characters with a polygenic mode of inheritance (i.e., where the phenotype depends on the joint action of several gene loci as well as nongenetic factors). This led Demant[6] to develop sets of "recombinant congenic" strains specifically for this task. These sets of about 20 strains typically differ from an inbred partner by about 12.5% of those loci originally poly-morphic between a donor and inbred partner strain. Though not yet widely used, they illustrate how specific strains can be developed as useful tools in biomedical research.

An important feature of this classical period was the development of flexible and adaptable nomen-clature rules for inbred strains, genetic loci, alleles, mutants, and chromosomes. These are administered by international nomenclature committees for mice and rats, which try to ensure that the same strain, mutant, etc., is not named differently by different investigators, or conversely, that different things do not end up with the same name. With rats, three competing nomenclature systems for the rat MHC managed to become established, and it took considerable effort for the rat nomenclature committed to reconcile them into a new system.

The maintenance of ever-increasing numbers of mouse and rat strains is expensive in terms of space and scientific resources. This has been alleviated to some extent by the development, toward the end of this classical period, of methods of freezing mouse embryos.[7] These can be maintained in liquid nitrogen for many years, saving considerably on maintenance costs and space. The development of associated methods of handling preimplantation embryos has also had important consequences for the development of transgenic strains.

The development of molecular techniques from the 1980s, largely driven by the Human Genome Project, has had a number of important consequences. The cloning and sequencing DNA taken together with the development of methods for handling early embryos led to the development of methods for producing transgenic strains following the injection of foreign DNA into the pronucleus of early embryos.[8] Later, embryonic stem cells lines were developed from cultured early embryos.[9] These could be maintained and manipulated as cell cultures, allowing gene targeting by homologous recombination in order to develop "knockout" mice, in which a specific gene was inactivated.[10] This has proved to be a powerful tool in finding out the functions of many genes, but it has also resulted in a remarkable proliferation of new strains.

Microsatellites are short repetitive DNA sequences with unique flanking sequences. These are widely distributed throughout the mouse and rat genomes and provide a large number of genetic markers, which have been used for genetic mapping and genetic quality control. These markers can also be used to map so-called "quantitative trait loci" (QTLs), which are loci controlling the inheritance of many complex characters, such as susceptibility to many toxic agents and diseases such as cancer, diabetes, and various aspects of behavior. Once these genes have been mapped to a general chromosomal location, they need to be identified. This usually requires extensive backcrossing programs using the methods used by Snell in developing his congenic-resistant strains. Unfortunately, such backcrossing takes two or three years, but by using an array of microsatellite markers, "speed-congenics" (see below) can be developed in about half the time, though at some cost in organization, testing, and reagents.

Keeping track of the vast amount of genetic information now being generated on gene sequences, polymorphic genetic markers, genetic maps, new strains, and phenotypes would have been impossible without the parallel development of informatics, which developed as a separate discipline from about the mid 1990s. The full impact of the World Wide Web has yet to be felt, though it is already substantially altering the way that science is done and communicated. Already, there are a large number of Web sites offering information and resources at locations throughout the world. Useful Web sites of potential value to mammalian geneticists include: Quantitative Genetics Resources (http://nitro.biosci.ari-zona.edu/zbook/book.html), World of Genetics Societies (http://www.faseb.org/genetics/), genet-ics-related Web sites (http://www.sidwell.edu/sidwell.resources/bio/VirtualLB/bioIweb.html), Genetics Education Center (http://www.kumc.edu/gec/), statistical genetics Web sites (http://www.rdg.ac.uk/~sns99kla/links.html), and animal behavior Web sites (http://www.societ-ies.ncl.ac.uk/asab/websites.html). A computer program, "MICE," which is used for automation of breed-ing records and is distributed free of charge to academic institutions, is also available through the Web (www.biomedcentral.com/1471–2156/2/4).

## THE MAJOR CLASSES OF GENETIC STOCKS

Table 9.1 classifies the different genetic types of laboratory mice and rats into a *genetically undefined* group, which includes outbred stocks, genetically heterogeneous stocks, and segregating hybrids and

**Table 9.1    The Major Classes of "Genetic" Stocks**

**Genetically Undefined**
Outbred stocks
Genetically heterogeneous stocks
Outbred selected stocks
Segregating hybrids
Advanced intercross lines
**Partially Genetically Defined**
Mutants on an outbred background
Transgenes on an outbred background
Inbred strains in development
**Isogenic Strains**
Inbred strains
Congenic strains
Recombinant inbred strains
Recombinant congenic strains
Consomic strains
F1 hybrids
Segregating inbred strains
Clones
Monozygous twins

outbred selected stocks. The genotype of any individual of this class of stock will be unknown unless the animal is individually genotyped at loci of interest. The *partially isogenic* group consists of stocks carrying mutants and transgenes that have a genetically heterogeneous background but in which the genotype at some loci is known, and of inbred strains in development. Finally, the *isogenic family* includes inbred strains, F1 hybrids, congenic strains, and others given in Table 9.1. The properties of these three main classes of stock are discussed below.

Most of the larger species of laboratory animals, such as primates, dogs, and cats, are only readily available as outbred stocks. However, there are a few colonies that carry mutations on an outbred background or have the genotype defined at some loci, such as the major histocompatibility complex. There are also some mutant stocks and inbred strains of rabbits, guinea pigs, hamsters, gerbils, and a few other species. However, with all these species, the characteristics, research uses, and maintenance are essentially the same as for mice.

## Genetically Undefined Stocks

### Outbred Stocks

An outbred stock is a breeding group of genetically heterogeneous animals that are usually maintained as a closed colony, without the introduction of animals from another stock or strain. The degree of genetic variation depends on the previous history of the colony. Colonies founded with a small number of pairs or that are maintained with relatively few breeding animals per generation will tend to lose genetic variation, and in some cases, the colony may become moderately or highly inbred. Outbred stocks are available for all species of laboratory animals.

### Research Uses

Outbred stocks continue to be widely used, even though for those species where there is a choice (mainly mice and rats), there is a compelling case for preferring isogenic strains.[11,12] For other species, there is generally no practical alternative. (The term "stock" is used for outbred colonies, with the term "strain" being reserved for isogenic strains, although in this chapter, in some cases, the term "strain" will be used collectively to mean all genetic types. The context should make it clear in which sense the term is being used.)

These animals have some advantages. They are cheaper to buy or breed, partly because they are more prolific than isogenic strains and partly because there are fewer outbred stocks, so they can be bred on a large scale with relatively less wastage. As a result, they tend to be more readily available in groups of a defined weight or age range.

The disadvantages are that each animal is genetically unique, so there is no information on the genotypes of individuals unless each is specifically genotyped. Phenotypic variation is usually greater than is found with isogenic strains, as individuals differ due to genetic and nongenetic factors. This means that more animals are usually needed to achieve a given level of statistical precision in a controlled experiment than if isogenic animals had been used. Also, stocks are subject to genetic change as a result of inbreeding and directional selection. Over a period of many generations, outbred mice and rats have tended to be selected for increased body size, so are often larger than their isogenic counterparts. Genetic change can occur rapidly over a period of a few generations, and this may mean that background data on the phenotypic characteristics of the stock may quickly lose its validity. In particular, there is good evidence that colonies with the same name such as "Wistar" rats are often genetically different. This can cause difficulties for people needing to replicate other people's work, as nominally identical animals may in fact be genetically distinct. Unfortunately, methods of genetic quality in outbred stocks are undeveloped. It is not even possible to distinguish between Wistar and Sprague-Dawley rats using known genetic markers.

Outbred stocks are mainly used in general research for studies using species in which there is no isogenic alternative, studies where genotype is judged to be of little importance, in noncritical studies, and in cases where the research worker is unaware of the advantages of isogenic strains. They are widely used in toxicological testing, though their use in such studies has been questioned.[13,14] Where animals of a broad range of genotypes are wanted, it is much better to use small numbers of isogenic animals of several strains rather than outbred animals of a single stock. However, they are suitable for within-animal experiments, such as crossover experimental designs or those comparing left and right sides, as in this case, precision is not affected by differences between animals. They may also be suitable for experiments requiring sources of live tissue and for the maintenance of parasites, assuming host–parasite relationships are not of interest. They continue to be widely used in neuroscience, though this may be due more to ignorance of the advantages of isogenic animals, rather than any good scientific argument for their continued use. A discussion of the relative merits of isogenic and nonisogenic stocks in aging research, which is relevant to many other fields of research, is given in Miller et al. (1999) and Festing (1999).[11,15]

## Nomenclature

Official rules for the nomenclature of outbred stocks were published by ICLA (the International Council on Laboratory Animals, now ICLAS, the International Council on Laboratory Animal Science) in 1972.[16] Stocks should be closed colonies for at least four generations and should be maintained with less than 1% inbreeding per generation (see below). They should be designated by a code consisting of uppercase letters or letters and numbers, starting with a letter. Where a stock was already known by a designation, which included other characters, it should retain its existing designation. This code should be preceded by a laboratory code that is the same as the one used for inbred strains (see below), with the code and strain designation being separated by a colon, e.g., Hsd:WIST, a stock of Wistar rats designated WIST and maintained by Harlan Sprague-Dawley.

In practice, the use of such standard designation is voluntary, as there is no international organization that currently takes a particular interest in the genetics and nomenclature of outbred stocks.

## Breeding and Maintenance

The aim in maintaining an outbred colony is usually to prevent change. In exceptional circumstances, the aim may be to change the colony in some way, and this can usually be done by selective breeding.

In any colony, genetic change occurs as a result of change in the frequency of alleles at the various genetic loci that are polymorphic in the colony. Within an outbred stock, individual animals will often differ at many thousands of loci. If the frequency of individual alleles changes, then the phenotype of the colony may also change. There are four main causes of genetic change in an outbred stock, as listed below.

*Genetic Contamination or Immigration* — An outcross is sometimes done deliberately if the breeding performance of a colony declines. However, the introduction of new genetic material may alter the characteristics of the colony. Genetic contamination may also occur by accident if the colony is maintained in physical proximity to another strain or stock.

*Mutation* — Mutations occur at the rate of about one in 100,000 to one in 1,000,000 per locus, and some of these will become established within the colony, depending on chance and whether or not they are deleterious or advantageous. Recessive genes may be maintained within the colony for many generations, even if they have large phenotypic effects. For example, the Rowett athymic nude rat mutation apparently occurred at some time prior to 1955 in a colony of outbred rats at the Rowett Research Institute in Scotland. It was maintained in the colony in a heterozygous state until 1978, when two heterozygotes were mated and produced hairless young.[17]

*Directional Selection* — If the stock is selected for some phenotype such as large body weight, high reproductive performance, or enhanced immune response to some antigen, then the frequency of alleles associated with these characters will change, thereby altering the characteristics of the colony. The rate of change will depend on the strength of selection and the extent to which the character is inherited. Many characters are correlated so that a change in body weight may lead to a change in other characters, such as life span and tumor incidence. Natural selection may also alter some characters. If there is an infection in the colony, those animals, which are most resistant, will tend to produce more offspring so that the frequency of "resistance" genes will increase. If there is a change in husbandry or environment, then those animals that thrive under the new conditions will tend to leave more offspring, and so on. Thus, in order to prevent genetic change, the colony should be maintained as far as possible without directional natural or artificial selection. This is usually achieved by using some procedure for selecting future breeding stock strictly at random, without taking any account of their phenotypic characteristics, though obviously abnormal animals would normally be excluded. If there are major deleterious genes present in the colony, such as blood clotting factors in dogs, then steps should be taken to eliminate affected animals and carriers using progeny testing methods described in many genetics textbooks.[18,19] Where single loci have been cloned and sequenced, it will often be possible to develop a PCR-based method for identifying carriers,[20] which should make it easy to eliminate them from the breeding stock.

*Random Drift* — Genetic segregation generates new combinations of genes in an essentially random manner, and in a large outbred stock colony, the frequency of an allele at any given genetic locus should remain constant, according to the Hardy–Weinburg law.[19] However, in smaller populations, gene frequency can change simply as a result of chance selection of breeding animals with certain alleles at any given genetic locus. Once an allele becomes fixed in the colony, with all animals being genetically identical at that locus, it will no longer be able to change. The rate of change due to random genetic drift depends on the level of inbreeding, which in turn, depends on the size of the breeding colony and on the amount of genetic variation already present. The coefficient of inbreeding ($F$) is the probability that the two alleles at a locus are identical by descent, i.e., that they are copies of the same gene at some previous period in the animal's ancestry. $F$ ranges from zero in a colony with no inbreeding to 1.0 or 100% in a fully inbred one. The inbreeding per generation, assuming random mating, is given by the following formula:

$$\Delta F = 1/8Nm + 1/8Nf$$

where $\Delta F$ is the increase in inbreeding per generation, and $Nm$ and $Nf$ are the numbers of males and females present in the breeding colony, which actually or potentially can leave offspring in the next generation. Note that the rate of inbreeding depends largely on the number of the most numerous sex. For example, if the colony has four breeding males, then the inbreeding with random mating will be

$1/32 = 3.1\%$ just from the males' side, however many females there are. This is important when considering colonies of larger animals, such as dogs, cats, and some species of primates, where a few stud males can be used for large numbers of females. In such circumstances, it is often necessary to maintain more males than would normally be needed in order to reduce the rate of inbreeding.

Inbreeding only reduces existing heterozygosity each generation. Over a period of time, the total inbreeding in the colony will be as follows:

$$\Delta F_t = \Delta F + (1-\Delta F)F_{t-1}$$

where $\Delta F_t$ is the inbreeding at generation $t$, and $\Delta F$ is the inbreeding for the current generation.

If there is a genetic "bottleneck" with a reduced number of breeding animals in one or a few generations, this will lead to an increase in inbreeding that cannot be undone, even if the colony is subsequently enlarged. Formulas taking into account variable sample sizes each generation are given by Falconer.[19]

Bottlenecks are common when a few breeding animals are used to found a new breeding colony, or when a colony is "cleaned up" following an outbreak of disease. Care must be taken to ensure that enough breeding animals are used.

As a general rule, it is recommended that inbreeding levels should not be more than about 1% per generation if the aim is to maintain the colony for long periods. Table 9.2 shows the inbreeding per generation from colonies maintained with various numbers of breeding individuals, and Figure 9.1 shows the coefficient of inbreeding over a period of 30 generations in colonies of various sizes.

Inbreeding over a period of time, such as five years, can be reduced by having a long generation interval. With mice and rats, it may be possible to have only about two generations per year by saving breeding stock from older females. Breeding from first litters can be avoided. With larger animals such as dogs and cats, there is even more scope for reducing the inbreeding by increasing the intergeneration interval.

A "maximum avoidance of inbreeding" system can also be used with small colonies, which approximately halves the rate of inbreeding. With random mating, on average, each breeding pair will contribute one breeding male and one breeding female to the next generation. However, some animals will contribute more and some less due to sampling variation. If each breeding female and each breeding male were to contribute exactly one female or male, respectively, to the next generation, this will halve the rate of inbreeding relative to random mating. There are various rotational breeding schemes that aim to ensure this happens. One of these is given in Table 9.3. However, in practice, there is little difference between the different rotational breeding schemes.[21] All rely on ensuring equal representation of the current generation of breeding animals in the next generation. It is not worthwhile to use a maximum avoidance of inbreeding system if the level of inbreeding is already well below the recommended 1% per generation. Halving something, which is already very small, may not be worthwhile. It is also usual to try to avoid

**Table 9.2  Inbreeding with Various Numbers of Breeding Males and Females**

| Number of Males | Number of Females | Inbreeding per Generation | Inbreeding after 10 Generations |
|---|---|---|---|
| 4 | 4 | 6.25 | 47.5 |
| 4 | 8 | 4.69 | 38.1 |
| 4 | 24 | 3.65 | 31.0 |
| 8 | 8 | 3.13 | 27.2 |
| 8 | 16 | 2.34 | 21.1 |
| 16 | 16 | 1.56 | 14.6 |
| 16 | 32 | 1.17 | 11.1 |
| 32 | 32 | 0.78 | 7.5 |
| 32 | 64 | 0.59 | 5.7 |
| 100 | 100 | 0.25 | 2.5 |
| 100 | 200 | 0.19 | 1.9 |

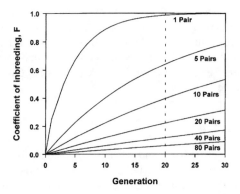

**Figure 9.1**    Coefficient of inbreeding over 30 generations with various numbers of breeding pairs.

Table 9.3   A Rotational System for Maximum Avoidance of Inbreeding[a]

| New Cage Number | Male from Old Cage Number | Female from Old Cage Number |
|:---:|:---:|:---:|
| 1 | 1 | 2 |
| 2 | 3 | 4 |
| 3 | 5 | 6 |
| 4 | 7 | 8 |
| 5 | 9 | 10 |
| 6 | 11 | 12 |
| 7 | 2 | 1 |
| 8 | 4 | 3 |
| 9 | 6 | 5 |
| 10 | 8 | 7 |
| 11 | 10 | 9 |
| 12 | 12 | 11 |

[a] It is assumed that the colony consists of 12 cages, and that these are to be replaced with a new generation of the same number of cages.

*Note:*   The breeding scheme ensures that each breeding male contributes one male, and each breeding female contributes one female to the next generation.

brother × sister mating, although in the long run, this does not matter, because any extra inbreeding that occurs as a result of mating closely-related individuals will be undone again in subsequent generations.

In summary, if the aim is to prevent genetic change, outbred stocks should be maintained as large randomly mated colonies with less than 1% inbreeding per generation, in which breeding individuals have been selected at random and without consideration of their phenotype, in order to avoid directional selection. Where it is not possible to maintain sufficiently large colonies, maximum avoidance of inbreeding should be used by ensuring that as far as possible, each breeding female contributes one female and each breeding male contributes one male to the next generation of breeding stock.

### Genetically Heterogeneous Stocks, Segregating Hybrids, and "Advanced Intercross Lines"

Genetically heterogeneous (GH) stocks have been synthesized by crossing several inbred strains and maintaining the offspring as an outbred stock. Such stocks have been used as a base population for selection experiments and in a range of behavioral studies.[22,23,24] They have also been used for mapping quantitative trait loci (QTLs), where over a period of many generations, there will have been considerable crossing over among the chromosomes, thereby substantially reducing linkage disequilibrium, i.e., the fragment of chromosome containing the QTLs and marker genes will be much shorter. This means that each QTL can be mapped more accurately than is possible using F2 hybrids.[25]

Segregating hybrids or advanced intercross lines (AILs)[26] is essentially similar, except that they tend to be produced by crossing just two inbred strains, followed by a random or semirandom mating system to avoid inbreeding. They are also used for genetic mapping.[27] These stocks should be maintained in the same way as outbred stocks, with large population sizes, and maximum avoidance of inbreeding procedures where necessary in order to minimize loss of heterozygosity.

### Outbred Selected Stocks

Genetic selection with the maintenance of genetic heterogeneity has been used by several investigators as a means of producing new animal models. Examples include the Biozzi mice selected for high and low immunological response to sheep red blood cells,[28] SENCAR mice selected for susceptibility to skin carcinogenesis,[29] the Dahl rats selectively bred for high or low blood pressure on a high salt diet,[30] and several stocks of mice selected for various aspects of response to alcohol.[31] This method of selection with the avoidance of inbreeding is more efficient and likely to result in a more substantial change in phenotype than the alternative method of selection with inbreeding that has sometimes been used.

In some cases, the original selected stock continues to be used in research, while in others, one or more inbred strains have been developed from the outbred selected stock. It is important for breeders to know whether a particular colony is of the original outbred stock, in which case, it should be maintained in the same way as other outbred stocks, described above, or whether it is an inbred strain, in which case, it should be maintained using methods described below.

### Partially Genetically Defined

Mutants may occur within an outbred stock, and with larger animals, transgenes will usually be developed in outbred stocks. Similarly, in rodents, outcrossing in order to obtain better breeding performance is quite common in the development of transgenic animals. In these situations, the breeder needs to consider how the stock is to be maintained in the future.

With large animals or species for which inbred strains are not available, there is little option but to maintain the mutation/transgene on the outbred background, taking account of the possible need to avoid genetic drift and inbreeding, which could reduce breeding performance. With mice and rats, a choice is possible. The stock can be maintained as an outbred stock, the colony could be inbred, or the mutation/transgene could be backcrossed to an existing inbred strain. Scientific considerations need to be taken into account. The expression of most mutants and transgenes is influenced by the genetic background. If this background is heterogeneous, the expression of the mutation will also be variable. Moreover, if the genetic background drifts because the colony could not be maintained so as to avoid inbreeding and genetic drift, or if outcrossing to another strain is used to boost breeding performance, then the expression of the mutation may vary over time. For this reason, most geneticists would recommend that the mutation/transgene be transferred to an inbred, or possibly hybrid, background by backcrossing, as described below in the development of a congenic strain.[32] Once a congenic strain has been developed, its characteristics will be fixed, and it can be maintained in small numbers without problems of genetic drift.

### The Isogenic Family of Strains and Their Derivatives

This family of strains includes inbred, segregating inbred, congenic, recombinant inbred, recombinant congenic, consomic, and F1 hybrid strains, as well as monozygous twins and genetic clones. These latter two classes are not discussed here. The most important feature of this class of strains is that at least two individuals, and usually many more, are genetically identical or nearly identical. All of these strains, with the exception of twins and clones, are based on inbred strains discussed below.

### Inbred Strains

These are produced by at least 20 generations of brother × sister mating or its genetic equivalent. For example, parent × offspring mating is one alternative, provided the mating is always to the younger of the two parents (several generations of mating female offspring to the same male is not the same thing, genetically). The effect of this mating scheme is to increase homozygosity within the strain to more than 98% (Figure 9.1), though the approach to complete homozygosity is asymptotic, and in theory, a strain never becomes fully inbred. All individuals within an inbred strain should also trace back to a single breeding pair in the 20th or a subsequent generation. This is to ensure that the strain is also isogenic (see below) by eliminating parallel substrains. Inbreeding also increases the total genetic and phenotypic variation, but all this variation is seen as differences between the substrains. Thus, if an outbred stock is inbred and all sublines are kept, the total phenotypic variation is substantially greater than in the original colony.

## *Properties of Inbred Strains*

Inbred strains have been described as "immortal clones of genetically identical individuals," and as such, have many useful properties that make them the animal of choice, where they are available, for many types of research. The main properties of these strains are as follows:

1. Isogenicity — All animals within a strain are virtually genetically identical. One consequence is that only a single individual needs to be genotyped at any locus in order to type the whole strain. Over a period of time, a catalog or "genetic profile" of the alleles carried by each inbred strain can be accumulated. This information can be used in planning and interpreting experiments and in mapping genes of interest. Isogenicity also implies that a single male and female taken from the colony should have all the alleles present in that colony. Hence, a daughter colony founded on a single breeding pair will, for most practical purposes, be genetically identical to the parent colony, at least until the colonies begin to diverge as a result of the accumulation of new mutations (see below). Isogenic individuals will also be immunologically histocompatible, so that skin, cell, and organ grafts exchanged between same-sex members of the same strain should not be immunologically rejected.

2. Homozygosity — Inbred strains are defined in terms of homozygosity. By the end of 20 generations of full sibling mating, the chance that any two alleles at a given locus are identical by descent (i.e., are copies of the same allele in a previous generation) is more than 98%. The most important practical consequence of this is that there should be no genetic segregation within the strain, so all genes will be expressed under appropriate conditions, and there will be no hidden recessive genes, which could cause confusion in breeding experiments.

3. Phenotypic uniformity — As there is no genetic variation within an inbred strain, the phenotype for highly inherited characters tends to be more uniform. The only variation between individuals will be due to nongenetic causes. One consequence is that, other things being equal, fewer inbred animals will be needed to achieve a given level of statistical precision than if outbred animals had been used. In some cases, the use of isogenic animals can lead to substantial reductions in the estimated numbers of animals needed to do a particular experiment. Table 9.4 shows the mean and standard deviation of sleeping time under hexobarbital anesthesia in five inbred strains and two outbred stocks of mice[33] and the estimated sample sizes that would be required in an experiment to detect a 4 min change in sleeping time as a result of some experimental manipulation. Note that this experiment could be done using an average of 18 inbred mice in each group or 244 outbred mice to achieve the same level of statistical precision. Clearly, for this reason alone, there is a strong case for using isogenic strains.

4. Long-term stability — In an outbred stock, change in allele frequency and, therefore, in phenotype, can be caused by directional selection, genetic drift due to inbreeding, and new mutations, assuming genetic contamination is avoided. An inbred strain is already fully inbred, so further inbreeding will have no effect. As there is no genetic variations within the strain, directional selection should be ineffective in changing the genotype and phenotype. Thus, the phenotype of an inbred strain will only change as a result of the fixation of new mutations or as a result of environmental changes (which can often be of great importance). New mutations are relatively rare, and only a quarter of them will normally be fixed with continued full sib mating, so inbred strains tend to stay genetically constant for quite long periods of time. Even the low amount of genetic drift due to new mutations can be eliminated by preserving frozen embryos. In a few cases, directional selection

**Table 9.4 Sleeping Time Under Barbiturate Anesthetic in Five Inbred Strains and Two Outbred Stocks of Mice, and Number of Mice Estimated to be Needed to Detect a Change in Sleeping Time in an Experiment Involving Control Mice and Those Treated with a Compound Thought to Alter Sleeping Time**

| Strain | n | Mean | SD | Number Needed[a] | Number Needed[b] |
|---|---|---|---|---|---|
| A/N | 25 | 48 | 4 | 16 | 23 |
| BALB/c | 63 | 41 | 2 | 7 | 7 |
| C57BL/HeN | 29 | 33 | 3 | 19 | 13 |
| C3HB/He | 30 | 22 | 3 | 41 | 13 |
| SWR/HeN | 38 | 18 | 4 | 105 | 23 |
| CFW | 47 | 48 | 12 | 144 | 191 |
| Swiss | 47 | 43 | 15 | 257 | 297 |

[a] Power analysis: number needed in a two-sample t-test to detect a 10% change in the mean (two-sided) with a 5% significance level and a power of 90%.
[b] Number needed to detect a 4 min change in sleeping time with a 5% significance level and a power of 90%.

may apparently alter strain characteristics, but this may be acting through associated microflora rather than through the genetics of the inbred strain. This long-term stability has important consequences. It means that genetic profiles will remain unaltered for long periods and phenotypic profiles of strain characteristics such as life span, types of spontaneous disease, immune functions, susceptibility to microorganisms, and biochemical and physiological characteristics can be built up by many research workers over a long period of time, provided the environment is also constant. However, some aspects of the environment, such as the associated microorganisms, diet, bedding, and social factors can alter strain characteristics, sometimes to quite a marked extent.

5. Individuality — Each strain is represented by a unique combination of alleles that confer a particular set of phenotypic characteristics. Sometimes these characteristics are useful in research, while at other times, they may preclude the use of a particular strain for a given research project. Many strains of mice were developed for use in cancer research and have high levels of a particular type of tumor. For example, strain AKR gets leukemia, SJL gets reticulum cell sarcoma, and C3H gets mammary tumors provided it carries the mammary tumor virus. However, other strains such as C57BL/6 have relatively low levels of cancer and tend to be relatively resistant to carcinogens. Differences in virtually any phenotypic characteristics are likely to be found among a group of independent strains. Many aspects of behavior such as open-field activity, wheel running, aggression, and learning ability differ between strains. Similarly, response to bacteria and viruses often varies dramatically between different strains. These strain differences have enormous importance, as strains can be chosen that have characteristics which make them useful for any given type of research. In extreme cases, whole areas of research have been opened up, based on the characteristics of a single mouse strain. For example, monoclonal antibody technology is all based on the plasmacytoma tumors induced in BALB/c mice by intraperitoneal injection of mineral oil, and the technology for gene targeting to produce "knockout" mice by homologous recombination, in which a single defined gene is inactivated, is based on the establishment of embryonic stem cell lines from strain 129 mice.

6. International distribution — The isogenic property of inbred strains means that daughter colonies can be set up using only a single breeding pair, and because of the long-term stability of inbred strains, these will remain similar to the parent colony for many generations. This means that work can be repeated using similar animals throughout the world. In contrast, although outbred stocks such as Wistar rats are available all over the world, genetic drift and directional selection mean that each colony is, to some extent, genetically different.

7. Identifiability — The isogenic property makes it possible to build up a genetic profile of each strain. This can then be used for genetic quality control purposes. Thus, if some white rats are supposed to be strain F344, this can be tested from a small sample of DNA using one of the methods discussed below (see genetic quality control). However, if the rats are supposed to be outbred Wistars, there is no known method of genetic quality control that will distinguish Wistars from Sprague-Dawleys.

8. Sensitivity — As a broad generalization, inbred strains tend to be more sensitive to environmental influences than outbred stocks or F1 hybrids (see below). This is a disadvantage in that it means that extra care is needed to ensure that they have the optimum environment. If not, they will tend to be more variable. However, it is an advantage in that they will also tend to be more sensitive to experimental treatments than other classes of stock.

## Nomenclature of Inbred Strains

The rules governing genetic nomenclature for mice and rats are formulated and maintained by international committees made up of research workers with an interest in the genetics of these two species. The rules are subject to constant revision as knowledge of genetics increases and as different types of animals become available. For example, transgenic strains were not envisioned at the time the rules were originally formulated. Full details of mouse genetic nomenclature are given on the Jackson Laboratory Web site (www.informatics.jax.org) and of rats and mice on the Rat Genome Database Web site (www.rgd.mcw.edu).

Briefly, inbred strains of mice and rats are designated by a code consisting of uppercase letters (e.g., SJL, LEW) or (less preferably) letters and numbers (e.g., C57BL, F344), except for strains that were already known by some other designation at the time that the rules were formulated. Short symbols are preferred, and duplicate designations must be avoided. Strains do not have names, only designations. Many people using rats seem to want to name them, sometimes with names that are quite inaccurate. For example, the DA strain of rats is sometimes called the "Dark Agouti" strain, though it is not dark, and the "D" in the designation actually stands for the D blood group (now an obsolete nomenclature), not for the word dark. A name is confusing, as it does not conveniently fit with the rest of the genetic nomenclature, particularly with the need to designate substrains. Fortunately, users of inbred mice seem to accept designation codes rather than names.

Where necessary, the level of inbreeding can be designated by an F followed by the number of generations of full sib mating, e.g., F87, or F?+50 in the case of a strain with unknown inbreeding followed by 50 known generations of sib mating.

Substrains can arise when branches of an inbred strain have been separated after 20 generations but before 40 generations of inbreeding, or when branches have been maintained separately for 100 or more generations from their common ancestor. In both cases, some divergence is expected. Substrains are also formed when genetic differences between branches have been found. A substrain is indicated by a slash and a number, a laboratory registration code, or a combination of the two. For example, FL/1Re, FL/2Re are two substrains of strain FL, both established in the laboratory with code Re (see below). Some exceptions, such as the "c" in BALB/c have been permitted for strains already known using a different designation.

Laboratory registration codes are administered by the Institute of Laboratory Animal Research (ILAR), and codes can be registered directly on their Web site (www4.nas.edu/cls/ilarhome.nsf). They are used, as above, to indicate substrain differences and also to remind people that genetically identical animals raised in different environments, such as found in different laboratories, may differ phenotypically due to nongenetic causes.

## Research Uses of Inbred Strains

Inbred strains are, or should be, the animal of choice for general research using mice or rats. They are the nearest thing to a pure reagent that is possible when using laboratory animals. As noted above, a strain represents a single genotype, which can be repeated by the thousands. It stays genetically constant for long periods, there is considerable background information on the genetic and phenotypic characteristics of the more common strains, the phenotype tends to be uniform, and genetic quality control is relatively easy using a wide range of DNA genetic markers. If the project requires a screen of animals with a wide range of potential genetic susceptibility to a toxic or pharmaceutical agent or microorganism, then small numbers of animals of several strains can be used. This is much more effective than using an outbred stock, which in fact, is often phenotypically quite uniform and does not necessarily have a wide range of susceptibility genotypes. There are compelling arguments for using this approach.[11,12,13,34]

**Table 9.5  Estimated "Top 10" Inbred Strains[a] of Mice and Rats**

| Rank | Mouse Strain | Rat Strain |
|------|--------------|------------|
| 1 | BALB/c | F344 |
| 2 | C3H | LEW |
| 3 | C57BL/6 | SHR |
| 4 | CBA | WKY |
| 5 | DBA/2 | DA |
| 6 | C57BL/10 | BN |
| 7 | AKR | WAG |
| 8 | A | PVG |
| 9 | 129 | BUF |
| 10 | SJL | WF |

[a] C57BL was separated into two major substrains.

Inbred strains are also of increasing importance as the "genetic background" for many mutants, "knockouts," and transgenic alterations. This is discussed in more detail below (see congenic strains). At least 17 Nobel prizes have been awarded for work that required inbred strains or would have been much more difficult without them.[35]

There are over 400 "straight" inbred strains of mice and about 200 inbred strains of rats, excluding congenic, recombinant, etc., strains discussed below, each of which is an inbred strain in its own right. The origin and phenotypic characteristics of these strains is available on the Jackson Laboratory and Rat Genome Database Web sites. Extensive genealogies of inbred mouse strains have also been developed.[36] However, probably 80% of research using inbred strains is done using about ten to 15 of the most common strains. The estimated "Top Ten" strains of mice and rats, according to the known number of laboratories maintaining each strain,[37] are listed in Table 9.5

There are two main reasons for choosing a strain such as C57BL/6 mice or F344 rats for a particular research project. The strain may be chosen because it is regarded as a good general-purpose strain, which can be used to replace the use of an outbred stock such as Swiss mice or Wistar rats. In this case, a strain would be chosen that has no known characteristics which would preclude its use for the project. Strains with a high incidence of a specific disease, which were highly aggressive, or which had some known immunological defects, for example, might not be suitable for the project. When choosing a suitable strain for a new project, several available strains could be screened to find one that showed an appropriate response to the treatment of interest. The project could then be continued with that strain, with occasional studies of other strains just to confirm that the results were not highly strain specific.

In contrast, a specific strain may be chosen because it has characteristics that make it useful for a particular type of research. For example, C57BL/6 mice maintained on a high-fat diet develop atherosclerosis, so they may be used as a model of that condition in humans.[36] C57BL/6 mice also like sweet tastes and alcohol and are highly active in an open field. A searchable database of these sorts of phenotypic characteristics of mouse and rat strains is available on the Web. Many strains are used for studying QTLs associated with phenotypes ranging from susceptibility to cancer to various aspects of behavior, growth, metabolic diseases, and immune function. A database of genetic markers in 55 mouse strains (http://www.cidr.jhmi.e.,du/mouse/mouse.html) developed at the Center for Inherited Disease Research makes it extremely easy to pick informative genetic markers in proposed crosses between any two strains.

### Breeding and Maintenance of Inbred Strains

The usual aim in maintaining a colony of laboratory animals is to prevent genetic change so that information on the characteristics of the strain can be accumulated and used in the planning and interpretation of future experiments. Provided genetic contamination due to a mating with an animal of another strain can be ruled out, the only way in which inbred strains can change is as a result of the accumulation of new mutations or from residual heterozygosity; the segregation polymorphisms still remaining in the strain after 20 generations of inbreeding. An economical and flexible breeding program that can rapidly respond to changes in demand is also wanted.

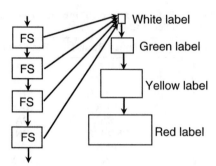

**Figure 9.2**    Diagram of the "traffic light" system for controlling the foundation stock (FS) and multiplication colonies of an inbred strain. The FS colony is self-perpetuating, and any surplus stock can be transferred to the multiplication colony in breeding cages with a white label. Their offspring can be used for a further three generations of multiplication, but the offspring of the red label colony are not used for breeding.

Unlike outbred stocks, inbred strains can be maintained with very small numbers of breeding stock. Inbred mice and rats are usually maintained as permanently mated monogamous pairs, and the minimum-sized breeding colony is one that just prevents the colony from dying out, given that breeding performance is often somewhat uncertain, particularly with some strains. Where large numbers of animals are needed for research purposes, an appropriate breeding scheme is to maintain a "stem-line" colony of about ten to 30 breeding pairs, with a multiplication colony of sufficient size to provide all the required experimental animals as a result of up to about four generations of breeding. The multiplication colony is used only to produce experimental animals and does not contribute to the long-term survival of the strain. This breeding scheme is shown diagrammatically in Figure 9.2.

The stem-line colony should be maintained by brother × sister mating, usually as pairs, and should, as far as possible, be kept physically separated from all other animals of the same species. Some breeders maintain such colonies in isolators or as frozen embryos. Genetic quality control methods (see below) should be used to authenticate the strain and monitor it over a period of time. Detailed records on each breeding pair should be maintained, as well as a pedigree chart showing the relationship between the pairs. These can be paper or computer-based records. Generally, all breeding pairs should be descended from a single breeding pair about five to seven generations back. Although directional selection within an inbred strain should have no effect, it is advisable to select for good breeding performance in order to try to prevent it from declining as a result of the accumulation of new deleterious mutations. Often, such mutations will have a very small effect, and the best method of selection is probably to base it on the average performance of the various sublines in the colony. For example, an index of productivity of each breeding pair, such as the number of young weaned per pair per week, can be recorded on the pedigree chart. Replacement breeding stock would then be chosen according to the average productivity of the different substrains, though a substantial rise in productivity could indicate genetic contamination with consequent hybrid vigor. A simplified example is given in Figure 9.3.

The multiplication colony should also be physically separated from other colonies of the same species as far as possible, in order to prevent genetic contamination with a nonstrain mating. The colony may be sib-mated if the aim is to identify any new mutations that may occur, as this will tend to reveal any recessive mutations. Alternatively, if the aim is simply to produce large numbers of experimental animals, the colony can be random mated, and trios or other mating systems can be used if these are more economical. Random mating for a few generations should have no adverse impact on the genetic quality of the animals. A few new mutations may accumulate, but in practice, these are unlikely to have any impact on research. However, as noted above, a maximum of about four generations of such mating should be used. A practical scheme for the multiplication colony is to use the "traffic light" scheme,[38] in which surplus offspring of the stem-line go into breeding cages with a white label, their offspring go into cages with a green label, their offspring go into cages with an amber label, and their offspring go into breeding cages with a red label. No offspring from red-label cages are used for breeding. Four generations of multiplication means that even a poorly breeding strain can produce large numbers of experimental animals.

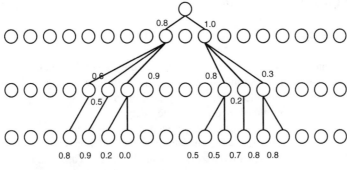

Mean subline 1= (0.8+ 0.6 + 0.5 + 0.9 +....+0.2+0.0)/8=0.59
Mean subline 2=(1.0+0.8+0.2+0.3+.......+ 0.8+0.8)/10=0.56

Conclusion: no significant difference, but probably best to select line 1

**Figure 9.3**    Diagram of a preprinted pedigree chart showing the output of individual breeding pairs (young weaned/female/week) and a scheme for selecting against any decline in breeding performance by comparing the average performance of each subline.

## F1-Hybrids

These are the first-generation cross between two inbred strains, and being isogenic, they have many of the more useful properties of the parental inbred strains. They also usually exhibit hybrid vigor, which means that they are more vigorous than inbred strains and are less influenced by adverse environmental conditions. As a result, they tend to be slightly more uniform than inbred strains.[39] However, they may also be less responsive to experimental treatments, so the increased uniformity does not necessarily mean that fewer animals are needed to attain a given level of statistical precision in an experiment. F1 hybrids are heterozygous at all the loci at which the parental strains differ, so they will not breed true. Mating two F1 hybrids gives an F2 hybrid, which is a genetically segregating generation. F2 hybrids are widely used for genetic mapping studies but are of less value as general research animals.

## Designation of F1 Hybrids

F1 hybrids are designated by giving the strain of the female parent first and the male parent second, with "F1" being appended. Thus, SJLDBA/2F1 would be the offspring of a female SJL and a male DBA/2 mouse. The strain designations are often abbreviated so that B6D2F1 is the offspring of a C57BL/6 female and a DBA/2 male.

## Research Uses of F1 Hybrids

F1 hybrids have one great advantage over pure inbred strains; they are much more robust. They tend to live longer, have fewer of the idiosyncrasies of the parental strains, and are less sensitive to adverse environmental conditions than inbred strains. They are particularly valuable as foster mothers in the production of transgenic strains. However, they should be used with caution in experiments involving breeding, as their offspring will be genetically segregating for the gene loci, which differ in the parental strains. Sometimes, they also have useful characteristics not normally found in the parental strains. For example, the NZBNZWF1 is widely studied as a model of autoimmune systemic lupus erythematosus, which is not found in the parental strains.[39] They can be of some value in immunological studies, as they will accept tissue, organ, tumor, or cell grafts from either of the parental strains without immunological rejection. One possible disadvantage of F1 hybrids is that they tend to be intermediate in phenotype, somewhere between the two parental strains. Thus, if an experiment requires several animals differing in their phenotypes, it is better to use several inbred strains rather than several F1 hybrids.

## Congenic and Coisogenic Strains

### Definition and Development

A pair of strains is said to be *coisogenic* if they differ at only a single genetic locus as a result of a mutation within one branch of the strain. Such strains are useful, because the effects of the mutation can be studied without the complication of genetic segregation in the genetic background. Unfortunately, coisogenic strains cannot be produced to order, as they depend on the mutation occurring within the inbred strain, which is uncontrollable. However, targeted mutations produced by homologous recombination using embryonic stem (ES) cells, which usually have the 129 inbred strain genotype, are basically coisogenic with strain 129 unless they are outcrossed to another strain. Thus, someone who has produced such mice should give careful consideration to maintaining it on the 129 genetic background in spite of its disadvantages. Outcrossing is usually done because strain 129 breeds poorly and is not well characterized. If or when ES cells of other strains become readily available, it will be possible to produce coisogenic strains for any locus, which can be manipulated in this way.

A pair of strains is said to be *congenic* if it approximates the coisogenic state as a result of backcrossing a gene (or, more strictly, an allele at a particular locus), known as the *differential allele*, to an inbred strain. Several methods of backcrossing can be used, depending on the mode of inheritance and method of identification of the gene. Basically, a donor strain is mated with the chosen background strain (often C57BL/6, but any strain may also be used depending on the aim of the study) to produce an F1 hybrid, designated N1. These animals are then mated to the inbred strain to produce the first backcross generation, designated N2. This will be segregating for the gene of interest, and carriers of the gene need to be identified. Many transgenes can be identified from a sample of DNA, and in this case, animals that carry the transgene would be selected for further backcrossing. If the gene can only be identified by its phenotypic effect, and if it is recessive, then it will be necessary to breed some of the N2 generation together in order to produce some homozygous mutant animals (the N is not incremented in this case as it does not count as a backcross generation) to continue with the backcrossing program. These procedures are repeated, ideally, until at least the N10 generation has been reached. At this stage, the strain carrying the mutation is said to be congenic with the background strain. If the differential allele can be made homozygous, it can be maintained just like an ordinary inbred strain (as above) with no further backcrossing.

One problem with this procedure is that it takes a lot of time. It is difficult to get even four generations of mice per year, so producing a congenic strain will often take as long as three years. Modern research requires instant results, so people are tempted not to go through the whole process before using the animals. One alternative is to produce "speed congenics."[40] From the offspring of the N2 generation onwards, about 20 males tested and known to carry the gene of interest are tested for their genotype at about 80 or more microsatellite genetic markers. The male who has the most alleles of the background strain summed across these loci is then chosen for further backcrossing. He will be mated to several females, as he needs to produce about 80 offspring. Half of these will be females, which will not be used, and half of the males will not carry the desired gene, leaving about 20 offspring to be tested for the next generation, from which the best male is again selected. Such a breeding program will approximately halve the time that it takes to produce a congenic strain, but it is expensive and needs careful organization.

### Nomenclature of Coisogenic and Congenic Strains

Coisogenic strains usually have the designation of the background strains followed by a hyphen, then the symbol of the differential allele, shown in italics (the nomenclature of genes is discussed below), e.g., C57BL/6J-*Lep$^{ob}$*. Where the gene is maintained in a heterozygous state, this is indicated by a + sign, a slash, and the gene symbol: C3H/N-+/*W.$^v$*

Congenic strains usually have the background strain designation (which is often abbreviated), a period, the donor strain designation, a dash, and the allele designation, e.g., B10.129-*H12$^b$*. B10 is an abbreviation for C57BL/10, and a full list of such abbreviations is given in the relevant nomenclature Web site. If the donor strain is not inbred or the genetic difference is complex, involving more than one allele, a less complete symbol may be used, such as the background strain and the gene symbols of the differential locus. Note that at present, gene symbols are changing quite frequently as the genes responsible for mutant phenotypes are identified, so the designation of the coisogenic and congenic strains also needs to change.

## Research Uses of Congenic Strains

Congenic strains have been used extensively by immunogeneticists to separate the genes responsible for the rejection of allografts, and subsequently, to study the genetics of the major histocompatibility complex without the additional complexity and "noise" created by segregation of the background.[41] Many sets of congenic strains, which differ at the MHC, are available. These have been used, for example, to study the effects of this complex locus on response to microorganisms such as the *Leishmania* parasite[42] and carcinogens such as urethane.[43] Such strains can be used without the necessity of genotyping individual animals. Many strains congenic for a mutation are also available, such as C57BL/6-*Lep^{db}*, the diabetes mutation maintained on the C57BL/6 genetic background.

More recently, geneticists have emphasized the great importance of transferring transgenes to an inbred background.[32] The expression of many mutants, knockouts, and transgenes depends on the rest of the genome. On a segregating background, the expression may be variable. Even more seriously, unless the strain is maintained in large numbers like a properly maintained outbred stock (see above), directional selection and genetic drift can drastically alter the expression of the gene over a period of a few generations. The directional selection in this case may be natural selection for reduced expression of the mutant. If the mutant is deleterious, then animals, which express it to a lesser extent, will have a reproductive advantage so that over a period of a few generations the expression may become much less severe. On the other hand, if the mutant or transgene can be backcrossed to an inbred background, then its characteristics will become largely fixed. If resources allow the transgene to be backcrossed to more than one strain, then the extent to which the background affects expression can also be studied. In some cases, genetic modifiers of transgene expression have been mapped.[44] Controlling such modifiers may provide a method of controlling the expression of some mutants in animals and in human medicine.[45]

## Maintenance of Congenic Strains

Once a strain is fully congenic, it can be maintained in the same way as any other inbred strain.

## Recombinant Inbred (RI) Strains

### Definition and Development

Sets of recombinant inbred (RI) strains are produced by crossing two standard inbred strains such as C57BL/6 and DBA/2 to produce an F1 hybrid, then brother × sister mating from these for at least 20 generations, but maintaining several separate lines. Ideally, at least 20 new lines are developed, though smaller sets can still be useful. In these strains, the genes for which the parental strains differ have been sorted into new combinations in each of the resulting RI strains. At any one genetic locus, such as the major histocompatibility complex, at which the parental strains differ, about half the strains should resemble each of the progenitor strains. There are many sets of RI strains available, developed from different pairs of progenitor strains.[46]

### Nomenclature of RI Strains

RI strains are known by an abbreviation of the two parental strains, with the female parent given first, separated by an upper case X, followed by numbers for the individual strains. For example, AXB2 is a strain derived from a cross between a strain A female and a strain C57BL/6 male, identified as strain 2 of this set. If the male parental strain already ends in a number, a hyphen may be used to distinguish the individual strains.

## Research Uses of RI Strains

Sets of RI strains are useful for investigating characters that differ between the two parental strains. If the parental strains differ for some phenotype, such as the activity of a particular enzyme or susceptibility to a pharmaceutical or toxic agent, it is often of interest to determine whether the difference is due to a single genetic locus, and if so, it would be desirable to genetically map that locus.

The procedure is to phenotype a full set of the RI strains developed from a cross between the strains for the character of interest, such as high or low activity of an enzyme. Obviously, this is only worthwhile if a suitable set of RI strains already exists, because it takes several years to develop and characterize each set. Thus, the first step would be to screen the parental strains of several sets of RI strains to see whether they differ. Assuming a suitable set of these strains can be found, small numbers of animals of each strain are phenotyped, and the distribution of strain means is studied. If the phenotypes fall into two distinct groups, each of which resembles one of the parental strains, then that would be evidence that the phenotype is controlled by a single genetic locus. In contrast, if there were a continuum of phenotypes covering many intermediate values, then this would suggest that the character has a more complex mode of inheritance. If there are sufficient numbers of individual RI strains in the set, it may be possible to resolve a character controlled by two loci, but it is rare to be able to distinguish more than about two loci.

If the phenotype indicates that the character is controlled by a single locus, then this can usually be genetically mapped to some chromosomal region. Suppose the study involves the BXD set of RI strains in which, in the original set, there are 26 individual strains, and the parental strains (C57BL/6 and DBA/2) differ in the activity of a particular enzyme. Suppose all 26 strains are typed, and in 16 of them, the enzyme activity resembles strain C57BL/6, and in the other ten, it resembles DBA/2, and all strains can unambiguously be classified into these two groups. The ratio of 16 to 10 is approximately a 50:50 ratio. The individual strains are designated BXD1, BXD2, BXD5, BXD6, BXD8, and so on (BXD3 and BXD4 were lost during inbreeding). The resemblance of each strain to one of the parental strains can be indicated by a B or a D, and a pattern of response such as DBBDD BBDBB, etc., can be built up. This is called the "strain distribution pattern" or SDP. Two genes tightly linked on the same chromosome will tend to have a similar or even identical SDP, because the loci will rarely have recombined during the inbreeding. The SDP of two unlinked genes will not be different from random. As an example, *Mod1* (an enzyme locus) and *D9Mit10* (a microsatellite locus) have identical SDPs of DBBDD BBDBB BDDBB DDBBB BBBDB B, and *D9Mit10* has been mapped to a particular location on mouse chromosome 9. Thus, *Mod1* can also be placed close to *D9Mit10* on chromosome 9 (in fact, Mod1 was mapped before D9Mit10, but this example illustrates the method).

In contrast, the SDP of *D1Mit5* on chromosome one is -BBDD BBBBB DDBDB BBBBD DDD (a "-" indicates no information). Of the 22 strains typed at both *Mod1* and *D1Mit5*, only 12 loci are concordant, and this ratio of 12/22 is not different from $^1/_2$, indicating no linkage between the two, as expected.

Most of the large sets of RI strains have been typed at many mapped loci, so if a single gene phenotype is indicated, then there is a high chance that it can be mapped simply by getting a computer to look for matches in the SDPs. No actual genotyping is required. Gene loci defined in terms of DNA can be mapped easily, because DNA samples from most of the RI strain sets are readily available.

Sets of RI strains can also sometimes be used to identify and map quantitative trait loci (QTLs) that are involved in the inheritance of polygenic characters. For example, there is a significant association between susceptibility to urethane-induced lung tumors and genotype at the *Kras2* locus in the AXB and BXA set of RI strains.[47] However, few sets of RI strains are sufficiently numerous to make it possible to resolve more than about two or three loci.

## Maintenance of RI Strains

Each RI strain is an inbred strain in its own right, and should be maintained as an inbred strain using methods described above.

## Recombinant Congenic (RC) Strains

### Definition and Development

Sets of RC strains are developed in a similar way to the development of RI strains, except that instead of inbreeding from the F2 hybrids, a few (generally about three) generations of backcrossing to one of the parental strains is done first, followed by sib mating of several independent lines for the equivalent of 20 generations.[6] Each backcross generation is regarded as being equivalent to two generations of sib mating. If the F1 hybrid is designated generation N1, then backcrossing to N3 is typically used so that each strain has, on average, a 12.5% contribution from the donor strain and 87.5% contribution from the background strain. The set would then be regarded as fully inbred at N3F14. Other levels of backcrossing may be used, and this will be indicated in the description of each RC set.

### Nomenclature of RC Strains

The individual RC strains are designated by uppercase abbreviations of the strain names, with the recipient (i.e., background) strain designated first, separated by a lowercase "c" with numbers to indicate the individual lines. For example, one of the CcS set of RC strains with BALB/c as the recipient strain and STS as the donor strain could be designated CcS6 if it was the 6th of such a set of strains.

### Research Uses of RC Strains

These strains are most useful for the identification and mapping of quantitative trait loci for characters, which differ between the donor and recipient strain. Assuming, for example, that a set of RC strains was developed from the N3 generation of a cross between two inbred strains, and that the character of interest (say a "high" or "low" response to a toxic agent in the donor and recipient strains, respectively) is controlled by several loci with different alleles in each of the two strains. For each locus, about 1/8 of the RC strains should contain the donor allele, and 7/8 of the strains should carry the background strain allele. The exact numbers will depend on chance sampling variation, so that there may be none, one, two, three, etc. of the strains with the donor allele, and some will have more than one. All strains are then phenotyped for the response to the toxic agent. If the response is due to a single locus, then most of the strains will have the response of the background strain, but a few may have that of the donor strain. If the character is controlled by several loci, then donor alleles will be present in a larger number of the strains, with a range of different phenotypes among the full set of strains. Most sets of RC strains have been genotyped at many marker loci, so that once a "high" strain has been identified, chromosomal regions, which are of donor origin, will already have been identified. Normally, one of the "high" RC strains will be crossed with the background strain. Segregation in the F2 generation will be studied for evidence that the difference is due to a single locus, resulting in Mendelian segregation. If this is the case, then one of the alleles for "high" response will have been trapped in the RI strain, and it can then be mapped and studied in more detail.

RC strains provide a powerful research tool, and they have been used in a wide range of studies, including susceptibility to cancer,[48] pathogens,[49] and bone strength,[50] as a few examples. However, so far their use has been limited to a few laboratories that made the considerable investment in developing and genotyping some useful sets. The main disadvantage is that they can usually only be used for studying characters that differ between the two parental strains. Thus, if the phenotype of interest is not seen in either of the parental strains, it is unlikely (though not impossible) that a particular set of RI strains will be of use in investigating that character.

### Maintenance of RC Strains

Once a set of RC strains is fully inbred, each strain is an inbred strain in its own right and should be maintained in the same way as any other inbred strain.

## Consomic or "Chromosome Substitution" Strains

### Definition, Development, and Research Uses

Consomic strains are developed by backcrossing a whole chromosome from one donor inbred strain into a recipient "background" strain. They are used for studying the genetics of complex traits and in the eventual identification of QTLs.[51] A minimum of ten backcrosses should be used, though ideally, all the background strain chromosomes should be present and uncontaminated with donor chromosomal fragments (with the exception of the donor chromosome).

The development of Y-chromosome consomic strains is relatively easy. A donor strain male is crossed with a recipient strain, and male offspring are again backcrossed to the recipient strain. This is repeated for ten generations, always using background strain females. Such strains have been used to study the phenotypic effects of genes on the Y chromosome.[52]

Strains consomic for autosomal chromosomes can now be developed using about five or six polymorphic genetic markers on the chosen chromosome, known to differ between the two strains. Offspring are then only chosen for backcrossing if they have all markers of the donor type for the chosen chromosome, i.e., if the chromosome has not recombined due to crossing over. A full set of consomic strains for the mouse would involve 21 strains, with 19 of them being the background strain with each of the 19 autosomes substituted and the other two being the background strain with either the X or Y chromosome from the donor strain. However, even a few of these strains can be used to answer questions about whether or not there are loci on a particular chromosome that influence a particular phenotype. For example, a number of consomic rat strains with chromosomes from BN strain rats substituted on an SS background have been developed. The background strain has high blood pressure when placed on a high-salt diet. These strains are now being characterized for 203 heart, lung, vascular, and blood function phenotypes. Strain SS.BN-13, for example, which has the BN chromosome 13 substituted, has lower mean arterial pressure than the SS strain, showing that there is a salt-related blood pressure gene on this chromosome.[53] In mice, these strains have been used to study susceptibility to germ-cell tumors in 129 strain mice, using a chromosome 19 substitution strain.[54]

Consomic strains will provide a powerful new set of genetic tools for studying quantitative trait loci as soon as they become more widely available. They provide the opportunity of mapping loci with quite small phenotypic effects, because sample size can be as large as needed to detect any effect thought to be of biological significance.[55] One note of caution is that QTLs may interact with one another, so that a difference seen on one genetic background may not be observed with a different background.

### Maintenance of Consomic Strains

Once backcrossing is completed, each consomic strain is an inbred strain in its own right and should be maintained in the same way as an inbred strain, as discussed above.

### Nomenclature of Consomic Strains

Consomic strains of mice are officially designated by RECIPIENT STRAIN-CHROMOSOME [DONOR STRAIN]. For example, C57BL/6J-Y[AKR] is a consomic strain with the C57BL/6 autosomes, but the Y chromosome from strain AKR.

## Mutants, Polymorphisms, Transgenic, and Gene-Targeted Mutations (Collectively Known as "Mutants")

Several thousand mutants, polymorphisms, and transgenes of various sorts are now used in biomedical research, and the numbers will increase substantially over the next few years as targeted mutations and mutants produced by chemical mutagenesis continue to be produced. These have a Mendelian mode of inheritance, though there may be complications due to variation in gene expression, penetrance, viability, and breeding performance. The spontaneous or induced mutations and polymorphisms may be caused by alteration of a single base pair, which may cause a mutation in a coding or regulatory region or be genetically silent, or by a deletion or insertion of a larger tract of DNA. In some cases, such as the dilute coat color allele in DBA/2 mice, the mutation has been caused by the insertion of a retrovirus within the gene.[56]

Transgenic (including knockout) strains are produced by incorporation of foreign DNA into the genome as a result of direct injection into early embryos or via embryonic stem cells. Such DNA is incorporated into the host chromosomes, and with the right regulatory sequences, it can be transcribed and translated by the host as though it was a host gene. So-called knockout mice are produced by inactivating an existing gene. Briefly, once a gene has been sequenced, an inactive copy of it can be synthesized *in vitro*, usually by splicing a gene for resistance to neomycin in the middle of the sequence. This construct is then incorporated into embryonic stem cell cultures *in vitro*. When these cells are cultured, a small amount of homologous recombination occurs, with the inactive gene replacing the host gene. Clones of cells in which this has occurred are identified by selective markers. These cells are then injected into blastocyst stage embryos, usually from an unrelated strain, and returned to a foster mother. Some of the resulting offspring are chimeras composed of cells from the host blastocyst and the cloned cells. Coat color markers are usually used so that chimeric offspring can easily be identified. In some cases, the gonads of the chimeric animals will also be chimeric, in which case offspring can be produced whose genomes contain the knocked-out gene. An overview of mutagenesis and transgenesis is given by Silver,[2] and more detailed descriptions of methods of producing transgenic animals of various species is given by Maclean.[57] The techniques are advancing rapidly and include the possibility of site- and time-specific mutagenesis by knocking out a gene only in a particular tissue at a time chosen by the scientist, and also "knock-in" methods for replacing a gene with an alternative, say from another species.

Transgenic strains are used in a number of ways:

1. They can be used to produce valuable proteins, in milk or other body fluids, that would otherwise be difficult to synthesize or extract. For example, transgenic sheep are currently being used to produce human alpha-1-antitrypsin, which will be used to treat patients with cystic fibrosis and emphysema.[58]
2. They can be used to produce animal models of human disease or to study the effects of environmental agents. For example, several strains of mice that are highly susceptible to cancer are being evaluated in the hope that through them, we will be better able to detect carcinogenic chemicals that would be a hazard to humans and the environment.[59]

The major use of these animals will be in studying the functions of genes. There appear to be about 30,000 genes in humans and laboratory mammals, and the function of only a few thousand of them is known. Finding the function of the remainder, and using the knowledge to combat human disease, are the challenges now facing biomedical research.

### Genetic Background of Mutants and Transgenes

Possibly the most important general point about this class of stock is that where a mutation causes a phenotype, this is virtually always subject to modification by other genes. A mutation that causes death or severe disability on one genetic background may be viable and have mild to nonexistent effects on another. Where a mutant or transgene is on a genetically segregating background, it is nearly always good research practice to backcross to one or more isogenic backgrounds before starting detailed investigation of the phenotype.[60] With an outbred background, genetic drift and the effects of natural selection for reduced expression of deleterious phenotypes in the first few generations may mean that results are not constant over a period of time.

### Characteristics and Research Uses of Mutants and Polymorphisms

Mutants may have occurred spontaneously or been induced by irradiation or chemicals. In the mouse, such mutants have been preserved for many years. Most mutants have some deleterious effect in the homozygous or heterozygous state, in contrast with polymorphisms discussed below. Many mutants have been identified as animal models of similar mutations in humans, while others such as the athymic nude mice and rats have also been widely used, because they have characteristics that are useful in a wide range of different disciplines. With the development of new molecular tools for studying mutants, it is now being realized that mutants who at first sight appear to be of little biomedical interest, such as coat color, circling, or curly tail mutants, often alter quite basic biological processes. Thus, they are a valuable resource for fundamental studies.

One of the advantages of mutants in research is that they are usually viable, and in many cases, there are several different mutant alleles at each locus, often with various gradations of severity. In some cases, this makes them more useful than knockouts. However, the obvious disadvantage is that if there are no mutations that affect a system of interest to a particular investigator, the only option may be to make an artificial one using transgenic methods. For people interested in basic research, a good strategy is often to find a new mutant, map it, identify the gene at the molecular level, and work out its mode of action. Many such studies are now reported in the literature, and more can be expected in the future. Although these studies will not lead to an immediate cure for human disease, in the long term, the knowledge accumulated in this way certainly will be useful.

Polymorphisms are genetic variants that are genetically silent or are present at a high frequency in a population. For example, several enzyme loci such as glucose phosphate isomerase (*Gpi1*) are polymorphic, with the alleles being recognizably following electrophoresis and appropriate staining. Inbred strains can differ at this locus. Similarly, there is extensive polymorphism at some repetitive loci. Microsatellite loci, which are short simple repetitive sequences of base pairs such as CA, repeated about 50 to 200 times are highly polymorphic in most species and are widely used as genetic markers for mapping and genetic quality control.[2] Minisatellite loci are also highly polymorphic, probably due to a high mutation rate.[61]

Recently, it has been found that there is extensive polymorphism of single nucleotides throughout the genome. These may occur within coding loci in which case they may or may not cause a mutation in introns and regulatory sequences where they may affect gene function, and in various types of noncoding DNA. These single nucleotide repeats (SNPs) are potentially valuable genetic markers once they are fully mapped and characterized,[62] as well as being of biological importance if they alter gene regulation or transcription.

## Nomenclature of Mutants and Polymorphisms

The nomenclature of mutants is reasonably straightforward. However, there are some complications, because gene symbols are changed frequently as a result of genetic advances. Thus, mutants such as obese were given a symbol, in this case *ob*, which indicated the allele and the locus. The wild type at the *ob* locus was then designated $+^{ob}$ or, when the context was clear, just +. However, when the gene was mapped and cloned, it was found to code for a protein named leptin, which was given the locus symbol, *Lep*, so the obese allele has now been renamed $Lep^{ob}$. Now the wild type is designated $+^{Lep}$ or $Lep^+$. Many mutants are undergoing such changes in their designations, with old symbols such as *c* for the albino locus now being redesignated $Tyr^c$. The nude mutation has been redesignated $Foxn1^{nu}$, because it is a mutation at a locus first described in *Drosophila*, which causes a forked head in that insect. As far as possible gene symbols for loci that are recognizable as identical in different species should be uniform across species.

Full details of the genetic nomenclature rules for mice are given on the Jackson Laboratory Web site (www.informatics.jax.org). In short, names for genes, loci, and alleles should be brief and, if possible, descriptive, e.g., "obese" or "congenital hydrocephalus." Genes are functional units, whereas a locus can be any distinct DNA sequence. Symbols for genes should be short abbreviations of one to four letters, starting with the same letter as the name. Arabic numbers can be included as part of the name where necessary, but the first symbol should always be a letter. Roman numerals and Greek letters should not be used. Hyphens are only used for clarity, such as when two numbers need to be separated. In published articles, gene symbols are given in italics.

Loci defined by anonymous DNA probes are given a symbol starting with a D followed by the chromosome assignment (i.e., the numbers 1 to 19, X or Y), a laboratory registration code (see above for nomenclature of inbred mouse strains), and a unique serial number. The laboratory methods for detection of the locus also need to be specified. Alleles are usually designated by a superscript. Where this is not possible (e.g., where superscripts are not accepted electronically), the symbol can be enclosed in chevrons, e.g., *Gpil^a* or *Gpil<a>*.

There are additional rules relating to things like pseudogenes, super-gene families or complexes, retroviruses, and for special classes of genes and gene complexes such as biochemical variants, lymphocyte antigens, histocompatibility loci, etc.

Transgenes are designated by a general formula:

$$TgX(YYYYY)\#\#\#\#\#Zzz$$

where *Tg* indicates a transgene, and *X* indicates the mode of insertion of the foreign DNA, with *N* for nonhomologous recombination, *R* for insertion via a viral vector, and *H* for homologous recombination. The *YYYYY* is a brief designation of the insert, with a range of standard abbreviations being available. The *#####* is a laboratory assignment number. Finally, *Zzz* is a laboratory registration code. As an example, given in the rules:

$$C57BL/6J\text{-}TgN(CD8GEN)23Jwg$$

This is the designation for inbred strain C57BL/6J carrying a transgene with the human CD8 genomic (GEN) clone. It is derived from the 23^rd mouse screened in a series of microinjection (N) in the laboratory of Jon W. Gordon (Jwg).

## Breeding and Maintenance of Mutant and Transgenic Strains

The breeding methods needed to maintain these strains will depend on the mode of inheritance of the mutant, including whether it is dominant, codominant, or recessive; whether it has a distinct phenotype; and whether all classes of stock are viable and fertile. The genetic background (inbred or outbred) also needs to be taken into account. A brief description of some of the more common situations is given below. However, more complex breeding systems may be needed to produce animals with a desired combination of mutant genes, or where identification of genotype is difficult. Also, genetic mapping studies may involve quite complex breeding schemes not discussed here.

*Both Sexes Fully Viable and Fertile* — Such a mutant can be maintained according to the genetic background. If this is inbred, then methods for maintaining inbred strains, described above, should be used. If the mutant or transgene is on a heterogeneous, outbred background and there are no plans to backcross to an inbred strain, and if the aim is to maintain the colony for some time, then it should be maintained as for an outbred stock. In order to prevent inbreeding and genetic drift, at least 25 breeding pairs should be used with random mating, or about 13 pairs if a maximum avoidance of inbreeding system is to be used. Smaller colonies can be maintained for short periods, although it may result in substantial genetic drift and change in expression of the mutation.

*A Recessive Mutation with One Sex Infertile* — In this case, the mutant is usually maintained by mating a homozygous animal of the viable sex with a heterozygous animal of the other sex. This is the common situation with nude mice, where the females often have poor breeding performance (depending on genetic background), but males are fully fertile. Half the offspring will be of the desired mutant phenotype, and the other half will carry the gene and the appropriate sex can be used for further breeding. The matings need to take account of the genetic background that will be inbred or outbred, and should be maintained as such.

*A Dominant Mutation with Homozygous Lethality* — Some dominant genes such as the yellow allele at the agouti locus are lethal in the homozygous state. In this case, matings are made between mutant animals and wild type ones, with approximately half of the offspring being of the mutant phenotype. Again, account needs to be taken of the genetic background.

*Recessive Mutation with Both Sexes Infertile* — This is quite common with mutations which have serious phenotypic effects, such as the obese and diabetic mutations in mice. The problem is that if two heterozygotes are mated, on average, $1/4$ of the offspring will be mutant, $1/2$ will be heterozygous carriers, and $1/4$ will be homozygous wild type. Unfortunately, the latter two classes will be phenotypically indistinguishable, but only the heterozygotes will be useful for further breeding.

The classical way of dealing with this situation is to set up test matings in order to identify heterozygotes. Known heterozygous animals can be crossed with individuals of unknown genotype. If mutant progeny are produced among a reasonably large number of offspring, then the animal of unknown genotype must be heterozygous. Unfortunately, this is time consuming and inefficient. Another alternative is to do random matings among the offspring of heterozygous matings. As $2/3$ of the animals are, on average, heterozygous, $2/3 \times 2/3 = 4/9$ of the matings will be between heterozygotes, but only $1/4$ of the offspring of such matings will be homozygotes. Generally, about 10 wild-type offspring will need to be produced to be reasonably confident that any particular mating is not between two heterozygotes, so can be discarded. Again, this is inefficient, though relatively simple.

If the mutant is on an inbred genetic background, an alternative method is to graft ovaries from a homozygous mutant animal into a wild-type female of the same inbred strain. This animal is then mated to a wild-type male to produce offspring known to be heterozygous for the mutation and suitable for breeding.

When a mutant locus has been sequenced and the nature of the mutation identified, it is often possible to develop a PCR-based method of genotyping individuals. All that is needed is a small sample of DNA and the laboratory capability of running the appropriate tests. Such a method has already been developed for the diabetes[63] and obese mutations[64] in mice, and is, or will be, available for many mutants. If heterozygotes can be identified in this way, then breeding mutant animals becomes much more efficient.

## Breeding Transgenes and Mutations Produced by Gene Targeting ("Knockouts")

Carriers of a transgene can be of three types: hemizygotes, with one copy of the transgene but no normal host allele; heterozygotes, in the case of a gene-targeted mutation, with one mutant and one normal wild-type allele; or homozygotes, with two copies of the transgene. Identification of carriers usually presents no problems as there will be a PCR-based test or a probe that can be used to identify them using Southern blots. Thus, all that is needed is a sample of DNA. This is often obtained from the tail tip, though if a PCR method is used, an ear punch, hair sample, saliva,[65] or even a fecal sample may provide sufficient DNA for the test. Once carriers have been identified, matings can be made to continue to maintain the transgene or backcross it to an inbred background or produce homozygous animals if the transgene is in a hemizygous or heterozygous state.

Problems may arise in differentiating between homozygous animals, i.e., with two copies of the transgene, and hemizygous or heterozygous animals, i.e., with one copy of the transgene. This will require a quantitative PCR[66] or Southern blot analysis or a progeny test. The latter involves mating the animal to some wild-type animals and testing the progeny for the presence of the transgene, assuming the animal of interest is fertile and viable. If all the progeny carry a copy of the transgene (and assuming about eight to ten progeny have been tested), then the animal of interest can be assumed to have been homozygous.

## GENETIC QUALITY CONTROL

### Aims

The aim of genetic quality control programs is to detect genetic contamination of one strain through an inadvertent mating with another strain. Currently, it is not possible to monitor strains for new mutations except by observing the phenotype. However, mutations that affect invisible characters such as minor changes in immune response, physiology, or susceptibility to infectious organisms may go undetected for many generations.

## Technical Methods

Historically, several methods of genetic quality control have ranged from the use of biochemical polymorphisms to the use of quantitative characters such as the shape of the skeleton and breeding performance.[67,68] Biochemical polymorphisms are often technically difficult to determine and somewhat limited in their distribution among strains of mice and rats. Skeletal morphology and breeding performance have the disadvantage of giving a statistical result, rather than a clear-cut positive or negative answer, though breeding performance should be routinely monitored for husbandry purposes so any change should be investigated.

The development of a large number of microsatellite and other DNA-based genetic markers detected using the polymerase chain reaction (PCR) has now completely changed the situation, although changes in phenotype noticed by animal technicians or scientific users of the animals continue to be an important way in which possible genetic contamination is first identified. The great advantage of DNA-based methods is that only a small sample of tissue is needed, it can be stored indefinitely in the deep freeze, and the techniques for genotyping are essentially the same for every locus. The only differences are in the sequences of the PCR primers and possibly some minor changes in the conditions for the PCR reaction.

Microsatellites, which are the most widely used markers, are short repetitive DNA sequences with unique flanking regions. They are highly polymorphic in the number of repeats. PCR primers usually consist of about 20 base pairs of the unique flanking DNA for each microsatellite. There are many thousands of microsatellites in the mouse and rat genomes, and primers are commercially available for many of them from Research Genetics (www.resgen.com). The basic technique involves taking a sample of tissue from the animals to be tested, preparing DNA, and amplifying one or more of the microsatellites using PCR with the appropriate primers. The resulting reaction mixture is run on an agarose or polyacrylamide gel with control samples from animals with a known genotype, where necessary. If agarose is used, alleles are usually visualized with UV light after staining with ethidium bromide, with different length alleles running different distances. If, with an inbred strain, the DNA bands are not aligned, this shows that the genotypes are not identical. Technical methods are given in many publications that use these markers, and by Litt (1991).[69]

There are several variants on this basic method. The system can be automated using DNA sequencing apparatus, though this is expensive and would normally only be economical if done on a large scale or if the apparatus is already available. As the bottleneck is usually running the gel, another alternative is to pool 5 or 10 samples of the PCR product in each well. Following electrophoresis, this will result in a strong band, with satellite bands if one or more of the samples has a different genotype.

The main difficulty with genetic quality control is in deciding the number of genetic loci to use, the sample size, and sample frequency. Many microsatellite loci have more than one allele at each locus, though sometimes these can only be identified using acrylomide rather than agarose gels. Based on about 7500 comparisons, there is about a 74% chance that two unrelated inbred mouse strains will be the same at any given microsatellite locus, assuming a resolution of six or more base pairs. This means that two unknown strains should be tested at ten loci to give a 95% chance of detecting one or more differences. However, any strain or stock will normally only be at risk of becoming genetically contaminated by other strains in the same animal house, so a critical set of markers can be chosen that will detect any contamination from these strains. Usually, loci should also be chosen that are on different chromosomes to increase their statistical independence, and known alleles should differ by more than about five base pairs so that they may be easily identified.

For routine monitoring, the sample size depends primarily on the presumed extent of any genetic contamination. A high level of contamination, say above 20%, can be detected with small sample sizes, but it is virtually impossible to detect a couple of wrong matings in a colony of a thousand breeding cages. Table 9.6 shows the sample size required to detect different levels of genetic contamination at a specified level of probability. From this, it is clear that the best approach is to do everything possible to avoid contamination in the first place.

**Table 9.6   Sample Size Needed to Give a 95% Chance of Detecting Genetic Contamination for Given Levels of Contaminated Animals in the Colony**

| Percent Contamination | Sample Size |
|:---:|:---:|
| 2 | 148 |
| 5 | 58 |
| 10 | 28 |
| 15 | 18 |
| 20 | 13 |
| 30 | 8 |
| 40 | 6 |
| 70 | 2 |

*Note:* The formula used is $S = \log(p)/\log(U)$, where $p$ is the chance of missing the contamination (in this case, $p = 0.05$), and $U$ is the proportion of animals in the colony that is uncontaminated.

## Genetic Monitoring of Isogenic Strains

Inbred strains are relatively easy to monitor because all individuals should be identical at all microsatellite loci, apart from any recent mutations. These will be quite rare. However, at present, there are no agreed upon standards on the number of loci to test or sample size and frequency. Generally, the effort expended in monitoring each strain should depend on the chance of genetic contamination, with account being taken of the importance of the colony and the likely damage to research or reputation of the breeder from a contamination. Danger arises when colonies are first established, because there may be no real assurance that they are what they are supposed to be, and maintaining animals of several strains in the same animal room clearly increases the chance of a mis-mating. Staff also needs to be well trained, with the avoidance of a culture of blame so that staff know that they will not get into trouble when reporting unexpected observations, should this turn out to be due to genetic contamination.

### Authentication of Newly Established Strains

Ideally, newly established colonies should be tested as soon as possible to ensure that they are of the correct genotype. With mice, control DNA from most strains is available from the Jackson Laboratory (www.jax.org). For rats, samples of DNA may be obtained from colleagues or known holders of the strains. If the colony is being established from a small breeding nucleus, it may be possible to test all animals. In this case, the main aim is to test the authenticity of the strain, though the possibility of contamination by one or more nonstrain animals should not be ruled out. A sample of five to ten animals is probably adequate at this stage, and they should probably be tested at about ten microsatellite loci.

### Existing Colonies

Breeding colonies of inbred strains will often be divided into a stem line and an expansion colony (see above). Ideally, the stem line colony will be physically separated from other colonies. If not, it should at least be kept with strains of a different coat color and microsatellite profile. With good physical separation and a small colony size, the chance of contamination is low, so the colony would not need to be monitored frequently. If it is maintained in an isolator with no other strain, then once it has been authenticated, it hardly needs any routine monitoring.

Expansion colonies may be large and at risk from other colonies in the same building. The colony might be monitored two to four times per year, with sample sizes of about ten animals, using a set of markers that will preclude contamination by all other strains in the building.

## Troubleshooting

In practice, genetic contamination is often picked up by the users of the animals, who obtain unexpected results. If DNA samples can be obtained from the abnormal animals (scientists should keep frozen tissue samples of their animals), then there is a good chance that contamination will be obvious using a few microsatellite markers. Conversely, if they match other animals in the colony and reference samples of DNA at 10 to 12 loci, then there is a good chance that the abnormal response is not due to genetic contamination. It may, however, be due to a new mutation and should be investigated as such. If live animals are still available, these should be outcrossed to an unrelated strain, and possible Mendelian segregation should be studied in the F2 and backcross generations.

## Genetic Monitoring of Outbred Stocks

Outbred stocks are subject to genetic drift as well as the possibility of genetic contamination. Any genetic monitoring scheme should aim to detect these. Also, in contrast to inbred strains, there can be no authentication stage, because there is no "correct" genotype for an outbred stock. Outbred animals with names like Wistar rats and Swiss mice from different colonies differ, and there is no international or national standard to say which is the "right" genotype. One further complication is that the outbred stock may turn out to be quite inbred if it has been maintained as a closed colony for many years. In this case, it may be quite difficult to find segregating markers suitable for monitoring genetic drift.

As with inbred strains, the preferred technical methods for monitoring outbred stocks is the use of microsatellite or other DNA-based genetic markers. However, the use of these markers is not so easy with outbred stocks, because when a locus is segregating within a colony, there may be several different alleles present. If there are small differences between alleles, classification of individual bands can become difficult. However, it should be possible to identify heterozygous animals, which have two bands.

There is no recent theoretical work on sample sizes, choice of markers, or frequency of monitoring in outbred stocks, so the suggestions given here may need to be modified according to individual experience.

## Choice of Marker Loci

Genetic drift will cause a change in gene (allele) frequency at loci, which are segregating within a colony, but is unable to change the frequency of "background" loci, which are fixed (i.e., the frequency of an allele is 100%). However, genetic contamination may introduce new alleles at such loci as well as changing frequency at segregating loci. Thus, some genetic markers can be chosen for nonsegregating background loci that will be sensitive to genetic contamination (taking into account other stocks in the animal house) or could be used to differentiate the stock from other stocks with the same name, and some for loci that are segregating to pick up drift and contamination. One strategy could be to do an initial survey of about 30 to 50 animals using about 20 or more genetic markers, and pick about four or five unlinked and apparently invariant background markers and four or five unlinked segregating markers for routine work. Segregating markers could be chosen that, as far as possible, had about 50% of heterozygous animals (i.e., with two bands) and 50% homozygous animals (i.e., one band), assuming that such markers can be found. If not, markers with the highest proportion of heterozygotes should be chosen. This should overcome the problem of trying to classify multiple alleles, as it is relatively easy to distinguish between one and two bands. In the absence of genetic drift, mutation, directional selection, or contamination, the frequency of heterozygotes should stay constant, according to the Hardy–Weinberg equilibrium, subject to sampling variation.

## Sample Size and Frequency and Data Evaluation

In theory, sample size and frequency will depend on the chance of genetic contamination, the probability of genetic drift, and the importance of any genetic changes to the breeder and the scientific community.

Initially, samples of about 20 animals could be taken about three to four times per year and evaluated at the five chosen invariant and five segregating loci. Sample size and frequency could be higher if automated genotyping is available. The invariant background loci should remain reasonably invariant, though small numbers of heterozygous animals may be found, because the initial sample of animals may have missed some of the genetic variation.

Any genetic drift due to inbreeding will tend to reduce the number of heterozygous animals. Thus, data on the number of heterozygous animals summed across all five of the segregating loci can be pooled to give a single estimate of the number of heterozygous loci in 20 animals at five loci, i.e., based on 100 observations. This could be calculated for each sample and might be graphed on a Shewart control chart of the type used in industrial quality control. Such charts are available for many statistical computer programs such as MINITAB. These charts provide rules for deciding whether the process is "out of control," which in this case would suggest genetic drift or contamination. Where this occurred, it would need to be investigated in more detail.

### Comparisons of Different Colonies

Breeders often want to maintain the same outbred stock at different locations and need some way of testing whether the colonies have drifted apart. The above protocol could be used to compare colonies. Assuming a set of background and segregating markers has been chosen, then samples of about 20 animals could be taken from each colony and tested at the same 10 loci. The invariant loci should be the same in both colonies. The segregating loci could be used to compare the total number of heterozygous animals in the two samples, using a contingency chi-squared test or comparing two proportions using the normal approximation of the binomial distribution. These tests are described in most statistical textbooks and are available in several computer statistical packages, such as MINITAB.

## REFERENCES

1. Malakoff, D., The Rise of the Mouse, Biomedicine's Model Mammal, *Science*, 288, 248, 2000.
2. Silver, L.M., *Mouse Genetics*, Oxford University Press, Oxford, 1995.
3. Little, C.C. and Tyzzer, E.E., Further studies on inheritance of susceptibility to a transplantable tumor of Japanese waltzing mice, *J. Med. Res.*, 33, 393, 1916.
4. Bailey, D.W., Recombinant inbred strains, an aid to finding identity, linkage, and function of histo-compatibility and other genes, *Transplant.*, 11, 325, 1971.
5. Taylor, B.A., Development of recombinant inbred lines of mice, *Behav. Genet.*, 6, 118, 1976.
6. Demant, P. and Hart, A.A.M., Recombinant congenic strains — a new tool for analyzing genetic traits determined by more than one gene, *Immunogenet.*, 24, 416, 1986.
7. Whittingham, D.G., Leibo, S.P., and Mazur, P., Survival of mouse embryos frozen to -196°C and -269°C, *Science*, 178, 411, 1972.
8. Gordon, J.W., Scangos, G.A., Plotkin, D.J., Barbosa, J.A., and Ruddle, F.H., Genetic transformation of mouse embryos by microinjection of purified DNA, *Proc. Natl. Acad. Sci. USA*, 77, 7380, 1980.
9. Evans, M.F. and Kaufman, M.H., Establishment in culture of pluripotential cells from mouse embryos, *Nature*, 292, 154, 1981.
10. Thomas, K.R. and Capecchi, M.R., Site-directed mutagenesis by gene targeting in mouse embryo-derived stem cells, *Cell*, 51, 503, 1987.
11. Festing, M.F.W., Warning: the use of genetically heterogeneous mice may seriously damage your research, *Neurobiology of Aging*, 20, 237, 1999.
12. Festing, M.F.W., Use of a multi-strain assay could improve the NTP carcinogenesis bioassay program, *Environ. Health Perspect.*, 103, 44, 1995.
13. Kacew, S. and Festing, M.F.W., Role of rat strain in the differential sensitivity to pharmaceutical agents and naturally occurring substances, *J. Toxicol. Environ. Health*, 47, 1, 1996.
14. Festing, M.F.W., Fat rats and carcinogen screening, *Nature*, 388, 321, 1997.

15. Miller, R.A., Austad, S., Burke, D., Chrisp, C., Dysco, R., Galecki, A., Jackson, A., and Monnier, V., Exotic mice as models for aging research: polemic and prospectus, *Neurobiology of Aging*, 20, 217, 1999.

16. Festing, M.F.W., Kondo, K., Loosli, R., Poiley, S.M., and Spiegel, A., International standardised nomenclature for outbred stocks of laboratory animals, *ICLA Bulletin*, 30, 4, 1972.

17. Festing, M.F.W., May, D., Connors, T.A., Lovell, D.P., and Sparrow, S., An athymic mutation in the rat, *Nature*, 274, 365, 1978.

18. Hutt, F.B., *Genetics for Dog Breeders*, W.H. Freeman, San Francisco, 1979.

19. Falconer, D.S., *Introduction to Quantitative Genetics*, 2nd ed., Longman, Chicago; New York, 1981.

20. Venta, P.J., Li, J., Yuzbasiyan-Gurkan, V., Brewer, G.J., and Schall, W.D., Mutation causing von Willbrand's disease in Scottish Terriers, *J. Vet. Intern. Med.*, 14, 10, 2000.

21. Nomura, T. and Yonezawa, K., A comparison of four systems of group mating for avoiding inbreeding, *Genet. Sel. Evol.*, 28, 141, 1996.

22. McClearn, G.E., Wilson, J.R., and Meredith, W., The use of isogenic and heterogenic mouse stocks in behavioral research, in *Contribution to Behavior Genetic Analysis. The Mouse as a Prototype*, Lindzey, G. and Thiessen, D.D., Eds., Appleton-Century-Crofts, New York, 1970, 3.

23. McClearn, G.E. and Hofer, S.M., Genes as gerontological variables: genetically heterogeneous stocks and complex systems, *Neurobiology of Aging*, 20, 147, 1999.

24. Heller, D.A., Ahern, F.M., Stout, J.T., and McClearn, G.E., Mortality and biomarkers of aging in heterogeneous stock (HS) mice, *J. Gerontol. Biol. Sci. Med. Sci.*, 53, B217, 1998.

25. Mott, R., Talbot, C.J., Turri, M.G., Collins, A.C., and Flint, J., From the cover: a method for fine mapping quantitative trait loci in outbred animal stocks, *Proc. Natl. Acad. Sci. USA*, 97, 12649, 2000.

26. Darvasi, A. and Soller, M., Advanced intercross lines, an experimental population for fine mapping, *Genetics*, 141, 1199, 1995.

27. Darvasi, A., Experimental strategies for the genetic dissection of complex traits in animal models, *Nature Genet.*, 18, 19, 1998.

28. Feingold, N., Feingold, J., Mouton, D., Bouthillier, Y., Stiffel, C., and Biozzi, G., Polygenic regulation of antibody synthesis to sheep erythrocytes in the mouse: a genetic analysis, *Eur. J. Immunol.*, 6, 43, 1976.

29. Updyke, L., Yoon, H.L., Chuthaputti, A., Pfeiffer, R.W., and Yim, G.K.W., Induction of interleukin-1 and tumor necrosis factor by 12-*O*-tetradecanoylphorbol-13-acetate in phorbol ester-sensitive (SENCAR) and resistant (B6C3F1) mice, *Carcinogenesis*, 10, 1107, 1989.

30. Rapp, J.R., Dahl salt-susceptible and salt-resistant rats. A review, *Hypertension*, 4, 753, 1982.

31. Collins, A.C. and Marks, M.J., Genetic studies of nicotine and nicotine/alcohol reactivity in humans and animals, in *Genetically Defined Animal Models of Neurobehavioral Dysfunctions*, Driscoll, P., Ed., Birkhauser, Boston, Basel, Berlin, 1992, p. 146.

32. Silva, A.J., Simpson, M.E., Takahashi, J.S., Lipp, H.-P., Nakanishi, S., Wehner, J.M., Giese, K.P., Tully, T., Able, T., Chapman, P.F., Fox, K., Grant, S., Itohara, S., Lathe, R., Mayford, M., McNamara, J.O., Morris, R.J., Picciotto, M., Roder, J., Shin, H.-S., Slesinger, P.A., Storm, D.R., Stryker, M.P., Tonegawa, S., Wang, Y., and Wolfer, D.P., Mutant mice and neuroscience: recommendations concerning genetic background, *Neuron*, 19, 755, 1997.

33. Jay, G.E., Variation in response of various mouse strains to hexobarbitol (Evpal), *Proc. Soc. Exp. Biol. Med.*, 90, 378, 1955.

34. Festing, M.F.W., Contemporary issues in toxicology: use of genetically heterogeneous rats and mice in toxicological research: a personal perspective, *Toxicol. Appl. Pharmacol.*, 102, 197, 1990.

35. Festing, M.F.W. and Fisher, E.M.C., Mighty mice, *Nature*, 404, 815, 2000.

36. Beck, J.A., Lloyd, S., Hafezparast, M., Lennon-Pierce, M., Eppig, J.T., Festing, M.F.W., and Fisher, E.M.C., Genealogies of mouse inbred strains, *Nature Genet.*, 24, 23, 2000.

37. Festing, M.F.W., *International Index of Laboratory Animals*, 6th ed., Michael F.W. Festing, Leicester, United Kingdom, 1993.

38. Lane-Petter, W. and Pearson, A.E.G., *The Laboratory Animal — Principles and Practice*, Academic Press, London, 1972.

39. Festing, M.F.W., Phenotypic variability of inbred and outbred mice, *Nature*, 263, 230, 1976.

40. Markel, P., Shu, P., Ebeling, C., Carlson, G.A., Nagle, D.L., Smutko, J.S., and Moore, K.J., Theoretical and empirical issues for marker-assisted breeding of congenic mouse strains, *Nature Genet.*, 17, 280, 1997.

41. Klein, J., *Biology of the Mouse Histocompatibility-2 Complex*, Springer-Verlag, Berlin, 1975.

42. Bradley, D.J., Models of complex host-parasite relationships: murine leishmaniasis, in *Animal Models in Parasitology*, Owen, D.G., Ed., MacMillan Press, Basingstoke, London, 1982, p. 69.

43. Miyashita, N. and Moriwaki, K., H-2 controlled genetic susceptibility to pulmonary adenomas induced by urethane and 4-nitroquinoline 1-oxide in A/Wy congenic strains, *Jpn. J. Cancer Res.*, 78, 494, 1987.

44. Cormier, R.T., Bilger, A., Lillich, A.J., Halberg, R.B., Hong, K.H., Gould, K., Bornstein, N., Lander, E.S., and Dove, W.F., The Mon1AKR intestinal tumor resistance region consists of Pla2g2a and a locus distal to D4Mit64, *Oncogene*, 19, 3182, 2000.

45. Nadeau, J., Modifier genes in mice and humans, *Nat. Rev. Genet.*, 2, 165, 2001.

46. Taylor, B.A., Recombinant inbred strains, in *Genetic Variants and Strains of the Laboratory Mouse*, Lyon, M.F., Rastan, S., and Brown, S.D.M., Eds., Oxford University Press, Oxford, 1996, p. 1597.

47. Malkinson, A.M., Genetic studies on lung tumor susceptibility and histogenesis in mice, *Environ. Health Perspect.*, 93, 149, 1991.

48. Tripodis, N. and Demant, P., Three-dimensional patterns of lung tumor growth: association with tumor heterogeneity, *Exp. Lung. Res.*, Sept. 2001, pp. 27, 521.

49. Fortin, A., Caradon, L.R., Tam, M., Skamene, E., Stevenson, M.M., and Gross, P., Identification of a new malaria susceptibility locus (Char4) in recombinant congenic strains of mice, *Proc. Natl. Acad. Sci. USA*, 98, 10793, 2001.

50. Yershov, Y., Baldini, T.H., Villagomez, S., Young, T., Martin, M.L., Bockman, R.S., Peterson, M.G., and Blank, R.D., Bone strength and related traits in HcB/Dem recombinant congenic mice, *J. Bone Miner. Res.*, 16, 992, 2001.

51. Nadeau, J.H., Singer, J.B., Martin, A., and Lander, E.S., Analysing complex genetic traits with chromosome substitution strains, *Nature Genet.*, 24, 221, 2000.

52. Kren, V., Qi, N., Krenova, D., Zidek, V., Sladka, M., Jachymova, M., Mikova, B., Horky, K., Bonne, A., Van Lith, H.A., Van Zutphen, B.F., Lau, Y.F., Pravenec, M., and St Lezin, E., Y-chromosome transfer induces change in blood pressure and blood lipids in SHR, *Hypertension*, 37, 1147, 2001.

53. Cowley, A.W., Roman, R.J., Kaldunski, M.L., Dumas, P., Dickhout, J.G., Green, A.S., and Jacob, H.J., Brown Norway chromosome 13 confers protection from high salt to consomic Dahl S rat, *Hypertension*, 37, 456, 2001.

54. Martin, A., Collin, G.B., Asada, Y., Varnum, D., and Nadeau, J.H., Susceptibility to testicular germ-cell tumours in a 129.MOLF-Chr 19 chromosome substitution strain, *Nature Genet.*, 23, 237, 1999.

55. Nadeau, J.H., Singer, J.B., Matin, A., and Lander, E.S., Analysing complex genetic traits with chromosome substitution strains, *Nature Genet.*, 24, 221, 2000.

56. Jenkins, N.A., Copeland, N.G., Taylor, B.A., and Lee, B.K., Dilute coat colour mutation of DBA/2J mice is associated with site of integration of an ecotropic MuLV genome, *Nature*, 293, 370, 1981.

57. Maclean, N., *Animals with Novel Genes*, Cambridge University Press, London; New York, 1994.

58. Brem, G. and Muller, M., Large transgenic mammals, in *Animals with Novel Genes*, Maclean, N., Ed., Cambridge University Press, London; New York, 1994, p. 266.

59. Gulezian, D., Jacobson-Ram, D., McCullough, C.B., Olson, H., Recio, L., Robinson, D., Storer, R., Tennant, R., Ward, J.M., and Neumann, D.A., Use of transgenic animals for carcinogenicity testing: considerations and implications for risk assessment, *Toxicologic Pathology*, 28, 482, 2000.

60. Linder, C.C., The influence of genetic background on spontaneous and genetically engineered mouse models of complex diseases, *Lab. Anim.*, 30, 34, 2001.

61. Jeffreys, A.J. and Neumann, R., Somatic mutation processes at a human minisatellite, *Hum. Mol. Genet.*, 6, 129, 1997.

62. Linblad-Toh, K., Winchester, E., Daly, M.J., Wang, D.G., Hirschhorn, J.N., Laviolette, J.P., Ardlie, K., Reich, D.E., Robinson, E., Sklar, P., Shah, N., Thomas, D., Fan, J.B., Gingeras, T., Warrington, J., Patil, N., Hudson, T.J., and Lander, E.S., Large-scale discovery and genotyping of single-nucleotide polymorphisms in the mouse, *Nature Genet.*, 24, 381, 2000.

63. Horvat, S. and Bunger, L., Polymerase chain reaction-restriction fragment length polymorphism (PCR-RFLP) assay for the mouse leptin receptor (Lep(db)) mutation, *Lab. Anim.*, 33, 380, 1999.

64. Namae, M., Mori, Y., Yasuda, K., Kadowaki, T., Kanazawa, Y., and Komeda, K., New method for genotyping the mouse Lep(ob) mutation, using polymerase chain reaction assay, *Lab. Animal Sci.*, 48, 103, 1998.

65. Irwin, M.H., Moffatt, R.J., and Pinkert, C.A., Identification of transgenic mice by PCR analysis of saliva, *Nature Biotech.*, 14, 1146, 1996.

66. McPherson, M.J., Quirke, P., and Taylor, G.R., *PCR — A Practical Approach*, IRL Press at Oxford University Press, Oxford; New York; Tokyo, 1991.

67. Nomura, T., Esaki, K., and Tomita, T., *ICLAS Manual for Genetic Monitoring of Inbred Mice*, University of Tokyo Press, Tokyo, 1984.

68. Hedrich, H.J., *Genetic Monitoring of Inbred Strains of Rats*, Gustav Fischer Verlag, Stuttgart, New York, 1990.

69. Litt, M., PCR of TG microsatellites, in *PCR — A Practical Approach*, McPherson, M.J., Quirke, P., and Taylor, G.R., Eds., IRL Press at Oxford University Press, Oxford; New York, 1991, p. 85.

CHAPTER 10

# Phenotypic Characterization of Genetically Engineered Mice

Charsa Rubin, Sue Knoblaugh, and Warren Ladiges

## CONTENTS

0-8493-1086-5/03/$0.00+$1.50
© 2003 by CRC Press LLC

# INTRODUCTION TO PHENOMICS

The powerful and wide-ranging genetic tools available in the laboratory mouse make it the major experimental model for studying mammalian gene function *in vivo*, and modeling human disease traits. Development and utilization of appropriate tools for assessing clinical phenotypes in mice is a crucial aspect of relevant model comparison in the post genome era.[1] The laboratory mouse is an excellent mammalian system for studying normal and disordered biological processes because of low cost and the technological ability to exploit genetic tools for investigation of mammalian gene function.[2] Many robust models of human disease may therefore be developed, and these in turn will provide critical clues to understanding gene function. Phenomics can be simply defined as the study of the phenotypic or biophysiological characteristics of mutant mice, especially genetically engineered mice.[3] An integrated, multidisciplinary approach is absolutely essential to fully exploit the power of mouse phenomics in molecular medicine. The establishment of an infrastructure for archiving and distributing the growing mutant mouse resource data will help assure the accessibility and utilization of newly created mutant mice lines. The objective of this chapter is to provide a brief summary of some of the pathobiological and physiological techniques currently being used in genetically engineered mice for the investigation of relevant human diseases and comparative mechanistic processes.

# TRANSGENIC TECHNOLOGY

The generation and characterization of genetically altered mice is intensive, time consuming, and technically demanding, and it includes preparation of the construct, selection of the mouse strain for embryo donation or ES (embryonic stem) clone, microinjection of DNA into pronuclei or targeted ES cells into blastocysts, identification of founder animals or chimeras, breeding and testing of transgenic progeny, and maintenance of colony records and health. It is essential to have the necessary expertise and resources to be successful in the development of new genetically altered animal models.

Popular mouse background strains used to generate mouse models include (C57Bl/6XC3H/He)F1 and (C57Bl/6XDBA/2)F1. However, the problems of genetic variation between F2 animals, genetic drift in subsequent generations, and the extensive backcrossing onto an inbred background required to regain genetic definition, are greatly minimized by using an inbred embryo strain. The disadvantages of inbred strains are decreased breeding efficiency, poor ovulators, eggs with small pronuclei (thus more challenge for microinjection), and increased susceptibility to lysis. The inefficiencies of C57BL/6 in superovulation, microinjection, and reproduction can be minimized by monitoring specific biological end points such as pronuclear egg formation and response to gonadotropin as well as diet, age, and light exposure. FVB and SWR female mice are highly suitable for the propagation of transgenes, as they have high ovulation rates, and oocytes are resistant to lysis.[4] For production of mutant mice via the microinjection of gene-targeted embryonic stem (ES) cells, inbred strains are routinely used as blastocyst donors to provide a constant genetic background for the production of chimeras. Generally, targeted ES cells (most commonly 129/SvJ and C57Bl/6 derived) are injected into BALB/c or C57BL/6 blastocysts, respectively. Pseudopregnant recipients can be any strain with high breeding efficiency. A popular outbred strain is Swiss Webster, which is an excellent breeder and exceptional foster mother (personal observations).

Procedures for generating transgenic mice have been described in detail.[5] In general, the transgene constructs for generating transgenic mice consist of a functional promoter, initiation codon, polyadenylation site, and full length cDNA or genomic DNA for a specific gene fused to the enhancer and promoter sequence and cloned into an appropriate vector. Because many vectors used in cloning can interfere with expression of the transgene, it is important that unique restriction sites at the 5' and 3' ends of the transgene are available to remove plasmid sequences prior to injection. Thermocycling with probes specific to the transgene is the most frequently used method for analyzing transgene integration, although Southern and dot blot analyses are also used. The choice of promoter depends on the target tissue of interest or the desire to direct the ubiquitous expression of a transgene. Promoters such as chick $\beta$-actin and cytomegalovirus are frequently used as broad-spectrum promoters by focusing the expression of genes to a broad array of tissues.[6] However, these promoters have limitations, because expression in liver is low or undetectable, and it is variable in other organs. Organ-specific promoters such as the rat albumin enhancer and promoter, which targets gene expression to the liver, and the lck promoter, which targets immature T cells, have been used successfully in various studies. Inducible promoters such as the tetracycline-inducible system can also be employed, because they are preferable in a number of situations in order to obtain maximum quantity of the gene product and to regulate its expression.

The gene of interest is excised from its vector sequences, purified by gel electrophoresis, and linearized. Linearized DNA of less than 10 kb in length is commonly used, because DNA with higher molecular weight becomes too viscous, making it difficult to load and inject through a 1 to 2 $\mu$m opening of a microinjection needle. The usual concentration of DNA for injection is 2 to 5 ng/uL. The foreign DNA is introduced into the murine germ line by microinjection of the pronuclei of one-cell fertilized eggs. Routinely, between 100 and 200 embryos are injected, which are then implanted into pseudopregnant females. Approximately 10 to 30% of the transferred eggs result in live births with litter sizes of five to eight per recipient. The pregnant females are monitored for phenotypic abnormalities during gestation, such as embryo reabsorption. Pups in the initial perinatal period are monitored for lack of suckling reflex, as well other abnormalities. At 2 to 4 weeks of age, the founder pups are screened for the presence of the transgene integration using tail or ear punch biopsies and PCR analysis. Southern blot analysis, while technically more cumbersome, provides an estimate of transgene copy number and is used to confirm PCR results. Once transgene positive pups are identified, they are raised to sexual maturity and mated to wild-type mice of the desired background strain. Transgene positive offspring will confirm that line as a permanent founder.

Embryonic stem cell procedures have been described in detail.[5] The ES cell lines used are of 129 derivation or C57Bl/6 origin. Homologous recombination vectors are transfected into ES cells via electroporation. The vectors are linearized and used at a concentration of about 1 $\mu$g/mL free of toxic contaminants, such as ethanol, ethidium bromide, and excess salt. Electroporation is performed on $2 \times 10^7$ ES cells in PBS (without $Ca^{++}$ and $Mg^{++}$) containing 25 $\mu$g vector. A short high-voltage electrical pulse is applied to the ES-vector mixture, which allows for pore development in the ES cell membranes and entry of the vector. A postelectroporation killing rate of 50%, as assessed by tryptan blue exclusion, is an early indication of a successful electroporation. Following electroporation, the cells are plated with selection media to isolate the targeted clones. When colonies of resistant cells are detected, they are tested by PCR or Southern blot for proper genetic manipulations. Because genetic drift and maintenance of totipotency are exquisitely sensitive to culture conditions, generating large pools of low-passage frozen stocks are required for successful thaw and chimera production.

Microinjection of the targeted ES cells into blastocyst (E3.5) stages is used to produce targeted genetic mutant mice. ES cells are injected into the space between the uncompacted eight-cell embryo and the zona pellucida, into the center of a compacted eight-cell/morula or into the blastocoel of a blastocyst, whether unhatched or hatched. Five to 15 ES cells are injected into each blastocyst, and nine to 15 of the microinjected blastocysts are transferred to the uterus of a pseudopregnant recipient. Pups are born approximately 18 days from transfer. Detection of chimerism is possible as early as 3 days later but is not confirmed and quantified until hair grows in, 5 to 7 days after birth. Only male chimeras, where ≥50% of the coat color is from the ES contribution, are selected for testing germ line transmission of the induced mutation. The male chimeras are test mated with C57BL/6 females. The resulting black offspring are derived only from a host embryo background, whereas black agouti offspring are derived from the ES cell contribution, due to coat color genetics of all the ES 129 substrains. These black agouti pups are further tested, using tail or ear biopsies, to determine whether the wild type or the targeted allele is transmitted. Pups determined to be heterozygous (+/-) for the mutations are retained for breeding.

## COLONY ASSESSMENT

It is essential for accurate phenotypic characterization that the integrity of the genetic alteration of a specific line be maintained over several generations of backcrossing using sound breeding practices and accurate record keeping. In addition, basic observational assessments at the colony level to detect any unusual characteristics of newly generated mouse lines are mandatory in research involving genetically engineered mice.

### Genotyping

An important strategy in validating and maintaining the integrity of genetically altered mutant mouse lines for distribution is to be able to accurately and efficiently identify the specific transgene or mutation. Because homozygosity may instill lethality or infertility, many imported transgenic and gene-targeted mutant lines are maintained as heterozygotes so must be individually genotyped. It is essential to obtain genotyping protocols from the scientific group originally characterizing the mouse line. Even then, a significant amount of time and resources may be necessary to optimize PCR protocols (or Southern blotting if necessary) for each mouse line. Oligonucleotide primers can be custom produced by a reliable commercial company. Genomic DNA for these assays is isolated from tissue biopsy samples using standard DNA isolation procedures. DNA probes for Southern blots can be labeled with $^{32}$P-dCTP using the random primer method. PCR and Southern blot analysis should be performed on tail or ear punch biopsies obtained from 2- to 4-week-old potential founder pups as a source of DNA for testing transgene integration or deletion mutation. Once mice integrating the transgene or exhibiting the deletion mutation have been identified, these mice are set up in breeding pairs. The goals of breeding the founder mice are to observe the inheritance and expression patterns of transgenic lines through several generations, to expand the colony to provide mice for experimental study, and to produce transgenic homozygosity for comparison with the heterozygous state. Breeding directly to homozygosity means +/- brother to +/- sister mating of offspring.

Accurate and user-friendly record keeping programs are vital for optimizing the expression of the mutation, maximizing the number of viable offspring, and minimizing the potential confounding influence of background genes from breeder parents. Record keeping systems should incorporate a combination of written and computer-based records. Either a commercial colony management software system or standard Excel spreadsheets can be used to maintain a master list of every mouse that is ear tagged or otherwise marked for identification using tattoos, ear punches, subcutaneous transponders, or toe clipping. Items recorded in the master list include birth date, transgene name, generation, line, sex, parents, phenotype and genotype, and information about the specific project. Establishing a pedigree for each founder mouse that is mated is a vital part of the recording system. Customized cage cards for breeder cages provide a summary of the breeding activity for each cage. Customized weaning cards record the line and ear tag numbers so weanlings can be readily located and tracked. A separate file on the Excel spreadsheet or within the database system for breeding records should contain birth dates of litters, number of pups born and weaned per litter, ear tag numbers of pups, breeder setup date, breeder retirement date, strain, generation, line, and comments.

Because many transgenic mice are developed utilizing strains that are not desirable genetically for use as a background strain, backcrossing to the desired congenic strain is necessary. As each consecutive generation is backcrossed to the desired strain, the proportion of the genome originating from that strain increases until at the tenth generation, the desired background strain is 99.9% of the genome and considered congenic. As backcrossing for 10 generations can take upwards of 2.5 to 3 years, a method to shorten this time can be highly cost-effective. In speed congenics, offspring are screened for those with higher percentages of the background strain-specific markers, using single sequence polymorphisms.[7] These tests can be performed by commercial laboratories. Those mice exhibiting higher percentages of the background strain in their genome are then used as breeders to produce the next generation. Using speed congenics can reduce the number of generations required to display a transgene or deleted mutation on a known, desired strain by one half to only five generations. In addition, embryo transfer can be used in an attempt to further shorten the time to have inbred mutant congenics to within a year.

Selection of controls for genetically altered mice is essential for valid scientific experiments. For stocks of mixed or segregated genetic background, littermate controls are best. In cases where lines are maintained by homozygous sibling matings and, therefore, do not have wild-type littermates available, determination of the appropriate controls can be based on Jackson Laboratory recommendations.

When transgenic or wild-type control or breeder mice are shipped into and received by a barrier facility, maintained in a specific pathogen-free environment, quarantine is essential to eliminate viruses or other pathogens that imported mice may be harboring. For example, it has been reported that up to 48% of the strains imported to the Jackson Laboratory in 1997 had evidence of prior infection with one or more viruses.[55] Genetic and microbiologic status of mouse lines should be assessed by the health surveillance staff (as described below).

## Colony Phenotyping

The objectives of phenotyping procedures are to cover the basics of observational assessment, gross necropsy examination, whole body and individual organ weights, clinical pathology assessments, and histological examination. As a first step in phenotyping mutant mouse lines, assessment of the colony as a whole unit is an important aspect. All mice should be observed at birth and at least once a week from birth until that time when the individual mouse is used for an experiment, euthanized, or dies naturally. Some phenotypes are readily seen or deducted, such as those mutations causing flaky skin (fsn)[8] or those that are homozygous lethal. Defects in development may be seen at or within a few days of birth or may not be seen until much later in the life span. Researchers and technicians should be aware that any deviation from normal may be a phenotype, not only in individual mice but also throughout the transgenic line. Reproductive problems such as small litter size, elongated gestation or gestation cycles, and increased cannibalization may be due to genetic engineering rather than to environmental problems. Differences in hair quality, growth rate, skeletal structure or behavior between transgenic strains should be recorded and statistically evaluated to determine the effect of the transgene on the deviation. Colony records of the parameters in Table 10.1 will help to track the effect of a transgene or deletion mutation within a colony of genetically engineered mice.

The effect of background strain in determining the phenotype of transgenic mice is at many times a large one. Researchers should have a good understanding of the specific influences the background strain of their mice will have on the effects they see in the colony and in individual mice. Jackson Laboratories has a thorough and informative Web site listing many of the common diseases of the most popular inbred strains (www.jax.org). For example, C57BL/6 mice commonly exhibit ulcerative dermatitis and hydrocephalus. Knowing that these diagnoses are a result of the background strain and not the transgene will assist in true characterization of the mutant mice.

Another complicating factor in colony phenotyping is infectious disease. In a specific pathogen-free environment, all rooms should be routinely examined for infectious processes through the use of sentinel cages. Despite that precaution, disease can and will break through the protective barrier and cause infectious processes in genetically engineered mice. A routine plan of specific diagnostic procedures should be in place so that any deviation from the norm can be evaluated and the effects of infectious processes ruled out.

**Table 10.1 Observational Assessments**

| Examine | To Determine |
|---|---|
| Embryonic death | Breeding efficiency |
| Litter size | |
| Fetal death | |
| Birth weight and rate of growth | Development |
| Hair growth | |
| Eyes and ears open | |
| Incisor eruption | |
| Stand and walk | |
| Physical exam for malformations | Various clinical parameters |
| Coat and skin condition | |
| Nasal or ocular discharge | |
| Hemogram, serum chemistry profiles | |
| Tumor development | |
| Eating, drinking, grooming | Simple behavioral patterns |
| Alertness, aggression | |
| Activity level, exploration | |
| Posture, climbing, locomotion | |
| Righting, twitches, tremors, reflexes | |
| Auditory startle, seizures | |
| Stereotypic behaviors | |

Observational assessments, summarized in Table 10.1, should include general appearance, posture, and mobility. These features can be monitored with the help of trained animal care and technical staff on a consistent and intense basis. Daily reports should be made and weekly assessments compiled. Specific aspects should be tailored to the individual research plan. For example, aging studies can include hair graying and alopecia, body weights, muscle atrophy, lordokyphosis, dermal thickness, and subcutaneous adipose. For hair regrowth assessment, hair is shaved from a 2-cm-square area at the dorsoventral back near the base of the tail. Regrowth is defined as the first appearance of hair in each of eight sections, designated by a transparent grid, in the shaved area. Aging ad lib-fed mice have been shown to have an extended hair regrowth time compared to calorically restricted mice.[9]

## Cryopreservation

Cryopreservation protects against catastrophic losses by such events as breeding cessation, genetic contamination, or disease outbreaks, stops genetic drift, and eliminates the necessity and expense of maintaining live breeders when specific lines are not in high demand. Cryopreserved stocks are pathogen free and can be recovered pathogen free. Cryopreservation of mouse lines can be approached using embryos, semen, or whole ovaries. Embryos take the most effort and expertise to freeze, but take little expertise to thaw and so, are especially suitable for transfer to outside facilities through shipment on dry ice. Semen preservation has the advantage that storage of just a few straws has the capability of generating the line many times over; however, the techniques have been most successful using hybrid animals.

## Health Surveillance

Surveillance for health quality is essential to detect unwanted and complicating spontaneously occurring mouse diseases. Health monitoring entails sampling of sentinel mice in the quarantine areas as well as any SPF mice in the colony. This can be done at intervals of 4 weeks in the quarantine area and every 12 weeks in the repository colony. Sentinel mice are often SW, 2-month-old females. These mice are free of all the agents that will be screened. In the barrier colony, sentinel mice are placed two per cage and placed on a cage rack so as to receive soiled bedding from about 70 cages each time cages are changed. In addition, foster mothers and retired breeders should also be sampled. Both types of surveillance are necessary, because not all infections in mice are spread by soiled bedding.

The skin should be examined for ectoparasites. The cecum and colon of mice should be examined for the presence of endoparasites. Mucosal scrapings from the duodenum, jejunum, cecum, and colon should be examined for pathogenic protozoa. In addition, every six weeks, a random sample of colony mice should be tested by anal tape test and fecal flotation for endoparasites, and the pelage examined by dorsal tape test for ectoparasites. Feces should be cultured to detect enteric pathogens (*Salmonella* sp., *Citrobacter rodentium,* enteropathogenic *Escherichia coli, Pseudomonas* sp., *Campylobacter* sp., and *Clostridium* sp.) PCR assay of the feces should be done for *Helicobacter* sp. (*H. hepaticus, H. bilis, H. rodentium*). A swab of the upper and lower respiratory tract should be cultured to detect respiratory pathogens (*Pasteurella* sp., *Bordetella* sp., *Streptococcus* sp., and *Staphylococcus* sp.). PCR assay of a smear from a nasopharyngeal swab should be done for CAR bacillus. Serology should be performed to detect the presence of various viral and mycoplasmal infections. Every six weeks, mice should be tested for MHV and MPV. On a quarterly basis, mice should be tested for MHV, MPV, TMEV, and rotavirus. In addition, mice should be tested semiannually for PVM, Sendai virus, reovirus-3, adenovirus, MTV, LCMV, MCMV, ectromelia and *Mycoplasma* sp.

Mice that die unexpectedly or spontaneously, or are identified through daily health screening in the quarantine or repository area as showing signs of poor health or disease, should be necropsied by a veterinary pathologist. This examination should consist of grossly visualizing major organ systems. This should be followed by a microscopic examination of all gross lesions within a subset of organs that include brain, heart, lungs, liver, kidney, adrenal glands, reproductive organs, gastrointestinal tract, pancreas, skin, and spleen. Where indicated, samples should be taken for microbial isolation, special diagnostic staining, or clinical chemistry determinations to aid in reaching a diagnosis. The lungs of immunodeficient mice can be silver stained and examined for *Pneumocystis* sp.

Helicobacter was initially recognized as a pathogen in rodents in 1994.[10] Several species of Helicobacter (*H. hepaticus, H. bilis, and H. rodentium*) are increasingly being recognized as a serious problem in mouse colonies because of their ability to induce hepatitis, inflammatory bowel disease, and rectal prolapse, and function as an intercurrent variable in biomedical research studies.[11] Also, some lines of mice generally do not breed well or thrive when infected with one or more Helicobacter species. A recent report detected *Helicobacter hepaticus* in 11 different genetically altered mouse lines.[12] Infection was associated with rectal prolapse and inflammatory bowel disease, and such animals would not be expected to reproduce optimally. *Helicobacter hepaticus* is apparently widespread in commercial mouse colonies and only a few large suppliers are currently certifying their mice to be free of certain *Helicobacter* spp. Helicobacter organisms are infectious and can easily be detected in sentinel mice using dirty bedding.[13]

## IMAGING

Recent advances in noninvasive imaging have greatly enhanced the ability to phenotypically characterize genetically engineered mice. Through the use of MRI, PET, and ultrasound, to name a few, the internal biochemical processes manipulated through transgene expression can be studied and visualized in minute detail. This capability has lead to many discoveries and developments in the world of genetic mutations. This section gives a general overview of a few of the advanced technologies now available for imaging transgenic mice.

In use by clinical practitioners for quite some time, ultrasonic imaging is becoming more widely used as a tool for the study of transgenic mice. Ultrasound can image the internal organs by detecting the echoes of ultrasound waves passed from a transducer through animals' bodies. Depending on the tissue, the image received will appear as a structure ranging in visualization on the gray scale from black to white with mineralized structure such as bone viewed as white and liquid structures such as urine and blood shown as black. Echocardiography is based on the same principles, applied to cardiac structure and function.

An emerging technology in advanced imaging is PET, or positron emission tomography.[14] Because disease is often expressed physiologically first, before it is anatomically observable, PET imaging provides a way to identify and characterize the nature of early onset, before it is clinically expressed. PET is an analytical nuclear medicine imaging technology that uses positron-labeled molecules in very low mass amounts to image and measure the function of biological processes, such as tumorigenesis, with minimal tissue and cellular disruption. Measuring, but not disturbing, the biological process is a fundamental and biologically important aspect of the tracer technique of PET. The assay depends on synthesizing a positron-labeled molecule that mimics a few steps of tumorigenesis so that kinetic analysis can estimate the concentration of reactants and products and the rates of reactions, using PET scanning. The PET scanner measures the changing regional tissue concentration of the labeled molecule and its labeled product over time.

Another category of noninvasive imaging useful in the study of transgenic mice, especially for cardiac and brain anatomy and function, is MRI, magnetic resonance imaging. Most MRI machines image hydrogen protons within the tissues. MRI is based on the property of the protons to have weak magnetic fields that will create an electrical current when exposed to a magnet. The "magnetic resonance" produced by the spinning, electrified nuclei registers as an electrical signal, determining the degree of brightness produced and creating an image of individual organs based on their specific amounts of hydrogen present. The differing degrees of brightness between separate organs provides contrast and leads to the ability to distinguish different areas of tissue. Adding a contrast agent, such as a gadolinium chelate, enhances the visualization of the tissues.[15]

Optical imaging can also be used to visualize the structure and function of tissues in genetically engineered mice. There are several methods in which the use of light is the basis of the imaging technique. Near-infrared spectroscopy is the transmission of light through tissue; absorption of one or more of the light wavelengths allows characterization of the region through which the light has passed. Infrared spectroscopy measures the reflection of light that has been differentially absorbed. Fluorescence optical imaging visualizes tissues that have taken up fluorescent dyes or natural chromaphores. Optical coherence tomography is analogous to B-mode ultrasonagrophy, but the acoustical waves used in ultrasound are replaced with infrared light.[15]

## PATHOLOGY TECHNIQUES

Pathological characterization of genetically engineered mice includes a complete necropsy for gross tissue evaluation and microscopic evaluation via histopathology. Specialized pathology methods such as immunohistochemistry and flow cytometry should also be utilized to investigate specific lesions, and the necropsy should be supported by clinical pathology findings. Genetically altered mice can and do have unexpected phenotypes. In addition, the phenotype of a particular gene alteration can be expressed in other tissues that may directly or indirectly influence the target tissue of interest. This is true of all genomic manipulations and underscores the importance of accurate and thorough phenotypic characterization and model validation by pathologists familiar with background mouse pathology and the human disease of interest. It is also important to understand that different strains of mice vary in normal anatomy, physiology, and behavior.

### Necropsy Procedure

Initial characterization should include necropsies performed on immature and mature mice of both genders as well as mice sacrificed in a moribund condition or found dead (autolytic state permitting). Embryos and fetuses should be examined in those mouse lines with high fetal loss or developmental anomalies. Blood for serum chemistry evaluation should be collected immediately before euthanasia. The necropsy is performed in a systematic and orderly fashion. Tissues in each system are evaluated grossly and microscopically (Table 10.2). Mice are weighed then humanely euthanized. The animal's weight, sex, age, and identification number are recorded. The gross necropsy begins with an external examination of the external body surface, all orifices, cranial vault, external surface of the brain, the nasal cavity and sinuses, the thoracic, abdominal, and pelvic cavities and viscera. It is important to keep in mind that all tissues should be examined *in situ* before being dissected from the body. Samples for microbiological culture are taken before tissue samples are collected. The entire body is palpated for superficial swellings, enlarged organs, or masses. If an abnormality of the hair coat is noted, a hair sample is manually plucked. Forceps are not recommended, as this may damage the hair shaft. Hair samples are collected from the same area in every mouse in a study in order to keep collection techniques standard. The skin should be sampled from an area on the mouse that was not plucked to prevent artifactual changes in the hair follicles being studied.

**Table 10.2  Tissues Collected from Genetically Engineered Mice for Potential Pathological Evaluation**

| | | | | |
|---|---|---|---|---|
| Adrenals | Esophagus | Liver | Pituitary | Spleen |
| Aorta | Eyes | Lungs | Prostate | Stomach |
| Bone marrow | Femur | Lymph nodes | Rectum | Thymus |
| Brain | Heart | Mammary gland | Salivary gland | Thyroid |
| Cecum | Ileum | Muscle (thigh) | Sciatic nerve | Tongue |
| Colon | Jejunum | Optic nerve | Seminal vesicle | Trachea |
| Duodenum | Kidneys | *Gonads | Skin | Urinary bladder |
| Epididymis | Lacrimal glands | Pancreas | Spinal cord | Uterus/vagina |

The following organs will be weighed prior to partitioning and fixation: brain, heart, lungs, liver, kidneys, spleen, uterus, gonads, and adrenal gland. The pituitary will be weighed postfixation, and paired organs will be weighed separately. Organ to body weight percentages and organ to brain weight ratios will be calculated. The intestinal tract will be collected and the cecum separated from the colon. Intestinal specimens will be gently inflated with fixative by intraluminal injection and prepared as an intestinal roll (by placing it on an index card and rolling it in a flat spiral around a central toothpick). Care must be taken when collecting the pancreas, as it is firmly adhered to both the small and large intestine. The stomach is removed and also inflated with fixative. The liver should be the last abdominal organ removed. One should enter the thorax so as to use the diaphragm as a handle while the liver is removed. Upon removal of the liver, the salivary glands can be removed, and the heart, lungs, and trachea can be removed as one. The lungs should also be inflated with fixative to ensure proper fixation for histology. Lastly, the brain is collected as well as the spinal column if needed. It is important to carefully remove the meninges and then the brain from the skull. Tissues should be preserved in 4F-1G (modified Karnovskys formalin-gluteraldehyde fixative) for one week then transferred to 10% neutral buffered formalin for storage. Primary fixation in 4F-1G will permit ultrastructural analysis if indicated.

## Histology

Proper trimming is important for proper orientation of tissue samples on the slides. This will ensure that the pathologist can identify normal tissue from any pathologic changes or lesions. Primary fixation in 4F-1G will permit ultrastructural analysis if indicated. Hematoxylin and Eosin (H&E) stain should be used on all histopathology samples. Special stains may be employed if needed to disclose certain pathological changes. Special stains include Congo red for amyloid, Trichrome for collagen, Von Kossa for calcium, and PAS for glycogen, especially in skeletal muscle and heart. Intracytochemistry for the different types of amyloid should be performed on the heart, if the data warrants it. Before trimming any bone samples, they must be decalcified in a hydrochloric acid decalcifying solution. After decalcifying, the samples must be rinsed with running water before being trimmed. The rinsing can take up to 3 to 4 h. Failure to properly rinse samples may result in inadequate staining.

## Immunohistochemistry

Immunohistochemisty (IHC) can be used to determine the presence of a particular protein within a tissue, the distribution in that tissue, and its precise location within a single cell. There are many IHC methodologies available. It is important that the methodology used be sensitive, produce little or no background, and give reliable and reproducible results. IHC is used at the light microscope level. It is an enzyme-based detection method that includes the use of horseradish peroxidase (HRP) or alkaline phosphatase (AP). Key points to keep in mind in IHC include choosing an enzyme system, preparing tissue properly, and using the correct tissue staining techniques. The two most commonly used enzyme detection systems are the horseradish reoxidase and the alkaline phosphatase systems. The HRP or peroxidase systems are most widely used and tend to produce a dense label. The HRP system provides better staining of nerve axons and projections in the CNS. The AP or alkaline phosphatase systems are more sensitive and allow for better cell morphology. The AP system is also used when endogenous peroxidase activity is a problem.[16]

Tissue preparation is crucial to any immunohistochemical procedure. The success of the application is based on antigen preservation and tissue morphology. These variables can be optimized if proper consideration is given to the fixative, method of fixation, tissue sample size, length of time in and temperature of the fixative, and whether the sample is frozen or embedded in paraffin blocks. The availability of primary antibodies that work well in mouse tissues is of primary concern. Some of the antibodies work better in frozen tissue, and some work better in paraffin-embedded tissues. Some antibodies are not commercially available. If specific antibodies are not available, individual investigators can be a good resource. One must consider tissue morphology when deciding on freezing versus paraffin embedding samples. It is faster to freeze the tissue, but the morphology is often compromised. It is time-consuming to embed samples, but the results are often enhanced due to improved morphology. There is not a universal fixative, so it is important or ideal to use a fixative that is known to work for the tissue type and target antigen. If IHC is to be performed on tissue that has been fixed in formalin, it is important to keep in mind that antigens may be masked. Antigen retrieval may be performed by steaming the slides in citrate buffer for 10 min. This method tends to work well and results in minimal destruction of cellular morphology. It is important to keep in mind that different antigen–antibody combinations may require different retrieval methods. Once the tissue is properly prepared, one can follow specific protocols for staining paraffin sections or frozen sections. The majority of the protocols involve blocking of various tissue components, antibody and detection reagent incubations, and buffer washes. For more detailed information, one should consult references discussing general techniques and specific applications.[16]

## Immunofluorescence Flow Cytometry

Although special pathology techniques such as immunohistochemistry can prove quite useful in evaluating tissues, flow cytometry is more sensitive and allows analysis of a large number of cells in tissues amenable to single-cell analysis. Flow cytometry provides the most basic measure of the cell cycle in G1, S, and G2 phase fractions, calculated from DNA histogram distributions. Additional information is obtained from Ki-67 staining simultaneously with DNA. BrdU pulse labeling of cells in S phase can also be used effectively to examine cell cycle kinetics in mice, followed by analysis with anti-BrdU antibody. Antibodies to cyclins A, B1, D, and E are analyzed simultaneously with DNA content by flow cytometry; tumors often exhibit dysregulated patterns of expression of cyclins throughout the cell cycle, and this is readily visualized by such analyses. Cell death, which often occurs through the apoptotic process, is just as important a parameter in understanding cellular behavior as is cell proliferation and control. Cell death and apoptosis can be readily visualized by flow cytometry.

DNA content, cell cycle, and cell kinetics measurement are used in the assessment of DNA content (ploidy), proliferative activities of cells and tissues, for observing evidence of cell-cycle checkpoint arrest following exposure to DNA damaging agents, and assessing cell survival and death. A broad spectrum of assays for these purposes is available, ranging from simple univariate DNA content measurements, through dual DNA/RNA staining. The BrdU-Hoechst method is based on continuous exposure to BrdU.[17] In addition, BrdU can be used for pulse labeling, and the incorporated BrdU can be detected by using an anti-BrdU antibody, or by UV exposure (BrdU containing DNA undergoes photolysis) followed by labeling of resultant DNA strand breaks by terminal transferase incorporation of FITC-dUTP. The cyclical expression of cyclins (e.g., cyclin A, B1, D, E) is believed to be central to the regulation of the cell cycle. Abnormal expression may be associated with conditions of altered cell proliferation. Although cyclin expression can be quantified by a variety of means, only flow cytometry can do so while identifying the expression as a function of the phase of the cell cycle and heterogeneity in expression in a population. The use of anticyclin antibodies simultaneously with DNA staining has emerged as a major area of current interest in cell cycle regulation.[18] Antibodies to cyclins B1 and A are especially easily used. Similarly, cell activation can be quantified with antibodies to proliferation-associated antibodies such as PCNA or Ki-67, and abnormal expression of proteins associated with transformation or immortalization, such as p53, can be quantitated. Susceptibility to apoptosis is thought to be related to oxidative states, as agents that are antioxidants can reduce apoptosis. In addition, induction of apoptosis is a cellular response to DNA damage. Thus, measurements of apoptosis are highly relevant to validating mouse cancer models. Flow cytometry has become a method of choice for quantitation of cell and nuclear features characteristic of apoptosis. Early cellular changes leading to apoptosis can be detected by altered membrane characteristics and light scatter.[19] The most characteristic feature of apoptosis is fragmentation of nuclear DNA by endonuclease. Loss of DNA fragments from permeabilized cells can be detected, using flow cytometry, by a reduction in cellular fluorescence leading to a subdiploid "apoptotic peak."[20] However, a more recent and, in many cases superior, detection method is based on fixation of cells and end labeling the apoptotic DNA fragments using terminal transferase to incorporate fluorescein labeled dUTP.[21]

## Confocal Microscopy

One of the most useful methods to characterize mouse phenotypes is microscopy, especially newer techniques like confocal microscopy. Confocal microscopy uses fluorescent antibodies or probes to illuminate the desired immunofluorescent image. This technique is advantageous for its production of fine-focused images that can be collected rapidly and stored to disk or hard drive, thus allowing for optimal interpretation. The immunofluorescent images are created through the use of lasers rather than mercury lamps, enabling the production of monochromatic light of a defined wavelength.[16] Immunofluorescence is most often used in evaluation of immunoglobin deposits in the skin and renal membranes, examination of neuropeptides in nerves, and antigen studies.[8] Immunofluorescence can be used with frozen sections or fresh whole cells. It should be noted that formalin-fixed tissues can autofluoresce and should not be used.

## Clinical Pathology

Clinical pathology assessment includes a complete blood count (red blood cell count, pack cell volume, hemoglobin, platelets, white blood cell count, white blood cell differential) and serum chemistries (albumin, alkaline phosphatase, alanine transaminase, urinary bilirubin, blood urea nitrogen, calcium, creatinine, cholesterol, glucose, phosphorus, total protein, Na, K, and Cl). Blood can be collected noninvasively by the retro-orbital technique. Urinalysis should also be performed. Urine collection will be described later in this chapter. Samples should be submitted to a lab experienced in the analysis of small volume rodent samples.

## TUMOR ASSESSMENT

Transgenic mice have long been an important resource for studying human neoplasms, both benign and malignant. In addition to the numerous genetically engineered mouse models developed over the years, many popular background strains have a genetic predisposition toward certain types of tumors. Knowing the effect of the background strain is important in research involving genetic manipulation with genes involved in tumor pathways. In addition, a detailed understanding of the biochemical cycles intrinsic in cell regulation is vitally important to the researcher dealing with transgenic mice in cancer research. As described in the pathology section, there are many assays in addition to normal necropsy procedures and techniques useful to the researcher experimenting with the cellular regulation details of cancer research. Flow cytometry, immunohistochemistry, Southern blot analysis, and fluorescence *in situ* hybridization can all be used to assist in evaluating the phenotype of specific cancers induced or discovered in genetically engineered mice.

The first step in phenotypic characterization of genetically engineered mice in cancer research is to develop a tumor tracking study specific to the type of cancer expected in the mouse model. For example, if the research involves hematopoietic tissues and neoplasms derived from them, an integral part of the tumor tracking would be periodic blood exams to evaluate complete blood counts and the status of hematopoiesis in mice involved in the study. In all types of studies, all mice expressing the cancer-related gene should be monitored at least weekly for signs of illness. A phenotypic endpoint should be determined so that no individual animal is left to suffer with a large tumor burden. That endpoint can be determined by reviewing IACUC requirements. In many cases, monitoring attitude, behavior, and simple biological parameters such as weight or body temperature is enough to fully maintain a colony of transgenic mice with potential tumors. In others, a weekly noninvasive evaluation of individual animals, including an assessment of attitude, measurement of weight, and performance of a basic physical exam including abdominal palpation of internal organs and all accessible lymph nodes will allow the researcher to track all developing neoplasms and comply with given tumor endpoint requirements. However, for transgenic mice that do not show a phenotype or tumor development as readily, such as those with intrathoracic, central or peripheral nervous system lesions, or for experiments in which the identification of the initial stages of tumor development is the research goal, there are other more invasive and more sensitive methods of tumor tracking available.

One method of additional phenotypic characterization is to attempt to induce tumorigenesis by the administration of a carcinogenic agent or radiation treatment to mice genetically engineered to be susceptible to carcinogenic insults. A popular chemical agent that is easy to administer and is a well-characterized mouse-specific carcinogen is ENU (*N*-ethyl-*N*-nitrosurea). An alkylating agent, ENU will provide a wide range of DNA damage and subsequent systemic tumorigenesis. Exposure to ENU commonly occurs through IP injection (5 umol per gm per mouse) in weanling mice, 3 weeks of age. Control mice followed for 12 months generally develop 5 to 10% tumors, so numbers above this in the transgenic group are indicative of tumor susceptibility. Latency to tumor development statistics of percent tumor-free survival can be performed using Kaplan Meier plots in addition to the use of log rank tests for determination of significant differences.

As gross observation and general physical exam measurements are lacking in sensitivity to the beginning stages of tumorigenesis, more advanced technologies to detect impending tumors are often required. One method of detecting early tumor growth is through the use of ultrasound imaging. Using ultrasound imaging, developing tumors can be seen as an unnatural enlargement of a previously normal size organ or as discrete nodules or extensions of a different grayness within or beside a normally colored tissue of origin. In this method, tumors can be visualized before they can be felt by palpation and long before the tumor develops to the point where illness occurs or the tumor burden becomes too great and euthanasia is required.

An emerging technology in advanced tumor imaging is PET, or positron emission tomography. It is known that tumorigenesis is associated with a high need for glucose for energy and to provide the carbon backbone to meet the high cell replication rates of tumors through activation of the hexose monophosphate shunt. This knowledge led to the development of the tracer $^{18}$F deoxyglucose (FDG), so that when injected systemically, high levels of FDG signal delineate neoplasms from surrounding tissues. In fact, whole-body FDG PET scanning is capable of detecting abnormal tumor metabolism before anatomic changes occur and is able to distinguish between malignant and benign anatomic abnormalities.

## NEUROLOGIC ASSESSMENT

Neurologic deficits in animals are manifested in loss of memory and normal behavior and in the expression of ataxia and loss of motor skills. In working with transgenic mice designed to explore neurologic function, both aspects of neurology should be examined.

### Learning and Memory Assessment

The Morris water task is one of the most widely used paradigms to evaluate learning and memory behavior. This is a spatial navigation task in which the mouse swims to find a hidden platform, using visual cues to locate the platform. Escape from the water is the positive reinforcement. The task is based on the principle that mice are highly motivated to escape from a water environment by the quickest, most direct route. The procedure requires a circular plastic tank about 30 cm in diameter and about 45 cm in height filled with tap water to a depth of about 20 cm. The tank is filled the day before so water can reach room temperature the next day for testing. Milk powder is mixed into the water so as to make an opaque liquid which makes the resting platform invisible to a mouse swimming on the surface of the water. Water is changed daily, the tank disinfected, and refilled for the next day of testing. Two platforms are used, one just below the surface of the water, and thus invisible, and one just above the surface of the water. It is essential that environmental cues and surroundings remain constant throughout the duration of testing. The same experimenter must handle the mice throughout an experiment, with no intermittent visitors. The first step of the procedure is pretraining, consisting of gently introducing the mouse to the pool and platform. Generally the mouse will jump off the platform and is allowed a swim up to 15 sec, before being guided back to the platform. Up to three pretraining sessions are conducted, measuring the visual ability of the mouse to see the room cues (latency to reach the visible platform), and the motor ability to swim in the pool (swim speed). The next step is training for the hidden platform, which starts by placing the mouse in the water at the edge of the pool, allowing 60 sec to reach the platform and climb out of the water. This is repeated for a variable number of predetermined trials over 3 to 10 days. A criterion of 10 sec to reach the platform is set, so that training is continued until controls reach the platform within this time frame. At the end of the training period, each mouse is tested on a trial probe that measures the ability of the mouse to identify the spatial location that previously contained the hidden platform within a 60 sec time period. Search time spent in the trained quadrant must be greater than search time spent in the other three quadrants of the pool. Normal performance on the visible platform task but impaired performance on the hidden platform task is interpreted as a deficit in learning and memory.[22]

A second learning and memory task that highly complements the Morris water test is the cued and contextual conditioning test. This trial is much easier to perform in terms of less elaborate equipment needed, less time for investigator and mouse, and less laboratory space required. The test takes only 2 days with 10 min per mouse per day. Cued and contextual conditioning is based on fear conditioning in that it measures the ability of a mouse to learn and remember an association between an aversive experience and certain environmental cues. This type of fear conditioning is highly intuitive to mice as a species, because it is in their nature to freeze at the hint of danger. Conditioning training occurs on Day 1, when a mouse is placed in a chamber and allowed to explore for 2 min. After the timed exploration period, the animal is subjected to 30 sec of an auditory cue, white noise from an 80 dB broad-band auditory clicker. Following cessation of the white noise, the unconditioned aversive stimulus, a 0.35 mA footshock, is administered for 2 sec through a grid in the floor of the chamber. The measurement of unconditioned fear is the amount of time spent freezing in the test chamber on Day 1. Twenty-four hours later, testing for Day 2 commences. The mouse is returned to the same chamber and scored for bouts of freezing behavior, in which there is no movement by the mouse except for respiration. Presence or absence of freezing behavior is recorded in cycles of every 10 sec for a continuous 5 min, after which the mouse is returned to its home cage. The number of seconds spent freezing is termed the contextually conditioned fear. The second phase of testing for Day 2 starts 1 h after the end of the first phase. The mouse is again placed in a chamber, though one that is slightly different in some way to provide an altered context. Freezing behavior is scored for 3 min. Contextual discrimination of fear conditioning is calculated by comparing the number of freezing bouts in the same chamber to the number of bouts in the altered contextual environment, the slightly different chamber. At the end of the second phase of 3 min scoring, the white noise auditory cue is presented in the altered context environment. Again, freezing behavior is scored for a time period of 3 min in the presence of the white noise. Cued conditioning is quantitated by comparing the number of freezing bouts in the altered environment without the presence of the auditory cue with the number of bouts in the altered environment with the presence of the auditory cue. Impairment or improvement of one of the components measured in this examination can generate information about neuroanatomy, neurotransmitters, and genes regulating emotional components of memory.[22]

There are various avoidance tasks, passive and active, that determine the ability of the mouse to remember adverse stimuli and learn to avoid their presentation. Passive avoidance tasks require the mouse to refrain from entering a chamber previously associated with an aversive stimulus. Active avoidance tasks require the mouse to leave a chamber where an aversive stimulus is present. All avoidance tasks are based on electrical shocks generated from a grid and producing a painful stimulus in the footpads. There are various intensity parameters available for use, with the minimum intensity being that required to produce vocalization and flinching. Latency to move to the required area, away from or toward the shock area dependent on whether the test is passive or active, is measured in seconds and compared between normal wild-type control mice and transgenic mice. Alterations in the avoidance test include step-down avoidance where a mouse is required to step off an elevated platform, and Y-maze and T-maze avoidance in which a shock is delivered in only one run of a three-pronged runway.[22]

## Motor Skills Assessment

Motor function evaluation is critical and often defines certain phenotypes. Mutant mouse models are often used to study ataxias and other neuromuscular disease. Thus, there is a need for tests that will enable investigators to examine and evaluate motor function. These tests are important for the evaluation of abnormalities in aging mice as well as specific muscular disease studies such as muscular dystrophy.

The first step in evaluating neuromuscular function is a complete neurologic exam including neurologic reflex tests. Easy tests for normal reflexes include the righting reflex, postural reflex, eye blink reflex, ear twitch reflex, and whisker-orienting reflex. In the righting reflex, the mouse is turned over onto its back; normal mice will immediately turn themselves over onto all four feet. The postural reflex involves placing the mouse in an empty cage and shaking the cage back and forth and up and down; normal mice will extend all four limbs to maintain the proper upright position. The eye blink and ear twitch reflexes are based on the normal avoidance action in response to a cotton swab nearing the eye or touching the ear. The whisker-orienting response is tested by lightly brushing the whiskers of a freely moving animal with a small brush; normal mice will stop moving their whiskers when touched and may turn their head to the side being brushed.[22]

The open field locomotion test can be used to evaluate motor function. The testing apparatus consists of a digiscan open field plexiglass box with a series of photocell receptors and emitters. When a mouse moves through a beam, the beam path is broken, and the photocell analyzer records the beam break. A computer-assisted analyzer tallies the number of beam breaks for each set of photocells, with an ability to distinguish between vertical and horizontal activity. Computer software can calculate a large number of relevant variables over a preset time period, including vertical and horizontal activities, total distance traversed, total number of movements, and time spent in the center of the open field versus time spent in the perimeter of the open field. A 5-min test session is sufficient to evaluate gross abnormalities in locomotion.[22]

Performance on the rotarod can be used as a way to detect cerebellar defects as well as other neuromotor deficits in mice. The rotarod is a machine with revolving wheels upon which the mice walk at constantly accelerating speeds. During experimentation, mice are placed on the wheels and the latency period for mice to fall off is measured. Mice with deficits in motor coordination or balance will fall off the rotarod well before the end of a 5-min test session.

In addition to the rotarod, the balance-beam test and the vertical pole test can be used to assess motor coordination and balance in transgenic mice. In the balance-beam exam, mice are trained to traverse a series of graded beams in order to reach an enclosed safety platform. A bright light illuminates the beginning of the beams to further encourage the mouse to walk across the beams toward the enclosed box at the other side. Common beam sizes are 28-mm,[2] 12-mm,[2] 5-mm,[2] 28-mm round, 17-mm round, and 11-mm round. Evaluations are based on latency to cross the beams and the number of times the hind feet slip off the beams. The vertical pole test also uses minimal amounts of equipment and is easy to perform. In this exam, a metal or plastic pole, 2 cm in diameter and 40 cm long, wrapped with cloth tape to increase traction, is held in a horizontal position with the mouse placed in the center. Gradually, the pole is lifted into a vertical position with latency to fall off the measured variable. Normal mice will stay on the pole even in the complete vertical position and may even walk up and down. Mice with deficits in motor coordination or balance will fall off before the pole hits 45°.[22]

Walking patterns or gait abnormalities can be evaluated with footprint analysis. The hindpaws of the mice are dipped in a nontoxic ink or paint, and the mice are allowed to walk on white paper within a dark tunnel. The mice walk down the tunnel and leave their footprints on the paper. The footprints can then be examined for any signs of ataxia or gait abnormalities. The ability of the mouse to walk a straight line is measured by variability in stride length, variability around a linear axis, hindbase width and frontbase width between left and right paws, overlap of paws, and the distance between each stride.[22]

Motor strength can be evaluated using the hanging wire exam. Mice use balance and grip strength to hang upside down on a wire. A standard wire cage lid can be used for this assessment, with duck tape around the edge so the animal will not walk off the edge. Simply place the mouse on the cage, shake the lid several times to cause the mouse to grip the wires, and turn the lid over so that the mouse is hanging by its feet to the underside of the lid. There should be enough of a drop from the lid that the mouse does not simply drop off, but not a distance so large that it will be injured should it fall during the exam; 20 cm from cage lid to ground is an adequate height. Latency to fall off is measured, as normal mice can hang upside down by their grip alone for several minutes. [22]

Some genetically engineered mouse models show neuromuscular dysfunction in more specific areas than these general neurologic exams can identify. For example, some transgenic mice will exhibit stereotypic behaviors, or repetitive, invariant, perseverative motor patterns that do not appear to be directed toward a specific goal. There are scoring systems to characterize such behaviors as chronic grooming, sniffing, or head dipping. Lesions on one side of the brain can cause circling, with the direction of the motion correlative with the side of the lesion. Circadian rhythms and sleep patterns can be disrupted in transgenic mice. Some transgenic mice will show extreme signs of neurologic dysfunction such as seizures, ataxia, immobility, or twitching that are indicative of loss of vestibular function.[22]

Diagnostic exams should be tailored toward the neurologic area most likely to be the leading cause of the visualized problems.

## CARDIOVASCULAR ASSESSMENT

### Cardiac Assessment

As heart disease in western society and around the world becomes more conspicuous as a source of illness and death, so do genetically engineered mice designed to facilitate the study of the cardiovascular system and the effects of specific genes on its various failings. Methods to assess the cardiovascular system are varied in their complexity and versatility. Investigators should carefully select those testing techniques best suited to the specific genes and questions being addressed. In all cases, evaluating cardiovascular parameters at rest and under stress is an important aspect of phenotypic characterization, because many abnormalities will not be discovered until the cardiovascular system is pushed beyond normal metabolic constraints.

One of the easiest systems for tracking and evaluating the cardiovascular system is telemetric monitoring. Implantable telemetric transmitters are placed inside the mouse while under general anesthesia. Depending on the type, they can be placed subcutaneously, intraperitoneally, or in specific blood vessels such as the abdominal aorta or carotid artery. Telemetry can be used to monitor body temperature, blood pressure, heart rate, and heart electrical activity through constant EKG readings. The major advantage with these systems is in the fact that 5 days postimplantation, cardiovascular parameters can be measured in freely moving awake mice for an extended duration. Mice with telemetry transmitters can function in all ways like normal mice, even reproductively. Female mice with telemetry devices implanted in the thoracic aorta are able to conceive, gestate, deliver, and provide postnatal care for pups without interference by the transmitter and with constant cardiovascular monitoring throughout.[23]

Echocardiography in mice has been more widely used in recent years to accurately assess cardiac structure and function as well as measure cardiac volume and output. At this point, two-dimensionally directed M-mode echocardiography is believed to be the leading imaging method for small animal cardiovascular work. To adequately visualize the heart and surrounding structures, the required linear array, broadband transducer should operate at no less than 12 to 15 MHz. Newer scanners operating at higher frame rates per second allow high resolution real-time imaging in multiple planes, allowing for calculations of two-dimensional left ventricular volume and mass. Mice are most frequently anesthetized for echocardiography as imaging conscious mice is difficult and prone to error. The investigator should be aware of the changes to the various cardiovascular parameters induced by anesthesia and factor these differences into any results obtained through the use of this procedure. Murine echocardiography is still a new technology, and there are limitations in its capacity to phenotype genetically engineered mice, such as decreased ability to view the right heart chambers and required operator expertise.[24]

MRI, or magnetic resonance imaging, is another form of imaging that can be used to assess cardiac structure and function. As with echocardiography, mice undergoing an MRI exam should be anesthetized. MRI provides noninvasive, highly accurate three-dimensional imaging. Unlike echo, the right chambers can be readily visualized, and operator expertise is inconsequential. In addition, serial assessment of cardiac structure and function in all age groups is reliable and accurate and provides reproducible serial measurements of cardiac output. However, MRI is costly, resource-intensive, and has limited availability to many investigators.[24]

A noninvasive exercise stress test can be used to assess mitochondrial physiology and cardiac function. Mice are exercised on a modular variable speed and angle treadmill enclosed in an airtight chamber, so that a known composition and constant flow atmosphere can be maintained. Paramagnetic sensors are used to monitor the $VO_2$ and $VCO_2$ of the outflow. It can readily be determined when the mouse begins making the transition from aerobic to anaerobic exercise by the rapid increase in exhaled $CO_2$. Anaerobic exercise produces excess lactate, which is buffered by $HCO_3^-$ as a result of the increased exhalation of $CO_2$. It is therefore suggested that the exercise stress test in the mouse provides a quantitative indication of the mitochondrial bioenergetic capacity.[25]

In addition to the aforementioned more generalized cardiovascular assessment techniques, there are numerous more specific procedures that can be used to characterize genetically engineered mice. Conductance volumetry catheterization permits *in vivo* determination of instantaneous pressure–volume relationships and calculation of load-independent indices of left ventricular function. Sonomicrometry measures immediate estimates of left ventricular cavity dimension with excellent resolution. X-ray contrast microangiography can be used to quantitate right ventricular dilatation and dysfunction as well as tricuspid regurgitation. Whole-animal elecrophysiologic studies with octapolar catheters can determine the molecular basis of conduction and arrhythmic disorders.[24] Quantitative fluorescence microscopy using individual artery perfusion of fluorescently labeled molecules measures microvessel and macrovessel permeability and reactivity. The interaction between macromolecules and the vascular wall can be determined using optical coherence tomography and 2-photon fluorescence. Other assays in vascular tissue include measurement of hydraulic conductivity and macromolecular flux across capillary endothelium, macromolecular localization in the vascular wall, adhesion molecule properties and signal transduction in leukocytes and platelets, atheroma quantification, and assessment of lipoprotein metabolism.

## Atherosclerosis

Transgenic mice have recently become an important model for the study of human atherosclerotic disease. This is an extremely important model, as cardiovascular disease is a leading cause of mortality in western society. Atherosclerosis involves occlusion of arteries by the formation of atherosclerotic plaques. Plaques are often composed of fibrin, platelets, endothelial cells, inflammatory cells, and smooth muscle cells. Occluded arteries are often opened via balloon angioplasty. Even after successful dilation, vessels can become reoccluded. Several mechanisms thought to be involved include thrombosis, smooth muscle proliferation, and extracellular matrix deposition. Thus, there are specific techniques used to assess this model. Histopathology is critical for examining the atherosclerotic plaques and the proposed mechanisms of their formation. Tissue preparation of vascular tissues includes sectioning the tissues longitudinally to assess intimal lesion development. Cross sections will enable wall thickness to be assessed. Special stains are employed to identify specific plaque components and to assess the integrity of the vessel walls. Routine histopath is performed with H&E staining. Van Gieson stains elastin fibers within the vessel wall. Oil red-O stains lipid and will allow one to visualize and measure the surface area of intimal lesions using a calibrated microscope.[26] Gomori's stain is used to demonstrate muscle fibers, collagen, and nuclei in vessel walls. Individual cell types are often identified on H&E. Immunostaining can be used to detect specific cell types within the intima and specific cell types that comprise atherosclerotic plaques. The proliferation of smooth muscle cells and the resulting intimal thickening is a major component in the pathogenesis of atherosclerotic lesions and is thought to contribute to restenosis following balloon angioplasty. The proliferation of smooth muscle cells can be measured on H&E sections via autoradiography by counting labeled nuclei of cells. Cross-sectional areas of vessel intima can be obtained by tracing images on a digitizing pad and using special software to calculate the area.[27]

## PULMONARY FUNCTION

There is increasing interest in the use of genetically engineered mice in inhalation toxicology studies involving host variables, including genetics, age, diet, and disease.[28] This interest derives from epidemiology that suggests an association with air pollution-related human mortality/morbidity, especially among individuals with cardiopulmonary disease. Several mouse models with genetically based cardiopulmonary diseases are now being incorporated into inhalation toxicology studies to investigate mechanisms that underlie host susceptibility. Current evidence indicates that mouse models of pulmonary hypertension, bronchitis, asthma, and cardiovascular disease, but not emphysema, appear to exhibit greater susceptibility to air pollution particulate matter. As in humans, host susceptibility appears to involve multiple genetic and environmental factors, but is poorly understood. As existing mouse models gain wider use, further development will encourage integration of genetic and environmental factors to better mimic the human conditions.

## Allergen-Induced Asthma

Mice are primed intraperitoneally with 10 μg ovalbumin (OVA) and 20 mg $Al(OH)_3$ in 0.2 mL of phosphate-buffered saline (PBS, pH 7.4) once per week for 3 weeks. Twenty-four hours after the last injection, the primed mice are placed individually in a 50-mL plastic tube and challenged by exposure to aerosolized 5% OVA/PBS, delivered by a nebulizer driven by compressed air at 5 L/min for 20 min. Challenge is carried out once a day for 6 consecutive days. Marked eosinophilic inflammation and AHR are generally induced by this treatment.[29]

## Lung Injury

Current aerosol irritant assays use mice in noxious atmospheres. For this reason, the Minimal Animal Stress Irritant Assay Chamber (MASIAC) has been developed based on the principle of avoidance.[30] The MASIAC can detect citric acid with more sensitivity than conventionally used assays. Responses were not significantly affected by the presence of other mice, and following multiple exposures to citric acid, the mice either sensitized to the irritant, or learned to avoid it. This new method could provide investigators with an alternative method of evaluating pulmonary irritants.

## Emphysema

Transgenic mouse models are being used to investigate cigarette smoke-induced emphysema.[31] The cigarette-smoking protocol consists of placing mice in a polypropylene whole-body inhalation chamber containing inlet pores and an exhaust pore. Smoke is generated through a 50-mL glass syringe equipped with a cigarette holder from one or multiple unfiltered cigarettes. The exposure period usually is long term on a daily basis. A typical protocol is exposure 6 days per week, for 6 months. Mice typically demonstrate progressive inflammatory cell recruitment beginning within the first month of smoking followed by airspace enlargement after 3 to 4 months of cigarette exposure.

## Specialized Histomorphic Techniques

Following euthanasia, lungs are inflation-fixed at 25 cm of water pressure with 4% paraformaldehyde in PBS for 1 min. The trachea is ligated, and the excised lungs and heart allowed to equilibrate in cold fixative. Lung and heart volumes are determined by fluid displacement. Each lobe is measured along its longest axis, bisected perpendicularly to the long axis, and processed into paraffin blocks. Five-micrometer sections are cut in series throughout the length of each lobe and stained with hematoxylin and eosin, Masson's trichrome stain for collagen, or orcein for elastin. Morphometric measurements are performed on mice at various ages. The overall proportion (% fractional area) of respiratory parenchyma and airspace is determined by using a point counting method, video images, and specialized software. Pressure–volume curves are generated by inflating lungs in 75-μL increments every 10 sec to a maximum pressure of 28 cm of water and then deflated. Pressure–volume curves are measured as lung volumes (mL/kg) at 10, 5, and 0 cm of water during the deflation curve.

## Bronchoalveolar Lavage

Bronchoalveolar lavage (BAL) is used to collect BAL cells, especially BAL macrophages. One milliliter of tissue culture medium is flushed into the cannulated trachea and aspirated back out for a total of three times. Up to eight to ten mice may need to be used and BAL cells pooled to provide sufficient numbers of macrophages for each determination.

# SPECIAL SENSORY ASSESSMENT

Assessment of the special senses allows for evaluation of cranial nerve function as well as normal structure and function of the eyes, ears, and brain. In some cases, a known phenotype for a specific transgene is the loss of visual or auditory acuity early in the life span. In others, a slight change in brain and cranial nerve structure or in development can have large affects on the ability of the genetically engineered mouse to hear and see normally from birth. Detection of changes in these special senses will provide a broader phenotype, especially as related to aging and degeneration studies.

## Auditory Assessment

There are several methods to test hearing function in mice. The most comprehensive approach is often the better method of phenotypic characterization, incorporating several types of testing. However, in mice genetically engineered to express mutations in a specific region or response, there may be one test in particular that is more highly sensitive to the area of interest. Researchers should be aware of the basis behind each testing system and coordinate the choice of test with the specific gene in question.

Exposure to an augmented acoustic environment (AAE) can be used to assess the amount of hearing loss evident in transgenic mice as compared to control mice. In AAE, mice are exposed beginning at approximately four weeks of age to a 70 dB SPL broadband noise. Exposure should be for at least 12 h per night and should continue for no less than 30 days sequentially. After AAE exposure, auditory performance is measured using one or more of several tests.[32]

The simplest auditory assessment exam is the startle reflex. In this test, the amplitude of response to immediate intense sounds is measured, with no prepulse present. The startle stimulus is a 100 dB 4 kHz tone burst, played in 10 ms duration with a 1 ms rise/fall time. Animals to be tested are placed in a startle chamber so that responses can be measured via transduction of movements into voltage units. Consequential voltage units are displayed on a digital storage oscilloscope that shows a spike-like voltage change in reaction to an acoustic startle response from the animal in the chamber. Amplitude of the response is defined as the largest peak-to-peak voltage deflection in the first 30 ms following onset of startle stimulus.[22]

By the addition of prepulse inhibition (PPI) to the simple startle reflex, researchers can evaluate behavioral responses to tones of moderate intensity. A 70 dB tone presented 100 ms prior to the intense startle stimulus results in a reduced startle amplitude. Using PPI, the neural pathways that involve the auditory brainstem and other central regions can be evaluated. The magnitude of the PPI response reflects the behavioral salience of the prepulse tones.[32]

The auditory-evoked brainstem response (ABR) can be used to measure auditory thresholds in mice by testing for cochlear sensitivity. The assay is based on the emission of synchronized neural discharges from the auditory nerve and brainstem in response to short acoustic stimuli. The procedure consists of anesthetizing mice and placing on a heating pad, then implanting with stainless steel electrodes below each ear superficial to the auditory nerve. Click stimuli tones, randomized over 8 to 32 kHz, are presented through earphones over both ears. The loudness of each tone is varied in 5 dB steps. The auditory threshold is defined as the lowest stimulus intensity at which a normalized ABR wave can be identified.

## VISION ASSESSMENT

While there are many genetically engineered mice produced specifically to study human ophthalmology, an important piece of knowledge to keep in mind is that with any form of systemic disease, pathology of the eye can follow along as a part of the spectrum. In addition to this, there are many background strains that have retinal degeneration and cataracts, among other ocular diseases, as part of the strain-specific lesions. Researchers should be aware of these complications and have measures in place to counteract or account for the effects of the statistical impediments.

To perform a general visual ability assessment, the visual cliff response (VCR) test is a simple and absolute test that can be conducted on many mice in a short period of time. The VCR is performed in a box with a horizontal surface and a vertical wall with a drop-off, extended with a piece of clear Plexiglass to give the appearance of a vertical cliff face. To emphasize the visual appearance of the box, black-and-white checkered paper can be layered on the horizontal and vertical drop-off surfaces. The mouse in question is placed on a platform at the border between the horizontal surface and the apparent drop-off, giving them a choice of stepping down on either surface. Mice with normal visual acuity will step down mostly on the horizontal surface, avoiding what they see as a vertical drop into space. Blind mice will step down equally on both sides since they cannot see the fake cliff.[22]

Another simple test of visual acuity is based on the tendency of mice to enter dark places, when placed into an area of high light. Given the choice, mice with normal light/dark perception will enter a darkened chamber placed in front of them, while blind mice will not. Fine tuning this assessment by timing the duration between placement in the light area and entering the dark chamber can help to generalize the amount of light/dark perception available in individual mice.[22]

A deeper examination of the ophthalmology of transgenic mice involves an examination of the eye. These exams are noninvasive, and the animals can be restrained by hand and without anesthetic, as no pain or major stress is involved. Response to light is evaluated by passing a small beam of light over the eye, causing pupil constriction when exposed to the light source, and pupil dilation as the light moves away. Following a general exam of the external eye structures and surrounding area, the left and right eye pupils are dilated using 1% tropicamide that produces full dilation in the mouse within 5 to 10 min. Direct examination of the left and right eyes is through the use of a Kowa SL-14 portable slit lamp. Using ophthalmoscopy, most of the major structures in the eye can be visualized, including the cornea, lens, retina, and optic nerve. Any deviation from normal structure may be a phenotype, though care must be taken not to confuse background strain affects on eye pathology with genetically engineered changes.

For more specific testing of ophthalmic function, greater effort and sophisticated equipment is required. Two of the more commonly tested areas of visual function are intraocular pressure to detect glaucoma and retinal function to test for visual acuity. Intraocular hydrostatic pressure can be measured using an electrophysiologic approach — the servo-null micropipette system (SNMS). For this procedure, all mice should be under general anesthesia supplemented with proparacaine or other local anesthetic topically applied to the eye. SNMS consists of a micropipette filled with 3M KCl solution to counteract fluid resistance of the extracellular fluid and carboxyfluorescein to enhance visualization, a ground reference placed on the conjunctiva, and a servo-null device. The micropipette tip is placed in the drop of topical anesthetic, overlaying the pupil at an angle of 60 to 70° relative to a tangent to the corneal surface, and then rapidly inserted into the anterior chamber through the cornea. This changes the hydrostatic pressure and forces aqueous humor into the pipette to displace the KCl solution. The resultant increase in electrical resistance is measured via a signal through a vacuum-pressure pump and is the value of the intraocular pressure (IOP). The IOP can be monitored at a rate of 3 to 5 measurements per second.[32] Testing for retinal function can be performed through a procedure called electroretinography. The instrument consists of a flashlamp with focusing and filtering optics and amplifiers so that light can be delivered to the eye in single pulses of 20 µsec duration. The electroretinogram (ERG) waveforms are recorded in triplicate and averaged, reflecting the amount of activity within the retina in response to the pulsing light beams. Mice undergoing ERG should be under general anesthesia and have topical anesthesia for the eye as well as a pupil dilator to allow for better light penetration. In preparation for a trial, mice should be dark-adapted overnight and placed under anesthesia in dim red light.[34]

## Tactile and Olfactory Assessment

For some genetically engineered mouse models, it may be important to differentiate between transgenic mice and wild-type controls in their ability to either feel or taste the world around them. There are simple "paperclip" assays designed for investigators interested in phenotyping and assessing the tactile and olfactory abilities of their mutant mice. Touch is evaluated through the use of the reflexive twitch response to Von Frey hairs exam in which fine wires of gradually increasing thickness are touched to the paw. Another part of the touch mechanism is the ability to feel pain through the sensory nerves of the body. The timed response to pick up or lick a paw placed on a hot plate or to move the tail out of the path of a high-intensity light beam are both indicators of the ability of an animal to feel and respond to pain.[16] A basic exam of olfactory ability is timing the retrieval of a buried food source. Another simple exam is measuring the amount of time spent sniffing a novel odor.[22] More extensive exams of smell involve learning in conditioned reward choice paradigms where a mouse must learn to press a lever in response to the smell of food or a female in heat.[16]

## RENAL FUNCTION

Phenotyping renal function in genetically engineered mice consists of clinical pathology assessment as well as gross and histopathology monitoring. Many of the parameters identified by urinalysis can give a good indication of the status of the kidneys and their ability to filter the blood supply and produce urine. Gross pathology and histopathology will allow a researcher to evaluate the anatomy of the kidney and urinary tract, in effect determining if a specific transgene creates a change in anatomical structure or in ultrastructure that could, in turn, lead to a change in physiology.

In order to perform a complete urinalysis, collection of urine from the mouse is required. Many mice will urinate due to the stress of handling, and a free catch specimen can be collected in a sterile blood or culture tube. Alternatively, mice can be placed on a square of Parafilm or can be voided by gently applying abdominal pressure in a sweeping motion from midabdomen down to above the bladder.[16] If more than a few drops of urine is required for analysis, multiple collections are needed from the same mouse over a series of several days. In this case, all urine collected should be kept in the refrigerator for storage. A microscopic exam of the urine can identify formed elements such as red and white blood cells and cellular casts. Multiple parameters of urine can be measured using a urine dipstick, namely urine protein, urine blood, and urine glucose. For experiments requiring daily urine sampling over an extended period of time or those in which large volumes are needed, the proper method of urine collection is with a metabolism cage. These cages have wire-bottom flooring and a funnel-shaped base. As the mice in the cage urinate and defecate, excretions drop into the funnel and are passed into a collection vessel at the base of the funnel.[35] Samples taken in this manner are not sterile and cannot be used for culture, though they can be used to determine the integrity of renal physiology. Urine volume can be used to evaluate renal concentrating ability, the loss of which is one of the first signs of renal disease. Less frequently measured urine elements can indicate certain types of renal pathology. Measurement of specific proteins in the urine via SDS-PAGE or nondenaturing gel electrophoresis can differentiate between glomerular proteinuria, where there is loss of large and small proteins, and tubular proteinuria in which only low-molecular weight proteins are lost. Presence in the urine of several proximal tubular cell enzymes, including $N$-acetylglucosaminidase and lysozyme, is indicative of tubular cell injury. Glomerular filtration rate can be assessed by measuring inulin clearance.[16]

A blood serum chemistry screen and complete blood count can be used to assist in evaluating renal function, namely the parameters BUN, creatinine, glucose, total protein, and bilirubin concentration. Elevations or depressions in these values can mean various changes in urinary metabolism and can indicate changes in ultrastructure best seen with histopathology evaluations. A loss in numbers of red blood cells, or anemia, can be due to chronic renal failure, as can abnormalities in blood electrolytes such as calcium and potassium.

# MUSCULOSKELETAL ASSESSMENT

## Lean Body Mass

Muscle wasting is a consequence of many systemic diseases such as diabetes, cancer, sepsis, hyperthyroidism, uremia, muscle disuse, and nerve injury. Thus, muscle mass determination is important to many systemic disease studies. Muscle-specific phenotyping procedures are important and consist of an intensive anatomic evaluation that involves muscle mass determination via gross pathology and histology, followed by biochemical techniques such as molecular flux spectroscopy. Such techniques are beyond the scope of this chapter. At the time of necropsy, the total mouse weight is obtained, and an analysis of muscle weights can be obtained by dissecting out individual muscles. An average of left and right sides should be used. Muscles are examined grossly, and samples are obtained for H&E staining. To better visualize muscle fibers, samples can be double stained with anticaveolin-3 IgG to help demarcate the sarcolemma or plasma membrane and propidium iodide to help demarcate nuclei. Electron microscopy is valuable in examining skeletal muscle disease at the cellular level. A muscle biopsy can also be utilized to investigate clinical manifestations of muscle disease such as muscle wasting. Serum creatinine kinase, a cytosolic muscle enzyme that elevates with lysis or necrosis of muscle fibers, is a good indicator of the extent of muscle wasting.[36]

## Bone Density Assessment

It is important to evaluate and compare bone mineral densities in various aging mouse cohorts and for any study that attempts to show the effect of any type of treatment on bone. Trabecular bone density can be measured by three established methods. These three methods include trabecular bone volume by histomorphometry (BV/TV%), trabecular bone density by peripheral quantitative computerized tomography (pQCT), and areal bone density of trabecular bone by duel-energy x-ray absorptiometry (DEXA). DEXA can be used to measure bone mineral content (BMC) and areal bone mineral density (BMD) (BMC/cm$^2$).[37]

Osteoporosis is a common bony disease of aging women. This disease process involves loss of bone mass with subsequent skeletal fragility. A mouse model has been developed and specific phenotyping techniques can be employed to study this model. Conventional radiography as well as microradiography can be performed to evaluate bone density. Biomechanical studies such as whole-bone three- or four-point bending tests to failure is conducted with a servohydraulic testing system, and microcomputed tomography can be used to determine cortical volumetric BMD. The bending strength is measured at mid-diaphysis. The bone is placed horizontally with the anterior surface upwards. A pressing force is directed vertically to the midshaft of the bone, and the bone is compressed with a constant speed until failure. Breaking force or maximal load is defined as bending load at failure.[38]

Histologic analysis starts with fixed tissues that are then decalcified and stained for H&E or prepared and histomorphometrically analyzed. Growth plate measurements can be made on tibial samples stained with alcian blue/Van Gieson stain and sectioned to 4 um thick. An image-processing system is used coupled to a microscope. Histologic analysis can help distinguish osteoporosis from osteomalacia. Osteomalacia is a failure to mineralize versus osteoporosis which is a reduction in bone mass. Bone histomorphometry is measured as the ratio of trabecular bone volume to total volume. Areas of trabecular bone within a reference area are stained with H&E and measured in sections. Measurements are made on printed copies by point counting with a square lattice.[38] Enzyme-histochemical staining can be done by staining for alkaline phosphatase activity or tartrate-resistant acid phosphatase. Serum assays are used to determine total protein, calcium, phosphorus, and creatinine. Serum osteocalcin, a marker of bone formation, is also helpful to determine. All these are important to distinguish osteoporosis from other osteopenic diseases such as primary hyperparathyroidism and renal osteodystrophy.[39]

## Experimental Arthritis

The ability to determine the severity of joint disease by gross physical examination is a useful clinical feature, because a numerical score is reflective of the degree of inflammation. The standard method reported in the literature assigns numerical values to digits and paws of each of the four limbs, with 0 = no signs, 1 = swelling or redness in one digit, 2 = swelling or redness in two or more digits, and 3 = swelling or redness in the entire paw. The scores from each of the four limbs are then added to obtain an arthritic score. Observations in T cell receptor transgenic mice, which develop an acute onset of collagen-induced joint disease,[40] suggested that when an entire paw was red or swollen, it was painful to the mouse. On the other hand, mice showed no apparent discomfort when joint disease was restricted to individual digits. Therefore, the discrepancy in correlation between arthritic score and distress became evident when comparing disease severity. For example, if a single digit on each of three limbs was affected, the mouse showed very little discomfort but had an arthritic score of 3. A mouse with an entire paw affected but on only one limb still had an arthritic score of 3 but was in obvious distress. Therefore, the scoring method was altered by multiplying each individual limb score by itself, thereby magnifying the scores associated with entire paw involvement, and then adding these multiplication grading scores from each limb to obtain a modified arthritic score.

## GASTROINTESTINAL ASSESSMENT

The gastrointestinal tract (GI) can have many abnormalities associated with it. There are various lesions occurring in the GI tract of genetically engineered mice. Some of these lesions are unique to the particular mouse line and some of them resemble human diseases. Thus, it is important to properly interpret these changes. Many of the GI diseases studied in transgenic mice involve the small intestinal tract and large intestinal tract or colon. In all cases, a thorough examination of the GI tract at necropsy is essential. The intestinal tract and cecum are first examined grossly from the serosal surface. One should check for enlargement of Peyer's patches or any masses within the lumen. Peyer's patches can be an early site of lymphoma. The entire length of the small and large intestine should be examined and opened to look for lesions such as tumors, polyps, adenomas, and carcinomas. Tumors should be described, counted, and placed in fixation. Later, they should be embedded and prepared for histopath. It is also important to section portions of the intestinal tract for histopathology.[16] The specimen can be taken as a cross section of the bowel or the preferred method of rolling. The intestinal "Swiss" rolls are a continuous segment of bowel. They are prepared by removing the intestines intact and gently inflating with fixative by intraluminal injection and preparing as an intestinal roll (by placing it on an index card and rolling it in a flat spiral around a central toothpick) and fixed by immersion.

Several strains of transgenic or gene-targeted mice develop chronic intestinal inflammation or tumors. These mice make an excellent model for the study of cellular and molecular makeup of preneoplastic stages of intestinal tumorigenesis. Thus, gross pathology and histopathology are critical for assessing preneoplastic and neoplastic lesions. Aberrant crypt foci (ACF) are possible precursors for colon cancer. They serve as excellent biomarkers for preneoplastic lesions. At necropsy, the small and large bowel should be longitudinally cut open and flushed with saline. The samples are fixed in formalin × 24 h then dipped in 0.2% solution of methylene blue in distilled water and rinsed. Using a light microscope at 40×, the samples are examined mucosal side up for ACF. ACF are distinguished by their increased size, prominent epithelial cells, and increased pericryptal space from surrounding normal crypts.[41] GEM are also important models for microbial-based diseases such as Helicobacter infection. Microbial infections are often linked to inflammatory diseases and tumorigenesis in the large and small bowel.

Mice are important models for the study of hepatic disease. There have been many models developed for use in detecting pathological and phenotypic expression of disease. They are especially useful for their susceptibility to developing liver cancers. Transgenic technology has had a tremendous impact on the study of human cancers, where aberrant gene expression is often found. As with any study, hepatic assessment starts with data about strain, mating, date of birth, and any clinical signs noted. It is important to keep careful records of individual mouse weights. One should specify if the weight is obtained before or after any type of fasting and before or after exsanguination. It is also important to perform abdominal palpation to assess for any outward signs of hepatomegaly or neoplasia. Clinical evaluation includes sampling blood for clinical pathology assessment. A clinical pathology assessment includes a complete blood count (red blood cell count, pack cell volume, hemoglobin, platelets, white blood cell count, white blood cell differential) and serum chemistries (albumin, alkaline phosphatase, alanine transaminase, urinary bilirubin, blood urea nitrogen, calcium, creatinine, cholesterol, glucose, phosphorus, total protein, Na, K, and Cl). Urine should be obtained for a urinalysis.

A complete necropsy should be performed. Each mouse should be weighed prior to starting the necropsy. The liver is accessed as previously described in the pathology techniques section. Again, it is important to remember that the liver is divided into four lobes and that each must be carefully removed. After complete excision, the entire liver should be weighed, and the relative liver weight can be calculated as a percent (liver weight × 100/body weight).[16] The liver should be examined and characterized grossly. The size, color, consistency, and cut surface appearance should be noted and recorded. Any nodular lesions should be described, measured, and recorded for each lobe. When nodules are found on the liver, it is also important to check for local invasion and for metastases elsewhere in the body. The most common sites for metastases include the lung and kidneys. As with any organ, the liver should also be characterized by histopathology. One should be careful to excise slices from each liver lobe and consistently sample from the same section or position of each lobe. This is especially important for DNA synthesis from histologic samples. Other techniques such as special stains or tests for the presence or absence of specific liver enzymes can be used to further identify and characterize lesions.

## DIABETES ASSESSMENT

Diabetes affects millions of people worldwide. It is a leading cause of death, and as a result, is a large area of study. Diabetes is actually a group of diseases characterized by aberrant glucose metabolism. Type I or insulin-dependent diabetes results from immune-mediated destruction of pancreatic beta-cells. Type II or noninsulin-dependent diabetes results in hyperglycemia without loss of endogenous insulin reserve or loss of pancreatic islets. Thus, Type II diabetes is characterized by the presence of insulin resistance, which is often associated with obesity and advancing age and accounts for most cases of diabetes. The use of mouse models and transgenic technology has been important in the understanding of this complex disease.[42]

Clinically, overt diabetic symptoms include polyuria, polydypsia, and weight loss. One easy phenotyping assay is monitoring for glucose in the urine. This test should be performed weekly on at-risk mice. Many other phenotyping procedures and tests have been adapted for evaluating diabetic mouse models. It is beyond the scope of this chapter to discuss all of them. Current procedures available to study diabetic models include urinary glucose and ketones, plasma glucose, lactate, ketones, serum lipids (triglycerides, nonesterified fatty acids, total and HDL cholesterol), plasma insulin, leptin, and corticosterone, assessment of insulin secretion and sensitivity, and glucose disposal by IVGTT with minimal model analysis, measurement of other circulating hormones, including C-peptide, glucagon, GLP-1 (active and total), pancreatic polypeptide, thyroxine and TSH, renal function parameters [urinary albumin excretion, serum creatinine and blood urea nitrogen, in vitro adipocyte metabolism and leptin production, in vitro assessment of insulin secretion from isolated islets, body composition (body weight, percent carcass lipid, and lean mass and percent, and dissected fat pad mass)].

Pathology is important for diabetes characterization and investigation, as it is useful in helping to differentiate true diabetes from other disease processes in the mouse. Islet failure (diabetes) can result from a number of causes, such as pancreatic developmental disorders (dysplasia, atresia, etc.), neoplasia, amyloid infiltration, lipidosis, infectious disease, etc., that must be differentiated from immune-mediated insulitis. Diabetes development is confirmed by histological examination showing destruction of at least 50% of islets.[43] By their very nature, diabetic mutants are likely to have immunologic perturbations. Therefore, immune function assays may be needed to thoroughly evaluate a particular mutant strain.

Recently, transgenic mice have become an important resource in the development of a nonobese diabetic model (NOD). This model consists of NOD mice carrying transgenes, gene knockout mutations, and small stretches of allotypic genetic material, enabling them to serve as mouse models for insulin-dependent (Type I) diabetes. Insulin-dependent diabetes is an autoimmune disease in which genetic factors and environmental influences play a role. The resulting immune reaction involves progressive lymphocyte infiltration into the islets and selective destruction of insulin-secreting pancreatic cells.[43] There are many important issues to consider when studying this particular mouse model. Dietary management is crucial for maintaining this model. An additional consideration, dependent on environmental conditions, is the onset of disease: 80% of females and 20% of males can be expected to develop diabetes by 12 to 30 weeks of age, depending on the NOD mouse strain. Thus, monitoring blood glucose and urine glucose is crucial. Histopathology involves staining pancreatic samples with H&E and evaluating for insulitis and diabetes. A histology grading scale is often used to evaluate the percentage of islets showing lymphoid infiltration within or around the islets.

Transgenic mice have also been used to explore the correlation between obesity and noninsulin-dependent diabetes. Obesity is often characterized by hyperinsulinemia and insulin resistance. This is often due to increased insulin secretion and reduced insulin clearance. The obese mouse is a good model to investigate this connection. Body condition and weight are important to monitor for this model. Weight can be measured weekly or even daily if indicated. It is critical to house the mice individually to measure body weight and food consumption. Food and water consumption can be most accurately measured via the use of metabolism cages, but estimations can be made using daily or weekly measures of a gram of food consumed subtracted from the amount provided.[16] Blood glucose and insulin levels as well as urine glucose should also be determined for these mice. Body weight and the weight of individual organs are also obtained at necropsy. The major organs as well as fat pads should be included. The ratio of each tissue or organ weight relative to body weight should also be calculated. The hepatic lipid content can be assessed on fresh frozen sections with the lipid-specific stain, Oil Red O. Glycogen content can be assessed on fixed paraffin-embedded tissues via periodic acid Schiff reaction.[44] The rest of the tissues should be stained with H&E for routine histopathology. This will allow the pathologist to assess for diabetes-related lesions.

## SKIN ASSESSMENT

One of the most common organs to study is the skin. It is relatively easy to examine and identify phenotypic variations which can become an important resource for gene-targeting studies. When evaluating the skin and coat, it is important to keep in mind that phenotypes can vary dramatically. Some mutations have no obvious effects, whereas others can dramatically change the appearance of the mice. Problems with the skin and coat can be due to environmental problems, parasites, autoimmune disease, nutritional disorders, or can be secondary to certain treatments or genetic changes.[16] Signs to watch for include dryness, scaling, alopecia, wounds, dermatitis, piloerection, matting, and excessive oiliness to the coat. It is important to recognize that mice do a lot of grooming to themselves and each other. Barbering is common and should not be confused with alopecia. Barbering appears to be a normal dominance behavior. It is important to keep in mind that mice can respond differently to epidermal injury than other mammals. Thus, there are some species-specific features that should not be misinterpreted pathologically.

General categories of cutaneous disease can be used for characterization studies and for identifying potential allelic mutations. These general categories of cutaneous disease include hair color mutations, eccrine gland defects, sebaceous gland defects, primary scarring disorders of the skin, hair shaft growth and structural defects, noninflammatory skin diseases, inflammatory skin diseases, papillomatous skin diseases, bullous and acantholytic skin diseases, and structural and growth defects in nails.[16] These disease processes can be identified and studied by clinical examination, skin biopsy procedures, and pathology techniques already discussed.

General necropsy procedures have already been discussed, but there are some specific techniques to keep in mind when obtaining skin samples. Skin should be collected from the dorsal and ventral trunk, eyelids, ears, muzzle, tail, and footpads.[8] For large study groups, skin samples should be obtained from the same location in all individual mice. This allows more effective comparison of individuals in the same and different groups. The orientation of the samples should be consistent. The recommended orientation is parallel to the long axis of the body.[45] Once obtained, the skin sample should be laid out flat with the hair side up. The sample should be placed on an unlined index card in a head-to-tail orientation and labeled with orientation and where on the body the sample was obtained.[16] Skin can then be fixed for histopathology or frozen for immunofluorescence or other biochemical and molecular studies. The epidermis and associated hair follicles are made of stratified squamous epithelia in which genes are turned on and off during growth and maturation. Thus, mice undergo major changes that may affect the phenotype of a mutation. Some of the most dramatic changes occur in the first 3 weeks of life. Thus, specific studies may warrant collecting samples at 2 to 3 day intervals during the first 3 weeks of life. Multiple biopsies can be collected from the same individual to help reduce the numbers of animals used.[16] Scanning electron microscopy can be used for skin punches, plucked hairs, and nails. The samples are carefully collected and placed in a gluteraldehyde solution. The front and rear feet are collected at necropsy by amputating at the carpus or tarsus, and then samples are prepared as above.

## IMMUNE FUNCTION ASSESSMENT

Phenotypes of genetically altered mice can be validated and enhanced by conducting basic immune function assays. These assays provide information to distinguish whether phenotypic characteristics are associated with secondary changes and ill health, or are the direct or indirect result of genetic manipulation. Basic immune analyses provide further evidence of the validity of these observations. In addition, novel phenotypes can be identified for understanding molecules and pathways involved in immunocyte development and immune function, including immunodeficiency, cancer, autoimmunity, biology of aging, and resistance to infectious diseases. The identification of lineage-specific developmental defects in mutant mice is also possible.

Complete blood counts (CBC) on each mutant line are performed to determine the total white blood cell counts and relative representation of each subpopulation (lymphocytes, neutrophils, bands, eosinophils, basophils, monocytes). In addition, hematocrits and other red blood cell indices, as well as platelet counts from each mutant line are obtained. Peripheral lymphocytes are further characterized using flow cytometry by staining lymphocytes with fluorochrome-labeled antibodies specific for CD4 helper T cells, CD8 cytotoxic T cells, CD3ε (T cell receptor component), B220 (pan-B cell marker), IgM, and IgD (separates immature from mature B cells). B and T lymphocyte function in mutant mice is assessed by stimulating splenocytes with T cell mitogens anti-CD3ε plus anti-CD28, and separately, B lymphocytes with anti-IgM or lipopolysaccharide (LPS).

Lymphocyte development in bone marrow and thymus is assessed, because transgenic and gene-targeted mutant mice with moderate to slight defects in hematopoietic development can compensate and show normal peripheral WBC indices. B cell development can be broken down into seven different stages by the expression of various surface markers using flow cytometry and the following combinations of antibodies: *B220 BIO, IgM FITC, CD43 PE combination*, which divides the B cell lineage into mature B cells (fraction F-B220$^{hi}$, IgM,$^+$ CD43$^-$), immature B cells (fraction E-B220$^{lo}$, IgM,$^+$ CD43$^-$), and pre-B cells (B220,$^+$ IgM,$^-$ CD43$^+$). If defects in B cell development are detected, further flow cytometry is done utilizing the following combination of antibodies: *Allophycocyanin-conjugated B220, PE-conjugated CD43, FITC-conjugated BP-1,* and *biotinylated HSA combination*, which divides B lineage into large pre-B (fraction C'-B220,$^+$ CD43,$^+$ HSA$^{hi}$, BP-1$^+$), pro-B (fraction C -B220,$^+$ CD43,$^+$ HSA$^{lo}$, BP-1$^+$), pro-B (fraction B-B220,$^+$ CD43,$^+$ HSA,$^+$ BP1$^-$), and pre-pro-B cells (fraction A-B220,$^+$ CD43,$^+$ HSA,$^-$ BP-1$^-$).

A similar approach is used to characterize T cell development by flow cytometry in mutant mice. Thymocytes are harvested and total thymus cellularity measured, as well as the relative representation of CD4$^-$CD8$^-$ (double-negative) CD4$^+$CD8$^+$ (double-positive) and CD4+8,$^-$ CD4–8$^+$ (single-positive) developmental populations. For this purpose, a combination of anti-CD4/anti-CD8/anti-CD3ε (T cell receptor) fluorochrome-labeled antibodies is used. CD69 levels, which are a measure of T cell activation, will also be measured. If any defects in thymocyte development are detected, further testing is warranted utilizing antibodies against other markers such as CD44, CD25, and Qa-2, a marker for positive selection (terminal differentiation of double-positive thymocytes into the single-positive compartment).

Myeloid and erythroid progenitors in total bone marrow are assessed by flow cytometry utilizing a combination of the following fluorochrome-labeled antibodies: *Gr-1/Mac1*, which identifies myeloid progenitors, and *CD61 and Ter119*, which identify megakaryocytes and erythroblasts, respectively.

A noninvasive assay to test for T cell function is the contact hypersenstivity assay (CHS). An exaggerated and sustained cutaneous swelling to the hapten dinotrofluoro-benzene (DNFB) is suggestive of a T cell defect. The procedure consists of sensitizing mice with DNFB on the abdominal skin with subsequent challenge with DNFB 4 days later on both ears. Cutaneous swelling is measured over time. The mean increase in ear thickness following DNFB challenge is calculated in units of $10^{-4}$ inches ± SEM.

## SPECIAL CONSIDERATIONS FOR AGING COLONIES

Aging is associated with the progressive decline in function of multiple organ systems. In order to identify and quantify these functional deficits in mice, a standardized, methodical, and consistent process must be in place.[46] Specialized phenotyping assays are necessary to distinguish phenotypic differences not easily discernible by gross observations, especially in aging cohorts. The specific assays to be used must be selected carefully to yield maximum amounts of information related to aging, while conserving resources of time, effort, and materials. Clinical and anatomical pathology assessments are standard measurements for comparing differences between aging cohorts and will cover a number of age-related conditions affecting multiple organ systems, including cardiovascular, kidney, pancreas, and skeletal muscle, as well as brain. Maximum life span can only be determined by allowing mice to live as long as they can. Euthanasia is necessary only if it is certain they would die without interference. The following reliable signs are used: sudden weight loss, failure to eat and drink, prominent appearing ribs and spine, and sunken hips; not responsive to being touched; slow or labored respiration; hunched up with matted fur; and cold to the touch. Mice up to 18 months of age are monitored 2 to 3 times per week, and mice older than 18 months, or mice with apparent health problems, are monitored daily, 7 days a week, or two to three times per day as needed.

Aging results in the progressive decline of the cardiovascular system, characterized in part by an increase in wall thickness of the ventricles. Aged rodents experience ventricular hypertrophy associated with an excess accumulation of collagen.[47] Systemic mitochondrial dysfunction will frequently compromise muscle and cardiac function. Therefore, it is useful to evaluate mitochondrial physiology and cardiac function. Many neurological deficits associated with aging are subtle and not grossly observable, especially learning and memory deficits. The Morris water task is presently the most frequently used paradigm to evaluate learning and memory abilities in genetically engineered mice.[22] Aged C57BL/6 mice show impairments in performance on this task.[48] Additional neurological assessments for locomotor function include open field activity and rotarod procedures.[49] Hearing is another neurosensory mechanism that exhibits an age-associated decline. The auditory-evoked brainstem response (ABR) is the most sensitive measure of auditory threshold that has been applied to mice.[22] Using this procedure, age-related hearing loss has been demonstrated in the C57/BL6 mouse strain beginning as early as 2 months of age.[50] The development of cataracts is an age-associated condition and can be readily evaluated in mice by slit lamp examination. The natural occurrence of age-related cataracts in mice and the protective effect of caloric restriction have recently been described.[51] Bone loss is associated with aging in man and mouse,[52] and bone scanning assays are frequently used to compare bone densities.

Caloric restriction (CR) is considered the "gold standard" against which other antiaging strategies are gauged.[53] There are a number of specific protocols, but the basic feature consists of offering CR rodents 60% of what the control animals ingest.[54] The amount of food consumed by the control group is measured on a weekly basis by weighing the food prior to offering it to the animals, and again at the end of the measuring period. Preweighed food allotments are provided three times a week, and adjusted each week according to the amount ingested by the control group. Open or closed diets are used. The NIH-31 Open Formula diet, which contains supplemental vitamins to provide the CR animals the same intake as that of the control animals, is a commonly used formula. The CR protocol for mice is started at 90% of the average control group beginning at 14 weeks of age, then reduced to 75% at 15 weeks of age, and further reduced to the full 60% at 16 weeks of age for the remainder of the study.

## REFERENCES

1. Rossant, J. and McKerlie, C., Mouse-based phenogenomics for modeling human disease, *Trends Mol. Med.*, 11, 502, 2001.
2. Moldin, S.O. et al., Trans-NIH neuroscience initiatives on mouse phenotyping and mutagenesis, *Mamm. Genome*, 8, 575, 2001.
3. Mahner, M. and Kary, M., What exactly are genomes, genotypes, and phenotypes? And what about phenomes?, *J. Theor. Biol.*, 1, 55, 1997.

4. Osman, G. et al., SWR: an inbred strain suitable for generating transgenic mice, *Lab. Anim. Sci.*, 2, 8, 1997.

5. Ladiges, W.C. and Ware, C.B., Transgenic animals in toxicology, in *Current Protocols in Toxicology,* Maines, M., Ed., John Wiley and Sons, New York, 1999, chap. 1.3.

6. Richardson, A. et al., Use of transgenic mice in aging research, *ILAR J.*, 3, 125, 1997.

7. Wakeland, E. et al., Speed congenics: a classic technique in the fast lane (relatively speaking), *Immunology Today*, 18, 472, 1997.

8. Sundberg, J.P. and Boggess D., Eds., *Systemic Approach to Evaluations of Mouse Mutations*, 1st ed., CRC Press, Boca Raton, FL, 2000.

9. Harrison, D.E. and Archer, J.R., Genetic differences in effects of food restriction on aging in mice, *J. Nutrition*, 117, 376, 1987.

10. Fox, J.G. et al., *Helicobacter hepaticus* sp. nov., a microaerophilic bacterium isolated from livers and intestinal mucosal scrapings from mice, *J. Clin. Microbiol.*, 32, 1238, 1994.

11. Ward, J.M. et al., Chronic active hepatitis and associated liver tumors in mice caused by a persistent bacterial infection with a novel Helicobacter species, *J. Natl. Cancer Inst.*, 86, 1222, 1994.

12. Foltz, C.J. et al., Spontaneous inflammatory bowel disease in multiple mutant mouse lines: association with colonization by *Helicobacter hepaticus*, *Helicobacter*, 3, 69, 1998.

13. Maggio-Price, L. et al., Diminished reproduction, failure to thrive, and altered immunologic function in a colony of T-cell receptor transgenic mice: possible role of *Citrobacter rodentium*, *Lab. Anim. Sci.*, 48, 145, 1998.

14. Czernin, J. and Phelps, M.E., Positron emission tomography scanning: current and future applications, *Annu. Rev. Med.*, 53, 89, 2002.

15. Budinger, T.F., Benaron, D.A., and Koretsky, A.P., Imaging transgenic animals, *Annu. Rev. Biomed. Eng.*, 1, 611, 1999.

16. Ward, J. M. et al., *Pathology of Genetically Engineered Mice*, 1st ed., Iowa State University Press, Ames, 2000.

17. Rabinovitch, P.S., June, C.H., and Kavanagh, T.J., Measurements of cell physiology: ionized calcium, pH, and glutathione, in *Clinical Flow Cytometry: Principles and Applications*, Bauer, K.D., Duque, R.E., and Shankey, T.V., Eds., Williams and Wilkins, Baltimore, 1992, p. 505.

18. Gong, J. et al., Unscheduled expression of cyclin B1 and cyclin E in several leukemic and solid tumor cell lines, *Cancer Res.*, 54, 4285, 1994.

19. Telford, W.G., King, L.E., and Fraker, P.J., Rapid quantitation of apoptosis in pure and heterogeneous cell populations using flow cytometry, *J. Immunol. Methods*, 172, 1, 1994.

20. Ormerod, M.G. et al., Quantification of apoptosis and necrosis by flow cytometry, *Acta Oncol.*, 32, 417, 1993.

21. Hotz, M.A. et al., Flow cytometric detection of apoptosis: comparison of the assays of *in situ* DNA degradation and chromatin changes, *Cytometry*, 15, 237, 1994.

22. Crawley, J.N., *What's Wrong With My Mouse?*, 1st ed., Wiley-Liss, New York, 2000.

23. Butz, G.M. and Davisson, R.L., Long-term telemetric measurement of cardiovascular parameters in awake mice: a physiological genomics tool, *Physiol. Genomics*, 5, 89, 2001.

24. Hoit, B.D., New approaches to phenotypic analysis in adult mice, *J. Mol. Cell Cardiol.*, 33, 27, 2001.

25. Wallace, D.C., Mouse models for mitochondrial disease, *A. J. Medical Gen.*, 106, 71, 2001.

26. Xiao, Q. et al., Plasminogen deficiency accelerates vessel wall disease in mice predisposed to atherosclerosis, *Proc. Natl. Acad. Sci.*, 94, 10335, 1997.

27. Lindner, V. and Reidy, M.A., Proliferation of smooth muscle cells after vascular injury is inhibited by an antibody against basic fibroblast growth factor, *Proc. Natl. Acad. Sci.*, 88, 3739, 1991.

28. Kodavanti, U.P. and Costa, D.L., Rodent models of susceptibility: what is their place in inhalation toxicology?, *Respir. Physiol.*, 128, 57, 2001.

29. Tanaka, H. et al., The effects of allergen-induced airway inflammation on airway remodeling in a murine model of allergic asthma, *Inflamm. Res.*, 50, 616, 2001.

30. Karwowski, A.S., MacLeod, B.A., and Quastel, D.M., The development and evaluation of a new aerosol irritant assay with minimal animal stress, *Pulm. Pharmacol. Ther.*, 14, 435, 2001.

31. Nikula, K.J. et al., A mouse model of cigarette smoke-induced emphysema, *Chest*, 117, 246S, 2000.

32. Willott, J.F., Turner, J.G., and Sundin, V.S., Effects of exposure to an augmented acoustic environment on auditory function in mice: roles of hearing loss and age during treatment, *Hear. Res.*, 142, 79, 2000.

33. Avila, M.Y. et al., Reliable measurement of mouse intraocular pressure by a servo-null micropipette system, *Invest. Ophthalmol. Vis. Sci.*, 42, 1841, 2001.

34. Li, J., Patil, R.V., and Verkman, A.S., Mildly abnormal retinal function in transgenic mice without Muller cell aquaporin-4 water channels, *Invest. Ophthalmol. Vis. Sci.*, 43, 573, 2002.

35. Suckow, M.A., Danneman, P., and Brayton, C., *The Laboratory Mouse*, 1st ed., CRC Press, Boca Raton, FL, 2001.

36. Galbiati, F. et al., Transgenic overexpression of caveolin-3 in skeletal muscle fibers induces a Duchenne-like muscular dystrophy phenotype, *PNAS*, 97, 9689, 2000.

37. Rosen, H.N. et al., Differentiating between orchiectomized rats and controls using measurements of trabecular bone density: a comparison among DXA, histomorphometry, and peripheral quantitative computerized tomography, *Calcif. Tissue Int.*, 57, 35, 1995.

38. Vidal, O. et al., Estrogen receptor specificity in the regulation of skeletal growth and maturation in male mice, *PNAS*, 97, 5474, 2000.

39. Lewis, D.B. et al., Osteoporosis induced in mice by overproduction of interleukin 4, *Proc. Natl. Acad. Sci.*, 90, 11618, 1993.

40. Cheunsuk, S. et al., Predictive parameters of joint disease in DBA/1 transgenic mice, *J. Gerontol.: Biol. Sciences*, 54A, B271, 1999.

41. Tanaka, T. et al., Chemoprevention of azoxymethane-induced rat colon carcinogenesis by the naturally occurring flavanoids, diosmin and hesperidin, *Carcinogenesis*, 18, 957, 1997.

42. Benecke, H. and Moller, D.E., Transgenic strategies used to study diabetes and obesity, in *Strategies in Transgenic Animal Science*, Monastarsky, G.M. and Robl, J.M., Eds., ASM Press, Washington, DC, 1995, chap. 8.

43. Jacob, C.O. et al., Prevention of diabetes in nonobese diabetic mice by tumor necrosis factor (TNF): similarities between TNF-alph and interleukin 1, *Proc. Natl. Acad. Sci.*, 87, 968, 1990.

44. Levin, N. et al., Decreased food intake does not completely account for adiposity reduction after ob protein infusion, *Proc. Natl. Acad. Sci.*, 93, 1726, 1996.

45. Peckham, J.C. and Heider, K., Skin and subcutis, in *Pathology of the Mouse*, Maronpot, R.R., Ed., Cache River Press, Vienna, 1999, chap. 22.

46. Miller, R.A. and Nadon, N.L., Principles of animal use for gerontological research, *J. Gerontology B.S.*, 55, 117, 2000.

47. Burgess, M.L., McCrea, J.C., and Hedrick, H.L., Age-associated changes in cardiac matrix and integrins, *Mech. Aging and Devel.*, 122, 1739, 2001.

48. Bellush, L.L. et al., Caloric restriction and spatial learning in old mice, *Physiology and Behavior*, 60, 541, 1996.

49. Ingram, D.K., Age-related decline in physical activity: generalization to nonhumans, *Medicine and Science in Sports and Exercise*, 32, 1623, 2000.

50. Willot, J. F. and Bross, L.S., Morphological changes in the anteroventral cochlear nucleus that accompany sensorineural hearing loss in DBA/2 and C57BL/6 mice, *Devel. Brain Res.*, 91, 218, 1996.

51. Wolf, N.S. et al., Normal mouse and rat strains as models for age related cataract and the effect of caloric restriction on its development, *Exp. Eye Res.*, 70, 683, 2000.

52. Ferguson, V.L. et al., The effects of age and dietary restriction without nutritional supplementation on whole structural properties in C57BL/6 mice, *Biomed. Sci. Instrum.*, 35, 85, 1999.

53. Bertrand, H.A. et al., Dietary restriction, in *Methods in Aging Research*, Byung, P.Y., Ed., CRC Press, Boca Raton, FL, 1999, p. 272.

54. Pugh, T.D., Klopp, R.G., and Weindruch, R., Controlling caloric consumption: protocols for rodents and rhesus monkeys, *Neurobiology of Aging*, 20, 157, 1999.

# CHAPTER 11

# Health Status and Health Monitoring

Axel Kornerup Hansen

## CONTENTS

0-8493-1086-5/03/$0.00+$1.50
© 2003 by CRC Press LLC

# INTRODUCTION

Health status may be defined as the actual status of an individual animal concerning its clinical, pathological, and physiological appearance. In more popular terms, the health status tells whether the animal is ill or not, but as no animal can be said to be only ill or healthy, health may be regarded more quantitatively than qualitatively. Infections, the environment, and genetic disorders may reduce the health of the animals and counteract the aim of receiving reproducible results in groups of animals with a low variation. Infections impose a current risk of irreversible health reductions in a high number of animals and, therefore, have to be dealt with on a daily basis. Therefore, many laboratory animal scientists automatically think of infection when hearing the term health, and laboratory animal health is often defined as being the same as laboratory animal microbiology. Environment and genetics are obviously equally important, but they have to be dealt with in the design of facilities, the optimal running of these, and the breeding procedures; therefore, the approaches are somewhat different. Environment and genetics are discussed elsewhere in this book. In this chapter, infections and the influence of microorganisms on experiments are discussed, as well as the precautions needed to reduce this impact.

# INFECTIOUS AGENTS IN LABORATORY RODENTS AND RABBITS

Infections are equally important, whether they occur in experimental mice or experimental pigs. However, it will go too far in this chapter to cover all animal species used in research, and, therefore, a short introduction to the infections of the most common laboratory animals, i.e., rodents and rabbits, is given, while information for other species must be found in textbooks dealing specifically with these. An excellent review of infections in rodents and rabbits has been given by Baker.[1]

## Bacterial Infections

Bacterial infections may cause disease as well as other negative consequences for research. It should also be noticed that most animals for research harbor a normal flora,[2] which normally does not interfere with research. Some bacteria even have a positive impact on the animal. So, laboratory animals not housed in isolators cannot, in the same way as it may be applied for viruses, be kept free of all bacteria, and for most research projects, this is not necessary. Specific bacteria of rodents and rabbits are listed in Table 11.1.

## Pasteurellaceae

"Pasteurella" pneumotropica is an important rodent bacterium formerly classified as Pasteurella, which probably should be classified by itself within Pasteurellaceae.[3] Most conventional rodent colonies are infected, but also barrier bred colonies of rats and mice may harbor this agent, mostly latently. Carrier prevalences in infected rodent colonies vary from a few percent up to 95%.[2,4] It may lead to upper respiratory disease or pyogenic infections such as subcutaneous abscesses or mastitis,[1] but generally, *P. pneumotropica* is a secondary pathogen in relation to a primary agent, such as Mycoplasma pulmonis or Sendai virus. Stress including experimental stress or immunosuppression may activate latent infections. The incidence of spontaneous deaths during inhalation anesthesia might be raised in infected animals.[5] Transmission is mainly horizontal by droplets, but newborn puppies may become infected during their gestational passage of the contaminated vagina.

*Pasteurella multocida* is a facultative pathogen of rabbits.[6] In conventional colonies, a high number of animals may be infected.[7] Infection is mostly subclinical, and epizootic disease is connected with environmental and host-related factors. Respiratory disease occurs as "snuffles," which may develop into conjunctivitis, abscessation, and acute septicemias as well as acute or chronic pneumonia. It is mostly observed during spring and fall. Direct contact is considered the chief means of spread. Suckling rabbits may be infected with *P. multocida* from carrier does within the first week of life. The infection does not seem to spread between rabbits not in close contact.[8–13] A barrier system is an efficient way of keeping rabbits free of the infection.[14] Transmission from other species, e.g., pigs and cattle, may occur.[15]

## Clostridium piliforme

*Clostridium piliforme* (formerly known as *Bacillus piliformis*) is the causative agent of Tyzzer's disease. In mice,[16] hamsters,[17–19] gerbils,[20,21] and rabbits,[22] this is a fatal disease characterized by multiple focal necrosis of the liver (Figure 11.1). Long slender bacteria are found in the cytoplasm of the hepatocytes at the periphery of the necrotic foci. These bacteria are also found in huge numbers in the alimentary tract, especially in the ileum and caecum, and especially in association with ileitis, caecitis, and colitis. It has long been known that different mutants infect different animal species[23,24] and that infection between species, therefore, is a rare event, with the exception that Mongolian gerbils under some circumstances may be sensitive to infection from other species, such as rats, mice, and rabbits.[25] In rats, it is a mild disease of weanlings connected with megaloileitis (Figure 11.2), multiple focal necrosis of the livers, and single necroses in the myocardium.[26] Resistance to development of Tyzzer's disease may be due to genetic traits.[27,28] The organisms probably persist in the intestinal epithelium of healthy animals. The prevalences of infected individuals in rat and mouse colonies vary, but it is often more than 50%.[29] The agent may cross the placenta.[30,31]

## Helicobacter spp.

*Helicobacter* spp. have been isolated from nearly all species of rodents,[32] but clinical significance has only been documented in relation to a few of these. *H. hepaticus* and maybe also *H. bilis* cause chronic hepatitis in mice.[32] *H. hepaticus* is probably also responsible for liver tumors in mice.[32] Susceptibility to disease seems to be genetically dependent, e.g., A/JCr[33] and B6C3F[34] mice seem to be highly susceptible, while C57BL[33] mice seem to be resistant. *H. cholecystus* causes hepatitis and pancreatitis in hamsters. Infection with *Helicobacter* spp. is probably rather common in rodent colonies.[32,35] A range of transgenic mice as well as some rats suffer from a syndrome consisting of gastric ulcers, colitis, proctitis, and rectal prolapses (Figure 11.3), which in its features resembles several aspects of human inflammatory bowel disease. The syndrome is primarily observed in strains in which transgenesis has disrupted the normal mucosal homeostasis by features such as cytokine imbalance, abrogation of oral

**Table 11.1  Important Bacterial and Fungal Infections Observed in Mice (M), Rats (R), Guinea Pigs (GP), Syrian or Chinese Hamsters (H), and Rabbits (RB)**

| Gram-Negative Bacteria | |
|---|---|
| *Bordetella bronchiseptica* | M, R, GP, H, RB |
| *Campylobacter coli/jejuni* | M, R, H, RB |
| *CAR Bacillus* | M, R, RB |
| *Citrobacter freundii* | GP |
| *Citrobacter rodentium* | M |
| *Eschericia coli* | M, R, GP, H, RB |
| *Francisella tularensis* | R, RB |
| *Fusobacterium necrophorum* | M, GP, RB |
| *Haemophilus* spp. | M, R, GP, H, RB |
| *Helicobacter bilis* | M |
| *Helicobacter cholecystus* | H |
| *Helicobacter cinnaedi* | H |
| *Helicobacter hepaticus* | M |
| *Helicobacter muridarum* | M, R |
| *Helicobacter rappini* | M |
| *Helicobacter rodentium* | M |
| *Helicobacter trogontum* | R |
| *Klebsiella pneumoniae* | M, R, GP, H, RB |
| *Leptospira* spp. | M, R |
| *"Pasteurella pneumotropica"* | M, R, GP, H, RB |
| *Pasteurella multocida* | M, R, GP, H, RB |
| *Pseudomonas aeruginosa* | M, R, GP, H, RB |
| *Salmonella* spp. | M, R, GP, H, RB |
| *Spirillum minus* | R |
| *Streptobacillus moniliformis* | M, R, GP |
| *Treponema paraluis-cuniculi* | RB |
| *Yersinia pseudotuberculosis* | M, R, GP, H, RB |
| **Gram-Positive Bacteria** | |
| *Clostridium perfringens* | M,RB |
| *Clostridium difficile* | GP, H, RB |
| *Clostridium piliforme* | M, R, H, RB |
| *Clostridium spiroforme* | RB |
| *Corynebacterium kutscheri* | M, R, GP, H |
| *Erysipelothrix rhusiopathiae* | R |
| *Listeria monocytogenes* | M, R, GP, H, RB |
| *Staphylococcus aureus* | M, R, GP, H, RB |
| Streptococcus group A/B/C/D/G | M, R, GP, H, RB |
| *Streptococcus pneumoniae* | M, R, GP, H, RB |
| **Chlamydiae and Mycoplasmae** | |
| *Chlamydia psitacci* | GP |
| *Mycoplasma pulmonis* | M, R |
| *Mycoplasma neurolyticum* | M |
| *Mycoplasma arthritidis* | M, R |
| *Mycoplasma caviae* | GP |
| *Mycoplasma cricetuli* | H |
| *Mycoplasma collis* | M |
| *Mycoplasma muris* | M |
| **Fungi** | |
| *Aspergillus* spp. | R |
| *Candida albicans* | M, GP |
| *Cryptococcus neoformans* | M, GP, H |
| *Microsporum canis* | GP, RB |
| *Pneumocystis carinii* | M, R, GP, H, RB |
| *Trichophyton mentagrophytes* | M, R, GP, RB |

*Note:*　Only infections which may influence research in some way is mentioned, and it should be kept in mind that animals not being gnotobiotic harbor a great number of other bacterial species which are not mentioned here.

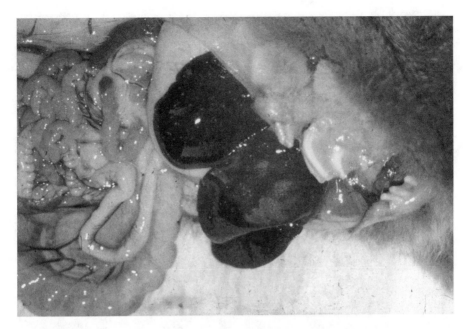

**Figure 11.1**  Tyzzer's disease in a rabbit. Note the white spots on the liver.

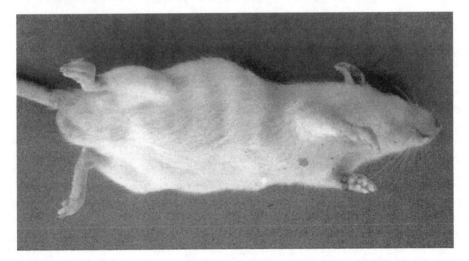

**Figure 11.2**  Megaloileitis in a post-weaned SPRD rat observed as a spiral on the right abdominal wall.

tolerance, alteration of epithelial barrier, and function or loss of immunoregulatory cells.[36] Members of the enteric flora are important factors in the development, as the syndrome may be prevented by antibiotic treatment[37] and germ-free conditions,[38] but while *Helicobacter pylori* has a key role in humans and also is involved in comparable experimental conditions in transgenic mice,[39] it is not clear whether spontaneous Helicobacter infections in transgenic mice have any impact on this syndrome.[40]

## Bordetella bronchiseptica

*Bordetella bronchiseptica* may be isolated from rabbits and guinea pigs, occasionally from rats, and seldom from mice, hamsters, and gerbils.[5] It may cause pneumonia, pleuritis, and pericarditis in guinea pigs, which might be fatal, especially if the animal carries other respiratory pathogens.[5] In rabbits, disease is mainly subclinical and characterized by focal chronic interstitial pneumonia.[41]

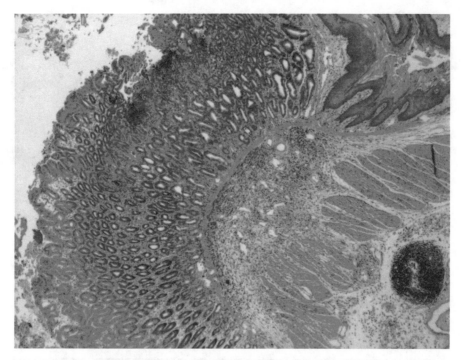

**Figure 11.3**  Certain transgenic knock-out mouse strains suffer from gastric ulcers, colitis, and rectal prolapses, as shown by this gastric ulcer in a plasminogen knock-out mouse. (Photo courtesy Kirsten Dahl.)

### Cilia-Associated Respiratory (CAR) Bacillus

CAR bacillus has been reported in mice, rats, and rabbits, but rat and mouse isolates differ from those of rabbits and should be regarded as different bacteria, the rat and mouse version being closely related to Flavobacterium,[42] and the rabbit version showing a higher similarity with Helicobacter.[43] Infected rats and mice are usually asymptomatic,[44,45] but CAR bacillus may be the cause of a highly contagious epizootic, slowly progressive and uncontrollable disease, called chronic respiratory disease (CRD), characterized by weight loss, rough hair coat, wheezing, rales, mucopurulent exudates, and severe peribronchial lymphoid cuffing.[46] In rabbits, no clinical signs of respiratory disease have been observed, although histopathological examination of the respiratory tree may reveal mild hyperplasia of lymphoid nodules subjacent to the respiratory mucosa with scattered bacilli in the lower respiratory system.[47] The infection is normally not transmitted to sentinels by the dirty bedding technique.[48]

### Corynebacterium spp.

*Corynebacterium kutscheri* is the cause of pseudotuberculosis in rats and mice. It has been found worldwide, but today it has become rather uncommon in laboratory animals as far as these are bred and kept in modern facilities. Infection is normally only observed in rats and mice,[49,50] although the organism also has been isolated from guinea pigs[51] and hamsters.[52] In immune-competent rats and mice, the agent may be subclinical,[49,50,53] but it may also cause abscessation in the superficial tissues and pulmonary emboli, while embolization in the mouse affects joints, liver, and kidney,[54] known as pseudotuberculosis. Genetics seem to be involved in the susceptibility of rats[55] and mice,[56] and therefore, mortality varies between infected colonies. Modes of excretion and spread of the agent are not fully known, but, probably, urine and feces from infected animals is contaminated. Transplacental infection has been demonstrated experimentally.[57] The prevalence of *C. kutscheri* within a colony may be less than 5%.

*C. renale* may cause urinary calculus in young rats.[58,59] *C. bovis* may be isolated from nude mice with scaly and crusty skin, often involving more than 80% of the animals.[60–64]

## Citrobacter rodentium

*Citrobacter rodentium,*[65] formerly known as *C. freundii* type 4280,[66] may in suckling and postweaned male mice be the cause of rectal prolapses, diarrhea, and dehydration. Feeding and genetic factors influence morbidity and mortality,[67] but mortality and prevalences are normally low. Transgenic and probably also nontransgenic mice of all ages may be affected by chronic debilitation, loss in reproductive efficiency, rectal prolapses, and death.[68] Also, alterations in immunological parameters may be observed, including outgrowth of an unusual population of cells in the spleen and blood, reduction in ascites production, loss of the capacity of peritoneal exudate cells to serve as feeders for the cloning of long-term T-cell lines, and inhibition of antigen-specific cytotoxic T-cell activity.[68] Antibiotic therapy may significantly reduce morbidity and mortality, increase litter size and frequency, and result in the normalization of many of the immunological assays.[68]

## Salmonellae

Salmonellae infect all species of warm-blooded animals, and until the introduction of barrier protected breeding systems, it was a common ruin of research projects involving rodents, especially mice. The prevalence of this organism has significantly diminished over the last 30 years. In mice and rats, *S. typhimurium*, and in guinea pigs, *S. enteritidis*, are the most common, causing various grades of diarrhea. The prevalences in infected colonies observed mostly range above 50% but may vary significantly.[69] Uterine infections (probably without passage of the placenta barrier) have been described for Salmonellae.[70]

## Streptobacillus moniliformis

*Streptobacillus moniliformis* may be isolated from mice,[71] rats,[72,73] and guinea pigs,[74,75] and it is transmissible to humans, in which it in rare cases causes rat bite fever, a purulent wound infection developing into endocarditis, petechial exanthema, polyarthritis, fever, and death. Recent cases have mostly been related to pets[76,77,78] and wild animals,[79] and have also been found in patients infected with human immunodeficiency virus.[80] In the mouse, disease begins as swelling of the cervical lymph nodes, which may turn into fatal septicemia. Chronic cases are characterized by arthritis in the distal parts of the legs and the tail. Abscessation and abortions may occur.[81] Genetic factors seem to be rather essential for the susceptibility to infection and disease. C57BL/6 mice seem to be highly susceptible.[71] In guinea pigs, the agent causes local abscesses, which do not spread.[82] In rats, clinical signs are uncommon, but cases of otitis media may occur.[72,73,74]

## Streptococcus spp.

Streptococcal infections are nonclinical and may spread between humans and animals, mostly by droplet infection through the intranasal route. Hemolytic streptococci and *S. pneumoniae* are, according to the FELASA guidelines for health monitoring,[48] the only ones to be reported, although their importance is probably questionable and some other types may be pathogenic as well. Prevalences within infected barrier bred colonies are generally around 10%.[2] Group C in guinea pigs as well as group G and A in rats and mice may be the cause of various pyogenic processes. *S. pneumoniae* may be found in guinea pigs, more seldom in rats and rabbits, and in rare cases in mice. The prevalence within infected colonies may vary from 15 to 55%. Disease is mostly related to stress, e.g., due to a poor environment or nutritional deficiencies. In rats, a mucopurulent discharge from the nose may initially be observed, and later the disease may progress into a noisy, abdominal respiration. Pathological changes are dominated by fibrin with various grades of focal bronchopneumonia developing into lobar fibrinous pneumonia. In guinea pigs, unexpected deaths are often the only visible signs of infection, while in rabbits, dyspnoea and depression often quickly turn into septicemia.[5]

## Staphylococci

Staphylococci are found worldwide in all species of animals and are spread between species, including animal to humans and vice versa. The majority of humans and animals are carriers of staphylococci. *Staphylococcus aureus* is found with a high prevalence in most colonies of laboratory rodents[2] as well as in most humans, while in wild mice, *S. aureus* is rather uncommon. Other types of Staphylococci common in laboratory rats and mice include *S. haemolyticus*, *S. xylosus*, *S. sciuri*, and *S. cohnii*.[83] The bacteria may be transmitted among hosts in various direct or indirect ways, including passive carriers among animal technicians. Staphylococcal disease in immune-competent animals is mainly secondary, e.g., due to trauma, stress, or the equivalent, and is characterized by pyogenic processes, such as abscesses in bite or surgical wounds, pneumonia in rodents kept in poorly ventilated units, and dermatitis in gerbils kept in too humid bedding. In immune-deficient animals, *S. aureus* may be a primary disease-causing agent, e.g., in the nude mouse, in which it causes multiple abscessation. Also, coagulase negative Staphylococci may cause disease in laboratory animals, e.g., *S. xylosus* is known to cause intestinal disease in mice,[84] dermatitis in gerbils,[85] and pneumonia in immune-suppressed rats.[86] Interference with research is mainly due to the activation of latent infection by stress or immunosuppression, but also the presence of abscesses in immune-deficient animals, typically nude mice, may be hazardous to research.

## Pseudomonas spp.

*Pseudomonas* spp. may be isolated from the respiratory, digestive, and genital systems of rats and mice, the more common ones being *P. aeruginosa* and *P. diminuta*. *P. aeruginosa* causes conjunctivitis and rhinitis and, under more severe or experimental conditions, pneumonia and septicemia in rats and guinea pigs.[5] In septicemic animals, abscessation of the liver, spleen, kidneys, and middle ears may be observed. Disease due to *P. aeruginosa* is mainly observed in immune-deficient, immune-suppressed, or stressed animals,[87] and, in general, it is secondary to something else. The prevalences in infected colonies of immune-competent animals kept in a high quality environment seldom reach more than 5 to 10%, but the prevalence of diseased animals in colonies of immune-deficient animals kept under poor environmental conditions may reach 100%, e.g., during ventilation breakdowns. Poor hygienic conditions, especially in relation to water used for drinking and cleaning may play an important role in the spread of *Pseudomonas* spp. *P. fluorescens* and probably also some other *Pseudomonas* spp. produce mucous in drinking nipples, which, however, is not known to have any impact on the animals. This condition is normally prevented by acidification of the drinking water with hydrochloric or citric acid.

## Mycoplasma spp.

Monoinfection with *Mycoplasma pulmonis* in rats causes mild symptoms. However, when complicated with other infectious agents, such as *P. pneumotropica*[88] or various viruses[89] as well as environmental inducers such as raised ammonia levels,[90] disease symptoms, such as snuffles, ruffled hair coat, bronchopneumonia, and arthritis, mostly in a mild form, occur. Additionally, it colonizes genitals of males and females, and at least in the latter, it may affect reproduction.[91,92,93] Even in the absence of clinical symptoms, *M. pulmonis* may raise the incidence of respiratory tract tumors,[94] decrease the cellular and humoral immune response,[95] decrease the severity of adjuvant arthritis,[96] and reduce the incidence of diabetes mellitus in BB rats.[97] The infection is far less common in mice, but symptoms are similar. Other *Mycoplasma* spp. infect rabbits and guinea pigs.

## Chlamydia spp.

*Chlamydia psitacci* may be the cause of conjunctivitis in guinea pigs. The diagnosis may be made by application of immunefluorescent antibodies to conjunctival scrapings, but routine screening is probably easier performed by PCR.[98]

## Dermatophytes

*Microsporum* and *Trichophyton* spp. may in rare cases be isolated from guinea pigs and rabbits. Clinical disease known as ringworm or dermatomycosis is rare, but it should be kept in mind that the infection is zoonotic.

## Viral Infections

Viruses should generally not be present in laboratory animals. This optimal condition has, however, not been reached, although surveys show that the number of virus infected rodent colonies has been declining over the last four decades. The most common infections are with corona-, parvo-, cardio-, and paramyxoviruses, and in that order.[35,99] Important viral infections in rodents and rabbits are listed in Table 11.2.

### DNA-Viruses

Several different types of DNA-viruses may infect rodents and rabbits, the major problem in rodents of today being parvoviruses. Except for the poxviruses, most DNA-viruses do not produce overt disease, but they may often have essential impact on research. As a rule of thumb, DNA-viruses cause persistent infection.

### Parvoviruses

Parvovirus infections in rats have traditionally been known to be caused especially by two viruses, Kilham rat virus (KRV) and Toolan's H1 virus (H1),[100] while in mice, they have been caused by minute virus of mice (MVM).[101] Antibodies are mostly detected by serology. Several antigenic types of parvoviruses are known, but KRV and H1 strains share common antigens and therefore cross-react in solid-phase serological assays. Orphan parvoviruses, a group of rodent parvoviruses distinct from MVM, KRV, and H1, were first discovered by the fact that antibodies to known rodent parvoviruses were detected by the immunofluorescence assay (IFA) but not by hemagglutination inhibition assay (HAI) in commercial breeding colonies of rats and mice.[102] Today, OPVs have been isolated from mice, rats, and hamsters, and they have further been divided into mouse parvovirus (MPV), rat parvovirus (RPV), and hamster parvovirus (HPV). RPV is assumed to be a variant of KRV,[103] while MPV resembles MVM in genome size, replication intermediates, and nonstructural proteins.[104] Cross-infection between species-specific strains does not seem to occur.[103] Horizontal transmission by fecal–oral contact is the most common. Vertical transmission is reported for some serotypes[105] but normally is not seen after the infection has balanced in the colony and the female breeders have developed protective immunity. Intrauterine infections may be observed in rare cases. The prevalence among adult animals is normally high, 50 to 80%, but lowers (sometimes even to zero) after a period of infection. MVM, H1, MPV, RPV, and HPV are not known to cause any clinical disease, while some KRV-serotypes have been reported to cause jaundice and ataxia in rats less than 10 days of age. Parvoviruses require a protein produced by the host cell during the S phase and, therefore, only replicate in rapidly dividing cells. In rats infected prior to the fourth day of life, intranuclear inclusions are present in the actively mitotic cells composing the external germinal layer of the cerebellum. This leads to necrosis, thereby preventing the normal development of the cerebellum and resulting in granuloprival cerebellar hypoplasia. Also, hepatitis with intranuclear inclusions in the hepatocytes has been reported. Parvovirus infections in rats were previously fairly common in Europe,[106] while it is far less common in mice. Results from Japan suggest that parvovirus infection in rats is most often caused by RPV.[107]

Rabbits may also harbor a parvovirus. Clinical signs in neonatal rabbits consist of anorexia and listlessness, while pathological signs are mostly located in the small intestines.[108,109,110]

**Table 11.2 Virus Infections Observed in Mice (M), Rats (R), Guinea Pigs (GP), Syrian or Chinese Hamsters (H), and Rabbits (RB)**

## DNA-viruses

### *Adenoviridae*

| | |
|---|---|
| Mouse adenovirus | M |
| Rat adenovirus | R |
| Guinea pig adenovirus | GP |

### *Herpetoviridae*

| | |
|---|---|
| Mouse cytomegalovirus | M |
| Rat cytomegalovirus | R |
| Guinea pig cytomegalovirus | GP |
| Virus III of rabbits | RB |
| Thymic virus | M |
| Guinea pig herpes-like virus | GP |
| Guinea pig X-virus | GP |

### *Papovaviridae*

| | |
|---|---|
| K virus | M |
| Mouse polyoma virus | M |
| Rat polyoma virus | R |
| Hamster papovavirus | H |
| Rabbit kidney vacuolating virus | RB |
| Virus of oral papillomatosis | RB |
| Rabbit papilloma virus | RB |

### *Parvoviridae*

| | |
|---|---|
| Kilham rat virus | R |
| Toolans H1 virus | R |
| Minute virus of mice | M |
| Hamster parvovirus | H |
| Mouse parvovirus | M |
| Rat parvovirus | R |

### *Poxviridae*

| | |
|---|---|
| Ectromelia virus | M |
| Mouse papule virus | M |
| Myxoma virus | RB |
| Shope's fibroma virus | RB |
| Rabbit pox virus | RB |
| Guinea pig pox-like virus | GP |

## RNA-Viruses

### *Arenaviridae*

| | |
|---|---|
| Lymphocytic choriomeningitis virus | M, GP, H |

### *Bunyaviridae*

| | |
|---|---|
| Hantavirus | R |

### *Caliciviridae*

| | |
|---|---|
| Rabbit hemorrhagic disease virus | RB |

**Table 11.2 Virus Infections Observed in Mice (M), Rats (R), Guinea Pigs (GP), Syrian or Chinese Hamsters (H), and Rabbits (RB) (continued)**

*Coronaviridae*

| | |
|---|---|
| Mouse hepatitis virus | M |
| Rat coronavirus | R |
| Sialodacryaodenitis virus | R |
| Guinea pig coronavirus | GP |
| Rabbit coronavirus | RB |

*Paramyxoviridae*

| | |
|---|---|
| Sendai virus | M, R, RB |
| Pneumonia virus of mice | M, R, H |
| Guinea pig parainfluenza type 3 | GP |

*Picornaviridae*

| | |
|---|---|
| Theiler's mouse encephalomyelitis virus | |
| Strain GDVII, FA, DA | M |
| Strain MHG | R |
| Guinea pig cardiovirus | GP |

*Reoviridae*

| | |
|---|---|
| Reovirus type 3 | M, R, H, GP |
| Mouse rotavirus | M |
| Rat rotavirus | R |
| Rabbit rotavirus | RB |

*Retroviridae*

| | |
|---|---|
| Type A viruses | M |
| Type B viruses | |
| Mouse mammary tumor virus | M |
| Type C viruses | |
| Leukemia viruses | M, R, H, GP |
| Sarcoma viruses | M, R |

*Togaviridae*

| | |
|---|---|
| Lactate dehydrogenase elevating virus | M |

*Unclassified Viruses*

| | |
|---|---|
| Grey lung virus | M, R |

## Adenoviruses

Host-specific strains of adenoviruses infect a range of species, including mice, rats, and guinea pigs, and in extremely rare cases, rabbits. The virus infects by oral or ocular transport, but close contact is required. In mice, two different substrains, MAD-FL[111] and MAD-K87[112], have been identified. In rats and mice, the infection is mostly clinically inapparent, although myocarditis, nephritis, adenitis, encephalitis, and mortality have been observed after experimental inoculation of neonatal mice.[113] In infected guinea pigs, necrotizing broncheoalveolitis is regularly observed, characterized by large basophilic intranuclear inclusion bodies in the desquamated bronchial epithelial.[114] The strain infecting guinea pigs has not yet been isolated *in vitro*, but the virus has been identified by polymerase chain reaction (PCR) in the upper airways on days six through 15 after inoculation, and, in addition, the virus has been spontaneously transmitted from an experimentally infected animal to immune-naive cage mates.[115]

Adenoviral disease in guinea pigs has been observed in Europe, the United States, and Canada. In most cases, no distinct clinical signs are observed, but occasionally, dyspnoea symptoms are discretely scattered among animals in a room. The diagnosis can be made by serology in all species. This includes guinea pigs,[114] but there is conflicting evidence on the degree of the cross-reactivity between guinea pig adenovirus and adenoviruses from other species.[115]

## Poxviruses

Ectromelia or mousepox is a fatal disease in mice induced by infection with ectromelia virus. The disease is most severe in DBA, C3H, and Balb/c mice, while black strains seem to be relatively resistant and even may harbor latent infections.[116,117,118] Prevalence of the overt disease may vary from few to all.[119] The virus infects through skin lesions, and after approximately 10 days of incubation, the infection causes edematous skin erosions and hyperplasias pathologically characterized by large eosinophilic cytoplasmatic inclusions in the epithelial cells. Extensive necrosis in the lymphatic organs and the liver are also observed. In mice surviving the infection, diagnosis may be made by serology, while diagnosis of acute infection is based upon polymerase chain reaction, immunohistochemistry, and virus isolation.[120] Today, the spontaneous disease is seldom in laboratory mice, although incidents have occurred, but major concern should be applied to biological materials of insecure origin.[121,120]

In rabbits, several poxviruses are known, i.e., myxoma, fibroma, and rabbit poxviruses. Of major concern is the myxoma virus, the cause of myxomatosis. In laboratory rabbits, this is a severe disease in which edema of the eyelids is the most dominant symptom. It may occur in a peracute version, which kills the rabbits within 1 week, or in an acute version with maybe 2 weeks of survival after edema also has developed around the anal, genital, oral, and nasal openings. Lethargy, hemorrhages, and convulsions occur just prior to death. The disease may be diagnosed by symptoms, but in countries in which the infection is under legal control, the diagnosis should be confirmed by virus isolation. Current screening may be performed by serology. Outbreaks in laboratory colonies are rare, as it mainly spreads through insects, but it should be kept in mind that the disease is endemic in the wild population of lagomorphs in Europe, North and South America, and Australia, even though it is under legal control in many countries.

## Herpesviruses

Cytomegaloviruses are also known as "salivary gland viruses." Several species-specific strains exist, among these, strains infecting mice, rats, and guinea pigs. Megalic cells and nuclear inclusions in the glandular epithelium of the salivary glands characterize infection. Infection is rare in laboratory colonies. Mouse thymic virus infects only mice, in which it may be found in the thymus and the salivary glands. Infection with mouse thymic virus is occasionally reported in laboratory mouse colonies, in which it may reach high prevalences. Acute herpesvirus-related disease may be induced experimentally in suckling mice,[122] but in general, infection in rodents is asymptomatic. But, in guinea pigs, clinical disease may be observed in breeding females, and transmission in utero cannot be excluded.[48,123] For all herpesviruses, diagnosis can easily be achieved by serology.

## RNA-Viruses

A range of different RNA-viruses infects rodents and rabbits. In immune-competent animals, RNA-virus infection is normally nonpersisting, except for those viruses able to incorporate themselves into the genome of the host, i.e., retroviruses or those viruses only generating an insufficient immune-response, e.g., lymphocytic choriomeningitis virus. The morbidity and mortality caused by these viruses vary greatly. Most of these viruses are host-specific with the major exception of lymphocytic choriomeningitis virus, but they are antigenically that close that the same virus is often stated to be able to infect more than one species, although infection in different species is probably caused by differing substrains.

## Coronaviruses

Coronavirus infection in mice or rats is the most important and the most common viral problem encountered in laboratory rodent facilities. Coronaviruses in mice, mouse hepatitis virus (MHV), and rats, sialodacryoadenitis virus (SDAV), and rat coronavirus (RCV), are antigenically close but are different and highly species-specific viruses. Clinical symptoms are not a major feature in relation to MHV infection in mice, but it depends on characteristics of the virus and the animal. Some strains are enterotropic, i.e., they infect through the gastrointestinal system, while others are pneumotropic, i.e., they infect through the respiratory system.[124] The most common symptom is diarrhea in suckling mice.[125] Several organs are affected by the infection, and the virus has a tropism for all tissues, which in connection with its immunomodulating effects makes it a significant modulator of research and an undesired organism in rodent facilities. The liver may be pale with multiple white, yellow, or hemorrhagic foci. The spleen may be enlarged, while the thymus may be reduced in size and there is widespread necrosis of lymphoid tissue. BALB/c and C57BL mice seem to be more sensitive than other mice.[126] In immune-deficient mice, such as SCID[127] and nude mice,[128] infection causes high mortality. Strains such as MHV-2, MHV-3, and MHV-A59 are more virulent than, e.g., MHV-1, MHV-S, MHV-Y, and MHV-Nu. A strain designated MHV-4 has a specific affinity for the nervous tissues.[129] Coronavirus infection in rats caused by one of several substrains of SDAV or RCV may often be asymptomatic, but clinical symptoms seem to be more common after corona viral infection in rats than in mice. A mild necrotizing rhinotracheitis develops into interstitial pneumonia with severe necrosis and swelling of salivary and lacrimal glands. In this phase, the ventral neck region of the rats is swollen, and red-brown porphyrin rings are observed around their eyes. Diseased rats fully recover within 5 weeks.[130] Corona viruses easily spread in rodent facilities, and the prevalence in a rat colony reaches 100% within 4 weeks.[131] Although unstable, coronaviruses may be transported passively between facilities by staff or equipment. Coronaviruses are nonpersisting in immune-competent animals but persisting in immune-deficient and certain transgenic animals. Current screening can easily be performed by serology.[132] Diagnosis of acute disease may be attempted by PCR[133] or tests for anticoronaviral IgA[131] on fecal samples.

Coronaviruses different from murine coronaviruses may infect guinea pigs or rabbits, and until the 1960s, pleural effusion as the result of infection with rabbit coronavirus was a common disease in laboratory rabbits.[134] Enteric infection with rabbit coronavirus may still be common.[48]

## Cardioviruses

Cardioviruses are picornaviruses producing enteric infection in a wide range of mammals. Similar to polioviruses, certain strains may, under specific conditions, invade the central nervous system and produce neurodegenerative disease. Theiler's mouse encephalomyelitis virus (TMEV) — occasionally in a more popular manner referred to as mouse polio — can be divided into three groups of substrains infecting mice, i.e., GDVII, FA, and DA. Some not well described strains infect rats.[135,136] Antibodies to TMEV are a common finding in rats,[99] however, without any clinical signs. GDVII and FA are far more virulent than DA, but in most cases, spontaneous infections are asymptomatic. GDVII may cause acute encephalitis. Paralysis develops when the virus leaves the gray matter and infects the white matter, thereby damaging the upper motor neuron system.[137] CD-1, DBA/2, SJL, and SWR mice seem to be far more susceptible.[138] DA causes a more long-term demyelinating disease. In infected colonies, prevalences are generally high, and the virus may be transmitted intrautero,[139] which may complicate rederivation.

Encephalomyocarditis virus (EMCV) is a virus commonly used for experimental infection of mice, especially in diabetes research.[140] However, spontaneous infections have not been found in mice.

Guinea pigs may also harbor a cardiovirus, the causative agent of the disease guinea pig lameness, which is a paralytic and mortal disease. Deficiency of vitamin C may be a predisposing factor, allowing spread to the central nervous system; therefore, in colonies, outbreaks are likely to occur if for accidental reasons the content of vitamin C in the diet is low.[141] Serological studies have revealed the presence of antibodies against TMEV in guinea pigs suffering from lameness,[142] but DNA technique shows that the virus is probably closely related to EMCV.[114] This virus is rather common in laboratory guinea pigs,[142] but it is seldom monitored by commercial breeders.

## Paramyxoviruses

Parainfluenzaviruses are unstable viruses producing nonpersisting infection in immunecompetent rodents and persisting infection in immune-deficient rodents.[143,144] They may be divided into type 1, type 2, type 3, and type 4, the latter, however, have only been found in humans. Sendai virus, a type 1 parainfluenzavirus, produces respiratory infections in rats and mice. Isolates are antigenically alike and show the same pathogenecity for both species. Clinical symptoms are rare, but a high mortality may be seen in young mice before or around weaning. DBA/2 and 129 mice may be more sensitive than other strains.[145] Pathological signs are catarrhal bronchitis eventually extending into the alveoli of the lung. Transmission is mainly respiratory.[146] Diagnosis can easily be made by serology, but the virus does not spread efficiently among sentinels by the dirty bedding technique.[147] Rabbits may be experimentally infected by Sendai virus,[148] and spontaneously occurring antibodies have also been found in rabbits.[149] Antibodies to Sendai virus have also been revealed in guinea pigs,[69] but as Sendai virus has not been isolated from guinea pigs and experimental infection has not been achieved, these antibodies have probably been cross-reactions from guinea pig parainfluenza virus. This type 3 parainfluenza virus was not isolated from guinea pigs until 1998,[150] but antibodies were found in guinea pigs already in the 1970s.[151] Guinea pig parainfluenza virus type 3 is a lineage of human parainfluenza virus type 3, probably introduced into guinea pig colonies via infected humans.[150] Serological screening for antibodies, e.g., enzyme-linked immunosorbent assay (ELISA), is at present the diagnostic method of choice, but as the guinea pig parainfluenza virus is not commonly available as antigen bovine parainfluenzavirus type 3 must be used for testing guinea pigs. In infected breeding colonies, all parainfluenzaviruses normally have high prevalences.

Infection with the pneumovirus, pneumonia virus of mice (PVM), causes a normally silent infection in mice, rats, hamsters, and gerbils. Serological prevalences are high in infected colonies, but pathological symptoms are absent in immune-competent mice. Weak pathological changes may be found in rats and in nude mice. In the latter, infection is also persistent,[143] in contrast to all immune-competent animals, which clear themselves of the infection. Current screening can be made by serology.

## Reoviruses

A number of wild-type and laboratory strains of reovirus type 3 have been recovered from vertebrates and nonvertebrates. So far, the virus has only been isolated from mice, but antibodies have been detected in rats, guinea pigs, and rabbits. Whether these antibodies are specific or are due to cross-reactions is actually not known. Previously, 15% of European guinea pig colonies were shown to have positive titers to reovirus type 3,[106] but clinical or pathological changes as a cause of reovirus infection in guinea pigs have never been reported, and the situation has been improved today. The virus is relatively heat stable, but temperature-sensitive mutants have also been developed. It is also resistant to some chemical disinfectants and can survive outside the body for longer periods. Transmission is mainly by the oral route, but air-borne contamination may occur. Intrauterine infection is described under experimental conditions but is not likely to occur in natural infection. Antibodies can be detected by serology.

## Rotaviruses

Rotaviruses have been isolated from numerous mammalian species. In general, they cause enteric infections leading to diarrhea, especially in newborn animals. These viruses are highly contagious, and prevalence in infected colonies is high. Infection with rabbit rotavirus is rather common in rabbits.[152,153] In mice, rotaviral disease is called epizootic diarrhea of infant mice (EDIM), and in rats, it is called infectious diarrhea of infant rats (IDIR).[154] EDIM and IDIR are different serotypes. Rodent rotaviral infections are less common today. Current screening can be performed in all species by serology, while diagnosis of acute disease can easily be achieved by capture-ELISA or latex-agglutination on feces. Different serogroups exist, and monitoring in mice and rabbits must be carried out using a serogroup A antigen.[48]

## Togaviruses

Lactate dehydrogenase elevating virus infects laboratory and wild mice worldwide. It is mostly known for its ability to elevate serum lactate dehydrogenase as well as a range of other serum enzymes in silently infected mice, but it may also produce paralysis after infection of the central nervous system. Especially the c strain, also called the Murphy strain, is likely to produce paralysis, and AKR and C58 mice seem to be more susceptible than other strains.[155] It is a common contaminant of transplantable tumor cell cultures.[156] It is not quite clear how it actually spreads. Cannibalism and fighting, especially among male mice, may be a way of transmission, but a parasitic vector may be needed, which makes epizootics in mouse colonies less likely. Transplacental infection may occur.[157] It is most easily diagnosed by testing mice for elevated levels of lactate dehydrogenase, while biological materials may be screened by PCR.[158]

## Arenaviruses

Lymphocytic choriomeningitis virus has been found naturally infecting hamsters, as the most frequent, and less frequently, mice and guinea pigs. Furthermore, it is zoonotic and has the capability of infecting man. In a few unlucky cases, this may lead to meningitis. Most cases have been associated with pet hamsters.[159,160,161] Infection may be persisting, in mice especially, if these are infected intrautero or within the first 7 days after birth. It has been widely used as experimental agent for studying viral immunology. It spreads slowly in the colony and prevalences seldom reach more than 10%, although the prevalence may be higher among animals in the lower racks.[162] Vertical spread from mother to foster is the principal way of infection.[163] Natural infections in mice and hamsters are normally silent, while more severe symptoms may be observed after infection with specific strains in guinea pigs.[164] Serology is the principal method for routine monitoring.

## Bunyaviruses

Hantaviruses are zoonotic viruses spreading from wild animals to humans, in which they mostly lead to silent infections but in unlucky cases produce hemorrhagic disease. Several types exist, but in relation to laboratory animals, the only type of interest is the Seoul Strain, producing inapparent infection in rats.[165] It is found mainly in the Far East and in the European Balkan region.[166] In other areas of the world, e.g., Scandinavia and United States, other types of Hantaviruses with specificity for other animal species may be found.[167] Infections in laboratory animals are rare, but they have occurred after housing wild rodents in laboratory animal facilities.[168] Current screening can easily be performed by serology.

## Caliciviruses

Rabbit haemorrhagic disease virus (RHDV) has over the last decades spread over Asia to Europe, where it is mostly found in slaughter rabbits. It is also found in Mexico. It causes sudden death, hemorrhages, especially from the nose. In acute cases, mortality will run up to 90%, but also more chronic cases with lower mortality, distress, and icterus may be seen. It is diagnosed by ELISA, but cross-reactions from nonpathogenic caliciviruses may occur.[169]

## Parasitological Infestations

Laboratory animals should be free of parasites. These may not have obvious clinical impact on the animals, but they have subclinical impact and may interfere with research in various ways. Also, parasitic infestations are symptomatic of low hygienic standards, and infestated animals may be suspected also to carry other types of infections. Important parasites in rodents and rabbits are listed in Table 11.3. The most common parasites found in laboratory rodents are the flagellates *Tritrichomonas* spp., but these also seem to have low impact on the animals. Pinworms and mites are also common findings. Encephalitozoon cuniculi has become more rare in colonies of rabbits and guinea pigs, but when it occurs, it does have a certain impact on research.

**Table 11.3  Parasitic Infestations Observed in Mice (M),
Rats (R), Guinea Pigs (GP), Syrian or Chinese
Hamsters (H), and Rabbits (RB)**

### Mastigophora (Flagellates)

| | |
|---|---|
| Chilomastix spp. | R,H,RB |
| Giardia spp. | M,R,GP,H,RB |
| Spironucleus muris | M,R,H |
| Tritrichomonas spp. | M,R,GP,H |
| Tetratrichomonas minuta | M,R,H |
| Pentatrichomas homonis | M,R,H |
| Trichomitis spp. | R |
| Hexamastix spp. | R,GP,H |
| Enteromonas spp. | R,GP |
| Retortamonas spp. | R,GP,RB |
| Monocercomonoides spp. | R,RB |
| Chilomitus spp. | GP |
| Octimitus spp. | R |

### Sarcodina (Amebas)

| | |
|---|---|
| Entamoeba muris | M,R,H |
| Entamoeba cuniculi | RB |

### Sporozoa

| | |
|---|---|
| Encephalitozoon cuniculi | M,R,GP,RB |
| Eimeria falciformis | M |
| Eimeria spp. | R |
| Eimeria caviae | GP |
| Eimeria spp.[a] | RB |
| Eimeria stiedae | RB |
| Cryptosporidium spp. | R |
| Toxoplasma gondii | M,R,GP,RB |
| Sarcocystis muris | M,R |
| Sarcocystis cuniculi | RB |
| Klossiella muris | M |
| Klossiella cobayae | GP |

### Ciliata

| | |
|---|---|
| Balantidium spp. | R,GP,H |

### Nematodes

*Stomach Worms*

| | |
|---|---|
| Graphidium strigosum | GP,RB |

*Intestinal and Cecal Worms*

| | |
|---|---|
| Trichostrongylus spp. | RB |
| Paraspidodera uncinata | GP |

*Pinworms*

| | |
|---|---|
| Aspiculuris tetraptera | M,R |
| Dermatoxys veligeria | RB |
| Passaluris ambiguus | RB |
| Syphacia muris | R,H |
| Syphacia obvelata | M,R,H |

**Table 11.3 Parasitic Infestations Observed in Mice (M), Rats (R), Guinea Pigs (GP), Syrian or Chinese Hamsters (H), and Rabbits (RB) (continued)**

*Bladder Worms*

| | |
|---|---|
| Trichosomoides crassicauda | R |

*Threadworms*

| | |
|---|---|
| Strongyloides ratti | M,R,H |
| Capillaria hepatica | M,R,RB |

*Lungworms*

| | |
|---|---|
| Protostrongylus spp. | RB |

**Cestodes**

*Adult Tapeworms*

| | |
|---|---|
| Cittotaenia variabilis | RB |
| Hymenolepis nana | M,R,H |
| Hymenolepis diminuta | M,R,H |

*Cysticerci of Tapeworms*

| | |
|---|---|
| Cysticercus pisiformis | RB |
| Coenurus serialis | R |
| Strobilicercus fasciolaris | M,R |

**Trematodes**

*Liver flukes*

| | |
|---|---|
| Fasciola hepatica | GP,RB |
| Dicrocoelium dendriticum | GP,RB |

**Hair Follicle Mites**

| | |
|---|---|
| Demodex aurata | H |
| Demodex caviae | GP |
| Demodex criceti | H |
| Demodex musculi | M |
| Demodex nanus | R |

**Ear Mange Mites**

| | |
|---|---|
| Notoedres muris | R,GP,H |
| Psoroptes cuniculi | RB |

**Body Mange Mites**

| | |
|---|---|
| Psorergates simplex | M |
| Notoedres cati | RB |
| Sarcoptes scabiei | M,R,GP,RB |

**Fur Mites**

| | |
|---|---|
| Cheyletiella parasitivorax | RB |
| Chirodiscoides caviae | gp |
| Myobia musculi | M |
| Myocoptes musculinus | M,GP |
| Listrophorus gibbus | RB |

**Table 11.3  Parasitic Infestations Observed in Mice (M), Rats (R), Guinea Pigs (GP), Syrian or Chinese Hamsters (H), and Rabbits (RB) (continued)**

| | |
|---|---|
| Radfordia affinis | M |
| Radfordia ensifera | R |
| Trichoecius romboutsi | M |

**Lice**

| | |
|---|---|
| Haemodipsus ventricosus | RB |
| Polyplax serrata | M |
| Polyplax spinulosa | R |
| Gliricola porcelli | GP |
| Gyropus ovalis | GP |

**Ticks**

| | |
|---|---|
| Haemaphysalis leporis-palustris | RB |

a  E. irresidua, E. magna, E. media, E. perforans, E. exigua, E. intestinalis, E. matsubayishii, E. nagpurensis, E. neoleporis, E. piriformsis.

## Pinworms

Laboratory rodents often harbor pinworms, i.e., ascarids of the family oxyuridae. In rodents, the most common species are Syphacia and Aspiculuris. *S. obvelata* is the most common in mice; *S. muris* is the most common in rats, while hamsters may occasionally harbor a third species, *S. mesocriceti*. In mice, *Aspiculuris tetraptera* is nearly as common as *Syphacia* spp. Up to 70% of U.S. laboratory animal facilities have recently been shown to carry pinworm infestations.[35] In rabbits, the most common species is *Passalurus ambiguus*. Pinworms live in the free lumen of the caecum, colon, and rectum. They are extremely infectious, as they easily spread with staff, equipment, etc., and the infectious dose is rather low. Clinical symptoms are rare. Diagnosis is, if not simply observing the worms during necropsy, made by detecting eggs in the tape or flotation test.

## Protozoans

*Tritrichomonas* spp. — The most common endoparasite in rodents is the flagellate *Tritrichomonas muris*. It is nearly the rule to find it in the caecum and colon of conventional rodents, but it may be found in barrier-protected rodents as well.[132] Transmission occurs by ingestion of pseudocysts. It seems to be apathogenic.

*Encephalitozoon cuniculi* — *Encephalitozoon cuniculi*, a protozoan parasite belonging to the subphylum Microspora, is the etiological agent of a spontaneous disease in rabbits, which should be called encephalitozoonosis but often is called nosematosis, as the agent previously belonged to the genus Nosema. The most common way of transmission seems to be the oral via infectious feces and urine. It is unclear whether vertical transmission may also occur.[170] The sporoplasm is extruded from a spore, which enters the host cells to multiply and mature into spores, which after cell rupture, restart the cycle. Infection has been described in many species, e.g., rabbits,[171] rats,[172] mice,[172] and guinea pigs,[170] but it is only common in rabbits and guinea pigs, while infection in other species is rare and probably the result of contact with contaminated rabbit colonies.[172] Infection is usually latent, but occasionally, rabbits exhibit various neurological signs, such as convulsions, tremors, torticollis, paresis, and coma. Lesions in the kidneys of infected rabbits are frequent, grossly manifested as multiple, pinpoint areas, randomly scattered over the surface, or more usual, as 2 to 4 mm indented gray areas on the cortical surface. In histopathology, granulomatous nephritis[173] and granulomatous encephalitis[174] are observed. Guinea pigs do not seem to develop disease. Infection is diagnosed by serology.[175] The disease is usually diagnosed by recognition of typical lesions and sometimes also the agent by histopathology of kidneys.

*Eimeria* spp. — Various Eimeria species are known to cause intestinal coccidiosis in rabbits. Infection is fairly common, but if observed, clinical symptoms are mostly found around weaning in breeding facilities and are rare in experimental facilities, although heavily infected rabbits may also develop clinical symptoms. Such rabbits show diarrhea in varying degrees, thirst, and dehydration. In subclinical cases, weight loss may be observed. Also, peracute cases with deaths prior to the presence of oocysts in the feces may be observed. The small and large intestines of heavily infected rabbits may show multiple white spots on the mucosa with a mixed mononuclear and polymorphnuclear exudate. Infection occurs orally with coccidial oocysts, which burst in the intestines, releasing sporozoites to invade the intestinal mucosa cells. Here they multiply into schizonts, which break, extruding a huge number of merozoites into the lumen, from which they invade new cells to repeat the process. After an unknown number of such asexual generations designated schizogony, the merozoites differentiate into female macrogametocytes or male microgametocytes. The microgametocytes leave the cells and unite with the macrogametocytes to form the oocysts. The most common enteric Eimeria species in the rabbit are *E. irresidua, E. magna, E. media,* and *E. perforans.*

    *E. stiedae* produces hepatic coccidiosis with multiplication in the bile ducts, but this infection is uncommon in laboratory rabbits. Clinical symptoms of this infection are mostly related to chronic debilitation of the liver, i.e., weight loss and chronic intestinal symptoms in rare cases leading to death.[1]

    The diagnosis of Eimeria infection is made by the observation of oocysts in the feces in the flotation test. The diagnosis coccidiosis is made by observation of the clinical and pathological symptoms combined with the observation of huge numbers of oocysts in the feces. Outbreaks of coccidiosis may be treated with sulfonamides, e.g., sulfadimidine as a 0.02% solution in the drinking water.

*Pneumocystis carinii* — The eucaryote *P. carinii,* occasionally referred to as a fungus, infects a wide range of mammal species, but it is difficult to diagnose and fully asymptomatic in immune-competent animals. However, in SCID mice and other immune-deficient animals, it may cause fatal pneumonia.[176,177] It is easily diagnosed in immune-deficient animals, e.g., by immunofluorescence staining of lung washings, and has shown, and it is quite widespread, even in barrier-protected colonies. Transmission seems to be airborne.[178]

*Ectoparasites* — In conventional rats and mice, ectoparasites are a common finding.[35] Mites such as *Myobia musculi, Radfordia affinis,* and *Myocoptes musculinus* in mice and *R. ensifera* in rats are the most common. This may cause no symptoms or a variety of skin lesions in the range from mild pruritus to serious pyoderma. Other ectoparasites are more rare, although lice infestations with *Polyplax* spp. have been observed in conventional rodent colonies. In barrier-protected rodent colonies, ectoparasites are generally absent.

## MICROBIAL INTERFERENCE WITH ANIMAL EXPERIMENTS

Clinical disease is only one of several ways that infectious agents may be hazardous to research, and the absence of disease symptoms should not be interpreted as the absence of hazardous infections.

### Pathological Changes, Clinical Disease, and Mortality

Many experiments have been ruined by disease or pathological changes caused by specific infections as described previously. Subclinical disease may also disturb essential parameters, e.g., subclinical viral infections may affect body weight in rats. Additionally, behavior will often be changed during subclinical disease, leading to disturbances in the open field test, etc. The presence of some microorganisms may leave changes in the organs, resulting in difficulties in the interpretation of the pathological diagnosis included in e.g., toxicological studies — a phenomenon often referred to as "background noise." Respiratory disease of any etiology may be responsible for deaths during anesthesia. Certain strains, inbred or transgenic, may be more prone to the development of specific pathological changes, e.g., certain inbred strains suffer from certain pathological changes or mice "knocked out" for various immunologically active genes suffer from a syndrome of gastric ulcers, colitis, and rectal prolapses, the causes of which are unkown[179] (Figure 11.3).

## Contamination of Biological Products

Microorganisms present in the animal may contaminate samples and tissue specimens, such as cells, sera, etc.[156] This may interfere with experiments performed on cell cultures or isolated organs. Furthermore, the introduction of such products into animal laboratories will impose a risk to those animals kept in that laboratory. In theory, any infection may contaminate certain products from the animal. However, viruses are known to represent the major risk, due to their ability to produce viremia in the animal and for some viruses to persist in specific organs, thereby increasing the time and the number of tissues at risk. The virus most commonly found is lactic dehydrogenase virus followed by reovirus type 3, lymphocytic choriomeningitis virus, minute virus of mice, mouse hepatitis virus, rat coronaviruses, Kilham rat virus, and Mycoplasma pulmonis.[156] Other *Mycoplasma* spp. may contaminate biological materials,[180] but often these are nonrelevant species derived from humans or unrelated species used as donors of serum for cell culture media. Recently, ectromelia virus has been brought into two U.S. laboratory animal facilities through contaminated cell lines, which have had enormous economic impact on these institutions.[121,120] Some protozoans, e.g., *Encephalitozoon cuniculi*, as well as bacteria also have the potential to contaminate transplantable tumors.

## Immunomodulation

Many experiments are based upon a functional immune system in the laboratory animal, e.g., in studies of autoimmune type 1 diabetes mellitus in the NOD mouse and the BB rat. Microorganisms may perform immunomodulation also in the absence of clinical disease, and the effect may be suppressing or activating or both at the same time but on different parts of the immune system. Viruses are normally described as the most frequent immune modulators, one of the reasons being the viremic phase in the pathogenesis of many virus infections during which cells of the immune system become infected, e.g., by adeno-, parvo-, or coronaviruses.[181] The thymus may become infected by thymic virus.[182] Leukemia viruses even integrate themselves into the leucocytes by reverse transcription. Through different mechanisms such infection of the immune cells may suppress the immune system. One of the most well-described viral infections suppressing the immune system is mouse cytomegalovirus infection,[183] an effect probably more specifically associated with the emission of cytokines from the macrophages resulting in the dominance of T-suppressor cells.[184–187] The immunosuppressive effect of Mycoplasma infections seems to differ from viral immunosuppression in the secretory stimulation of the macrophages, as interferon production during Mycoplasma infection seems to be impaired.[188] *Mycoplasma* spp. have generalized effects on the specific response to some antigens[189] as well as influence on the unspecific host defense mechanisms, especially in the respiratory system.[190–195] Immunomodulatory bacteria include group A Streptococci, *Pseudomonas aeruginosa*, *Eschericia coli*, and *Salmonella* spp., the effect which can be suppressing as well as stimulating and often is mediated through endotoxin production, the active component probably being lipid A.[196] Parasites with impact on the immune system are protozoans such as Toxoplasma gondii[197–201] or helminths such as *Syphacia* spp.[202]

## Physiological Modulation

Some microorganisms have a specific effect on enzymatic, hematological, and other parameters monitored in the animal during an experiment, lactate dehydrogenase (LDH) elevating virus in mice inhibiting the clearance of LDH and a number of other enzymes[203] being the classical example. Also, mouse hepatitis virus alters the hepatic enzyme activity.[204] Such organic function disturbances may change the outcome of the experiment, e.g., the altered function of the liver and macrophages in mice infected with mouse hepatitis virus, may lead to an altered response in toxicology, nutrition, and other items in which the liver is involved.[205] Such disturbances may be irreversible for some drugs, while reversible for other drugs, as it has been described in mice infected with *Clostridium piliforme*.[206]

## Interference with Reproduction

In breeding colonies, changes in fertility may be the cause of serious trouble. Infections giving rise to clinical disease in a major part of the population are likely to reduce fertility. It is also quite obvious that infections such as rotaviruses[207] or mouse hepatitis virus[208] causing high mortality in newborn animals will reduce the outcome of breeding or disturb experiments with newborn animals. Also, direct effects of the infection on reproduction such as the change in sex hormones, pathological anatomical changes in the reproductive tract, or infection of the embryo causing abortion and stillbirths are observed, as, for example, may be the case for parvoviruses.[100] Many microorganisms, e.g., retroviruses, lymphocytic choriomeningitis virus,[163] parvoviruses,[105] ectromelia virus,[209] cardioviruses,[139] and *Clostridium piliforme*[30,31] posses the ability to cross the placenta barrier. Uterine infections — probably without passage of the placenta barrier — have been observed for many bacteria, e.g., *Salmonella* spp.[70] and *Pasteurella pneumotropica*.[210] However, with the exception of the retroviruses and lymphocytic chorio-meningitis virus, spontaneous infections of fetuses are rare. Mycoplasma pulmonis is known to be harmful to reproduction, altering a number of reproductive parameters,[211] and infection with *Mycoplasma* spp. and certain purulent bacteria is a risk in embryo transfer, for which the application of aseptic principles is essential.[212]

## Competition Between Microorganisms within the Animal

In some studies, experimental infection is the main detail of the study, or inactivated bacteria are used for the induction of some immunological response, so obviously, the animal should not already carry that infection. If an animal is used for propagation of viruses, it may experimentally be immuno-suppressed to make it susceptible to the inoculated virus. However, some of the organisms already present in the animal may propagate instead. Some infections, such as *Corynebacterium kutscheri*, Kilham rat virus, Sendai virus and sialodacryoadenitis virus[213] reduces the severity of disease caused by other agents, thereby destroying infectious disease models.

## Modulation of Oncogenesis

Infectious agents may induce cancer, enhance the oncogenic effect of certain oncogens, or reduce the incidence of cancer in laboratory animals. Spontaneous tumors represent the most important implication in studies with aging animals or other long-term studies. Retroviruses[214] and *Helicobacter* spp.[32] may be oncogenic. Some microorganisms, e.g., *Mycoplasma pulmonis*[215] and *Citrobacter rodentium*,[216] that are not primary oncogens may increase, while other organisms, e.g., *Salmonella*[217,218,219] and H1 parvovirus,[220,221,222] may decrease the incidence of specific tumors.

# ERADICATION METHODS

To produce animals free of hazardous infection, uninfected breeding animals for the upstart of a colony are essential. These are produced by so-called rederivation, e.g., hysterectomy or embryo transfer. Alternatives to rederivation are cessation of breeding, stamping out, antibiotic treatment, or vaccination, although these methods are less recommendable.

## Cesarian Section

Offspring removed from the uterus of their mother by sterile techniques will be free of infections, not having the capability of transplacental infection. Cesarian sections may eliminate even infections, which do pass the placenta, as not every fetus of an infected mother will harbor the infection. Therefore, upstart of breeding colonies on the basis of breeding animals born by caesarian section have been applied as rederivation principle for years. The cesarian section procedure for rederivation consists of several steps.

The exact date of expected birth of the mother of those fetuses to become the future breeding animals needs to be known for planning the further parts of the procedure. Therefore, mating has to be done as a so-called time mating, i.e., the female is given 24 h with the male, after which the male is removed. Mating needs to be checked by observation of a vaginal plug or by the microscopic observation of sperms in a vaginal smear. If the female reproductive cycle can be synchronized, which is actually possible for some larger farm animals, this step may be made rather efficient with a high mating rate. In rabbits, in which ovulation is induced by mating and not by a cycle, time mating is also a quite simple matter, and artificial insemination may be applied. Rodents are, however, difficult to synchronize. Mice may be synchronized by placing the females in the same cage as the male, however separated by a grid, for two days. When the grid is removed on the third day, a mating rate of 50% may be achieved. In rats, five times as many females need to be set up to compensate for those 80%, which from a statistical point of view will not reach estrus within 24 h. The number of females to be mated may be reduced by up to 50% by performing a vaginal smear prior to mating and only mating proestral females. As female rats often lose their plugs into the bedding, it is advisable to place a rat pair to mate on a grid and search for the plug under the grid.

To avoid spontaneous birth of the litters, it may be advantageous to treat the mother with a gestagen, e.g., medroxyprogesteron (15 mg per kg body weight) 3 days prior to expected birth. Females treated this way must be released by caesarian section or euthanized, as they are no longer able to give birth by themselves. The section should be performed as precisely dated as possible. Exact duration of the pregnancy differs between different strains, and the closer the section is performed to the real time of birth, the higher the rate of success. To avoid contamination, strictly aseptic principles should be obeyed, which may be achieved in different ways (Figure 11.4). The uterus should never be introduced into the isolator, as this increases the risk of infection with *Pasteurella pneumotropica*. Preferably, the donor mother is euthanized rather than anesthetized. The method of euthanasia should be mechanical. The uterine blood supply in rats and mice is sufficient enough to allow 15 min for the puppies to be released, while the entire procedure should not last more than 4 min in rabbits or guinea pigs.

Standard procedure is to place the puppies with a foster mother, whose own offspring have been removed. For rodents, the highest success rate is achieved if the foster mother has given birth not later than 2 days and not earlier than 10 days prior to receiving the sectioned puppies. Before placing these with the foster mother, they should be rubbed and heated to achieve a normal blood circulation, as the foster mother may readily cannibalize cold and pale puppies.

If germ-free animals are needed, the foster mother should be germ-free and kept in an isolator. Germ-free conditions are also the most attractive for the upstart of microbiologically defined animals, as it eases health monitoring procedures. Germ-free status is not possible for hamsters and guinea pigs, and for these species, a mother of the microbiological status to be obtained should be used. Guinea pig puppies may be fed a fluid guinea pig diet, and therefore, the foster mother may not be needed at all, but she is hardly recommended, as she will supply the puppies with colostrum and, as feeding the puppies is difficult and not as successful as natural feeding. Offspring of all species may be fed with irradiated milk from a mother of their own species instead of using a foster mother. This technique is of course the only possible way if germ-free animals are to be produced for the first time. In all other cases, it is only recommended for rabbits, as these only need to be fed once per 24 h. If done, it is of the utmost importance to calculate the dose of milk, weigh the animals before and after, and give them only the calculated dose to prevent them from developing aspiration pneumonias. The puppies are weaned from the foster mother according to normal weaning procedures of the species in question.

At least 8 weeks should pass before the success of the rederivation is evaluated by health monitoring. A few of the offspring have to be sacrificed along with the foster mother. Before breeding the animals outside the isolator, it is advisable to supply them with a basic microflora. Breeding gnotobiotic animals of the same species with a defined flora in another isolator most easily solves this problem. If one of these is transferred to the isolator with the germ-free animals, these will automatically get the defined flora. This flora should at least contain some anaerobes and some Gram positives. Transfer from the isolator to the barrier unit may be done differently, but it is most easily done if the barrier unit has a port that fits the isolator port.

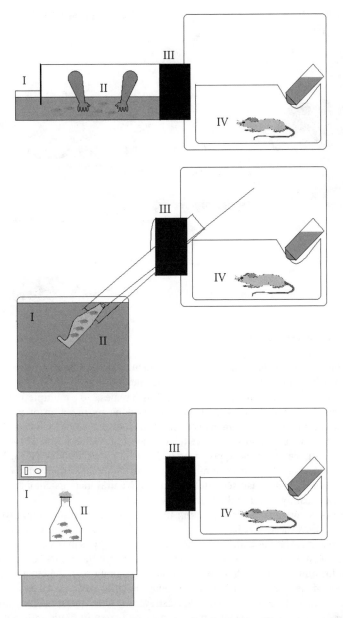

**Figure 11.4** Three ways of performing caesarian section for rederivation of laboratory animals. In A the uterus removed from the donor mother is transferred to a surgical isolator (I) and the offspring are released through gloves underneath a desinfective solution (II) and transferred via the isolator port (III) to a positive pressure isolator with a foster mother (IV). In B the uterus is placed in the desifinfective solution (I) and the offspring are placed in a bottle cut open (II), which can be lifted into the high pressure isolator through a tube passing through the isolator port (III). In C the section is performed in a laminar air flow bench (I) and the offspring are placed in an Ehrlen-Meyer bottle (II) which enters through the isolator port (III) by standard procedures.

## Embryo Transfer

As an alternative to cesarian section, embryo transfer has been used over the last 20 years. The advantages to cesarian section are that the embryos may be kept for long periods in liquid nitrogen and used for further rederivation whenever needed. The technique is also a must for the production of transgenic

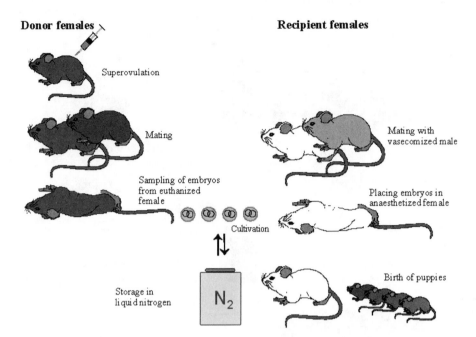

**Figure 11.5**  Principles of embryo transfer.

animals. Disadvantageous is that more specialized equipment is needed for this method than for cesarian section. Embryo transfer rederivation is also a multistep method (Figure 11.5)

Embryos are normally harvested from superovulated females, i.e., multiple ovulations have been induced by hormonal treatment. In mice and rats, this is done by injection of pregnant mare serum gonadotropin (PMSG) combined with human chorionic gonadotropin (HCG) and best results are achieved in premature animals. HCG is normally given 48 h later than PMSG and has to be correlated to the light cycle. Females are then mated with the appropriate male the night after giving HCG. Next morning, the females are checked for plugs. In premature rats, these more frequently remain in the vagina than is the case for adult rats as used for cesarian section (see above). Natural matings compared to superovulated matings lead to a significantly lower embryo recovery.

The donor female is euthanized for harvesting the embryos. For rederivation procedures, two-cell stages are mostly used, as these are the most appropriate for freezing. These are recovered from the salpinx, 36 h after mating. Having euthanized the donor female, the salpinx is excised and transferred into an appropriate medium, in which a needle is used to flush out the embryos. Dependent on the risk of collecting hazardous infections, such as *Mycoplasma* spp., along with the embryos, the embryos are submitted to a series of rinsing steps. The embryos may be cultivated *in vitro* in order to synchronize with mating of the recipient female. When placed in straws in some appropriate medium supplemented with sucrose and some freezing protectant, embryos may be frozen in liquid nitrogen. The straws may simply be dropped into the nitrogen; a procedure known as vitrification. However, survival may be increased by the use of a so-called cryostat, i.e., equipment that slowly reduces the temperature. For thawing, the embryos are placed in a 37° water bath and the straws are emptied into a petri dish containing a dilution medium. The embryos are then submitted to a number of elution steps. There are no known limits for how long the embryos can be stored in liquid nitrogen and in professional hands the viability is normally above 90%.

If the outcome needs to be germ-free, transfer of embryos and all related procedures should be performed in isolators. The recipient mother needs to be mated with a vasectomized male to induce pseudopregnancy. This is, in principle, done and checked as described for time mating of donors for cesarian section. The mating is synchronized with production and cultivation of embryos and the stage in which these are to be transferred. The pseudopregnant female is placed in a surgical isolator and anaesthetized and the abdomen is entered through the back skin and both sides of the flank muscles as described for ovariectomy. The ovary with the salpinx is carefully grasped from the abdominal cavity, and while the bursa is opened with great care, the embryos are collected with a transfer pipette. The embryos are placed in the salpinx through the infundibulum. Equal amounts of embryos are placed in both salpinces. The wound is closed with sutures, and the recipient female is returned to develop her pregnancy in the isolator. From this point, the steps are similar to those of cesarian section.

## Cessation of Breeding and Burnout

If a virus possesses the ability to induce an immune response in the host animal strong enough to eliminate the virus and protect against reinfection as well as an infectivity that high, that a prevalence of 100% can be reached in the colony, the virus may be eliminated from breeding colonies by a time break in all breeding procedures. The principle is that all animals gets infected and develop protective immunity, and until this has happened, no naïve animals are introduced. This has been successful for especially coronaviruses in rats[223] and mice.[224] The period from the last births before the break until the first matings after the break should be at least 6 weeks. This method is definitely not applicable for immune-deficient rodents, such as nude or SCID mice, and in relation to transgenic animals, caution should be taken, as viruses may behave abnormally in these animals and, e.g., develop persistent states or immunity may not develop.[225] The same principles may eventually be applied in experimental facilities by not taking in new animals for a period of 6 weeks. This is called burnout. None of these methods are applicable for bacterial infections, as these are generally persistent.

## Stamping Out

Removal of infected animals from the colony, a method known as stamping out, has been found to be efficient for some infections. In practice, this has most successfully been applied as a tool against infection with *Encephalitozoon cuniculi* in rabbits,[226] in which the method is performed by current serological testing of all breeding animals and consequent removal of all positive responders. The success for this agent is probably related to its complicated life cycle, while more simple infections easily spreading among the offspring may not become totally eliminated by this method.

## Antibiotic Treatment

Antibiotics have been used for the total eradication of specific pathogenic bacteria in immune-competent animals. This method, known as selective decontamination, has been more or less successfully applied to mice,[227] guinea pigs,[228] rabbits,[229] and dogs.[230] Mice may even be made germ-free by decontamination.[231] However, care should be taken, as the effect of the antibiotic treatment may simply be that the microorganism can no longer be detected, and not that it has been eliminated forever.[232]

## Vaccination

If all other measures are unsuccessful or impossible to apply, the use of animals protected by vaccination may be considered. If vaccination is unavoidable, one should in the first place consider whether it might be sufficient to vaccinate the female breeding animals only, as passive immunization through maternal antibodies in most cases must be considered as less interfering with research than active immunization of the experimental animal. Apart from this, it is easier and cheaper. In this way, pneumonia due to *B. bronchiseptica* may be almost totally eliminated from breeding colonies of guinea pigs, although the agent still persists in the colony.[233] Vaccination of the experimental animals has been used against Sendai virus pneumonia,[234,235,236] ectromelia,[237] mycoplasmosis,[238,239] and the various effects of infection with cytomegalovirus.[240] Especially for larger laboratory animals, such as pigs, this is a common way of dealing with infectious problems.[241] However, in regard to infections, which the animal colony can actually be kept free of, vaccination should be considered bad practice. Some of the unwanted microbial effects on research may be seen even in vaccinated animals, e.g., the immunosuppressive effect of Sendai virus.[242]

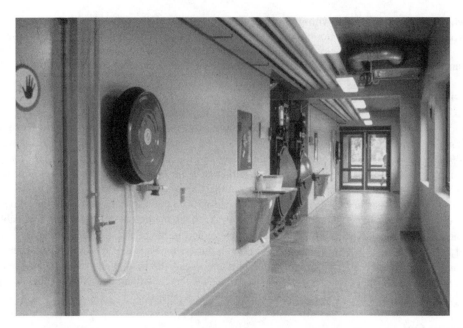

**Figure 11.6**   The outside of two separated barrier units. The shower entrance of barrier 1 is closest followed by a diptank for chemical disinfection into barrier 1, autoclave for barrier 1, autoclave for barrier 2, diptank for barrier 1, shower entrance and chemical disinfection lock for barrier 2 (M&B Ltd, Denmark).

## CONTAINMENT FACILITIES

To prevent infections in laboratory animals, these should be housed in facilities, in which certain protective measures reduce the risk of infections. This is especially important for breeding colonies, as infections in these will spread to a number of studies, and reestablishment of a breeding colony is a time-consuming, expensive and occasionally an impossible procedure. However, as experimental facilities normally house a number of different studies, some of which would be as time-consuming and expensive to restart, these should also be protected in an appropriate way. Animals deriving from facilities in which no protective measures are applied are called conventional.

### Barrier Housing

In animal units, in which the staff is allowed to move freely around, protective measures are used for decontamination of the staff, materials, and fresh air entering the unit, i.e., a barrier is physically as well as mentally in front of the unit (Figure 11.6). Such a barrier may be run at different levels. Basically, materials and diets are autoclaved or chemically decontaminated at entry, and the staff should not be allowed contact to animals of the same species within a certain period, e.g., 48 h. In the same way, animals should only be introduced if health monitoring has documented the absence of unwanted infections. In breeding units, this normally means that new breeding animals are only introduced by rederivation. In breeding units, staff members are only allowed to enter through a 3-room shower, while in experimental units, the staff in some facilities are allowed to enter after changing their clothes, only. Ingoing and preferably also outgoing air is filtered, and the air pressure in the facility is maintained at approximately 15 mmHg above the surrounding pressure to prevent air-borne infection. In some facilities, protection is further enhanced by the use of facemasks and gloves by the staff (Figure 11.7). However animals kept in barrier-protected facilities may (even if the staff is equipped with protective clothing) only be kept free of certain species-specific infections. They will not be free of infections shared between

**Figure 11.7**  An animal technician dressed for working in a barrier-protected animal unit (Ellegaard Göttingen Minipigs, Denmark).

their own species and humans.[2] Therefore, they will have a bacterial flora in their gastrointestinal, respiratory, and genital systems and on their skin, which can never be fully defined.

Animals bred in barrier-protected facilities are sold under different terms. Preferable, the term microbiologically defined should be used to indicate that the animals have been protected and health monitored. Occasionally, the term specific pathogen free (SPF) is used, however, without always clearly defining which specific pathogens are actually considered. Some commercial breeders have their own registered trademarks, such as Virus Antibody Free (VAF™) or Cesarian-Originated Barrier-Sustained (COBS™).

If the aim is to protect the surroundings, e.g., because animals of an unknown or a known but unacceptable microbiological quality are housed, all containment facilities may be run in the opposite direction, i.e., with a negative air pressure and all the decontamination procedures applied for staff and materials leaving the facility.

## Cubicles and Filter Cabinets

To enhance the protection of the animals inside a facility, all animals may be placed behind a glass wall in a number of built-in boxes containing only a limited number of cages. Such "boxes" are called cubicles (Figure 11.8.) Ventilation inlets and outlets are placed inside each cubicle. The advantages of cubicles are that they reduce spread of infection among individual animals as well as the spread of allergens in the room.

As an alternative to a cubicle, which is a stationary part of an animal unit, a transportable filter cabinet may be used (Figure 11.9). This is a closed cabinet supplied with its own ventilation motor or individually plugged into the central ventilation. These cabinets should imply the same advantages as cubicles in relation to reduced spread of infections and allergens. If they are supplied with their own

**Figure 11.8**   Animals placed behind glass walls in so-called cubicles (University of Washington, Photo Gavin
                 Sisk).

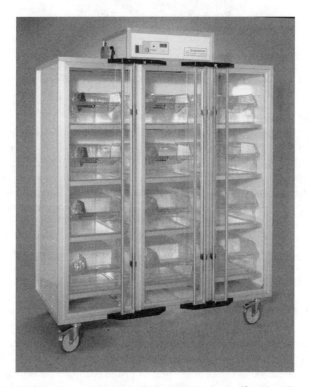

**Figure 11.9**   A ventilated cabinet for maintenance of laboratory animals (Scanbur, Denmark).

ventilation motor, they may also be advantageous in facilitating setup of animal facilities without
expensive investments in central ventilation. They may also be used for protecting animals ,and staff in
facilities far from central facilities, e.g., when animals for a 1-day acute study are to be housed in the
research laboratory without exposing the animals to stressful noises or the staff to allergens.

**Figure 11.10** A cage protected by a lid with a filter, a so-called filter-top (Tecniplast, Italy).

**Figure 11.11** An individually ventilated cage (IVC) rack (Tecniplast, Italy).

## Filter-Top Cages and Individually Ventilated Cage Systems (IVC)

In some cases, it may be practical or economical to place the barrier around the individual cage rather than the entire unit. This principle may be used inside a barrier-protected unit to enhance the protection of the individual animal or as an alternative to running such a facility. The simplest principle is to place a lid on the cage. This lid, called a filter-top, is provided with a filter to allow the passage of the air (Figure 11.10). The system is also called a "microisolator," but although these filter-tops may be more or less sealed to the cage, they cannot be regarded as isolation. The risk for spreading infections as well as allergens may be reduced, but certainly not eliminated. Occasionally, concentrations of trace gases in a filter-topped cage may be elevated,[224] which may be hazardous to the animals.[243] In individually ventilated cage (IVC) systems (Figure 11.11), each filter-topped cage is supplied with its own ventilation, and all cages are isolated from one another as well as the surroundings. Principles and degrees of sealing depend on the brand of the system. This allows the microbiological separation of each individual cage,

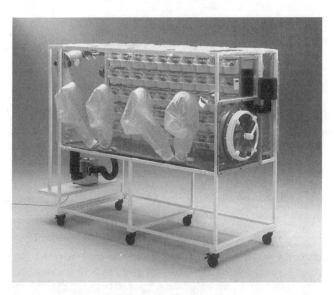

**Figure 11.12** A flexible film isolator for laboratory animals (Isotec, UK).

thereby isolating the animals not only from the surroundings, but also from one another. To allow the continued separation, cages are changed in a specially equipped laminar airflow system. Some IVC systems may be run at positive and negative pressures.

## Isolators

In some studies, it is necessary to have a full knowledge of the microflora of the animals used. For such studies, animals are kept in isolators (Figure 11.12). These are fully closed systems in which the animals are only handed through gloves tightly sealed to the isolator wall. Ingoing air is filtered through an absolute filter, and outgoing air is blown through a silicone lock. All materials and diets are first introduced after autoclaving or irradiation. In the isolator-lock chemicals, such as peracetic acid or potassium monopersulfate, are used to sterilize all surfaces. Animals are transported from one isolator to another in a closed cylinder, which fits exactly with the port of the isolator. The pressure in the isolator is positive to the surroundings.

Animals kept in isolators may be kept without any germs at all; at least as far as no such germs can be shown by any of the methods available. Such animals are called germ-free or axenic. However, these animals may also be supplied with a fully defined flora of which every microorganism is known. Such animals are called gnotobiotic. This term may be used for germ-free animals as well.

Isolators are necessary for certain types of studies. Within microbiology and nutrition, it may be essential to know the role of specific microorganisms, which may be achieved by running the animal model of formerly germ-free animals monoinfected with the organism to be studied. This will allow the separation of characteristics of the physiology into germ-free associated characteristics (GAC) or microflora-associated characteristics (MAC), i.e., characteristics that are related to the animal or the microorganism, respectively.[244] In cancer research, isolators are also commonly applied as strong-acting pharmacological immunosuppressors, which is likely to induce problems from opportunistic pathogens. Also, isolators are essential for rederivation of laboratory animals.

## HEALTH MONITORING

### Scope

Animals maintained as microbiologically defined according to rederivation and containment principles described above, should be regularly controlled to confirm this status. Therefore, a number of animals

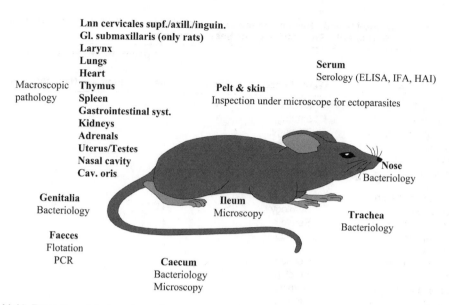

**Lnn cervicales supf./axill./inguin.**
**Gl. submaxillaris (only rats)**
**Larynx**
**Lungs**
**Heart**
Macroscopic **Thymus**
pathology **Spleen**
**Gastrointestinal syst.**
**Kidneys**
**Adrenals**
**Uterus/Testes**
**Nasal cavity**
**Cav. oris**

**Serum**
Serology (ELISA, IFA, HAI)

**Pelt & skin**
Inspection under microscope for ectoparasites

**Nose**
Bacteriology

**Genitalia**
Bacteriology

**Ileum**
Microscopy

**Trachea**
Bacteriology

**Faeces**
Flotation
PCR

**Caecum**
Bacteriology
Microscopy

**Figure 11.13** Examples of test performed on each sampled animal in health (microbiological) monitoring of rodents.

are sampled from the colony at frequent intervals and are subjected to a range of tests (Figure 11.13). This practice is called health monitoring, although microbiological monitoring would be a more appropriate term.

As quite a number of examinations are necessary for health monitoring, the procedures will normally be performed on animals sampled only for this purpose, and the status will be used as a picture of the population and not the individual. e.g., a mouse colony may be infected with *Helicobacter hepaticus*, i.e., mice from this colony may or may not have hepatic changes. The individual animal used in a research project may not even harbor the agent, and hence, health monitoring of laboratory animal colonies is based upon the principle that a few animals can be sampled for examination, but the results can be used to describe the entire colony. Therefore, if one animal is infected with a certain organism, the entire colony is considered infected with that particular organism, and if the infection is not found in any of the animals sampled, the entire colony is considered free of that organism.

It is essential to define the microbiological entity, i.e., a definition of the group of animals for which a sample is predictive, which is a complicated matter. One isolator, one individually ventilated cage, or a simple barrier-protected one-room unit used for breeding at a commercial vendor may be defined as one microbiological entity each, as the idea of the system is to keep the animals out of contact with the surroundings and due to the limited space, there is close contact between the animals, but in experimental facilities, it is often difficult to define the microbiological entity. Some bacteria easily spread from one room to another, while others do not. Under such circumstances, it would be safest but also impossible to define each cage as a microbiological entity. Therefore, individual judgments are necessary.

When the estimated number of animals have been sampled, these are subjected to a number of clinical examinations and laboratory assays, after which the microbiological status is given as a list of agents stating whether or not that agent has been found.

The commercial breeders without whom the historical process would probably have been much slower have traditionally developed most health monitoring programs. However, reports from different breeders have been difficult to compare, and, therefore, there is a need for standardization of health monitoring programs. To fulfill this, the Federation of European Laboratory Animal Science Associations (FELASA) has issued guidelines for health monitoring of various species, i.e., rodents and rabbits under breeding as well as experimental conditions,[48] pigs, dogs and cats,[245] primates,[246] and ruminants.[247] These papers set standards for which agents to test for, which methods to use, how many animals to test, how frequently this should be done and how this should be reported. Agents and methods recommended for rodents and rabbits are shown in Table 11.4. Some of these recommendations are pragmatic. If a more

**Table 11.4 Agents and Methods Recommended Being Included in Health Monitoring Programs According to FELASA Guidelines for Health Monitoring**

| Viruses | Test Method | Mice | Rats | Guinea Pigs | Hamsters | Rabbits |
|---|---|:---:|:---:|:---:|:---:|:---:|
| Coronaviruses | Serology | + | + | | | |
| Ectromeliaavirus | Serology | + | | | | |
| Guinea pig adenovirus | Serology | | | + | | |
| Guinea pig cytomegalovirus | Serology | | | + | | |
| Hantaviruses | Serology | | + | | | |
| Lymphocytic choriomeningitis virus | Serology | + | | | + | |
| Parvoviruses | Serology | + | + | | + | |
| Pneumonia virus of mice | Serology | + | + | | | |
| Rabbit hemorrhagic disease virus | Serology | | | | | + |
| Rabbit pox viruses | Serology | | | | | + |
| Rabbit rotavirus | Serology | | | | | + |
| Reovirus type 3 | Serology | + | + | | | |
| Sendai virus | Serology | + | + | + | + | |
| Theiler's mouse encehalomyelitis virus | Serology | + | + | | | |

**Bacteria and Fungi**

| | | | | | | |
|---|---|:---:|:---:|:---:|:---:|:---:|
| Bordetella bronchiseptica | Cultivation | | + | + | | + |
| Chlamydia psitacci | No recommendation | | | + | | |
| Citrobacter rodentium | Cultivation | + | | | | |
| Clostridium piliforme | Pathology, Serology | + | + | | + | + |
| Corynebacterium kutscheri | Cultivation | + | + | | | |
| Dermatophytes | Cultivation | | | + | | |
| Helicobacter spp. | Cultivation, PCR | + | + | + | | |
| Mycoplasmae spp. | Serology | + | + | | | |
| Pasteurellaceae | Cultivation | + | + | + | + | + |
| Salmonellae | Cultivation | + | + | + | + | + |
| Streptobacillus moniliformis | Cultivation | | + | + | | |
| Streptococci (β-hemolytic) | Cultivation | + | + | + | | + |
| Streptococcus pneumoniae | Cultivation | + | + | + | | |

**Parasites**

| | | | | | | |
|---|---|:---:|:---:|:---:|:---:|:---:|
| Ectoparasites | Microscopy | + | + | + | + | + |
| Endoparasites | Microscopy | + | + | + | + | + |
| Encephalitozoon cuniculi | Serology | | | + | | + |

thorough health monitoring is wanted, some of the principles described below may be used in addition to the FELASA recommendations. Furthermore, FELASA issued a set of guidelines for the accreditation of health monitoring laboratories.[248]

## Sampling Strategies

How many animals to sample, how often to do it, what to look for, and which methods to use may be based upon scientific judgment rather than strictly obeying international guidelines.

**Table 11.5 Calculation of Sample Sizes for Health Monitoring of Laboratory Animals.**

*Nosografic sensitivity ($N_1$)*

$$(N_1) = \frac{\text{Infected animals reacting in the assay}}{\text{Infected animals whether reacting or not}}$$

*The estimated prevalence (p) in the colony*

$$p = \frac{\text{Infected animals}}{\text{Total number of animals}}$$

*The risk of a false negative result in an infected colony (C)*

$$C = \frac{\text{Number of infected colonies tested with a negative result}}{\text{Total number of infected colonies tested}}$$

*Sample size (S) for colonies with more than 1000 animals*

$$S \geq \frac{\log C}{\log (1-(p*N_1))}$$

*Sample size (S) for colonies with less than 1000 animals*

$$S \geq (1 - C^{1/D})* (T- ((D-1)/2))$$

D=Number of infected animals     T=Total number of animals

---

The prevalence that a certain infection reaches depends on many factors, e.g., the contact between the animals, the resistance of the animals, but characteristics of the agent itself play a major role. The sample size needed to detect a specific agent in a colony therefore depends upon the acceptable risk of a false negative result, the sensitivity of the method applied and the estimated prevalence, that the infection as a minimum would reach if present in the colony (Table 11.5).[249] As soon as a sample has been taken, it becomes historical. Curiosity only will dictate when to take the next sample. The result of one sampling visualizes whether changes have occurred since the last sampling.

## Choice of Method

Some microorganisms are easily cultivated from easily accessible sites of the animal, while others can only be cultivated if a range of specific conditions are fulfilled. A few rodent organisms, e.g., Spirillum minus, the cause of the Japanese variant of rat bite fever, sodoku, cannot even be grown in vitro. Therefore, a health program necessarily consists of different types of assays.

Enteric helminths and protozoans may be diagnosed by direct microscopy after flotation test, on smears from the caecum and ileum, or on tapes used for sampling around the anus. Inspection under a stereomicroscope may reveal ectoparasites.

A range of organs should be inspected macroscopically to decide whether further investigation by histopathology is needed, while histopathological examination is seldom the method of choice for current screening and is more efficient as a tool for a final diagnosis and validation of the impacts of a certain microorganism.

Cultivation is the diagnostic method of choice for microorganisms readily grown on artificial media, e.g., bacteria and fungi. Samples for bacteriological cultivation are typically taken from organs such as the nose, trachea, genitals, liver, and cecum inoculating on selective and nonselective media. Further procedures are then designed to grow pure cultures and to identify exactly those organisms searched for.[5]

For microorganisms not easily cultivated serology, such as immunofluorescence assay (IFA), enzyme-linked immunosorbent assay (ELISA) and western immunoblotting is widely used, although the latter is rather laborious and, therefore, not recommended for routine use in the FELASA guidelines.[48] Serology is applied for all viruses, but lactate dehydrogenate (LDH) elevating virus, which is simply monitored by testing for LDH activity. Also a few bacteriae, such as *Clostridium piliforme,*[26] and protozoans, such as *Encephalitozoon cuniculi*[175] are screened by serological methods. Some organisms like *Mycoplasma* spp. may be diagnosed by both serological and cultural methods.[238] Parvoviruses are typically diagnosed by a combination of both solid-phase assays, such as ELISA or IFA, to detect common parvoviral antigens and hemagglutination inhibition assays to differentiate between the different types.[250] Serological results are historical and do not give any evidence whether the animal actually harbors that particular microorganism. The main pitfalls are the lack of sensitivity due to a pure antibody response to the infection or a low specificity due to cross-reactions, which mostly occurs with bacteria, as these contain more than 2000 antibody-inducing proteins, while viruses are far more simple.

Molecular biological techniques represent another attractive choice for noncultivable microorganisms. These methods are generally divided into two: those based upon the DNA (occasionally, RNA) already being present in detectable amounts, e.g., *in situ* hybridization and southern blotting, and those in which the DNA or RNA must be amplified before detection, e.g., the polymerase chain reaction (PCR). To be able to apply molecular biological methods for detection or identification of an infectious agent one or several specific sequences from that organism must be known. *Helicobacter hepaticus* is an example of an infection for which PCR may be the method of choice.[251] Figure 11.13 shows a typical range of methods applied for health monitoring of rodents.

## Reporting

When all examinations have been performed and the results are known, a report is issued (Table 11.6). These reports are made available by the vendor for all purchasers of the animals, e.g., by publishing them on the Internet (Table 11.7). No animals should be allowed into an animal facility without a careful study of their health report prior to their arrival. The veterinarians at experimental facilities should in a similar way make the results of the current monitoring of the animals in experiments available for the animal users.

## Sentinels

In experimental facilities it is often not possible to sample animals directly from the experiments, and therefore, health monitoring is performed on animals placed in the unit only for health monitoring purposes, so-called sentinels.[132] These animals are sampled and replaced at regular intervals and subjected to a range of tests in the same way as it is done on animals from breeding facilities. To facilitate the occurrence of the same infections in these animals as in the experimental animals, these animals should be of the same species. Certain precautions in relation to age and genetics of the sentinels may have to be considered specifically for the infections to be monitored.[26] They should be placed on the lower shelves. Furthermore, it is a general principle to supply these animals with some bedding sampled from the dirty bedding of the other animals in the room. This might spread some but certainly not all infections to the sentinels, e.g., it is a safe technique for mouse hepatitis virus[252] but unsafe for Sendai virus.[147] It is probably also unsafe for pinworm infestation, but for such infections, the diagnosis may be made directly on fecal samples from the experimental animals.

## Screening of Biological Materials

As described above, biological materials, such as cells and sera, are as likely as the animals or even more to introduce hazardous infections in an animal facility. There are basically three ways of protecting the facility against this risk. One way is by only allowing studies including such materials in negative pressure isolators. Another way is by only allowing those materials that derive directly from facilities in which the microbiological status is clearly defined as described above. This reduces the risk to a level

**Table 11.6 An Example of a Health Monitoring Report from a Commercial Vendor**

### FELASA-APPROVED HEALTH MONITORING REPORT

**Name and address of the breeder:** The Breeding Company Ltd.

**Date of issue:** May 22 2001     **Unit N°:** 100-12     **Latest test date:** 08-04-01

**Rederivation:** 1999     **Protocol No:** 000397-000398

**Species:** Mice     **Strains:** C57Bl/6J//XXX-*pmn*, B10M/*XXX*, XXX:NMRI, DBA1/J/XXX

| | HISTORICAL results | LATEST TEST results | LABORATORY | METHOD |
|---|---|---|---|---|
| **VIRAL INFECTIONS** | | | | |
| Minute virus of mice | Neg. | 0/10 | Lab. Ltd. | ELISA |
| Mouse hepatitis virus | Neg. | 0/10 | Lab. Ltd. | ELISA |
| Pneumonia virus of mice | Neg. | 0/10 | Lab. Ltd. | ELISA |
| Reovirus type 3 | Neg. | 0/10 | Lab. Ltd. | ELISA |
| Sendai virus | Neg. | 0/10 | Lab. Ltd. | ELISA |
| Theilers encephalomyelitis virus | Neg. | 0/10 | Lab. Ltd. | ELISA |
| Ectromelia virus | Neg. | N.T. | Lab. Ltd. | IFA |
| Hantaviruses | Neg. | N.T. | Lab. Ltd. | IFA |
| Lymphocytic choriomeningitis virus | Neg. | 0/10 | Lab. Ltd. | IFA |
| Lactic dehydrogenase virus | N.T. | N.T. | | |
| | | | | |
| **BACTERIAL AND FUNGAL INFECTIONS** | | | | |
| Bordetella bronchiseptica | Neg. | 0/10 | Lab. Ltd. | Culture |
| Campylobacter spp. | Neg. | 0/10 | Lab. Ltd. | Culture |
| Citrobacter rodentium | Neg. | 0/10 | Lab. Ltd. | Culture |
| Clostridium piliforme | Neg. | 0/10 | Lab. Ltd. | ELISA |
| Corynebacterium kutscheri | Neg. | 0/10 | Lab. Ltd. | Culture |
| Helicobacter spp. | Neg. | 0/10 | Lab. Ltd. | Culture |
| Leptospira spp | N.T. | N.T. | | |
| Mycoplasma spp | Neg. | 0/10 | Lab. Ltd. | ELISA |
| Pasteurella spp. | | | | |
|  Pasteurellaceae | Pos. | 8/10 | Lab. Ltd. | ELISA |
|  Pasteurella pneumotropica | Pos. | 4/10 | Lab. Ltd. | Culture |
|  Other Pasteurella spp | Neg. | 0/10 | Lab. Ltd. | Culture |
| Salmonellae | Neg. | 0/10 | Lab. Ltd. | Culture |
| Streptobacillus moniliformis | Neg. | 0/10 | Lab. Ltd. | Culture |
|  -hemolytic streptococci | Neg. | 0/10 | Lab. Ltd. | Culture |
| Streptococcus pneumoniae | Neg. | 0/10 | Lab. Ltd. | Culture |

*Other species associated with lesions:*
None

| | HISTORICAL | LATEST TEST | LABORATORY | METHOD |
|---|---|---|---|---|
| **PARASITOLOGICAL INFECTIONS** | | | | |
| Arthropods | Neg. | 0/10 | Lab. Ltd. | Inspection |
| Helminths | Neg. | 0/10 | Lab. Ltd. | Flotation |
| Eimeria spp | Neg. | 0/10 | Lab. Ltd. | Flotation |
| Giardia spp | Neg. | 0/10 | Lab. Ltd. | Microscopy |
| Spironucleus spp | Neg. | 0/10 | Lab. Ltd. | Microscopy |
| Other flagellates | Neg. | 0/10 | Lab. Ltd. | Microscopy |
| Klossiella spp | N.T. | N.T. | | |
| Encephalitozoon cuniculi | N.T. | N.T. | | |
| Toxoplasma gondii | N.T. | N.T. | | |

### PATHOLOGICAL LESIONS OBSERVED
**Stock:** XXX:NMRI    **Lesions:** None

### ABBREVIATIONS FOR LABORATORIES
Lab. Ltd.   Laboratory Company Ltd, X Boulevard, X-Town, X-Country
Pos.       Positive results previously observed
Neg.   Positive results never observed
0/10 No positives out of 10 samples
N.T.      Not tested

**Table 11.7 Internet Addresses at which Health Monitoring Reports May Be Found**

| Breeder | Origin | URL |
|---|---|---|
| B&K Universal | Worldwide | www.bku.com |
| Charles River Laboratories | Worldwide | www.criver.com |
| Clea Japan | Japan | www.clea-japan.co.jp |
| Ellegaard Göttingen Minipigs | Denmark & U.S. | www.minipigs.dk |
| Harlan | Worldwide | www.harlan.com |
| Jackson Laboratories | U.S. | www.jax.org |
| M&B | Denmark | www.m-b.dk |
| RCC | Switzerland | www.rcc.ch |
| Taconic | U.S. | www.taconic.com |

comparable to introducing animals from that facility. A third way is to test all biological materials before allowing them into the animal facility.

As viruses impose the highest risk, it is not possible to test biological materials by simple cultivation. Today, PCR[253] is applicable for most laboratory animal viruses, and therefore, this method is the fastest, cheapest, and from an ethical point of view, most desirable method. However, if PCR is not available for everyone of those agents to be searched, for it may be necessary to test the materials by the mouse antibody production (MAP) test.[254] In this test, germ-free rodents of the appropriate species are exposed to the material in an isolator. To increase the risk of infection, the material is given by the intranasal as well as the intraperitoneal route. Four days later, serum is sampled for an enzymatic test for lactate dehydrogenase elevating virus. Twenty-eight days later, the animals are exsanguinated, and serological assays are made for all agents searched for.

## QUARANTINE HOUSING OF ANIMALS IMPORTED FROM NONVENDOR SOURCES

If the supplier of animals for an experimental facility is unable to supply a health report or the containment facilities at this supplier are regarded unsatisfactory, animals should not be introduced directly into an experimental facility. This problem has increased with the increased use of transgenic animals, as many of these animals, mostly mice, are not produced by commercial breeders but are produced by a range of different research groups, who globally exchange these animals between them. Therefore, especially university animal facilities currently receive requests from their animal users to receive transgenic mice from different sources, the health status of which can be difficult to judge. Taking into account that these transgenic animals are often costly and difficult to get for restocking if they are accidentally lost, animal facilities need to take precautions to reduce the risk of infections in their animals, while still allowing the intake of mice from a number of different sources. This is most efficiently done by breeding animals on site in a barrier-protected unit into which animals are only introduced by rederivation.

To get a smaller number of animals into experiments on a current basis, it may become necessary to create a facility in which it is possible to quarantine animals before entry into the main experimental facility. Here the animals should be maintained for at least 4 weeks before health monitoring is performed to define the status of these animals. If sentinels are used, and this will often be the case, 4 weeks of quarantine may be insufficient. If this health monitoring states the absence of unwanted microorganisms, the animals may be transferred to the experimental facility. In larger institutional units, animals are delivered currently and, therefore, each delivery needs to be kept separate inside the quarantine facility, which makes the use of negative pressure isolators or IVC systems the most appropriate choice for quarantine housing. Furthermore, one sampling for health monitoring can never be made 100% safe, and it may, therefore, be the safest for units dealing with many minor deliveries from different sources, if all delivered animals are routinely rederived to the institution's own barrier breeding facility and only from a breeding stock inside this unit delivered to individual projects in the experimental facilities. This may, of course, prove to be impossible for all animals, and, in practice, it may become necessary to differentiate deliveries into safe and unsafe deliveries. Animals from known commercial breeders with high standard barrier facilities and efficient health monitoring procedures may be considered safe deliveries, and therefore allowed direct access to the experimental facility, eventually through a quarantine unit, while all other animal deliveries are regarded unsafe and subjected to rederivation. This method would not take into consideration that animals may get infected during transport. Another security measure would be to divide the experimental unit into small separate units, which are protected from one another in a way, that will secure that they each can be regarded as their own microbiological entity, i.e., they should have their own barrier and staff. Inside each separation all animals should derive from the same source.

Which system to set up must be judged independently by each institution, which needs to weigh the costs of the protective measures against the economic value of their experiments.

## REFERENCES

1. Baker, D.G., Natural pathogens of laboratory mice, rats, and rabbits and their effects on research, *Clinical Microbiol. Rev.*, 11, 231, 1998.

2. Hansen, A.K., The aerobic bacterial flora of laboratory rats from a Danish breeding centre, *Scandinavian J. Lab. Anim. Sci.*, 19, 59, 1992.

3. Mutters, R., Ihm, P., Pohl, S., Frederiksen, W., and Mannheim, W., Reclassification of the genus Pasteurella Trevisan 1887 on the basis of DNA homology with proposals for the new species Pasteurella Dagmatis, Pasteurella Canis, Pasteurella Stomatis and Pasteurella Langaa., *Int. J. Syst. Bacteriol.*, 35, 309, 1985.

4. Nakagawa, M., Saito, M., Suzuki, E., Nakayama, K., Matsubara, J., and Muto, T., Ten years-long survey on pathogen status of mouse and rat breeding colonies, *Jikken Dobutsu*, 33, 115, 1984.

5. Hansen, A.K., *Handbook of Laboratory Animal Bacteriology*, CRC Press, Boca Raton, FL, 1999.

6. Loliger, H.C. and Matthes, S., Infectious factor diseases in domestic small animals (carnivorous and herbivorous fur animals, wool and meat rabbits), *Berl Munch. Tierarztl. Wochenschr.*, 102, 364, 1989.

7. Donnio, P.Y., Legoff, C., Avril, J.L., Pouedras, P., and Grasrouzet, S., Pasteurella-multocida — oropharyngeal carriage and antibody-response in breeders, *Vet. Res.*, 25, 8, 1994.

8. Deeb, B.J. and DiGiacomo, R.F., Respiratory diseases of rabbits, *Vet. Clin. North Am. Exotic Anim. Pract.*, 3, 465, 2000.

9. Dillehay, D.L., Paul, K.S., DiGiacomo, R.F., and Chengappa, M.M., Pathogenicity of Pasteurella Multocida A:3 in Flemish Giant and New Zealand White Rabbits, *Lab. Anim.*, 25, 337, 1991.

10. DiGiacomo, R.F., Allen, V., and Hinton, M.H., Naturally acquired Pasteurella Multocida Subsp. Multocida infection in a closed colony of rabbits: characteristics of isolates, *Lab. Anim.*, 25, 236, 1991.

11. DiGiacomo, R.F., Xu, Y.M., Allen, V., Hinton, M.H., and Pearson, G.R., Naturally acquired Pasteurella Multocida infection in rabbits: clinicopathological aspects, *Can. J. Vet. Res.*, 55, 234, 1991.

12. DiGiacomo, R.F., Taylor, F.G., Allen, V., and Hinton, M.H., Naturally acquired Pasteurella Multocida infection in rabbits: immunological aspects, *Lab. Anim. Sci.*, 40, 289, 1990.

13. DiGiacomo, R.F., Jones, C.D., and Wathes, C.M., Transmission of Pasteurella Multocida in rabbits, *Lab. Anim. Sci.*, 37, 621, 1987.

14. Scharf, R.A., Monteleone, S.A., and Stark, D.M., A modified barrier system for maintenance of Pasteurella-free rabbits, *Lab. Anim. Sci.*, 31, 513, 1981.

15. al Lebban, Z.S., Corbeil, L.B., and Coles, E.H., Rabbit Pasteurellosis: induced disease and vaccination, *Am. J. Vet. Res.*, 49, 312, 1988.

16. Tyzzer, E.E., A fatal disease of the Japanese Waltzing Mouse caused by a spore-bearing Bacillus (Bacillus Piliformis N. Sp.), *J. Med. Res.*, 37, 307, 1917.

17. Zook, B.C., Huang, K., and Rhorer, R.G., Tyzzer's disease in Syrian hamsters, *J. Am. Vet. Med. Assoc.*, 171, 833, 1977.

18. Zook, B.C., Albert, E.N., and Rhorer, R.G., Tyzzer's disease in the Chinese hamster (Cricetulus Griseus), *Lab. Anim. Sci.*, 27, 1033, 1977.

19. Zook, B.C., Huang, K., and Rhorer, R.G., Tyzzer's disease in Syrian hamsters, *J. Am. Vet. Med. Assoc.*, 171, 833, 1977.

20. White, D.J. and Waldron, M.M., Naturally-occurring Tyzzer's disease in the gerbil, *Vet. Rec.*, 85, 111, 1969.

21. Port, C.D. Richter, W.R., and Moise, S.M., Tyzzer's disease in the gerbil (Meriones unguiculatus), *Lab. Anim. Care*, 20, 109, 1970.

22. Allen, A.M., Ganaway, J.R., Moore, T.D., and Kinard, R.F., Tyzzer's disease syndrome in laboratory rabbits, *Am. J. Pathol.*, 46, 859, 1981.

23. Fujiwara, K., Nakayama, M., Nakayama, H., Toriumi, W., Oguihara, S., and Thunert, A., Antigenic relatedness of Bacillus-piliformis from Tyzzers disease occurring in Japan and other regions, *Jpn. J. Vet. Sci.*, 47, 9, 1985.

24. Fujiwara, K., Nakayama, M., and Nakayama, H., Antigenic relatedness of Bacillus-piliformis (Tyzzer) from different sources, *Lab. Anim. Sci.*, 33, 490, 1983.

25. Hansen, A.K., The Use of the Mongolian Gerbil as Sentinel for Infection with Bacillus Piliformis in Laboratory Rats, in Proceedings of the 4th FELASA Symposium, 449, Federation of European Laboratory Animal Science Associations, Lyon, France, 1990.

26. Hansen, A.K., Andersen, H.V., and Svendsen, O., Studies on the diagnosis of Tyzzer's disease in laboratory rat colonies with antibodies against Bacillus Piliformis (Clostridium Piliforme), *Lab. Anim. Sci.*, 44, 424, 1994.

27. Hansen, A.K., Svendsen, O., and Mollegaard-Hansen, K.E., Epidemiological studies of Bacillus Piliformis infection and Tyzzer's disease in laboratory rats, *Z Versuchstierkd*, 33, 163, 1990.

28. Hansen, A.K., Dagnaes-Hansen, F., and Mollegaard-Hansen, K.E., Correlation between Megaloileitis and antibodies to Bacillus Piliformis in laboratory rat colonies, *Lab. Anim. Sci.*, 42, 449, 1992.

29. Hansen, A.K., Skovgaard-Jensen, H.J., Thomsen, P., Svendsen, O., Dagnaes-Hansen, F., and Mollegaard-Hansen, K.E., Rederivation of rat colonies seropositive for Bacillus Piliformis and the subsequent screening for antibodies [Published erratum appears in *Lab. Anim. Sci.*, February, 43, 1, 114, 1993], *Lab. Anim. Sci.*, 42, 444, 1992.

30. Friis, A.S., Demonstration of antibodies to Bacillus Piliformis in SPF colonies and experimental transplacental infection by Bacillus Piliformis in mice, *Lab. Anim.*, 12, 23, 1978.

31. Friis, A.S., Studies on Tyzzer's disease: transplacental transmission by Bacillus Piliformis in rats, *Lab. Anim.*, 13, 43, 1979.

32. Fox, J.G. and Lee, A., The role of Helicobacter species in newly recognized gastrointestinal tract diseases of animals, *Lab. Anim. Sci.*, 47, 222, 1997.

33. Ward, J.M., Anver, M.R., Haines, D.C., and Benveniste, R.E., Chronic active hepatitis in mice caused by Helicobacter Hepaticus, *Am. J. Pathol.*, 145, 959, 1994.

34. Hailey, J.R., Haseman, J.K., Bucher, J.R., Radovsky, A.E., Malarkey, D.E., Miller, R.T., Nyska, A., and Maronpot, R.R., Impact of Helicobacter Hepaticus infection in B6C3F(1) mice from twelve national toxicology program two-year carcinogenesis studies, *Toxicol. Pathol.*, 26, 602, 1998.

35. Jacoby, R.O. and Russell Lindsey, J., Risks of infection among laboratory rats and mice at major biomedical research institutions, *ILAR J.*, 266, 1998.

36. Bhan, A.K., Mizoguchi, E., Smith, R.N., and Mizoguchi, A., Colitis in transgenic and knockout animals as models of human inflammatory bowel disease, *Immunol. Rev.*, 169, 195, 1999.

37. Rath, H.C., Schultz, M., Freitag, R., Dieleman, L.A., Li, F.L., Linde, H.J., Scholmerich, J., and Sartor, B., Different subsets of enteric bacteria induce and perpetuate experimental colitis in rats and mice, *Infect. Immun.*, 69, 2277, 2001.

38. Kawaguchi-Miyashita, M., Shimada, S., Kurosu, H., Kato-Nagaoka, N., Matsuoka, Y., Ohwaki, M., Ishikawa, H., and Nanno, M., An accessory role of TCR gamma delta(+) cells in the exacerbation of inflammatory bowel disease in TCR alpha mutant mice, *Eur. J. Immunol.*, 31, 980, 2001.

39. Smythies, L.E., Waites, K.B., Lindsey, J.R., Harris, P.R., Ghiara, P., and Smith, P.D., Helicobacter pylori-induced mucosal inflammation is Th1 mediated and exacerbated in IL-4, but not IFN-gamma, gene-deficient mice, *J. Immunol.*, 165, 1022, 2000.

40. Dieleman, L.A., Arends, A., Tonkonogy, S.L., Goerres, M.S., Craft, D.W., Grenther, W., Sellon, R.K., Balish, E., and Sartor, R.B., Helicobacter hepaticus does not induce or potentiate colitis in interleukin-10-deficient mice, *Infect. Immun.*, 68, 5107, 2000.

41. Uzal, F.A., Feinstein, R.E., Rehbinder, C., and Persson, L., A study of lung lesions in asymptomatic rabbits naturally infected with *B. Bronchiseptica*, *Scand. J. Lab. Anim. Sci.*, 16, 3, 1989.

42. Wei, Q., Tsuji, M., Takahashi, T., Ishihara, C., and Itoh, T., Taxonomic status of CAR bacillus based on the small subunit ribosomal RNA sequences, *Chin. Med. Sci. J.*, 10, 195, 1995.

43. Cundiff, D.D., Besch-Williford, C.L., Hook, R.R., Jr., Franklin, C.L., and Riley, L.K., Characterization of cilia-associated respiratory bacillus in rabbits and analysis of the 16S RRNA gene sequence, *Lab. Anim. Sci.*, 45, 22, 1995.

44. Shoji-Darkye, Y., Itoh, T., and Kagiyama, N., Pathogenesis of CAR bacillus in rabbits, guinea pigs, Syrian hamsters, and mice, *Lab. Anim. Sci.*, 41, 567, 1991.

45. Shoji, Y., Itoh, T., and Kagiyama, N., Pathogenicities of two CAR bacillus strains in mice and rats, *Jikken Dobutsu*, 37, 447, 1988.

46. Itoh, T., Kohyama, K., Takakura, A., Takenouchi, T., and Kagiyama, N., Naturally occurring CAR bacillus infection in a laboratory rat colony and epizootiological observations, *Jikken Dobutsu*, 36, 387, 1987.

47. Caniatti, M., Crippa, L., Giusti, M., Mattiello, S., Grilli, G., Orsenigo, R., and Scanziani, E., Cilia-associated respiratory (CAR) bacillus infection in conventionally reared rabbits, *Zentralbl. Veterinarmed.* [B], 45, 363, 1998.

48. Nicklas, W., Baneux, P., Boot, R., Decelle, T., Deeny, A.A., Fumanelli, M., and Illgen-Wilcke, B., Recommendations for the health monitoring of rodent and rabbit colonies in breeding and experimental units, *Lab. Anim.*, 36, 20, 2002.

49. Amao, H., Komukai, Y., Sugiyama, M., Takahashi, K.W., Sawada, T., and Saito, M., Natural habitats of Corynebacterium kutscheri in subclinically infected ICGN and DBA/2 strains of mice, *Lab. Anim. Sci.*, 45, 6, 1995.

50. Amao, H., Komukai, Y., Akimoto, T., Sugiyama, M., Takahashi, K.W., Sawada, T., and Saito, M., Natural and subclinical Corynebacterium Kutscheri infection in rats, *Lab. Anim. Sci.*, 45, 11, 1995.

51. Vallee, A., Guillon, J.C., and Cayeux, P., Isolation of a strain of Corynebacterium Kutscheri in a guinea pig, *Bull. Acad. Vet. Fr.*, 42, 797, 1969.

52. Amano, H., Akimoto, T., Takahashi, K.W., Nakagawa, M., and Saito, M., Isolation of Corynebacterium Kutscheri from aged Syrian hamsters (Mesocricetus Auratus), *Lab. Anim. Sci.*, 41, 265, 1991.

53. Amao, H., Komukai, Y., Sugiyama, M., Takahashi, K.W., Sawada, T., and Saito, M., Natural habitats of Corynebacterium Kutscheri in subclinically infected ICGN and DBA/2 strains of mice, *Lab. Anim. Sci.*, 45, 6, 1995.

54. Weisbroth, S.H. and Scher, S., Corynebacterium Kutscheri infection in the mouse. II. Diagnostic serology, *Lab. Anim. Care*, 18, 459, 1968.

55. Suzuki, E., Mochida, K., and Nakagawa, M., Naturally occurring subclinical Corynebacterium Kutscheri infection in laboratory rats: strain and age related antibody response, *Lab. Anim. Sci.*, 38, 42, 1988.

56. Hirst, R.G. and Wallace, M.E., Inherited resistance to Corynebacterium Kutscheri in mice, *Infect. Immun.*, 14, 475, 1976.

57. Juhr, N.C. and Horn, J., Model infection with Corynebacterium Kutscheri in the mouse, *Z. Versuchstierkd.*, 17, 129, 1975.

58. Osanai, T., Ohyama, T., Kikuchi, N., Takahashi, T., Kasai, N., and Hiramune, T., Distribution of Corynebacterium Renale among apparently healthy rats, *Vet. Microbiol.*, 52, 313, 1996.

59. Takahashi, T., Tsuji, M., Kikuchi, N., Ishihara, C., Osanai, T., Kasai, N., Yanagawa, R., and Hiramune, T., Assignment of the bacterial agent of urinary calculus in young rats by the comparative sequence analysis of the 16S RRNA genes of Corynebacteria, *J. Vet. Med. Sci.*, 57, 515, 1995.

60. Gobbi, A., Crippa, L., and Scanziani, E., Corynebacterium Bovis infection in immunocompetent Hirsute mice, *Lab. Anim. Sci.*, 49, 209, 1999.

61. Gobbi, A., Crippa, L., and Scanziani, E., Corynebacterium Bovis infection in Waltzing mice, *Lab. Anim. Sci.*, 49, 132, 1999.

62. Duga, S., Gobbi, A., Asselta, R., Crippa, L., Tenchini, M.L., Simonic, T., and Scanziani, E., Analysis of the 16S RRNA gene sequence of the Coryneform bacterium associated with hyperkeratotic dermatitis of Athymic nude mice and development of a PCR-based detection assay, *Mol. Cell Probes*, 12, 191, 1998.

63. Scanziani, E., Gobbi, A., Crippa, L., Giusti, A.M., Pesenti, E., Cavalletti, E., and Luini, M., Hyperkeratosis-associated Coryneform infection in severe combined immunodeficient mice, *Lab. Anim.*, 32, 330, 1998.

64. Scanziani, E., Gobbi, A., Crippa, L., Giusti, A.M., Giavazzi, R., Cavalletti, E., and Luini, M., Outbreaks of hyperkeratotic dermatitis of Athymic nude mice in Northern Italy, *Lab. Anim.*, 31, 206, 1997.

65. Schauer, D.B., Zabel, B.A., Pedraza, I.F., Ohara, C.M., Steigerwalt, A.G., and Brenner, D.J., Genetic and biochemical-characterization of Citrobacter-rodentium sp-nov, *J. Clinical Microbiol.*, 33, 2064, 1995.

66. Barthold, S.W., Coleman, G.L., Bhatt, P.N., Osbaldiston, G.W., and Jonas, A.M., The etiology of transmissible murine colonic hyperplasia, *Lab. Anim. Sci.*, 26, 889, 1976.

67. Barthold, S.W., Osbaldiston, G.W., and Jonas, A.M., Dietary, bacterial, and host genetic interactions in the pathogenesis of transmissible murine colonic hyperplasia, *Lab. Anim. Sci.*, 27, 938, 1977.

68. Maggio-Price, L., Nicholson, K.L., Kline, K.M., Birkebak, T., Suzuki, I., Wilson, D.L., Schauer, D., and Fink, P.J., Diminished reproduction, failure to thrive, and altered immunologic function in a colony of T-cell receptor transgenic mice: possible role of Citrobacter rodentium, *Lab. Anim. Sci.*, 48, 145, 1998.

69. Nakagawa, M., Saito, M., Suzuki, E., Nakayama, K., Matsubara, J., and Matsuno, K., A survey of Streptococcus pneumoniae, Streptococcus zooepidemicus, Salmonella spp., Bordetella bronchiseptica and Sendai virus in guinea pig colonies in Japan, *Jikken Dobutsu*, 35, 517, 1986.

70. Okewole, P.A., Uche, E.M., Oyetunde, I.L., Odeyemi, P.S., and Dawul, P.B., Uterine involvement in guinea pig Salmonellosis, *Lab. Anim.*, 23, 275, 1989.

71. Wullenweber, M., Kaspareit-Rittinghausen, J., and Farouq, M., Streptobacillus moniliformis epizootic in barrier-maintained C57BL/6J mice and susceptibility to infection of different strains of mice, *Lab. Anim. Sci.*, 40, 608, 1990.

72. Wullenweber, M., Jonas, C., and Kunstyr, I., Streptobacillus moniliformis isolated from otitis media of conventionally kept laboratory rats, *J. Exp. Anim. Sci.*, 35, 49, 1992.

73. Koopman, J.P., Van den Brink, M.E., Vennix, P.P., Kuypers, W., Boot, R., and Bakker, R.H., Isolation of Streptobacillus moniliformis from the middle ear of rats, *Lab. Anim.*, 25, 35, 1991.

74. Fleming, M.P., Streptobacillus moniliformis isolations from cervical abscesses of guinea pigs, *Vet. Rec.*, 99, 256, 1976.

75. Kirchner, B.K., Lake, S.G., and Wightman, S.R., Isolation of Streptobacillus moniliformis from a guinea pig with granulomatous pneumonia, *Lab. Anim. Sci.*, 42, 519, 1992.

76. Cunningham, B.B., Paller, A.S., and Katz, B.Z., Rat bite fever in a pet lover, *J. Am. Acad. Dermatol.*, 38, 330, 1998.

77. Peel, M.M., Dog-associated bacterial infections in humans: isolates submitted to an Australian reference laboratory, 1981–1992, *Pathology*, 25, 379, 1993.

78. Rygg, M. and Bruun, C.F., Rat bite fever (Streptobacillus moniliformis) with septicemia in a child, *Scand. J. Infect. Dis.*, 24, 535, 1992.

79. Mathiasen, T. and Rix, M., Rat-bite — an infant bitten by a rat, *Ugeskr. Laeger*, 155, 1475, 1993.

80. Rordorf, T., Zuger, C., Zbinden, R., von Graevenitz, A., and Pirovino, M., Streptobacillus moniliformis endocarditis in an HIV-positive patient, *Infection*, 28, 393, 2000.

81. Kaspareit-Rittinghausen, J., Wullenweber, M., Deerberg, F., and Farouq, M., Pathological changes in Streptobacillus moniliformis infection of C57bl/6J mice, *Berl Munch. Tierarztl. Wochenschr.*, 103, 84, 1990.

82. Fleming, M.P., Streptobacillus moniliformis isolations from cervical abscesses of guinea-pigs, *Vet. Rec.*, 99, 256, 1976.

83. Vogelbacher, M. and Bohnet, W., Distribution of Staphylococcus species in laboratory mice and laboratory rats compared with those found in house mice and Norway rats, in *Harmonization of Laboratory Animal Husbandry*, Proceeding of the Sixth Symposium of the Federation of the European Laboratory Animal Science Associations, 19–21 June, Basel, Switzerland, O'Donoghue, P.N., Eds., Royal Society of Medicine, London, 1997.

84. Rozengurt, N. and Sanchez, S., Enteropathogenic catalase-negative cocci, in *Welfare and Science*, Proceedings of the Fifth Symposium of the Federation of the European Laboratory Animal Science Associations, 8–11 June, Brighton UK, Bunyan, J., Ed., 402, Royal Society of Medicine Press, London, 1994.

85. Solomon, H.F., Dixon, D.M., and Pouch, W., A survey of Staphylococci isolated from the laboratory gerbil, *Lab. Anim. Sci.*, 40, 316, 1990.

86. Detmer, A., Hansen, A.K., Dieperink, H., and Svendsen, P., Xylose-positive Staphylococci as a cause of respiratory disease in immunosuppressed rats, *Scand. J. Lab. Anim. Sci.*, 18, 13, 1991.

87. Urano, T. and Maejima, K., Provocation of pseudomoniasis with cyclophosphamide in mice, *Lab. Anim.*, 12, 159, 1978.

88. Brennan, P.C., Fritz, T.E., and Flynn, R.J., Role of Pasteurella Pneumotropica and Mycoplasma Pulmonis in murine pneumonia, *J. Bacteriol.*, 97, 337, 1969.

89. Schoeb, T.R., Kervin, K.C., and Lindsey, J.R., Exacerbation of murine respiratory mycoplasmosis in Gnotobiotic F344/N rats by Sendai virus infection, *Vet. Pathol.*, 22, 272, 1985.

90. Broderson, J.R., Lindsey, J.R., and Crawford, J.E., The role of environmental ammonia in respiratory mycoplasmosis of rats, *Am. J. Pathol.*, 85, 115, 1976.

91. Cassell, G.H., Lindsey, J.R., and Davis, J.K., Respiratory and genital mycoplasmosis of laboratory rodents: implications for biomedical research, *Isr. J. Med. Sci.*, 17, 548, 1981.

92. Cassell, G.H., Wilborn, W.H., Silvers, S.H., and Minion, F.C., Adherence and colonization of mycoplasma pulmonis to genital epithelium and spermatozoa in rats, *Isr. J. Med. Sci.*, 17, 593, 1981.

93. Busch, K. and Naglic, T., Natural uterine Mycoplasma pulmonis infection in female rats, *Vet. Med. (Praha)*, 40, 253, 1995.

94. Kimbrough, R. and Gaines, T.B., Toxicity of hexamethylphosphoramide in rats, *Nature*, 211, 146, 1966.

95. Lai, W.C., Pakes, S.P., Owusu, I., and Wang, S., Mycoplasma pulmonis depresses humoral and cell-mediated responses in mice, *Lab. Anim. Sci.*, 39, 11, 1989.

96. Taurog, J.D., Leary, S.L., Cremer, M.A., Mahowald, M.L., Sandberg, G.P., and Manning, P.J., Infection with mycoplasma pulmonis modulates adjuvant- and collagen-induced arthritis in Lewis rats, *Arthritis Rheumatol.*, 27, 943, 1984.

97. Voot, L., Sadewasser, S., and Kloeting, I., The development of BB-rat diabetes is delayed or prevented by infections or applications of immunogens, *Zeitschrift Fur Versuchstierkunde*, 31, 197, 1988.

98. Freymuth, F., Vabret, A., Brouard, J., Duhamel, J.F., Guillois, B., Petitjean, J., Gennetay, E., Gouarin, S., and Proust, C., Epidemiology of viral infection and asthma, *Revue Francaise d'Allergologie et d'Immunologie Clinique*, 38, 319, 1998.

99. Zenner, L. and Regnault, J.P., Ten-year long monitoring of laboratory mouse and rat colonies in French facilities: a retrospective study, *Lab. Anim.*, 34, 76, 2000.

100. Kilham, L. and Ferm, H.V., Rat virus (RV) infections of pregnant, fetal, and newborn rats, *Proc. of the Soc. of Exp. Biol. and Med.*, 106, 825, 1961.

101. Crawford, L.V., A minute virus of mice, *Virology*, 29, 605, 1966.

102. McKisic, M.D., Lancki, D.W., Otto, G., Padrid, P., Snook, S., Cronin, D.C., Lohmar, P.D., Wong, T., and Fitch, F.W., Identification and propagation of a putative immunosuppressive orphan parvovirus in cloned T cells, *J. Immunol.*, 150, 419, 1993.

103. Ueno, Y., Sugiyama, F., Sugiyama, Y., Ohsawa, K., Sato, H., and Yagami, K., Epidemiological characterization of newly recognized rat parvovirus, "Rat Orphan Parvovirus," *J. Vet. Med. Sci.*, 59, 265, 1997.

104. Ball-Goodrich, L.J. and Johnson, E., Molecular characterization of a newly recognized mouse parvovirus, *J. Virol.*, 68, 6476, 1994.

105. Kilham, L. and Margolis, G., Transplacental infection of rats and hamsters induced by oral and parenteral inoculations of H-1 and rat viruses (RV), *Teratology*, 2, 111, 1969.

106. Kraft, V. and Meyer, B., Seromonitoring in small laboratory animal colonies. A five year survey: 1984–1988, *Z. Versuchstierkd.*, 33, 29, 1990.

107. Ueno, Y., Iwama, M., Ohshima, T., Sugiyama, F., Takakura, A., Itoh, T., and Yagami, K., Prevalence of "orphan" parvovirus infections in mice and rats, *Exp. Anim.*, 47, 207, 1998.

108. Metcalf, J.B., Lederman, M., Stout, E.R., and Bates, R.C., Natural parvovirus infection in laboratory rabbits, *Am. J. Vet. Res.*, 50, 1048, 1989.

109. Matsunaga, Y. and Chino, F., Experimental-infection of young-rabbits with rabbit parvovirus, *Archives of Virology*, 68, 257, 1981.

110. Matsunaga, Y., Matsuno, S., and Mukoyama, J., Isolation and characterization of a parvovirus of rabbits, *Infect. Immun.*, 18, 495, 1977.

111. Larsen, S.H. and Nathans, D., Mouse adenovirus: growth of plaque-purified FL virus in cell lines and characterization of viral DNA, *Virology*, 82, 182, 1977.

112. Takeuchi, A. and Hashimoto, K., Electron microscope study of experimental enteric adenovirus infection in mice, *Infect. Immun.*, 13, 569, 1976.

113. Heck, F.C., Jr., Sheldon, W.G., and Gleiser, C.A., Pathogenesis of experimentally produced mouse adenovirus infection in mice, *Am. J. Vet. Res.*, 33, 841, 1972.

114. Hansen, A.K., Velschow, S., Clausen, F.B., Svendsen, O., Amtoft-Neubauer, H., Kristensen, K., and Jorgensen, P.H., New infections to be considered in health monitoring of laboratory rodents, *Scand. J. Lab. Anim. Sci.*, 27, 65, 2000.

115. Butz, N., Ossent, P., and Homberger, F.R., Pathogenesis of guinea pig adenovirus infection, *Lab. Anim. Sci.*, 49, 600, 1999.

116. Bhatt, P.N. and Jacoby, R.O., Mousepox in inbred mice innately resistant or susceptible to lethal infection with Ectromelia virus. III. Experimental transmission of infection and derivation of virus-free progeny from previously infected dams, *Lab. Anim. Sci.*, 37, 23, 1987.

117. Jacoby, R.O. and Bhatt, P.N., Mousepox in inbred mice innately resistant or susceptible to lethal infection with Ectromelia virus. II. Pathogenesis, *Lab. Anim. Sci.*, 37, 16, 1987.

118. Bhatt, P.N. and Jacoby, R.O., Mousepox in inbred mice innately resistant or susceptible to lethal infection with Ectromelia virus. I. Clinical responses, *Lab. Anim. Sci.*, 37, 11, 1987.

119. Werner, R.M., Allen, A.M., Small, J.D., and New, A.E., Clinical manifestations of mousepox in an experimental animal holding room, *Lab. Anim. Sci.*, 31, 590, 1981.

120. Lipman, N.S., Perkins, S., Nguyen, H., Pfeffer, M., and Meyer, H., Mousepox resulting from use of Ectromelia virus-contaminated, imported mouse serum, *Comp. Med.*, 50, 426, 2000.

121. Dick, E.J., Kittell, C.L., Meyer, H., Farrar, P.L., Ropp, S.L., Esposito, J.J., Buller, R.M.L., Neubauer, H., Kang, Y.H., and McKee, A.E., Mousepox outbreak in a laboratory mouse colony, *Lab. Anim. Sci.*, 46, 602, 1996.

122. Mayo, D., Armstrong, J.A., and Ho, M., Activation of latent murine cytomegalovirus infection: cocultivation, cell transfer, and the effect of immunosuppression, *J. Infect. Dis.*, 138, 890, 1978.

123. Chatterjee, A., Harrison, C.J., Britt, W.J., and Bewtra, C., Modification of maternal and congenital cytomegalovirus infection by anti-glycoprotein B antibody transfer in guinea pigs, *J. Infect. Dis.*, 183, 1547, 2001.

124. Homberger, F.R., Enterotropic mouse hepatitis virus, *Lab. Anim.*, 31, 97, 1997.

125. Broderson, J.R., Murphy, F.A., and Hierholzer, J.C., Lethal enteritis in infant mice caused by mouse hepatitis virus, *Lab. Anim. Sci.*, 26, 824, 1976.

126. Taguchi, F., Hirano, N., Kiuchi, Y., and Fujiwara, K., Difference in response to mouse hepatitis virus among susceptible mouse strains, *Jpn. J. Microbiol.*, 20, 293, 1976.

127. Percy, D.H. and Barta, J.R., Spontaneous and experimental infections in Scid and Scid/Beige mice, *Lab. Anim. Sci.*, 43, 127, 1993.

128. Sebesteny, A. and Hill, A.C., Hepatitis and brain lesions due to mouse hepatitis virus accompanied by wasting in Nude mice, *Lab. Anim.*, 8, 317, 1974.

129. Barthold, S.W. and Smith, A.L., Mouse hepatitis virus strain — related patterns of tissue tropism in suckling mice, *Arch. Virol.*, 81, 103, 1984.

130. Jacoby, R.O., Bhatt, P.N., and Jonas, A.M., Pathogenesis of Sialodacryoadenitis in gnotobiotic rats, *Vet. Pathol.*, 12, 196, 1975.

131. Hansen, A.K. and Jensen, H.J.S., Mice infected with mouse hepatitis virus shed diagnostically detectable amounts of IgA in feces, *Scand. J. Lab. Anim. Sci.*, 24, 66, 1997.

132. Hansen, A.K. and Jensen, H.J.S., Experience from sentinel health monitoring in units containing rats and mice in experiments, *Scand. J. Lab. Anim. Sci.*, 22, 1, 1995.

133. Casebolt, D.B., Qian, B., and Stephensen, C.B., Detection of enterotropic mouse hepatitis virus fecal excretion by polymerase chain reaction, *Lab. Anim. Sci.*, 47, 6, 1997.

134. Jensen, H.J., Elimination of intercurrent death among rabbits inoculated with Treponema Pallidum, Acta Pathol. Microbiol. Scand.[B], *Microbiol. Immunol.*, 79, 124, 1971.

135. McConnell, S.J., Garner, F.L., Spertzel, R.O., Warner Jr., A.R., and Yager, R.H., Isolation and characterization of a neurotropic agent from adult rats, Proc. Soc. Exp. Biol. Med., 115, 362, 1964.

136. Ohsawa, K., Watanabe, Y., Miyata, H., and Sato, H., Genetic analysis of TMEV-like virus isolated from rats: nucleic acid charcterisation of three-dimensional protein region, *Lab. Anim. Sci.*, 48, 418, 1998.

137. Lipton, H.L. and Dal Canto, M.C., Chronic neurologic disease in Theiler's virus infection of SJL/J mice, *J. Neurol. Sci.*, 30, 201, 1976.

138. Lipton, H.L. and Melvold, R., Genetic analysis of susceptibility to Theiler's virus-induced demyelinating disease in mice, *J. Immunol.*, 132, 1821, 1984.

139. Abzug, M.J. and Tyson, R.W., Picornavirus infection in early murine gestation: significance of maternal illness, *Placenta*, 21, 840, 2000.

140. Buschard, K., Hastrup, N., and Rygaard, J., Virus-induced diabetes mellitus in mice and the thymus-dependent immune system, *Diabetologia*, 24, 42, 1983.

141. Clausen, F.B., Lykkesfeldt, J., and Hansen, A.K., On the possible connection between vitamin C, cardiovirus and guinea pig lameness, in Proc. of the 7th FELASA meeting, 101, Royal Society of Medicine, London, 2001.

142. Hansen, A.K., Thomsen, P., and Jensen, H.J., A serological indication of the existence of a guinea pig poliovirus, *Lab. Anim.*, 31, 212, 1997.

143. Carthew, P. and Sparrow, S., Persistence of pneumonia virus of mice and Sendai virus in germ-free (Nu/Nu) mice, *Br. J. Exp. Pathol.*, 61, 172, 1980.

144. Carthew, P. and Sparrow, S., Sendai virus in nude and germ-free rats, *Res. Vet. Sci.*, 29, 289, 1980.

145. Parker, J.C., Whiteman, M.D., and Richter, C.B., Susceptibility of inbred and outbred mouse strains to Sendai virus and prevalence of infection in laboratory rodents, *Infect. Immun.*, 19, 123, 1978.

146. Iida, T., Experimental study on the transmission of Sendai virus in specific pathogen-free mice, *J. Gen. Virol.*, 14, 69, 1972.

147. Artwohl, J.E., Cera, L.M., Wright, M.F., Medina, L.V., and Kim, L.J., The efficacy of a dirty bedding sentinel system for detecting Sendai virus infection in mice: a comparison of clinical signs and seroconversion, *Lab. Anim. Sci.*, 44, 73, 1994.

148. Machii, K., Otsuka, Y., Iwai, H., and Ueda, K., Infection of rabbits with Sendai virus, *Lab. Anim. Sci.*, 39, 334, 1989.

149. Iwai, H., Machii, K., Ohtsuka, Y., Ueda, K., Inoue, S., Matsumoto, T., and Satoh, Z., Prevalence of antibodies to Sendai virus and rotavirus in laboratory rabbits, *Jikken Dobutsu*, 35, 491, 1986.

150. Ohsawa, K., Yamada, A., Takeuchi, K., Watanabe, Y., Miyata, H., and Sato, H., Genetic characterization of parainfluenza virus 3 derived from guinea pigs, *J. Vet. Med. Sci.*, 60, 919, 1998.

151. Welch, B.G., Snow, E.J., Jr., Hegner, J.R., Adams, S.R., Jr., and Quist, K.D., Development of a guinea pig colony free of complement-fixing antibodies to parainfluenza virus, *Lab. Anim. Sci.*, 27, 976, 1977.

152. Rizzi, V., Legrottaglie, R., Cini, A., and Agrimi, P., Electrophoretic typing of some strains of enteric viruses isolated in rabbits suffering from diarrhea, *Microbiologica*, 18, 77, 1995.

153. Percy, D.H., Muckle, C.A., Hampson, R.J., and Brash, M.L., The enteritis complex in domestic rabbits — a field-study, *Can. Vet. J. — Revue Veterinaire Canadienne*, 34, 95, 1993.

154. Vonderfecht, S.L., Huber, A.C., Yolken, R.H., and Strandberg, J.D., Infectious diarrhea of infant rats — clinicopathologic observations and characterization of a viral agent, *Lab. Anim. Sci.*, 33, 502, 1983.

155. Martinez, D., Wolanski, B., Tytell, A.A., and Devlin, R.G., Viral etiology of age-dependent polioencephalomyelitis in C58 mice, *Infect. Immun.*, 23, 133, 1979.

156. Nicklas, W., Kraft, V., and Meyer, B., Contamination of transplantable tumors, cell lines, and monoclonal antibodies with rodent viruses, *Lab. Anim. Sci.*, 43, 296, 1993.

157. Crispens, C.G., Jr., Lactate dehydrogenase virus and mouse embryos, *Nature*, 214, 819, 1967.

158. Chen, Z.Y. and Plagemann, P.G.W., Detection of lactate dehydrogenase-elevating virus in transplantable mouse tumors by biological assay and RT-PCR assays and its removal from the tumor cell, *J. Virol. Methods*, 65, 227, 1997.

159. Biggar, R.J., Woodall, J.P., Walter, P.D., and Haughie, G.E., Lymphocytic choriomeningitis outbreak associated with pet hamsters. Fifty-seven cases from New York state, *JAMA*, 232, 494, 1975.

160. Deibel, R., Woodall, J.P., Decher, W.J., and Schryver, G.D., Lymphocytic choriomeningitis virus in man. Serologic evidence of association with pet hamsters, *JAMA*, 232, 501, 1975.

161. Biggar, R.J., Douglas, R.G., and Hotchin, J., Letter: Lymphocytic choriomeningitis associated with hamsters, *Lancet*, 1, 856, 1975.

162. Smith, A.L., Paturzo, F.X., Gardner, E.P., Morgenstern, S., Cameron, G., and Wadley, H., Two epizootics of lymphocytic choriomeningitis virus occurring in laboratory mice despite intensive monitoring programs, *Can. J. Comp. Med.*, 48, 335, 1984.

163. Parker, J.C., Igel, H.J., Reynolds, R.K., Lewis, A.M., Jr., and Rowe, W.P., Lymphocytic choriomeningitis virus infection in fetal, newborn, and young adult Syrian hamsters (Mesocricetus Auratus), *Infect. Immun.*, 13, 967081, 1976.

164. Dutko, F.J. and Oldstone, M.B., Genomic and biological variation among commonly used lymphocytic choriomeningitis virus strains, *J. Gen. Virol.*, 64 (Pt 8), 1689, 1983.

165. Meyer, B.J. and Schmaljohn, C.S., Persistent hantavirus infections: characteristics and mechanisms, *Trends in Microbiol.*, 8, 61, 2000.

166. Lee, H.W., Lee, P.W., and Johnson, K.M., Isolation of the etiologic agent of Korean hemorrhagic fever, *J. Infect. Dis.*, 137, 298, 1978.

167. LeDuc, J.W., Epidemiology of Hantaan and related viruses, *Lab. Anim. Sci.*, 37, 413, 1987.

168. Desmyter, J., LeDuc, J.W., Johnson, K.M., Brasseur, F., Deckers, C., and van Ypersele, D.S., Laboratory rat associated outbreak of haemorrhagic rever with renal syndrome due to Hantaan-like virus in Belgium, *Lancet,* 2, 1445, 1983.

169. Capucci, L., Fusi, P., Lavazza, A., Pacciarini, M.L., and Rossi, C., Detection and preliminary characterization of a new rabbit calicivirus related to rabbit hemorrhagic disease virus but nonpathogenic, *J. Virology,* 70, 8614, 1996.

170. Boot, R., van Knapen, F., Kruijt, B.C., and Walvoort, H.C., Serological evidence for encephalitozoon Cuniculi infection (Nosemiasis) in gnotobiotic guinea pigs, *Lab. Anim.*, 22, 337, 1988.

171. Waller, T., Morein, B., and Fabiansson, E., Humoral immune response to infection with encephalitozoon cuniculi in rabbits, *Lab. Anim.*, 12, 145, 1978.

172. Gannon, J., A survey of encephalitozoon cuniculi in laboratory animal colonies in the United Kingdom, *Lab. Anim.*, 14, 91, 1980.

173. Flatt, R.E. and Jackson, S.J., Renal nosematosis in young rabbits, *Pathol. Vet.*, 7, 492, 1970.

174. Koller, L.D., Spontaneous nosema cuniculi infection in laboratory rabbits, *J. Am. Vet. Med. Assoc.*, 155, 1108, 1969.

175. Boot, R., Hansen, A.K., Hansen, C.K., Nozari, N., and Thuis, H.C.W., Comparison of assays for antibodies to Encephalitozoon cuniculi in rabbits, *Lab. Anim.*, 34, 281, 2000.

176. Shultz, L.D., Schweitzer, P.A., Hall, E.J., Sundberg, J.P., Taylor, S., and Walzer, P.D., Pneumocystis-carinii pneumonia in SCID mice, *Current Topics in Microbiol. and Immunol.*, 152, 243, 1989.

177. Walzer, P.D., Kim, C.K., Linke, M.J., Pogue, C.L., Huerkamp, M.J., Chrisp, C.E., Lerro, A.V., Wixson, S.K., Hall, E., and Shultz, L.D., Outbreaks of pneumocystis-carinii pneumonia in colonies of immunodeficient mice, *Infect. Immun.*, 57, 62, 1989.

178. Hughes, W.T., Natural-mode of acquisition for de novo infection with pneumocystis-carinii, *J. Infect. Dis.*, 145, 842, 1982.

179. Bhan, A.K., Mizoguchi, E., Smith, R.N., and Mizoguchi, A., Colitis in transgenic and knockout animals as models of human inflammatory bowel disease, *Immunol. Rev.*, 169, 195, 1999.

180. Hopert, A., Uphoff, C.C., Wirth, M., Hauser, H., and Drexler, H.G., Mycoplasma detection by PER analysis, *In Vitro Cellular & Developmental Biology-Animal*, 29A, 819, 1993.

181. Mims, C., Virus-related immunomodulation, in *Viral and Mycoplasmal Infections of Laboratory Rodents, Effects on Biomedical Research*, Bhatt, P.N., Jacoby, R.O., Morse III, H.C., and New, A.E., Eds., Academic Press, New York, 1986.

182. Cross, S.S., Parker, J.C., Rowe, W.P., and Robbins, M.L., Biology of mouse thymic virus, a herpes virus of mice, and the antigenic relationship to mouse cytomegalovirus, *Infect. Immun.*, 26, 1186, 1979.

183. Hamilton, J.D., Fitzwilliam, J.F., Cheung, K.S., and Lang, D.J., Effects of murine cytomegalovirus infection on the immune response to a tumor allograft, *Rev. Infect. Dis.*, 1, 976, 1979.

184. Loh, L. and Hudson, J.B., Murine cytomegalovirus-induced immunosuppression, *Infect. Immun.*, 36, 89, 1982.

185. Loh, L. and Hudson, J.B., Murine cytomegalovirus infection in the spleen and its relationship to immunosuppression, *Infect. Immun.*, 32, 1067, 1981.

186. Shanley, J.D. and Pesanti, E.L., Effects of antiviral agents on murine cytomegalovirus-induced macrophage dysfunction, *Infect. Immun.*, 36, 918, 1982.

187. Shanley, J.D. and Pesanti, E.L., Replication of murine cytomegalovirus in lung macrophages: effect of phagocytosis of bacteria, *Infect. Immun.*, 29, 1152, 1980.

188. Kaklamanis, E. and Pavlatos, M., The immunosuppressive effect of mycoplasma infection. I. Effect on the humoral and cellular response, *Immunology*, 22, 695, 1972.

189. Specter, S.C., Bendinelli, M., Ceglowski, W.S., and Friedman, H., Macrophage-induced reversal of immunosuppression by leukemia viruses, *Fed. Proc.*, 37, 97, 1978.

190. Romero-Rojas, A., Ponce-Hernandez, C., Ciprian, A., Estrada-Parra, S., and Hadden, J.W., Immuno-modulatory properties of Mycoplasma pulmonis. I. Characterization of the immunomodulatory activity, *Intl. Immunopharmacol.*, 1, 1679, 2001.

191. Westerberg, S.C., Smith, C.B., Wiley, B.B., and Jensen, C., Mycoplasma-virus interrelationships in mouse tracheal organ cultures, *Infect. Immun.*, 5, 840, 1972.

192. Pollack, J.D., Weiss, H.S., and Somerson, N.L., Lecithin changes in murine mycoplasma pulmonis respiratory infection, *Infect. Immun.*, 24, 94, 1979.

193. Laubach, H.E., Kocan, A.A., and Sartain, K.E., Lung lysophospholipase activity in specific-pathogen-free rats infected with Pasteurella pneumotropica or Mycoplasma pulmonis, *Infect. Immun.*, 22, 295, 1978.

194. Wells, A.B., The kinetics of cell proliferation in the tracheobronchial epithelia of rats with and without chronic respiratory disease, *Cell Tissue Kinet.*, 3, 185, 1970.

195. Ventura, J. and Domaradzki, M., Role of mycoplasma infection in the development of experimental bronchiectasis in the rat, *J. Pathol. Bacteriol.*, 93, 342, 1967.

196. Thomsen, A.C. and Heron, I., Effect of mycoplasmas on phagocytosis and immunocompetence in rats, *Acta Pathol. Microbiol. Scand.* [C.], 87C, 67, 1979.

197. Ruskin, J. and Remington, J.S., Toxoplasmosis in the compromised host, *Ann. Intern. Med.*, 84, 193, 1976.

198. Krahenbuhl, J.L., Ruskin, J., and Remington, J.S., The use of killed vaccines in immunization against an intracellular parasite: Toxoplasma gondii, *J. Immunol.*, 108, 425, 1972.

199. Ruskin, J. and Remington, J.S., Resistance to intracellular infection in mice immunized with toxoplasma vaccine and adjuvant, *J. Reticuloendothel. Soc.*, 9, 465, 1971.

200. Ruskin, J. and Remington, J.S., Toxoplasmosis in the compromised host, *Ann. Intern. Med.*, 84, 193, 1976.

201. Swartzberg, J.E., Krahenbuhl, J.L., and Remington, J.S., Dichotomy between macrophage activation and degree of protection against Listeria monocytogenes and Toxoplasma gondii in mice stimulated with Corynebacterium parvum, *Infect. Immun.*, 12, 1037, 1975.

202. Sato, Y., Ooi, H.K., Nonaka, N., Oku, Y., and Kamiya, M., Antibody production in Syphacia Obvelata infected mice, *J. Parasitol.*, 81, 559, 1995.

203. Notkins, A.L., Enzymatic and immunologic alterations in mice infected with lactic dehydrogenase virus, *Am. J. Pathol.*, 64, 733, 1971.

204. Ruebner, B.H. and Hirano, T., Viral hepatitis in mice. Changes in oxidative enzymes and phosphatases after murine hepatis virus (MHV-3) infection, *Lab. Invest.*, 14, 157, 1965.

205. Tiensiwakul, P. and Husain, S.S., Effect of mouse hepatitis virus infection on iron retention in the mouse liver, *Br. J. Exp. Pathol.*, 60, 161, 1979.

206. Friis, A.S. and Ladefoged, O., The influence of Bacillus Piliformis (Tyzzer) infections on the reliability of pharmacokinetic experiments in mice, *Lab. Anim.*, 13, 257, 1979.

207. Vonderfecht, S.L., Huber, A.C., Eiden, J., Mader, L.C., and Yolken, R.H., Infectious diarrhea of infant rats produced by a rotavirus-like agent, *J. Virol.*, 52, 94, 1984.

208. Gustafsson, E., Blomqvist, G., Bellman, A., Holmdahl, R., Mattsson, A., and Mattsson, R., Maternal antibodies protect immunoglobulin deficient neonatal mice from mouse hepatitis virus (MHV)-associated wasting syndrome, *Am. J. Reprod. Immunol.*, 36, 33, 1996.

209. Schwanzer, V., Deerberg, F., Frost, J., Liess, B., Schwanzerova, I., and Pittermann, W., Intrauterine infection of mice with ectromelia virus, *Z. Versuchstierkd.*, 17, 110, 1975.

210. Blackmore, D.K. and Cassillo, S., Experimental investigation of uterine infections of mice due to Pasteurella pneumotropica, *J. Comp. Pathol.*, 82, 471, 1972.

211. Brown, M.B. and Steiner, D.A., Experimental genital mycoplasmosis: time of infection influences pregnancy outcome, *Infect. Immun.*, 64, 2315, 1996.

212. Hill, A.C. and Stalley, G.P., Mycoplasma pulmonis infection with regard to embryo freezing and hysterectomy derivation, *Lab. Anim. Sci.*, 41, 563, 1991.

213. Barthold, S.W. and Brownstein, D.G., The effect of selected viruses on Corynebacterium Kutscheri infection in rats, *Lab. Anim. Sci.*, 38, 580, 1988.

214. Schramlova, J., Otova, B., Cerny, J., and Blazek, K., Electron-microscopic demonstration of virus particles in acute lymphoblastic leukaemia in Sprague-Dawley rats, *Folia Biol.* (Praha), 40, 113, 1994.

215. Lynch, D.W., Lewis, T.R., Moorman, W.J., Burg, J.R., Groth, D.H., Khan, A., Ackerman, L.J., and Cockrell, B.Y., Carcinogenic and toxicologic effects of inhaled ethylene oxide and propylene oxide in F344 rats, *Toxicol. Appl. Pharmacol.*, 76, 69, 1984.

216. Barthold, S.W. and Jonas, A.M., Morphogenesis of early 1, 2-dimethylhydrazine-induced lesions and latent period reduction of colon carcinogenesis in mice by a variant of Citrobacter freundii, *Cancer Res.*, 37, 4352, 1977.

217. Ashley, M.P., Neoh, S.H., Kotlarski, I., and Hardy, D., Local and systemic effects in the non-specific tumour resistance induced by attenuated Salmonella enteritidis 11RX in mice, *Aust. J. Exp. Biol. Med. Sci.*, 54, 157, 1976.

218. Tindle, R.W., Neoh, S.H., Ashley, M.P., Hardy, D., and Kotlarski, I., Resistance of mice to Krebs Ascites tumour, Sarcoma S180 and PC6 plasmacytoma after immunisation with Salmonella enteritidis 11RX, *Aust. J. Exp. Biol. Med. Sci.*, 54, 149, 1976.

219. Tindle, R.W., Neoh, S.H., Ashley, M.P., Hardy, D., and Kotlarski, I., Resistance of mice to Krebs Ascites tumour, Sarcoma S180 and PC6 plasmacytoma after immunisation with Salmonella enteritidis 11RX, *Aust. J. Exp. Biol. Med. Sci.*, 54, 149, 1976.

220. Toolan, H.W. and Ledinko, N., Inhibition by H-1 virus of the incidence of tumors produced by adenovirus 12 in hamsters, *Virology*, 35, 475, 1968.

221. Toolan, H.W., Lack of oncogenic effect of the H-viruses for hamsters, *Nature*, 214, 1036, 1967.

222. Toolan, H.W., Rhode, S.L., III, and Gierthy, J. F., Inhibition of 7,12-dimethylbenz(a)anthracene-induced tumors in Syrian hamsters by prior infection with H-1 parvovirus, *Cancer Res.*, 42, 2552, 1982.

223. Brammer, D.W., Dysko, R.C., Spilman, S.C., and Oskar, P.A., Elimination of sialodacryoadenitis virus from a rat production colony by using seropositive breeding animals, *Lab. Anim. Sci.*, 43, 633, 1993.

224. Krohn, T.C. and Hansen, A.K., Carbon dioxide concentrations in unventilated IVC cages, *Lab. Anim.*, 36, 209, 2002.

225. Rehg, J.E., Blackman, M.A., and Toth, L.A., Persistent transmission of mouse hepatitis virus by transgenic mice, *Comp. Med.*, 51, 369, 2001.

226. Cox, J.C., Gallichio, H.A., Pye, D., and Walden, N.B., Application of immunofluorescence to the establishment of an encephalitozoon cuniculi-free rabbit colony, *Lab. Anim. Sci.*, 27, 204, 1977.

227. van der Waaij, D. and Berghuis-de Vries, J.M., Selective elimination of enterobacteriaceae species from the digestive tract in conventional and antibiotic treated mice, *J. Hygiene* (Cambridge), 69, 405, 1974.

228. van der Waaij, D., Cohen, B.J., and Anver, M.R., Mitigation of experimental bowel disease in guinea pigs by selective decontamination of the aerobic gram negative flora, *Gastroenterology*, 67, 460, 1974.

229. Heidt, P.J. and Timmermanns, C.P.J., Selective decontamination of the digestive tracts of pregnant rabbits: a method for producing enterobacteriaceae-free rabbits, *Lab. Anim. Sci.*, 25, 594, 1975.

230. Walker, R.I., MacWittie, T.J., Sinha, B.L., Egan, J. E., Ewald, P.E., and McClung, G., Antibiotic decontamination of the dog and its consequences, *Lab. Anim. Sci.*, 28, 55, 1978.

231. Srivastava, K.K., Pollard, M., and Wagner, M., Bacterial decontamination and antileukemic therapy of AKR mice, *Infect. Immun.*, 14, 1179, 1976.

232. Hansen, A.K., Antibiotic treatment of nude rats and its impact on the aerobic bacterial flora, *Lab. Anim.*, 29, 37, 1995.

233. Stephenson, E.H., Trahan, C.J., Ezzell, J.W., Mitchell, W.C., Abshire, T.G., Oland, D.D., and Nelson, G.O., Efficacy of a commercial bacterin in protecting strain 13 guinea pigs against Bordetella bronchiseptica pneumonia, *Lab. Anim.*, 23, 261, 1989.

234. Tagaya, M., Mori, I., Miyadai, T., Kimura, Y., Ito, H., and Nakakuki, K., Efficacy of a temperature-sensitive Sendai virus vaccine in hamsters, *Lab. Anim. Sci.*, 45, 233, 1995.

235. Iwata, H., Tagaya, M., Matsumoto, K., Miyadai, T., Yokochi, T., and Kimura, Y., Aerosol vaccination with a Sendai virus temperature-sensitive mutant (HVJ-PB) derived from persistently infected cells, *J. Infect. Dis.*, 162, 402, 1990.

236. Kimura, Y., Aoki, H., Shimokata, K., Ito, Y., Takano, M., Hirabayashi, N., and Norrby, E., Protection of mice against virulent virus infection by a temperature-sensitive mutant derived from an HVJ (Sendai virus) carrier culture, *Arch. Virol.*, 61, 297, 1979.

237. Bhatt, P.N. and Jacoby, R.O., Effect of vaccination on the clinical response, pathogenesis and transmission of mousepox, *Lab. Anim. Sci.*, 37, 610, 1987.

238. Cassell, G.H., Davis, J.K., and Lindsey, J.R., Control of mycoplasma pulmonis infection in rats and mice: detection and elimination vs. vaccination, *Isr. J. Med. Sci.*, 17, 674, 1981.

239. Cassell, G.H. and Davis, J.K., Protective effect of vaccination against mycoplasma pulmonis respiratory disease in rats, *Infect. Immun.*, 21, 69, 1978.

240. Howard, R.J. and Balfour, H.H., Jr., Prevention of morbidity and mortality of wild murine cytomegalovirus by vaccination with attenuated cytomegalovirus, Proc. Soc. Exp. Biol. Med., 156, 365, 1977.

241. Hansen, A.K., Microbiological quality of laboratory pigs, *Scand. J. Lab. Anim. Sci.*, 25, 145, 1998.

242. van Hoosier, G.L., Answer on a question conc. Sendai virus vaccination, in Viral *and Mycoplasmal Infections of Laboratory Rodents, Effects on Biomedical Research*, Bhatt, P.N., Jacoby, R.O., Morse III, H.C., and New, A.E., Eds., 61, Academic Press, New York, 1986.

243. Corning, B.F. and Lipman, N.S., A comparison of rodent caging systems based on microenvironmental parameters, *Lab. Anim. Sci.*, 41, 498, 1991.

244. Gustafsson, B.E. and Norin, K.E., Development of germ-free animal characteristics in conventional rats in antibiotics, *Acta Pathol. Microbiol. Scand.*[B], 85B, 1, 1977.

245. Rehbinder, C., Baneux, P., Forbes, D., van Herck, H., Nicklas, W., Rugaya, Z., and Winkler, G., FELASA recommendations for the health monitoring of breeding colonies and experimental units of cats, dogs and pigs — report of the Federation of European Laboratory Animal Science Associations (FELASA) Working Group on Animal Health, *Lab. Anim.*, 32, 1, 1998.

246. Weber, H., Berge, E., Finch, J., Heidt, P., Kaup, J.P.P.G., Verschuere, B., and Wolfensohn, S., Health monitoring of non-human primate colonies. Recommendations of the Federation of European Laboratory Animal Science Associations (FELASA) Working Group on Non-Human Primate Health accepted by the FELASA Board of Management, 21 November 1998, *Lab. Anim.*, 33 Suppl 1, S1, 1999.

247. Rehbinder, C., Alenius, S., Bures, J., las Heras, M.L., Greko, C., Kroon, P.S., and Gutzwiller, A., FELASA recommendations for the health monitoring of experimental units of calves, sheep, and goats. Report of the Federation of European Laboratory Animal Science Associations (FELASA) Working Group on Animal Health, *Lab. Anim.*, 34, 329, 2000.

248. Homberger, F.R., Boot, R., Feinstein, R.E., Hansen A.K, and Van-der-Logt, J., FELASA Guidance Paper for the Accreditation of Laboratory Animal Diagnostic Laboratories. Report of the Federation of European Laboratory Animal Science Associations (FELASA) Working Group on Accreditation of Diagnostic Laboratories, *Lab. Anim.*, 33 Suppl 1, S19, 1999.

249. Hansen, A.K., Statistical aspects of health monitoring of laboratory animal colonies, *Scand. J. Lab. Anim. Sci.*, 20, 11, 1993.

250. Riley, L.K., Knowles, R., Purdy, G., Salome, N., Pintel, D., Hook, R.R.J., Franklin, C.L., and Besch-Williford, C.L., Expression of recombinant parvovirus NS1 protein by a baculovirus and application to serologic testing of rodents, *J. Clin. Microbiol.*, 34, 440, 1996.

251. Riley, L.K., Franklin, C.L., Hook, R.R.J., and Besch-Williford, C., Identification of murine helicobacters by PCR and restriction enzyme analyses, *J. Clin. Microbiol.*, 34, 942, 1996.

252. Homberger, F.R. and Thomann, P.E., Transmission of murine viruses and mycoplasma in laboratory mouse colonies with respect to housing conditions, *Lab. Anim.*, 28, 113, 1994.

253. Homberger, F.R., Smith, A.L., and Barthold, S.W., Detection of rodent coronaviruses in tissues and cell cultures by using polymerase chain reaction, *J. Clin. Microbiol.*, 29, 2789, 1991.

254. Lewis, V.J. and Clayton, D.M., An evaluation of the mouse antibody production test for detecting three murine viruses, *Lab. Anim. Sci.*, 21, 203, 1971.

# CHAPTER 12

# Nutrient Requirements, Experimental Design, and Feeding Schedules in Animal Experimentation

Merel Ritskes-Hoitinga and André Chwalibog

## CONTENTS

0-8493-1086-5/03/$0.00+$1.50
© 2003 by CRC Press LLC

# INTRODUCTION

This chapter on nutrition of laboratory animals discusses the influence of diets, dietary composition, and feeding schedules on experimental results and animal health and welfare. Food is consumed every day, and many processes in the body are dependent upon and affected continuously by what is eaten and when and how it is ingested. The chapter provides scientific information and examples of designing sound experiments in the field of nutrition. This is not only important when doing nutritional studies, but also in other studies in which nutritional interference is undesirable. A good experimental design will contribute to reliable and reproducible results without unnecessarily compromising animal welfare. As many species spend a large part of their day foraging, suggestions concerning dietary enrichment are presented. An increased effort in the field of dietary enrichment is expected to contribute to the improvement of the welfare of laboratory animals and more reliable experimental results. The possible conflict that might arise between the need for standardization and the need for enrichment is a current research challenge for laboratory animal scientists and nutritionists.

# NUTRIENT REQUIREMENTS

Nutrients and energy are required for a number of different functions in the body, for the vital processes necessary for an animal to survive and for the productive processes, such as reproduction, growth, and lactation. As stated by Fuller and Wang,[1] in attempting to estimate requirements, one must ask three questions: Requirements of what animal or population of animals? Requirements for what? How are requirements to be specified? The first question arises from great differences between animals in requirements. Animals of different age, live weight, and sex have different requirements. Furthermore, even when these differences are excluded, the individual animals will still differ in their requirements due to genetic constitution, which is more difficult to quantify. It is often observed from animal experiments that even individuals of the same age and sex, belonging to the same genetic line, and being kept in similar environments show an in-between individual coefficient of variation up to 15 to 20% for many nutritional parameters. The second question relates to the criteria we choose to define requirement. For example, whether it is defined in terms of animal performance as body weight gain and feed conversion for gain or in terms of metabolic responses. The third question requires a specification of terms, whether requirements are, for example, for metabolizable energy or for net energy, or for digestible protein or for amino acids, or for what combination of energy and protein is preferred. In general, the nutritional requirements of farm animals and laboratory animals are similar. All require energy, protein, carbohydrate, lipid, minerals, and vitamins supplied in diets that should be palatable and free from chemical and biological contaminations. Therefore, much information regarding farm animal nutrition may be applied to laboratory animals. However, we have to keep in mind that the aim of laboratory animal nutrition is not the highest production, but optimum performance and nutritionally unbiased response to biomedical treatments.

It is well established that a diet deficient in composition or quantity may influence not only animal growth or reproductive performance but also immune response and resistance to diseases. Furthermore, prior to changes in animal performance, a number of physiological and anatomical changes caused by a deficiency or an excess of nutrients may interact with the action of biomedical treatments. Therefore, this chapter will outline some of the important effects of nutrition on the performance of laboratory animals, concomitantly emphasizing possible interactions between nutrition and experimental results.

Most biomedical experiments are, nowadays, performed with small rodents, and because the existing data are most abundant for rats and mice, the focus will be on these species. Because of the diversity of aspects associated with the supply of nutrients and energy for laboratory animals, it is simply impossible to debate all possible effects of nutrition on experimental results, so only some of the major nutritional characteristics of macronutrients and energy will be discussed in the following sections.

## Food Intake

Laboratory rodents are usually fed *ad libitum*. It is generally accepted that the voluntary food intake is related to the energy requirement of the animal. The classical experiments of Adolph in 1947[2] demonstrated that when the rat diets were diluted with inert materials to produce a wide range of energy concentration, the animals were able to adjust the amount of food eaten so that their energy intake remained constant.[2] However, the concept that "animals eat for energy" has several aspects. In case of extensive dilution of the diet with materials of low digestibility, the ability to adjust the intake may be impaired by the gastrointestinal capacity being a limiting factor. In this case, food intake may be insufficient to cover energy and often also nutrient requirements. On the other hand, when the dietary energy density is high enough to cover the energy requirements, increase in energy density by supplementing the diet with extra fat or carbohydrate may result in nutrient deficiencies. This occurs because the animal usually stops eating when its energy requirement has been met. Furthermore, not only the energy density of a diet and the capacity of the gastrointestinal tract influence food intake, but reduced intake is commonly observed in deficiency states, especially with diets low in protein[3] or unbalanced in amino acids, or in case of deficiency or excess of some trace minerals and vitamins. Diets that taste well, e.g., "hamburger diets," have the consequence that rats eat more than meeting the energy requirement, which makes them obese.

The physiological state of the animal also plays an essential role in food intake; several reports show increased intake with the onset of pregnancy, but other reports suggest little or no changes. Lactation is usually associated with a marked increase in food intake, which may in a rat at peak lactation be nearly three times that of a nonlactating rat. Considering that the voluntary food intake may be subjected to marked variation depending on nutritional and physiological factors, it is difficult to specify an expected daily consumption; consequently, Table 12.1 demonstrates the approximate voluntary daily food intake in the most common laboratory animals when fed commercial pelleted diets. The metabolizable energy content of these diets usually lies within the range of 8 to 12 MJ/kg diet.

For growing animals, the wide range of daily food intake is directly related to the age and live weight (LW), increasing during the growth period. For example, the voluntary food intake of growing rats, reported by Thorbek et al.[4] was 15 g/day at the age of 5 weeks and a LW of 100 g, while from 7 to 8 weeks (200 g LW) to the 18th week of age (400 g LW), it was remarkably constant at about 25 g/day. With a dietary gross energy concentration of 16.5 MJ/kg food, the values correspond to about 1400 kJ/LW per kg metabolic live weight (LW, kg$^{0.75}$) at 100 and 200 g LW, but only to 820 kJ/LW, kg$^{0.75}$ at 400 g LW. (The metabolic live weight is a parameter used to compare species. It is used in order to correct for differences in the metabolic rate per kg.) Comparable values can be calculated  from the

**Table 12.1  Estimated Average Food Intake (g/day)**

| Species | Growing | Adult | Pregnant | Lactating |
|---|---|---|---|---|
| Mouse | 3–5 | 5–7 | 6–8 | 7–15 |
| Rat | 8–25 | 25–30 | 25–35 | 35–65 |
| Hamster | 6–12 | 10–12 | 12–15 | 20–25 |
| Guinea pig | 35–45 | 45–70 | 70–80 | 100–130 |
| Rabbit | 120–200 | 200–300 | 300 | 300–400 |

experiments of Pullar and Webster[5] with "lean" rats, indicating a pattern of decreasing energy consumption and thereby food consumption in relation to metabolic live weight. Kleiber[6] has suggested that the maximum food intake, for all species, is proportionally related to basal metabolic rate (BMR) with the ratio 4:1. Assuming a constant BMR of 320 kJ/LW, $kg^{0.75}$ the maximum intake in growing laboratory animals would be about 1.3 MJ/LW, $kg^{0.75}$. In correspondence with Kleiber's principle, Clarke et al.[7] presented general equations for growth, pregnancy, and lactation in all laboratory animals, suggesting constant proportions between energy supply for different life processes and metabolic live weight. Despite the fact that such approaches may give some indication of the level of food intake, the accuracy of the predicted values is doubtful because energy intake in relation to metabolic live weight is not constant at maintenance level[8] or during growth.[4] Nevertheless, it is practical to use when doing nutritional studies, as one compares groups of animals of the same strain, age, sex, etc., at the same time and makes sure that the intake of food and nutrients is similar, except for those the researcher wants to differ.

## Digestion

The supply of nutrients required for the body functions depends on the transformation of the dietary constituents into simpler elements (amino acids, glucose, fatty acids) before they can pass through the mucous membrane of the gastrointestinal tract into the blood and lymph. The process of digestion results from muscular contraction of the alimentary canal, microbial fermentation, and action of digestive enzymes secreted in digestive juices. In monogastric animals like rats and mice, microbial activity in the large intestine is low; these animals mainly process food compounds by means of the digestive enzymes and acids.

In suckling rats or mice, the action of the digestive system and the secretion of enzymes are restricted to hydrolysis of the components of maternal milk. The serous glands of the tongue produce a lingual lipase which is important for the digestion of milk triglycerides.[9] In contrast to the serous glands, the salivary glands of neonatal rats and mice are functionally immature, and amylase activity is negligible during the first 2 postnatal weeks. In the pancreas, amylase activity does not begin to increase until 2 weeks after birth.[10] The gastric secretion of HCl, pepsinogen, and pepsin is minimal, thus allowing intact protein to pass into the small intestine, where it is absorbed as intact macromolecules by the process of pinocytosis.[9] Weaning of rats and mice, which normally begins at 17 days of age and is completed by day 26, constitutes a significant change in dietary composition from milk, with a high content of lipids (9%) and a low content of carbohydrate (4%) to a diet low in fat and high in carbohydrate. The change in dietary composition necessitates changes in digestive function. The intestinal hydrolases (maltase, sucrase, isomaltase, trehalase) that are involved in digestion of carbohydrate from solid food cannot be detected in the intestines of rats during the first 2 postnatal weeks, but their activities rise rapidly later.[11] The activities of amylase, chymotrypsin, trypsinogen, and lipase change little before weaning but increase dramatically at the time of weaning.[12] The gastric secretion of acids and pepsinogen rises to adult levels during the third and fourth postnatal weeks, coincident with the transition to solid food.[9] Although the enzymatic changes that occur in the gastrointestinal tract about weaning time seem to be directly related to the change of diet from milk to solid food, there is evidence that the primary cause of the enzymatic development is not a change of diet.[13] Among other regulatory factors that have been suggested, glucocorticoids, thyroxine, glucagon, gastrin, cholecystokinin, prostaglandins, and insulin play an important role as potential regulators of postnatal development of the gastrointestinal tract.[9] In older rats, the digestive capacity is stabilized and remains almost constant under normal feeding conditions.

The presence of food in the stomach may have a significant influence on the bioavailability and pharmacokinetics of certain drugs.[14] Rats are night-eaters, and usually they are fasted overnight prior to dosing with different drugs, assuming that overnight fasting will result in the postabsorptive state.[15] However, there is substantial evidence that the rate of gastric emptying, e.g., the rate of passage in the alimentary tract, depends on diet composition. In balance experiments with rats at about 140 g LW, using glass beads as a marker, it was demonstrated[16] that the highest amount of marker was recovered in faeces about 30 h after the beginning of eating, and only marginal amounts could be detected in the digestive tract at 72 h after feeding. Protein and fat levels in the diet did not affect the rate of passage. However, crude fiber strongly increased the rate of passage with the highest recovery of the marker about 20 h from feeding. It was also demonstrated that reduced level of microbial activity in the hind-gut decreased the rate of passage time by about 15 h. It is interesting to note that in these experiments, the level of microbial activity was regulated by the administration of the antibiotic Nebacitin, thereby indicating that compounds that alter microbial metabolism might affect the rate of food passage in the alimentary tract and, subsequently, the digestibility of nutrients.

## Metabolism

The starting point of metabolism is the substances produced by the digestion of food. Digested carbohydrate (DCHO), fat (DFAT), and protein (DPROT) are the main groups of nutrients involved in a variety of catabolic and anabolic processes in the body. The general relations between the intake of digested nutrients and the end products of their metabolism in the body are presented in Figure 12.1.

The soluble part of DCHO is mainly absorbed as glucose, which can be stored as glycogen. It can be oxidized and used as a source of energy and utilized in the process of *de novo* lipogenesis. The insoluble carbohydrates (fibers) are fermented and transformed into short-chain fatty acids (SCFA) which finally are sources of energy and precursors in the synthesis of glucose and body fat. Although SCFA are a significant source of energy for herbivorous animals (in the guinea pig and the rabbit, approximately 30% of the energy is supplied from SCFA metabolism), the microbial capacity for SCFA production in monogastrics such as the rat and the mouse is limited. The free fatty acids (FFA) and triglycerides (TG) from DFAT are transformed into body fat, and they can be oxidized, becoming an efficient energy source. In case of inadequate energy supply from a diet, body fat can be mobilized as an additional energy source. Amino acids from DPROT are synthesized and retained in the body or milk and partly deaminated and oxidized with concomitant transfer of their carbon skeletons into gluconeogenesis and ketogenesis conversion of ammonia into urea.

## Carbohydrates

Carbohydrates (CHO) constitute the largest proportion of food consumed by laboratory animals, except carnivores. They are the most important components of plants, constituting up to 75% of the dry matter present in feeds of plant origin. Dietary CHO, consisting of $\alpha$-monosaccharide units (soluble CHO), are readily digested by endogenous enzymes and constitute the major energy source for laboratory animals. A number of CHO can be used by the rat, and as reviewed by the National Research Council,[17] glucose, sucrose, maltose, fructose, and starch support similar levels of performance. However, contents of lactose or galactose in the diet that are too high may cause diarrhea and poor performance.

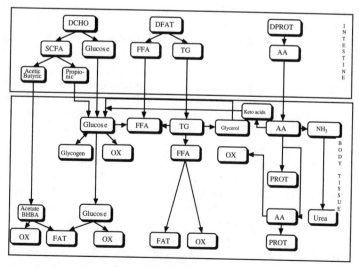

**Figure 12.1**   Nutrient partition in the body; digested carbohydrate (DCHO), fat (DFAT), and protein (DPROT); short-chain fatty acids (SCFA), free fatty acids (FFA), triacylglycerols (TG), amino acids (AA), oxidized substrates (OX), and $\beta$-hydroxybutyric acid (BHBA).

Diets for rats and mice are mainly based on starch, which is a relatively inexpensive energy source, yielding 17.6 kJ/g. There is a direct relation between CHO intake and the level of fatness of the animals, because CHO, exceeding the amount needed to meet the energy requirement, will be stored as body fat.[18,19] Although rats and mice are able to regulate food intake, depending on the energy density of the diet, an extensive supply of soluble CHO may result in obesity, especially in animals with genetic predispositions to obesity, like Zucker "fatty" rats.[20] Obesity is also more likely to occur when the diet is high in sugar because sugar is utilized more efficiently than starch.[21] All rodents exhibiting obesity or obesity-related diabetes syndrome are characterized by diminished glucose tolerance, which may result in hyperinsulinemia and inappropriate hyperglycemia,[22] and thereby in reduced longevity.

The CHO composed of ß-monosaccharide units are called insoluble CHO, collectively referred to as "dietary fiber." They resist the action of digestive enzymes but can be utilized by microorganisms in the large intestine. Plant cell wall material, which is the major source of dietary fiber, is composed chiefly of cellulose, hemicellulose, and pectins. The microbial metabolism is limited in rats and mice compared with herbivorous animals, and fiber digestion is almost entirely confined to microbial activity in the hind-gut.[23] A small amount of dietary fiber (2 to 5% of the diet) should be included in the diets. This can be processed by the microorganisms. Fiber is important because of its water-holding capacity and its influence on the peristaltic mobility of the intestines by which food components are driven through the alimentary tract. Furthermore, dietary fiber components influence the composition and function of the intestinal microflora and activate the microbial synthesis of several vitamins. It has been demonstrated that pectin stimulates the microbial synthesis of thiamin, riboflavin, and niacin.[24] Pectin can absorb various antimetabolites and thus reduce their degree of absorption.[25] There are also indications that fiber increases peripheral sensitivity to insulin, perhaps by increasing the number of insulin receptors.[26] On the other hand, an excess of dietary fiber has a negative influence on nutrient and energy digestibility because it increases the rate of passage of food components in the digestive tract and subsequently reduces the time of action of digestive enzymes. Total glucose, lipid, and protein absorption is decreased when high levels of dietary fiber are consumed.[27,28] Furthermore, an association between dietary fiber and periodontitis and oronasal fistulation in rats has been reported in several investigations,[29] and is presumed to relate to the presence of long sharp fibers of oats or barley.[30]

## Fat

Dietary fat is required as a source of essential fatty acids (EFA), for the absorption of the fat-soluble vitamins, and to enhance the palatability of the food. Lipids are an excellent source of energy providing 2.5 times more energy (39.8 kJ/g) than carbohydrates (17.6 kJ/g) and protein (18.4 kJ/g). However, if a diet contains adequate CHO and protein, fat is not used as a source of energy, but is stored as body lipids.[31] Both the amount and composition of dietary fat are important in laboratory animal nutrition. A high fat level in the diet increases cholesterol synthesis. Especially high inputs of saturated fat furnish acetyl-CoA in excess of that required for energy production and body fat synthesis, and the excess acetyl-CoA is used for cholesterol formation. The fatty acid composition of the dietary fat affects antioxidant mechanisms in the colon mucosa, presumably because the composition of cell membranes reflects the fatty acid composition of the diet.[32] There is evidence that high-fat diets elevate the toxic effects of nuclear-damaging agents and carcinogens.[33] It has been demonstrated that increase in fat level may alter the acute genotoxic effects of carcinogens, a phenomenon associated with the initiation of colon cancer.[34] Furthermore, mammary tumor incidence is related to the fatty acid composition of the diet, with a greater incidence of tumors in rats fed diets containing polyunsaturated fatty acids (PUFA) when compared to diets with saturated fat. In the rat, increasing levels of linoleic acid were correlated with increasing chemically induced mammary tumor incidence up to maximum at 4.5% of dietary linoleic acid.[35,36] The same effect was observed in the mouse, but at a higher level of linoleic acid (8.4%).[37] Feeding high levels of fat, particularly PUFA, may depress immune responsiveness,[38] which may lead to an increased susceptibility to infections. One should, however, realize that the minimum recommended level for linoleic acid in rats and mice is only about 0.6%, so the chosen experimental concentrations were high. Furthermore, the effect of linoleic acid on mammary tumor development depends on the animal model used.[39]

The early works of Burr and Burr[40,41] first established that the rat does not thrive on diets rigidly devoid of fat but develops a number of deficiency syndromes. The linoleic acid family of PUFA was shown to reverse effects of fat-free diets. The linoleic, linolenic, and arachidonic acids are usually referred to as essential fatty acids (EFA). Mammals lack the enzymes that introduce double bonds at carbon atoms beyond C-6 in the fatty acid chain. This makes the double bond at the 12th carbon atom of linoleic acid "essential." After absorption, linoleic acid can be oxidized, accumulated in the adipose tissue, and converted to PUFA and incorporated into structural lipids.[42] The list of symptoms ascribed to EFA deficiency ranges from classical signs such as reduced growth rate, dermal lesions, increased water permeability of the skin, increased susceptibility to bacteria, decreased prostaglandin synthesis and reproductive failure, reduced myocardial contractility, abnormal thrombocyte aggregation, swelling of liver mitochondria, and increased heat production.

Dietary requirements of EFA are usually stated in terms of linolate. An amount equivalent to 1 to 1.5% of the metabolizable energy (ME) of the diet has been found adequate for most monogastric animals. However, studies on growing pigs indicate an even lower requirement of 0.26% of the ME.[43] This level is likely to be present in all natural food compounds used for laboratory animals, but not in highly refined or purified diets, which must be fortified with EFA-rich sources like soybean oil. Also, diets containing a high level of saturated fatty acids (>5%) may require a greater supply of EFA; hence, EFA enhance the utilization of saturated fatty acids.[44]

## Protein

The nutritive effects of a protein depend on the amino acids, which are released from the protein by digestive processes. For nutritional purposes, amino acids are classified into two groups: nonessential (NEAA) and essential (EAA) amino acids. The NEAA are not necessary as dietary components, because they can be synthesized in the body via intermediates of carbohydrate metabolism or by transformation of some EAA into certain NEAA. EAA, however, cannot be synthesized in the body, at least not at a rate adequate to meet physiological requirements, and they must therefore be supplied with the diet (Table 12.2).

Protein synthesis can only take place when all the amino acids required to form a certain protein are present together, thus, a relative inadequacy of one amino acid impairs utilization of the rest. The amino acid in lowest concentration in relation to the requirement will therefore determine the rate at which protein can be synthesized in the body. Subsequently, amino acids present in excess of the requirement for protein synthesis will not be used for synthesis, but their nitrogen-free components will be oxidized or used in gluconeogenesis and ketogenesis, while the nitrogenous component (ammonia) is converted by the liver to urea and excreted by the kidneys with concomitant energy loss. It is evident that EAA must be present in the diet in correct quantities and proportions in order to be synthesized into animal protein. However, the animal must also receive a sufficient amount of NEAA as nitrogen source for protein synthesis. If an inadequate amount of NEAA is absorbed (and produced from body protein turnover), they will be resynthesised from dietary EAA.

**Table 12.2 Essential and Nonessential Amino Acids for Growing Rats and Mice**

| Essential | Nonessential |
|---|---|
| Arginine | Alanine |
| Histidine | Asparatic acid |
| Isoleucine | Cystine |
| Leucine | Glutamic acid |
| Lysine | Glycine |
| Methionine | Hydroxyproline |
| Phenylalanine | Proline |
| Threonine | Serine |
| Tryptophan | Tyrosine |
| Valine | |

Inadequate amino acid supply is the most common of all nutrient deficiencies. Signs of protein deficiency include reduced protein concentration in the blood, reduced protein synthesis rate in the tissues and synthesis of certain enzymes and hormones, decreased food intake, reduced growth rate, and infertility. On the other hand, an excess supply of one or more amino acids may cause amino acid imbalance and, consequently, decreased protein utilization. The classical experiments with growing rats fed a rice diet with lysine and threonine as limiting amino acids demonstrated that if the lysine content of the diet was held constant and the threonine content was increased stepwise, a point was reached at which the growth of rats fed on a threonine-supplemented diet was retarded unless the lysine content of the diet was also increased.[45] The same phenomenon was seen in the reverse situation, with threonine being held constant and lysine increased. Furthermore, the sites in the brain that regulate food intake are sensitive to an alteration in the proportion of amino acids in the blood plasma, and an imbalance of amino acids may cause a reduced food intake.[46] There is also evidence that a surplus of arginine and histidine may depress protein utilization, while the ingestion of a large amount of methionine or tyrosine (20 to 50 g/kg food) is followed by serious metabolic and histopathological changes in addition to depressed food intake and retarded growth of the rat.[47] Excess methionine inhibits ATP synthesis, causing irregularities in energy metabolism, and excess tyrosine causes a specific toxic syndrome with histopathological changes in the skin, pancreas, and liver, and severe eye lesions. For some amino acids, negative effects of an excess supply may only be prevented by the addition of other amino acids that are structurally similar. Growth depression in rats caused by surplus isoleucine and valine can be prevented by addition of their "antagonistic" amino acid leucine. It is also interesting to note that a high-quality protein is much less affected by an excess of a single amino acid than is a protein source of poorer quality, as demonstrated for egg protein versus barley and potato protein.[48]

High dietary protein levels should be avoided in case of kidney disease, as the relatively high urea excretion will negatively influence kidney function. Dietary protein can also interfere with tumor studies. Results from experiments with rats indicate that supplementation of methionine to soybean protein isolates increased mammary tumor progression in rat.[49] It was also suggested that a decreased level of dietary methionine may decrease tumor cell proliferation.

When discussing the role of dietary protein in laboratory animal nutrition, it has to be emphasized that the utilization of protein for different life processes is dependent on energy supply. The relationship between protein balance and protein and energy intake has been recognized for years[50,51,52] (Figure 12.2).

For a low protein diet, an increase of energy intake to the point A (Figure 12.2) will increase protein retention, while the extended energy supply from A to B will not have any effect on protein retention. For a high protein diet, the pattern is the same, however, because of higher protein supply, protein retention is elevated. For such a diet, the increase from A to B will stimulate protein retention until the limit is reached at point B. The presented relationship between energy and protein is, of course, a simplification, and other factors should also be considered, such as the extent to which energy is supplied from carbohydrate or fat,[53,54] the supply of other nutrients,[55] amino acid profile in the diet,[54] and endocrinological regulation of anabolic and catabolic processes.

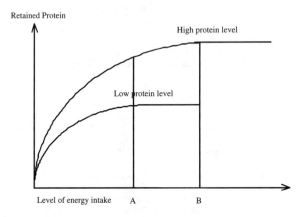

**Figure 12.2**   Relation between protein retention and energy intake.

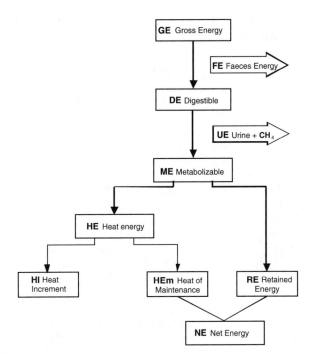

**Figure 12.3** The partition and losses of food energy in the body.

## Energy

All functions of the body require energy, which is supplied from carbohydrate, fat, and protein. The results of inadequate energy supply are obvious, but an energy intake that is too high can be harmful as well. Excess energy intake produces obesity, several obesity-associated diseases, and reproductive failure, and reduces longevity.[56–61] There is evidence that rapid growth rates and obesity are associated with an increased occurrence of spontaneous and induced tumors in laboratory animals.[62,63]

The total energy, gross energy (GE), available in food can be determined by complete combustion in the calorimetric bomb. As shown in Figure 12.3 the animals cannot use all GE, as some energy is lost in feces, urine, methane, and hydrogen. The remaining energy, called metabolizable energy (ME), is the energy available in the animal for maintenance, growth, reproduction, lactation, and work (Figure 12.3).

The ME value of a food varies according to the species of animal to which it is given. For rats and mice, ME values are almost similar, as these animals digest foods to much the same extent, and losses in the form of methane are negligible. However, for herbivorous animals like guinea pigs and rabbits, the same foods are digested to a lower extent, and due to the fermentation processes, more methane is lost, consequently reducing the ME value. Caused by energetic expenses of the digestion and absorption of nutrients and due to energetic inefficiency of the reactions by which absorbed nutrients are metabolized, part of the ME is lost as heat increment. The deduction of the heat increment of a food (dietary-induced thermogenesis) from its ME gives the net energy (NE) value of the food. The NE is the remaining part of food energy used for different life processes, and is, therefore, the unique measure of the energetic value of the food.

An energy system based on NE is the most accurate system to evaluate energetic values of foods, but it is difficult to access, and from a practical point of view, it is advisable to use ME for the evaluation of foods for laboratory animals. Relatively few types of feedstuffs are used for laboratory animals, primarily grain and oilseed cakes, where the utilization of ME does not vary much. Extensive studies on different animals have shown that ME can be calculated with reasonable certainty on the basis of the digested quantities of nutrients as follows:[64]

|  |  | Rats | Rabbits |  |
|---|---|---|---|---|
| ME, kJ = |  | 18.4 | 18.2 | × digest. crude protein, g |
|  | + | 39.4 | 39.5 | × digest. fat, g |
|  | + | 15.2 | 18.8 | × digest. crude fiber, g |
|  | + | 17.5 | 17.1 | × digest. nitrogen-free extract, g |

The calculation of ME will always be subject to some uncertainty concerning the energy constants for the respective nutrients. In practice, the same accuracy can be obtained by using only the content of organic matter (OM) as a basis. With growing rabbits, Thorbek and Chwalibog[65] found the following relation between ME and digestible OM as ME, kJ = 18.7 x digest. OM, g. The factor 18.7 corresponds to the factors 17.0, 18.4, and 20.6 found for calves, pigs, and poultry, respectively. Because the digestibility of organic matter generally is about 65%, ME in food for rabbits and guinea pigs could be calculated as ME, kJ = OM, g x 0.65 x 18.7. For rats, mice, and hamsters, a factor 19 kJ/g digestible OM can be applied. Assuming that the digestibility of organic matter is 85%, the following calculation could be used: ME, kJ = OM, g x 0.85 x 19.[66] Usually, the ME value of a certain diet is given by the manufacturer, but it is often unclear whether this is an analyzed value or calculated by using fixed, theoretical constants.

If the energy value of foods is expressed in ME units, the requirements must be expressed in the same units, thus we need to know the animal's maintenance requirement for metabolizable energy (MEm) as well as the efficiency of ME utilization for growth, pregnancy, and lactation. Theoretically, the maintenance requirement is defined as the amount of energy necessary to balance anabolism and catabolism, giving an energy retention around zero.[67] For laboratory animals, there are few empirical results about MEm.[17,68] MEm values for rats and mice are usually suggested to be around 100 kcal (420 kJ) per LW,kg$^{0.75}$.[69,70] Although this constant value has been used as the measure of MEm in laboratory animals, it is debatable whether there is any constant value of MEm independent of nutritional, genetic, and environmental conditions.[67,71] There is substantial information about ME utilization for growth, pregnancy, and lactation in farm animals,[69] but surprisingly, these values are not used for laboratory animals. The standards for energy requirements of laboratory animals are based on the measurements of the voluntary food intake at different physiological states[17] or on equations produced by Clarke et al.[7] According to these equations, requirements of ME for maintenance of rats are 450 kJ/LW,kg$^{0.75}$ and requirements for the other physiological states (inclusive of maintenance) are calculated as growth = 1200 kJ/LW,kg$^{0.75}$, pregnancy = 600 kJ/LW, kg$^{0.75}$, and lactation = 1300 kJ/LW, kg$^{0.75}$.

## Requirements

The primary element in the evaluation of nutritional requirements is knowledge of the ability of animals to transform nutrients and energy obtained from a diet into body components and products. In spite of much information available on different aspects of laboratory animal nutrition, there are only few methodical investigations concerning nutrient and energy balances during growth,[72] pregnancy,[73] and lactation.[69]

A series of methodical studies concerning protein and energy metabolism in growing rats during 5 months after weaning was carried out by Thorbek et al.[4] The results of these studies furnish valuable data concerning nutrient and energy utilization and accretion during growth, providing the necessary basis for calculation of the requirements. The experiments were performed with male albino rats fed *ad libitum* on nonpurified commercial diets.

During the growth period from 5 to 18 weeks of age, 56% of consumed nitrogen was excreted in urine and 18% in feces, while 26% was retained in the body. During the same period, energy excreted in urine constituted only 2% of dietary energy, and 18% of the energy intake was excreted in feces, but the major part, 53%, was measured as the total heat energy, and subsequently, the remaining 27% was retained in the body protein and fat. These relations clearly indicate that about 20 to 30% of the consumed protein or energy is finally retained in the growing laboratory rat kept under ordinary dietary and housing conditions.

In the same experiment, protein retention (RP) followed the pattern for growing farm animals,[74] increasing RP to about 1.7 g/day per animal at the age between 7 and 8 weeks and then decreasing to a constant plateau of 0.5 g/day at 16 to 18 weeks of age. Fat retention (RF) gradually increased above 2 g/day until 10 to 14 weeks of age, after which a slow decrease was observed. The amount of retained energy (RE), which is a sum of energy retained in protein and fat, reached the highest level of 125 kJ/d at the age of 8 to 10 weeks, then it gradually decreased to about 80 kJ/d because of decreasing protein and fat retention.

Based on the presented values, the requirements for protein and energy in the growing rat are calculated by the factorial approach as demonstrated in Tables 12.3 and 12.4 The calculations are performed for weeks 5, 8 (around maximum RP), 10 (around maximum RF), and 18 (mature animals). Protein requirement is estimated, assuming an average protein digestibility of 80%[52] and that endogenous nitrogen excretion in urine and feces is related to the live weight of animals.[75]

Metabolizable energy required for maintenance (MEm) is calculated from the following function: MEm, kJ/d = $32.5 + 251 \times LW, kg^{0.75}$, while the amount of ME required for energy retention in protein (RPE) and in fat (RFE), i.e., for growth, is calculated with the efficiencies of ME utilization as 0.50 and 0.77 for RPE and RFE, respectively[76] (Tables 12.3 and 12.4).

The pattern of protein retention showed that the highest requirement for digestible protein was at 8 weeks of age, with relatively low fat retention at 5 weeks of age. The total requirement for ME was markedly lower than in the later part of the growth period. The presented values demonstrate the pattern of requirement during growth, but for practical diet formulation, it is interesting to note the changes in the required concentration of DP per 100 kJ ME. It decreased linearly from 1.2 g/100 kJ at 5 weeks to 0.7 g/100 kJ at 18 weeks of age, thus indicating the necessity to provide diets with different concentrations of protein during the growth period. As this would give many practical problems, usually only one dietary composition for the entire growth period is used for laboratory animals.

**Table 12.3  Requirement for Digestible Protein for Growing Rats in Relation to Live Weight (LW) and Retained Protein (RP)**

| Age (Week) | LW (g) | RP (g/d) | Requirement (g/day) | | |
|---|---|---|---|---|---|
| | | | Maintenance | Growth | Total |
| 5 | 100 | 1.3 | 0.4 | 1.7 | 2.1 |
| 8 | 220 | 1.7 | 0.8 | 2.3 | 3.1 |
| 10 | 290 | 1.3 | 1.0 | 1.7 | 2.7 |
| 18 | 400 | 0.5 | 1.3 | 0.7 | 2.0 |

*Note:*  Mean values based on the data from Thorbek, G., Chwalibog, A., Eggum, B.O., and Christensen, K., Studies on growth, nitrogen, and energy metabolism in rats, *Arch. Tierernähr.*, 32, 827, 1982.

**Table 12.4  Requirement for Metabolizable Energy for Growing Rats in Relation to Live Weight (LW), Retained Energy in Protein (RPE), and Retained Energy in Fat (RFE)**

| Age (Week) | LW (g) | RPE (kJ/day) | RFE (kJ/day) | Requirement (kJ/day) | | |
|---|---|---|---|---|---|---|
| | | | | Maintenace | Growth | Total |
| 5 | 100 | 30 | 20 | 80 | 90 | 170 |
| 8 | 220 | 40 | 80 | 110 | 180 | 290 |
| 10 | 290 | 30 | 95 | 130 | 180 | 310 |
| 18 | 400 | 12 | 70 | 160 | 120 | 280 |

*Note:*  Mean values based on the data from Thorbek, G., Chwalibog, A., Eggum, B.O., and Christensen, K., Studies on growth, nitrogen, and energy metabolism in rats, *Arch. Tierernähr.*, 32, 827, 1982.

## Allowances

The knowledge of the amount of nutrients and energy required by animals is necessary to establish nutritional allowances for laboratory animals. There is a clear distinction between the term requirement and allowance. The requirement is a statement of what animals on average require for a particular function, but the allowance is greater than this amount by a safety margin designed principally to allow for variations in requirements between individual animals and to account for possible variations of nutrient content in the same foods or diets. As a result of the large number of factors that can influence the dietary requirements, manufacturers add nutrients that are in excess of the estimated requirements.[77] However, there are inconsistencies in the magnitude of these safety margins, and unfortunately, this distinction between requirements and allowance terms is not strictly defined for laboratory animals. In many publications, it is not clear whether the "requirements" or so-called nutritional standards refer to requirements or to allowances. This is one of the main reasons for many of the discrepancies between different recommendations. Keeping in mind the possible confusion caused by the use of inconsequent terminology, the tables in this section present only a general outline of nutritional allowances recommended for common laboratory animals. For detailed description of nutritional standards, the publications by Coates,[78] Clarke et al.,[7] The National Research Council,[17] and Eggum and Beames[68] are suggested.

Nutrient and energy allowances for laboratory animals are rarely expressed in terms of quantity per day but by means of the concentrations of nutrients and energy in the diet. In practice, most laboratories use only two different diets for each animal species; one during growth and maintenance (adult, nonproducing animals), and the other for pregnant and lactating animals. Maintenance diets can often be used during growth as well, as the safety margins are such that no deficiency in nutrients will arise under *ad libitum* conditions.

Knowledge of the requirements for the individual amino acids for different laboratory animals and different life processes is limited. The most abundant information exists for laboratory rats, although, depending on the source of information, there is a considerably broad range of recommended values, as demonstrated in Table 12.5.

**Table 12.5  The Range of Recommended Levels for Essential Amino Acids for Laboratory Rats as g/100 g Protein[a], as g/kg Diet[b], as mg/kg[o.75] per Day[c], and as g/MJ, ME per Day[d]**

| Amino Acid | g/100g Protein | g/kg Diet | mg/kg[o.75] | g/MJ |
|---|---|---|---|---|
| Arginine | 5.0–6.0 | 4.3 | 0–10 | 0.39 |
| Histidine | 2.5–3.0 | 2.8 | 0–17 | 0.19 |
| Isoleucine | 5.0–6.0 | 6.2 | 30–49 | 0.32 |
| Leucine | 7.5–8.0 | 10.7 | 16–64 | 0.49 |
| Lysine | 6.0–9.0 | 9.2 | 10–33 | 0.45 |
| Methionine+Cystine | 5.0–10.0 | 9.8 | 20–43 | 0.30 |
| Phenylalanine+Tyrosine | 6.0–10.0 | 10.2 | 16–52 | 0.52 |
| Threonine | 4.0–4.5 | 6.2 | 20–54 | 0.32 |
| Tryptophan | 1.2–1.5 | 2.0 | 5–10 | 0.10 |
| Valine | 5.0–6.0 | 7.4 | 18–47 | 0.39 |

[a] From Chwalibog, A., Jakobsen, K., Tauson, A-H., and Thorbek, G., Heat production and substrate oxidation in rats fed at maintenance level and during fasting, *Comp. Biochem. Phys.*, 121, 423, 1998; and Solleveld, H.A., McAnulty, P., Ford, J., Peters, P.W.J., and Tesh, J., Breeding, housing, and care for laboratory animals, in *Laboratory Animals*, Ruitenberg, E.J. and Peters, P.W.J., Eds., Elsevier Science Publishers B.V., Amsterdam, 1986, chap. 1.

[b] From National Research Council, *Nutrient Requirements of Laboratory Animals*, Fourth ed., National Academy of Sciences, Washington, DC, 1995.

[c] From Owens, F.N. and Pettigrew, J.E., Subdividing amino acid requirements into portions for maintenance and growth, in *Absorption and Utilization of Amino Acids*, Friedman, M., Ed., CRC Press, Inc., Boca Raton, FL, 1989, chap. 2; and Shin, I-S., Subdividing amino acid requirements for maintenance from requirements for growth, Thesis, Oklahoma State University, 1990.

[d] From Eggum, B.O. and Beames, R.M., Use of laboratory animals as models for studies on nutrition of domestic animals, in *Laboratory Animals*, Ruitenberg, E.J. and Peters, P.W.J., Eds., Elsevier Science Publishers B.V., Amsterdam, 1986, chap. 9.

Concerning protein supply in general, adult mice, rats, and hamsters require 70 to 120 g crude protein per kg diet with 90% of dry matter. Depending on protein digestibility and biological value, this is equivalent, in the natural diets, to a supply of about 50 to 70 g digestible protein per kg food. For the other body functions, a supply of 200 to 240 g crude protein per kg food may be recommended, corresponding to 120 to 140 g digestible protein for maintenance and productive functions.

The level of crude fat in the diet for adult animals is recommended at about 20 g/kg diet, but at about 50 g/kg for productive animals. For the mouse, rat, and hamster, the crude fiber in the diet should preferably not exceed 80 g/kg. On the other hand, in the guinea pig and rabbit, considerable amounts of cellulose, hemicellulose, and pectin are broken down in the large intestine (50 to 75% of the digestion capacity of ruminants). For these animals, a content of 100 to 200 g crude fiber/kg is recommended.

As mentioned before, the nutritional standards for laboratory animals are not comprehensive. A lot of work is required to establish nutrient and energy requirements and, consequently, nutritional allowances for different species and life processes, as well as for animals kept under different conditions and used for different research purposes. The presented values of nutritional allowances are therefore to be used as general recommendations that should be revised when new scientific data become available. In order to improve standardization, the use of the National Research Council Guidelines[17] for defining the minimum requirements per species is recommended.

## TYPES OF DIETS

### Natural-Ingredient Diets

The pelleted diets that are usually the standard diets used in most laboratory animal facilities are in most instances made from natural ingredients. Manufacturers can produce diets according to variable or fixed formulas. In a variable formula, the final product levels are kept as constant as possible, i.e., aimed at keeping the level of nutrients in the end product as constant as possible. This means that the amount of individual ingredients is adjusted for variation in nutrient levels of the raw material.[81] In a fixed formula, the recipe does not change for a particular type of diet, i.e., the same proportions of raw material ingredients are used each time a batch is produced.[81] As natural ingredients can differ, depending on the weather conditions or the soil where these have been grown, also the commercially available pelleted natural-ingredient diets are subject to variation. This has to be taken into consideration when one uses so-called "standard" diets.

Natural-ingredient diets can also be divided in open- and closed-formula diets.[82] In open-formula diets, all dietary ingredients and their concentrations are reported and should not vary from batch to batch. In closed-formula diets, the dietary ingredients used are reported, however, the concentration of each dietary ingredient is not stated by the manufacturing company. The concentration of dietary ingredients may vary from batch to batch or with availability of ingredients.[82]

Natural-ingredient diets can be offered in several physical forms, e.g., meal form, as extruded or expanded pellets. The extruded form is the conventional pelleted form. The expanded form is produced by forcing a wet mixture of the basal ingredients through a die at high pressure, accompanied by injection of superheated steam. This way the pellets become more voluminous. It is claimed that expanded diets reduce food intake (higher availability of nutrients), increase food conversion, reduce microbial counts, and reduce wastage as compared to the conventional extruded pellets.

Manufacturers provide information on the dietary products in their catalogs and on their Web sites. The amount of information provided differs, but generally, information is provided on which ingredients have been used and what nutrient and contaminant concentrations are to be expected in these diets. Also, microbiological quality is often presented. The way the catalog values for nutrient concentrations have been established is not standardized. The most common way is to analyze a few production batches of the same diet and then present the average value in the catalog, with no information on the batch-to-batch variation. It is a well-known fact that dietary batches produced by different manufacturers can differ (between-brand variation), but also batches produced at the same production unit can differ on the basis of variation in natural ingredients (between-batch variation).[83] In case a production firm produces a similar type of diet at different production sites, variation must be expected as well, as natural ingredients used at the different sites are usually not similar. In case the natural ingredients used are exactly the same, a variation in storage and production methods may still result in variation of the final product.

## Between-Brand Variation

As natural ingredients are, by definition, not constant in their composition, and manufacturing companies use different natural ingredients in varying amounts, the so-called "standard" diets for, e.g., growth and maintenance, show variation. In Table 12.6 catalog values for a number of nutrients from minipig maintenance diets from nine different manufacturing companies are given, which illustrates the variation in composition between brands.[84] A varying dietary composition will result in different experimental results. By feeding female Wistar rats ten brands of standard maintenance rodent diets, varying experimental results were obtained (Table 12.7).[85] This demonstrates that historical control groups can never be used reliably and that there is a need for a concurrent control group at all times.

**Table 12.6  Between-Brand Variation: Variation in Catalog Values of Selected Nutrients of Nine Brands of Minipig Maintenance Diets**

| Nutrient | Lowest Value | Highest Value |
|---|---|---|
| Metabolizable energy (MJ/kg) | 9.5 | 11.2 |
| Crude protein (%) | 13.2 | 15.4 |
| Lysine (%) | 0.57 | 0.94 |
| Isoleucine (%) | 0.40 | 0.74 |
| Cu (mg/kg) | 10.0 | 38.0 |
| Mn (mg/kg) | 50.0 | 160.0 |
| Fe (mg/kg) | 130.0 | 770.0 |
| Se (ug/kg) | 150.0 | 520.0 |
| Vitamin B (mg/kg) | 3.0 | 18.0 |
| Vitamin C (mg/kg) | 0.0 | 100.0 |

*Source:* From Ritskes-Hoitinga, J. and Bollen, P., The Göttingen minipig, a refined animal model?, *Der Tierschutzbeauftragte*, 2, 87, 1999.

**Table 12.7  Between-Brand Variation: Variation in Seven Parameters After Feeding Ten Different Brands of Rodent Maintenance Diets *ad libitum* to Female Outbred Wistar Rats for a 4-Week Period (Six Animals per Group)**

| Parameter | Lowest Mean Group Value | Highest Mean Group Value |
|---|---|---|
| Body weight (g) | 137.0 | 163.0 |
| Food intake (g/day) | 11.0 | 15.6 |
| Water intake (mL/day) | 18.2 | 24.8 |
| Kidney calcification score (0–3) | 0.0 | 1.3 |
| Urine production (mL/day) | 8.6 | 17.0 |
| Urine Ph | 6.3 | 8.4 |
| Caecal weight (g/100 g BW) | 1.1 | 3.9 |

*Source:* From Ritskes-Hoitinga, J., Mathot, J.N.J.J., Danse, L.H.J.C., and Beynen, A.C., Commercial rodent diets and nephrocalcinosis in weanling female rats, *Lab. Anim.*, 25, 126, 1991.

**Table 12.8 Within-Brand Variation: Comparison of Catalog Values of a Maintenance Minipig Diet with Analyzed Values from a Number of Nutrients from Five Dietary Batches from the Same Manufacturing Company**

| Parameter | Catalog Value | Lowest Value | Highest Value |
|---|---|---|---|
| Crude oil (%) | 2.4 | 2.2 | 3.4 |
| Crude protein (%) | 13.9 | 14.1 | 16.5 |
| Crude fiber (%) | 11.6 | 7.1 | 13.2 |
| Calcium (%) | 1.02 | 0.89 | 1.13 |
| Phosphorus (%) | 0.68 | 0.53 | 0.73 |
| Iron (mg/kg) | 130.0 | 130.0 | 299.0 |
| Zinc (mg/kg) | 110.0 | 109.0 | 200.0 |
| Vitamin E (mg/kg) | 53.0 | 32.0 | 216.0 |

*Source:* From Ritskes-Hoitinga, J. and Bollen, P., Nutrition of (Göttingen) minipigs: facts, assumptions and mysteries, *Pharm. and Tox.*, 80, 5, 1997.

## Within-Brand Variation

If one sticks to the same type of diet from the same manufacturing company, one will be inclined to think that the dietary composition will remain the same over time. However, as mentioned before,[5] natural ingredients vary in their composition. As a consequence, dietary batches produced from "the same" natural ingredients will also show variation in nutrient analyses over time. Table 12.8 illustrates the variation in five batches of minipig maintenance diets from the same manufacturing company.[86]

Generally speaking, nutrient levels in commercial diets fulfil requirements of all essential nutrients more than sufficiently, at least when diets have been transported and stored under proper conditions. If one looks at the vitamin E levels in Table 12.8, a batch level of 32 mg/kg will fulfil the needed minimum requirement. However, as vitamin E is unstable under higher environmental temperatures, inappropriate storage may quickly reduce the vitamin E level, becoming deficient at the time of feeding. In another batch, the vitamin E level was as high as 216 mg/kg. As vitamin E is an antioxidant, such a high level can theoretically provide protection if one works with, e.g., oxidative stress models. Due to this batch-to-batch variation, it is advisable to buy diets with a batch analysis certificate at all times. Although this will increase costs, one gets at least basic knowledge on nutrient content, microbiology, and contaminants of a particular batch. This will also give the opportunity to reject a batch before the start of a particular experiment or to exclude the dietary composition as a possible interfering factor, when unexpected results have been obtained. In the case of studies performed under Good Laboratory Practice Guidelines, it is mandatory to buy diets with a batch-analysis certificate, in order to document all necessary details of the experiment performed. Table 12.9 gives an example of a typical batch analysis certificate.

## Purified Diets

### Standardization

Purified or semipurified diets (also named synthetic or semisynthetic diets) are defined as being formulated with a combination of natural ingredients, pure chemicals, and ingredients of varying degrees of refinement.[83] This results in diets having a much more standardized composition than natural-ingredient diets, consequently leading to (more) reproducible results and thereby more responsible use of laboratory animals. By using purified diets in studies evaluating the influence of dietary P level on the induction of nephrocalcinosis (kidney calcification) in female Wistar rats, a 0.2% P level prevented the occurrence of kidney calcification in each study, wheres a 0.5 to 0.6% P level would induce a severe degree of nephrocalcinosis.[87] By making specially prepared purified diets, it is possible to approximate the desired nutrient levels closely and reproducibly. Table 12.10 shows the between-batch variation for eight batches of purified diets used in rat experiments in which the dietary etiology of nephrocalcinosis was studied. The levels obtained are close to the desired concentrations, and between-batch variation is relatively small. These highly standardized and reproducible levels can usually only be obtainable when using purified diets.

**Table 12.9 An Example of a Batch Analysis Certificate**

SPECIAL QUALITY CONTROL OF
SMALL ANIMAL DIETS

# SDS

CERTIFICATE OF ANALYSIS      Special Diets Services

PRODUCT: STANRAB SQC
BATCH NO: 4630      PREMIX BATCH NO: P764
DATE OF MANUFACTURE: 08-DEC-89

| Nutrient | Found Analysis | | Contaminant | Found Analysis | | Limit of Detection |
|---|---|---|---|---|---|---|
| Moisture | 6.5 | % | Fluoride | 15 | mg/kg | 1.0 mg/kg |
| Crude Fat | 2.8 | % | Nitrate as NaNO3 | 1413 | mg/kg | 1.0 mg/kg |
| Crude Protein | 18.4 | % | Nitrite as NaNO2 | 2.5 | mg/kg | 1.0 mg/kg |
| Crude Fibre | 10.1 | % | Lead | 0.85 | mg/kg | 0.25 mg/kg |
| Ash | 6.6 | % | Arsenic | Non Detected | mg/kg | 0.2 mg/kg |
| Calcium | 0.84 | % | Cadmium | 0.11 | mg/kg | 0.05 mg/kg |
| Phosphorus | 0.69 | % | Mercury | Non Detected | mg/kg | 0.01 mg/kg |
| Sodium | 0.24 | % | Selenium | 0.05 | mg/kg | 0.05 mg/kg |
| Chloride | 0.51 | % | | | | |
| Potassium | 1.20 | % | | | | |
| Magnesium | 0.22 | % | Total Aflatoxins | Non Detected | mcg/kg | 1 mcg/kg each of B1,B2,G1,G2 |
| Iron | 216 | mg/kg | | | | |
| Copper | 11 | mg/kg | | | | |
| Manganese | 75 | mg/kg | Total P.C.B | Non Detected | mcg/kg | 10.0 mcg/kg |
| Zinc | 68 | mg/kg | Total D.D.T | Non Detected | mcg/kg | 1.0 mcg/kg |
| | | | Dieldrin | Non Detected | mcg/kg | 1.0 mcg/kg |
| | | | Lindane | Non Detected | mcg/kg | 1.0 mcg/kg |
| | | | Heptachlor | 2 | mcg/kg | 1.0 mcg/kg |
| | | | Malathion | Non Detected | mcg/kg | 20.0 mcg/kg |
| Vitamin A | 5.2 | iu/g | Total Viable Organisms x 1000 | 44.75 | per grm | 1000/g |
| Vitamin E | 84 | mg/kg | | | | |
| Vitamin C | | mg/kg | Mesophilic Spores x 100 | 490.00 | per grm | 100/g |
| | | | Salmonellae Species | Non Detected | per grm | Absent in 20 grm |
| | | | Presumptive E.coli | Non Detected | per grm | Absent in 20 grm |
| | | | E.coli Type 1 | Non Detected | per grm | Absent in 20 grm |
| | | | Fungal Units | 25 | per grm | Absent in 20 grm |
| | | | Antibiotic Activity | Non Detected | | |

Signed ....R S F Field....
Dated ....9/1/90....

**Table 12.10 Between-Batch Variation in Eight Batches of Purified Diets**

| Nutrient | Targeted Concentration | Lowest Value | Highest Value |
|---|---|---|---|
| Calcium (%) | 0.50 | 0.52 | 0.57 |
| Phosphorus (%) | 0.40 | 0.40 | 0.43 |
| Magnesium (%) | 0.05 | 0.06 | 0.06 |

*Source:* From Ritskes-Hoitinga, J., Nephrocalcinosis in the Laboratory Rat, Thesis Utrecht University, The Netherlands, 1992.

## Achieving the Aimed Nutrient Levels

In case one wants to achieve (very) low levels of certain nutrients, it is often necessary to use purified diets, because due to contamination of natural ingredients, it is not possible to achieve these low concentrations. Especially in toxicity studies in rodents, it is advisable to use purified diets composed according to the American Insitute of Nutrition[88] guidelines: the AIN93 G (G = growth) or AIN 93 M (M = Maintenance) diets. This will lead to a constant dietary composition for each study. These guidelines are a "cookbook receipe" that ensures that all the nutrient requirements for rodents are fulfilled.[17] However, the vitamin B12 level is lower in the AIN-93 diet than that advised by the National Research Council,[17] and it is advised to double the vitamin B12 level in the AIN diet, in order to live up to the minimum requirements. By using these purified diets, more reproducible results will be obtained that will reduce interlaboratory variation. Hasemann, Huff, and Boorman[89] and Roe[90] clearly demonstrated the great variation there is to be found in control groups in 2-year toxicological studies (Table 12.11). The mammary tumor incidence in control groups could vary from 2 to 44%. Part of this large variation is the result of the variation in dietary composition when using natural-ingredient diets.

## Fulfilling Essential Needs — Avoiding Toxic Levels

In order to fulfill the essential needs for each species, the National Research Council documents provide the best documented scientific basis.[91] For minipigs, the scientific documentation for nutrient requirements is uncertain and insufficient, which is therefore under investigation.[84,86,92]

When reducing specific nutrient levels, one must be aware of the essentiality of these nutrients and the minimum necessary levels, in order to avoid (sub)deficiency problems, unless this is the goal of the study. One should consult the literature to make a responsible choice of how far a certain nutrient level can be reduced without causing serious health or welfare problems or premature death of the animals, as this will also compromise experimental results.

In a study by DeWille et al.[93] the influence of dietary linoleic acid on mammary tumor development in transgenic mice was studied. Three levels of dietary linoleic acid were given, 0, 1.2, and 6.7%. There was a significant reduction of mammary tumor development on the 0 level of linoleic acid, as compared to the other two dietary groups (Figure 12.4). As linoleic acid is an essential fatty acid necessary for the development of cell membranes,[17] this cannot be considered a reliable control group. In case the diet does not contain linoleic acid, general health is compromised. There were initially 25 animals in the 0 level of linoleic acid at the start of the study; however, in the results section, only data of 15 animals were presented.[93] In the case that cell membranes cannot develop, it is questionable whether tumors can arise.

What is to be said about lowering nutrient levels also applies to increasing nutrient levels. One needs to examine the literature to find the toxic levels for a certain nutrient for a certain species, because toxic levels negatively influence animal health and welfare and, thereby, experimental results. Some nutrients,

**Table 12.11 Variation in Results among Control Groups in 2-Year Toxicological Studies in Five Different Laboratories**

| Fischer 344 Rats | Results in Males | Results in Females |
|---|---|---|
| Number of studies | 41 | 42 |
| Percent survival to 2 years | 44–78 | 50–86 |
| Liver neoplastic nodules (%) | 0–12 | 0–12[a] |
| Mammary fibroadenomas (%) | 0–8 | 2–44[a] |

[a] Significant interlaboratory variation ($P < 0.05$).

*Source:* Haseman, J.K., Huff, J., and Boorman, G.A., Use of historical control data in carcinogenicity studies in rodents, *Toxicol. Path.*, 12, 126, 1984; and Roe, F.C.J., Historical histopathological control data for laboratory rodents: valuable treasure or worthless trash?, *Lab. Anim.*, 28, 148, 1994.

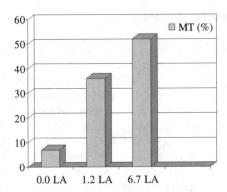

**Figure 12.4**   Relationship between dietary linoleic acid levels and mammary tumor frequency in transgenic mice. MT: mammary tumor frequency (%: number of animals with mammary tumors); LA: dietary linoleic acid level (%); mammary tumor frequency on the two highest levels of linoleic acid was significantly higher than on the 0 level linoleic acid. The frequencies on the two highest dietary linoleic acid levels were not significantly different from each other.

like selenium and methionine, have a narrow "safety margin," and relatively small increases in dietary levels will lead to toxic effects. Information on toxicity levels of nutrients is available in documents published by the National Research Council.[91,94]

## Purified Diets Versus Species

Disadvantages of purified diets are that the palatibility is often lower than that of natural-ingredient diets. This may induce the need for an acclimatization phase, in which a natural-ingredient diet becomes gradually replaced by the purified diet. It is the authors' experience that this is species dependent: rats will usually accept purified diets readily, and therefore, a 1-week acclimatization to this diet is sufficient. In mice, one can register a drop in body weight in the first week and irregular food intake over 2 to 4 weeks, when an acute change from a natural-ingredient diet to purified diet has been made. Feeding purified diets containing "unnatural" components, like, e.g., fish oil to the herbivorous rabbit, made a gradual adaptation phase of 6 months necessary, in which the purified diet gradually replaced the natural-ingredient diet.[95] When rabbits stop eating, one should give them some grassmeal on top of the diet to increase palatibility in order to keep them eating, otherwise they will readily die within 2 to 3 days due to fasting-induced hyperlipemia. When choosing dietary ingredients and animal model, one must be aware of possible species-specific characteristics. Increasing concentrations of fish oil fed to the herbivorous rabbit caused liver pathology, which coincided with a higher degree of aortic atherosclerosis (Figure 12.5), and therefore, other animal species, like, e.g., the omnivorous pigs, are considered better animal models in which to study the effects of fish oil.

## Points to Consider when Preparing Purified Diets

When using purified diets, one must make sure that the species-specific requirements are covered. However, there is a risk of creating shortages of unknown essential nutrients by using purified diets as compared to natural-ingredient diets. This is because these unknown nutrients are present as "natural contaminants" in natural-ingredient diets. Chromium and vanadium are examples of substances that are possibly essential nutrients for rodents, which were not described until the latest revision of the recommendations for rodents by the National Research Council in 1995.[17]

The selection of certain refined ingredients can be critical. An example of this is the use of a short-type cellulose fiber (Arbocel R B-00) in a purified diet, which caused intestinal obstruction and death in rats.[96] The fiber content of the diet was 10.5%. By replacing this short-type fiber by a longer type (Arbocel R B-200), the intestinal problems disappeared. When using oils like fish oil, which oxidize readily due to their high content of polyunsarated fatty acids, one needs to prepare the diet mixture fresh each day. Oils should be stored under liquid nitrogen until the day of mixing into the diet and feeding it. Antioxidants can be added to the diet as well.

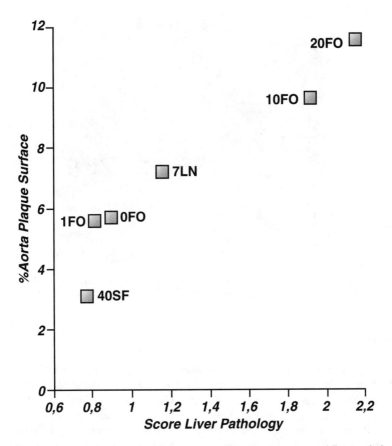

**Figure 12.5** Relationship between the amount of atherosclerosis in the aorta and liver pathology score in a rabbit study examining the influence of fish oil on atherosclerosis: correlation between group mean scores for liver pathology (X-axis) and relative aorta plaque area (Y-axis). N = 6, correlation coefficient = 0.96, P = 0.003. 20FO = 20 energy% fish oil diet; 10FO = 10 energy% fish oil diet; 1FO = 1 energy% fish oil diet; 0FO = 0 energy% fish oil diet; 7LN = 7 energy% linseed oil; 40 SF = 40 energy% sunflower seed oil diet. All diets contained 40 energy% total fat. (From Ritskes-Hoitinga, J., Verschuren, P.M., Meijer, G.W., Wiersma, A., van de Kooij, A.J., Timmer, W.G., Blonk, C.G., and Weststrate, J.A., The association of increasing dietary concentrations of fish oil with hepatotoxic effects and a higher degree of aorta atherosclerosis in the ad lib.-fed rabbit, *Food and Chem. Tox.*, 36, 663, 1998. With permission.)

## *Pelletability and Feeding Devices*

Due to the composition of the purified diet, it is often difficult to pellet these diets, producing pellets of a loose structure, leading to increased spillage. In most cases, purified diets are given in a powdered form. Special feeding devices (Figure 12.6) have been developed in order to be able to feed purified diets, also in groups of mice and rats in Macrolon cages, without having too much spillage. Feeding the purified diets in open food hoppers without restricted access will make it possible for animals to play with the food and create much wastage. The same occurs when using metabolic cages for collecting feces and urine. If the animal is relatively small in comparison to the hopper, the animal can hide inside the food hopper. As rodents like dark environments, they will use the food hopper as a sort of hiding place, as the rest of the metabolic cage is usually a brighter environment. The food sticks to the fur, thereby making food intake measurements unreliable. Also, the food that sticks to the fur will contaminate the urine and fecal collections. The least spillage is achieved when using the proper size metabolic cage for a certain species and age and size.

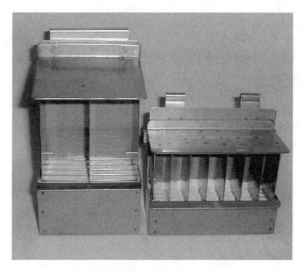

**Figure 12.6**   Examples of feeding devices suitable for feeding powdered diets to rats (left) and mice (right) housed in Macrolon cages (feeding devices kindly provided by Scanbur A/S, Denmark).

## Contaminants

Contaminants may be defined as undesirable substances (usually foreign) which when present at a sufficiently high concentration in the food may affect the animal and therefore the outcome of the experiments.[81] Possible contaminants include industrial chemicals, pesticides (e.g., DDT), plant toxins, mycotoxins (e.g., aflatoxin), heavy metals, nitrates, nitrites, bacteria, and bacterial toxins.[81] The specific parameters that are critical may vary from user to user and depend upon the objectives of the study in which the diet is to be used. The user must determine what is critical to the particular study, as it is not feasible to identify and analyze for every possible contaminant. On the basis of this information, a researcher can establish which contaminants at what maximum levels can be allowed in the diet. There are several documents stating general guidelines for maximum allowed concentrations of contaminants.[81,97] One of the guidelines that gives maximum limits, to which toxicologists all over the world are referring to, was issued by the Environmental Protection Agency.[98] Dietary production firms have often developed their own maximum limits. Besides contaminants, several nutrients can be toxic at concentrations not far above the dietary requirements. This information can be found in the NRC documents on nutrient requirements.

As different guidelines state different levels, what should be chosen as the "correct" maximum tolerated levels? First, one has to decide which guidelines are most appropriate in the experimental setting one is working in and then choose these guidelines as the institutional policy. One may also consider developing specific institutional guidelines. Second, for each group of experiments of similar type, one can do a literature search to figure out whether there are contaminants that will interfere with the specific purpose of the study. That way, specific maximum levels of specific contaminants can be established and requested for when ordering diets for specific types of experiments. It is clear that one needs a batch analysis certificate report in order to be able to judge the levels of contaminants in a specific batch of diet. In some cases, one may need to request for separate analyses of specifically specified contaminants, in case they are not included in the routine analyses. Should any contaminant be above an acceptable level, then either a different diet should be used or a different acceptance criteria adopted for the diet used in that particular study.[81] In the last case, this needs good argumentation, especially in regulatory studies, as governmental inspectors will likely investigate this argumentation.

By using purified ingredients and diets, contaminants can be virtually avoided. In natural-ingredient diets, more contaminants in higher levels are to be found.

## Quality Considerations

### Storage Conditions

Diets must be kept under suitable storage conditions (cool, dry place without access by wild rodents) at all times at the manufacturer and the user, to ensure they remain within the specifications until the recommended expiry date. Information on proper storage conditions is provided by the manufacturer. Providing diets *ad libitum*, means that diets are exposed to room temperature 24 h/day, which may cause a decrease of some nutrients. Therefore, diet present in the cages at room temperature must be discarded regularly and replaced by fresh food (e.g., one to two times per week for "regular" natural-ingredient diets — the guidelines from the manufacturer should be followed). In case of using. e.g., highly unsaturated fats like fish oil, it may be necessary to provide fresh food every day. Preservatives like butyl hydroxytoluene (BHT) may be added in order to prevent oxidation. Usually BHT is only added when highly oxidative oils are used. During transport, the diets must remain dry and protected from damage, heat, and wild rodents. No chemicals should be transported or stored with the diets.[81] When diets arrive at the user´s facility, they have to be inspected for damage or contamination and discarded if unfit for use.

### Sterilization

Diets can be sterilized by gamma-irradiation or autoclaving in order to further reduce the microbiological contamination levels. This is especially requested for barrier units. Besides reducing the microbiological contamination, these processes reduce the contents of unstable nutrients like vitamins A and E as well. Manufacturers offer special diets that are meant to be sterilized: in these diets, the content of unstable nutrients has been increased, to make sure that after the sterilization process, these levels still live up to the requirements.

Serious problems in a breeding colony of rats arose in a barrier unit, due to malfunctioning of a valve of the autoclave at the entrance of the barrier unit, in which all materials entering the barrier unit were sterilized.[99] Rats developed alopecia and had reduced weight gain and a decreased production rate. By moving the animals from the barrier unit to the conventional unit, the problems disappeared. Food analyses revealed that vitamins A, B1, and B6 were only present in traces in the diet after autoclaving, and were therefore at a deficient level. After mending the autoclave, the problems disappeared.

### Pellet Hardness

Feeding of (too) hard pellets was found to reduce the growth of preweaned mice.[100] Hardness of pellets is measured as the amount of pressure, in kp weight, that is required to crush a pellet. In one type of diet, the measured pellet hardness could vary between 4 and 50 kp.[100] A value higher than 20 is considered problematic. A pellet hardness that is too high will make it difficult for the young preweaned mice to obtain enough food, which will reduce growth. Part of this effect is suggested to be mediated through the effect on the mothers.[101]

## FEEDING SCHEDULES

### Ad libitum Intake and Isocaloric Exchange

When offering diets *ad libitum*, the diet is available at all times. As discussed earlier, the voluntary food intake is in principle determined by the energy need of the animals. As fat has a much higher energy content than protein and carbohydrates, adding fat cannot be done without thinking. Only adding fat to a control diet, which thereby becomes the test diet, will lead to an increased metabolizable energy content of the test diet as compared to the control diet. As animals eat according to energy need, this means that the food intake of the test diet in grams under *ad libitum* conditions becomes lower than on the control diet. This implies that intake of all nutrients in control and test group will differ, which makes it impossible to make a reliable comparison of these two groups, as one cannot conclude if effects are the result of the fat addition alone. To interpret the effect of fat addition reliably, one has to execute isocaloric exchange. When wanting to add a certain amount of fat to a diet, it is calculated what amount of energy this represents. Then the same amount of energy is substracted by withdrawing that in the form of carbohydrates. The dietary composition will change in such a way that this will result in a similar nutrient intake (grams) in control and test animals, except for the fat and carbohydrate intake (Table 12.12).[83]

## Restricted Feeding

Restricted feeding refers to restricting the amount of food as compared to *ad libitum* feeding, while still insuring nutritional adequacy.[102] This implies that the amount of energy is restricted. When feeding restrictedly, it must be secured that the restricted feeding level provides enough essential nutrients. By feeding rodents restrictedly instead of *ad libitum*, remarkable improvement of health is achieved. Feeding restrictedly has more positive effects on health than changing dietary composition to, e.g., a higher fiber content under *ad libitum* conditions.[102]

   *Ad libitum* feeding is still considered "normal" practice for rodents, however, it is considered bad veterinary practice for pigs, monkeys, rabbits, and dogs, as they become obese.[102] The fact that rodents are still being fed *ad libitum*, has probably more to do with economical and practical aspects, than scientific reasons. *Ad libitum* feeding as opposed to moderate restricted feeding (75% of *ad libitum* intake) has a clear negative impact on rodent health, as it shortens survival time, increases cancer incidence, shortens cancer latency period, and increases the incidence of degenerative diseases in kidney and heart.[102] These effects have been found to be very reproducible. Besides impairing the health, the number of animals at the start of long-term toxicological studies needs to be increased drastically, as it is necessary that 25 animals per sex will survive a 2-year period. This conflicts with the idea of the 3 Rs, where one strives for Reduction, Replacement, and Refinement. A bad health of the rodents conflicts with Refinement, and the increase in the number of animals conflicts with Reduction. Therefore, moderate food restriction should become the new standard in laboratory facilities.

**Table 12.12 Examples of Expected Results of Low and High-Fat Diet Formulations when Fed to Rats**

| | Diet 1 | Diet 2 | Diet 3 | Diet 4 |
|---|---|---|---|---|
| | **Low-fat** | **High-fat** | **High-fat, adjusted** | **High-fat, adjusted** |
| *Diet Ingredient* | | | | |
| Protein (g) | 20 | 20 | 20 | 20 |
| Carbohydrate (g) | 60 | 40 | 15 | 15 |
| Fat (g) | 10 | 30 | 30 | 30 |
| Fiber (g) | 4 | 4 | 4 | 4 |
| Mineral mix (g) | 4 | 4 | 4 | 4 |
| Vitamin mix (g) | 1 | 1 | 1 | 1 |
| Test compounds (g) | 1 | 1 | 1 | 1 |
| "Inert" compound (g) | – | – | – | 25 |
| TOTAL (g) | 100 | 100 | 75 | 100 |
| Energy value (kcal/g) | 4.10 | 5.10 | 5.47 | 4.10 |
| *Expected Intake* | | | | |
| Energy (kcal/day) | 82 | 82 | 82 | 82 |
| Food (g/day) | 20 | 16 | 15 | 20 |
| Protein (g/day) | 4 | 3.2 | 4 | 4 |
| Carbohydrate (g/day) | 12 | 6.4 | 3 | 3 |
| Fat (g/day) | 2 | 4.8 | 6 | 6 |
| Fiber (g) | 0.8 | 0.64 | 0.8 | 0.8 |
| Mineral mix (g/day) | 0.8 | 0.64 | 0.8 | 0.8 |
| Vitamin mix (g/day) | 0.2 | 0.16 | 0.2 | 0.2 |
| Test compound (g/day) | 0.2 | 0.16 | 0.2 | 0.2 |
| "Inert" compound (g/day) | – | – | – | 5 |

Keenan et al.[103] stated that *ad libitum* overfeeding of rodents is at present one of the most poorly controlled variables affecting the current rodent bioassay. In the past decades, the variation in results have even increased under *ad libitum* conditions, which will lead to the use of an increased number of animals. Moderate dietary restriction (70 to 75% of adult *ad libitum* food intake) is advised as a method to improve uniformity, increase exposure time, and increase statistical sensitivity of chronic bioassays to detect true treatment effects.[103] However, moderate dietary restriction will only improve uniformity in individually housed animals, where there is control of individual food intake. A restricted amount of food in group-housed animals is expected to increase variation due to differences in individual food intakes, based on the hierarchy in the group. It will be a challenge to develop restricted feeding methods in group-housed rodents, in order to fulfil the animals social needs as well.

## Meal Feeding

For some animal species, it is normal practice to meal-feed. e.g., Göttingen minipigs usually get a meal in the morning and in the afternoon. Adult dogs are often fed one meal per day at a time point that is chosen by the researcher, i.e., when the least interference with experimental results is expected.

In order to mimic postprandial studies in man, rats can be trained to eat meals.[104] Rats are fast learners and know within a week that they are fed meals, and they will be able to eat what they need during these periods. This of course depends on sufficient time available in relation to the dietary composition (in case the palatability is low or the composition is such that it needs more time for chewing, food intake can be limited when food is presented during a relatively short period). Crossover studies cannot be done reliably in rats, as a small change in dietary composition will be noticed by the rats, who will consequently reduce food intake immediately.[104]

## Pair Feeding

In case a test substance has a bad taste or gives negative health effects, the consequence usually is that food intake is reduced as compared to the control group. In order to be able to judge the effects of the test substance, this necessitates that food intake in the control group be reduced to the same level as that in the test group. Through measuring the amount of food the test group eats, the control group receives a similar amount of food the next day or the next week. One can do this on an individual basis or on a group basis. These decisions are to be made by the responsible scientist, judging the specific circumstances of a particular study. Although measuring food intake takes time, it is a necessary step in order to obtain reliable results.

## Influence of Feeding Schedules in Pharmacological Studies

The effect and pharmacokinetics of pharmacological substances (e.g., oral antibiotics) are largely dependent on the time of administration in relation to the time of feeding. An empty stomach is often required when applying substances per gavage, as e.g., in toxicological or pharmacological studies. This practice is considered necessary in order to avoid mixing with food, which could dilute the test substance or interact with it. How long animals need to be fasted before the "bare" effect of pharmacological substances tested can be judged, is an important animal welfare issue.[105] A rat will have an empty stomach after 6 h.[106] Fasting for longer periods leads to increased locomotory and grooming behavior, and 18 h fasting caused a body weight loss of at least 10%.[106] Fasting for 2, 3, and 4 days led to a body weight loss of 11, 16, and 13%, respectively, in rats with an initial body weight of 75, 112, and 225 g.[70] In case an empty intestinal tract is required, fasting up to 22 h is required in the rat.[107] In that case, it is advised, at least if it does not interfere with the experiment, to give the rats sucrose cubes or a 10% glucose or maltose solution for the time required to empty the gastrointestinal tract. To avoid the intake of bedding, animals shall be housed on grid floors during fasting. Depending on the type of study or which part of the intestinal tract is studied, duration of fasting should be adapted accordingly. In nutritional studies, fasting before blood collections should be avoided, as one studies the effects of dietary composition, and not the effects of fasting. In some cases, short-term fasting may be necessary, as, e.g., triacylglycerols in high concentrations may interfere with certain (colorimetrical) measurements.

Feeding by gavage is considered a stressful event that may influence metabolism. Vachon et al.[108] compared feeding a similar meal by gavage versus voluntary intake, which gave different results. The voluntary intake of the meal gave results similar to human studies, whereas giving the same meal by gavage did not.[108] Feeding by gavage leads to a reduced gastrointestinal passage time. Blood glucose and insulin peak faster after eating a meal voluntarily as compared to giving a similar meal by gavage, even though in the last case, the meal arrives in the stomach faster. Gavage bypasses the first part of the digestion in the mouth, which therefore avoids physical processing (chewing) and adding of saliva and enzymes.

### Influence of Feeding Schedules on Circadian Rhythms

Circadian rhythms are biological functions with a certain periodicity. The typical diurnal frequency is one cycle per 24 h (+/- 4 h). Circadian rhythms are generated endogenously in the brain in a part that is called the circadian oscillator, located in the suprachiasmatic nuclei in the hypothalamus.

In order to prevent obesity, rabbits, pigs, and dogs are fed restrictedly. Usually, the restricted amount of food is provided during the normal working hours of the personnel. In certain nutritional studies, e.g., postprandial studies, a restricted amount of food is given to rats, usually during the daytime as well. When fed *ad libitum*, nocturnally active animal species like the mouse, rat, hamster, and rabbit consume almost all their food during the hours of darkness. The natural behavior of these species in the wild also indicates that the dark period is used for foraging and eating. When a restricted amount of food is given during the normal working hours of the personnel, food-restricted animals start eating immediately, as they are hungry, and as a consequence, many biochemical and physiological functions become related to this event. This implies that feeding during daylight will lead to changes in natural rhythms in, e.g., nocturnal species like rabbits and rats.

Some metabolic parameters (e.g., blood glucose and insulin) that are directly linked to the time of food ingestion will directly shift to the time of food access. Other activity rhythms that do not seem directly related to food intake, e.g., the 24 h rhythm of locomotor activity and core body temperature, will also be changed and influenced by the altered feeding schedule.[109] Although some activity rhythms are influenced by periodically restricted food access, this can coincide with an unaffected circadian oscillator at the same time: the effect is called "masking." This implicates that when the restricted feeding schedule is stopped again, the rhythms are dictated solely by the internal oscillator again, which has not been influenced by the feeding schedule. It is also possible that external variables influence the circadian oscillator. This process is defined as "entrainment." The time needed in order to reach a "stable" phase again can last up to 50 to 60 days.[110] Thus, by using periodic food access, some functions are rearranged immediately. Simultaneously, entrainment and masking are taking place. The masking of certain physiological functions is probably needed for maintaining vital functions during the time-consuming process of achieving a new homeostatic state for functions implying complete circadian reorganization.

Many metabolic functions are brought out of phase when restricting food access to some hours during the day. This accounts especially for nocturnal animals being fed during the light period. The process of reentrainment can require 50 to 60 days, and during that time, physiological functions like locomotor activity, digestive functions, and urine excretion will be affected.[111] Possibly the feeding of a restricted amount of food at a more "natural" time point (during the dark hours) may prevent the phase-shifting of metabolic processes. As it is outside the scope of this chapter to discuss this issue in depth, further details and references can be found in Ritskes-Hoitinga and Jilge (2001).[112]

## DIET AND WELFARE

### Transport and Acclimatization

Knowledge of the species is important when transporting animals. Before transport, getting specialist advice for each particular species is needed, e.g., (mini)pigs will vomit when being fed just before transport; rabbits may develop stomach rupture when transported with a full stomach; rats and mice will acclimatize faster after transport when food and water has been provided during the transport.[113] After transport of rats within one continent, i.e., without a shift in light-dark rhythm, it was reported that an acclimatization period of at least 3 days was considered sufficient for nutritional studies.[114] As a general rule, a 1-week acclimatization period is advised. Whether or not transport stress has an affect on some individuals, will usually become clear during the first week after arrival. When transporting animals between continents, a longer acclimatization period of up to 3 to 4 weeks may be necessary, due to the shift in the light-dark schedule.[113] However, as mentioned under circadian rhythms, a 50- to 60-day period may even be necessary.[111]

## Enrichment and Variation Versus Need for Standardization

From preference testing, it is known that rats prefer to work for food instead of having it available at all times. For each species, there are certain species-specific essential basic needs connected to searching and finding food (e.g., rooting of pigs). If these essential needs are not fulfilled, abnormal behavior like stereotypies can occur. Pigs can develop sham chewing, mice can develop rotating behavior inside the cage, rabbits can make digging movements in the cage, dogs can pace back and forth, etc. The fact that some animals develop stereotyped behavior, will increase the variation in results, as animals performing stereotypies related to movement, will have a lower body weight than the animals that have not developed stereotypies. The fact that stereotypies develop implies that an essential need of an animal has not been fulfilled. Once an animal has developed stereotypies, one should not try to prevent this behavior as it is a "suitable" adaptation to a difficult situation. Possible ways of enriching the environment include letting the animals work and search for food. This aspect is well known and used in zoos, but not so much yet in laboratory animal facilities. Probably this is due to the fear that this will compromise standardization. However, by frustrating the animals' basic needs, one is also compromising animal welfare and standardization. Therefore, the animals should be given the benefit of the doubt, and fulfilling species-specific needs around the feeding process should be introduced whenever possible. Knowledge of the natural feeding time and behavior are important factors to consider. The time of day at which a restricted amount of food is given can be an important tool for improving welfare: e.g., Krohn, Ritskes-Hoitinga, and Svendsen[115] reported a significantly reduced frequency of stereotypies by feeding rabbits a restricted amount of food just before the dark period instead of providing an *ad libitum* or restricted amount in the morning.

Giving food rewards is an important tool for teaching and training animals. Monkeys have been trained to present their arm voluntarily for taking blood samples by giving them a banana as reward. These social activities with human beings are also important enrichment tools and will improve trust toward people handling them. That way, animals will be less stressed during experiments and the pain threshold goes up. Which food rewards are chosen and in what amounts need careful consideration to avoid interference with the experimental results and health of the animal.

Certain dietary schedules require individual housing. As individual housing compromises the well-being of social species, alternative ways of feeding need to be considered whenever possible. For example, the animals can be individually fed for a certain period each day and then socially housed for the remaining part of the 24-h period. A current challenge of modern laboratory animal science is to develop enrichment related to feeding, as this has so far not been exploited. Further research will prove whether nutritional enrichment may compromise standardization, and if so, how and when. However, it may also prove that by enriching the feeding process, animal welfare increases, thereby reducing variation and improving standardization.

## REFERENCES

1. Fuller, M.F., and Wang, T.C., Amino acid requirements of the growing pig, in *Manipulation Pig Production*, Barnett J.L. et al., Eds., Werribee: Australasian Pig Science Association, 97–111, 1987.
2. McDonald, P., Edwards, R.A., and Greenhalgh, J.F.D., *Animal Nutrition*, Longman, Chicago; 1981, chap. 16.

3. Forebs, J.M., *Voluntary Food Intake and Diet Selection in Farm Animals*, CAB International, Wallingford, 1995, chap. 11.

4. Thorbek, G., Chwalibog, A., Eggum, B.O., and Christensen, K., Studies on growth, nitrogen, and energy metabolism in rats, *Arch. Tierernähr.*, 32, 827, 1982.

5. Pullar, J.D. and Webster, A.J.F., Heat loss and energy retention during growth in congenitally obese and lean rats, *Br. J. Nutr.*, 31, 377, 1974.

6. Kleiber, M., *The Fire of Life*, John Wiley and Sons, New York, 1961, chap. 18.

7. Clarke, H.E., Coates, M.W., Eva, J.K., Ford, D.J., Milner, C.K., O'Donoghue, P.N., Scott, P.O., and Ward, R.J., Dietary standards for laboratory animals: Report of the Laboratory Animal Centre Diets Advisory Committee, *Lab. Anim.*, 11, 1, 1977.

8. Eggum, B.O. and Chwalibog, A., A study on requirement for maintenance and growth in rats with normal or reduced gut flora, *Z. Tierphysiol., Tierernährg.u. Futtermittelkde.*, 49, 104, 1983.

9. Henning, S.J., Functional development of the gastrointestinal tract, in *Physiology of the Gastrointestinal Tract*, Johnson, R., Ed., Raven Press, New York, 1987, chap.9.

10. Prochazka, P., Hahn, P., Koldovsky, O., Noh Ynek, M., and Rokos, J., The activity of alpha-amylaze in homogenates of the pancreas of rats during early post-natal development, *Physiol. Bohemoslov.*, 13, 288, 1964.

11. Rubino, A., Zimbalatti, F., and Auricchio, S., Intestinal disaccharidase activities in adult and suckling rats, *Biochem. Biophys. Acta*, 92, 305, 1964.

12. Descholdt-Lanckman, M., Robberecht, P., Camus, J., Baya, C., and Christophe, J., Hormonal and dietary adaptation of rat pancreatic hydrolases before and after weaning, *Am. J. Physiol.*, 226, 39, 1974.

13. Henning, S.J., Postnatal development: coordination of feeding, digestion, and metabolism, *Am. J. Physiol.*, 241, G199, 1981.

14. Melander, A., Influence of food on the bioavailability of drugs, *Clin. Pharmacokinet.*, 3, 337, 1978.

15. Jefery, P., Burrows, M., and Bye, A., Does the rat have an empty stomach after an overnight fast? *Lab. Anim.*, 21, 330, 1987.

16. Raczynski, G., Eggum, B.O., and Chwalibog, A., The effect of dietary composition on transit time in rats, *Z. Tierphysiol., Tierernährg.u. Futtermittelkde.*, 47, 160, 1982.

17. National Research Council, *Nutrient Requirements of Laboratory Animals*, Fourth ed., National Academy of Sciences, Washington, DC, 1995.

18. Chwalibog, A., Jakobsen, K., Henckel, S., and Thorbek, G., Oxidation and fat retention from carbohydrate, protein, and fat in growing pigs, *Z. Tierphysiol., Tierernährg.u. Futtermittelkde., J. Anim. Physiol. a. Anim. Nutr.*, 68, 123, 1992.

19. Chwalibog, A., Tauson, A-H., Fink, R., and Thorbek, G., Oxidation of substrates and lipogenesis in pigs (*Sus scrofa*), mink (*Mustela vison*), and rats (*Ratus norvegicus*), *Thermochim. Acta*, 309, 49, 1998.

20. Rafecas, I., Esteve, M., Remesar, X., and Alemany, M., Plasma amino acids of lean and obese Zucker rats subjected to a cafeteria diet after weaning, *Biochem. Int.*, 25, 797, 1991.

21. Glick, Z., Bray, A., and Teague, R.J., Effect of prandial glucose on brown fat thermogenesis in rats: possible implications for dietary obesity, *J. Nutr.*, 114, 1934, 1984.

22. Herberg, L., Interrelationships between obesity and diabetes, *Proc. Nutr. Soc.*, 50, 605, 1991.

23. Eggum, B.O., Andersen, J.O., and Rotenberg, S., The effect of dietary fibre level and microbial activity in the digestive tract on fat metabolism in rats and pigs, *Acta Agric. Scand.*, 32, 145, 1982.

24. Rotenberg, S., Eggum, B.O., Hegedüs, M, and Jacobsen, I., The effect of pectin and microbial activity in the digestive tract on faecal excretion of amino acids, fatty acids, thiamin, riboflavin and niacin in young rats, *Acta Agric. Scand.*, 32, 310, 1982.

25. Rotenberg, S. and Andersen, J.O., The effect of antibiotica on some lipid parameters in rats receiving cornstarch, potato flour or pectin in the diet, *Acta Agric. Scand.*, 32, 151, 1982.

26. Anderson, J.W., Physiological and metabolic effects of dietary fiber, *Fed. Proc.*, 44, 2902, 1985.

27. Cummings, J.H., Nutritional implications of dietary fiber, *Am. J. Clin. Nutr.*, 31, 521, 1978.

28. Zhao, X., Jørgensen, H., and Eggum, B.O., The influence of dietary fibre on body composition, visceral organ weight, digestibility and energy balance in rats housed in different thermal environments, *Br. J. Nutr.*, 73, 5, 687, 1995.

29. Robinson, M., Hart, D., and Pigott, G.H., The effects of diet on the incidence of periodontitis in rats, *Lab. Anim.*, 25, 247, 1991.

30. Madsen, C., Squamous-cell carcinoma and oral, pharyngeal and nasal lesions caused by foreign bodies in feed. Cases from a long-term study in rats, *Lab. Anim.*, 23, 241, 1989.

31. Chwalibog, A. and Thorbek, G., Estimation of net nutrient oxidation and lipogenesis in growing pigs, *Arch. Anim. Nutr.*, 53, 253, 2000.

32. Kuratko, C. and Pence, B., Rat colonic antioxidant status: interaction of dietary fats with 1,2 dimethylhydrazine challenge, *J. Nutr.*, 122, 278, 1992.

33. Bird, R. and Bruce, R., Effect of dietary fat levels on the susceptibility of colonic cells to nuclear-damaging agents, *Nutr. Cancer*, 8, 93, 1986.

34. Bull, A., Bronstein, J., and Nigro, N., The essential fatty acid requirement for azoxymethane-induced intestinal carcinogenesis in rats, *Lipids*, 24, 340, 1989.

35. Ip, C., Carter, C.A., and Ip, M.M., Requirement of essential fatty acid for mammary tumorgenesis in the rat, *Cancer Res.*, 45, 155, 1985.

36. Ip, C., Fat and essential fatty acid in mammary carcinogenesis, *Am. J. Clin. Nutr.*, 45, 218, 1987.

37. Fisher, M.S., Claudio, J.C., Locniskar, M., Belury, M.A., Maldve, R.E., Lee, M.L., Leyton, J., Slaga, T.J., and Bechtel, D.H., The effect of dietary fat on the rapid development of mammary tumors induced by 7,12-dimethylbenz(a)anthracene in SENCAR mice, *Cancer Res.*, 52, 662, 1992.

38. Crevel, R.W.R., Friend, J.V., Goodwin, B.F., and Parish, W.E., High-fat diets and the immune response of C57 B1 mice, *Br. J. Nutr.*, 67, 17, 1992.

39. Ritskes-Hoitinga, J., Meijers, M., Meijer, G.W., and Weststrate, J.A., The influence of dietary linoleic acid on mammary tumour development in various animal models, *Scand. J. LAS*, 23, 1, 463–468, 1996.

40. Burr, G.O. and Burr, M.M., A new deficiency disease produced by the rigid exclusion of fat from the diet, *J. Biol. Chem.*, 82, 345, 1929.

41. Burr, G.O. and Burr, M.M., On the nature and role of the fatty acids essential in nutrition, *J. Biol. Chem.*, 86, 587, 1930.

42. Innis, S.M., Essential fatty acids in growth and development, *Prog. Lipid Res.*, 30, 39, 1991.

43. Christensen, K., Determination of linoleic acid requirements in slaughter pigs, *Nat. Inst. Anim. Sci. Rep.*, 577, 1, 1985.

44. Holman, R.T., Essential fatty acid deficiency, in *Progress in the Chemistry of Fats and Other Lipids*, Vol. 9, Pergamon Press, Elmsford, New York, 1968, p. 619.

45. Eggum, B.O., Biochemical and methodological principles, in *Protein Metabolism in Farm Animals*, Bock, H.D., Eggum, B.O., Low, A.G., Simon, O., and Zebrowska, T., Eds., Oxford Scientific Publications, Deutscher Landwirtschaftsverlag, Berlin, 1989, chap. 1.

46. Rogers, Q.R. and Leung, P.M.B., The influence of amino acids on the neuroregulation of food intake, *Fed. Proc. Fed. Am. Soc. Exp. Biol.*, 32, 1709, 1973.

47. Harper, A.E., Benevenga, N.J., and Wohlhueter, R.M., Effects of ingestion of disproportionate amounts of amino acids, *Physiol. Rev.*, 50, 428, 1970.

48. Eggum, B.O., Bach Knudsen, K.E., and Jacobsen, I., The effect of amino acid imbalance on nitrogen retention (biological value) in rats, *Br. J. Nutr.*, 45, 175, 1981.

49. Hawrylewicz, E.J., Huang, H.H., and Blair, W.H., Dietary soybean isolate and methionine supplementation affect mammary tumor progression in rats, *J. Nutr.*, 121, 1693, 1991.

50. Dean, J. and Edwars, D.G., The nutritional value of rat diets of differing energy and protein levels when subjected to physical processing, *Lab. Anim.*, 19, 311, 1985.

51. Edwars, D.G., Porter, P.D., and Dean, J., The responses of rats to various combinations of energy and protein. I. Diets made from purified ingredients. *Lab. Anim.*, 19, 328, 1985.

52. Edwars, D.G. and Dean, J., The responses of rats to various combinations of energy and protein. II. Diets made from natural ingredients, *Lab. Anim.*, 19, 336, 1985.

53. Yoshida, A., Harper, A.E., and Elvehjem, C.A., Effects of protein per calorie ratio and dietary level of fat on calorie and protein utilization, *J. Nutr.*, 63, 555, 1957.

54. Eggum, B.O., A study of certain factors influencing protein utilization in rats and pigs, *Nat. Inst. Anim. Sci. Rap.*, 406, 1, 1973.

55. Szelényi-Galàntai, M., Jacobsen, I., and Eggum, B.O., The influence of dietary energy density on protein utilization in rats, *Acta Agric. Scand.*, 31, 204, 1981.

56. Solleveld, H.A., McAnulty, P., Ford, J., Peters, P.W.J., and Tesh, J., Breeding, housing and care for laboratory animals, in *Laboratory Animals*, Ruitenberg, E.J. and Peters, P.W.J., Eds., Elsevier Science Publishers B.V., Amsterdam, 1986, chap. 1.

57. Keenan, K.P., Smith, P.E., Hertzog, P., Soper, K., Ballam, G.C., and Clark, R.L., The effects of overfeeding and dietary restriction on Sprague-Dawley rat survival and early pathology biomarkers of aging, *Toxicol. Pathol.*, 22, 300, 1994.

58. Keenan, K.P., Soper, K., Smith, P.F., Ballam, G.C., and Clark, R.L., Diet, overfeeding and moderate dietary restriction in control Sprague-Dawley rats. I. Effects on spontaneous neoplasmas, *Toxicol. Pathol.*, 23, 269, 1995.

59. Keenan, K.P., Soper, K., Hertzog, P.R., Gumprecht, L.A., Smith, P.F., Mattson, B.A., Ballam, G.C., and Clark, R.L., Diet, overfeeding and moderate dietary restriction in control Sprague-Dawley rats. II. Effects on age-related proliferative and degenerative lesions, *Toxicol. Pathol.*, 23, 287, 1995.

60. Keenan, K.P., Laroque, P., Soper, K., Morrissey, R.E., and Dixit, R., The effects of overfeeding and moderate dietary restriction on Sprague-Dawley rat survival, pathology, carcinogenity, and the toxicity of pharmaceutical agents, *Exp. Toxicol. Pathol.*, 48, 139, 1996.

61. Keenan, K.P., Ballam, G.C., Dixit, R., Soper, K., Laroque, P., Mattson, B.A., Adams, S.P., and Coleman, J.B., The effects of diet, overfeeding and moderate dietary restriction on Sprague-Dawley rat survival, disease and toxicology, *J. Nutr.*, 127, 851S, 1997.

62. Suzuki, T., Tana-Ami, S., Fujiwara, H., and Ishibashi, T., Effect of the energy density of non-purified diets on reproduction, obesity, alopecia and aging in mice, *Exp. Anim.*, 40, 499, 1991.

63. Ross, M.H. and Brass, G., Influence of protein under- and overnutrition on spontaneous tumour prevalence in the rat, *J. Nutr.*, 103, 944, 1965.

64. Schiemann, R., Nehring, K., Hoffmann, L., Jentsch, W., and Chudy, A., *Energetische Futterbewertung und Energinormen*, VEB Deutsch. Landwirtschaftsverlag, Berlin, 1971.

65. Thorbek, G. and Chwalibog, A., Tilvækst, fordøjelighed, kvælstof- og energiomsætning hos voksende kaniner målt ved forskellige foderkombinationer (Growth, nitrogen and energy metabolism in growing rabbits measured at different feed combinations), *Nat. Inst. Anim. Sci. Rep.*, 510, 1, 1981.

66. Chwalibog, A., *Ernæring af laboratoriedyr (Nutrition of laboratory animals)*, DSR Forlag, The Royal Veterinary and Agricultural University, Copenhagen, 1989.

67. Chwalibog, A., Energetics of animal production, *Acta Agric. Scand.*, 41, 147, 1991.

68. Eggum, B.O. and Beames, R.M., Use of laboratory animals as models for studies on nutrition of domestic animals, in *Laboratory Animals*, Ruitenberg, E.J. and Peters, P.W.J., Eds., Elsevier Science Publishers B.V., Amsterdam, 1986, chap. 9.

69. Canas, R., Romero, J.J., and Baldwin, R.L., Maintenance energy requirements during lactation in rats, *J. Nutr.*, 112, 1876, 1982.

70. Chwalibog, A., Jakobsen, K., Tauson, A-H., and Thorbek, G., Heat production and substrate oxidation in rats fed at maintenance level and during fasting, *Comp. Biochem. Phys.*, 121, 423, 1998.

71. Blaxter, K., *Energy Metabolism in Farm Animals*, Cambridge University Press, Cambridge, 1989, chap. 8.

72. Klein, M. and Hoffmann, L., Bioenergetics of protein retention, in *Protein Metabolism in Farm Animals*, Bock, H.D., Eggum, B.O., Low, A.G., Simon, O., and Zebrowska, T., Eds., Oxford Scientific Publications, Deutscher Landwirtschaftsverlag, Berlin, 1989, chap. 11.

73. Imai, K., Ohnaka, M., Ondani, M., and Niiyama, Y., Maintenance energy requirement in pregnant rats and net energetic efficiency for fetal growth during late pregnancy, *J. Nutr. Sci. Vitaminol.*, 32, 527, 1986.

74. Tauson, A-H., Chwalibog, A., Jakobsen, K., and Thorbek, G., Pattern of protein retention in growing boars of different breeds, and estimation of maximum protein retention, *Arch. Anim. Nutr.*, 51, 253, 1998.

75. Chwalibog, A., *Husdyrernæring (Animal Nutrition)*, DSR Forlag, The Royal Veterinary and Agricultural University, Copenhagen, 2000, chap. 4.

76. Thorbek, G., Chwalibog, A., and Henckel, S., Energetics of growth in pigs from 20 to 120 kg live weight, *Z. Tierphysiol., Tierernährg. u. Futtermittelkde.*, 49, 238, 1983.

77. Lang, C.M. and Harrell, G.T., Laboratory animal science in the future, *Scand. J. Lab. Anim. Sci.*, 3, 166, 2000.

78. Coates, M.E., The nutrition of laboratory animals, in *The UFAW Handbook on the Care and Management of Laboratory Animals*, Hume, C.W., Ed., UFAW, Churchill Livingstone, Edinburg, London and New York, 1976, chap. 3.

79. Owens, F.N. and Pettigrew, J.E., Subdividing amino acid requirements into portions for maintenance and growth, in *Absorption and Utilization of Amino Acids*, Friedman, M., Ed., CRC Press, Inc., Boca Raton, FL, 1989, chap. 2.

80. Shin, I-S., Subdividing amino acid requirements for maintenance from requirements for growth, Thesis, Oklahoma State University, 1990.

81. British Association of Research Quality Assurance (BARQA), Guidelines for the Manufacture and Supply of GLP Animal Diets, 1992.

82. Thigpen, J.E., Setchell, K.D.R., Alhmark, K.B., Locklear, J., Spahr, T., Caviness, G.F., Goelz, M.F., Haseman, J.K., Newbold, R.R., and Forsythe, D.B., Phytooestrogen content of purified, open- and closed-formula laboratory animal diets, *Lab. Anim. Sci.*, 49, 530, 1999.

83. Beynen, A.C. and Coates, M.E., Nutrition and experimental results, in *Principles of Laboratory Animal Science*, Van Zutphen, L.F.M., Baumans, V., and Beynen, A.C., Eds., Elsevier Scientific Publishers, Amsterdam; New York, 2001, p. 111.

84. Ritskes-Hoitinga, J. and Bollen, P., The Göttingen minipig, a refined animal model?, *Der Tierschutzbeauftragte*, 2, 87, 1999.

85. Ritskes-Hoitinga, J., Mathot, J.N.J.J., Danse, L.H.J.C., and Beynen, A.C., Commercial rodent diets and nephrocalcinosis in weanling female rats, *Lab. Anim.*, 25, 126, 1991.

86. Ritskes-Hoitinga, J. and Bollen, P., Nutrition of (Göttingen) minipigs: facts, assumptions and mysteries, *Pharm. and Tox.*, 80, 5, 1997.

87. Ritskes-Hoitinga, J., Nephrocalcinosis in the Laboratory Rat, Thesis Utrecht University, The Netherlands, 1992.

88. Reeves, Ph., G., Nielsen, F.H., and Fahey, Jr., G.C., AIN-93 Purified Diets for Laboratory Rodents: Final Report of the American Institute of Nutrition Ad Hoc Writing Committee on the Reformulation of the AIN-76A Rodent Diet, *J. Nutrition*, 123, 1939, 1993.

89. Haseman, J.K., Huff, J., and Boorman, G.A., Use of historical control data in carcinogenicity studies in rodents, *Toxicol. Path.*, 12, 126, 1984.

90. Roe, F.C.J., Historical histopathological control data for laboratory rodents: valuable treasure or worthless trash?, *Lab. Anim.*, 28, 148, 1994.

91. National Research Council (NRC), Nutrient Requirements of Sheep 1985; of Dogs 1985; of Beef Cattle 1984; of Mink and Foxes, 1982; of Laboratory Animals (Rat, Mouse, Guinea Pig, Hamster, Gerbil, Vole), 1995; of Poultry 1994; of Fish 1993; of Horses 1989; of Dairy Cattle 1989; of Swine 1998; of Cats 1986; of Goats 1981; of Nonhuman Primates 1978; of Rabbits 1977, National Academy Press, Washington, DC.

92. Ritskes-Hoitinga, J. and Bollen, P., The formulation of a test diet in establishing the nutrient requirements and optimum feeding schedules for minipigs, *Scand. J. Lab. Anim. Sci.*, 25, 27, 1998.

93. De Wille, J.W., Waddell, K., Steinmeyer, C., and Farmer, S.T., Dietary fat promotes mammary tumorigenesis in MMTV/v-Ha-*ras* transgenic mice, *Cancer Letters*, 69, 59, 1993.

94. National Research Council, *Vitamin Tolerance of Animals*, National Academy Press, Washington, DC, 1987.

95. Ritskes-Hoitinga, J., Verschuren, P.M., Meijer, G.W., Wiersma, A., van de Kooij, A.J., Timmer, W.G., Blonk, C.G., and Weststrate, J.A., The association of increasing dietary concentrations of fish oil with hepatotoxic effects and a higher degree of aorta atherosclerosis in the ad lib.-fed rabbit, *Food and Chem. Tox.*, 36, 663, 1998.

96. Speijers, G.J.A., Voedingsvezel en haarballen, NVP Symposium Proceedings "voeding en kwaliteit van proef en dier," 1987.

97. GV-Solas publication no. 9, Definition of Terms and Designations in Laboratory Animal Nutrition (Part 1); Use of Feedstuffs and Bedding Materials for Nonclinical Laboratory Studies (Part 2), 1980.

98. Environmental Protection Agency, Proposed health effects test standards for toxic substances control act test rules, *Good Laboratory Standards for Health Effects*, in *Federal Register*, 44, 91, 1979.

99. Wyss-Spillmann, S.K., Homberger, F.R., Lott-Stolz, G., Jörg, R., and Thomann, P.E., Alopecia in rats due to nutritional deficiency, in *Proceedings of the Sixth Felasa Symposium, 19–21 June 1996*, O'Donoghue, Ph.N., Ed., Basel, Switzerland, 1997.

100. Koopman, J. P., Scholten, P.M., Roeleveld, P.C., Velthuizen, Y.W.M., and Beynen, A.C., Hardness of diet pellets and its influence on growth of pre-weaned and weaned mice, *Z. Versuchstierkunde*, 32, 71, 1989.

101. Koopman, J.P., Scholten, P.M., and Beynen, A.C., Hardness of diet pellets and growth of pre-weaned mice: separation of direct effects on the young and indirect effects mediated by the lactating females, *Z. Versuchstierkunde*, 32, 257, 1989.

102. Hart, R.W., Neumann, D.A., and Robertson, R.T., Eds., *Dietary Restriction: Implications for the Design and Interpretation of Toxicity and Carcinogenicity Studies*, ILSI Press, Washington, DC, 1995.

103. Keenan, K.P., Ballam, G.C., Soper, K.A., Laroque, P., Coleman, J.B., and Dixit, R., Diet, caloric restriction, and the rodent bioassay, *Toxicol. Sci.*, 52, 2 Suppl, 24, 1999.

104. Ritskes-Hoitinga, J., van het Hof, K.H., Kloots, W.J., de Deckere, E.A.M., van Amelsvoort, J. M.M., and Weststrate, J.A., Rat as a model to study postprandial effects in man. Alternative methods in Toxicology and the Life Sciences Series volume 11, in *Proceedings of the World Congress on Alternatives and Animal Use in the Life Sciences: Education, Research, Testing*, Goldberg, A.M. and van Zutphen, L.F.M., Eds., 1995, p. 403.

105. Claassen, Ed., *Neglected Factors in Pharmacology and Neuroscience Research*, Elsevier, Amsterdam, 1994.

106. Vermeulen, J.K., de Vries, A., Schlingmann, F., and Remie, R., Food deprivation: common sense or nonsense?, *Anim. Tech.*, 48, 45, 1997.

107. SGV Newsletter, no. 24 (Spring 2001): http://www.sgv.org/Newsletter/news-24.htm.

108. Vachon, C., Jones, J.D., Nadeau, A., and Savoie, L., A rat model to study postprandial glucose and insulin responses to dietary fibers, *Nutr. Rep. Int.*, 37, 1339, 1988.

109. Jilge, B. and Hudson, R., Diversity and development of circadian rhythms in the European rabbit: a review, *Chronobiol. Int.*, 18, 1, 2001.

110. Jilge, B., Hörnicke, H., and Stähle, H., Circadian rhythms of rabbits during restrictive feeding, *Am. J. Physiol.*, 253, R46, 1987.

111. Jilge, B. and Stähle, H., Restricted food access and light-dark: impact of conflicting zeitgebers upon circadian rhythms of the rabbit, *Am. J. Physiol.*, 264, R708, 1993.

112. Ritskes-Hoitinga, J. and Jilge, B., FELASA — Quick reference paper on laboratory animal feeding and nutrition, www.felasa.org, 2001.

113. Van Ruiven, R., Meijer, G.W., van Zutphen, L.F.M., and Ritskes-Hoitinga, J., Adaptation period of laboratory animals after transport: a review, *Scand. J. Lab. Anim. Sci.*, 23, 185, 1996.

114. Van Ruiven, R., Meijer, G.W., Wiersma, A., Baumans, V., van Zutphen, L.F.M., and Ritskes-Hoitinga, J., The influence of transportation stress on selected nutritional parameters to establish the necessary minimum period for adaptation in rat feeding studies, *Lab. Anim.*, 32, 446, 1998.

115. Krohn, T.C., Ritskes-Hoitinga, J., and Svendsen, P., The effects of feeding and housing on the behaviour of the laboratory rabbit, *Lab. Anim.*, 33, 2, 101, 1999.

# CHAPTER 13

# Impact of the Biotic and Abiotic Environment on Animal Experiments

Nancy A. Johnston and Timo Nevalainen

## CONTENTS

## INTRODUCTION

Laboratory animal science is concerned with standardization of all the factors that may have an impact on animals and, consequently, on experimental results, as nicely presented almost half a century ago by Biggers and coworkers.[1] Unnecessarily large variation is the enemy of scientists; hence, all approaches and methods to control variation should be utilized. This is a key element in animal experiments with the aim to operate with best practice and with relatively low numbers of animals.

Recently, the original definition of alternative methods by Russell and Burch[2] has been reiterated and updated in the Declaration of Bologna. A closer look at refinement and reduction alternative methods shows that we are dealing with the same topics. In essence, the refinement alternative is any activity for improvement of animal welfare through housing or procedure practices, and the reduction alternative is any activity toward lowest possible number of animals. In practical terms, this means that every scientist can and should apply these alternative methods.

How to define the optimal number of laboratory animals in terms of refinement and reduction? Refinement and reduction are not independent of each other; hence, an understanding of their interplay is crucial. Figure 13.1 is an attempt to accomplish this. In the reduction axis, there is a window of appropriate numbers of animals, below and above which the experiment becomes meaningless and

0-8493-1086-5/03/$0.00+$1.50
© 2003 by CRC Press LLC

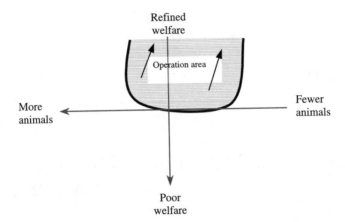

**Figure 13.1**  Schematic presentation of refinement (y-axis) and reduction (x-axis) interplay illustrating the preferred operation area (shaded) for animal studies, and the preferred direction (arrows) the "loop" should be drawn to. The refinement axis is straightforward, while the reduction axis has a window for appropriate number of animals, below which the statistical power is too low to reach significance, and above which animals are "wasted."

unethical. This is because the first scenario fails to draw conclusion due to too few animals being used and consequently poor statistical power, and the latter is guilty of unnecessarily large group size. The refinement axis is more straightforward, the higher the refinement value, the better. These two axes establish the operation area in animal studies, and arrows show the preferred direction.

Environmental biotic and abiotic factors can interfere with a study and study results essentially in two ways. If the factor changes the mean of determination, it may be the lesser of the evils, because it should do so in experimental and control groups. Change in variation is far more troublesome, because this is bound to increase the appropriate number of animals with all the consequences of this. Unfortunately, there are scarce data to assess refinement and reduction dimensions of all the factors.

There are lists of factors that have an impact on animals, and they should be considered thoroughly while designing and planning an animal study. These lists provide a summary of items of variable environmental parameters, which should be carefully described in scientific articles. This is fundamentally important for the reader to understand the message, and for other scientists to accept and, if deemed necessary, to successfully repeat the study.[3]

## BIOTIC FACTORS

One of the major sources of variation in laboratory animal science is contamination or infection of the research animals with microbial agents. Controlling or eliminating the agents of infection contributes to the standardization of experiments using animals. The renewed call for searching for alternative methods in laboratory animal science necessitates an appropriate control of these biological agents. Refinement of techniques requires the elimination of unknown variables that biotic agents cause. Reduction of the number of animals is only possible if unexpected deaths of animals and clinical illnesses are eliminated.

Whether or not these infectious agents produce clinical disease, the exposure and infection of laboratory animals with infectious agents can influence the outcome of experiments. The effects of these agents may lead to false conclusions or misinterpretation of data. The introduction of this biological variability may also cause a failure to create reproducible studies.[4–6] The benefits of eliminating clinical disease seem obvious, and the benefits of eliminating subclinical infections through creating a biologically controlled environment are becoming more widely recognized.

The environment of laboratory animals may be classified based on the presence of biotic factors: "conventional," specific-pathogen-free (SPF), or sterile (axenic). Many research animals are housed in conventional facilities, where there are opportunities for animals to come into contact with many microbes — bacteria, fungi, viruses, and parasites — known and unknown with little or no control. These interactions with microbes may produce a myriad of host responses, including clinical or subclinical disease, immune responses, behavioral or physiological changes, decrease in reproductive capabilities, and many others. These microbes create variability in the experimental environment and add a layer of uncertainty to any data collected. The conventional facility with widespread disease and infection is becoming obsolete, as researchers understand how these biotic factors may influence research parameters.

In SPF facilities, a select number of bacteria, viruses, and parasites are excluded from the facility. The animals are not sterile, and they are not free from all infections or exposure to microbes, but they are kept healthy by removing the specific microbes that cause the greatest illness or harm to them as determined by the individual facility. In this environment, animals are tested for infection or exposure to the selected viruses, bacteria, and parasites, but carry an unknown microbial flora that may differ substantially among facilities or even among individual animals in the same facility.[7]

The animals housed in the most controlled environments are classified as axenic or gnotobiotic. Axenic animals are kept totally free from association with microbes, and gnotobiotic animals have only a few known, nonpathogenic microorganisms as normal flora.[8] These animals, although free from the variability that is common among the conventionally reared animals, may have different nutritional needs or demonstrate different responses to drug metabolism than animals with a normal flora.[9]

Many of the microbes that infect laboratory animals result in clinical illness. Due to modern husbandry techniques, more controlled environments, and health monitoring programs, the incidence of many of these diseases is low. Clinical illness in rodents may manifest itself as poor body condition, weight loss, poor hair coat, decreases in reproductive measures, or death. Larger animals may display signs of illness as anorexia, lethargy, among others. Clinical illness often alters the physiologic and behavioral parameters of the affected animals. The effect on research due to clinical illness in a colony of animals may be devastating. Fortunately, most large laboratory animals, such as dogs, cats, and pigs, have been vaccinated against the major causes of illness.[10] Rodent diseases with high morbidity and mortality rates have not been controlled by vaccines but rather by strict control of husbandry, sanitation, animal purchase, and health surveillance. Clinical illness may cause other serious problems for research groups, resulting in unreliable data, loss of study animals, or poor reproducibility. In some immunodeficient animals or genetically modified animals, the disease may persist in the colony, because the individual animals are incapable of clearing the infectious agent.

Subclinical infections pose a more serious risk to disruption of research than clinical infections because there are no visible manifestations or warning signs of disease. Researchers may not even be aware that the colony is infected and what serious complications may occur in the research due to the infection. Immune effects of disease may persist for months after the infection has occurred.[4,8,11] Gross pathological or histopathological changes that are discovered at the end point of a study may have been caused by illness or infection, not by the experimental procedure or compound being tested, leading to false conclusions. One example may be the lesions associated with subclinical *Helicobacter* spp. infection in mice. Infection with this bacteria may produce chronic proliferative hepatitis[12] or inflammatory large bowel disease,[4,5,13-15] both of which may confound research objectives.[16,17] Changes in behavior, activity level, memory, or learning are known to occur in rodents with endoparasite infections.[18-22] Endoparasite infection has been reported to alter nonopioid mediated analgesia,[23] inhibit growth rate,[24] and change the immune response.[25]

With many strains of rodents, the environment in which they live can influence the expression of phenotype. Several studies compared strains of laboratory rodents for differing susceptibility to various pathogens, such as *Mycoplasma pulmonis*[26] or Sendai virus.[27] The differences are attributed to host-specific factors and immune system differences among the strains. The nonobese diabetic mouse (NOD) strain develops insulin-dependent diabetes mellitus, but at different frequencies, depending on the microbial environment in which they live. After Mouse Hepatitis Virus exposure, the diabetes incidence in one colony decreased significantly.[28] Other parasitic and microbial factors have influenced the incidence of diabetes in this model.[29,30] The microbial flora of an individual animal can influence development of autoimmunity. Several studies have compared the incidence of autoimmune diseases in conventionally reared animals to specific pathogen-free reared animals and have found significant differences between the two groups with all other factors remaining constant.[31,32] Stimulation of the immune system, chronic infection, or chronic inflammation may contribute to the variation documented. With expanding populations of transgenic animals in the laboratory animal facilities, and often untested and unknown phenotypes of these animals, the standardization of the microbial flora is crucial to determine the true characteristics of any transgenic strain.

**Table 13.1  References to Selected Murine Pathogens Causing Research Complications**

| Virus | Produces Signs of Lesions | Alters Immunity | Alters Neoplasia | Alters Metabolism |
|---|---|---|---|---|
| Reo-3 | 39–41 | 41 | 42 | 39, 40 |
| MHV | 43 | 64–66 | 42, 43, 63 | 43 |
| LCM | 44, 45 | 46, 47 | 42, 44, 67 | 46 |
| LDH-E | | 48–50, 68, 69 | 42, 48, 50 | 48–50, 70 |
| MVM | | 70, 71 | 42, 51 | |
| Sendai | 52, 53 | 54 | 42 | 73 |
| CMV | | 55, 56 | | 56 |
| Polyoma | | | 42 | |
| Thymic agent | 57 | 57 | | |
| Mycoplasma | 58 | | | 59, 60 |
| Mycoplasma arthritidis | 61, 62 | | | |

*Source:* Adapted from Jacoby, R.O. and Barthold, S.W., Quality assurance for rodents used in toxicological research and testing, in *Scientific Considerations in Monitoring and Evaluating Toxicologic Research*, Gralla, E.J., Ed., Hemisphere, Washington, DC, 1981, pp. 27–55; and Loew, F.M. and Fox, J.G., Animal health surveillance and health delivery systems, in *The Mouse in Biomedical Research, Volume III*, Foster, H.L., Small, J.D., and Fox, J.G., Eds., Academic Press, New York, 1983, pp. 69–82.

The elimination of biological variation may become even more important as toxicological studies, immunological studies, or aging studies become more common. These experimental categories rely heavily on standardized animal models for valid results.[33,34] The end point assays often measure more specialized functions of cells or organs of laboratory animals, requiring more uniformity among the animals. Any long-term study depends on the consistency of the animal and experimental variables throughout the entire length of the study. Any change in the health status of animals used in a study may disrupt the final interpretation of the data and may render the study not reproducible. Table 13.1 illustrates some of these research complications associated with certain murine pathogens. This table demonstrates only a small fraction of the possible ways in which microbial factors can alter research data in mice.

There are at least two other major benefits for controlling the microbial status of laboratory animals: reducing the risk of zoonotic disease exposure to humans and increasing the welfare of the animals. Several animal pathogens have zoonotic potential. Some, like Herpes B virus or lymphocytic choriomeningitis virus, have devastating effects in people infected with these agents.[35,36] When the risk of zoonotic disease exposure is high, the quality and scope of research is compromised due to concern for human safety. With the increasing concerns for animal welfare and adoption of refinement and reduction techniques, eliminating animal disease will benefit the quality of life for the laboratory animals as well as the quality of research.

## ABIOTIC FACTORS

The main emphasis is on the abiotic factors operant inside the cage or the pen. The basic elements there are wall-, bottom-, and top-material(s), diet, bedding, and water bottle or nipple. The optimal situation would, of course, be that anything introduced into the cage would be made of material already present or of inert material. The question is not whether the item is toxic or not, but rather whether it includes potential to interfere with the study. It may not be a problem in all studies, but one can easily foresee that there are sensitive studies, and for the sake of certainty, we may need to adhere to a "no-new-materials" approach. This approach has often not been followed when designing environmental enrichment programs.

### Bedding

Bedding properties that may have an impact on animals and results fall into two categories:

1. Endogenous factors such as enzyme induction, carcinogenicity, and ureolytic factors
2. Exogenous factors such as chemical residues, microbes, and dirty environment

A study by Vesell[74] clearly indicates causal relationship of softwood bedding and increased liver microsomal enzyme activity. This has later been attributed to cedrol or alpha-cedrene in cedar wood[75] and to alpha-pinene in spruce and pine wood.[76] Despite ample time for implementation, a recent study on volatile compounds in selected commonly used European beddings showed that many are virtually loaded with harmful compounds, and autoclaving is the second best thing to not using these types of bedding[77] (Figure 13.2). Because heat treatments like drying used by the manufacturer partially lower these concentrations, facilities may be using beddings with a highly variable profile of volatile substances. There should be no doubt that this will cause interference in all relevant studies.

It has been suggested that certain bedding materials used may contain carcinogenic compounds. Some studies demonstrate an association between the use of cedar material and an increased incidence of spontaneous tumors in rodents.[78,79] More specifically, Sabine[80] showed that cedar bedding was associated with a higher frequency of hepatic and mammary tumors in some mouse strains, yet there are studies in mice where no effect could be seen.[81–83]

Urease is an enzyme commonly found in plants, and hence, also in bedding materials of plant origin. A study comparing various beddings found three- to sevenfold amounts of activity in heat-treated hardwood as compared to pelleted alfalfa or pelleted corncobs, with large variation between batches. Furthermore, a heat-stable activator of bacterial urease was found in hardwood chips and crushed corncobs.[84] Because enteric bacteria contain urease, capable of increasing cage ammonia levels, it is unclear whether the urease activity in bedding has practical consequences.

When beddings of plant origin are used, residues such as heavy metals and herbicides accumulate to the bulk during growth. Although there are no widely used recommendations for maximum allowed concentrations of these residues, it appears feasible to apply the respective recommendations, which apply to diets.[85] In countries with timber industry, it must be remembered that antifungal agents are commonly used to prevent discoloring of wood. Practical experience has shown that rodent pups do not thrive on bedding containing these agents, yet effects on adult animals are yet to be assessed. Furthermore, low concentrations of N-nitrosoamines have been detected in bedding, but the practical validity and significance of this finding is unclear.[86]

Microbial accumulation in bedding material before use seems to be rare. Apparently, this requires poor drying or damage to packages during transport and consequent introduction of excessive moisture. Adverse health effects, such as fungal rhinitis[87] and tracheobronchial disease[88] have been reported in rats. One would assume that commonly used filter top cages with wood bedding would provide an optimal microenvironment for fungi in the cage, but a study to assess this with rats, with and without Penicillium added, shows the opposite.[89]

The criteria used for determining when bedding change is needed are variable. Some facilities change bedding before the fur of the animals becomes wet or soiled or when the bedding looks dirty or is smelly. A possible criterion to be used for the purpose could be bacterial and fungal accumulation in bedding during use. An exponential increase is the point at which bedding should be changed.[90]

New ventilation solutions, like individually ventilated cages (IVC), enhance ventilation inside the cage to high numbers of air changes per hour. Obviously, through the drying effect, this has led facilities to considerably reduce cage change frequency. A study focusing on cage microenvironment and animal health suggested that cage change every 2 weeks with 60 air changes per hour provides optimal conditions for mice in IVCs.[91] Yet it remains to be seen whether the environment in IVCs becomes dirty enough to cause impairment of hepatic microsomal enzymes, as reported in standard cages by Vesell.[92]

## Enrichment

Environmental enrichment deals with animal groups, housing enclosures, animal care procedures, and in-the-cage "furniture." It is not enough if enrichment does not cause mortality or morbidity or result in fighting or cause stress in the animals. It should not, as stated by Council of Europe expert group, interfere with the study or interpretation of the results.[93]

The Council of Europe expert group emphasizes the need for enrichment for all laboratory species unless there is scientific or veterinary reason not to use it.[93] One may assume that interference with the

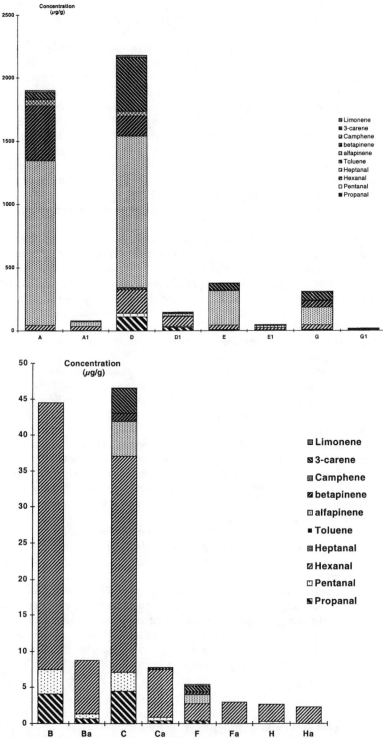

**Figure 13.2** Volatile compounds in commonly used beddings before and after autoclaving. (A) illustrates summative concentrations of ten volatile compounds of beddings with the highest sum; and (B) illustrates those with the lowest sum. Capital letters by x-axis depict bedding code, when followed by a, the same when autoclaved. (Reprinted with permission from *Scand. J. Lab. Anim. Sci.,* 23,101–104, 1996.)

experimental outcome could be an example of a scientific reason, and fighting between incompatible animals a veterinary reason.

Beneficial effects of environmental enrichment were shown half a century ago by Hebb,[94] who reported changes in physiology, behavior, and brain anatomy. Since then, two comprehensive reports have been published on refining mouse[95] and rabbit husbandry,[96] both containing numerous and detailed examples.

Two opposite approaches to enrichment are expressed in the views that state the following:

- Laboratory rodent environments can be effectively enriched using standard husbandry.
- The rodent cage is barren and lacks the necessary elements for physiological and psychological well-being.

Some enrichment forms have been associated with an increase in aggression, particularly in mice,[97] which made the facility in question abandon the approach. Assessment of the effects of enrichment is laborious and requires carefully executed studies. Furthermore, there are differences in response to enrichment between mouse strains.[98]

Whether environmental enrichment has a "reduction" outcome remains a controversial issue, and most likely, the last word has not been said on this topic. Many enrichment regimes are originally tailored to provide animals with better capability to cope with various challenges, which is assumed to lead to less variation in study results.[99–101] Yet, there are a few studies showing that enrichment may increase variation.[102,103]

Weighing "refinement" and "reduction" outcomes simultaneously has the best potential for assessing enrichment. A study by Eskola et al.[104] looked at arbitrarily chosen physiological and serum chemistry parameters from animals housed with and without two enrichment items made from aspen wood and thus using a no-new-materials approach. Rats used one of the items extensively, which can be considered verification of its enrichment and refinement value. The group applied power analyses to yield estimates of the number of animals needed to obtain significant results. The study showed increased and decreased numbers depending on which parameter was studied. Combining results, it was shown that enrichment as such is a statistically significant determinant of the appropriate number of animals needed in a study.[105,106]

If environmental enrichment changes variation, the scientists — some of whom may be totally unaware of which enrichment regimen is used in the study — may end up discovering or losing significances. Enrichment may not be the first thing they suspect as having an added contribution to between-animal variation.

Implementation of environmental enrichment should not be everyone's hobby, but a facility's choice based on scientific data. Enrichment should not be changed within a study or between studies meant to be compared or to be a continuum of each other. This makes it a difficult task to manage unless the same enrichment regimen is kept steady over a long period of time, just like the diet.

## Cage Material

Pertinent recommendations and guidelines on laboratory animal housing commonly contain specifications on space allocation and enrichment approaches.[93,94,107–110] Only a few of these deal with the cage materials as such.[94,107]

Traditionally, only two cage materials have been widely used in laboratory rodent housing: stainless steel and polycarbonate. Because stainless steel cages have mostly been grid floor and polycarbonate cages solid bottom enclosures, all possible choices have not been utilized. The real valid comparison of cage materials should be made with the same floor type.

So far, material choices have been based on practical aspects. Steel cages are perceived as durable and autoclaveable with nontransparent walls, making animal observation more laborious. Polycarbonate cages provide insulation, are lighter to handle, and make observations easier. The Berlin report and Rodent Refinement Working Party recommend plastic as the preferred rodent cage material.[95,106] Yet, especially to albino animals, the light in polycarbonate cages may be excessive.[111] Even though polycarbonate cages are more popular, the true refinement outcome of these housing alternatives necessitates assessment of the animals' preference.

Comparisons of solid bottom to grid floor have been made[112–114] and mice preference to different cage types has been studied by Baumans, Stafleu, and Bouv.[115] The effect of cage type has also been combined with the use of certain enrichment objects.[116] In another study by the same group, rats' preference to polycarbonate or stainless steel was compared, but there seemed to be no clear preference to either of the materials.[117]

In a recent study, rats could choose between stainless steel and polycarbonate cage halves. In most cage variations, the rats seemed to prefer stainless steel, irrespective of the cage material they were born and raised in. However, the position of a food hopper and illumination seemed to have significant effect on the choice.[118] Contrary to the general opinion, it appears that rats do not find steel less attractive than polycarbonate as cage material. Furthermore, animals in transparent cages are usually exposed to more variable lighting, a possible cause for increased variation.

## Humidity

For relative humidity (RH) in the cage, there is an optimal range, where no adverse effects can be seen in rodents and rabbits. ILAR gives a range between 30 to 70%,[110] but European guidelines call for a more narrow range of 45 to 65%.[108,119] Both ranges may seem wide, but because RH is closely related to temperature, together with strict temperature control, these ranges are not always easy to achieve.

The RH can be excessive in filter top cages,[120–124] mainly due to decreased ventilation.[125,126] Consequently, waste gases like carbon dioxide and ammonia may accumulate inside the cage.[121–123,126] These conditions may cause negative effects on the animals and their health.

Too low RH, which can easily occur during the heating season in cold climates, is associated with a specific disease called ringtail in rats, where necrotic belts are seen around the tails, and it may result in partial tail amputation.[127] From an occupational health and animal health view, it is noteworthy that at about 50% RH, viability of microbes in the air is lowest.[128] And, low RH increases concentration of dust particles and thus animal allergens in the animal room.[129]

## Light

Due to major difference between human and rodent vision, animal room lighting designed for personnel may not be suitable for the animals. Furthermore, most of the common laboratory animals are nocturnal or crepuscular, and they are in resting phase during our working hours. Some of the animals are albinos, which make them more sensitive to light than pigmented animals. The light intensity for albino animals to thrive well is quite low, about 25 lux. If one increases illumination to 60 lux, albino rats may develop retinal degeneration in 3 months.[111]

To accomplish a pleasant light intensity in the cage, the animals must be provided with shade, and overall illumination in the animal room should not be too bright. Introduction of cage "furniture," such as shelters and tubes, provide shaded place, where animals can retrieve. This is especially true in transparent cages, where the animals' preferred location often is under the hopper containing food and water bottle.[118]

Well below intensities leading to retinal damage, light intensity has been associated with reproductive and behavioral changes.[130] In this respect, the length of the photoperiod seems decisive. When lights are on less than 8 h a day, the rat estrous cycle ceases. Similarly, they may show constant estrus, devoid of ovulation, when photoperiod exceeds 14 h a day.[131] In general, equal length of dark and light period during 24 h is recommended for laboratory animals.

Distance and angle to light source, cage material and type, and presence or absence of nontransparent shelves in the cage rack may yield quite variable illumination between cage rows. All measures should be taken to make the illumination as even as possible, because only then will any possible reaction to light show least variation. If this cannot be accomplished, then the next best thing is to randomize cage positions in the rack, and simply live with the possibility of increased variation in the results.

## Sound

There are two philosophical approaches to sounds and laboratory animals. The "silence is golden" attitude aims at exclusion of all sounds. The opposite attitude regards total exclusion of sounds as deprivation of an acoustic environment for the animals, and that they should be provided with purposely made sounds, e.g., a radio music program. In addition to this refinement dimension, sounds cause a wide variety of physiological responses, with a possibility to interfere with the study results.

The reaction to sounds depends on hearing capabilities and on quality and type of sound. Unlike humans, many animals hear high-frequency sounds, so-called ultrasounds (>20 kHz). Albino rats can hear sounds ranging from 250 Hz to 80 kHz at 70 db, and the most sensitive area is between 8 to 38 kHz.[132,133]

Because of differences in rodent and human hearing ranges, the human ear may not be the right tool to assess sound disturbance to the animals. Björk et al.[134] calculated a new R-weighing to illustrate how rats hear, and analogically calculated H-weighing for humans, and compared these to commonly used A-weighing for humans. The comparison shows that studies dealing with rats and sound studies require different instrumentation than used in humans, and that R-weighing provides a new and important tool for rat hearing assessment (Figure 13.3).

Different sound types are combined with different behavioral responses in rats. Noise types result in fear reactions like startle, flight, and freezing, even when the sound pressure is low. Wave-type (like whistling, resembling rats' own vocalization) sounds cause movement and listening or no reaction at low-pressure sound levels. Adaptation to sounds occurs rapidly, but memory is short.[135]

Animal care routines and equipment, study apparatus, doors, sinks, and ventilation systems are the principle sources of external sounds in an animal facility. More than half of sources screened in an animal facility were shown to have ultrasound components. Overall frequencies and sound pressure varied considerably.[136]

A logical time to carry out experiments would be the time at which there is least noise activity, i.e., during weekends. Then any sound disturbance, like cage change and adding diet to hoppers, would be absent. If this is not possible, simply finding out which day(s) major animal care routines in the room are done, and then allowing the longest possible time to elapse before commencing the procedures or sampling of the study, is advisable.

## Temperature

The effects of ambient temperature on animals and on experimental results have been assessed thoroughly, and consequently, temperature control is in use in most facilities. There are guidelines for animal room

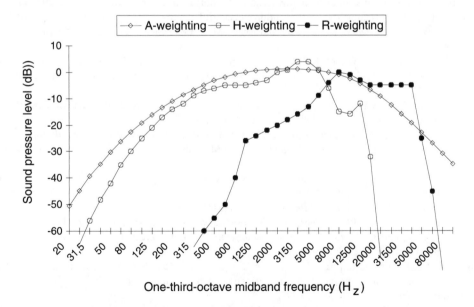

**Figure 13.3**  Illustration of novel sound pressure weighing, called R-weighing (rat), which takes account of rat's hearing sensitivity, with comparison to similarly processed, new H-weighing (human) and most commonly used A-weighing (human). (Reprinted with permission from *Lab. Anim.*, 34, 136–144, 2000.)

temperature,[108,110] giving species-specific ranges. Within these ranges, it is necessary to maintain tight temperature control. Yet, room values are not necessarily those inside the cage, where animals may enjoy warmer temperatures depending on a variety of factors like cage size and type, animal and group size, effective ventilation inside the cage, cage change interval, and type and amount of bedding.

Ideally, the temperature should be maintained at the set value ±1°C. If this is successful, temperature is unlikely to change metabolic rate in rodents, which as such may cause considerable variation in results.[131] Conversely, 4°C change in ambient temperature, possible within recommended temperature ranges, may result in a magnitude change of toxicity.[137]

As compared to standard caging, newer housing solutions like cubicles, ventilated cabinets, isolators, and individually ventilated cages increase ventilation efficiency inside the cage, especially so when no recirculation is allowed. This increases the chance for more uniform temperature between the cages. If ambient temperature is important in the study, this should result in less variation of results.

## REFERENCES

1. Biggers, J.D., McLaren, A., and Michie D., Variance control in the animal house, *Nature,* 182, 77–80, 1958.
2. Russell, W.M.S. and Burch, R.L., *The Principles of Humane Experimental Techniques*, Methuen, London, 1959.
3. GV-SOLAS, Working committee for the biological characterization of laboratory animals, Guidelines for specification of animals and husbandry methods while reporting the results of animal experiments, *Lab. Anim.,* 19, 106–108, 1985.
4. GV-SOLAS, Implications of infectious agents on results of animal experiments, *Lab. Anim.,* 33(Suppl.1), S1:39–S1:87, 1999.
5. FELASA, FELASA recommendations for the health monitoring of rodent and rabbit breeding colonies and experimental units, *Lab. Anim.,* 36, 20, 2002.
6. Baker, D.G., Natural pathogens of laboratory mice, rats, and rabbits and their effects on research, *Clin. Microbiol. Rev.,* 11, 2, 231–266, 1998.
7. O'Rourke, J., Lee, A., and McNeill, J., Differences in the gastrointestinal microbiota of specific pathogen free mice: an often unknown variable in biomedical research, *Lab. Anim.,* 22, 297–303, 1988.
8. Frost, W.W. and Hamm, T.E. Jr., Prevention and control of animal disease, in *The Experimental Animal in Biomedical Research*, Rollin, B.E. and Kesel, M.L., Eds., CRC Press, Boca Raton, FL, 1990, pp. 133–152.
9. Schaedler, R.W. and Orcutt, R.P., Gastrointestinal microflora, in *The Mouse in Biomedical Reseach, Volume III*, Foster, H.L., Small, J.D., and Fox, J.G., Eds., Academic Press, New York, 1983, pp. 327–345.
10. FELASA, FELASA recommendations for the health monitoring of breeding colonies and experimental units of cats, dogs, and pigs, *Lab. Anim.,* 32, 1–17, 1998.
11. Pakes, S.P., Lu, Y-S., and Meunier, P.C., Factors that complicate animal research, in *Laboratory Animal Medicine*, Fox, J.G., Cohen, B.J., and Loew, F.M., Eds., Academic Press, New York, 1984, pp. 649–665.
12. Fox, J.G., Yan, L., Cahill, R.J., Hurley, R., Lewis, R., and Murphy, J.C., Chronic proliferative hepatitis in A/JCr mice associated with persistent *Helicobacter hepaticus* infection: a model of helicobacter-induced carcinogenesis, *Infect. Immun.,* 64, 1548–1558, 1996.
13. Ward, J.M., Anver, M.R., Haines, D.C., Melhorn, J.M., Gorelick, P., Yan, L., and Fox, J.G., Inflammatory large bowel disease in immunodeficient mice naturally infected with *Helicobacter hepaticus*, *Lab. Anim. Sci.,* 46, 15–20, 1996.
14. Cahill, R.J., Foltz, C.J., Fox, J.G., Dangler, C.A., Powrie, F., and Schauer, D.B., Inflammatory bowel disease: an immunity-mediated condition triggered by bacterial infection with *Helicobacter hepaticus*, *Infect. Immun.,* 65, 3126–3131, 1997.
15. Shomer, N.H., Dangler, C.A., Schrenzel, M.D., and Fox, J.G., *Helicobacter bilis*-induced inflammatory bowel disease in scid mice with defined flora, *Infect. Immun.,* 65, 4858–4864, 1997.
16. Truett, G.E., Walker, J.A., and Baker, D.G., Eradication of infection with *Helicobacter* spp. by use of neonatal transfer, *Comp. Med.,* 50, 4, 444–451, 2000.

17. Whary M.T., Cline J.H., King A.E., Corcoran C.A., Xu S., and Fox J.G., Containment of *Helicobacter hepaticus* by use of husbandry practices, *Comp. Med.,* 50, 1, 78–81, 2000.

18. McNair, D.M. and Timmons, E.H., Effects of *Aspiculuris tetraptera* and *Syphacia obvelata* on exploratory behavior of an inbred mouse strain, *Lab. Anim. Sci.,* 27, 1, 38–42, 1977.

19. Webster, J.P., The effect of *Toxoplasma gondii* and other parasites on activity levels in wild and hybrid *Rattus norvegicus, Parasitology,* 109, Pt 5, 583–589, 1994.

20. Braithwaite, V.A., Salkeld, D.J., McAdam, H.M., Hockings, C.G., Ludlow, A.M., and Read, A.F., Spatial and discrimination learning in rodents infected with the nematode *Strongyloides ratti, Parasitology,* 117, 145–154, 1998.

21. Mohn, G. and Philipp, E.-M., Effects of *Syphacia muris* and the anthelmintic fenbendazole on the microsomal monooxygenase system in mouse liver, *Lab. Anim.,* 15, 98–95, 1981.

22. Agersborg, S.S., Garza, K.M., and Tung, K.S.K., Intestinal parasitism terminates self tolerance and enhances neonatal induction of autoimmune disease and memory, *Eur. J. Immunol.,* 31, 851–859, 2001.

23. Kavaliers, M. and Colwell, D.D., Parasite infection attenuates nonopiod mediated predator-induced analgesia in mice, *Physiol. Behav.,* 55, 3, 505–510, 1994.

24. Wagner, M., The effect of infection with the rat pinworm (*Syphacia muris*) on rat growth, *Lab. Anim. Sci.,* 38, 4, 476–478, 1988.

25. Sato, Y., Ooi, H.K., Nonaka, N., Oku, Y., and Kamiya, M., Antibody production in *Syphacia obvelata* infected mice, *J. Parasitol.,* 81, 4, 559–562, 1995.

26. Reyes, L., Steiner, D.A., Hutchison, J., Crenshaw, B., and Brown, M.B., *Mycoplasma pulmonis* genital disease: effect of rat strain on pregnancy outcome, *Comp. Med.,* 50, 6, 622–627, 2000.

27. Parker, J.C., Whiteman, M.D., and Richter, C.B., Susceptibility of inbred and outbred mouse strains to Sendai virus and prevalence of infection in laboratory rodents, *Infect. Immun.,* 19, 1, 123–130, 1978.

28. Wilberz, S., Partke, H.J., Dagnaes-Hansen, F., and Herberg, L., Persistent MHV (mouse hepatitis virus) infection reduces the incidence of diabetes mellitus in non-obese diabetic mice, *Diabetologia,* 34, 2–5, 1991.

29. Cooke, A., Tonks, P., Jones, F.M., O'Shea, H., Hutchings, P., Fulford, A.J.C., and Dunne, D.W., Infection with *Schistosoma mansoni* prevents insulin dependent diabetes mellitus in non-obese diabetic mice, *Parasite Immunol.,* 21, 169–176, 1999.

30. Oldstone, M.B.A., Viruses as therapeutic agents. I. Treatment of nonobese insulin-dependent diabetes mice with virus prevents insulin-dependent diabetes mellitus while maintaining general immune competence, *J. Exp. Med.,* 171, 2077–2089, 1990.

31. Brabb, T., Goldrath, A.W., von Dassow, P., Paez, A., Liggitt, H.D., and Goverman, J., Triggers of autoimmune disease in a murine TCR-transgenic model for multiple sclerosis, *J. Immunol.,* 159, 497–507, 1997.

32. Penhale, W.J. and Young, P.R., The influence of the normal microbial flora on the susceptibility of rats to experimental autoimmune thyroiditis, *Clin. Exp. Immonol.,* 72, 288–292, 1988.

33. Van der Loght, J.T.M., Necessity of a more standardized virological characterization of rodents for aging studies, *Neurobiol. Aging,* 12, 6, 669–672, 1991.

34. Sebesteny, A., Necessity of a more standardized microbiological characterization of rodents for aging studies, *Neurobiol. Aging,* 12, 6, 663–668, 1991.

35. Muchmore, E., An overview of biohazards associated with nonhuman primates, *J. Med. Primatol.,* 16, 55–82, 1987.

36. Fox, J.G., Newcomer, C.E., and Rozmiarek, H., Selected zoonoses and other health hazards, in *Laboratory Animal Medicine,* Fox, J.G., Cohen, B.J., and Loew, F.M., Eds., Academic Press, New York, 1984, pp. 613–648.

37. Jacoby, R.O. and Barthold, S.W., Quality assurance for rodents used in toxicological research and testing, in *Scientific Considerations in Monitoring and Evaluating Toxicologic Research,* Gralla, E.J., Ed., Hemisphere, Washington, DC, 1981, pp. 27–55.

38. Loew, F.M. and Fox, J.G., Animal health surveillance and health delivery systems, in *The Mouse in Biomedical Research, Volume III,* Foster, H.L., Small, J.D., and Fox, J.G., Eds., Academic Press, New York, 1983, pp. 69–82.

39. Stanley, N.F., Dorman, D.C., and Ponsford, J., Studies on the pathogenesis of a hitherto undescribed virus (hepatoencephalomyelitis) producing unusual symptoms in suckling mice, *Aust. J. Expt. Med. Sci.,* 31, 147, 1953.

40. Walters, M.N.I., Joske, R.A., Leak, P.J., and Stanley, N.F., Murine infection with retrovirus. I. Pathology of the acute phase, *Br. J. Exp. Pathol.*, 44, 427, 1963.

41. Walters, M.L., Stanley, N.F., Dawlins, R.L., and Alpers, M.P., Immunologic assessment of mice with chronic jaundice and runting induced by reovirus 3, *Br. J. Exp. Pathol.*, 54, 329-345, 1973.

42. Cross, S.S. and Parker, J.C., Viral contaminants of mouse tumor systems, *Bacteriol. Proc.*, 163, (abstract), 1967.

43. Piazza, M., Hepatitis in mice, in *Experimental Viral Hepatitis*, Thomas, Springfield, IL, 1969, pp. 13–140.

44. Hotchin, J., The contamination of laboratory animals with lymphocytic choriomeningitis virus, *Am. J. Pathol.*, 64, 747–769, 1971.

45. Mims, C.A. and Tosolini, F.A., Pathogenesis of lesions in lymphoid tissue of mice infected with lymphocytic choriomeningitis (LCM) virus, *Br. J. Exp. Pathol.*, 50, 584–592, 1969.

46. Bro-Jorgensen, K. and Volkert, M., Defects in the immune system of mice infected with lymphocytic choriomeningitis virus, *Infect. Immun.*, 9, 605–614, 1974.

47. Jacobs, R.P. and Cole, G.A., Lymphocytic-choriomeningitis virus-induced immunosupression: a virus-induced macrophage defect, *J. Immunol.*, 117, 1004–1009, 1976.

48. Rowson, K.E.K. and Mahy, B.W. J., Lactic dehydrogenase virus, *Virol. Monogr.*, 13, 1–121, 1975.

49. Notkins, A.L., Enzymatic and immunologic alterations in mice infected with lactic dehydrogenase virus, *Am. J. Pathol.*, 64, 733–746, 1971.

50. Notkins, A.L., Lactic dehydrogenase virus, *Bacteriol. Rev.*, 29, 143, 1965.

51. Parker, J.C., Collins, M.J., Cross, S.S., and Rowe, W.P., Minute virus of mice. II. Prevalence, epidemiology, and occurrence as a contaminant of transplantable tumors, *J. Natl. Cancer Inst.*, 45, 305–310, 1970.

52. Appell, L.H., Kovatch, R.M., Reddecliff, J.M., and Gerone, P.J., Pathogenesis of Sendai virus infection in mice, *Am. J. Vet. Res.*, 32, 1935–1941, 1971.

53. Robinson, T.W.E., Cureton, R.J.R., and Heath, R.B., The pathogenesis of Sendai virus infection in the mouse lung, *J. Med. Microbiol.*, 1, 89–95, 1968.

54. Kay, M.M.B., Long-term subclinical effects of parainfluenza (Sendai) infection on immune cells of aging mice, *Proc. Soc. Exp. Biol. Med.*, 158, 326–331, 1978.

55. Selgrade, M.K., Ahamed, A., Sell, K.W., Gershwin, M.E., and Steinberg, A.D., Effect of murine cytomegalovirus on the *in vitro* responses of T and B cells to mitogens, *J. Immunol.*, 116, 1459–1465, 1976.

56. Lussier, G., Murine cytomegalovirus (MCMV), *Adv. Vet. Sci. Comp. Med.*, 19, 223–247, 1975.

57. Cohen, P.L., Cross, S.S., and Mosier, D.E., Immunologic effects of neonatal infection with mouse thymic virus, *J. Immunol.*, 115, 706–710, 1975.

58. Lindsey, J.R., Baker, H.J., Overeash, R.G., Cassell, G.H., and Hunt, C.E., Murine chonic respiratory disease, *Am. J. Pathol.*, 64, 675–716, 1971.

59. Ventura, J. and Domaradzki, M., Role of mycoplasma infection in the development of experimental bronchiectasis in the rat, *J. Pathol. Bacteriol.*, 93, 342–348, 1967.

60. Green, G.M., The Burns Amberson lectures in defense of the lung, *Am. Rev. Respir. Dis.*, 102, 691–703, 1980.

61. Cole, B.C. and Cassell, G.H., Mycoplasma infections as models of chronic joint inflammation, *Arthritis Rheum.*, 22, 1375–1381, 1979.

62. Eckner, R.J., Han, G., and Kumar, V., Immuno-supression of cell-mediated immunity but not humoral immunity, *Fed. Proc. Fed. Am. Soc. Exp. Biol.*, 33, 769, 1974.

63. Barthold, S.W., Research complications and state of knowledge of rodent coronaviruses, in *Complications of Viral and Mycoplasmal Infections in Rodents to Toxicology Research Testing*, Hamm, T.F., Ed., Hemisphere, Washington, 1986, pp 53–89.

64. Lardans, V., Godfraind, C., van der Logt, J.T., Heessen, W.A., Gonzalez, M.D., and Coutelier, J. P., Polyclonal B lymphocyte activation induced by mouse hepatitis virus A59 infection, *J. Gen. Vir.*, 77, 1005–1009, 1996.

65. Schijns, V.E., Haagmans, B.L., Wierda, C.M., Kruithof, B., Heijnen, I.A., Alber, G., and Horzinek, M.C., Mice lacking IL-12 develop polarized Th1 cells during viral infection, *J. Immunol.*, 160, 3958–3964, 1998.

66. De Souza, M.S. and Smith, A.L., Characterization of accessory cell function during acute infection of BALB/cByJ mice with mouse hepatitis virus (MHV) strain JHM, *Lab. Anim. Sci.,* 41, 112–118, 1991.
67. Koler, M., Ruttner, B., Cooper, S., Hengartner, H., and Zinkernagel, R.M., Enhanced tumor susceptibility of immunocompetent mice infected with lymphocytic choriomeningitis virus, *Cancer Immunol. Imunother.,* 32, 117–24, 1990.
68. Hayashi, T., Hashimoto, S., and Kawashima, A., Effect of infection by lactic dehydrogenase virus on expression of intercellular adhesion molecule-1 on vascular endothelial cells of pancreatic islets in streptozotocin-induced insulitis of CD-1 mice, *Int. J. Exp. Pathol.,* 75, 211–217, 1984.
69. Verdonck, E., Pfau, C.J., Gonzalez, M.D., Masson, P.L., and Coutelier, J.P., Influence of viral infection on anti-erythrocyte autoantibody response after immunization of mice with rat red blood cells, *Autoimmunity,* 17, 73–81, 1994.
70. Takei, I., Asaba, Y., Kasatani, T., Maruyama,, T. Watanabe, K., Yanagawa, T., Saruta, T., and Ishii, T., Suppression of development of diabetes in NOD mice by lactate dehydogenase virus infection, *J. Autoimmun.,* 5, 665–673, 1992.
71. Tattersall, P. and Cotmore, S.F., The rodent parvoviruses, in *Viral and Mycoplasmal Infections of Laboratory Rodents: Effects on Biomedical Research,* Bhatt, P.N., Jacoby, R.O., Morse, III, A.C., and New, A.E., Eds., Academic Press, New York, 1986, pp.305–348.
72. Kimsey, P.B., Engers, H.D., Hirt, B., and Jongeneel, V., Pathogenicity of fibroblast- and lymphocte-specific variants of minute virus of mice, *J. Virol.,* 59, 8–13, 1986.
73. Yunis, E.J. and Salazar, M., Genetics of life span in mice, *Genetica,* 91, 211–223, 1993.
74. Vesell, E., Induction of drug-metabolizing enzymes in liver microsomes of mice and rats by softwood bedding, *Science,* 157, 1057–1058, 1967.
75. Bang, L. and Ourisson, G., Hydroxylation of cedrol by rabbits, *Tetrahedron Letters,* 1881–1884, 1975.
76. Nielsen, J.B., Andersen, O., and Svendsen, P., Effekt af stroelse på leverens Cytochrome P-450 i mus, *ScandLAS nyt,* 11, 7–13, 1984.
77. Nevalainen, T. and Vartiainen, T., Volatile organic compounds in commonly used beddings before and after autoclaving, *Scand. J. Lab. Anim. Sci.,* 23,101–104, 1996.
78. Schoental, R., Carcinogenicity of wood shavings, *Lab. Anim.,* 7, 47–49, 1973.
79. Schoental, R., Role of podophyllotoxin in the bedding and dietary Zealarone on incidence of spontaneous tumors in laboratory animals, *Cancer Res.,* 34, 2419–2420, 1974.
80. Sabine, J.R., Exposure to an environment containing the aromatic red cedar, *Juniperus virginiana*: procarcinogenic, enzyme-inducing, and insecticidal effects, *Toxicol.,* 5, 221–235, 1975.
81. Vlahakis, G., Brief communication: Possible carcinogenic effects of cedar shavings in bedding of C3H-A$^{vy}$fB mice, *J. Natnl. Cancer Inst.,* 58, 149–150, 1977.
82. Jacobs, B.B. and Dieter, D.K., Spontaneous hepatomas in mice inbred from Ha:ICR Swiss stock: Effects of sex, cedar shavings in bedding, and immunization with fetal liver or hepatoma cells, *J. Natl. Cancer Inst.,* 61, 1531–1534, 1978.
83. Tennekes, H.A., Wright, A.S., Dix, K.M., and Koeman, J.H., Effects of dieldrin, diet, and bedding on enzyme function and tumor incidence in livers of male CF-1 mice, *Cancer Res.,* 41, 3615–3620, 1981.
84. Gale, R.G. and Smith, A.B., Ureolytic and urease activating properties of commercial laboratory animal bedding, *Lab. Anim. Sci.,* 31, 56–59, 1981.
85. Fox, J., Selected aspects of animal husbandry and good laboratory practises, *Clinical Toxicol.* 15, 539–553, 1973.
86. Silverman, J. and Adams, J.D., N-Nitrosoamines in laboratory animal diet and bedding, *Lab. Anim. Sci.,* 33, 161–164, 1983.
87. Royals, M.A., Getzy, D.M., and Vandewoude, S., High fungal spore load in corncob bedding associated with fungal-induced rhinitis in two rats, *Contemp. Topics,* 38, 64–66, 1999.
88. Hubbs, A.F., Hahn, F.F., and Lundgren, D.I., Invasive tracheo-bronchial aspergillosis in an F344/N rat, *Lab. Anim. Sci.,* 41, 521–524, 1991.
89. Pernu, N., Hyvärinen, A., Toivola, M., Nevalainen, A., Harri, M., and Nevalainen, T., Filter top cages: moldy homes for rodents?, *Scand. J. Lab. Anim. Sci.,* 27, 1–12, 2000.
90. Haataja, H., Voipio, H-M., and Nevalainen, T., Bedding bacterial counts as indicators of hygiene in rats, *Scand. J. Lab. Anim. Sci.,* 16, 123–126, 1989.

91. Reeb-Whitaker, C.K., Paigen, B., Beamer, W.G., Bronson, R.T., Churchill, G.A., Schweitzer, I.B., and Myers, D.D., The impact of reduced frequency of cage changes on the health of mice housed in ventilated cages, *Lab. Anim.*, 35, 58–73, 2000.

92. Vesell, E., Lang, C.M., and White, W.J., Hepatic drug metabolism in rats: impairment of a dirty environment, *Science,* 179, 896–897, 1973.

93. Hansen, A.K., Baumans, V., Elliot, H., Francis, R., Holgate, B., Hubrecht, R., Jennings, M., Peters, A., and Stauffacher, M., Future principles for housing and care of laboratory rodents and rabbits. Report concerning revision of the Council of Europe Convention ETS 123 Appendix A concerning questions related to rodents and rabbits issued by the Council's working group for rodents and rabbits. Part A, Actions and proposals of the working group, 1999.

94. Hebb, D.O., The effects of early experience on problem-solving at maturity, *Am. Psychol.*, 2, 306–307, 1947.

95. Jennings, M., Batchelor, G.R., Brain, P.F., Dick, A., Elliot, H., Francis, R.J., Hubrecht, R.C., Hurst, J.L., Morton, D.B., Peters, A.G., Raymond, R., Sales, G.D., Sherwin, C.M., and West, C., Refining rodent husbandry: the mouse, Report of the Rodent Refinement Working Party, *Lab. Anim.*, 32, 233–259, 1998. See also: www.LAL.org.uk.

96. Second Report of the BVAAWF/FRAME/RSPCA/UFAW joint working group on refinement, Refinements in rabbit husbandry, *Lab. Anim.,* 27, 301–329, 1993.

97. Haemisch, A., Voss, T., and Gärtner, K., Effects of environmental enrichment on aggressive behaviour, dominance hierarchies and endocrine states in male DBA/2J mice, *Physiol. and Behav.*, 56, 1041–1048, 1994

98. van de Weerd, H.A., Baumans, V., Koolhaas, J.M., and van Zupthen, L.F., Strain specific behavioural response to environmental enrichment in the mouse, *J. Exp. Anim. Sci.*, 36, 117–127, 1994.

99. Broom, D.M., Indicators of poor welfare, *Brit. Vet. J.,* 142, 524–526, 1986

100. Baumans, V., Environmental enrichment: practical applications, in *Animal Alternatives, Welfare and Ethics*, van Zupthen, L.F.M. and Balls, M., Eds., Elsevier, Amsterdam; New York, 1997, pp. 187–197.

101. Stauffacher, M., Comparative studies in housing conditions, in *Harmonization of Laboratory Animal Husbandry*, O'Donoghue, P.N., Ed., Royal Soc. Med. Press, London, 1997.

102. Haemisch, A. and Gärtner, K., The cage design affects intermale aggression in small groups of male mice: strain specific consequences on social organization, and endocrine activations in two inbred strains, *J. Exp. Anim. Sci.*, 36, 101–116, 1994.

103. Tsai, P.P. and Hackbarth, H., Environmental enrichment in mice; Is it suitable for every experiment?, Abstracts of Scientific Papers, ICLAS, FELASA, 26–28 May, 1999, Palma de Mallorca, Balearic Islands, Spain.

104. Eskola, S., Lauhikari, M., Voipio, H.-M., and Nevalainen, T., The use of aspen blocks and tubes to enrich the cage environment of laboratory rats, *Scand. J. Lab. Anim. Sci.,* 26, 1–10, 1999.

105. Eskola, S., Lauhikari, M., Voipio, H.-M., Laitinen, M., and Nevalainen, T., Environmental enrichment may alter the number of rats needed to achieve statistical significance, *Scand. J. Lab. Anim. Sci.*, 26, 3, 134–144, 1999.

106. Mering, S., Kaliste-Korhonen, E., and Nevalainen, T., Estimates of appropriate number of rats: interaction with housing environment, *Lab. Anim.,* 35, 80–90, 2001.

107. Brain, P.F., Buttner, D., Costa, P., Gregory, J.A., Heine, W.O.P., Koolhaas, J., Militzer, K., Ödberg, F.O., Scharmann, W., and Stauffacher, M., Rodents, in *The Accommodation of Laboratory Animals in Accordance with Animal Welfare Requirements*, O'Donoghue, P., Ed., Proceedings of an International Workshop held at the Bundesgesundheitsamt, Berlin 17–19 May 1993, 1–14.

108. European convention for the protection of vertebrate animals used for experimental and other scientific purposes, Council of Europe, Publications and Documents Division, Strasbourg, 1986.

109. Multilateral Consultation of Parties to the Convention for the protection of vertebrate animals used for experimental or other scientific purposes (ETS 123): Resolution on the accommodation and care of laboratory animals, May 1997.

110. Institute of Laboratory Animal Resources (ILAR), *Guide for the Care and Use of Laboratory Animals*, National Academy Press, Washington, DC, 1996.

111. Schlingmann, F., Pereboom, W.J., and Remie, R., The sensitivity of albino and pigmented rats to light: a mini review, *Anim. Techn.,* 44, 71–86, 1993.

112. Manser, C.E., Morris, T.H., and Broom, D.M., An investigation into the effects of solid or grid flooring on the welfare of laboratory rats, *Lab. Anim.*, 29, 353–363, 1995.

113. Manser, C.E., Elliott, H., Morris, T.H., and Broom, D.M., The use of a novel operant test to determine the strength of preference for flooring in laboratory rats, *Lab. Anim.*, 30, 1–6, 1996.

114. Van de Weerd, H.A., van den Broek, F.A.R., and Baumans, V., Preference for different types of flooring in two rat strains, *Appl. Anim. Behav. Sci.*, 46, 251–261, 1996.

115. Baumans, V., Stafleu, F.R., and Bouv, J., Testing housing system for mice — the value of a preference test, *Z. Versuchstierk.*, 29, 9–14, 1987.

116. Eskola, S. and Kaliste-Korhonen, E., Effects of cage type and gnawing blocks on weight gain, organ weights and open-field behaviour in Wistar rats, *Scand. J. Lab. Anim. Sci.*, 25, 180–193, 1998.

117. Kaliste-Korhonen, E., Kelloniemi, J., and Harri, M., Cage material and rat behaviour, *Scand. J. Lab. Anim. Sci.*, 23, 125–128, 1996.

118. Heikkilä, M., Sarkanen, R., Voipio, H.-M., Mering, S., and Nevalainen, T., Cage position preferences of rats, *Scand. J. Lab. Anim. Sci.*, 28, 65–74, 2001.

119. Home Office, Animals (Scientific Procedures) Act 1986: Code of Practice for the Housing and Care of Animals Used in Scientific Procedures (HC 107), London: Her Majesty's Stationary Office, 1986.

120. Baer, L.A., Corbin, B.J., Vasques, M.F., and Grindeland, R.E., Effects of the use of filtered microisolator tops on cage microenvironment and growth rate of mice, *Lab. Anim. Sci.*, 47, 327–329, 1997.

121. Corning, B.F. and Lipman, A., A comparison of rodent caging systems based on microenvironmental parameters, *Lab. Anim. Sci.*, 41, 498–503, 1991.

122. Lipman, N.S., Corning, B.F., and Coiro, Sr., M.A., The effects of intracage ventilation on microenvironmental conditions in filter-top cages, *Lab. Anim.*, 26, 206–210, 1992.

123. Perkins, S.E. and Lipman, A., Characterisation and quantification of microenvironmental contaminants in isolator cages with a variety of contact beddings, *Contemp. Top.*, 34, 93–97, 1995.

124. Simmons, M.I., Robie, D.M., Jones, J.B., and Serrano, L.J., Effect of filter cover on temperature and humidity in a mouse cage, *Lab. Anim.*, 2, 113–120, 1968.

125. Keller, L.S., White, W.J., Snider, M.T., and Lang, C.M., An evaluation of intracage ventilation in three animal caging systems, *Lab Anim. Sci.* 39, 237–242, 1989.

126. Serrano, L.J., Carbon dioxide and ammonia in mouse cages: effect of cage covers, population, and activity, *Lab. Anim. Sci.*, 21, 75–85, 1971.

127. Njaa L.R., Utne F., and Braekken O.R., Effect of humidity on rat breeding and ringtail, *Nature* (London), 180, 290, 1957.

128. Anderson, J.D. and Cox, C.S., Microbial survival, in *Airborne Microbes*, 17th Symposium of the Society of General Microbiology, Gregory P.H. and Monteith, J.L., Cambridge University Press, Cambridge, 1967, pp. 203–226.

129. Jones, R.B., Kacergis, J.B., MacDonald, M.R., McKnight, F.T., Turner, W.A., Ohman, J.L., and Paigen, B., The effect of relative humidity on mouse allergen levels in an environmentally controlled mouse room, *Am. Ind. Hyg. Assoc.*, 56, 398–401, 1995.

130. Donnelly, H. and Saibaba, P., Light intensity and the oestrous cycle in albino and normally pigmented mice, *Lab. Anim.*, 27, 385–390, 1993.

131. Clough, G., Environmental effects on animals used in biomedical research, *Biological Reviews*, 57, 487–523, 1982.

132. Gourevitch, G. and Hack, M.H., Audibility in the rat, *J. Comp. Physiol. Psychol.*, 62, 289–291, 1966.

133. Kelly, J.B. and Masterton, B., Auditory sensitivity of the albino rat, *J. Comp. Physiol. Psychol.*, 91, 930–936, 1977.

134. Björk, E., Nevalainen, T., Hakumäki, M., and Voipio, H-M., R-weighting provides better estimation for rat hearing sensitivity, *Lab. Anim.*, 34, 136–144, 2000.

135. Voipio, H.-M., How do rats react to sound?, *Scand. J. Lab. Anim. Sci.*, 24 (Supplement 1), 1–80, 1997.

136. Sales, G.D., Wilson, K.J., Spencer, K.E.V., and Milligan, S.R., Environmental ultrasound in laboratories and animal houses: a possible cause for concern in the welfare and use of laboratory animals, *Lab. Anim.*, 22, 369–375, 1988.

137. Harri, M.N.E., Effect of body temperature on cardiotoxicity of isoprenaline in rats, *Acta Pharmacol. Toxicol.*, 39, 214–224, 1976.

CHAPTER **14**

# Experimental Design and Statistical Analysis

**Michael F.W. Festing and Benjamin J. Weigler**

## CONTENTS

0-8493-1086-5/03/$0.00+$1.50
© 2003 by CRC Press LLC

## INTRODUCTION

It is difficult to design a perfect animal experiment. It needs to be just the right size; too small and it may not be able to detect important biological effects; too large and scientific resources and animals will be wasted. Decisions need to be taken about the choice of animals, including species, strain, and sex. Diet, housing, and other environmental factors need to be considered. Treatments must be chosen, which may involve deciding on suitable dose levels, routes of administration, and time factors. This often involves a certain amount of guess work. Possible interactions between these treatments and the strain and sex of the animals must be considered. Is it important to know whether males and females or different strains respond in the same way? The characters to be measured must be chosen, and the importance of measurement errors must be taken into account. Ways of reducing variability need to be considered, as heterogeneity will reduce the power of the experiment to demonstrate, on statistical grounds, any differences in treatment effects that truly exist. And finally, the methods used for statistical analysis of the resulting data must be considered in order to maximize the effective use of all of the information available to the scientist.

One result of having to make these choices is that well-designed experiments are relatively rare in the real world of biomedical research. In practice, experiments can range from those deemed very good and properly address the research questions economically and efficiently, to those that provide the intended answers but are quite inefficient, to those that are so awful that they lead to erroneous conclusions and waste resources. On ethical grounds, it is difficult to justify experiments that are poorly designed if the mistakes have arisen from poor understanding of statistical principles by the investigator.

Fortunately, most experiments give the intended results if only because the effects of many treatments are so dramatic that they are obvious even within the context of a badly designed experiment. However, when the effects are more subtle, there is a danger that real biological differences will be obscured by "noise" or uncontrolled variation in the study that can lead to false negative findings (known as type II error). In other cases, experiment-related noise can act in such a way as to produce false positive results (or type I error). Both scenarios can be minimized through proper planning in the design and analysis phase of laboratory animal research. The aim of this chapter is to give a brief introduction to the principles of experimental design and the statistical analysis of animal experiments in hopes of guiding research workers in planning and executing their own simpler studies and to facilitate their communications with statisticians for more advanced and complicated ones. It is a chapter *about*, rather than a description *of* experimental design and statistics. The concepts are illustrated with numerical examples of real data that have been analyzed using the MINITAB statistical package (version 13, MINITAB Inc., State College, PA), though other statistical software would have given similar results. It should, therefore, be used in conjunction with a textbook such as one of those listed in the reference section and with statistical software manuals. Readers are encouraged to work with persons knowledgeable in this area whenever necessary to ensure proper usage of the methods presented in this chapter.

## THE NEED FOR IMPROVED EXPERIMENTAL DESIGN

A basis exists for concern in the misuse of statistical methods within the literature of laboratory animal science.[1,2] A survey of papers published in two academic toxicology journals found that roughly a third of the experiments were about twice as large as they needed to be, none of them used blocking, a technique for improving precision, and more than 60% used incorrect statistical methods.[3] A similar survey of papers published in a veterinary journal, commissioned by its editors, found mistakes that included failure to use randomization, potential bias, inappropriate statistical methods, and a few cases where the conclusions were not supported by the data.[4] These problems are not unique to animal experiments. Surveys of published papers in human medical journals also show that there is room for improvement,[5,6] with some authors suggesting that the misuse of statistics can be viewed as a breach of ethical principles in science.

# BASIC STATISTICAL CONSIDERATIONS

## Observational and Epidemiologic Studies

Data used to explore a scientific hypothesis can be acquired in several different ways. Population-based investigations involving surveys, cross-sectional studies, case-control studies, and cohort studies are widely used for epidemiologic investigations and have recently been reviewed elsewhere.[7] Epidemiologic work also includes methods to obtain unbiased measures of disease frequency, such as incidence and prevalence, that are valuable for comparing trends and the impact of prevention efforts. Epidemiologic studies, whether prospective or retrospective, can be designed to explore exposure–disease relationships in situations where formal controlled experiments are impossible, unethical, or impractical, such as in assessing whether smoking causes lung cancer in human beings. Hypothesis-based observational investigations can shed light onto possible risk factors for disease, and thereby aid in the development of intervention strategies. Relative risks, odds ratios, and various other measures of association are used to quantify the importance of study factors on disease development in groups of individuals or animals. The study factors can include natural events and exposures that are not controlled by the researcher and can address agent variability, dose, route, host genetics, and important covariables more typical of the spectrum of real-world conditions.

Epidemiologic evidence is used in conjunction with findings from controlled experiments in laboratory animals to strengthen conclusions about causality and to dissect mechanistic pathways. A logic-based analysis using nine criteria provide for the critical evaluation of putative causes of effect obtained from observational datasets.[8] They differ from the traditional Henle–Koch postulates in part because they allow for conditions where infectious agents play no or only partial roles and also where lack of animal models makes establishment of causation difficult. For example, the biological plausibility reasoned from knowledge that tobacco smoke contains carcinogens and consistency of the association in different populations adds strength to conclusions regarding tobacco smoke as a cause of lung cancer. The contributions of genetic, social, and environmental factors in disease pathways can also be addressed through observational methods.[9]

Clinical trials are often discussed in epidemiology texts and nearly always represent one form of prospective randomized controlled experiment, as discussed below. These typically involve vaccines, drugs, nutrients, or procedures, with the objective being to determine safety and efficacy for prevention or treatment of a disease or condition. Losses to follow-up and ensuring that individuals remain compliant throughout the test period are problems that must be appropriately addressed through the design and statistical analysis.[10]

## Controlled, Randomized Experiments

In contrast to purely observational studies, randomized, controlled, animal experiments offer scientific opportunities to investigate possible causal relationships between two or more treatments and one or more observed responses, with the scientist in control of many sources of variation in the data. Basically, the aim of the experiment is to obtain data directed toward some clearly stated hypothesis that can be evaluated through statistically valid tests. The most important features of the controlled experiment are that there will be one or more independent variables or treatments, such as the dose of a compound (usually including a zero dose — designated the control), which can be manipulated by the experimenter, and one or more dependent or response variables that can be measured, counted, or otherwise assessed to determine whether they are altered by the treatment. The experiment must be replicated among animals (or other subjects) in order to assess the extent of the variation between individuals treated alike. Experimental subjects must be assigned to the treatment groups using some form of randomization procedure in order to minimize the chance of bias occurring as a result of the treatment groups differing before the experiment starts, or of having different environments during the course of the experiment.

The purpose of the experiment may be to test a specific hypothesis, or it may simply be to estimate the means of the different treatment groups, or some correlation or other statistically appropriate relationship among the variables. For example, a study involving different dose levels of a compound may aim at testing the hypothesis that there is no relationship between dose and response. Alternatively, if it is obvious that some type of association is likely, the aim of the experiment may be to quantify the relationship with a specified level of precision.

## Basic Principles of Statistical Inference

Once the data from an experiment have been collected, the scientist is faced with the problem of deciding whether there is sufficient evidence to conclude that the treatment(s) have had some effect, such as altering the measured characteristics of subjects within the treated groups compared with the controls. The means of two groups treated identically will differ simply as a result of chance factors, so how large does a difference have to be in order to reasonably conclude that it has been caused by the treatment? In order to answer this question, statisticians will usually estimate the probability that a difference at least as large as the one observed could have arisen by chance, in the absence of a treatment effect, as a result of variation among the experimental subjects.

In order to calculate these probabilities, statisticians usually assume that the observed set of experimental subjects really represent a sample from a larger hypothetical population of similar subjects, and the aim is to make inferences about these populations based on the observations of the individuals in the sample. In particular, the aim is often to estimate the probability that the observed sample data from each treatment group could instead have been derived from the same source population, regardless of their exposure to the treatment(s). It is usual to set up the so-called *"null hypothesis"* that the two (or more) samples are derived from the same source population, and the *"alternative hypothesis"* that they are from different populations. The null hypothesis will then be rejected if it is "very unlikely" that the observed difference could have arisen by chance, as evidenced by the data. If not sufficiently compelling, the null hypothesis is not rejected, and the inference is made of no evidence for a difference due to the treatment(s). It is up to the scientist to decide what "very unlikely" means, but in practice, a probability of 0.05 (5%) or less is usually accepted as the cut-off point. This implies that in about 5% of all comparisons, the null hypothesis will be rejected when in fact it is true (a *type I error*)

Experiments can also make the other kind of mistake by failing to detect a true treatment effect. This false negative result is known as a *type II error*, and it can arise if the experimental material is extremely variable or if the sample size for the experiment is too small, so that quite large differences between group responses can frequently result by chance. The *power* of an experiment is 1-(type II error rate), and is usually expressed as a percentage. Clearly, it is important to devise powerful experiments capable of detecting treatment effects of magnitudes sufficiently large so as to be of biological importance or clinical relevance, as dictated by the area of study. This is discussed in more detail when considering sample sizes, below. Note that comparisons of, say, the differences between two group means, can be statistically significant but of little biological interest. Thus, statistical significance and biological significance are not synonymous.

So how are these probabilities calculated? The answer depends on the type of data collected. Computer packages are now readily available to do the actual calculations, but in order to use them, it is necessary to know something about the type of data, the experimental design, and the purpose of the study.

## Types of Data

It is necessary to consider the types of data available before embarking on any type of statistical analysis. Two major classifications of data are as follows:

- Categorical data such as dead/alive, male/female, etc., which can usually be summarized by counts, proportions, and percentages
- Measurement data (also known as interval or continuous-type data) which will usually be summarized using means or medians, sometimes with some estimate of the variation among subjects

In order to calculate the required probabilities, it is also necessary to know something about the distribution of the data for the outcome variable(s) of interest. Counts could be the number of abnormal cells in a microscope field, where it is assumed that there are a large and uncountable number of normal cells, with approximately the same number expected in each field. If the mean number of abnormal cells is low (say commonly 0, 1, 2, but rarely more than 5 or 6), then these data may follow a *"Poisson"* distribution. If the counts are much higher, say with a mean of 20 or more, then although the data are still discrete, the distribution will often resemble that of a *normal* or *Gaussian* distribution. In contrast, the outcome data may be the proportion of animals in a test group showing some adverse effect. In this case, the data may follow a *"binomial"* distribution, where, like all proportions, the outcome is bounded by zero and one.

With measurement data, the most common distribution encountered is the normal distribution, which has a bell-shaped curve that is symmetrical about the mean. However, other distributions are commonly encountered, and scatter plots of the raw data can be used to explore this feature during the analysis phase of experiments. Most biological characteristics are not actually normally distributed, but the *central limit theorem* of statistics shows that mean values of samples drawn from populations of any form, if the sample size is sufficiently large, will tend toward normality. Moreover, it is not the data as such that needs to have a normal distribution for many types of statistical analysis, but it is rather that the "residuals" or deviations from the group means must be approximately normally distributed. As this is reasonably common, or can be made so by an appropriate data transformation, this accounts for the widespread use of statistical methods based on the assumption that the data have a normal distribution. When the data is the concentration of some substance, the underlying distribution may instead be *log-normal*, which is asymmetrical with a long tail to the right due to a small number of very high values.

## Computer Software

The actual calculations necessary to work out the desired probabilities and associated statistical information are at best tedious, and at worst extremely complicated and difficult. Fortunately, a wide range of statistical software is available for doing the calculations. Readers are strongly urged to use software such as MINITAB, SAS, SPSS, STATISTICA, StatsDirect, Stata, Genstat, GLIM, BMDP (now maintained by SPSS), or other dedicated statistical packages available for various computer platforms. A listing of many such packages with brief comments and sources for further information is given by http://www.execpc.com/~helberg/statistics.html. Spreadsheet software, though useful for many purposes, does not have the complete range of statistical methods needed for most analysts, its output is often nonstandard, and in some cases, the algorithms may even be inaccurate. It would be extremely tedious, for example, to produce the graph shown in Figure 14.5 using a spreadsheet program.

## EXPERIMENTAL DESIGN

### The "Experimental Unit"

The *experimental unit* is the entity that can be assigned at random to one of the treatments, independently of all other experimental units. It is also the unit of statistical analysis. Any two experimental units must be able to be assigned to different treatments. For example, in a teratogenesis experiment, it is the pregnant female that can be assigned to a treatment group, not the individual *in utero* pups. So, the data from individual pups within the litter will need to be aggregated, and some form of aggregate score will be the metric used in the statistical analysis.

If there are several mice in a cage, and the cage is assigned to a treatment group, such as one of a number of possible diets, then the group of mice in the cage is the experimental unit. However, if mice within a cage can be assigned to different treatment groups, such as when a compound is administered by injection to each animal, then the individual mouse can be the experimental unit. In a crossover experiment, an animal may be assigned to a treatment for one time interval, then rested, and then assigned to a different treatment for a second time interval. In this case, the experimental unit is the animal for a period of time. Similarly, if the animal can have its back shaved, and different compounds can be applied topically to different patches of skin chosen at random, with some measurement being made on each patch during the experiment, then the patch is considered the experimental unit.

## Formal Experimental Designs

Experiments can range in size and complexity all the way from the use of a single animal in an uncontrolled experiment where no statistical analysis is possible, up to large formal controlled experiments involving hundreds of animals. A *controlled experiment* is one in which different treatment groups are compared, with the treatments being under the control of the experimenter. Virtually all controlled experiments should conform to one of the standard types of design such as "completely randomized," "randomized block," "Latin square," "split-plot," "repeated measures," "sequential," "factorial," or other more advanced design. Some of these are considered separately, below. The main aim of these designs is to control some of the variation so as to make the experiment more powerful, i.e., capable of detecting the differences between treatments deemed to be important, though the aim with factorial designs is to obtain more information for the same input of resources. This is discussed below.

Some designs can be used to take account of a natural structure among the experimental units, such as litters of offspring, a lack of availability of the required number of animals all at one time, or some logistical bottleneck in making the observations. For example, the experiment might involve sacrificing some of the animals, taking blood samples, and determining hematological and blood biochemical parameters. This will take time, and conditions may change during data collection. In this type of situation, it may be wisest to use an experimental design that splits the experiment into a number of smaller mini-experiments, the results of which are combined in the final statistical analysis.

## Factorial Treatment Arrangements

Superimposed on these design options, there may be a specific structure to the treatments. For example, a single factor design will vary one independent variable, such as administration of compounds A, B, C, or D, or various levels of one of these. In contrast, a *factorial* arrangement of treatments might vary two independent variables such as the administration of, say, four compounds to both sexes resulting in a 4 (compounds) × 2 (sexes) "factorial" design. Factorial designs are often highly efficient, as they can produce extra information at little or no additional cost.

## A Well-Designed Experiment

The aim of the experiment needs to be clearly specified, and it needs to be designed so as to do the following.

### Avoid Bias

Bias (systematic error) can arise, for example, when subjects in different treatment groups have different environments or when they are otherwise noncomparable due to features of the experimental conditions. In this case, any environmental difference could be mistaken for an effect of the treatment. Bias can be avoided first by ensuring that the experimental subjects are allocated at random to the different treatment groups, and all subsequent manipulations, housing, etc., are done in random order. It would not be sufficient to allocate the animals to the treatments at random and then process all the controls on the first day and all those in the treated group on the next day, because there may be environmental factors that influence results differently on different days.

If there is any subjective element in collecting the data, this should be done "blind" using coded samples, so that the person collecting the data does not know the treatment group of an individual subject. With human experiments, double blinding should be used wherever possible, with neither the patient nor the doctor knowing the group to which the patient belongs.

Other types of bias due to sampling problems, measurement error due to clinical judgements, laboratory devices and assays, and reporting problems can also occur, potentially resulting in misleading conclusions about the treatment effect. When bias is known or suspected after a study has been completed, the results should include discussions of the implications and provide caveats to its interpretation and generalizability.

### Have High Power

The power of an experiment is the probability (usually expressed as a percentage) of being able to detect a specified treatment effect and call it statistically significant given certain other assumptions (discussed below when considering sample size). Generally, experiments should be designed to have the highest possible power. This can be achieved by using uniform material such as isogenic strains of mice or rats free of intercurrent disease, housed under optimum conditions, and of a narrow weight and age range. If measurement error is likely to be large, it might be possible to reduce it by using multiple determinations of things like enzyme activity or behavior such as open field activity. Other sources of variation can often be controlled by using a randomized block or Latin square design. If within-subjects designs, such as crossover designs, are possible, then they will often lead to a useful increase in the power of the experiment.

Power can also be increased by increasing the sample size (see below). However, that will cost time, money, and animals, so it is best to concentrate on the other things first, and then use the minimum number of subjects consistent with achieving the scientific objectives of the study.

### Have a Wide Range of Applicability

There are a large number of variables that need to be fixed by the scientist in order to do a sensible experiment. These include the age, sex, strain, diet, caging, physical and social environments, and methods of measuring the characters of interest. It is often important to know the extent to which an observed result is dependent on any of these variables. Every experiment is done using a particular set of these fixed effects, and biomedical research would be impossible if the results could never be generalized to a broader range of conditions. Thus, if the ambient temperature in the animal facility where the experiment was done was maintained at 23°C, it would usually be reasonable to assume that similar results would have been obtained if it had instead been maintained at 21°C (plus or minus the same daily variation).

On the other hand, there are circumstances where it really is important to find out whether males and females or strain X and strain Y animals differ in their response. In this circumstance, a factorial design can be used. What is sometimes not appreciated is that in many cases, both sexes can be used without any increase in the total number of animals required. Thus, suppose an experiment is designed to find out, say, whether a compound affects mean blood pressure in rats, and it has initially been decided to use ten male rats in the treated and control groups, respectively. However, there may be serious concern that males and females will respond differently. An alternative design would be to use five male and five female rats in the control, and the same numbers in the treated group. The same total number of animals will be used, so this modification to the experiment costs nothing in terms of additional resources and animals. The statistical analysis of these data would then require a more complex two-way analysis of variance (ANOVA) rather than a t-test or one-way ANOVA. However, it would also provide more information, because it would indicate whether males and females respond in the same way to the treatment. If there is no evidence that males and females respond differently, then the effect of the drug treatment can be assessed on the means of all ten animals in each group, so there is little loss in precision, and extra information has been obtained. In contrast, if males and females respond differently, then the mean response of the males will be based on only five individuals in each group, so precision will be lower than if only males had been used. However, this is not as serious a problem as it would seem, because the precision of a mean depends partly on the number in the group and partly on the precision with which the standard deviation is estimated. With the factorial design, the standard deviation will be estimated from the whole experiment including both males and females, so it will be estimated much more precisely than if the experiment had only consisted of five males in each group.

In general, factorial designs provide extra biological information at little extra cost. Thus, provided they are properly used and correctly analyzed, they can be strongly recommended as a way of increasing the amount of information from each experiment, and thereby reducing animal use and saving scientific resources. However, problems may arise with such designs if there are substantial numbers of missing values due, for example, to the death of animals before they are measured.

## Be Simple

Clearly, an experiment should not be so complex that mistakes are made in its execution. Detailed protocols should be prepared showing exactly what needs to be done at each stage of the experiment, and great care needs to be taken to ensure that there are no serious bottlenecks that make it impossible to carry out the assigned tasks in the time provided. A pilot experiment is recommended as a first step if there is any doubt about the logistics of the study.

## Be Capable of Being Analyzed Statistically

Most experiments will need to be analyzed statistically to determine the probability that the observed results could simply have arisen by chance. The method of statistical analysis needs to be planned at the same time as the experiment is planned, and it is often a good idea to generate some simulated results and do a statistical analysis at the planning stage. A scientist should not start an animal experiment without knowing how the data are to be handled. There have been examples of scientists who have done a series of experiments without knowing how to analyze the data, only to find out that the results from the first experiment clearly showed that modifications to the design were necessary.

## Randomization

Randomization throughout the whole experiment is fundamental to all controlled studies. Scientists should welcome the need to randomize, as it protects them from making false claims about the results of the experiment. Picking animals from a box is not an adequate method of randomization, as those caught first may be different from the other animals with respect to their responses in the experiment. There are several practical ways of doing the randomization. Physical randomization can be done by writing numbers or treatment designations on pieces of paper and drawing them out of a bag. For example, if the aim is to randomize 20 rats to four different groups A–D, then five bits of paper labeled "A," five labeled "B," etc., can be placed in a bag. The first rat is caught, and one of the pieces of paper is drawn from the bag to see the group to which it is assigned. This is repeated for each rat, with the pieces of paper not being returned to the bag.

Many computer programs will generate random numbers, or place a group of numbers in random order. So, in the above example, the numbers 1 through 20 (or alternatively five ones, five twos, etc.) could be placed in a column of a software package such as MINITAB, and a command can be used to randomize their order. If numbers are used, then the first five numbers would represent the rats assigned to treatment A, the next five to B, etc. The first rat caught would be rat one in the column of values, and the group to which it was assigned would have been already determined by the computer randomization.

Tables of random numbers can also be used, and this is described in most statistical texts. In general, these tables can be a bit tedious to use, though undoubtedly they produce good randomizations. A system for enumerating each subject or cage of animals is required for use of many randomization procedures, thereby adding to practical difficulties.

With randomized block experimental designs (see below), the whole experiment is divided into a number of "mini" experiments, and randomization is only done within a block. Typically, if there are four treatments, then block one would consist of four animals matched so as to be as similar as possible. Four bits of paper can be placed in the bag labeled A, B, C, and D. The first rat is caught, and one of the bits of paper is drawn from the bag to assign it to one of the four treatments, and so on. The same thing is then done for block two, etc., for all remaining blocks.

## Sample Size

It is important for experiments to be just the right size to address the specified research questions. Very small experiments will lack power and may be incapable of detecting biologically important responses to the treatments. Very large experiments may well waste time, laboratory reagents, and valued animals. Unfortunately, there is no singular way of determining the most appropriate sample size for a given study design. The two methods available, namely *power analysis* and the *resource equation* methods, though helpful, each have inherent limitations. However, both are better than relying entirely on past experience or intuition.

### *Power Analysis*

This method is recommended in particular for large, relatively simple, expensive experiments such as clinical trials. It can also be used for any other experiment where the requisite information is available. It would be well worthwhile using power analysis if a series of similar experiments are planned, but it is more difficult to use for "one-off experiments" that are entirely new, have no preexisting information regarding the expected variability, and for which future repetitions are considered unlikely.

The method depends on a mathematical relationship between the following:

- The sample size
- The effect size of biological interest or clinical relevance
- The standard deviation among the experimental subjects
- The desired power of the experiment
- The significance level (type I error rate) to be used in the analysis
- The alternative hypothesis

The sample size is usually what is being calculated, given knowledge or estimates of the other parameters. However, in some cases, sample size may be fixed, and the aim may be to determine the power of the experiment or the effect size that could be detected in the context of a proposed experiment.

The effect size is the change in mean, median, or proportion between treatment groups which is thought to be of biological importance or clinical relevance. For example, in an experiment designed to study the effect of a test compound on blood pressure in rats, it might be decided that unless the compound reduces blood pressure by, say, 20% or more, the compound is unlikely to be of much interest. So the effect size would be specified as a 20% change in the mean value of blood pressure between treatment groups.

Other things being equal, the larger the effect size specified, the smaller the experiment needs to be to detect it. In experiments where there are multiple dependent variables such as panels of hematology and blood biochemistry values, it is necessary to decide which of these is most important, and the experiment will need to be designed primarily to detect the change in that variable. Deciding which variable is most important in a study involving gene microarrays, where there may be several thousand dependent variables, is not likely to be easy. The standard deviation (for quantitative characters) must also be estimated. This is a problem because the experiment has not yet been done, so the estimate must come from previously related experiments or from the literature. This presents difficulties for one-off-type experiments, as defined above. Pilot studies are sometimes suggested as a way of getting estimates for the standard deviation in advance of the experiment, but such studies are usually very small, so the standard deviation is not reliably estimated in that way. Unfortunately, small differences in the standard deviation can make a large difference to the estimate of the numbers of subjects required.

The desired power is usually set somewhat arbitrarily between about 80 and 90%. Higher power requires larger experiments. Similarly, the significance level is usually set at 5% (0.05), as this is generally accepted as the critical value for statistical significance.

The alternative hypothesis is either that there is a difference between the means (or other parameter being studied), in which case a *two-tailed* test is used, or for some biological reason, it may be known that the difference can only occur in one direction, in which case a *one-tailed* test will need to be used. Most laboratory animal experiments will use two-tailed tests, because the need is to detect *any* difference (beneficial or detrimental) among treatments. One-sided tests should be used only if there is compelling evidence that the response can only occur in one direction.

A power analysis is usually used to determine sample size. However, if the sample size is fixed, say due to limited facilities, then it may be of interest to estimate the power or effect size of a proposed experiment.

Having specified five of the above variables, the sixth is then estimated, using appropriate formulae. However, even for a simple comparison of two means, the formulae may appear daunting to persons untrained in these methods. Fortunately, several modern statistical software packages include power calculation algorithms, and several stand-alone packages also exist.[11] There are a number of Web-based resources that can be used for relatively simple calculations without cost to the users. For example, DSS Research (http://www.dssresearch.com/SampleSize/default.asp) offers a power and sample size calculator for one and two sample *t*-tests and for comparing two proportions. In that case, users are asked to specify the type II error rate (100 - the power), rather than the power itself. So, if an 80% power is wanted, the type II error rate to be entered would be 20%. Note that when comparing two proportions, the standard deviation is not required because this is implicit in the proportion and sample size. Several other sites offer power calculations. As some of these seem to be transient, a Web search using terms such as "statistical power" should be used to find those that are currently active. Note also, that different programs will often give slightly different answers, because they use different algorithms and assumptions for the calculations. Finally, if more than one outcome is being studied, and it is difficult to decide which is the most important, the power calculations may need to be done separately for each, and the final size of the experiment will depend on finding a suitable compromise between the estimates for each outcome.

As an example, suppose that an experiment is proposed to explore the effect of some compound incorporated into the diet on litter size at birth in BALB/c mice. Data is available to show that the average litter size per cage (i.e., averaging all litters produced by one cage) when these mice were housed in trios (as specified for the experiment) was 5.3 pups with a standard deviation of 0.9 pups. How large an experiment will be required to detect a 20% change in litter size with a power of 80% and using a significance level of 0.05?

The results will be analyzed using a two-sample *t*-test. It is debatable whether a one-sided or a two-sided test should be used. A one-sided test might be appropriate if the only outcome of interest is a reduced litter size, as might be the case in some drug toxicity experiments. On the other hand, if there is a possibility that it could instead increase litter size, then a two-sided test would be most appropriate. The alternate hypothesis should be specified before doing any statistical analysis and should not be changed following the analysis. When these parameters were entered into MINITAB version 13, it was estimated that 13 cages per group would be needed using a two-tailed test. The program nQuery Advisor (Version 2.0, Statistical Solutions, Cork, Ireland) gave identical results. The requirements for sample size estimations grow in complexity in the case of more than two treatment groups requiring hypothesis testing via analysis of variance methods, in which case, help from professional statisticians should be sought.

## Sample Size Using the Resource Equation Method

There are situations where it is difficult to apply a power analysis, such as when there is no estimate of the standard deviation, there are many dependent and independent variables, or when it is difficult to specify an effect size of interest. The Resource Equation method[12] is useful in these situations. It is based on the law of diminishing returns. Increasing the size of a small experiment gives good returns, but increasing sample size when the experiment is already large gives poor returns. Mead[12] suggests that, in this context, for experiments producing quantitative data that are to be analyzed by the ANOVA, the error degrees of freedom, E, should be somewhere between 10 and 20. The ANOVA is discussed below. However, for the very simplest design the following formula can be used:

$$E = \text{(Total number of experimental units)} - \text{(total number of treatments)}$$

In the above example of the effects of a compound on litter size in BALB/c mice, the power analysis method suggested the experiment would need 13 cages per group (with the assumptions stated above). Using the resource equation method, if there were 11 cages per group, 22 cages in total, with two treatments, E would be 22 - 2 = 20. Thus, this method would give roughly comparable results to the power analysis method if the aim is to detect a 20% change in the mean with an 80% power. However, if the effect size to be detected was quite small, say a 10% change in the mean, then the resource equation method would underestimate the required sample size. In general, the resource equation method is likely to be appropriate for complex biological experiments where relatively large treatment effects are expected. The power analysis method is preferred, but where it is difficult or impossible to use, the resource equation method is often useful.

## Formal Experimental Designs

### Completely Randomized Design

With this design, subjects are allocated to treatments strictly at random using one of the randomization methods described above. If the treatment is something like an injection that can be given at the start of the experiment, the subjects can then be coded, and the rest of the experiment can be done "blind," without the investigator knowing to which treatment an individual subject belongs until the data are decoded at the time of the statistical analysis. This would not, of course, be possible, if the treatment is a diet that has to be fed over a period of time, though even in this case, the diet can be coded so that it is not known to the staff which group received which diet.

This design can have any number of treatment groups, and unequal numbers in each group usually present no problems, provided the treatments do not have a factorial structure, such as males and females with three dose levels. However, in most cases, the experiment has highest precision if there are equal numbers in each group, except in the situation where a control group is being compared with several different treatment groups. In this case, the control group might be increased in size to benefit study power. In this situation, ideally, if there are t treatment groups, the ratio of numbers in the control group to those in the other groups should be the square root of t. Thus, with five treated groups, including the control, the numbers in the control group should be approximately the square root of 5 (i.e., 2.2) times that in each of the treated groups (assuming these numbers are equal). This would normally be rounded down to twice the numbers in the treated groups. If the resulting data is quantitative, then it will often be analyzed by ANOVA methods or via nonparametric methods such as the Mann–Whitney rank-sum test or Kruskal–Wallis test.[16,17] Table 14.1 shows the results of an experiment using a completely randomized design in which adult C57BL/6 mice were randomly allocated to one of four dose levels of a hormone compound, and the uterus weight was measured after an appropriate time interval. The method of analysis depends to some extent on the distribution of the data. These data are analyzed in the next section.

Completely randomized designs may also produce categorical data. Table 14.2 comes from a study of the effect of the nonsteroidal antiestrogen agent Tamoxifen on mutations in BigBlue transgenic mice. Mice were assigned to a control group or a group treated with Tamoxifen, and the number of genetic mutations occurring in a region of the transgene in the mice was recorded. Then about 50 such mutations from each group were chosen at random, and their nucleotide sequence was determined. The table shows the number of mutations of three types. The question is whether the type of mutation differed between the two groups. The analysis of these data is discussed below.

The main problem with the completely randomized design is that where the experimental material is heterogeneous, such as if the body weight of the animals is quite variable, or if the experiment has to be split and done at different times, it may be inefficient. In such cases, a randomized block or Latin square design may be substantially more powerful.

**Table 14.1  Uterus and Body Weight (g) in C57BL/6 Mice Treated with Various Doses of Estrogen (Arbitrary Units)**

| Dose | Body Weight | Uterus Weight |
|------|-------------|---------------|
| 0    | 12.69       | 0.0118        |
| 0    | 11.58       | 0.0088        |
| 0    | 10.39       | 0.0069        |
| 0    | 11.89       | 0.0090        |
| 1    | 11.21       | 0.0295        |
| 1    | 10.97       | 0.0264        |
| 1    | 10.03       | 0.0189        |
| 1    | 10.64       | 0.0242        |
| 2.5  | 11.28       | 0.0515        |
| 2.5  | 11.97       | 0.0560        |
| 2.5  | 10.67       | 0.0449        |
| 2.5  | 12.16       | 0.0514        |
| 7.5  | 12.64       | 0.0833        |
| 7.5  | 11.92       | 0.0948        |
| 7.5  | 12.69       | 0.1017        |
| 7.5  | 13.57       | 0.0780        |
| 50   | 13.29       | 0.1130        |
| 50   | 13.24       | 0.0623        |
| 50   | 12.78       | 0.0802        |
| 50   | 13.26       | 0.0912        |

*Note:* It is debatable whether body weight should have been recorded to two decimal places.

**Table 14.2  Type of Mutation in Control Transgenic Mice and Mice Treated with Tamoxifen**

| Type of Mutation | Treatment Control | Tamoxifen | Totals |
|------------------|-------------------|-----------|--------|
| C:C to T:A       | 7                 | 20        | 27     |
| G:C to A:T       | 15                | 6         | 21     |
| Other            | 31                | 26        | 57     |
| Totals           | 53                | 52        | 105    |

## Randomized Block Design

This design can be used to take account of heterogeneous features of the study or subjects, such as animals differing in their initial body weight, and time and space variables that result when the measurements cannot all be done at the same time, or when the animals have to be housed in more than one room or on different shelves in the same room. Essentially, a randomized block design involves splitting the experimental material into a series of "mini-experiments" each involving one or a few subjects on each treatment. For example, with an experiment involving four treatments and five animals per treatment, block 1 might consist of the four heaviest animals assigned at random to one of the four treatments. These animals would be started on the experiment at the same time, housed on the same shelf, and measurements would be made at the same time. Block 2 might have the next four heaviest animals, possibly started at a later date, but again housed together, possibly in a different room, and terminated at a different time from block 1, and so on, with the next three blocks.

Randomized block designs can be particularly useful when the experimental material (e.g., animals) has some sort of natural structure. For example, it may be difficult to collect enough animals of a transgenic strain of mouse all at once to do an experiment with the desired sample size. However, it may be possible to do the experiment as a randomized block design using small numbers as they become available. The only limitation would be the need to collect enough animals to have one representative for each treatment group.

In Table 14.3, an experiment done as a randomized block is shown. This was an *in vitro* experiment (emphasizing the point that the methods discussed here are applicable to all experiments, whether or

**Table 14.3 Cell Counts in Tissue Culture Dishes**[a]

| Treatment | Block 1 | Block 2 | Block 3 | Block 4 | Means |
|-----------|---------|---------|---------|---------|-------|
| con | 100 | 81 | 62 | 128 | 92.8 |
| tpa | 514 | 187 | 294 | 558 | 388.3 |
| gen | 35 | 82 | 148 | 241 | 126.5 |
| gen + tpa | 120 | 84 | 134 | 1011 | 337.3 |

[a] All received media. The controls (con) had no additives, some dishes had tpa (tpa) added, some genistine (gen), and some genistine + tpa. For further details, see text.

not they involve animals) in which dishes of cells were treated with a vehicle (the control), TPA (a phorbol ester), Genistine, or TPA+Genistine. Data were the number of cells following further incubation under specified conditions. Notice that there are four treatments, but they have a factorial structure in that half the dishes had Genistine and half did not and half had TPA and half did not. The miniexperiment using just four dishes was repeated four times. Thus, this was a randomized block experiment with four treatments arranged as a 2 × 2 factorial layout. The statistical analysis of this experiment is explained below.

## Within-Subjects Designs

Sometimes it is possible to do a "within animal" design such that the animal receives the treatments in sequential order, usually with a rest period between them. This is generally a powerful design, because the variation within an individual is usually less than that between individuals, but it is only appropriate for certain types of relatively minor treatments that do not permanently alter the animal. As an example, Table 14.4 shows an experiment to study taste preferences in mice. Cages containing two mice were set up with two drinking bottles each, one containing distilled water and the other the test compound (alcohol, sucrose etc.). The control was two bottles of distilled water, one arbitrarily designated as the "treated" one. The position of the bottles on the cages was rotated each day, and the percentage of test fluid consumed as a percentage of total fluid consumption was recorded over 1-week periods, with a 1-week rest between treatments. Treatments were applied to cages in a random order. Notice that with this experimental design, the experimental unit is a cage of two mice for a period of time, and it is assumed that any one treatment does not permanently alter the behavior (taste preference) of the mice. The aim was to determine whether the mice preferred any of the test solutions over others.

This design is similar to a randomized block design, with the "block" in this case being the cage of mice, and treatments (the different test fluids) being assigned to time periods at random. The analysis of these data is described below.

## Other Designs

There are several other designs that can be used in appropriate circumstances, but which cannot be discussed in detail here. A Latin square design can be used to control two sources of heterogeneity that cannot be included together in one block of a randomized block design. For example, a replicated 3 × 3 Latin square was used to compare the behavioral effects of bleeding from the orbital sinus, diethyl-ether anaesthesia and sham anaesthesia in rats.[13] Each of three rats received all three treatments, with a rest between them, but balanced in such a way as to cancel out any possible trend in behavior over the period of the experiment.

An *incomplete block design* can be used when there is a natural block size due to the nature of the experimental units, but more treatments are to be compared than can fit into a block. These designs can be quite complex, so professional advice is usually necessary.

*Sequential designs* can be used when the response of an individual subject to some treatment can be obtained quickly, and a decision can be made on whether to proceed to the next individual or to terminate the experiment. This may follow because the accumulated information has provided sufficient evidence of a positive effect, or the power of the experiment is now high enough to preclude an effect of a specified magnitude. The "up-and-down" method for determining the LD50 of a compound is an

**Table 14.4  Percent of Test Fluid Consumed by C57BL Mice Offered Distilled Water and One of the Test Fluids in a Two-Bottle Choice Experiment Using a Repeated Measures Design[a]**

| Cage | Treatment and Percent of Test Fluid Consumed | | | | |
|------|------|------|------|------|------|
| 1 | B | C | A | E | D |
|   | 69.6 | 61.9 | 54.9 | 69.4 | 78.3 |
| 2 | A | D | E | B | C |
|   | 48.2 | 81.5 | 60.9 | 61.1 | 43.9 |
| 3 | C | D | A | B | E |
|   | 53.4 | 74.7 | 49.9 | 58.2 | 68.5 |
| 4 | E | A | D | B | C |
|   | 64.5 | 50.4 | 73.6 | 55.3 | 50.7 |

[a] Mice were housed two per cage. The control group had two water bottles, one of which was arbitrarily designated as the "treatment."

Treatments were as follows: A — control, distilled water in both bottles; B — 0.02% saccharin; C — 0.05 M sodium chloride; D — 0.04 M sucrose; and E — 10% ethanol.

example of this type of design.[14] Such designs are usually efficient and use small numbers of animals, but they are only applicable in some circumstances.

*Split plot designs* are ones where there are two types of experimental units. For example, several cages, each containing two mice, might be assigned to different dietary treatments. Within each cage, one of the mice might be given an injection of vitamins, while the other is given an injection of the vehicle as a control. In this case, the cage of mice is the experimental unit for comparing diets, but the mouse is the unit for comparing the effects of the vitamin injection. The statistical analysis of such designs is quite complicated.

*Fractional factorial designs* can be used to compare many different treatments simultaneously using relatively few experimental subjects.[15] They have been used, for example, in screening potentially active compounds in drug discovery. Animals may receive ten or more treatments (presence or absence of compounds A, B, C, etc.) in such a way that if one or more of the compounds is biologically active for a particular end point, it can be demonstrated following the statistical analysis. This is a useful way of reducing the use of animals and costs, but requires more advanced statistical methods than can be described here.

## STATISTICAL ANALYSIS

### Data Screening

The first step in any statistical analysis is to screen the data for obvious entry, transcription, or measurement errors, and to obtain a general feel or impression of the results. Graphical methods available in all good modern statistical packages are invaluable for this. Figure 14.1 shows a dotplot produced by the MINITAB statistical package of the data shown in Table 14.1. It shows individual observations, with clear evidence that the hormone compound treatments are increasing uterus weight in mice, but with the variation also increasing. Figure 14.2 is a similar dotplot for the data in Table 14.3. Treatment groups have been labeled 1–4 in this case, though the final statistical analysis will take account of the factorial nature of the treatments, and the blocking has been ignored. Figure 14.2 shows that one point in treatment group 4 appears to be an outlier. When such an outlier is found, it should be checked to ensure that it is not simply a typographical or other type of data entry error. Assuming that this is not the case (as is the situation here), the presence of the outlier will need to be taken into account when analyzing the data. This is discussed below.

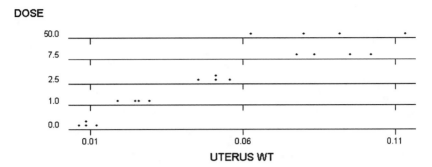

**Figure 14.1**   Dotplot (MINITAB) of the raw uterus weight data in Table 14.1, by dose level. Note that uterus weight certainly increases as the dose level increases, but that the variability also increases.

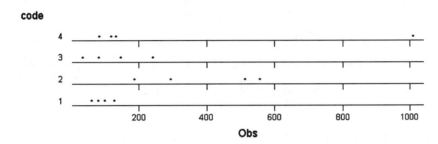

**Figure 14.2**   Dotplot for the cell count data in Table 14.3. Treatment groups have been coded 1–4, with 1 being the control group (medium alone), 2 being TPA, 3 being Genistine, and 4 being TPA and Genistine. The blocking has been ignored. Note that there is some evidence, particularly in group 2, that variation increases as the mean increases. Note also the outlier in group 4.

## The Analysis of Variance (ANOVA)

Most designed experiments producing measurement data can be analyzed by the ANOVA. This is a highly versatile statistical method which can be used for experiments with several treatment groups and with all types of experimental designs. It is much more versatile than the $t$-test, which can only be used for comparing two groups. In fact, when there are only two groups, the $t$-test and the ANOVA give mathematically equivalent results. Thus, there is really no need to use the $t$-test for this type of analysis.

The ANOVA quantifies the total variation in the data, and partitions it into components associated with treatments, error and in some cases other components. With modern computer packages, it is extremely easy to do a one-way ANOVA. The data from Table 14.1 were analyzed using MINITAB. With this and most other software packages, the observations are placed in one column of a datasheet with codes representing the treatment groups in another column. A menu command is then used to do the analysis. However, the ANOVA assumes 1) that the variation in each group is approximately the same, 2) that the deviations from the group means (known as the residuals) have an approximately normal distribution, and 3) that there is independent replication of the observations. Serious departures from these three conditions will lead to results which are not reliable. The first two conditions can be assessed by doing a normal probability plot of the residuals, which should produce an approximately straight line, and by plotting the "fits" (group means) versus the residuals. In MINITAB and many other good packages, these graphs are produced easily as part of the ANOVA. The two plots for the data in Table 14.1 are shown in Figure 14.3 and Figure 14.4. Clearly, there is some evidence for nonnormality of residuals, as the points in Figure 14.3 do not lie on a straight line, and there is evidence from Figure 14.4 that there is more variation in some groups than in others. The practical importance of these two departures from the assumptions underlying the analysis is to reduce the ability of the study to discern true differences that might really exist between the treatment groups, giving an increased chance of a type II error. The third assumption depends on the correct design of the experiment. Any two

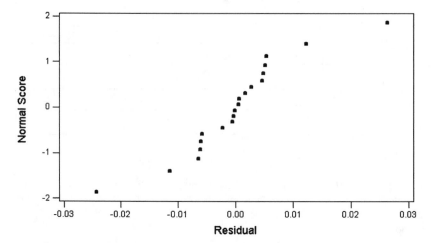

**Figure 14.3**   Normal probability plot of the residuals from the analysis of the uterus weight data from Table 14.1
(MINITAB). In judging whether the points fit a reasonably straight line, not too much attention
should be given to a few apparent outliers, so in this case, it is debatable whether a transformation
is necessary. However, the variation certainly increases with the mean, as shown in Figure 14.4,
suggesting that a transformation would be advisable.

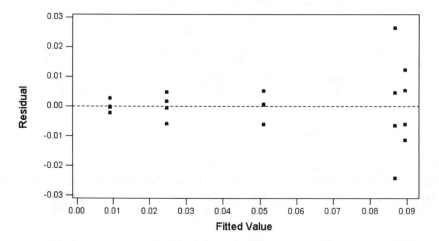

**Figure 14.4**   Plot of fitted values versus residuals for the raw uterus weight of Table 14.1. Note the increase in
variation as the fitted values (group means) increase.

experimental units must be capable of being assigned to different treatments, and there must be no
systematic biases that could affect treatment groups differently. Unfortunately, there is no easy way of
checking that this assumption has been met.

The output of an ANOVA shows a number of calculated values and statistical findings, as shown in
Table 14.5, which is the analysis of the log (see below) of the uterus weights ($\times100$) from the data shown
in Table 14.1. Note that coding the raw data by multiplying it by a constant such as 100 does not alter
the statistical analysis, and is advisable in cases such as this one in order to avoid rounding errors and
excessive numbers of decimal places in the output. The first column headed "Source" lists the sources
of variation present in the data, in this case Dose, Error, and Total. The next column headed "DF" lists
the *degrees of freedom* for each source of variation. These can be regarded as the number of units of
variation associated with each source of variation. There are five dose levels, and the degrees of freedom
for the dose is shown as four. This is because two observations will give one estimate of the variation
among dose levels (the difference between them), three observations will give two estimates of variation,
and n observations will give n - 1 estimates of variation. Thus, the DF are given by n - 1, where n is
the number of objects.

**Table 14.5 ANOVA Analysis of the Uterus Weight Data Given in Table 14.1, Using the Log of Uterus Weight (×100)**

**Analysis of Variance for LogWt**

| Source | DF | SS | MS | F | P |
|--------|----|----|----|----|----|
| Dose | 4 | 2.81567 | 0.70392 | 110.54 | 0.000 |
| Error | 15 | 0.09552 | 0.00637 | | |
| Total | 19 | 2.91120 | | | |

| Level | N | Mean | StDev | Individual 95% CIs For Mean Based on Pooled StDev | | | |
|-------|---|------|-------|------|------|------|------|
| | | | | ------+-------------+-------------+-------------+---- | | | |
| 0.0 | 4 | -0.0476 | 0.0953 | (--*-) | | | |
| 1.0 | 4 | 0.3879 | 0.0822 | | (-*--) | | |
| 2.5 | 4 | 0.7058 | 0.0397 | | | (-*--) | |
| 7.5 | 4 | 0.9492 | 0.0523 | | | | (-*--) |
| 50.0 | 4 | 0.9279 | 0.1081 | | | | (--*-) |
| | | | | ------+-------------+-------------+-------------+---- | | | |
| Pooled StDev = | 0.0798 | | | 0.00 | 0.35 | 0.70 | 1.05 |

The next column, headed "SS" is the sums of squares associated with each source of variation. This can be taken as a quantitative estimate of each source of variation. The column headed "MS" (for mean square) puts this on a per-unit basis by dividing the SS by the DF column. The column headed F shows the F-statistic associated with the relevant sources of variation (only dose in this case). This can be used to produce the value for "p" shown in the last column. Before computers were available, these F-values had to be looked up in statistical tables, but nowadays, the computer produces the p-value, which in this case is 0.000, shown to three decimal places. Because F-statistics derive from probability distributions with a very long tail, the true p-value is actually slightly larger than zero, and the literature would report the result as <0.001. This p-value is the probability that differences as large as, or larger than those obtained in this experiment could have arisen by chance sampling variation, in the absence of any true difference between treatment groups.

Note that the error mean square is the pooled estimate of the variance. Its square root is the pooled standard deviation. This is used in calculating *post-hoc* comparisons, and can be used in presenting the results.

## Transformations

When the data do not fit the assumptions necessary for an ANOVA, they can be transformed to another scale which may be more appropriate, or a nonparametric test can be used. A log transformation is often useful when the standard deviation increases with the mean, an angular transformation[16] is often used for percentage data when many of the values are less than about 20% or more than about 80%, and a square root transformation can be used for data with a Poisson distribution, particularly where the mean count is quite low. A log transformation (to any base such as e or 10) of the uterus weight (×100 to avoid negative numbers) data in Table 14.1 corrects the deviations from normality, and makes the variation in each group about equal as judged by the two graphs. In each case, residuals plots should be used to assess the success of any transformation. Note however that transformations of the raw data can sometimes complicate interpretation of findings, because they change the units of measurement and summary values. Nonetheless, if the null hypothesis is rejected following an appropriate scale transformation, this implies that the samples are unlikely to come from the same population, even if this is not always apparent on the untransformed scale.

Another powerful and useful transformation when there is nonnormality and outliers is to replace the observations by their ranks. An ANOVA and F-test applied to ranks is equivalent to the Kruskal–Wallis test. This transformation, with the resulting data being analyzed by the ANOVA, can be used in many cases where no nonparametric method is available.[15]

## *Post-hoc* Comparisons and Orthogonal Contrasts

When an experiment has more than two treatment groups, as above, the first step is to conduct an ANOVA to test the null hypothesis. However, if the ANOVA results are significant (say $p < 0.05$) the scientist may want to know which group means differ from one another, or which differ from the control group. There are several methods for making such comparisons. Using $t$-tests to compare all possible combinations of paired means is not advisable because the resulting type I error rate then becomes artificially inflated above the specified value. Most computer packages offer several *post-hoc* comparison methods that differ in their most appropriate use and in their characteristics. Among the most common of these, Dunnett's test is used when the aim is only to compare each group with a control group, but where the treatments are dose levels, it may be more appropriate to study the dose–response relationship using regression analysis or orthogonal contrasts in an ANOVA. There are several different tests for comparing all means. These include Tukey's test, Fisher's least significant difference, Duncan's multiple range test, and the Newman–Keuls test.[15–17] Each of these has slightly different properties, and it is not possible to state which is most appropriate in any given situation. Other tests will be offered in some computer software packages. Scientists should consult a statistical text to find out which tests are most appropriate for their particular analysis.

For the hormone compound data in Table 14.1, there was one control group and three treatment groups in a study that demonstrated a clear dose–response relationship on uterus weights, including one very high dose level. Presumably the highest dose was chosen to try to gauge the maximum possible response under these test conditions. With this type of data, Dunnett's test might be used to see which groups differed from the controls. Further comparisons would really depend on the purpose of the experiment. With the four lower dose levels, it may be of interest to quantitatively explore the form of the dose–response relationship. These methods will be discussed in the section on linear regression.

When categorical data is produced in a "contingency table" (Table 14.2), then an analysis appropriate to counts is needed, such as a chi-square ($\chi^2$) test. In this case, the hypothesis to be tested is that the proportions of each type of mutation are the same in the two groups. This is equivalent to testing the hypothesis that the rows and columns are independent. The formulae for doing a $\chi^2$ test is given in most statistical textbooks, and it is available in most software packages. For the genetic mutation experiment of Table 14.2, the calculated $\chi^2$ statistic = 10.5 with two degrees of freedom [calculated as $(r - 1) \times (c - 1)$, where r and c are the number of rows and columns, respectively] and a corresponding $p = 0.005$, so the null hypothesis of no difference in mutation type by treatment group would be rejected. The results could be presented as percentages or proportions, in which case a confidence interval should be stated, as described below (presentation of data). Thus, in the controls, 13% (5 to 25%) of the mutations were C:C to T:A, 28% (17 to 42%) G:C to A:T and 58% (44 to 72%) "other," with the numbers in parentheses being the 95% confidence interval for the percentage, as calculated using the MINITAB statistical package. The main assumptions required for chi-square tests are that the observations be independent, and it is best if the expected values in each cell of the table (by way of the calculations) are not less than five. Most good computer packages will list the expected values, but they can be calculated manually as the product of the row and column totals for a particular cell in the table, divided by the grand total number of observations. If there are small expected numbers in some of the cells of a contingency table, Fisher's exact test can be used instead. However, that test is tedious to do by hand even for a $2 \times 2$ table, so published tables[17] or software packages such as StatXact (version 3 or later for Windows, Cytel Software Corp., Cambridge, MA) can be used for this purpose.

Another way of comparing groups, often favored by statisticians, is to use a set of what are called "orthogonal contrasts."[15] Where the treatments are different levels of a variable such as dose of a test compound, and these are equally spaced on some scale, then orthogonal contrasts can be used to test whether there is a linear or nonlinear trend. However, a discussion of the use of these methods is beyond the scope of this chapter.

### Two- or Three-Way ANOVA

The analysis of the data in Table 14.3 is a bit more complicated, though still easy given an appropriate computer software package. The experiment used a randomized block design, and the block effect needs to be removed in the analysis. The treatments had a factorial structure, so this needs to be built into the analysis, and there is one outlier value in the data to think about (Figure 14.2). In the MINITAB statistical package, the observations are put in one column of the datasheet, the block number in the next, the TPA treatment (coded one or two) in a third column, and the coded Genistine treatment in the fourth column. An ANOVA is then done taking account of the block, the TPA treatment, the Genistine treatment, and the interaction between TPA and Genistine. Details of exactly how this is done are given in the software manuals. The usual plots of the type shown in Figures 14.1 and 14.2 are produced (not shown here). From these, it is clear that the data needs to be transformed prior to the final analysis. A log transformation was made, and the data were reanalyzed. This time, the two graphs showed no evidence of nonnormality or heterogeneous variances. Moreover, on this scale, the outlier disappeared. So this analysis would probably be the final one necessary to address the hypothesis (Table 14.6). However, what if the outlier did not disappear after the transformation? One approach would be to do the analysis with and without the outlier to see what difference it makes in the analysis. In this experiment, elimination of the data point representing the outlier on the log scale (where as noted it no longer appeared to be an outlying value) did not make any meaningful difference to the interpretation of results (data not shown). The main change was that the TPA*Genistine interaction now approached significance at $p = 0.06$, so it might be concluded that had the experiment been larger, it might have shown that TPA and Genistine applied together reduced the number of cells more than their combined individual effects.

Looking at the ANOVA results in Table 14.6a (with the outlier), the "main effect" (i.e., averaged over both levels of Genistine) of TPA was to significantly increase the counts from a mean of 110 to 363 cells (Table 14.6b on the untransformed scale), the main effect of Genistine was not significant and there was no interaction between them.

Analysis of the drinking taste preference data in Table 14.4 presents no problems. It is analyzed as a two-way ANOVA without interaction. This eliminates the effect of differences between cages so that the effects of the treatments can be assessed with high precision. The graphs to explore normality of the residuals and heterogeneity of variances give no evidence that would preclude the use of the raw data in the analysis. However, the data in this case are the percentages of test fluid consumed. Had these percentages been higher (say many above 80%) or much lower (below 20%), then an angular transformation may have been necessary. This is again an example where Dunnett's test seems most appropriate in order to compare each solution with the control. A potential criticism of the design might be that as

**Table 14.6a Analysis of Variance of the Log of the Data in Table 14.3**

**Analysis of Variance for LogScore**

| Source | DF | SS | MS | F | P |
|---|---|---|---|---|---|
| replicat | 3 | 0.74028 | 0.24676 | 3.91 | 0.049 |
| TPA | 1 | 0.77213 | 0.77213 | 12.24 | 0.007 |
| Gen | 1 | 0.04627 | 0.04627 | 0.73 | 0.414 |
| TPA*Gen | 1 | 0.09994 | 0.09994 | 1.58 | 0.240 |
| Error | 9 | 0.56775 | 0.06308 | | |
| Total | 15 | 2.22636 | | | |

**Table 14.6b Means (on the Untransformed Scale) for the Main Effects of TPA and Gen and the TPA*Gen Interactions**

| Main Effect,TPA | Mean | Main effect, Gen | Mean | Interaction, TPA*Gen | Mean |
|---|---|---|---|---|---|
| None | 109.63 | None | 240.50 | None, None | 92.75 |
| TPA | 362.75 | Gen | 231.88 | None, Gen | 126.50 |
| | | | | TPA, None | 388.25 |
| | | | | TPA, Gen | 337.25 |

**Table 14.7  Analysis of Variance with Dunnett's Posthoc Comparisons of the Data from Table 14.4**

| Factor | Type | Levels | Values | | | | |
|---|---|---|---|---|---|---|---|
| Cage | Random | 4 | 1 | 2 | 3 | 4 | |
| Treats | Fixed | 5 | 1 | 2 | 3 | 4 | 5 |

**Analysis of Variance for Percent**

| Source | DF | SS | MS | F | P |
|---|---|---|---|---|---|
| Cage | 3 | 205.14 | 68.38 | 4.44 | 0.026 |
| Treats | 4 | 1819.17 | 454.79 | 29.53 | 0.000 |
| Error | 12 | 184.78 | 15.40 | | |
| Total | 19 | 2209.09 | | | |

**Means**

| Treats | N | Percent |
|---|---|---|
| 1 | 4 | 50.850 |
| 2 | 4 | 61.050 |
| 3 | 4 | 52.475 |
| 4 | 4 | 77.025 |
| 5 | 4 | 65.825 |

Dunnett 95.0% Simultaneous Confidence Intervals
Response Variable Percent
Comparisons with Control Level
Treats = 1 subtracted from:

| Treats | Lower | Center | Upper |
|---|---|---|---|
| 2 | 2.411 | 10.200 | 17.989 |
| 3 | -6.164 | 1.625 | 9.414 |
| 4 | 18.386 | 26.175 | 33.964 |
| 5 | 7.186 | 14.975 | 22.764 |

Dunnett Simultaneous Tests
Response Variable Percent
Comparisons with Control Level
Treats = 1 subtracted from:

| Level Treats | Difference of Means | SE of Difference | T-Value | Adjusted P-Value |
|---|---|---|---|---|
| 2 | 10.200 | 2.775 | 3.6760 | 0.0106 |
| 3 | 1.625 | 2.775 | 0.5856 | 0.9360 |
| 4 | 26.175 | 2.775 | 9.4333 | 0.0000 |
| 5 | 14.975 | 2.775 | 5.3969 | 0.0006 |

four treatments are being compared with the control, the latter should probably have been assessed more accurately using another set of control measurements on each cage. The results of the analysis are presented in Table 14.7 Dunnett's test was used for the *post-hoc* comparisons, and this indicated that the controls did not differ significantly from group 3 (sodium chloride), but did differ significantly ($p < 0.001$) from each of the other three groups.

Note that the cages differed significantly, implying that the blocking was worthwhile. With the analysis shown in Table 14.7, the pooled within-group variance (given by the error mean square) was 15.4, giving a standard deviation of 3.9. When the data were reanalyzed ignoring the blocking (not shown), the error mean square was 25.99, giving a standard deviation of 5.1. Thus, by decreasing the standard deviation from 5.1 to 3.9, the blocking increased the power of the experiment to a useful extent.

## Multiple Dependent Variables

In many experiments, there is more than one dependent variable (character) being measured. Commonly, each variable will be analyzed separately. However, if the variables are correlated, the tests will not be independent of each other. Thus, if a false positive or negative result occurs with one of the variables, it may also occur with the other. There are various "multivariate" methods such as principal components analysis that can sometimes be used in such cases. In extreme cases, there may be hundreds or even thousands of measurements done on each individual, such as in experiments involving gene microarrays. Specialist advice is needed in such situations.

Another common situation is when an experimental subject is measured several times over a period of minutes, hours, or days. Growth curves are a typical example. This is sometimes analyzed as a "repeated measures" design. However, there are objections to such an analysis, as it treats each observation measurement over time as an independent variable, when in fact, they are correlated. It is usually better to analyze some function of the measurements such as their average, or the difference between the first and last, or the slope of the best fitting line (estimated separately for each individual using regression, discussed below), or the area under the curve. If the repeated measures type response variable is of the categorical type, other sophisticated approaches (e.g., the general mixed model of SAS) may be required.

## Linear Regression and Correlation

In some experiments, the aim is to see whether there is any linear association or possibly causal relationship between two or more variables. For example, Table 14.1 shows body and uterus weight in the mice. The "product-moment correlation," also sometimes called the Pearson correlation, can be used to quantify the linear association between these on a scale of -1, which implies a perfect inverse relationship, to +1, which implies a perfect positive relationship. The correlation between body and uterus weight in this case is 0.685, between body weight and log uterus weight is 0.592, and between uterus weight and log uterus weight is 0.945. Note that the last of these is less than 1.0, because the relationship between the two is not linear. The value of a linear correlation may depend on the scale of measurement. Correlation coefficients are often tested to see whether they differ significantly from zero. This can be done using tables, available in some text books. Alternatively, p-values are often given in computer output. All these correlations differ from 0 at $p < 0.01$. There are a number of other types of correlation that might be appropriate if one or more of the variables is categorical. These are discussed in most statistical text books.[15–17]

A linear association between an independent or predictor variable usually designated "X" and a dependent or outcome variable designated "Y" can be quantified using linear regression analysis. This gives the best fitting straight line $Y = a + bX$, where Y is the estimated Y-value, a and b are the intercept and slope constants estimated from the data, respectively, and X is the value of the X-variable. "Best fitting" in this case means the line that minimizes the sum of squared deviations from the line. As an example, the data in Table 14.1, excluding the top dose, can be used to explore the relationship between the dose of an estrogen, and uterus weight. The output of a regression analysis of these data is given in Table 14.8. The output includes an ANOVA table, in addition to the parameters for the best-fitting

**Table 14.8 Regression Analysis of Uterus Weight Versus Dose, Omitting the Highest Dose Level**

Regression Analysis: UTERUS WT versus DOSE
The regression equation is
Uterus Weight = 0.0151388 + 0.0103382 Dose
  S = 0.0087338        R-Sq = 93.0%              R-Sq(adj) = 92.5 %

| | | Analysis of Variance | | | |
|---|---|---|---|---|---|
| Source | DF | SS | MS | F | P |
| Regression | 1 | 0.0142147 | 0.0142147 | 186.352 | 0.000 |
| Error | 14 | 0.0010679 | 0.0000763 | | |
| Total | 15 | 0.0152826 | | | |

*Note:* It would probably have been better to multiply the uterus weights by, say, 100, in order to avoid the large number of decimal places in the analysis. This can lead to serious arithmetical errors, particularly when using spreadsheets, which do not always use double precision arithmetic.

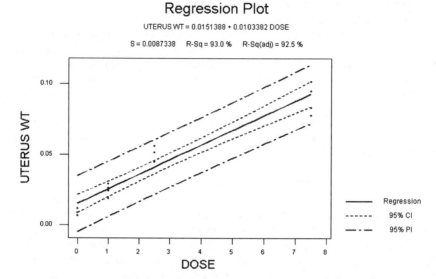

**Figure 14.5**   Regression plot for the uterus weight data of Table 14.1 versus dose of estrogen (omitting the highest dose level, because it was well outside the range of the other doses). The best fitting line is shown (solid). The inner dashed lines show the 95% confidence interval for the mean (the solid line). The outer dashed lines show the 95% prediction interval for individual points, i.e., 95% of individual values should fall within these lines. Note that the straight line fit is not ideal in this case; other transformations may result in a better fit.

regression equation, which indicates whether the slope of the line differs significantly from zero. Note that this analysis gives the best fitting straight line, though the true relationship may not be linear.

Many software packages will also produce a graph like the one shown in Figure 14.5. This shows the best fitting straight line, which is clearly not an ideal fit because all of the 2.5 dose group values lie clearly above the line. This plot also shows an inner set of confidence bands, which are the 95% confidence interval for the mean value of Y, given any value of X (i.e., for any value of X we can be 95% confident that the true value of Y falls within the inner bands shown). The outer set of bands represent the 95% prediction interval for individual values of Y, given any value of X, i.e., there is a 95% confidence that individual points will fall within these outer bands. The plot also includes the formula for estimating any value of Y given any value of X, and also shows the $R^2$ value; i.e., the proportion of the variation in Y that is accounted for by variation in X. In this case, approximately 93% of the variation in Y is accounted for by variation in X. However, the plot also shows that the relationship between dose and uterus weight is apparently not linear but instead somewhat curved, as noted. A plot of log (dose + 1) versus uterus weight provides a slightly better fit ($R^2$ = 94.4%).

Some software, such as MINITAB, also makes it relatively easy to fit a second degree polynomial curve to the data, necessary for the proper evaluation of some data sets, though discussion of curve fitting is beyond the scope of this chapter.

## PRESENTATION OF DATA

The issues and methods presented in this chapter have thus far been focused on the design and analysis of animal experiments to encourage the fruitful expenditure of resources. Unfortunately, even well-guided studies can be wasteful if they fail to convey the essential findings in a way that has potential to benefit the field of study. Studies must be presented in a way that allows others to understand their purpose, including the hypothesis being tested and the basis for selecting the clinical or biological outcomes of effect. Details of the source population available for study, including genetic nomenclature of the animals and their husbandry conditions, should be described, making clear the target population to which the results are to be generalized. The sampling technique, methods of randomization to treatment group, rationale for blocking and/or pairing along with any criteria for exclusion from treatment, should be included as part of the methods section. The type I and type II error rates (or the study power) should be specified as the basis for making judgements about the null hypothesis in advance of the data collection phase. The decision to use one-tailed or two-tailed significance tests should be noted and justified with respect to the hypothesis. Since the algorithms used for computations and proper interpretation of outputs may vary, the brand and version number of statistical software used for analysis must be stated.

A thorough set of annotated guidelines for reporting statistics in biomedical research has recently been published[18] including appendices showing style preferences for presenting numbers and checklists for reporting on clinical trials in the peer-reviewed literature. Similar, though less extensive, guidelines and checklists have been published by others.[19] These authors emphasise that numbers should be rounded when presented, but not when analyzed (due to potential loss of information), and that two significant digits are generally sufficient. Percentages should normally be reported to one decimal place at maximum, unless the sample size is less than 100 in which case whole numbers should be used. Numerators and denominators should always be shown. Continuous data should be summarized with the mean and standard deviation (sometimes including the use of "±" symbols) if they are approximately normally distributed, otherwise the median and interquartile range (i.e., the values representing the $25^{th}$ to $75^{th}$ percentiles of the distribution) is preferred.

Standard errors of the mean (SEM) are sometimes mistakenly used to summarize the variability of data in lieu of the standard deviation. Their proper use is to provide a measure of precision for estimates of a mean drawn from a (possibly hypothetical) population, and they therefore can promote misleading conclusions about the actual dispersion of the observations. The use of unlabeled error bars on charts can likewise present confusion, since it is left unclear to the reader as to whether they represent the sample standard deviation, SEM, or confidence interval. Where possible scatter plots should be used to show the individual observations. In any event, the sample size used to calculate SEM values must be included for them to be of any value, since they are calculated as:

$$\text{SEM} = (\text{standard deviation}) / \sqrt{(\text{sample size})}$$

Scientists should use confidence intervals to describe the precision of estimates, including the upper and lower limits that form the bounds on the value of interest. Consistent with the logic in establishing the type I error rate at 5%, most would report the 95% confidence intervals for all primary comparisons of effect. Many statistical programs use confidence intervals in presenting the results of an analysis. For example, the results of Dunnett's test given in Table 14.7 are given in the form of the difference of each group mean from the control and 95% confidence interval for that difference. A 95% confidence interval constructed around difference measures will not include zero if the results are significant at the 5% level. Confidence intervals add value beyond that of p-values, since they span the range of estimates for the true (typically unknowable) population difference or effect, as well as showing whether the difference is statistically significant. Therefore they can be used to aid in judgements about importance of the findings, particularly where biological importance may differ markedly from statistical significance.

Tables or figures should be used to present the main study findings wherever possible, constructed in a way that emphasizes important trends or comparisons. When it is of interest to compare several mean values by treatment group, they should be listed as columns in tables instead of rows to facilitate visual comparisons. Means should be reported to no more than one decimal place beyond the data they summarize, and standard deviations to no more than two places. Categorical data can often be presented in the text for space savings, unless the numbers of categories warrant a table or chart.

Reports should include methods used to provide assurance that statistical assumptions for the chosen hypothesis tests were satisfied. This can include normal probability plots of residuals as shown (Figures 14.3 and 14.4) and design considerations to demonstrate independence of the observations.

Test statistics (such as the "F" value or "$\chi^2$-statistic") should be reported along with the corresponding degrees of freedom, allowing readers to verify the p-value for each explanatory variable. Actual *p*-values should be shown to two significant digits, and the smallest value that needs to be reported is $p < 0.001$. ANOVA results should be reported as a table where possible, including the "Source," "SS," and "MS" values for exploring between-group differences. Any efforts to test for interaction and their subsequent treatment in the analysis should be described, along with how any outliers were handled in the analysis.

Outliers should not be deleted unless there is independent evidence suggesting that the data are incorrect, and the reasons for any deletions should be stated. If data have been transformed to fulfil any statistical assumptions, the results of the analysis should be converted back to the original units of measurement in the report. Finally, the clinical or biological implications of the conclusions should be thoughtfully developed to aid in any judgements or decisions that might emerge from the work and thereby promote utility of the findings for the biomedical research community.

## REFERENCES

1. Festing, M.F.W., The scope for improving the design of laboratory animal experiments, *Lab. Anim.*, 26, 256, 1992.
2. Festing, M.F.W., Reduction of animal use: experimental design and quality of experiments, *Lab. Anim.*, 28, 212, 1994.
3. Festing, M.F.W., Are animal experiments in toxicological research the "right" size?, in *Statistics in Toxicology*, Morgan, B.J.T., Ed., Clarendon Press, Oxford, 1996, p. 3.
4. McCance, I., Assessment of statistical procedures used in papers in the *Australian Veterinary Journal*, *Aust. Vet J.*, 72, 322, 1995.
5. Altman, D.G., Statistics in medical journals, *Statistics in Medicine*, 1, 59, 1982.
6. Altman, D.G., Misuse of statistics is unethical, in *Statistics in Practice*, Gore, S.M. and Altman, D.G., Eds., British Medical Association, London, 1982, p. 1.
7. Weigler, B.J., A primer in epidemiologic methodology, *Comp. Med.*, 51, 208, 2001.
8. Hill, A.B., The environment and disease: association or causation?, *Proc. R. Soc. Med.*, 58, 295, 1965.
9. Gordis, L., *Epidemiology*, 2nd ed., W.B. Saunders Co., Philadelphia, PA, 2000.
10. Meinert, C.L., *Clinical Trials: Design,Conduct, and Analysis*, Oxford University Press, New York, 1986.
11. Thomas, L., A review of statistical power analysis software, *Bull. Eco. Soc. Amer.*, 78, 126, 1997.
12. Mead, R., *The Design of Experiments*, Cambridge University Press, Cambridge, New York, 1988.
13. van Herck, H., Baumans, V., Boere, H.A.G., Hesp, A.M.P., and van Lith, H.A., Orbital sinus blood sampling in rats: effects upon selected behavioural variables, *Lab. Anim.*, 34, 10, 2000.
14. Lipnick, R.L., Cotruvo, J.A., Hill, R.N., Bruce, R.D., Stitzel, K.A., Walker, A.P., Chu, I., Goddard, M., Segal, L., Springer, J.A., and Myers, R.C., Comparison of the up-and-down, conventional LD50, and fixed-dose acute toxicity procedures, *Food and Chemical Toxicology*, 33, 223, 1995.
15. Montgomery, D.C., *Design and Analysis of Experiments*, 5th ed., John Wiley & Sons, New York, 2001.
16. Snedecor, G.W. and Cochran, W.G., *Statistical Methods*, Iowa State University Press, Ames, IA, 1980.
17. Fisher, L.D. and Van Belle, G., *Biostatistics. A methodology for the Health Sciences*, John Wiley & Sons, Inc., New York, 1993.
18. Lang, T.A. and Secis, M., *How to Report Statistics in Medicine*, American College of Physicians, Philadelphia, PA, 1997.
19. Altman, D.G., Gore, S.M., Gardner, M.J., and Pocock, S.J., Statistical guidelines for contributors to medical journals, in *Statistics with Confidence*, Gardner, M.J. and Altman, D.G., Eds., British Medical Journal, London, 1989, p. 83.

CHAPTER **15**

# Common Nonsurgical Techniques and Procedures

**Cynthia A. Pekow and Vera Baumans**

## CONTENTS

0-8493-1086-5/03/$0.00+$1.50
© 2003 by CRC Press LLC

## INTRODUCTION

Humane experimental technique minimizes distress to research animals and research personnel alike. This chapter outlines procedures for restraint, sampling, and dosing, with suggested humane refinements. Primary species covered are the mouse, rat, guinea pig, rabbit, and pig.

Successful, efficient, humane technique requires practice and skill development. Adequate time must be budgeted to train and work alongside experienced instructors, to develop comfort and proficiency with specific procedures. Appropriate use of sedatives, analgesics, and anesthetics in research animals can minimize distress. Use of such medication is specified or suggested with a number of the techniques described in this chapter, and in the chapter on anesthetics.

In addition, stress may be minimized if animals are first adapted to the research environment and are familiar with caretakers and research staff. Acclimatization and quarantine periods vary with the species and their place of origin. Many animals can be conditioned to readily accept common handling and restraint techniques and procedures.

The techniques described here are certainly not the only methods available. In selecting a particular method, an individual should consider his proficiency, the distress to the animal of the restraint as well as of the procedure, and the history of success. Each person who is privileged to work with research animals must have the integrity to know when a technique or method is not right in his hands, and to seek assistance or employ another method.

## HANDLING AND PHYSICAL RESTRAINT

Safe, firm, gentle handling and restraint is key to humane technique. A soft voice and demeanor plus a calm, consistent laboratory atmosphere assist in minimizing tension for animals and researchers. Animals communicate with smells and sounds that are often undetected by humans, and they should be protected from exposure to what they may perceive as distressful situations in other animals. Ideally, potentially distressful procedures are not performed in the animal housing area.[1,2]

### Rats

Rats respond positively to gentle handling. Their inclination to hide and enter small spaces can be used to assist with restraint. Use of leather or wire mesh gloves is not necessary. Heavy gloves decrease dexterity of the handler, and rats may tear toenails on wire mesh gloves. A nervous or aggressive animal can be grasped with the aid of a towel or cloth drape.

Rats may be handled by the tail for initiation of restraint, but care must be taken to grasp the tail at the base, near to the body. The skin of the tail may be torn if handled near the tail tip. In general, rats are uncomfortable restrained by the tail alone, and this technique is used only for quick transfer of animals, for example, from cage to cage. Preferably, the tail is held only to keep the animal in place, and the weight of the rat's body is supported from underneath (Figure 15.1).

Rats unaccustomed to being handled may be placed in a towel or drape and allowed to hide in folds of material.

There are three basic grip techniques, useful for restraining rats for procedures such as injection. In general, rats placed on a solid surface such as a countertop or handler's forearm, can be readily grasped

**Figure 15.1**    To remove a rat from the cage, the handler gently grasps the tail near the base, and supports the body with the other hand.

for restraint. Do not place rats onto wire bar cage lids, as they will grasp the bars, and may damage toenails as their grip is pried loose. Selection of restraint method will vary with the size and docility of the rat, and the size of hands, comfort level, and procedure to be performed by the handler.

For a secure two-handed grasp, one hand is placed over the rat's shoulders, with thumb and index finger under the forepaws, gently pushing the forepaws toward the rat's head. The forepaws may cross under the rat's chin. The other hand encircles the body at mid-abdomen, to keep the rear legs restrained. Care is taken not compress the chest or restrict the rat's breathing (Figure 15.2).

In a similar method using a single hand, the rat's head is restrained by the index and middle fingers pushing up from behind on each side of the rat's lower jaw. The thumb and fourth finger encircle the thorax behind the rat's forepaws (Figure 15.3).

When procedures are performed that require direct control of the head, rats may be grasped by the scruff. The skin over the neck, including the skin at the base of the ears, is firmly held by the index finger and thumb to control the head, and the skin of the back is held by the other three fingers. Care is taken that the grasp does not restrict the rat's ability to breathe (Figure 15.4).

Use of terry towels, drapes, or stockinette provide secure and rapid restraint. A rat may be quickly rolled into or held beneath a towel, leaving access to the body part needed (Figure 15.5). Cotton tube stockinette material is an adjunct to other restraint methods and provides a hiding place for the rat, yet easy access for the handler. The rat is gently guided into the bunched material, much like putting on a sweater or jumper. A band of tape at the neck area prevents the rat from exiting the cloth tube (Figure 15.6).

A variety of commercially available restraint devices come in a range of sizes and are useful for permitting access to tail vessels. Most rats will voluntarily enter the openings of these rigid restrainers (Figure 15.7). Rats can be trained to sit quietly in the restrainer. Cone-shaped plastic bags ("decapicones") are available for restraint of animals for decapitation by guillotine. These bags can also be used to briefly restrain rats or mice for access to blood vessels in the tail or rear legs (Figure 15.8).

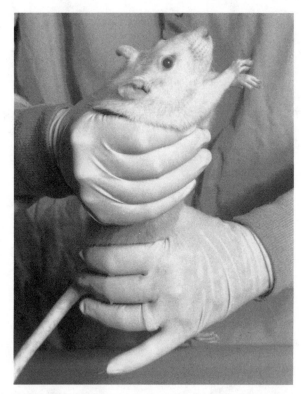

**Figure 15.2**  A secure two-handed rat restraint technique. Care is taken not to compress the chest.

**Figure 15.3**  Restraint of the rat using one hand.

**Figure 15.4** Restraint of the rat by the scruff.

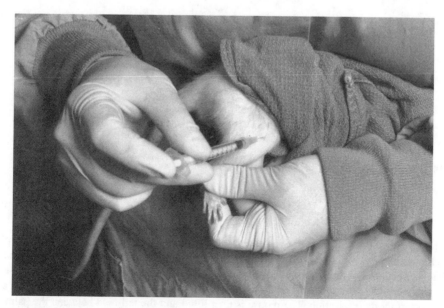

**Figure 15.5** Restraint of the rat in a towel. The body part needed extends from the towel. In this example, an intramuscular injection is being made in the quadriceps muscle of a rear leg.

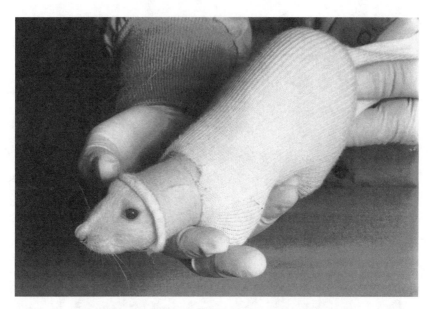

**Figure 15.6**   Restraint of the rat in stockinette. Injections may be made through the material, or the material may be cut to allow access.

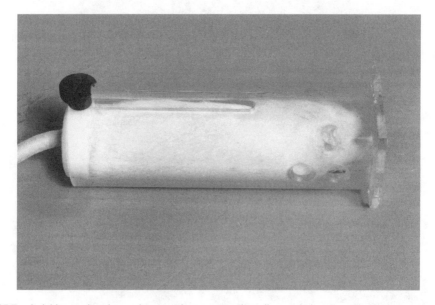

**Figure 15.7**   A rigid restrainer is used to permit access to the tail vessels in rats and mice.

## Mice

Mice can be removed from the cage by grasping the tail near to the tail base. For restraint, place the animal on a surface on which it can take a firm grip, such as a wire bar cage lid. A confident, gentle grasp that keeps control of the head will counter the mouse's common bite-defense response to restraint. The skin *including the base of the ears* must be held in the grip. The mouse should not be grasped so tightly as to restrict its respiration. The mouse is then lifted and the tail held between the palm and the ring finger or little finger (Figure 15.9).

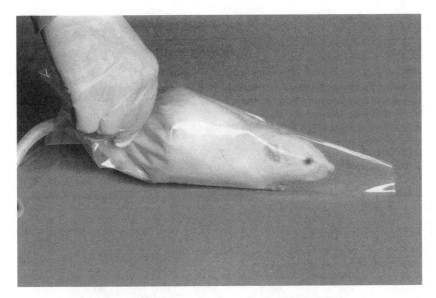

**Figure 15.8**   A plastic bag restrainer allows access to tail vessels in rats and mice and can be cut to permit access to rear legs. Animals become hot in these bags and must not be left in for more than a few minutes. Note the breathing hole at the front end of the bag.

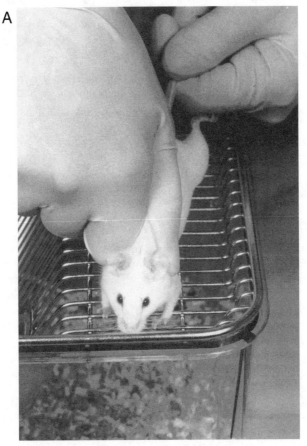

**Figure 15.9**   Restraint of the mouse. (A) The skin including the base of the ears must be held in the grasp to keep control of the mouse's head.                                                                 (continued)

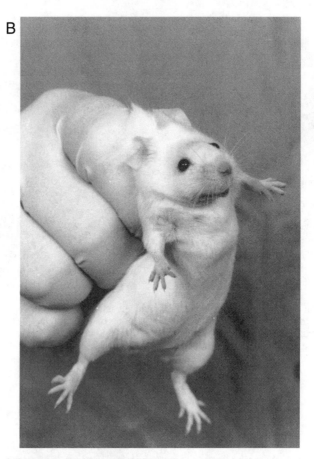

**Figure 15.9 (continued)** Restraint of the mouse. (B) The tail is tucked between the palm and last fingers.

Rigid plastic and metal restraint devices are available to permit access to tail vessels. A plastic cone-shaped bag ("decapicone") may be used to restrain mice for access to the tail or a rear leg (Figure 15.8).

## Guinea Pigs

Guinea pigs must be approached calmly and surely to decrease the chance of a panic response, in which frightened animals race about the cage. Guinea pigs are handled with one hand encircling the thorax from in front or behind, and the other hand supporting the hindquarters (Figure 15.10). Holding the thorax too tightly may induce shock. Guinea pigs tend to hide in terry towels or drapes, which may be used to assist restraint. A firmly wrapped towel may be used as a primary method of restraint.

## Rabbits

As an adaptation for rapid escape from predators, rabbits have powerful hind leg muscles, coupled with a light skeleton. This combination predisposes rabbits to fracture or subluxation of the lumbar spinal vertebrae (broken back), likely to occur if an animal attempts to leap or kicks while being restrained. Restraint methods therefore emphasize secure control of the front and hind end.

To remove a rabbit from a cage, the scruff or loose skin over the shoulders is grasped. The rabbit is turned to face the handler. The other hand is placed over or under the rabbit's hindquarters, and the animal is lifted out (Figure 15.11). Rabbits can be securely carried close to the handler's body with a two-handed restraint, with the animal's head tucked under the handler's arm (Figure 15.12). Rabbits

**Figure 15.10** Restraint of the guinea pig. The body weight is supported with one hand, and the thorax is encircled from the front or rear.

should never be handled or lifted by the ears or by the scruff alone, or allowed to struggle while restrained. If a rabbit struggles while being carried, the animal should be placed on the floor and temporarily released until a more secure grasp can be achieved.

Use of tranquilizing medications such as phenothiazine drugs (for example, acepromazine maleate at 0.8 mg/kg, s.c.,) or opioid combinations (such as fentanyl/fluanisone [Hypnorm®] at 0.2 mL/kg i.m.) can make restraint less stressful. About 15 min prior to handling, the medication can be given to an animal, held by the scruff, without removal from its cage. Aggressive or markedly frightened animals may first be covered with a drape while in the cage, then grasped for lifting, or given tranquilizing medication.

A variety of restraint devices can be used to permit access to blood vessels, for oral dosing, or other procedures. Ideally, rabbits are acclimated to restraint in these devices prior to experimental use. Rigid metal or plastic units (Figure 15.13), cloth bags (Figure 15.14), and towels (Figure 15.15) can be used. Once secured, rabbits may be less likely to struggle if their eyes are covered by a lightweight cloth. Struggling while restrained in a device can easily lead to lumbar spinal damage.

## Pigs

Pigs can be trained to cooperate in a variety of procedures, with the use of positive reinforcement, such as food treats. Untrained or frightened animals can be extremely noisy and may injure themselves or handlers in attempts to flee restraint. Pig boards or similar large flat objects may be used to herd a pig into a corner for a rapid procedure such as a quick injection. Soft snares or loop twitches may be also be used around the upper jaw, for such short procedures, although this method may be stressful (Figure 15.16). Longer procedures require use of sedation. V-shaped troughs may be used to restrain small pigs in dorsal recumbency.

Use of slings is a humane way to restrain pigs for many procedures. Pigs are trained with treats to readily enter a sling and to remain calm while in the sling. For heavier animals, slings are available that use cranks to raise the pig to table height, once the pig has walked into the sling at floor level (Figure 15.17).

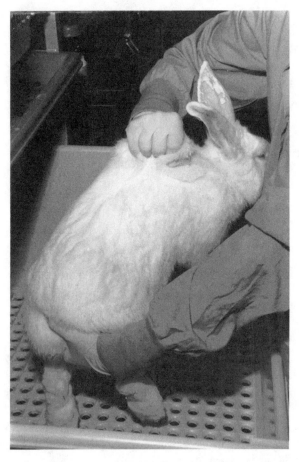

**Figure 15.11** Restraint of a rabbit for removal from a cage. The scruff and hindquarters are securely held to decrease chance of injury.

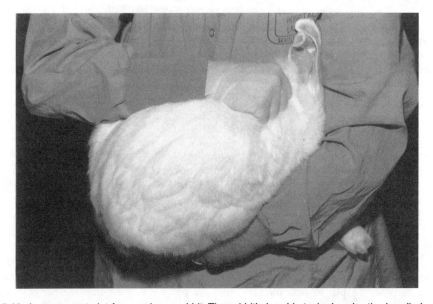

**Figure 15.12** A secure restraint for carrying a rabbit. The rabbit's head is tucked under the handler's arm.

**Figure 15.13** A plastic rabbit restrainer.

**Figure 15.14** A rabbit held in a cloth restraint bag designed for cats. Zippered openings in the bag allow access to the body.

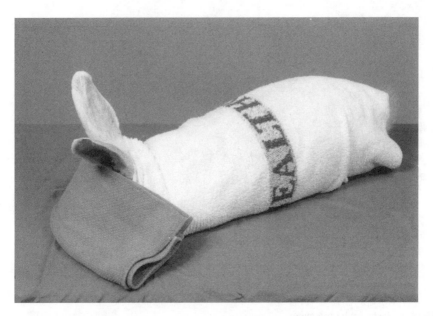

**Figure 15.15** A terry towel restraint permits access to the ears. The rabbit's weight secures the towel. With any method of restraint, covering the eyes may calm the animal.

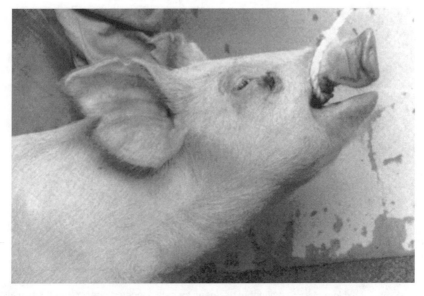

**Figure 15.16** Restraint of a large pig with a soft loop twitch.

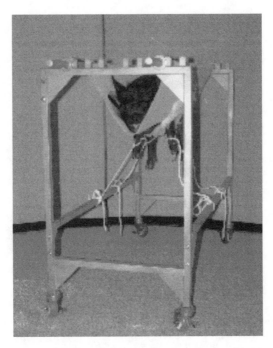

**Figure 15.17** A pig resting in a sling. (Photo courtesy of Panepinto and Associates, Colorado.)

## METHODS OF IDENTIFICATION

Temporary marking methods are those that are typically groomed off, rubbed off, or shed by an animal. These methods include use of colored pens on albino animals or light-colored parts of an animal, livestock marking crayons, and some dyes. In rats, colored ink rings marked on the tail provide ready identification lasting several days, and the color can be reapplied as it fades. Shaving a patch of hair can also serve as a temporary identification. Color photographs may be used for identification of animals with distinctive coat marking patterns.

Tattoos, made with clamp tongs (Figure 15.18) or tattoo pencils (Figure 15.19) permanently place dye into the skin. Tattooing is commonly used in rodents (ears or tails), rabbits (ears), and pigs (ears). Local anesthetic such as prilocaine-lidocaine cream, possibly supplemented with general sedation, or general anesthesia, is a humane consideration during tattoo placement. Tattooing is of advantage in marking neonatal animals, as the technique is permanent yet minimally invasive. A tattoo placed with correct technique using a tattoo pencil in a neonatal rodent tail can be clearly read throughout the animal's life. Correct technique involves placing the ink into the dermal layer of the skin, and some initial practice is required. Removal of toes from neonates as a method of identification is difficult to justify, with use of tattoos as a preferred method.

Ear tags, used in rodents, rabbits, and pigs, are generally placed at weaning. Correct placement decreases chances that the tag will cause pain or irritation to the animal. Tags placed too close to the ear margins are readily torn out, while those placed too deeply crimp the ear. Tags used in rodents should be placed hanging on the lower aspect of the ear, so that the ear is not bent from the weight of the tag when the animal's head is held in a naturally erect position. In rabbits, tags are placed in the fold at the lower medial aspect of the pinna.

Ear punches and ear snips may be used to identify pigs, mice, rats, and other rodents. Ear snips are made in neonatal piglets. In rodents, punches or snips may be made at weaning or later. A numbering system is based upon the number and location of the holes or snips (top, middle, or bottom, and right or left ear).[3]

Implantable microchips are of approximately 1 cm size. In rodents and rabbits, these chips are inserted subcutaneously using a trocar in the dorsal neck. Surgical glue may be used to seal the hole

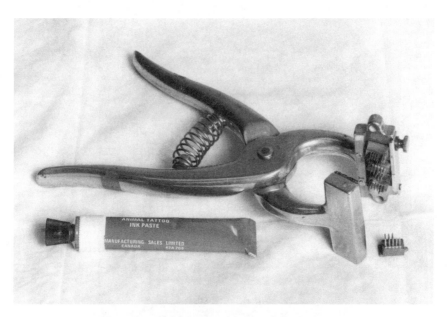

**Figure 15.18** Tattoo clamp, numbers, and tattoo ink.

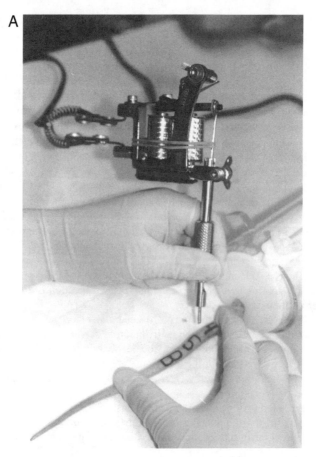

**Figure 15.19** (A) Tattoo pencil used on a rat tail. (Photos courtesy of AIMS, Inc., www.aims@animalid.com.)

(continued)

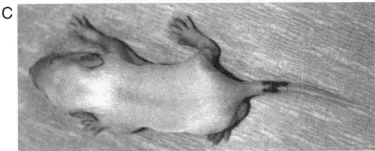

**Figure 15.19 (continued)** (B) Adult rat with tail tattoo. (C) Neonatal rat with tail tattoo. (Photos courtesy of AIMS, Inc., www.aims@animalid.com.)

left by the trocar in small animals such as mice. An electronic scanner is used to read the identification number unique to each chip. Though more expensive than other methods of identification, microchip systems can be integrated with data recording and census programs.

## ADMINISTRATION OF SUBSTANCES BY INJECTION

Selection of needle diameter and length is made with consideration of several factors. A smaller needle causes less damage and pain when inserted into tissue. Larger diameter needles are advantageous in permitting large volumes to be given rapidly, and they facilitate passage of viscous substances. When administering blood or cells, cell damage is less likely if a large bore needle is used. However, diameter is also limited by the size of the animal and, in particular for intravenous injections, by the size of the selected vein. Recommended needle sizes for specific species and routes of injection are given in Table 15.3.

**Table 15.2 Catheter Dimensions in Relation to French Size**

| External Diameter | |
|---|---|
| Metric Gauge (mm) | French Gauge/Charrièr no. |
| 0.33 | 1 |
| 0.67 | 2 |
| 1.00 | 3 |
| 1.33 | 4 |
| 1.67 | 5 |
| 2.00 | 6 |
| 2.33 | 7 |
| 2.67 | 8 |
| 3.00 | 9 |
| 3.33 | 10 |
| 3.67 | 11 |
| 4.00 | 12 |
| 4.33 | 13 |
| 4.67 | 14 |
| 5.00 | 15 |
| 5.33 | 16 |
| 5.67 | 17 |
| 6.00 | 18 |
| 6.33 | 19 |
| 6.67 | 20 |
| 7.00 | 21 |
| 7.33 | 22 |
| 7.67 | 23 |
| 8.00 | 24 |
| 8.33 | 25 |
| 8.67 | 26 |
| 9.00 | 27 |
| 9.33 | 28 |
| 9.67 | 29 |
| 10.00 | 30 |
| 10.33 | 31 |
| 10.67 | 32 |
| 11.00 | 33 |
| 11.33 | 34 |

**Table 15.1 Needle Sizes in Relation to Standard Wire Gauge**

| External Diameter | |
|---|---|
| Metric Gauge (mm) | Standard Wire Gauge |
| 0.25 | 30 |
| 0.35 | 28 |
| 0.40 | 27 |
| 0.45 | 26 |
| 0.50 | 25 |
| 0.55 | 24 |
| 0.65 | 23 |
| 0.70 | 22 |
| 0.80 | 21 |
| 0.90 | 20 |
| 1.10 | 19 |
| 1.25 | 18 |
| 1.45 | 17 |
| 1.65 | 16 |
| 1.80 | 15 |
| 2.10 | 14 |
| 2.40 | 13 |
| 2.80 | 12 |
| 3.00 | 11 |
| 3.25 | 10 |
| 3.65 | 9 |

Other injection considerations include the pH, tonicity, and temperature of the substance being administered.[4] Extremes in pH cause local tissue damage. Ideally, solutions should be at pH = 7.4. Tonicity refers to the concentration of solutes that affects the movement of water in or out of cells. Isotonic solutions such as 0.9% saline or 5% dextrose are often used for injection. Hypertonic solutions are most commonly administered by intravenous route in a large vessel to promote rapid dilution and avoid irritation to the vessel wall. Solutions for rehydration or any large volume to be administered should be warmed to prevent inducing hypothermia, particularly in small animals. A review of best practices for substance administration has been made.[5]

Warming an animal or the extremity to be injected induces vasodilation, making the blood vessels easier to see and to access. Rate of intravenous infusion can vary from bolus, to slowly over several minutes, to continuous over several hours or longer. Solutions of high or low pH and nonisotonic solutions are better tolerated as slow or continuous infusions.

Adequate restraint may need to be augmented with local or general anesthesia to ensure accurate needle placement with minimal tissue trauma and pain for the animal. Butterfly needles (needles with attached flexible tubing) are advantageous in many situations. Once a butterfly needle has been seated in a blood vessel and taped in place, the flexible tubing permits the animal some freedom from complete restraint during the injection period.

**Table 15.3 Recommended Needle Gauge and Dosing Volume for Different Routes of Administration**

| Species | Subcutaneous | | Intraperitoneal | | Intramuscular | | Intravenous | | | Oral | |
|---|---|---|---|---|---|---|---|---|---|---|---|
| | Needle Gauge | Volume (mL) | Needle Gauge | Volume (mL) | Needle Gauge | Volume (mL) | Needle Gauge | Bolus Volume (mL) | Slow Infusion Volume (mL) | Needle Gauge or French Tube Size | Volume (mL) |
| Mouse, 25g | 25 | 0.25 | 25 | 0.5 | 27 | 0.05 per site | 26 | 0.125 | 0.3 (25 mL/kg) | 20 | 0.25 |
| Rat, 200g | 25 | 1 | 25 | 2.0 | 25 | 0.1 per site | 25 | 1 | 4 (20 mL/kg) | 18 | 2 |
| Guinea pig, 200g | 23 | 1 | 25 | 2 | 25 | 0.1 per site | 26 | 1 | 4 | 18 | 2 |
| Rabbit, 4 kg | 23 | 4 | 23 | 20 | 25 | 1 | 23 | 8 | 40 (10 mL/kg) | 13 or 8.Fr | 15 |
| Pig, 25 kg | 22 | 25 | 20 | 25 | 23 | 6.25 | 23 | 62.5 | 125 (5 mL/kg) | 22 | 150 |

Volume refers to good practice for single or multiple doses. Prior to redosing, consider time for absorption of prior dose, particularly for substances in nonaqueous base, or irritating substances. Multiple sites may be used.

Bolus IV injection rate for rodents and rabbits is about 3 mL/min. Slow IV infusion rate is over the course of 5 to 10 min or longer. Tolerance of IV injection is highly dependent on the vehicle used.

Oral dosing volume is dependent on whether the stomach is empty. Oil-based substances are less well tolerated in large volume than are aqueous-based substances.

## Rats

Subcutaneous injections are made in the scruff or flank. Rats may be restrained manually or wrapped in a towel. The skin is gently raised or "tented." The needle is inserted at a shallow angle, parallel to the animal, and is directed away from the fingers of the handler. Gently lifting the inserted needle to raise the skin confirms correct placement in the subcutis.

Intravenous injection sites include the lateral tail veins, lateral saphenous veins, dorsal metatarsal veins, tongue veins, and penis vein. For access to the tail and leg veins, restraint is achieved with the use of a rigid restraint tube, towel, cloth, or plastic cone (Figure 15.20). Topical anesthetic cream has *not* been found to be effective in decreasing pain response to tail stick in the rat.[6] General anesthesia is necessary for access to the veins of the tongue or penis. Gently warming the rat or its tail with a heating pad, heat lamp, or hot-water bag will dilate the vessels. Some rats have a build-up of reddish-brown scale stain on the tail, which may obscure the blood vessels. The stain can be cleaned off by gently rubbing the tail with a gauze sponge moistened with rubbing alcohol. The needle is inserted at a very shallow angle. Correct placement can be confirmed by aspiration of blood. For repeat intravenous injections over a short period, catheters can be placed into the tail vein. For repeated access over several days, surgical implantation of a jugular catheter is recommended.

Intramuscular injections are made in the quadriceps muscles (anterior aspect) of the thigh. The posterior muscles of the thigh may be used with caution, as the sciatic nerve passes thru this muscle group. Firm restraint or anesthesia is necessary to preclude the leg from kicking during the injection, to prevent the needle from tearing and bruising the muscles. Care is taken to visualize the depth at which the needle is placed, ideally centering the tip near the middle of the muscle group. Use of a short needle gives better control. Restraint may be achieved with the use of a cloth or towel (Figure 15.5).

Intraperitoneal injections are made in the lower half of the abdomen, off the midline. Animals can be restrained manually, beneath or in a towel, or with the upper half of the body in a rigid tube (Figure 15.21). The needle is directed in a shallow angle. Once the needle is in the abdomen, but prior to injection, aspiration may be done to help confirm correct needle placement. The plunger is pulled back gently to check for influx of urine, blood, or intestinal contents into the syringe. Should aspiration confirm that the needle has entered the bladder, or a blood vessel, or the gut, the needle is removed from the animal, and the entire procedure is begun again with a fresh needle and syringe.

For intradermal (intracutaneous) injection, the site of injection must first be shaved. The skin is held between two fingers, and the needle delicately advanced into the cutis. A bleb will be formed on the skin as the injectate enters. Local or general anesthesia is generally required for accurate intradermal injection technique and for humane considerations.

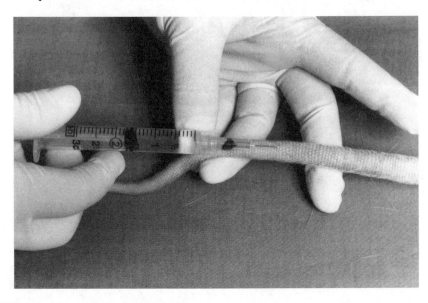

**Figure 15.20** Intravenous injection in the rat tail.

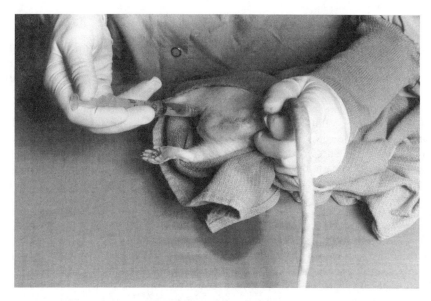

**Figure 15.21** Use of a towel for restraint of the rat for intraperitoneal injection in the lower half of the abdomen.

Intracardiac injection is performed when the rat is in a deep plane of anesthesia. This technique is most commonly performed as a terminal (nonrecovery) procedure. The heartbeat can be palpated by placing fingers on either side of the chest, at the level of the rat's elbows. The needle is directed into the heart from one of several possible sites. With the animal positioned in dorsal recumbency, a longer needle may be directed on the midline from below the sternum, through the diaphragm, at about a 45 degree angle toward the back. Alternatively, the animal is placed on its side, and the needle directed at a 90 degree angle from the side of the chest just behind the elbow, between the ribs (Figure 15.22). With any approach, a single well-controlled, quick thrust of the needle is needed to pierce the heart. Rapid flow of blood into the syringe confirms correct placement. Repeated attempts to place the needle increase the likelihood of causing severe damage to the lungs, heart, or blood vessels within the chest.

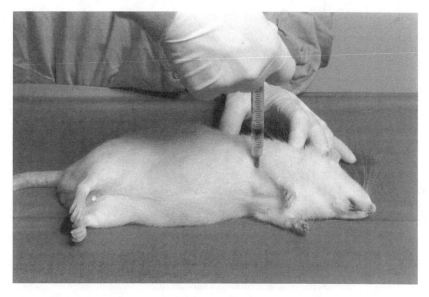

**Figure 15.22** One technique for intracardiac injection in the rat.

## Mice

Subcutaneous injections are made along the flank, scruff, or behind the elbow. The mouse is restrained by the scruff and tail, and the needle is inserted at a shallow angle (Figure 15.23). If the scruff is used, the needle is directed parallel to the animal to avoid the cervical vertebrae and away from the handler's fingers. Gently lifting the inserted needle to raise the skin confirms correct placement in the subcutis.

Intravenous injections can be given via the lateral tail veins. Injection via the retro-orbital sinuses is controversial, as the retro-orbital tissue may be readily damaged, leading to formation of a hematoma. Gently warming the mouse with a heat lamp or pad will cause vasodilation of the tail vessels. Use of lidocaine-prilocaine cream does *not* decrease reaction to tail stick in the mouse.[6] For tail vein injection, mice are held in a rigid restraint device (Figure 15.24). The needle is inserted at a very shallow angle, almost parallel to the tail, just far enough into the tissue to seat the needle past the bevel. Aspiration of blood to confirm needle placement is not productive in mice. Pushing a small volume of injectate can confirm correct placement. If the needle is correctly seated in the vein, the vessel will appear to clear as the injectate flows through. If the needle is outside the vein, a bleb will appear. Accessing the vessel is easiest about 2/3 of the distance down the tail (1/3 up from the tail tip).

Retro-orbital injection is made in the anesthetized mouse. The needle is directed under the lid and behind the eye at the dorsal aspect, or at the medial canthus, into the retrobulbar sinus. Aspiration of a small volume of blood confirms correct placement. After the needle is removed, the lids are held shut with a dry gauze sponge for several seconds, to achieve hemostasis. There is attendant risk of hematoma or ocular damage with this technique.

Intramuscular injection, a delicate procedure in the mouse, is made in the quadriceps muscle. Restraint technique should include holding the leg to be injected to prevent it from kicking during the injection. The foot may be secured alongside the tail beneath the fingers holding the tail. Care is taken to visualize the depth at which the needle is placed, ideally centering the tip near the middle of the muscle group.

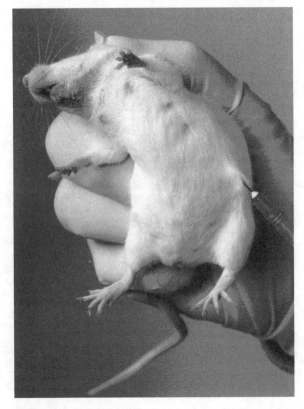

**Figure 15.23** Subcutaneous injection in the mouse.

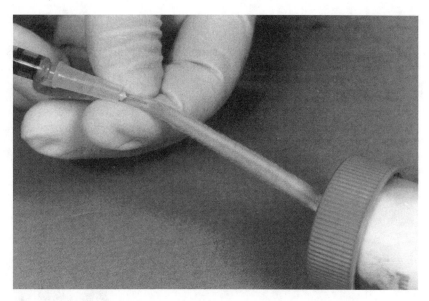

**Figure 15.24** A technique for restraint of the mouse for tail vein injection.

Intraperitoneal injection is made in the lower half of the abdomen, off the midline. The mouse is restrained manually (Figure 15.25). The technique is as described for the rat.

Intradermal injections follow the technique as described for the rat.

Intracardiac injection is performed with the mouse in a deep plane of anesthesia. This technique is most commonly performed as a terminal (nonrecovery) procedure. The needle is directed into the heart from one of several possible sites. With the animal positioned in dorsal recumbency, a needle may be directed on the midline from below the sternum, through the diaphragm, at about a 45 degree angle toward the back. Alternatively, the animal is placed on its side, and the needle directed at a 90 degree angle from the side of the chest just behind the elbow, between the ribs, as described for the rat. With any approach, a single well-controlled, quick thrust of the needle is needed to pierce the heart. Rapid flow of blood into the syringe confirms correct placement. Repeated attempts to place the needle increase the likelihood of causing severe damage to the lungs, heart, or blood vessels within the chest.

**Figure 15.25** Intraperitoneal injection in the mouse.

## Guinea Pigs

The scruff or flank is generally used as the site for subcutaneous injections. The guinea pig is held in a towel, or on its abdomen, gently pressed on a firm surface. A tent of skin is made, and the needle is directed away from the handler's fingers (Figure 15.26).

Intravenous access can be challenging in the guinea pig. Sites include the vessels of the ear (Figure 15.27), the lateral pedal veins, dorsal metatarsal veins, and penis vein. General anesthesia is necessary for access to the penis vein. Warming the animal gently with a heating pad or warm towel will dilate blood vessels.

The quadriceps muscles are the preferred site for intramuscular injection. The guinea pig can be restrained in a towel with the rear leg extended, or the animal can be held between the handler's arm

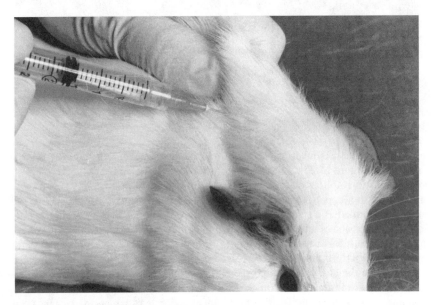

**Figure 15.26** Subcutaneous injection in the guinea pig.

**Figure 15.27** The ear vessels of the guinea pig.

and chest. The leg must be held firmly to prevent kicking during the injection. Care is taken to visualize the depth at which the needle is placed, ideally centering the tip near the middle of the muscle group.

Intraperitoneal injection is accomplished as described for the rat. The guinea pig is restrained manually or wrapped in a towel. Intradermal and intracardiac injections are as described for the rat.

## Rabbits

The scruff is the most common site for subcutaneous injection in the rabbit, although any area of loose skin over the back or flanks may be used. If the substance to be injected is an irritant, for example, antigen plus adjuvant for antibody production, the scruff should be avoided as an injection site, as this area must routinely be grasped for restraint of the animal. The skin is raised or tented, and the needle inserted pointing away from the handler's fingers.

The lateral ear veins are the most readily accessible site for intravenous injection in the rabbit. Warming the rabbit or the ears will cause vasodilation. A subcutaneous injection of phenothiazine tranquilizer (acepromazine maleate at 0.8 mg/kg) or an opioid combination (fentanyl/fluanisone [Hyp-norm®] at 0.2ml/kg given i.m.) 15 min prior to injection will calm the rabbit and cause dilation of the blood vessels. Topical anesthetic (lidocaine-prilocaine cream) generously applied 15 min prior to injection will effectively numb the ear vessels.[6]

Rabbits may be restrained for injection in any number of rigid or cloth restraint devices or wrapped in a towel (Figures 15.13, 15.14, and 15.15) The ear veins are very superficial to the skin. Use of a butterfly needle assists in needle insertion at a very shallow angle, almost parallel to the ear surface (Figure 15.28). Small catheters may be inserted and taped in place for repeat injections or infusions.

Intramuscular injection sites include the quadriceps muscles of the anterior thigh and lumbar muscles (Figure 15.29). The rabbit is restrained on a tabletop with its head tucked under the elbow of the handler. The quadriceps are readily palpated above the femur, and the leg is held steady with the hand cupped around the rabbit's stifle. The lumbar muscle sites are on either side of the spinal column. The needle is placed parallel to the spinal column, in the muscles between the caudal-most rib and the anterior aspect of the ilium (point of the hip).

Intraperitoneal injections are made in the lower half of the abdomen, off the midline. The rabbit is held by the scruff and raised to an upright position on the handler's lap. If the rabbit is agitated, an assistant may give the injection while the handler grasps the rabbit's feet and scruff. Alternatively, a cloth bag or towel may be used to restrain the rabbit. Aspiration is recommended to assist in confirming correct needle placement.

Intradermal injections follow the technique as described for the rat.

Intracardiac injections are made when the rabbit is in a deep plane of anesthesia. This is most commonly a terminal (nonsurvival) procedure for the rabbit. The rabbit is placed in dorsal recumbency. The needle is directed from the lateral aspect of the chest at the level of the rabbit's elbow, between the ribs. The needle is advanced with a smooth thrust to ensure the heart is pierced. Blood will pulsate back into the needle once it enters the heart. Use of a butterfly needle facilitates the procedure.

## Pigs

The moderately loose skin over the dorsal neck area, caudal to the ears is the primary site for subcutaneous injection in the pig. Depending on the volume and "sting" of the injectate, the animal can be completely restrained or the injection made via a small butterfly needle, while the pig is confined in a relatively small space.

The vessels of the ear are the primary site for intravenous injection in the pig. Use of a small butterfly needle is recommended, as is the use of topical anesthetic cream (lidocaine-prilocaine) applied 15 min prior to injection to numb the site. Gently warming the ear can dilate the vessels. Butterfly needles or catheters can be taped in place for repeat injections or longer infusions.[7]

Intramuscular injection sites in pigs include the quadriceps muscles of the hind leg and the muscles of the dorsal lateral neck. The pig can be restrained in a sling or with use of a soft snare or twitch. If a butterfly needle is used, the pig may be allowed to move about in a small area, while the person giving the injection seats the needle then walks beside the pig while giving the injection.[8]

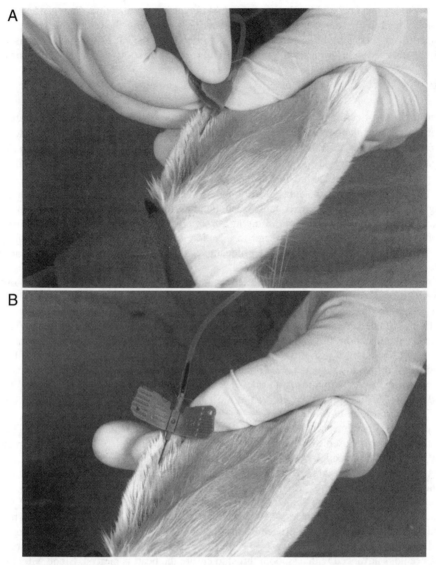

**Figure 15.28**  The ear vessels of the rabbit. The central ear artery can be seen as it runs from base to tip down the center of the ear. (A) Injection into the lateral ear vein. A 25 gauge butterfly needle is positioned flat along the ear. (B) Blood flow into the tubing confirms proper needle placement.

Intraperitoneal injections are made in the lower half of the abdomen. The pig can be restrained in a sling or with a snare. Aspiration is used to assess needle placement prior to the injection.

## ORAL DOSING

Oral dosing can be performed by gastric intubation (oral gavage) or by voluntary ingestion of test compounds in feed or water. Consideration of how the compound may alter the palatability of feed or water must be made and intake monitored accordingly. Moreover, voluntary ingestion may be less accurate.

Fasting an animal prior to oral dosing may be appropriate, depending on the feeding pattern of the species, and the time of day or light cycle. Rodents and rabbits practice cecotrophy, and may not have an empty stomach even after fasting.

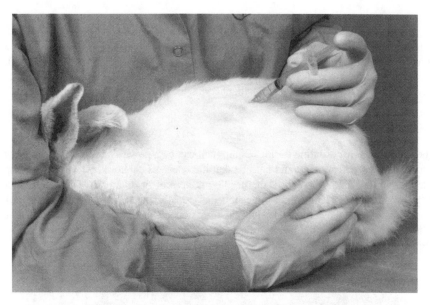

**Figure 15.29** Intramuscular injection in the lumbar muscles of the rabbit.

Gavage tubes come in flexible and rigid forms. Metal feeding needles with bulbed tips for rodents are available in straight, curved, and malleable forms, and do not require use of a mouth gag (Figure 15.30). Rubber or vinyl tubes often require a mouth gag be used to prevent the animal from chewing on the tube.

The length of needle or tube is determined by measuring from the animal's lips to its last rib. This distance should be marked on a long tube to guide correct placement depth. Once the tube or needle is in place, confirmation that the tube has not entered the respiratory tract is essential prior administering the dose. With proper tube placement, the animal is observed to breathe, but there is an absence of air passage through the tube as the animal respires.

Administration of large volumes can overload the stomach and may reflux into the esophagus or trachea, causing acute pneumonia, or may pass immediately into the small bowel. Table 15.3 gives suggested volumes for gavage. In planning a dosing protocol, the frequency of dosing, volume, pH, tonicity, and vehicle (aqueous

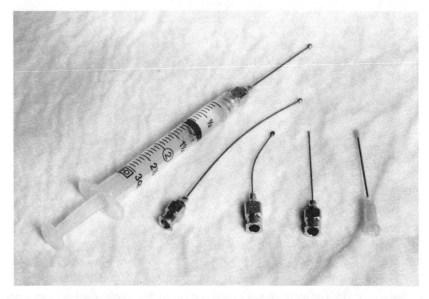

**Figure 15.30** Oral gavage needles for rodents come in a range of gauges and lengths. All feature a probe-ended tip to reduce trauma and decrease likelihood of insertion into the respiratory tract.

or oil) must be considered. Larger volumes of aqueous vehicle are better tolerated than oil vehicle. Depending on the length of the needle or tube inserted, substances may be administered into the esophagus instead of the stomach. Animals are not anesthetized prior to dosing, because maintenance of gag and swallow reflexes are important to detect proper tube placement and decrease chances of reflux and aspiration after dosing. However, animals resisting the procedure may be lightly sedated. Gentle technique is essential, as severe trauma may occur to the soft tissues of the mouth, pharynx, trachea, or esophagus if the technique is forced.

## Rats

Rats can be trained to accept routine dosing. Equipment can be a probe-ended gavage needle or a rubber tube. The rat is manually restrained and held so that the rear legs cannot kick. The head is tipped back with gentle pressure by the probe-end of the needle on the roof of its mouth. The needle should easily slip its entire (premeasured) length into the esophagus. Ideally, the handler should observe the animal swallowing the tube or needle (Figure 15.31). If any resistance is met, the tube may be in the respiratory tract and should be retracted to the level of the mouth and reinserted. A technique for dosing rats using rubber tubing held in a ring-stand has been described.[9]

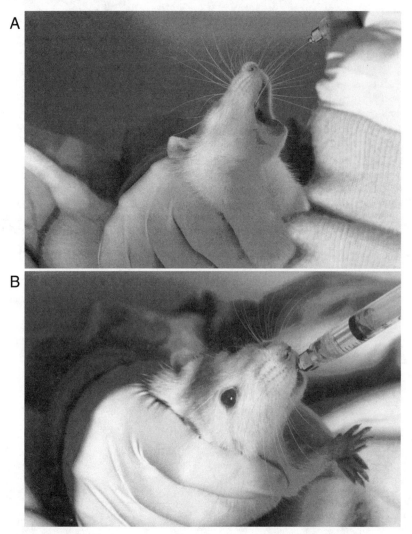

**Figure 15.31** Oral gavage in the rat using a metal gavage needle. (A) Tipping the head back with gentle pressure on the roof of the mouth. (B) The needle should slip its entire length without meeting any resistance. The needle must never be forced.

Capsule dosing of rats can be used for materials that are insoluble, or form poor suspensions. A mini gel capsule (size 9, approximately 8.4 mm length, 2.6 mm diameter, with a capacity of 0.025 mL) is used for rats of 150+ gm weight. A capsule-dosing syringe is used to eject the capsule into the distal esophagus. Insertion of the dosing syringe follows the same technique described for dosing with liquids.[10]

## Mice

Straight or curved probe-ended metal feeding needles are used for oral gavage in mice (Figure 15.32). The technique follows that described for rats.

## Guinea Pigs

Guinea pigs can be dosed using metal probe-ended feeding needles, or with a mouth gag and rubber tube, size 5 or 6 F. If the guinea pigs become agitated by gavage procedure, they may require light sedation (but not general anesthesia) before the procedure. The technique is similar to that described for the rat but is considered more difficult in the guinea pig due to the position of the molars.

A technique for capsule dosing of guinea pigs has been described. A mini gel capsule (size 9, approximately 8.4 mm length, 2.6 mm diameter, with a capacity of 0.025 mL) is used for guinea pigs of 300+ gm weight. A capsule-dosing syringe is used to place the capsule at the esophagus entrance. The capsule-dosing syringe is placed into the mouth using the same technique described for dosing with liquids, but the animal does not swallow the end of the syringe. The syringe is removed from the mouth, and the animal's neck is gently stroked to encourage swallowing.[10]

A technique for obtaining gastric juice samples has been described in which the guinea pig is held in a vertical position by an assistant, who uses a strip of gauze looped under the top incisor teeth to pull the animal's head into a position of dorsoflexion. A second strip of gauze is used to control the lower jaw. A flexible feeding tube is used.[11]

## Rabbits

Rabbits may be dosed using a rubber tube and mouth gag. The animal is restrained in a towel, cloth bag, or restraint box. Tipping the rabbit back may facilitate the procedure.

## Pigs

Pigs can often be induced to take fluids from a syringe inserted into the side of the mouth. The pig can be held in a sling or restrained with a soft snare. Well-socialized pigs under 25 kg can be restrained in a "bear hug," held with the pig's back against the handler's chest. Pigs can be trained to accept passage of a rubber stomach tube, while restrained in a sling or held in a hug. A 30 French size tube is appropriate for pigs of 15 kg. The soft tube is inserted until the pig swallows the end. Listening for stomach noises at the free end of the tube can help confirm proper placement. A mouth gag is not used unless the pig is resistant to the procedure. Alternatively, a long metal feeding needle designed for large rats can be used in small pigs.

## INTRATRACHEAL AND INTRABRONCHIAL INSTILLATION

Intratracheal and intrabronchial instillation are accomplished in anesthetized animals. Small volumes are injected directly into the trachea via the pharynx, with the aid of an otoscope and tracheal transillumination. Alternatively, but with more risk for the animal, the trachea may be pierced from the ventral aspect of the neck.

Intrabronchial administration is accomplished after passing a small gauge flexible plastic tube through an endotracheal tube, into the lung.[12] In larger species, a sterile fiberoptic bronchoscope passed through the endotracheal tube may be used to guide catheter placement.

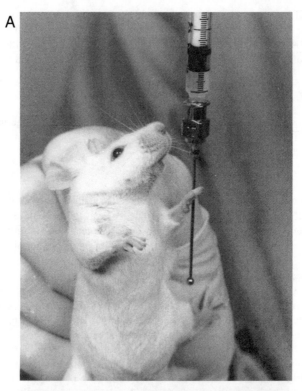

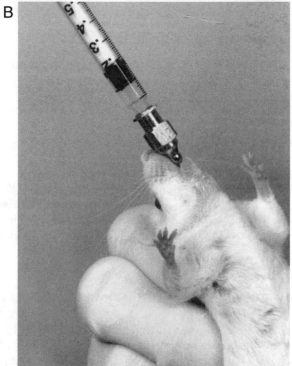

**Figure 15.32** Oral gavage in the mouse using a metal gavage needle. (A) The correct needle length is the distance from lips to the first rib. (B) Correctly placed, the needle is advanced without meeting resistance.

## BLOOD SAMPLING

Blood samples can be obtained from various sites of the body, using a variety of methods: from the veins, from the arteries, or by puncturing the orbital vessels, by cardiac puncture, or by decapitation. The choice of method depends on several factors, including the purpose of the blood collection, need for arterial or venous sample, duration and frequency of sampling, and whether the experiment is terminal for the animal. In small species, blood is often collected with the animal under anesthesia, to assist in immobilization and to decrease distress to the animal. When selecting a method for blood sampling in the conscious animal, consideration must be given to the potential for stress-induced effects on biochemical parameters.

Blood volume is estimated based on body weight. Several papers have reviewed recommendations on how much blood can be removed over a given time interval, without causing abnormal physiologic response and distress in animals.[13,14] A typical rule is not to exceed removal of 10% of blood volume (roughly 1% of body weight or 8 mL/kg) in any given 2-week period. For example, this would equate to removal of 3 mL from a 300 g rat, or 0.2 mL from a 20 g mouse, or 35 mL from a 3.5 kg rabbit. Removal of larger volumes necessitates longer recovery periods for the animal or may cause distress or even death.

Exsanguination is only performed on deeply anesthetized animals. One can expect to obtain approximately half the animal's total blood volume at exsanguination via cardiac puncture. Blood can also be collected after decapitation using a guillotine or scissors. For example, one can obtain approximately 9 mL from a 300 g rat, or 0.6 mL from a 20 g mouse, or 115 mL from a 3.5 kg rabbit. More precise techniques, such as a cut-down and cannulation of the carotid artery in the neck with infusion of fluids via the contralateral jugular vein or an ear vein, or aorta puncture, can maximize the blood volume collected at exsanguination. Infusion of fluids may influence blood parameters.

Local anesthesia, sedation, or general anesthesia are refinements that improve success and decrease stress for many blood-draw techniques. For repeated sampling over long periods, surgical placement of indwelling catheters is recommended. For multiple sampling at short intervals, catheters or butterfly needles can be taped in place. Once the needle has been removed from the blood vessel, light pressure with a dry gauze sponge is applied until hemostasis is confirmed.

### Rats

For blood withdrawal from the lateral tail vein, the rat is held in a restraint device (rigid tube, towel, or bag). If the tail has a buildup of reddish brown scale stain, a gauze sponge moistened with rubbing alcohol may be used to gently rub away the stain, to permit visualization of the blood vessels. The tail may be dipped in 45 degree water for about 1 min to induce vasodilation. The vein is most readily accessed about 1/3 the length up from the tip. Subsequent venipunctures are made ascending the tail. A 25 gauge needle or butterfly needle is inserted at a very shallow angle. Blood flow may be augmented by very gently milking with the fingers from the base of the tail toward the tip. Typically this technique is used for small volumes of blood, up to about 1 mL.

The ventral tail artery is accessed with the rat placed in dorsal recumbency. Using a finger, pressure is applied about 5 cm from the tail tip to raise the vessel. The artery is then punctured near to the base of the tail, and blood is collected. The technique may be performed in awake animals if an assistant is available to hold the rat firmly wrapped in a towel. However, anesthesia of the rat assists with this technique, to eliminate struggling against restraint and ease accurate blood vessel access and hemostasis.

The dorsal metatarsal or lateral saphenous veins on the lateral aspect of the rear leg may be punctured for blood collection. The rat is held wrapped in a towel or plastic cone with the leg extended. Bland ointment (petrolatum) applied to the leg will slick back the hair to permit visualization of the vessel, and also will prevent blood from soaking into the hair. Alternatively, the site may be shaved. The loose skin of the thigh is gently gathered to raise the blood vessel (Figure 15.33). The vessel is punctured, and blood is collected in a capillary tube or test tube.[15]

A technique for blood draw from the jugular vein in the unanesthetized rat requires participation of two people. A handler places the animal in dorsal recumbency on a board designed with a flat level surface that ends in a 45 degree slant, so that the rat's head can be tipped below the body. The forepaws

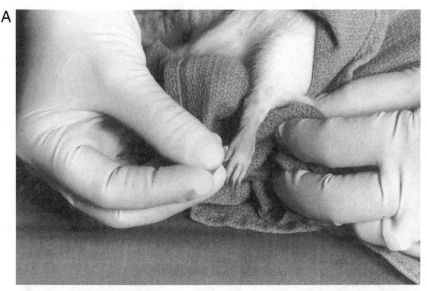

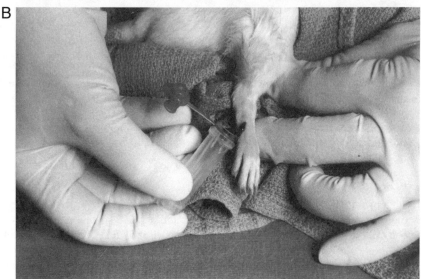

**Figure 15.33** Lateral saphenous venipuncture in the rat. (A) The vein is punctured. (B) Blood is collected in a tube or pipet.

are secured straight out to the sides with soft rope ties. The person taking the blood sample manipulates the rat's head downward and turned to the side, using a plastic cup held over the head. A needle with syringe attached is inserted 1 cm lateral to the midline, to about a 1 cm depth, and the sample is collected. Digital pressure is applied immediately after the needle is removed to aid in hemostasis. Restraint in this manner can be distressful to the animal, and the team performing the technique must be well practiced and coordinated to complete the procedure rapidly.

In the anesthetized rat, the jugular vein is accessed with the rat placed in dorsal recumbency on a flat surface. The syringe is kept parallel to the surface, and inserted about 1 cm lateral to the midline to a depth of about 1cm. Gentle suction is maintained in the syringe, which is slowly withdrawn from the animal until blood is seen to flow into the syringe. Alternatively, a small incision in the neck permits direct visualization of the jugular vein.

Tail tip removal has been justified in some protocols where multiple short-interval, small-volume samples are required. The last 0.5 to 1 mm of the tail is snipped off with iris scissors. Blood is gently milked from the cut, and then pressure is applied with gauze to achieve hemostasis. Repeat samples are obtained by removing the clot. Volumes of 0.1 to 0.2 mL are obtained.

Blood may be obtained from the retro-orbital plexus in the anesthetized rat. The rat is placed in lateral or ventral recumbency. A Pasteur pipette or microhematocrit capillary tube is passed beneath the upper lid at the medial canthus (inside corner) of the eye. The tube is gently pushed and twisted until it penetrates the conjunctiva. Without removing the tube from beneath the lid, the tube is gently retracted until blood flows into the tube. Once the flow is started, tilting a Pasteur pipette can allow gravity to assist. When the sample has been obtained, the tube is removed from the eye, and the lids are immediately held closed with a dry gauze sponge for several seconds, to achieve hemostasis. This technique is used to obtain amounts of blood from 0.5 to 3 mL. There is risk of corneal or ocular damage with this technique.[16]

Cardiac blood withdrawal is performed when the rat is in a deep plane of anesthesia, and is generally a terminal (nonrecovery) procedure. The technique is as described for intracardiac injection in the rat. Alternatively, the chest may be opened and the heart directly visualized for accurate direction of the needle.

## Mice

Blood may be collected from the lateral saphenous vein of the unanesthetized mouse. The mouse is held head-first in a restraint device, so that only the rear legs and tail are free. A rolled paper towel may be used for restraint, or a 50 cc plastic centrifuge tube may be used if holes are cut to permit air flow into the tube. The skin on the upper thigh is gently but firmly squeezed by the handler, using the same hand that is holding the tube. This serves to secure the mouse in the tube, and raises the vein. Bland ointment (petrolatum) applied to the leg will slick back the hair to permit visualization of the vessel, and also will prevent blood from soaking into the hair. Alternatively, the site may be shaved. Using a 25 gauge needle, the vessel is punctured at the most proximal visible aspect, and blood is collected as it wells up. A dry gauze sponge is used to apply pressure to the puncture site, and the pressure on the upper thigh is released. For repeat samples, the scab may be brushed off with a dry gauze sponge. A volume of 200 mL may be readily collected with this technique (Figure 15.34).[15]

Blood may be collected from the retro-orbital sinus of the mouse. This technique is performed with the mouse under anesthesia. The technique is as described for retro-orbital plexus sampling in the rat. Volumes of up to 1% of body weight are readily and rapidly obtained with this method (Figure 15.35). There is danger of corneal or ocular damage with this technique.

Cardiac blood withdrawal is performed when the mouse is in a deep plane of anesthesia and is generally a terminal (nonrecovery) procedure. The technique is as described for intracardiac injection in the rat. Alternatively, the chest may be opened and the heart directly visualized for accurate direction of the needle.

## Guinea Pigs

Small amounts of blood can be obtained from the unanesthetized guinea pig. With the animal restrained in a towel, the lateral pedal vein on a rear foot can be punctured with a small (25 gauge) needle, and blood can be collected into a microhematocrit tube or small test tube (Figure 15.36). In larger guinea pigs, small samples can be obtained from the lateral ear vessels (Figure 15.27). These vessels are quite small, and warming the ear may help vasodilation. A technique has been described for collection of blood from the jugular vein of mesmerized guinea pigs.[11] In the anesthetized guinea pig, blood collection from the jugular vein or via cardiac puncture is as described for the rat.

## Rabbits

The rabbit ear provides an easily accessible site for blood collection. Blood may be removed from the lateral ear vein or central ear artery. The artery is easy to cannulate and is the suggested site for volumes of more than 1 mL. Warming the rabbit or the ears will cause vasodilation. A subcutaneous injection of phenothiazine tranquilizer (acepromazine maleate at 0.8 mg/kg) or an opioid combination (fentanly/flu-anisone [Hypnorm®] at 0.2 mL/kg i.m.) 15 min prior to injection will calm the rabbit and cause dilation of the blood vessels. Topical anesthetic (lidocaine-prilocaine cream) generously applied 15 min prior to blood withdrawal will effectively numb the ear vessels.[6]

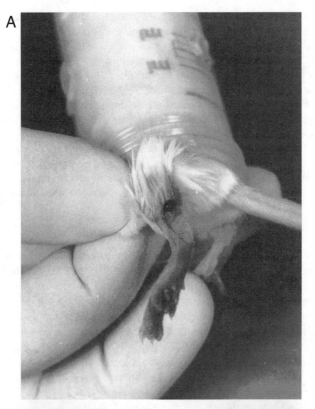

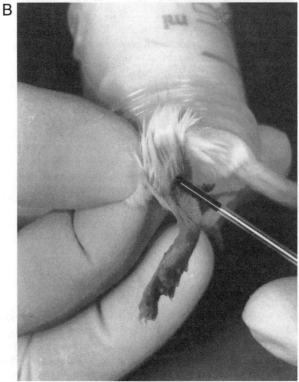

**Figure 15.34** Lateral saphenous venipuncture in the mouse. The vein is visible on the lateral thigh when the hair is slicked with petrolatum. (A) The vein is punctured. (B) Blood is collected in a tube or pipette.

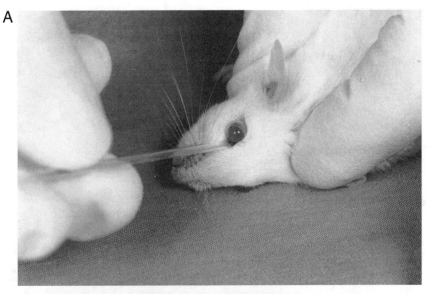

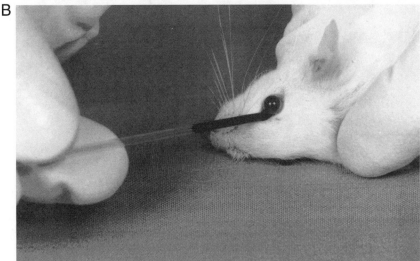

**Figure 15.35** Retro-orbital sinus puncture in the mouse. (A) The microhematocrit tube is placed under the lid and gently twisted until it penetrates the conjunctiva. (B) Blood is collected by capillary action.

Rabbits may be restrained in any number of rigid or cloth restraint devices or wrapped in a towel. The ear vessels are superficial to the skin. Use of butterfly needles assists in needle insertion at a very shallow angle, almost parallel to the ear surface (Figure 15.37). Correct placement of the needle into the artery is confirmed by a pulsing blood flow back into the tubing. Small catheters may be inserted and taped in place for repeat sampling. Hemostasis after arterial stick must be carefully attended to and may require pressure with a dry gauze sponge for 60 sec or longer.

Cardiac puncture, typically as a terminal (nonsurvival) procedure for exsanguination, is performed with the rabbit in a surgical plane of anesthesia. A large needle attached to flexible tubing is used to pierce the heart. With the rabbit in dorsal recumbency, the needle is directed from the level of the elbow, between the ribs, to the center of the chest. The needle is inserted with a firm thrust, so as to pierce the heart. A pulsing flow of blood into the attached tubing confirms correct placement. If blood does not immediately flow, slight redirection of the needle may assist. However, repeated probing in the chest increases the likelihood of killing the animal prior to obtaining the sample. If blood does not appear, completely remove the needle and make a new attempt with a fresh needle. An alternative method

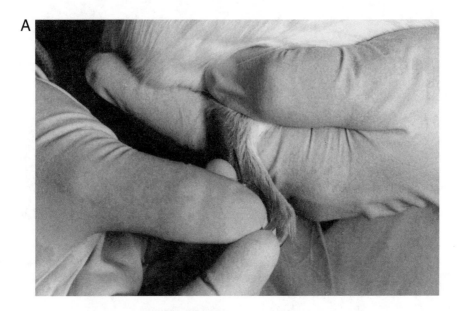

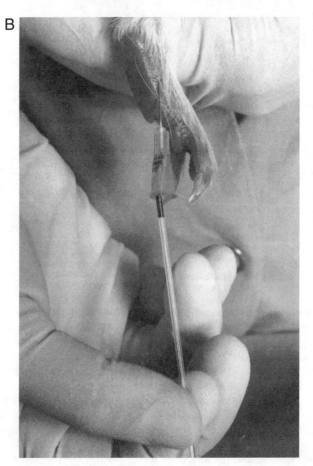

**Figure 15.36** Blood collection from the lateral pedal vein of the guinea pig. The guinea pig is restrained in a towel. (A) A 25 gauge needle is threaded into the vein. (B) Blood is collected into a hematocrit tube placed into the needle hub.

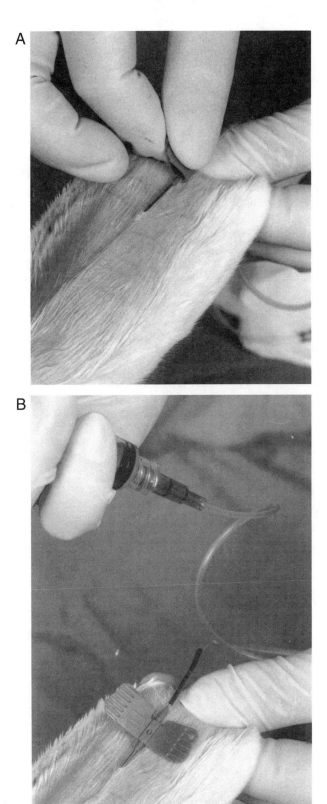

**Figure 15.37** Blood collection from the central ear artery of the rabbit, using a butterfly needle. (A) The needle is inserted at a very shallow angle. (B) Pulsing blood into the tubing confirms correct placement.

involves opening the skin of the ventral neck to visualize the carotid artery, which is then cannulated for blood removal. High-rate flow of saline into the contralateral jugular vein or an aterial ear vein during the exsanguination procedure maintains blood pressure for a longer time and will allow more blood to be collected from the animal. Fluid administration may affect blood parameters. When blood flow ceases, the animal should be euthanatized with a barbiturate injection or by another approved method.

## Pigs

Common sites for blood removal from the pig include the ear veins, jugular veins, and anterior vena cava. Occasionally, the cephalic vein is used, most often in smaller pigs. Often, untrained pigs must be sedated or anesthetized to permit ear vessel cannulation. Sedation has the advantage of promoting vasodilation of the vessels, depending on the medication used. However, pigs trained to sit in a sling may permit vessel cannulation, particularly if the site is first numbed with topical anesthetic (lidocaine-prilocaine cream). Small catheters or butterfly needles may be taped into place for long-term or repeat sampling in the restrained pig.

Puncture of the jugular vein may be done in the anesthetized pig or in an unanesthetized pig restrained in a sling or held by a soft rope snare around the upper jaw.[7] The pig's head is pulled up and back. Most pigs will pull backward from a snare so that the assistant holding the snare stands in front of the animal. The needle is directed caudodorsally in the deepest point of the jugular groove, formed between the medial sternocephalic and brachiocephalic muscles. Use of a vacuum blood collection tube facilitates the technique. The collection tube is advanced onto the needle once the needle has been placed into the neck. Good restraint is essential to prevent laceration of the vessel.[7,8]

## MILK COLLECTION

A milk collection apparatus can be made for use in rodents (Figure 15.38). Size of the equipment will vary with the size of the animal. A large glass tube is fitted with a two-hole rubber stopper. A T-shaped

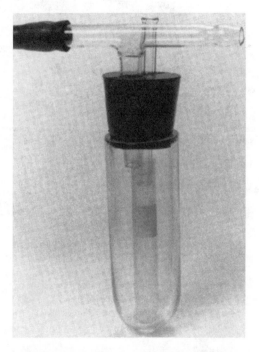

**Figure 15.38** Milk collection equipment for rats. The T-shaped glass tube is connected to a vacuum pump via the rubber tubing. The open limb is operated with a fingertip to create intermittent vacuum, simulating sucking. The open straight glass pipe through the stopper ends in a small collecting tube inside the large glass tube, and its outer end is applied to the teat.

glass tube and a straight glass tube are placed through the stopper. The straight tube is placed into a collection tube within the larger glass tube, and the other end of the straight tube is placed over a teat. One end of the T-shaped tube is connected to a vacuum, with between 250 to 300 mm mercury suction pressure. The other end of the T-shaped tube is intermittently occluded with the operator's finger to obtain an interrupted vacuum on the teat, to mimic sucking action. Gently massaging the mammary glands toward the teat during milking can assist flow. In rats, up to 7 cc of milk can be obtained per day. A discussion of practical aspects of milk collection in rats has been published.[17]

## BONE MARROW COLLECTION

Bone marrow collection is accomplished while the animal is under anesthesia. For survival procedures, the site is shaved and prepped as for surgery. In rodents and rabbits, the tibia is the most common collection site. The knee joint is flexed, and a needle is inserted through the trochlea into the marrow cavity, parallel to the long axis of the bone. For mice, a 25 gauge needle is gently yet firmly pushed and twisted into the marrow cavity. Somewhat larger needles may be used in other species, depending on the size of the bone. A low speed dental drill may be used to create a hole for needle insertion. Up to 1 mL of marrow may be obtained from the tibia of a rabbit. Pre-emptive and postprocedural analgesia is recommended. For pigs, the long bones, pelvis, or sternum are the sites for marrow collection. Consult a standard veterinary text for technique in this species.[8]

## URINE AND FECES COLLECTION

Rodents, particularly mice, tend to void samples of urine and feces when they are picked up and restrained. Gentle manual compression of the abdomen or placing the animal in an empty cage may provoke urination.

To collect uncontaminated urine samples, percutaneous cystocentesis (aspiration of urine via needle through the abdominal wall) may be used. The anesthetized or sedated animal is restrained in an upright position or in dorsal recumbency. A needle is directed into the abdomen toward the midline, just anterior to the pubis, and gentle aspiration is used to collect the urine. In some species, such as rabbits, the full bladder may be palpated to confirm location prior to centesis. The bladder will decrease in size as urine is aspirated, and may move away from the needle. However, the bladder should not be squeezed during the procedure. If urine does not flow into the syringe, the needle should be withdrawn and an attempt made with a clean needle. Probing in the abdomen may damage gut or other tissues and contaminate the bladder. If the bladder cannot be palpated with the animal in dorsal recumbency, a blind attempt may be made, directing the needle on the midline, midway between the umbilicus and brim of the pelvis. Blind technique carries increased risk of internal damage or contamination from the gut.

For longer-term collection, metabolism cages are available to separate and collect urine and feces (Figure 15.39).

Catheterization of the urethra is possible in some species but requires experience. Care is taken to avoid contamination of the catheter, which can introduce pathogenic bacteria into the bladder. Once the catheter has entered the bladder, urine should flow readily with gentle compression of the abdomen or light aspiration with a syringe on the catheter.

In female rats, a 22 gauge flexible over the needle catheter is used, with the needle removed. The anesthetized rat is held in dorsal recumbency, with the head toward the handler. The rat's tail is held by the handler's index finger, and the thumb applies gentle traction cranially on the abdomen to open the genital papilla. The catheter is advanced first in a caudal direction over the pelvis, then cranially into the bladder.

In unanesthetized rabbits, a flexible size 9 F catheter may be passed through the urethra. Sedation with subcutaneous acepromazine maleate at 0.8 mg/kg, will calm a struggling animal and will encourage male rabbits to protrude the penis, which may facilitate catheterization. The male is restrained in dorsal recumbency, and the catheter is passed into the urethra and directed caudally. The catheter will follow the urethra first caudally, then curve up and ventrally into the bladder. A rabbit doe is placed in ventral recumbency.

**Figure 15.39** Sampling urine and feces from a rat using a metabolism cage. (Techniplast Gazzada S.A.R.L., Italy.)

The catheter must be directed first vertically into the caudal part of the vagina, then brought into a horizontal position and into the urethral opening located in the ventral vagina. Urine collection may be facilitated by placing the rabbit in a box some hours before collection, to discourage premature voiding.

## CEREBROSPINAL FLUID

For all species, collection of cerebrospinal fluid (CSF) is conducted with the animal in a surgical plane of anesthesia. The area over the dorsal atlanto-occipital area is shaved, and cleaned with a presurgical scrub. The animal is placed in lateral recumbency, and the head is held firmly flexed forward, taking care not to obstruct respiration. A needle is introduced through the atlanto-occipital membrane, on the midline. The needle is kept parallel to the table surface. The needle is advanced slowly and carefully, keeping in mind that the target area is usually quite small and shallow. A decrease in resistance to advancing the needle usually indicates that the needle has entered the CSF space. The sample may be contaminated with blood, particularly if repeated samples are taken. Catheters may be placed for repeated sampling in larger rodents, rabbits, and larger species. Catheters may be placed in the lumbar subarachnoid space, or through the skull into the third intracerebroventricular space. Clinical textbooks on veterinary medicine should be consulted prior to CSF collection in the pig.[8]

### Rats

Between 0.1 to 0.5 mL may be obtained from the cisterna magna. A 22 or 23 gauge needle is used. A technique for collection from the lumbar region has also been described.[18,19]

## Mice

To reveal the puncture site, the skin is incised from a point 4 mm cranial to the external occipital protuberance to a point 1 cm cranial to the shoulder. About 0.025 mL of CSF can be collected with a 22 gauge needle. [20]

## Guinea Pigs

A 23 gauge needle is introduced through the atlanto-occipital membrane, at a 20 to 30 degree angle between the needle and the axis of the head. A volume of up to 0.33 mL can be obtained.

## Rabbits

A 22 or 23 gauge 1.5 in spinal needle is used. The needle is inserted about 2 mm caudal to the occipital protuberance and advanced slowly toward the midline. A volume of 1.5 to 2.0 mL can be obtained.

## BLOOD PRESSURE MEASUREMENT

The most precise blood pressure measurements are obtained from instruments implanted in central vessels, typically the femoral artery or abdominal aorta. Several commercial implantable telemetric blood pressure monitors are available for use in animals as small as mice. A surgical procedure is required for their placement. Use of telemetry allows collection of data that is free from the physiological effects of restraint on the animal.

Indirect blood pressure measurement can be made with adapted versions of the sphygmomanometer used in humans. In rats and mice, a pneumatic cuff can be fitted over the animal's tail and inflated to occlude blood flow. A detector and pressure transducer in the cuff provide a signal that is converted into analog voltage for recording. Because mice and rats alter blood circulation in their tails depending on ambient temperature, room and restrainer conditions must be held constant from one reading to the next. In addition, animals unconditioned to handling and being held in a restraint device will not provide reliable readings. Tail cuff blood pressure monitors are available commercially.

In large species, noninvasive methods (Doppler ultrasonography and oscillometry) are available for detecting blood pressure but are not considered as accurate as direct arterial puncture measures. The Doppler method only measures systolic pressure, while the oscillometric system measures systolic and diastolic pressures. A pressure cuff can be placed around a forepaw or hindpaw or tail. Allowing an animal at least 15 min to become used to the pressure cuff each time it is applied may improve consistency of the readings.

## REFERENCES

1. Olfert, E.D., Cross, B.M., and McWilliam, A.A., Eds., *Guide to the Care and Use of Experimental Animals*, Vol. 1, Canadian Council on Animal Care, Ottawa, Ontario, 1993, http://www.ccac.ca/guides/english/toc_v1.htm.
2. van Zutphen, L.F.M., Baumans, V., and Beynen, A.C., Eds., *Principles of Laboratory Animal Science,* rev. ed., Elsevier, Amsterdam, 2001.
3. Lawson, P.T., Ed., *Assistant Laboratory Animal Technician*, American Association for Laboratory Animal Science, Sheridan Books, Chelsea, MI, 1999, p. 45.
4. Dhein, C.R., VM551, Fluid, Electrolyte, and Acid-Base Abnormalities, http://www.vet-med.wsu.edu/courses_vm551_crd/fluidrx_text.html.
5. Diehl, K.-H., Hull, R., Morton, D., et al., A good practice guide to the administration of substances and removal of blood, including routes and volumes, *J. Applied Toxicol.* 21, 15, 2001, http://www3.interscience.wiley.com/cgi-bin/abstract/76510682/START.

6. Flecknell, P.A., Liles, J.H., and Williamson, H.A., The use of lignocaine-prilocaine local anaesthetic cream for pain-free venepuncture in laboratory animals, *Lab. Anim.*, 24, 142, 1990.
7. Framstad, T., Oystein, S., and Aass, R., Bleeding and intravenous techniques in pigs, Norwegian School of Veterinary Science, http://www.oslovet.veths.no/teaching/pig/pigbleed.
8. Swindle, M.M. *Surgery, Anesthesia, and Experimental Technique in Swine*, Iowa State University Press, Ames, IA, 1998, chap. 1.
9. Svendsen, P. and Hau, J., Eds. *Handbook of Laboratory Animal Science, Vol. I, Selection and Handling of Animals in Biomedical Research*, CRC Press, Boca Raton, FL, 1994, pp. 254, 255.
10. Guide to oral capsule dosing of the rat and guinea pig, http://www.torpac.com.
11. Shomer, N.H., Aasrofsky, K.M., Dangler, C.A., et al., Biomethod for obtaining gastric juice and serum from the unanesthetized guinea pig (porcellus), *Contemporary Topics in Lab. Anim. Sci.*, 38, 5, 32, 1999.
12. Brown, R.H., Walters, D.M., Greenberg, R.S., et al., A method of endotracheal intubation and pulmonary functional assessment for repeated studies in mice, *J. Appl. Phys.*, 87, 2362, 1999, http://www.jap.org.
13. Joint Working Group on Refinement: British Veterinary Association, Fund for the Replacement of Animals in Medical Experiments, Royal Society for the Prevention of Cruelty to Animals, Universities Federation for Animal Welfare, Removal of blood from laboratory animals, *Lab. Anim.*, 27, 1, 1993, http://www.lal.org.uk/laban.htm.
14. McGuill, M.W. and Rowan, A.N., Biological effects of blood loss: implications for sampling volumes and techniques, *ILAR News*, 31, 4, 5, 1989, http://www4.nas.edu/cls/ijhome.nsf/web/McGuill3104.
15. Hem, A., Smith, A.J., and Solberg, P., Saphenous vein puncture for blood sampling of the mouse, rat, hamster, gerbil, guinea pig, ferret and mink, *Lab. Anim.*, 32, 364, 1998, http://www.uib.no/vivariet/mou_blood/Blood_coll_mice.html.
16. vanHerck, H., Baumans, V., Brandt, C.J.W.M., et al., Blood sampling from the retro-orbital plexus, the saphenous vein and the tail vein in rats: comparative effects on selected behavioural and blood variables, *Lab. Anim.*, 35, 2, 131, 2001.
17. Rodgers, C.T., Practical aspects of milk collection in the rat, *Lab. Anim.*, 29, 4, 450–455, 1995.
18. Strake, J.G, Mitten, M.J., Ewing, P.J., et al. Model of *Streptococcus pneumoniae* meningitis in adult rats, *Lab. Anim. Sci.*, 45, 5, 1996.
19. Petty, C., *Research Techniques in the Rat*, Charles C. Thomas, Springfield, IL, 1982.
20. Vogelweid, C.M. and Kier, A.B., A technique for the collection of cerebrospinal fluid from mice, *Lab. Anim. Sci.*, 38, 1, 9102, 1988.

# CHAPTER 16

# Production of Polyclonal and Monoclonal Antibodies

**Coenraad Hendriksen and Jann Hau**

## CONTENTS

0-8493-1086-5/03/$0.00+$1.50
© 2003 by CRC Press LLC

## INTRODUCTION

Many breakthroughs in biomedical science have been achieved by the use of polyclonal and monoclonal antibodies. One of the earliest, and probably best-known examples, is the discovery by Behring and Kitasato in the 1890s[1] of the therapeutic effects of diphtheria antiserum. The history of monoclonal antibodies, however, is much younger and dates to the pioneering work of Köhler and Milstein in 1975.[2] Nowadays, polyclonal antibodies (Pabs) and monoclonal antibodies (Mabs) are indispensable tools in the laboratory. They are used for immunoassays (e.g., as a diagnostic tool), for affinity chromatography, as immunomarkers (e.g., in pathology), and in basic research (e.g., to discover new proteins and to characterize complex antigenic structures). Furthermore, they are of crucial value in the clinic. Although vaccines have replaced most therapeutic polyclonal antisera, some, such as rabies antiserum and snake antivenom are still important in third world countries. Mabs are increasingly being used as carriers in drug-targeted therapy.

The production of Pabs as well as Mabs requires the use of substantial numbers of animals, and because of the invasiveness of the experimental procedure, the animals are protected by legislation regulating the use of laboratory animals. In the case of Pabs, animals are immunized to induce an optimal humoral immune response, and in the case of Mabs, animals are needed to obtain antigen-specific B-lymphocytes and to produce ascites if *in vitro* production is not feasible.

The procedures employed to generate Pabs and Mabs are frequently under discussion for animal welfare reasons. In particular, certain aggressive immunostimulatory products (adjuvants) used to enhance the immune response and some routes and locations of immunization and the ascites technique are believed to be associated with pain and distress. In addition, animals used for production of Pabs may spend a long time, being regularly boosted and bled, in the animal house.

The aims of this chapter are to discuss the factors that influence the production of Pabs and Mabs and to provide best practice principles for antibody production methods. Furthermore, approaches to reduce, replace, and refine the use of animals will be presented and discussed. But first, an overview of the immune response is given to provide a basis for the understanding of the mechanism underlying the induction of antibody synthesis. Additional information may be obtained from any of several immunology text books.[3,4] After presenting and discussing the many parameters in immunization protocols and the production of antibodies, a summary will be presented of existing guidelines for *in vivo* Pab and Mab production. An overview of existing literature on Pab and Mab production as well as useful addresses and Web sites are given by Smith et al.[5]

## THE IMMUNE RESPONSE

One of the most complex and fine-tuned activities in the intact vertebrate organism is the immune response. An immune response is triggered by contact of the organism with an antigen,* which is a structure recognized by the immune system as foreign (non-self). Antigens can be presented to the immune system as complex (particulate) multiantigens, such as bacteria, viruses, parasites, and artificial particles or as single antigens such as proteins or polysaccharides.[6] The immune system, responsible for the immune response, can be broadly divided into the innate system and the acquired (adaptive) system, each composed of cellular and humoral elements. Interactions between the two are important for the functioning of both systems. The innate system, which appears to be the evolutionary oldest, lacks a high level of specificity and efficiency but responds rapidly. Elements that are involved include macrophages, dendritic cells, and humoral factors such as complement activation system. The adaptive — antigen specific — immune system, on the other hand, has a high level of specificity and efficiency, but it responds more slowly.[7] The principle actors in the antigen-specific immune response are the B-lymphocytes, responsible for antibody production, and the T cells (thymus-derived lymphocytes), responsible for cytotoxic responses (cytotoxic T cells, $T_c$ cells) and T helper cells ($T_h$–lymphocytes, $T_h$–cells) for B cells and $T_c$ cells. Finally, the maintenance of the immune system requires intercellular communication

---

* In this text, the term antigen is used instead of the term immunogen. Immunogens are defined as substances that are able to induce an immune response, whereas an antigen is a substance that can react with an antibody. The two are not always synonymous, but often, no distinction is used in the literature.

mediated through cell-to-cell contact (e.g., by the production of cytokines) and the support of so-called accessory cells such as fibroblasts and endothelial cells.

The antigen-specific immune response can be divided into three phases: the inductive phase, the effector phase, and the establishment of immunological memory.[8] In the inductive phase, an antigen surviving the primary defenses (e.g., skin, mucosa) of the organism is recognized as "foreign" by so-called "antigen-presenting-cells" (APCs: monocytes, dendritic cells, and B-lymphocytes). Dendritic cells recognize all antigens, while B-lymphocyte APCs employ a specific receptor-dependent form of antigen binding. The APC binds and internalizes the antigen and processes it into peptides that, together with MHC class molecules, are then exposed on the surface of the APC for presentation to the antigen-receptors on the T-cells. This contact leads to activation of the T cells.

After antigenic contact, only a small percentage of the lymphoid cells will become involved in the immune response; initially less than 10 per 1,000,000 cells, increasing to 1 in 1000 cells within 8 days of antigen-dependent activation, if the antigen is a potent immunogen. The rest of the cells remain dormant.

The next step in the immune response is the effector phase. Depending on the type of antigen, activation of T cells leads to cell-mediated responses (mainly cytotoxic, such as is the case for viral antigens) or to T-cell help for B cells (i.e., in case of protein antigens). Once activated, $T_h$-cells become more sensitive to the action of growth factor cytokines, such as IL-1 and IL-2. In the mouse, at least three kinds of $T_h$-cell subsets can be recognized; $T_{h0}$, $T_{h1}$, and $T_{h2}$ cells. These cells also produce cytokines, each subpopulation generating different sets of cytokines, which modulate the immune response. A $T_{h1}$ response is predominantly cell mediated, whereas a $T_{h2}$ response is dominated by humoral factors. Cytokines induce the proliferation and ultimately differentiation of B-lymphocytes into antibody-producing plasma cells. Each plasma cell is genetically encoded to produce an exactly defined antibody with specificity against a single antigen epitope.

The final phase of the antigen-specific immune response is the induction of memory after primary contact with the antigen. Memory, which is based on the differentiation of B cells, gives the secondary immune response its characteristics of speed, higher magnitude, and avidity compared to the primary response, and dominance of the IgG antibody subclasses over IgM.[8] In the mouse, $T_{h1}$ cells bias the secondary immune response toward cell-mediated immunity and the IgG2a subclass of antibody, while $T_{h2}$ cells bias the response toward humoral immunity and IgE plus the IgG1 subclass of antibodies.

## ANTIBODIES

Antibodies, also called immunoglobulins or gamma globulins, are of fundamental importance for the identification and elimination of foreign material that enters the organism. In mammals, antibodies belong to one of five immunoglobulin (Ig) classes: IgA, IgD, IgE, IgG, or IgM, each with a specific range of activities and characteristics. In avian species, immunoglobulins belong to one of three classes: IgA, IgM, or IgY. The IgY molecule is functionally similar to but differs structurally from the mammalian IgG molecule. IgA and IgG may be further subdivided into subclasses in some species. The core structure of all Ig molecules comprises two identical, covalently linked heavy (H) chains and two identical light (L) chains. Each chain has a variable (V) and a constant (C) region. Antibody molecules are bifunctional, having an antigen binding site and a biological effector function site. The heavy- and light-chain V regions' domains form the antigen-binding sites and differ in structure from one antibody specificity to the next (between clone variation). The heavy-chain C regions (Fc segment) are relatively constant in structure within each subclass of antibodies and contain the biological effector functions, such as complement activation.

## POLYCLONAL ANTIBODIES VERSUS MONOCLONAL ANTIBODIES

The basic difference between Mabs and Pabs is implied in the terminology. Mabs are antibodies obtained from one plasma cell clone. By definition, antibodies produced by a single clone are monospecific, having specificity directed against only one antigen determinant. Generally, however, an antigen epitope

**Table 16.1  Characteristics of Polyclonal and Monoclonal Antibodies**

| Characteristic | Polyclonal Antibodies | Monoclonal Antibodies |
|---|---|---|
| Specificity | Low to high | High |
| Sensitivity for antigenic modification (e.g., genetic polymorphism, heterogeneity of glycosylation, denaturation) | Low | High |
| Affinity/avidity | Avidity equals sum of affinity of  individual specificities; increases with  time | Avidity = affinity; constant |
| Cross reactivity with other antigens | Unlikely after absorption | Likely |
| Production time | Usually 4 to 8 weeks | Up to 3 to 6 months |
| Batch-to-batch variability | Yes | In principle not |

*Source:* From Leenaars, M., Claassen, E., and Boersma, W.J.A., Modulation of the humoral immune response; antigens and antigen presentation, in *Immunology Methods Manual. I.* Lefkovits, Ed., Academic Press, New York, 1996, p. 989.

consists of perhaps five to ten amino acids. And, because most antigens are much larger, they normally constitute a number of different epitopes, each inducing a plasma cell clone (polyclonal) producing antibodies against a particular epitope. Therefore, the antibody response against such antigens normally consists of a combination of a number of monoclonal antibodies (sometimes hundreds to thousands). Some characteristics of Pabs and Mabs are given in Table 16.1

When making a choice between the generation of Pabs or Mabs, the desired application of the antibody and the time and money available for its production should be taken into consideration. In research, while many questions can be answered utilizing Pabs, Mabs are preferred for diagnostic purposes and large-scale production systems. Single Mabs are not usable in precipitation assays, but this problem can be overcome by mixing different Mabs with specificities against different epitopes raised against the antigen

# POLYCLONAL ANTIBODIES

## Production of Pabs

Scientists performing an immunization procedure often want a number of conditions to be met, such as the antibody titer should be high and be obtained within a reasonable time, the antibody response should be specific, the antibodies induced should be of high avidity, and the amount of antibody produced should be sufficient for the intended purpose. The aim of this section is to identify these factors and to ensure proper immunization procedures, which combine acceptable immunological results with minimal pain and suffering for the animals.

Generally speaking, an immunization procedure includes a number of critical factors that may influence the outcome of the immunization procedure and the welfare of the animals. They include the following:

- The antigen
- The use of an adjuvant
- The choice of animal
- Immunization protocol (route of injection, volume, boosting schedule)
- Blood collection

### The Antigen

Antigens can be single molecules like proteins or particulate complex multiantigens. The group of particulate complex antigens includes intact microorganisms like bacteria, viruses, parasites,  protozoa,

**Table 16.2 General Features Influencing Immunogenicity of Antigens**

| Parameter | Immunogenicity |
|---|---|
| Source | Xenogenic > allogenic > syngenic > autologous |
| Chemical nature | Protein > polysaccharide > lipid |
| Size | High molecular weight > low molecular weight |
| Complexity | Complex > simple |
| Stability | Stable > unstable |
| Degradability | Degradable > undegradable |

*Source:* From Leenaars, M., Claassen, E., and Boersma, W.J.A., Modulation of the humoral immune response; antigens and antigen presentation, in *Immunology Methods Manual. I.* Lefkovits, Ed., Academic Press, New York, 1996, p. 989.

and mammalian cells, but also artificial particles. The single antigens include proteins, peptides, and polysaccharides. Some antigen features that influence immunogenicity are shown in Table 16.2. The intensity of the immune response is directly related to the size of the antigen and the degree of foreignness. Particulate antigens are generally very immunogenic, while single antigens are less immunogenic. Small antigens (<3 to 4 kDa) are not able to elicit an antibody response in themselves[9,10] and need a protein-carrier molecule to activate T cells. A rule of thumb is the bigger the molecule, the better the antibody response. The molecule should be in a form that it can be easily processed by the cells of the immune system, but on the other hand, have a certain structural stability and be present in more or less the same form in booster injections. In order to be a good immunogen, the molecule must possess at least one epitope that can be recognized by the cell surface antibody found on B cells, and it must have at least one surface structure that can be recognized simultaneously by a class II protein and a T-cell receptor. Furthermore, the greater the phylogenetic difference between the antigen donor and the animal to be immunized, the better the immune response that is evoked.

An important aspect to consider is the quality of the antigen. When raising Pabs, as opposed to Mabs, the purity of the antigen is of major importance. Minute impurities may prove to be immunodominant and will lower the immune response wanted.[11] It can also be time consuming and laborious to purify the antigen, but usually the resources are well spent, as it is much easier to render an antiserum only weakly contaminated with antibodies against impurities functionally monospecific, than to remove a lot of unwanted specificities by extensive absorption procedures. Antigen preparations can be toxic due to contamination with endotoxins or chemical residues used to inactivate microorganisms (for example, formaldehyde or propiolactone) or an extreme pH.[12] Antigen preparations should ideally be prepared and stored according to the rules of Good Laboratory Practice.

## The Adjuvant

Adjuvants (from the Latin word *adjuvare* = help) are substances used to enhance and modulate the immune response. They have been used since 1926,[13] and the ideal adjuvant can be characterized as a product that stimulates high and sustainable antibody titers, is efficient in a variety of species, is applicable to a broad range of antigens, is easily and reproducibly prepared in an antigen mixture, is easily injectable, is effective in a small number of injections, has a low toxicity for the immunized subject, and is not hazardous to the investigator. Unfortunately, the adjuvant that meets all of these criteria remains to be identified.

There are more than 100 known adjuvants, but only some of these are routinely utilized. Depending on their mode of action or composition, adjuvants can be categorized in a number of groups (Table 16.3). The adjuvant products most frequently used are the water-in-oil (W/O) emulsions, in particular, Freund's adjuvant[14] as can be seen from a survey performed in the Netherlands (Figure 16.1). Freund's Incomplete Adjuvant (FIA) is an oil-in-water emulsion composed of mineral oil (Bayol F) at 85 to 90% and a detergent (Arlacel A; mannide monooleate) at 10 to 15%. Freund's Complete Adjuvant (FCA) is FIA to which is added heat-killed, dried Mycobacterium species, usually *M. tuberculosis*, *M. smegmatis*, or *M. butyricum*. The use of FCA is favored, because it gives rise to long-term persistence of the immune responses, and although this is not common practice in research laboratories, booster doses[15] would not always be needed because of the depot effect of FCA. When antigenic material in a water solution is

**Table 16.3  Overview of Categories of Adjuvants that may be Used for Routine Polyclonal Antibody Production**

| Category | Examples | Mode of Action |
|---|---|---|
| Mineral salts | Al(OH)3, AlPO4 | Vehicle, depot effect |
| Oil emulsions (W/O, O/W, W/O/W) | FIA, Specol, Montanide | Vehicle, depot effect, activation of macrophages |
| Microbial (like) products | LPS, MDP, MPL, TDM | Stimulation B or T cells, activation of macrophages, enhanced antigen uptake |
| Saponins | Quil-A | Facilitate cell-cell interaction, aggregation of antigen |
| Synthetic products | DDA | Activation of macrophages and complement |
|  | Iscoms | Facilitate cell-cell interaction, stimulation of T cells, aggregation of antigen |
|  | Liposomes | Vehicle, enhanced antigen uptake |
|  | NBP | Activation of macrophages and complement |
| Cytokines | IL-2, IL-1, IFN-γ | Growth and differentiation of B and T cells |
| Adjuvant formulations | FCA, TiterMaxTM, RIBI, Gerbu, Softigen | Combinations of the above |

W/O = water-in-oil; O/W = oil-in-water; W/O/W = water-in-oil-in-water.

FIA =   Freunds incomplete adjuvant; LPS = lipopolysaccharide, MDP = muramyl dipeptide; MPL = mono-phosphoryl lipid A; TDM = trehalose dimycolate; DDA = dimethyldioctadecylammonium bromide; ISCOMs = immunostimulating complexes; NBP = nonionic block polymer; FCA = Freunds complete adjuvant.

*Source:* Adapted from Leenaars, P.P.A.M., Hendriksen, C.F.M., De Leeuw, W.A., Carat, F., Delahaut, P., Fischer, R., Halder, M., Hanley, W.C., Hartinger, J., Hau, J., Lindblad, E.B., Nicklas, W., Outschoorn, I.M., and Stewart-Tull, E.S., The production of polyclonal antibodies in laboratory animals, *ATLA*, 27, 79, 1999.

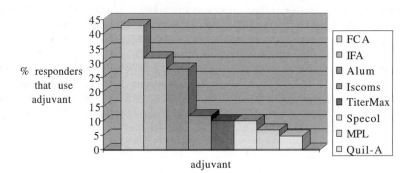

**Figure 16.1**   The use of adjuvant. Percentage of responders using a specific adjuvant product in mice. (From Leenaars, P.P.A.M. and Hendriksen, C.F.M., Ontwikkeling van "COMParative Adjuvant Selection System" (Development of "COMParative Selection System"), RIVM report 623860006, 1999. With permission.)

injected into an animal, a prompt and rapid dissemination of the injected material occurs. Alum precip-itation or adsorbtion of antigens, as in many vaccines used in humans, causes only a slight retention of the antigen. However, if the same material is injected as a water-in-oil emulsion, with or without mycobacteria, the oil vesicles can be retained at the site of injection for many months, while the material undergoes slow degradation. The rate of antigen elimination from the injection site has been reported to have a half-life of approximately 14 days. The antibody response to emulsified protein antigens seems to be relatively constant for up to a year.[16] However, the persistence of cell-mediated reactions, partic-ularly at the site of the original depot, may lead to tissue inflammation and granulomas, which may become necrotic.[12,17] Moreover, an additional finding might be the presence of granulomas in lungs, kidneys, liver, heart, lymph nodes, and skeletal muscle due to adjuvant/antigen dissemination.[17,18] This phenomenon might be explained by uptake of the emulsion by macrophages and transport by the lymphatics.[11] Also, the induction of a number of different autoimmune diseases might occur. In chickens, the use of FCA should be discouraged, because it results in a reduction in the egg laying frequency, and thus, antibody productivity, as compared to chickens immunized with other adjuvants including FIA.[19] Freund's adjuvant-induced arthritis constitutes the only animal model of chronic pain that has been validated to a significant extent,[20,21] and the side effects of adjuvants, in particular FCA, have led to serious concern for the welfare of the animals. A number of studies have been undertaken to replace FCA by adjuvant products with less irritant properties. Unfortunately, only a few products have been demonstrated to have less undesirable effects while being as immunogenic as FCA. Efficient alternative adjuvants are now available not only for stimulating humoral immunity but also for stimulating cell-mediated immunity[22,23] and for immunomodulation (Table 16.3). These include the RIBI adjuvant system containing squalene; monophosphoryl lipid A; trehalose dimycolate; saponins; dextran polymer particles; ethylene-vinyl acetate polymers; Muramyl dipeptide (MDP); pegylated C8/C19 mono-/diglycerides; Cholera toxin B subunit; Montanide; Specol; and ISCOMs (immunostimulating complexes). Adjuvant products like the mineral salts (e.g., $AlPO_4$), the only adjuvants allowed for human application, are almost without adverse effects, but the immunostimulatory effects are not as impressive as for many of the adjuvants used for experimental purposes. More detailed information on adjuvants can be obtained from several review articles and books.[24–27]

There is evidence to suggest that FCA is a human health hazard because of unpleasant complications for sensitized personnel who accidentally injure themselves while immunizing the animals. Even if the skin is pricked accidentally and no injection made, this can result in a painful and dramatic swelling of the area, and the wound may heal slowly and discharge up to 2 years after the incident.[28]

One of the first questions that might be asked when setting up an immunization procedure is whether an adjuvant is needed. In case of particulate antigens, it might be possible to obtain a high antibody titer without the use of an adjuvant. When an adjuvant is required, it is up to the investigator to decide which type of adjuvant is the most appropriate. When specific immunomodulation is part of the immunization objectives, ISCOMs, Quil-A and other modern adjuvants should be considered. However, the overall welfare of the laboratory animal to be immunized must be taken into consideration. As it is known that oil-adjuvants combined with bacterial components (e.g., FCA) may induce considerable side effects, the use of these adjuvants should be discouraged. If they are used, then the administration of the adjuvant should comply with certain restrictive conditions, such as a limitation to the volume injected, the location of injection, and the number of boost injections. FCA should never be injected twice, and normally, it would be administered with the first immunization of the animal. Repeated injections of FCA may lead to severe tissue reactions and may result in anaphylactic shock if the animal has become sensitized to any of the bacterial immunogens. The antibodies against the bacterial components of complete adjuvants can disturb the use of the antiserum, which is an additional reason for avoiding the use of these types of adjuvants.

The preparation of the antigen and adjuvant mixture should occur *lege artis*. First, the mixture should be prepared aseptically, to minimize the risk of possible contamination. When an oil emulsion is used, the stability and quality of the emulsion should be checked. In case of a water-in-oil emulsion, drops of the mixture placed on the surface of water in a dish will remain as discrete white drops on or just below the surface, indicating that the water phase containing the antigen is entirely enclosed within the oil.[24] If, on the other hand, the emulsions form a cloud of tiny particles when dropped in the water, it is an oil-in-water emulsion. Water-in-oil emulsions that are not properly prepared are ineffective as adjuvants.[29] The time between the preparation of the mixture and its administration should be short. If stored, the mixture should be stored in a freezer and the emulsion checked after thawing before injection.

## The Animal

### Species

An essential issue to consider when producing Pabs is the selection of the animal species and animal strain. The rabbit is by tradition the most widely used species, but the choice of species should depend on the purpose of the immunization. There are many factors that may influence which species will be the most appropriate for a specific purpose, e.g., the volume of antibody or antiserum needed and the ease to obtain large blood samples, the phylogenetic relationship between the recipient and the donor of the antigen, the character of the antibody synthesized by the recipient species, and the intended use of the antibody.[7]

When large amounts of antibody are needed, farm animals such as sheep or goats are preferred. Mice and rabbits are most frequently used when small amounts are required, particularly because these animal species are easy to bleed compared to, for example, guinea pigs and hamsters. Because of the continuous transovarian passage of antibodies from blood to egg yolk in birds,[30] it is convenient to harvest and purify antibodies from the egg yolk of the domestic fowl, and several fairly simple methods have been described.[31–35] Chicken egg antibodies (IgYs), which are phylogenetically a progenitor of IgG antibodies, can be extracted from the egg to concentrations of approximately 100 mg IgY/egg. Compared to the IgG productivity of the rabbit (approximately 200 mg Ig/bleeding), a chicken produces about ten times as much IgY. In addition, IgY does not cross-react with mammalian immunoglobulins, thereby reducing the risk for false-positive results in, e.g., an ELISA.[35] Antibody production against highly conserved antigens (such as intracellular proteins) requires a wide phylogenetic distance between the recipient and the donor animal, e.g., the use of chickens for producing antibodies to mammalian proteins. Chicken antibodies raised against a protein in one mammalian species will often react against the analogous protein in other mammalian species.[36,37]

The use of chickens for production of antibodies is attractive from an ethical viewpoint and with respect to Russell and Burch's principle of the three Rs,[39] the principle to replace, reduce, and refine the use of laboratory animals when possible. Mammals can thus be replaced by a species with a lower degree of neurophysiological sensitivity, and the number of animals needed can be considerably reduced. Oral immunization techniques are being developed,[38] thus eliminating restraint and distress associated with administration of antigen and harvest of antibodies and refining the methodology so that this production of antibodies may no longer be considered an experimental procedure from a legal point of view.

However, the use of IgY technology is not widespread. This is probably due to a number of factors, such as tradition, the limited availability of conjugated antibodies directed against IgY, and the infrequent use of the chicken as a laboratory animal, in general, and their specific requirements for housing.[12] More information about the IgY technology can be obtained from Schade et al.[35]

## Species/Strain-Stock

The antibody response to most antigens seems to be genetically determined and under polygenic control involving histocompatibility antigens.[40,41] A species and also often strain difference is thus to be expected.[42] Minimal interindividual variation in antibody response is observed if inbred strains of animals are used in immunization series using the same antigen and immunization procedure. If this minimal variation is important, mice, rats or chickens, which are all readily available as inbred strains, may be preferred. Outbred stocks like rabbits, although selected through generations for high antibody response, have been documented to exhibit a remarkable interindividual variation in their immune response.[43]

Even animal strains may differ in their immune response. For example, BALB/c mice tend to be Th2-like responders, while C57BL/6 mice are Th1 responders. In certain instances, the genetic control of the immune response seems to be dependent on antigen dose, and increasing doses can transform a poorly responding mouse strain to a good responder.[44]

## Sex

Although there are no scientific reasons for not using male animals, traditionally, female animals are preferred. These animals are generally more docile and less aggressive in social interactions, and can therefore be group-housed more successfully. Group housing has also been reported for castrated male rabbits, but there may be ethical problems associated with castrating animals in order to make them easier to group house. Modern guidelines for the care of laboratory animals like the CoE (ETS 123) [Appendix A] presently being revised, advocate group-housing whenever possible. However, group housing of male animals is not unproblematic and may be associated with stress and elevated corticosteroid levels. A number of reports have demonstrated that social stress may impair the initial immune response in immunized animals.[45]

A sex difference in the magnitude of antibody response within inbred strains has been observed,[46] and the humoral immune response is suppressed during pregnancy in the mouse.[47]

## Age

The outcome of an immunization procedure depends on the immune status of the animal. In general, young adult animals are better responders, as they are naive to immunostimulatory agents in their environment. The immune response declines with advancing age after young adulthood. Chickens should be of egg-laying age by the time antibody is harvested, but there is no great difference in titers obtained in young egg-layers compared with older hens about to cease egg-laying.[48] The following minimum ages are recommended for Pab production: mice and rats at six weeks; rabbits and guinea pigs at three months; chickens at three to five months; goats at six to seven months; and sheep at seven to nine months.

## The Immunization Protocol

### The Injection Route

The location at which the antigen is deposited in part determines the lymphoid organs activated and the type of antibody response that will be induced. However, there are other aspects taken into consideration when selecting the route of injection: the species used, the quantity of the antigen and adjuvant mixture, the choice of the adjuvant, and the welfare of the animals. However, often the choice seems mostly determined by tradition. Frequently used routes are subcutaneous (s.c.), intramuscular (i.m.), intraperitoneal (i.p.), intravenous (i.v.), and intradermal (i.d.). Some characteristics of these routes as well as advantages and disadvantages are summarized in Table 16.4. Alternative routes that are being used include oral administration, intranasal inoculation, footpad injection, and intrasplenic injection. Oral administration is preferred when using specific adjuvants, such as *lactobacillus*, and when the requirement is a peripheral as well as a mucosal immune response. Intranasal administration, with antigen in aqueous solution, is used for the induction of tolerance. Animals usually have to be anesthetized and the material be given by pipette or microcanula.[49] Footpad injection (particularly when an oil adjuvant is used) and intrasplenic injection give rise to serious animal welfare concern. The intrasplenic route can be used when only small amounts of antigen are available and a direct delivery of antigen to lymphoid tissue is needed. Footpad injection is used when high numbers of B cells are required from a local lymph node. However, generally, these routes are not necessary for routine Pab production and should be justified on a case-by-case basis. If a (subcutaneous) footpad injection is given, only one hind foot should be used, and the animals should be housed on soft bedding.

### The Volume of Injection

As a principle, injection volumes should be as small as possible to limit the level of side effects, but also, because the immune response is generally stronger against an antigen administered in high concentration in a small volume than a low concentration in a large volume of vehicle. An important aspect to consider is the adjuvant used. Oil adjuvants, in particular in combination with microbial products and viscous gel adjuvants, form a depot at the injection site and induce a local inflammation, potentially inducing sterile abscesses. Therefore, emulsion antigens should be administrated in smaller volumes than aqueous antigens. Information on recommended volumes for emulsions and aqueous products is given in Tables 16.5a and 16.5b, respectively.

In case of s.c. injection, multiple small volumes can be given in multiple sites, instead of one large volume. This may reduce the intensity of side effects (e.g.,[50]). However, the number of injection sites should be limited.[12]

**Table 16.4  Injection Routes: Advantages and Disadvantages**

| Injection Route | Details | Advantage | Disadvantage |
|---|---|---|---|
| s.c. | Most frequently used route<br>Route preferred<br>Do not inject material in part of animal used for restraint<br>Limit locations to <5 | Relatively large volumes can be administered<br>Inflammatory processes can be easily monitored | Slow absorption |
| i.m. | Skeletal muscles are well vascularized<br>Use *B.femoris* in small animals<br>Not recommended for injection of oil adjuvant in rodents | Rapid absorption, in particular, with muscular activity<br>In large animals, relatively large volumes can be administrated | Antigen and adjuvant can spread along interfacial planes and nerve bundles and may damage sciatic nerve and have other serious side effects<br>Local reactions can be easily overlooked |
| i.p. | An efficient route for antigen delivery (easy access to lymphatics, large intercellular clefts in lymphatics, transport help by respiratory activity) | Relatively large volumes of inoculums can be accommodated | Relative high percentage of injection failures<br>Oil adjuvant induces peritonitis<br>Risk for anaphylactic shocks at booster injection of aqueous antigen |
| i.v. | Antigen is delivered primarily to spleen and secondary to lymph nodes<br>Route of choice for particulate antigen | Rapid distribution of antigen | No oil or viscous gel adjuvant can be used<br>High risk for anaphylactic shock at booster immunization |
| i.d. | Limit number of injections to <5 per animal | Efficient processing of antigen due to high density of Langerhans dendritic cells<br>Small quantities of antigen already effective | Use of oil adjuvants leads to ulcerative processes |

**Table 16.5a  Maximum Volume of Injection (in mL) Used for Injection of Emulsions per Route of Injection for Different Animal Species**

| Species | Vol | | | | |
|---|---|---|---|---|---|
| | s.c. | i.m. | i.p. | i.v. | i.d. |
| Mice | 0.1 | 0.05 | N.R. | N.A. | 0.05 |
| Rats | 0.2 | 0.1 | N.R. | N.A. | 0.05 |
| Guinea pigs | 0.2 | 0.2 | N.R. | N.A. | 0.05 |
| Rabbits | 0.25 | 0.25 | N.R. | N.A. | 0.05 |
| Sheeps/goats | 0.5 | 0.5 | N.R. | N.A. | 0.05 |
| Cattle | 0.5 | 0.5 | N.R. | N.A. | 0.05 |
| Poultry | 0.25 | 0.5 | N.R. | N.A. | 0.05 |

N.R. = not recommended; N.A. = not acceptable.

*Source:* Adapted from Leenaars, P.P.A.M., Hendriksen, C.F.M., De Leeuw, W.A., Carat, F., Delahaut, P., Fischer, R., Halder, M., Hanley, W.C., Hartinger, J., Hau, J., Lindblad, E.B., Nicklas, W., Outschoorn, I.M., and Stewart-Tull, E.S., The production of polyclonal antibodies in laboratory animals, *ATLA*, 27, 79, 1999; Hohmann, A.W., Freund's adjuvant and immune responses, in *The Use of Immuno-adjuvants in Animals in Australia and New-Zealand*, ANZCCART, Glen Osmond, SA, Australia, 1998; and Dutch Inspectorate 2000.

## The Quantity of Antigen

Guidelines on the quantity of antigen that should be used are difficult to give. The quantity depends on the inherent properties of the antigen, whether the antigen is purified or a component in a mixture of

**Table 16.5b Maximum Volume of Injection (in mL) Used for Injection of Aqueous Antigens per Route of Injection for Different Animal Species**

| Species | Vol | | | | |
|---|---|---|---|---|---|
| | s.c. | i.m,. | i.p. | i.v. | i.d. |
| Mice | 0.5 | 0.05 | 1.0 | 0.2 | 0.05 |
| Rats | 0.5–1.0 | 0.1 | 5.0 | 0.5 | 0.05 |
| Guinea pigs | 1.0 | 0.2 | 5–10 | 0.5–1.0 | 0.05 |
| Rabbits | 1.5 | 0.2–0.5 | 10–20 | 1–5 | 0.05 |
| Sheeps/goats | 2.0 | 2.0 | N.A. | N.G. | 0.05 |
| Cattle | 2.0 | 2.0 | N.A. | N.G. | 0.05 |
| Poultry | 0.5 | 1.0 | N.A.[a] | N.G. | 0.05 |

[a] Air sacs and liver can be easily damaged.
N.A. = not acceptable; N.G. = not given.

*Source:* Adapted from Leenaars, P.P.A.M., Hendriksen, C.F.M., De Leeuw, W.A., Carat, F., Delahaut, P., Fischer, R., Halder, M., Hanley, W.C., Hartinger, J., Hau, J., Lindblad, E.B., Nicklas, W., Outschoorn, I.M., and Stewart-Tull, E.S., The production of polyclonal antibodies in laboratory animals, *ATLA*, 27, 79, 1999; Hohmann, A.W., Freund's adjuvant and immune responses, in *The Use of Immuno-adjuvants in Animals in Australia and New-Zealand*, ANZCCART, Glen Osmond, SA, Australia, 1998; and Dutch Inspectorate, 2000.

antigens, the adjuvant used, the route and frequency of injection, etc. In general terms, >25 to 50 ug up to mg quantities of protein antigen in conjunction with an adjuvant are needed to ascertain a high-titer antibody response. Although smaller doses may be used for smaller animals, the antigen dose is not increased or decreased in proportion to body weight. A better way is to think in terms of the number of lymphoid follicles to which the antigen will be distributed. Recommended antigen doses in combination with FCA are 10 to 200 ug for mouse and 250 to 5000 ug for goat and sheep. For each antigen, there is a dose range called the "window of immunogenicity." Too much or too little antigen may induce suppression, sensitization, tolerance, or other unwanted immunomodulation.[7] Very low doses (<1 to 5 ug) are used to induce hypersensitivity (allergy),[9,49] and should be avoided in immunization animals, particularly because booster injections may result in an anaphylactic shock in the animals.

Based on the finding that high antigen doses activate a large number of low-affinity B-cell clones, while a low dose activates only the high-affinity B-cell clones, it is recommended that a lower dose be used at booster immunizations.

## Booster Immunization

Generally, it is not sufficient to immunize only once, even with the help of an adjuvant. The time interval for booster immunization is critical, but fixed periods are difficult to give. As a rule of thumb, a booster can be considered after the antibody titer has plateaued or begins to decline. Usually, this is 3 weeks after the primary immunization in case of an aqueous antigen and at least 4 weeks when a depot-forming adjuvant is used. The number of booster immunizations should be limited to a maximum of two or three. If antibody responses are still insufficient, the experiment should, in general, be terminated. However, in case of antigens of low molecular weight, more boosters might be needed. Also, when animals are used as antibody donors over a long period of time, frequent booster immunizations might be necessary. If a water-in-oil emulsion has been administered in the initial immunization, a depot of antigen has been established in the animal, and booster immunizations may be carried out without the use of an adjuvant. The depot will remain in the animal for many months. Too frequent immunizations with very low amounts of antigen might lead to tolerance. Intermittent bleeding of a hyperimmunized animal or plasmaphoresis appear to help maintain a high serum antibody titer.

Booster antigen mixtures should never be inoculated at the site of the previous injections, and certainly not in granulomas or swellings. Furthermore, booster immunizations might advantageously be given by a different route than the primary immunization, because the object of this procedure is to stimulate as many memory cells as possible.

## Blood Collection

A number of conditions should be met when collecting blood from an animal. The technician should be experienced with the technique, and the procedure has to be performed in a quiet surrounding. If the animal is not stressed during bleeding, the use of sedative to facilitate blood sampling is usually unnecessary. However, in case of exsanguination, general anesthesia is a prerequisite, and at the end of the procedure, the animal should be subjected to euthanasia following appropriate standards.

Blood should normally be collected only from the sites recommended in guidelines, e.g., by the BVA/FRAME/RSPCA/UFAW Joint Working Group on Refinement.[51] These guidelines also specify the maximum amount of blood that can be collected. When maximum volumes are taken, the frequency should be less than once a fortnight. Table 16.6 specifies maximum amounts of blood to be collected for common immunization species. As a general rule, a maximum of approximately 8 mL/kg body weight can be removed once a fortnight.[52]

The needle used should be matched for the vessel size and should preferably be of relative large bore to facilitate rapid collection and to prevent hemolysis. The volume to be removed per bleeding should not exceed 15% of the total blood volume, which equals about 1% of the animal's body weight.

## General Aspects

During the entire experiment, immunized animals should be closely monitored for general appearance, for food and water intake, and for local reactions at the injection site. In case no information is available about the effects of the antigen or adjuvant, as well as of the immunization period and booster interval, it might make sense to perform a pilot study. Macroscopic and microscopic examination should be part of such a study.[53] A postmortem analysis of animals dying spontaneously during an immunization period should be performed.

## Existing Guidelines on the Production of Polyclonal Antibodies

Several guidelines on the production of Pabs have been established by various national control authorities, organizations, and institutions, such as those of the Canadian Council on Animal Care,[54] the NIH,[55] and the Danish (1990) and Dutch (1993/2000) Inspectorates. An overview was prepared by Leenaars et al.[12] Others can be found on the Internet.

Table 16.6 Maximum Volumes of Blood that Should be Collected

| Animal Species | Volume (mL) |
| --- | --- |
| Mouse | 0.3 |
| Hamster | 0.3 |
| Rat | 2.0 |
| Guinea pig | 5.0 |
| Rabbit | 15a |
| Monkey | 20–200a |
| Sheep | 200–600a |
| Horse | 500–700a |

a Depending on body weight.

Source: From Leenaars, P.P.A.M., Hendriksen, C.F.M., De Leeuw, W.A., Carat, F., Delahaut, P., Fischer, R., Halder, M., Hanley, W.C., Hartinger, J., Hau, J., Lindblad, E.B., Nicklas, W., Outschoorn, I.M., and Stewart-Tull, E.S., The production of polyclonal antibodies in laboratory animals, ATLA, 27, 79, 1999.

The general aim of the guidelines is to ensure appropriate procedures, which combine the scientific interest in producing satisfactory results and the animal welfare aspect. The guidelines can be used as a tool for scientists, animal technicians, animal welfare officers, and ethical review committees. Most of the guidelines are not mandatory, but they can be used as a reference to assure proper treatment of animals. Differences between the guidelines occur, but most agree about a number of aspects. Thus, there is general consensus that footpad injection is not necessary; the use of FCA is discouraged, while the use of alternative adjuvant products is recommended. Furthermore, most guidelines set limitations to the volumes to be injected and the blood samples taken.

## New Developments in Polyclonal Antibody Production

In one of the previous sections, information was already provided about the production of egg-yolk (IgY) antibodies. This technology can be seen as both a reduction and a refinement alternative. Due to the high IgY concentration in the egg yolk and the egg-laying time, large amounts of antibodies can be produced in one animal. Consequently, the number of animals can be reduced. It is a refinement alternative, because animals do not have to be bled. For isolation of IgY from egg yolk, easy-to-use commercial kits are now available, and also the number of secondary antichicken conjugate antibodies has increased. This makes the IgY technology an attractive alternative to mammalian antibody production.

Another approach to conventional Pab production is the induction of a systemic humoral response by transcutaneous immunization. By using, e.g., Cholera toxin as an adjuvant, the antigen can be applied with a skin plaster, facilitating transcutaneous uptake. Oral immunization eliminates invasive injections and is also a novel trend that should be welcomed from an animal welfare aspect.

The available *in vitro* techniques for Pab production as routine methodology are still limited. Cloning and production of immunoglobulins with specificity against a particular antigen by nonvertebrate hosts are indeed a possibility, but only time will tell whether this technology develops into a practical adaptable technique for smaller-scale antibody production and experimental antibody production. Unfortunately, it appears to be extremely difficult to culture plasma cells *in vitro* and to produce detectable amounts of Pabs or antibodies of required specificity. Usually, IgM antibodies are produced. However, novel rDNA technologies, such as phage display libraries, which are now being used for production of monoclonal antibodies, might in the future be adapted for Pab production.[56] The phage display technique permits the selection of antibody fragments, which are expressed on the surface of filamentous phage particles, from large libraries constructed from the B cells of individuals or assembled *in vitro* from the genetic elements encoding antibodies.

# MONOCLONAL ANTIBODIES

## Introduction

Monoclonal antibodies (Mabs) are antibodies produced by a single B-cell clone when an antigen epitope interacts with the immune system. Normally, B cells are not able to survive outside the body. However, in 1975, Köhler and Milstein[2] discovered that by fusion of a B cell with a myeloma cell, they were able to immortalize the cell. These cells, called hybridoma cells, could produce virtually unlimited quantities of (monoclonal) antibodies. Mabs have become important tools in the laboratory and in the clinic and are used as therapeutic agents, components of kits for immunoassay and affinity chromatography, as carrier in drug-targeted therapy, and in basic research. At present, more than 100,000 different Mabs are available.[57] Most are produced in small quantities (<0.1 g) solely for bench-related purposes. Larger amounts are often required for use in diagnostic kits and reagents (0.1 to 0.5 g), for routine diagnostic procedures, and in preclinical evaluation studies (0.5 to 10 g), and for prophylactic and therapeutic purposes (>10 g).

## Animal Models in the Development and Production of Mabs

The classical procedure of Mab production includes two separate procedures in laboratory animals: immunization for the generation of B-cell clones and production of Mabs by the ascites induction method. A schematic outline of Mab production is given in Figure 16.2.

### Immunization

The choice of animal species and strains as immune spleen cell donors for fusion is largely dependent on the myeloma cell line available and the origin of the antigen. The mouse is the animal most commonly used for immunization, because there are a variety of mouse myeloma cell lines available. BALB/c mice are generally used for immunization because many of the myeloma cells available for fusion are of BALB/c origin.[58] In case BALB/c mice are unable to produce the B-cell clone of interest or in specific cases, B cells can also be obtained from other species, such as the rat, hamster, and human. The immunization protocols are quite similar to those for the production of Pabs. However, at the end of the immunization period (generally three days after the booster immunization), the animal is killed, and the spleen is removed for isolation of B cells. Instead of the spleen, other lymphoid tissue, such as mesenteric or peripheral intestinal lymph nodes can also be used.

### Fusion and Selection

The isolated B cells are fused with the myeloma cells. Fusion procedures include electrofusion, Epstein-Barr transformation, and, most frequently, chemical fusion by polyethylene glycol. The fused (hybridoma) cells are then cultured *in vitro* in microtiter plates containing HAT (hypoxanthine, aminopterin, thymidine) medium. Selection in this medium allows hybridoma cells to survive. Hybridomas positive for the specific antigen epitope are further selected and subcloned *in vitro*. This selection procedure is essential in further Mab production, as an inadequate selection procedure might lead to

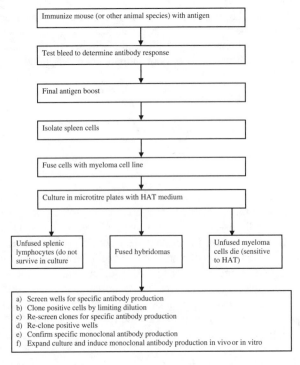

**Figure 16.2** Schematic outline of the production of monoclonal antibodies in mice.

contamination of the desired hybridoma clone with other hybridoma clones. Usually, it is because of poor subcloning and selection that Mab production fails because of unspecificity.[59] Finally, Mab production is scaled up by injecting (i.p.) the hybridoma cells into mice (the ascites method) or by growing them in *in vitro* culture.

## The Production of Mabs in the Mouse by the Ascites Method

From the existing literature, it can be concluded that the *in vivo* production of Mabs involves a number of steps that are believed to substantially impair animal welfare.[60] Regulatory authorities that categorize ascites production as "severe distress" recognize this, and in various countries, strong policies exist to discourage *in vivo* Mab production in favor of *in vitro* production.[61]

The production of Mabs includes a number of steps[60] that might affect the total volume of ascites or the Mab concentration: the i.p. injection of a primer, the effects of the primer after administration, the i.p inoculation of hybridoma cells, the growth of tumor cells in the peritoneal cavity, the production of ascites, and finally, the tapping of the animals to yield the ascites. Some of these steps are also believed to substantially interfere with animal welfare. The parameters affecting ascites production are given in Table 16.7.

### Intraperitoneal (i.p.) Injection of Primer

Intraperitoneal injection always introduces a certain risk for incorrect administration.[62] As a result, the primer might accidentally be injected in one of the abdominal organs or in the inoculation site. This might result in additional pathological effects and in a reduced priming effect.

### The Effects of the Primer after Administration

The administration of a primer is almost always a prerequisite for ascites production. Not using a primer leads to tumor formation without ascites production. The primer is believed to have several effects: it suppresses the immune system so that the growth of the hybridoma cells is not strongly impaired; it causes toxic irritation that leads to peritonitis and the secretion of serous fluid; and it retards the clearance of hybridoma cells and ascites from the peritoneal cavity by obstructing the lymphatic outflow.[63–65] The primer most frequently used is pristane (2,6,10,14-tetramethyl pentadecane), but FIA has also been used for this purpose.[66]

**Table 16.7 Parameters Affecting Ascites Production**

| Parameter | Specification | Comment |
|---|---|---|
| Animal | Age | Highest Mab concentration in mice between 43 to 75 days of age |
| Sex | | Males have a longer period of secretion, higher conc./mL and greater volume of ascites, probably due to testosterone |
| Strain | | Nude mice (irradiated mice) and SCID mice can be used in case of nonsyngeneic hybridoma cells |
| | | Yields can be increased by use of BALB/c-derived cross-bred F1 hybrids |
| Hybridoma cells | Type of cell | Some cells show a large variety in growth pattern in animals and tend to grow poorly in some animals |
| | Number of cells | Cell number affects the duration of secretion |
| | | Maximum number ranges from $3.2 \times 10^6$ to $5 \times 10^6$; Optimum: $10^6$ Minimum number: $6 \times 10^5$ per mouse |
| Primer | Product | By the use of IFA the interval between priming and hybridoma cell inoculation can be shortened |
| Volume | | Optimum volume is 0.5 ml, but also lower volume (0.1–0.2 mL) has been shown to be effective while causing less distress |

*Source:* From Hendriksen, C.F.M. and De Leeuw, W., Production of monoclonal antibodies by the ascites method in laboratory animals, *Research in Immunology*, 6, 149, 535, 1998.

As can be expected, the primer induces diffuse inflammation in the peritoneal cavity. In what way this effects animal welfare is not clear.[67]

### The i.p. Injection of Hybridoma Cells

Incorrect i.p. injection of hybridoma cells may lead to growth of tumor cells in the injection canal or in the intestinal wall.

### Growth of Hybridoma Cells in the Peritoneal Cavity

Within a few days after i.p. administration, hybridoma cells start to form solid and soft tissue masses throughout the peritoneal cavity.[67] The duration of the Mab production and the survival time of the animal depend on the number and the aggressiveness of the hybridoma cells injected.[68]

### The Production of Ascites

Production of ascites is believed to occur due to a number of mechanisms: absorption of intraperitoneal fluid is prevented by obstruction of subdiaphragmatic lymphatics; vascular permeability is altered, causing edema and fluid accumulation within the abdomen; and different cytokines contribute to imbalance of influx and efflux of lymphatic and vascular fluid.[64,69] The volume of ascites might range between 0 to 10 mL, but generally, 2 to 4 mL is obtained. Due to the formation of ascites, animals might gain weight and get dehydrated. In what way ascites formation effects animal welfare is not clear, but from the human clinic, it is well known that tumor growth, in particular when these tumors invade the abdominal organs, ascites' fluid production and stretching of the peritoneum during a short time period is painful.

A summary of clinical, pathophysiological, and pathological effects of ascites production is given in Table 16.8.

### Tapping of Ascites

Ascites can be obtained in a relatively short interval between the onset of ascites production (generally about 7 days after i.p. inoculation of hybridoma cells, but hybridoma dependent) and the death of the animal due to, e.g., respiratory distress, tumor growth or metastasis, at about day 14. In some countries, only one terminal tap under anesthesia is allowed, but in many other countries, no regulations exist with regard to the number of taps. Frequently, ascites are harvested more times, generally every 1 to 3 days on the rate of ascites production as assessed by the degree of abdominal distension and clinical condition of the animal. In the United States, the maximum number of taps permitted differs based on institutional policies. The maximum number of taps is generally considered to be three, as lethality due to ascites production increases, and the volume of ascites decreases after the third tap. The argument being used

**Table 16.8 Clinical, Pathophysiological, and Pathological Effects of Ascites Production**

| Clinical | Pathophysiological | Pathological |
|---|---|---|
| Abdominal distension | Anorexia | Peritonitis |
| Decreased activity | Anemia | Infiltrative tumor growth |
| Decreased body mass | Dehydration | Adhesions in the abdomen |
| Sunken eyes | Tachypnoe | Enlarged abdominal organs |
| Difficulty in walking | Decreased venous and arterial | Blood in the peritoneal cavity |
| Hunched posture | blood flow | Hydrothorax |
| Apathy | Ascites production | |
| Death | Immunosuppression | |

*Source:* From Hendriksen, C.F.M., Production of monoclonal antibodies by the ascites method: animal welfare implications, in *Proceedings of the Production of Monoclonal Antibodies Workshop,* McCardle, J.E. and Lund, C.J., Eds., August 29, 1999, Bologna, Italy, Alternatives Research and Development Foundation, Eden Prairie, 2000, p. 6.

for re-tapping is to increase the volume of ascites obtained per animal, thus resulting in a reduction of the number of animals used. Others, however, believe that it would be better to use more animals, but to reduce the level of pain and suffering per animal. As a general statement, animals with significant clinical abnormalities of pain or distress should be euthanized.

## The Production of Mabs by in vitro Methods

Although *in vitro* procedures for Mab production have been available for many years (in fact *in vitro* production has always been needed for subcloning and selection), the use of these methods has been limited, due to inadequacy and high costs. Traditionally, *in vitro* methods were expensive, labor intensive, or produced low Mab concentrations. However, these methods have been substantially improved, even for small-scale production. For an overview of these methods, the reader is referred to De Geus and Hendriksen[70] and Jackson, Trudel, and Lipman.[71]

The *in vitro* methods can be categorized according to the principles underlying their culture systems. Generally, they are divided in one-compartment and two compartment systems. The one-compartment technologies, in which hybridoma cells and medium are in the same compartment, are generally used for small-scale production. They include static cultures such as tissue flask cultures and gas-permeable bags and suspension culture such as roller bottles and spinner flasks. These one-compartment technologies allow the growth of cells in low densities only (not exceeding several millions cells per mL). As a consequence, the resulting quantities of Mab are small (the Mab concentration ranges between 10 to 200 ug/mL), and a concentration and purification step is often needed before the Mabs can be used. Cultures also have a finite life, as the supply of $O_2$ and nutrients is suboptimal, and metabolites and $CO_2$ lead to intoxication. However, these methods are easy to use and do not require specific training in most laboratories experienced in cell culture.[72]

The other group of production methods are based on two-compartment technologies. Hybridoma cells and secreted Mab are retained in small volume compartments, which are separated by semipermeable membranes from the larger volume of the basal medium. The semipermeable membranes allow an effective diffusion of nutrients, oxygen, and metabolites. High cell densities can be obtained in these methods ($10^7$ to $10^9$ cells per mL), resulting in large quantities of Mab per mL (1 to 10 mg per mL), almost comparable to ascites concentrations. High cell density methods include the Integra CELLine culture systems, the miniPERM bioreactors, the Tecnomouse or the hollow fiber bioreactors. The disadvantage of these methods is that they require moderately expensive equipment and well-trained technicians. However, from costs comparison studies,[73] it has become clear that if all costs related to production are included (equipment, materials, staff, production time, etc), *in vitro* produced Mabs can compete with ascites produced Mabs in terms of costs per mg Mab (Table 16.9) From a recent hearing, the NRC concluded[74] that *in vitro* methods are able to replace ascites production in >90% of all hybridomas. For some reason, some hybridoma cell-lines do not adapt to *in vitro* culture conditions (<4%), although many of the cases are due to suboptimal selection. Based on this information, several European countries issued restrictive regulations toward ascites production. *In vivo* production is only permitted in exceptional cases, for instance, where verifiable efforts have failed

**Table 16.9 Monoclonal Antibody Production: *in vivo* Versus *in vitro*[a]**

|  | Flasks and Roller Bottles | Gas-Permeable Bags | Mini Fermentor Chambers | Hollow Fiber Filters | Ascites |
|---|---|---|---|---|---|
| Rate | 300–900 mL/flask | 600 mL/bag | 7 mg/day | 3–6 mg/day | 1–3 mL/tap |
| Purity | 75–90% | 75–90% | 60–65% | 35–75% | 25–65% |
| Conc. | 0.01–0.3 mg/mL | 0.01–0.2 mg/mL | 2–3 mg/mL | 0.5–4 mg/mL | 1–5 mg/mL |
| Cost | $0.50–2/mg | $0.50–2/mg | $1–3/mg | $1–10/mg | $1–2/mg |

[a] Values differ with different hybridomas.

*Source:* From Peterson, N., A comprehensive cost comparison of *in vitro* and ascites approaches to Mab production, in *Proceedings of the Production of Monoclonal Antibodies Workshop*, McCardle, J.E. and Lund, C.J., Eds., August 29, 1999, Bologna, Italy, Alternatives Research and Development Foundation, Eden Prairie, 2000, p. 48.

to produce the Mab *in vitro*. NIH has stated that *in vitro* production should be the first choice. If *in vivo* production must be performed, animal welfare aspects should be taken into consideration and receive high priority.

# REFERENCES

1. Behring, E. von and Kitasato, Uber das Zustandekommen der Diphtherie-Immunität und der Tetanus-Immunität bei Tieren (Production of diphtheria immunity and tetanus immunity in animals), *Deutsche Medizinische Wochenschrift*, 49, 113, 1890.
2. Köhler, G. and Milstein, C., Continuous cultures of fused cells secreting antibody of predefined specificity, *Nature*, 256, 495, 1975.
3. Roitt, J., Brostoff, J., and Male, D., *Immunology*, 6th ed., Mosby, Baltimore, MD, 2001.
4. Austyn, J.M., and Wood, K.J., *Principles of Cellular and Molecular Immunology*. Oxford University Press, Oxford, 1994.
5. Smith, C.P., Jensen, D., Allen, T., and Kreger, M., Eds., Information resources for adjuvants and antibody production: comparisons and alternative technologies 1990–1997, *AWIC Resources Series*, 3, 1, 1997.
6. Leenaars, M., Claassen, E., and Boersma, W.J.A., Modulation of the humoral immune response; antigens and antigen presentation, in *Immunology Methods Manual. I.* Lefkovits, Ed., Academic Press, New York, 1996, p. 989.
7. Hanly, W.C., Artwohl, J.E., and Bennett, B.T., Review of polyclonal antibody production procedures in mammals and poultry, *ILAR News*, 37, 93, 1995.
8. McCullough, K.C., Hendriksen, C.F.M., and Seeback, T., *In vitro* methods in vaccinology, in *Veterinary Vaccinology*, Pastoret P.-P., Blancou, J., Vannier, P., and Verschueren, C., Eds., Elsevier, Amsterdam; New York, 1997, p. 69.
9. Poulsen, O.M., Hau, J., and Kollerup, J., Effect of homogenization and pasteurization on the allergenicity of bovine milk analysed by a murine anaphylactic shock model, *Clin. Allergy*, 17, 5, 449–458, 1987.
10. Poulsen, O.M., Nielsen, B.R., Basse, A., and Hau, J., Comparison of intestinal anaphylactic reactions in sensitized mice challenged with untreated bovine milk and homogenized bovine milk, *Allergy*, 45, 321, 1990.
11. Leenaars, M., Adjuvants in Laboratory Animals, Thesis, Erasmus University Rotterdam, Ponsen & Looijen BV., Wageningen, The Netherlands, 1997.
12. Leenaars, P.P.A.M., Hendriksen, C.F.M., De Leeuw, W.A., Carat, F., Delahaut, P., Fischer, R., Halder, M., Hanley, W.C., Hartinger, J., Hau, J., Lindblad, E.B., Nicklas, W., Outschoorn, I.M., and Stewart-Tull, E.S., The production of polyclonal antibodies in laboratory animals, *ATLA*, 27, 79, 1999.
13. Ramon, G., Sur l'augmentation anormale de l'antitoxine chez les chevaux producteurs de serum antidiphterique, *Bull. Soc. Centr. Med. Vet.*, 101, 227, 1925.
14. Freund, J., Casals, J., and Hosmer, E.P., Sensitization and antibody formation after injection of tubercle bacilli and parafin oil, *Proc. Soc. Exp. Biol. Med.*, 37, 509, 1937.
15. Hohmann, A.W., Freund's adjuvant and immune responses, in *The Use of Immuno-adjuvants in Animals in Australia and New-Zealand*, ANZCCART, Glen Osmond, SA, Australia, 1998.
16. Talmage, J. and Dixon, F.J., The influence of adjuvants on the elimination of soluble protein antigens and the associated antibody responses, *J. Infect. Dis.*, 93, 176, 1953.
17. Kittell, C.L., Banks, R.E., and Hadick, C.L., Raised skin lesions in rabbits after immunization, *Lab. Anim.*, 25, 16, 1991.
18. Hau, J., The rabbit as antibody producer — advantages and disadvantages, in Symposium über Zucht, Haltung und Ernährung des Kaninches. Das Kanin als Modell in der Biomed. Forschung. Altromin & Akademi für tierärztliche Fortbildung, 1988, p. 68.
19. Bollen, L.S., Hau, J., Freund's complete adjuvant has a negative impact on egg laying frequency in immunised chickens, *In Vivo*, 13, 107, 1999.
20. Colpaert, F.C., Evidence that adjuvant arthritis in the rat is associated with chronic pain, *Pain,* 28, 201, 1987.

21. Erb, K. and Hau, J., Monoclonal and polyclonal antibodies, in *Handbook of Laboratory Animal Science* Vol. I, 99, Svendsen, P. and Hau, J., Eds., CRC Press, Boca Raton, FL, 1994, pp. 293–309.

22. Lindblad, E.B. and Hau, J., Escaping from the use of Freund's complete adjuvant, in *Progress in the reduction, refinement and replacement of animal experimentation*, Proceedings from the 3rd world congress on alternatives and animal use in the life sciences 1999, Balls, M., van Zeller, A.-M., and Halder, M.E., Eds., Elsevier, Amsterdam, 2000, 1681–1685.

23. Herbert, W.J., Methods for preparation of water-in-oil, and multiple, emulsions for use as antigen adjuvants; and notes on their use in immunization procedures, in *Handbook of Experimental Immunology*, Weir, D.M., Ed., Blackwell Scientific, Oxford, 1967.

24. Lindblad, E.B. and Sparck, J.V., Basic concepts in the application of immunological adjuvants, *Scand. J. Lab. Anim. Sci.*, 14, 1, 1987.

25. Claassen, E. and Boersma, W.J.A., Characteristics and practical use of new generation adjuvants as an acceptable alternative for Freund's complete adjuvant, *Res. in Immunol.*, 143, 475, 1992.

26. Stewart-Tull, D.E.S., *The Theory and Practical Application of Adjuvants*, John Wiley & Sons, New York, 1995.

27. Jennings, V.M., Review of selected adjuvants used in antibody production, *ILAR J.*, 37, 119, 1995.

28. Chapel, H.M. and August, P.J., Report on nine cases of accidental injury to man with Freund's complete adjuvant, *Clin. Exp. Immunol.*, 24, 558, 1976.

29. Freund, J., Some aspects of active immunization, *Ann. Rev. Microbiol.*, 1, 291, 1947.

30. Bollen, L.S. and Hau, J., Immunoglobulin G in the developing oocytes of the domestic hen and immunospecific antibody response in serum and corresponding egg yolk, *In Vivo*, 11, 395, 1997.

31. Jensenius, J.C., Andersen, I., Hau, J., Crone, M., and Koch, C., Eggs: conveniently packaged antibodies. Methods for purification of yolk IgG, *J. Immunol. Methods,* 46, 63, 1981.

32. Svendsen, L., Crowley, A., Ostergaard, L.H., Stodulski, G., and Hau, J., Development and comparison of purification strategies for chicken antibodies from egg yolk, *Lab. Anim. Sci.*, 45, 89, 1995.

33. Svendsen Bollen, L., Crowley, A., Stodulski, G., and Hau, J., Antibody production in rabbits and chickens immunized with human IgG. A comparison of titre and avidity development in rabbit serum, chicken serum and egg yolk using three different adjuvants, *J. Immunol. Meth.*, 27, 191, 113, 1996.

34. Schade, R., Staak, C., Hendriksen, C., Erhard, M., Hugl, H., Koch, G., Larsson, A., Pollmann, W., van Regenmortel, M., Rijke, E., Spielmann, H., Steinbusch, H., and Straughan, D., The production of avian (egg yolk) antibodies: IgY. ECVAM Workshop Report 21, *ATLA*, 24, 925, 1996.

35. Schade, R., Behn, I., Erhard, M., Hlinak, A. and Staak, C., Eds., *Chicken Egg Yolk Antibodies, Production and Application. IgY-Technology.* Springer Lab Manual, Berlin, 2001.

36. Hau, J., Westergaard, J.G., Svendsen, P., Bach, A., and Teisner, B., Comparison between pregnancy-associated murine protein-2 (PAMP-2) and human pregnancy-specific beta 1-glycoprotein (SP-1), *J. Reprod. Fertil.*, 60, 115, 1980.

37. Hau, J., Westergaard, J.G., Svendsen, P., Bach, A. and Teisner, B., Comparison of the pregnancy-associated murine protein-1 and human pregnancy zone protein, *J. Reprod. Immunol.*, 3, 341, 1981.

38. Persdotter Hedlund, G. and Hau, J., Oral immunization of chickens using cholera toxin B subunit and Softigen as adjuvants results in high antibody titre in the egg yolk, *In Vivo*, 15, 384, 2001.

39. Russell, W.M.S. and Burch, R.L., *The Principles of Humane Experimental Technique*, Methuen & Co. LTD, London, 1959.

40. Cannat, A., Feingold, N., Caffin, J.C., and Serre, A., Studies on the genetic control of murine humoral response to immunization with a peptidoglycan-containing fraction extracted from *Brucella melitensis*, *Ann. Immunol.*, 130c, 675, 1979.

41. Lifschitz, R., Schwartz, M., and Mozes, E., Specificity of genes controlling immune responsiveness to (T,G)-A-L and (Phe, G)-A-L, *Immunol.*, 41, 339, 1980.

42. Long, D.A., The influence of the thyroid gland upon immune responses of different species to bacterial infection, CIBA Found, Colloq., *Endocrinol.*, 10, 287, 1957.

43. Harboe, N.M.G. and Ingild, A., Immunization, isolation of immunoglobulins, and antibody titre determination, *Scand. J. Immunol.*, 17 Suppl. 10, 245, 1983.

44. Young, C.R. and Atassi, M.Z., Genetic control of antibody response to antigenic sites by increasing the dose of antigen used in immunization, *J. Immunogen.*, 9, 343, 1982.

45. Abraham, L., O'Brien, D., Poulsen, O.M., and Hau, J., The effect of social environment on the production of specific immunoglobulins against an immunogen (human IgG) in mice, in *Welfare and Science*, Bunyan, J., Ed., 1994, p. 165.

46. Kaplan, P.J., Caperna, T.J., and Garvey, J.S., Bovine serum albumin humoral immune response in aged Fischer 344 rats, *Mech. Ageing Dev.*, 16, 61, 1981.

47. Poulsen, O.M. and Hau, J., Suppression of humoral antibody response during pregnancy in mice, *In Vivo*, 4, 381, 1990.

48. Bollen, L.S. and Hau, J., Comparison of immunospecific antibody response in young and old chickens immunized with human IgG, *Lab. Anim.*, 33, 71, 1996.

49. Nielsen, B.R., Poulsen, O.M., and Hau, J., Reagin production in mice: effect of subcutaneous and oral sensitization with untreated bovine milk and homogenized bovine milk, *In Vivo*, 3, 271, 1989.

50. Halliday, L.C. et al., Physiologic and behavioral assessment of rabbits immunized with Freund's complete adjuvant, *Cont. Topics Lab. Anim. Sci.*, 39, 8, 2000.

51. BVA/FRAME/RSPCA/UFAW Joint Working Group on Refinement: removal of blood from laboratory animals and birds, *Lab. Anim.*, 27, 1–22, 1993.

52. Zutphen van, L.F.M., Baumans, V., and Beynen, A.C., Eds., *Principles of Laboratory Animal Science*, Elsevier, Amsterdam; New York, 2001, p. 322.

53. Leenaars, P.P.A.M., Koedam, M.A., Wester, P.W., Baumans, V., Claassen, E., and Hendriksen, C.F.M., Assessment of side effects induced by injection of different adjuvant/antigen combinations in rabbits and mice, *Lab. Anim.*, 32, 387, 1998.

54. CCAC, Guidelines on Antibody Production, Ottawa, Ontario, Canada, 2002.

55. Grumstrupp-Scott, J. and Greenhouse, D.D., NIH intramural recommendations for the research use of Freund's complete adjuvant, *ILAR News*, 2, 9, 1988.

56. De Kruijf, J. et al., New perspectives on recombinant human antibodies, *Immunology Today*, 17, 453, 1996.

57. Marx, U., Embleton, M.J., Fischer, R., Gruber, F.P., Hansson, U., Heuer, J., de Leeuw, W.A., Logtenberg, T., Merz, W., Portelle, D., Romette, J.-L., and Straughan, D.W., Monoclonal antibody production. The report and recommendations of ECVAM workshop 23, *ATLA*, 25, 121, 1997.

58. Johnson, D.R., Murine monoclonal antibody development, in *Methods in Molecular Biology*, Vol. 51, Antibody Engineering Protocols, Paul, S., Ed., Humane Press, Totowa, NJ, 1995, p. 123.

59. Hendriksen, C.F.M., A call for a European prohibition of monoclonal antibody production by ascites in laboratory animals, *ATLA*, 26, 523, 1998.

60. Hendriksen, C.F.M., Production of monoclonal antibodies by the ascites method: animal welfare implications, in *Proceedings of the Production of Monoclonal Antibodies Workshop*, McCardle, J.E. and Lund, C.J., Eds., August 29, 1999, Bologna, Italy, Alternatives Research and Development Foundation, Eden Prairie, 2000, p. 6.

61. Hendriksen, C.F.M. and De Leeuw, W., Production of monoclonal antibodies by the ascites method in laboratory animals, *Research in Immunology*, 6, 149, 535, 1998.

62. Walvoort, N.C., Assessment of distress through pathological examination, in *Replacement, Reduction and Refinement: Present Possibilities and Future Prospects*, Hendriksen, C.F.M. and Köeter, H.B.W.M., Eds., Elsevier, Amsterdam; New York, 1991, p. 265.

63. Kuhlmann, I., Kurth, W., and Ruhdel, I., Monoclonal antibodies: *in vivo* and *in vitro* production on a laboratory scale, with consideration of the legal aspects of animal protection, *ATLA*, 17, 73, 1989.

64. Moore, J.M. and Rajan T.V., Pristane retards the clearance of particulate materials from the peritoneal cavity of laboratory mice, *J. Immunol. Meth.*, 173, 273, 1994.

65. O'Driscoll, C.M., Anatomy and physiology of the lymphatics, in *Lymphatic Transport of Drugs*, Charman, W.N. and Stella, V.J., Eds., CRC Press, Boca Raton, FL, 1992.

66. Gilette, R.W., Alternatives to pristane priming for ascites fluid and monoclonal antibody production, *J. Immunol. Meth.*, 99, 21, 1987.

67. Jackson, L.R., Trudel, L.J., Fox, J.G., and Lipman, N.S., Monoclonal antibody production in murine ascites. I. Clinical and pathological features, *Lab. Anim. Sci.*, 49, 70, 1999.

68. Brodeur, B.R., Tsang, P., and Larose, Y., Parameters affecting ascites tumour formation in mice and monoclonal antibody production, *J. Immunol. Meth.*, 71, 265, 1984.

69. Behammer, W., Kluge, M., Ruschoff, J., and Mannel, D.N., Tumor necrosis factor effects on ascites formation in an experimental tumor model, *J. Interferon and Cytokine Research*, 14, 403, 1996.

70. De Geus, B. and Hendriksen, C.F.M., Eds., *In vivo* and *in vitro* production of monoclonal antibodies: current possibilities and future perspectives, *74th Forum in Immunology, Research in Immunology*, 6, 149, 533, 1998.

71. Jackson, L.R., Trudel, L.J., and Lipman, N.S., Small-scale monoclonal antibody production *in vitro*: methods and resources, in *Proceedings of the Production of Monoclonal Antibodies Workshop*, McCardle, J.E. and Lund, C.J., Eds., August 29, 1999, Bologna, Italy, Alternatives Research and Development Foundation, Eden Prairie, 2000, p. 16.

72. Falkenberg, F.W., Monoclonal antibody production: problems and solutions, *Research in Immunology*, 149, 542, 1998.

73. Peterson, N., A comprehensive cost comparison of *in vitro* and ascites approaches to Mab production, in *Proceedings of the Production of Monoclonal Antibodies Workshop*, McCardle, J.E. and Lund, C.J., Eds., August 29, 1999, Bologna, Italy, Alternatives Research and Development Foundation, Eden Prairie, 2000, p. 48.

74. National Research Council (NRC), Monoclonal Antibody Production, National Academy Press, Washington, DC, 1999.

# CHAPTER 17

# Laboratory Animal Analgesia, Anesthesia, and Euthanasia

Patricia Hedenqvist and Ludo J. Hellebrekers

## CONTENTS

0-8493-1086-5/03/$0.00+$1.50
© 2003 by CRC Press LLC

# INTRODUCTION

Legislation and humane use of live vertebrate animals in scientific research require that experiments must be performed in a way that minimizes pain and suffering. As a rule, painful experiments should be performed under local or general anesthesia.

One of the major problems in laboratory animal anesthesia and pain control lies in the fact that, under many circumstances, it is difficult to ascertain adequacy of anesthesia or analgesia. The major obstacles one faces are the recognition and quantification of pain, less so in the animal under anesthesia than in the wake animal, potentially experiencing pain.

A plethora of parameters of physiological or behavioral background have been named for their relevance for recognizing and quantifying pain in animals. Although many of these parameters will change under painful circumstances, their discriminative power is limited to the extent that they cannot at all times differentiate between pain and other causes involving a stress response. Complicating matters further, some species are known to mask any signs of pain, based upon their potential role as victims or bait of predatory animals.

Under many circumstances involving laboratory animals, the Principle of Analogy is employed. This approach ascribes pain to an animal in the case where it may be expected based on human experience, if pain cannot definitely be excluded based on findings from the animal involved. Although this introduces the risk of anthropomorphism, it provides the "benefit of the doubt" to the animal, which will consequently receive analgesic treatment when indicated.

Our present day understanding of the physiology of pain clearly demonstrates the relevance of a proactive, i.e., preventative, approach to pain. Through early detection and installation of effective treatment, prevention of the development of peripheral and central sensitization is pursued. With this approach the development of pain (intensity and duration) will be reduced and thus easier to treat. The distress to the animal will be reduced and a full and fast recovery promoted.

In discussing laboratory animal anesthesia, a differentiation can be made between those animals that are anesthetized for reasons of instrumentation and those animals that are anesthetized for allowing the execution of an experiment. Under the latter circumstances, scientific data collection is performed during anaesthesia. This automatically necessitates an in-depth evaluation of potential interactions between the data to be collected and the anesthetic protocol. Of major importance under these circumstances is the precise definition of the actual parameters under investigation. Thereby, the evaluation of the potential, direct or indirect effects of the anaesthetics used can be achieved based upon available physiologic and pharmacological knowledge.

The scope of this chapter is to give an overview of the present day's views on laboratory animal analgesia and anesthesia. For more detailed information, the reader is advised to consult the different laboratory animal or veterinary anesthesia handbooks.

## ADMINISTRATION OF ANALGESICS AND ANESTHETICS

Refinement of techniques for administration of substances has recently been reported by Morton et al.[1]

### Topical Administration

Local anesthetics can be applied as gel, cream, solution, or aerosol and are used for blocking pain in the eye and ear canal, on mucous membranes, and on the skin. Local anesthetics can be used for analgesic supplementation during general anesthesia in the form of regional or epidural block, or be deposited directly around exposed nerves during surgery.

Lidocaine gel or spray is used for mucous membranes of the mouth, nose, pharynx, larynx, vagina, urethra, and rectum for analgesia during surgery. A 2 to 4% solution produces effect in 5 min and lasts for 30 min.[2] Topical local anesthetic solution can be used during nasal, aural, or extraocular surgery.

EMLA-cream, a combination of lidocaine and prilocaine, is effective in reducing pain from venipuncture or skin biopsy in larger species (cat, dog, pig, and rabbit).[3] After clipping the fur or coat, the cream is applied for 60 min under an adhesive plastic bandage, resulting in 1 h of full skin anesthesia (Figure 17.1).

Patches for transdermal administration of the opioid fentanyl (25, 50 µg/h) can be used in larger species (cat, dog, pig, sheep, and goat) for control of postoperative or chronic pain.[3] Even though this method involves topical administration, the aim is to achieve a systemic analgesic effect. Because fentanyl plasma concentrations vary considerably with patches, they are best used to deliver a background level of analgesia. Skin temperature and perfusion are among the factors that will influence fentanyl absorption rate.

General anesthetics and analgesics may sometimes be applied on the mucus membranes of the mouth or rectum, in order to avoid painful injections or immediate hepatic drug metabolism, which occurs after oral and intraperitoneal administration. Examples are rectal administration of ketamine or buccal administration of buprenorphine in cats.[4,5]

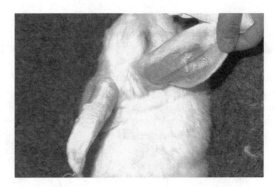

**Figure 17.1**    EMLA cream on rabbit ears for cannulation of central ear artery.

Before limb amputation, exposed nerves are bathed in local anesthetic for 2 to 3 min prior to transection.[6] This will significantly decrease postoperative pain and possibly reduce the development of phantom limb pain.

## Administration by Injection

Intravenous injection of anesthetics is the safest injection route because it allows for dose adjustment in the individual animal. Repeated injections (boluses) may be used to prolong duration of anesthesia, but to achieve a stable plane of anesthesia, and consequently, more stable hemodynamics, continuous intravenous infusion is preferred.

A precise control over anesthetic depth can be achieved by using continuous infusion of drugs that are rapidly metabolized (e.g., propofol). By changing the infusion rate, the level of anesthesia is rapidly changed, which gives a control comparable to that of inhalation anesthesia.

Another benefit of continuous intravenous access is that fluids, specific antagonists, or other emergency drugs can be administered with immediate effect. Peripheral veins for IV injections may be difficult to access in several rodent species, like the hamster and guinea pig.

Injection of anesthetics by the intraperitoneal, intramuscular, or subcutaneous route does not allow for the same dose adjustment as intravenous administration. Therefore, only substances with a broad safety margin should be chosen for these routes. Intraperitoneal administration is commonly used in rodent anesthesia. It should not be used in pregnant rodents or animals larger than rodents. Rapid absorption due to the large uptake area is considered an advantage, but care must be taken not to inject into abdominal organs. A short needle stop, which can be made from the protective needle sheath, prevents the needle from entering the peritoneal cavity to a depth more than 4 or 5 mm.[7]

Drug that is absorbed from the peritoneal cavity by the portal system is subjected to hepatic first-pass elimination. This stands in contrast to drugs administered by the intramuscular or the subcutaneous route, which reach the systemic circulation without previous metabolism. Textbooks on veterinary anesthesia often recommend the same drug dose for IM, SC, and IP injection in rodents, not taking into consideration the possible effects of first-pass hepatic metabolism. When changing routes of administration, from intraperitoneal to intramuscular, or subcutaneous injection, drug doses should be reduced by approximately 50 to 75%.

Intramuscular injections should be avoided in rodents because they can cause unnecessary muscle damage and pain.[8] Irritant substances such as ketamine may cause muscle necrosis and should be diluted and injected at several sites to limit this effect. When IM injection is indicated, the cranial thigh muscle mass (quadriceps) is most suitable. The epaxial muscles along the spine may also be used. Injections into the caudal thigh muscle mass incorporate the risk of damaging the ischiadic nerve.

Subcutaneous injection is often an alternative because it is easy to perform and less painful than IM injection. The rate of absorption is often comparable with that after IM injection. Irritant substances will cause damage when injected SC, which can be avoided by dilution with saline or water for injection.[1]

## Administration by Inhalation

Inhalation anesthesia is useful for animals of most species and ages. Advantages are fast induction and recovery, greater control than with most injectable anesthetics, and stable physiological effects. Under most circumstances, the concurrent oxygen supplementation contributes to limiting the overall anesthetic risk. Due to minimal metabolism of most modern inhalants, they are less likely to be a significant variable in the research project.[9] Disadvantages are high costs and need for scavenging systems, as well as for greater technical competence.

The safest way to administer inhalants is by using an anesthetic machine. With help of a calibrated vaporizer, the volatile agent can be delivered at a precise and accurate concentration. For brief procedures in small animals, a simple bell jar may be used (see rodent anesthesia). This technique necessitates great caution because dangerously high vapor concentrations may be produced with modern volatile agents. The volatile anesthetic is delivered to the animal by means of a carrier gas. If oxygen alone is used as carrier, the result will be a high arterial blood $O_2$ levels. Oxygen mixed with air or nitrous oxide ($N_2O$), will produce more physiological arterial $O_2$ levels. Nitrous oxide has the advantage of contributing to the analgesic effect without causing any respiratory depression and of reducing the concentration of the volatile needed by approximately 20%. This limits the cardiovascular depression of the volatile agent is limited.

If the animal is allowed to breathe spontaneously, arterial blood carbon dioxide partial pressure may increase. By mechanical ventilation, more physiological $CO_2$ levels can be maintained during anesthesia.

Induction of inhalation anesthesia can be achieved by means of an induction chamber or a face mask. Chamber induction is useful for rodents, kittens, pups, piglets, birds, and other small animals not accustomed to manual restraint. Mask induction should only be used in animals that are accustomed to handling. Anesthesia is maintained by keeping the animal on the mask, or by gas delivery via an endotracheal tube. Effective scavenging systems are necessary in order to prevent exposure of staff to waste anesthetic gas (see Operator Safety).

Induction of anesthesia with inhalants should not be performed in nonsedated rabbits, because they are reluctant to inhaling vapors and respond with extended breath holding and violent struggling.[10,11,12] Instead, a short-acting injectable agent can be used for induction, followed by intubation and inhalation anesthesia.

## ANALGESIC AND ANESTHETIC DRUGS

### Opioids

Opioids are used for treating moderate to severe pain, with morphine being the most potent drug for treatment of severe pain. Synthetic opioids act in a similar manner to morphine and can likewise induce some degree of sedation in high doses. The analgesic effect is mainly mediated via μ-receptors, located in the central and peripheral nervous system, and on inflammatory cells. Analgesia is also mediated by kappa receptors, which are abundant in the spinal cord. Opioids are mostly used for their central effects but have been shown to have additional local analgesic effects, for example in mice, after intraplantar injection or topical application on the tail.[13,14] Opioids reduce the amount of anesthetic needed for surgery and have been shown to have a preemptive effect in dogs and rodents.[15,16]

Side effects seen with higher dosages may include respiratory depression, as a result of a decreased responsiveness of the respiratory center to carbon dioxide. This effect seems to be less significant in animals compared with humans. Older opioids such as morphine and pethidine may induce vasodilation and inhibition of baroreceptor reflexes. When using the most potent agents IV, atropine can be given to prevent or treat severe bradycardia.[3] All opioids decrease hydrochloric acid secretion in the stomach and prolong gastric emptying time. They also decrease propulsive activity in the gastrointestinal tract and may induce urinary retention. This does not seem to be clinically significant in rodents and rabbits.

Opioid drugs are generally well absorbed from the gut and after IP, SC, and IM injection. The bioavailability after oral dosing is poor due to substantial first-pass hepatic metabolism, which also reduces bioavailability after IP administration. Repeated use may induce tolerance, necessitating increased dosages to maintain efficacy. All opioid drugs can be reversed with naloxone, which has antagonistic actions on all opioid receptors. Its duration of action is short (30 to 60 min), and it may be used to reverse the opioid effects after overdosing or to speed up recovery. Because no analgesia remains after naloxone reversal, analgesia must be provided by other means (e.g., NSAIDs) if pain is expected.

*Morphine* has a short elimination half-life (1 to 2 h). Due to longer persistence in the cerebrospinal fluid, dogs only need readministration after approximately 4 h. In cats, metabolism of morphine is slow, and duration of action is approximately 6 h. Morphine has no ceiling effect, meaning that the higher the dose, the greater the analgesic effect. Slow IV injection is necessary to avoid histamine release and consequent hypotension. Morphine may cause vomiting in dogs and cats, but not when it is given after surgery or to a traumatized cat or dog.[3] Morphine has been shown to reduce natural killer cells and phagocytic activity.[17]

*Fentanyl, alfentanil, and sufentanil* are μ-agonists suitable for providing intraoperative analgesia because of their high potency and short duration of action. Preadministration with atropine or glycopyrrolate is recommended depending on the species and pre-existing vagal tone. The onset of action is quick (2 to 5 min) and duration between 5 to 10 min (alfentanil) and 20 to 30 min (fentanyl and sufentanil).[3] Depending on the dose administered, respiratory support may be needed. One advantage of opioids is that they do not reduce myocardial contractility or coronary blood flow, which makes them useful for cardiac surgery and for taking physiologic measurements in cardiovascular protocols.[18] During general surgery, they need to be combined with other anesthetics, such as inhalants, to provide full anesthesia.

Fentanyl is available in combination with droperidol (Innovar®) for use in dogs, cats, rabbits, and rodents, for 30 min of potent analgesia. Duration of sedation is considerably longer. Another fentanyl combination is with fluanisone in Hypnorm®, which is commonly used for sedation and analgesia in rodents and rabbits. Hypnorm in combination with a benzodiazepine, such as midazolam, is used to attain 20 to 40 min of surgical anesthesia.[3]

Fentanyl or sufentanil can be combined with medetomidine for the production of completely reversible anesthesia in rats.[19,20] Due to the severe hypoxia produced by these combinations, oxygen supplementation is strongly recommended. Anesthesia with infusion of sufentanil in combination with inhalation of nitrous oxide anesthesia has been described in dogs.[21]

*Buprenorphine and butorphanol* are mixed agonists used for providing postoperative pain relief. Although buprenorphine is 25 times as potent as morphine, the maximum analgesic effect is less.[17] Peak plasma levels are reached after 30 to 60 min and duration is up to 8 h in most species. Sheep need readministration more often (approximately every 3 h). Buprenorphine can be used to reverse pure μ-agonists such as morphine or fentanyl, while providing long-lasting analgesia. Side effects are minimal when compared to morphine. Buprenorphine can safely be given before induction with inhalation anesthesia.

Butorphanol has effects on μ- and κ-receptors and is useful to control mild to moderate pain.[6] Its duration of action is short to moderate (1 to 2.5 h in dogs and cats). Like buprenorphine, it may be used to reverse the effects of μ-agonists. It is four times as potent as morphine, and peak plasma levels are reached after 45 to 60 min.

## NSAIDs

Nonsteroidal anti-inflammatory drugs are a group of weak organic acids with anti-inflammatory, analgesic, antipyretic, and antihyperalgesic properties.[3] They act by inhibiting enzymes (cyclooxygenases) and are well absorbed after oral, SC, and IM injection. Metabolism occurs in the liver. The most common side effects of NSAIDs are gastrointestinal ulceration and renal function disturbances, but they are less frequently seen with the more modern drugs. At therapeutic doses, most NSAIDs do not impair clotting time or prolong bleeding time. With the exception of carprofen, NSAIDs should be avoided in animals with impaired renal blood flow, e.g., during hypovolemia. Due to their embryotoxicity, NSAIDs are contraindicated in pregnant animals.

*Salicylates* (e.g., aspirin) stimulate the respiratory center directly (medullary effect) and indirectly (production of $CO_2$).[17] Salicylates interfere with platelet aggregation, thereby prolonging bleeding time. Aspirin is given by mouth and may be used to treat mild pain. Metabolism includes conjugation with glucoronic acid. In cats, which lack the enzyme for conjugation, half-life is considerably prolonged.

*Ketorolac* (acetic acid derivate) is pharmacologically similar to aspirin, except it does not interfere with platelets.[3] It is readily absorbed following oral or intramuscular administration.

*Ketoprofen* (proprionic acid derivate) has stronger analgesic properties than aspirin and can be used in dogs, cats, swine, rabbits, and rodents.[17] It is available in tablet and injection preparations and may be used alone for postoperative analgesia or in combination with an opioid.

*Carprofen* is the most commonly used NSAID for perioperative analgesia, and is often effective for controlling postoperative pain.[3] It may be combined with an opioid to achieve the most efficient pain relief. Half-life varies considerably in different species (dog, 8 h; sheep, 24 h). It is suitable for administration once or twice daily by mouth or by SC injection, and is safe to administer preoperatively in a number of species (e.g., rat, dog).

## Local Anesthetics

Local anesthetics are weak bases, which in the ion form stop nerve transmission by blocking $Na^+$ channels.[2] Duration of action is dependent on the drug's lipid solubility. Bupivacaine, for example, is more lipid than lidocaine and therefore of longer duration. Toxicity is primarily systemic and related to absorption from the injection site to the circulation. By adding a vasoconstrictor like adrenaline, the risk of toxicity is reduced, and intensity as well as duration of analgesia extended. Adrenaline must not be included if local anesthetics are used on extremities (e.g., toe, ear, or penis) because necrosis and loss of the extremity may result from vasoconstriction.

Not only sensory nerves, but also sympathetic and motor nerves, will be blocked by local anesthetics, with the results of muscle relaxation. When local anesthetics are used during surgical procedures, the level of general anesthesia can be kept at a lighter plane, and side effects reduced. The combination of a sedative and local anesthesia may even replace general anesthesia during minor procedures. With the use of local anesthetics, analgesia can be provided during and after surgery, and thus prevent severe physiological effects of surgery. The need for other analgesics will also be reduced.

Local anesthetics can be used topically, by local infiltration, in the area of a nerve, intrapleurally, and epidurally.[3] Intravenous injection of high doses may result in convulsions, hypotension, ventricular arrhythmia, and myocardial depression. Gels may be used for application in the urethra with catheterization, and aerosols for spraying the airways before intubation or bronchoscopy. Intrapleural administration is used after thorax surgery or in animals with rib fractures, and subcutaneous infiltration before skin suturing. The use of local anesthesia is essential before limb amputation to avoid development of phantom pain.

The most commonly used drugs are lidocaine and bupivacaine. The duration of action for lidocaine is 60 to 90 min. If adrenaline is included in a 1:200,000 concentration, duration will be increased by 50%. Due to a high rate of protein binding, which prevents absorption, bupivacaine has a longer duration of action (2 to 6 h). The long duration makes bupivacaine suitable for control of postoperative pain. Addition of adrenaline has little effect on duration.

*Ropivacaine* has similar properties to bupivacaine but is less cardiotoxic. Metabolism of all three drugs occurs in the liver with cytochrome P-450. Safe maximum doses in most species are 4 mg/kg for lidocaine and 1 to 2 mg/kg for bupivacaine.[6] The LD50 for bupivacaine in adult rats is 30 mg/kg.[21] For epidural analgesia or anesthesia, local anesthetics as well as opioids, ketamine, or xylazine may be used.

EMLA-cream is a mixture of *prilocaine* and lidocaine that will induce full skin anesthesia after 60 min of application.[3] Its main use is in the prevention of pain with venipuncture in cats, dogs, pigs, and rabbits.

MS-222 (*tricaine*) is a soluble local anesthetic used for sedation, immobility, and anesthesia in fish and amphibians.[22] The drug is delivered via the water or respiratory system and has local as well as systemic effects. Onset and recovery is rapid.

## Anticholinergics

Anticholinergics are parasympathetic antagonists that reduce salivation, limit or correct a decrease in heart rate, and dilate bronchi.[23] Their use is contraindicated in animals with an elevated heart rate, and they should be used with caution in species that normally have a very high heart rate (e.g., birds).

Anticholinergics are used to block the vagal reflex, which slows the heart rate during induction of anesthesia and intubation. Routine incorporation in every anesthetic protocol is not warranted, instead they are recommended in combination with drugs that cause vagal bradycardia (opioids) or increased salivation (ketamine). An overdose of anticholinergics will precipitate seizures.

*Atropine* has a faster onset and a shorter duration than *glycopyrrolate*.[9] In the dog, the duration of action is 30 to 60 min for atropine and between 2 to 3 h (reduction of heart rate) and 7 h (reduction of secretions) for glycopyrrolate. Glycopyrrolate is a large polar quaternary ammonium molecule that will not diffuse across the blood-brain barrier or the placenta, and thus has minimal effects on the CNS and fetus. Administration routes are SC, IM, or IV. Several animal species (rats, rabbits, cats) can destroy large amounts of atropine with atropine esterase in the liver.

## Sedatives and Tranquilizers

These agents are administered to relieve anxiety, decrease stress, and ensure animal and staff safety during restraint. They are often given before induction of anesthesia to allow a smoother induction and recovery and to reduce the need for anesthetics, with the result that side effects are minimized. Upon administration, the animal should be left in a quiet area for 10 to 30 min to allow the treatment to take effect. The group includes benzodiazepines, phenothiazine, and butyrophenone derivates, and they are often used in conjunction with a dissociative or opioid drug to provide general anesthesia.

*Benzodiazepines* (diazepam, midazolam, and zolazepam) are weak bases with sedative, muscle relaxant, and anticonvulsant properties.[9] They exert their effect via potentiation of GABA (gamma-aminobutyric acid), through binding on specific GABA-receptor sites. The result is a postsynaptic inhibition with modulation of the release of norepinephrine, dopamine, and serotonin. Cardiovascular as well as respiratory systems are minimally affected and no analgesia is provided. Sometimes when a benzodiazepine is administered as the first or only drug, the animal reacts by getting more difficult to handle. By adding an opioid, sedation is increased and handling facilitated. The effects of benzodiazepines can be reversed with the specific antagonist flumazenil.

The most common formulation of *diazepam* is in a propylene glycol base, making it insoluble in water. Uptake after SC or IM administration is unpredictable, making IV injection preferable. The half-life of diazepam varies between 1 to 7 h in the rat, guinea pig, rabbit, and dog. *Midazolam* is slightly more potent than diazepam and is water soluble, which makes it suitable for intravenous, subcutaneous, and intramuscular injection. It is compatible with atropine, fentanyl, Hypnorm, and ketamine. Combinations of *zolazepam* and the dissociative agent tiletamine (Zoletil® or Telazol®) are useful for attaining surgical anesthesia in rats, ferrets, gerbils, dogs, and cats. They have been reported to cause nephrotoxicity in rabbits and tachycardia in dogs. Due to respiratory depression, oxygen supplementation and assisted ventilation may be necessary in dogs and cats.[23]

*Acepromazine* is a phenothiazine derivate acting as a sedative by central dopamine blockade, and has peripheral alpha-adrenergic antagonistic effects.[9] At lower doses, it causes sedation and drowsiness, at higher doses, ataxia and somnolence. It has no analgesic effects. Acepromazine reduces the occurrence of [nor]adrenaline-induced ventricular arrhythmias. Side effects include vasodilation, hypotension, hypothermia, and a lowered CNS seizure threshold. It should not be used in animals with hypotension or in pediatric animals. It can be administered IM or SC.

*Butyrophenone derivates* include droperidol, fluanisone, and azaperone, with similar properties to the phenothiazine derivates.[9] Droperidol is combined with the opioid fentanyl in Innovar-Vet®, and fluanisone with fentanyl in Hypnorm®.

In dogs, Innovar-Vet causes profound analgesia for about 30 min and is useful for minor procedures, but in cats, undesirable CNS stimulation may occur. In rats, muscle relaxation is poor. Adverse effects may include bradycardia, hypotension, respiratory depression, hypoxia, hypercapnia, and acidosis. Anticholinergic premedication eliminates some of these side effects. Hypnorm is commonly used in rodents and rabbits for its sedative and analgesic effects and may be administered SC, IM, or IV. It is often used in combination with a benzodiazepine such as midazolam, for surgical anesthesia. During Hypnorm + midazolam anesthesia in rats, cardiac output is almost double that of pentobarbital anesthetized rats.[24]

Azaperone has minimal cardiovascular effects and produces immobilization in pigs lasting about 20 to 40 min.[18] It is not recommended for use in other species.

## Volatiles and Gases

Inhalants are halogenated hydrocarbons (halothane), halogenated ethers (iso-, des-, en-, sevoflurane), or inorganic gases ($N_2O$). Volatile anesthetics (halo-, iso-, des-, en-, sevoflurane) enhance the inhibitory neurotransmitter GABA (gamma-aminobutyric acid), by interacting with CNS $GABA_A$ receptors. Each volatile has a unique range of temperature-dependent vapor pressures, usually giving rise to concentrations much higher than those needed for anesthesia. Thus, a calibrated vaporizer is necessary for controlled delivery (Figure 17.2).

Minimum alveolar concentration (MAC), a measure of inhalation anesthetic potency, is the concentration that produces immobility in 50% of subjects during noxious stimulation and corresponds to a light level of anesthesia. MAC is similar between animals of one species and only varies 10 to 20% between animal species.[25] To reliably reach an adequate depth of anesthesia, with 95% of the subjects unresponsive to a noxious stimulus, 1.5 MAC is needed. A deep level of anesthesia is represented by 2.0 MAC and may constitute an overdose.[26] MAC decreases with age, and the highest values are found in neonates. Premedication with opioids, alpha-2-agonists or tranquilizers will reduce MAC values. The high MAC value for $N_2O$ in animals indicates that anesthesia cannot be maintained with $N_2O$ alone because more than 100% is needed to suppress reaction to a noxious stimulus. Nitrous oxide is more effective in humans (MAC 104%).

**Figure 17.2** Isoflurane vaporizer.

**Table 17.1 MAC Values in Some Laboratory Animal Species**

| Species | Halothane | Isoflurane | Sevoflurane | $N_2O$ |
|---------|-----------|------------|-------------|--------|
| Mouse | 0.96–1.0 | 1.35–1.41 | — | 150–275 |
| Rat | 0.81–1.23 | 1.17–1.52 | 2.4–2.5 | 155–235 |
| Rabbit | 0.8–1.56 | 2.05 | 3.7 | — |
| Cat | 0.82–1.19 | 1.61–1.63 | 2.58 | 255 |
| Dog | 0.86–0.93 | 1.28–1.39 | 2.10–2.36 | 188–297 |
| Primate | 0.89–1.15 | 1.28–1.46 | 2 | 200 |

Source: From Steffey, E.P., Inhalation anesthetics, in *Lumb & Jones' Veterinary Anesthesia*, third ed., Thurmon, J.C., Tranquilli, W.J., and Benson, G.J., Eds., Williams & Wilkins, Baltimore, MD, 1996.

By measuring end-tidal gas concentrations in intubated animals during inhalation anesthesia, anesthetic depth can be compared throughout the procedure, and multiple animals can be studied under the same anesthetic conditions.[9] This is more difficult to achieve with injection anesthesia.

Chronic exposure to trace anesthetic gases is a health hazard, and scavenging of excess gas is essential. Human exposure to trace amounts of halothane is more hazardous than exposure to isoflurane, due to halothane's high degree of metabolism. Especially with mask and box induction, the risk of exposure is high, and isoflurane a better choice. Most countries have regulations on the use of inhalants to ensure staff safety (see section on Operator Safety).

Onset, depth and recovery of anesthesia are dependent on the concentration of the inhalant in the brain and spinal cord. The speed of induction is related to the solubility of the inhalant in the blood. Isoflurane for example, is less soluble than halothane and produces a quicker induction. By increasing the concentration of a less-soluble agent, the speed of induction can be increased. With halothane, stable anesthesia is reached within 10 min in most laboratory species.[9] Likewise, recovery is slower with the more lipid-soluble agents. Halothane is stored in fatty tissue, and upon recovery, the storage will leave the fat and slow the return to consciousness.

Inhalants are mainly eliminated by the lungs. The degree of hepatic metabolism is highest for halothane (20%), followed by enflurane (2%), isoflurane (0.2%), and desflurane (0.02%). All inhalants depress cardiopulmonary function and renal blood flow in a dose-dependent manner.[26]

*Nitrous oxide* ($N_2O$) is poorly soluble in blood, oil, and fat; therefore, uptake and equilibration throughout the body is rapid. During induction with $N_2O$ and a volatile, the rate of uptake of the inhalant increases, which is especially valuable during mask induction. By adding 60 to 70% $N_2O$ to the inspired $O_2$, the MAC of halothane or isoflurane is reduced by approximately 20 to 25%, and consequently, cardiopulmonary function is less depressed.[9] The lowest $O_2$: $N_2O$ ratio should be 1:2, so that the $O_2$ concentration is at least 33%.[25] At the end of anesthesia, $N_2O$ delivery has to be discontinued first, and 100% oxygen given for 5 to 10 min to prevent the development of diffusion hypoxia. Nitrous oxide is most useful in animals with a body weight over 2 kg because in smaller animals volatile anesthetics equilibrate rapidly without $N_2O$.[9] Its use is not recommended in herbivores because it readily diffuses into gas-filled intestinal segments.

*Halothane* induces a pronounced dose-dependent depression of myocardial contractility and sensitizes the heart for the arrhythmogenic effects of [nor]adrenaline.[9] Cardiac output and total peripheral resistance are decreased, and cerebral blood flow is increased. At deep anesthetic levels, ventilation becomes inadequate, and blood pressure falls. Hepatic metabolism of halothane during hypoxia can result in the formation of radicals, which give rise to hepatotoxicity, by reaction with cell membrane proteins.[25] Radical formation is less likely with isoflurane or desflurane, due to their low degrees of metabolism. Because halothane is degraded to toxic end products by sunlight, it is contained in dark bottles containing a preservative (thymol).

*Isoflurane* maintains overall cardiovascular functions slightly better than halothane but is a stronger respiratory depressant and has greater vasodilatory effects.[9] It is a more potent coronary vasodilator than halothane or enflurane, and does not sensitize the heart to cathecolamines. In all, the safety margin is greater than with halothane. The insignificant amount of metabolism reduces the potential for hepatic and renal injury. Induction and recovery are faster than with halothane. Because deep anesthesia is rapidly achieved, care must be taken not to overdose the animal.

The solubility in blood is even lower for *sevoflurane* than for isoflurane, which results in faster induction and recovery. Cardiovascular and respiratory effects are similar to those of isoflurane.

The properties of *desflurane* are equal to those of sevoflurane, with the exception that the very low boiling point of desflurane makes the use of a heated vaporizer necessary for controlled delivery. Toxicity is very low.

*Enflurane* has the potential of causing nephrotoxicity, and in dogs, it may induce seizures.[9]

## Hypnotics

By definition, hypnotics are drugs that induce sleep and they have no intrinsic analgesic properties.[9] Low doses are sedative, whereas higher doses can produce a light plane of anesthesia. They are used for immobilization, minor superficial surgery (in conjunction with analgesic treatment), or induction of anesthesia. They can be administered IV by injection or infusion, or IP (in rodents). Barbiturates, propofol, and chloral hydrate belong to this group.

Barbiturates enhance GABA-mediated inhibition of synaptic transmission and can be classified as ultrashort-acting (thiopental, methohexital), short-acting (pentobarbital), or long-acting (Inactin). They are poor analgesics. Sleep time is affected by sex, age, strain, time of day, room temperature, and nutritional status, as well as other factors. Administration of glucose or adrenergic agents to animals recovering from barbiturate anesthesia may result in reanesthetization.[9] Barbiturates decrease cerebral metabolism and may induce development of tolerance, due to induction of hepatic enzymes.

*Thiopental and methohexital* are ultrashort acting barbiturates that are highly lipid soluble and rapidly cross the blood–brain barrier, inducing loss of consciousness in 10 to 20 sec upon IV injection.[27] Due to redistribution from the CNS to other tissues, duration of action is only 5 to 30 min. Recovery is prolonged with repeated doses of thiopental due to storage in body fat and slow hepatic clearance. The rate of metabolism is higher for methohexital, recovery from anesthesia faster and with less hangover effects, even after repeated boluses. Both agents are used for short procedures or to enable endotracheal intubation followed by inhalation anesthesia. Side effects include hypotension and respiratory depression, whereby apnea may follow induction and necessitate mechanical ventilation. Thiopental is available in a 1 to 2.5% solution and is highly irritating if not injected intravenously. Methohexital is less irritating.

*Pentobarbital* is rapidly absorbed from most sites and metabolized by the liver cytochrome P450-enzyme system. Metabolism is slow, with a half-life of 38 min in the mouse and 200 min in the dog.[9] The 6% alkaline solution is irritating and should be diluted with water for injection, except when administered IV. Pentobarbital has been shown to produce inadequate and inconsistent anesthesia in several species. It is a respiratory and cardiovascular depressant, potentially inducing hypoxemia, hypercapnia, and prolonged hypotension. Side effects include reduced blood pressure, stroke volume, pulse pressure, and central venous pressure. It impairs myocardial contractility, has a narrow margin of safety, and is especially hazardous to use in rabbits.[9] There are no antagonists available, and recovery is slow.

*Propofol* is an alkylphenol that is formulated as an emulsion in soybean oil and glycerol. It produces a dose-dependent level of sleep by enhancement of the GABA-receptor function. The distribution half-life is 1 to 6 min, and the elimination half-life is 16 to 55 min.[9] Upon IV injection, consciousness is lost in 30 to 60 sec, with a duration of 5 to 10 min. IV induction often results in respiratory depression and a fall in blood pressure. Apnea can be avoided by slow injection. Intramuscular injection results in sedation only.

Due to propofol's minimal accumulation in body tissues and rapid metabolism in the liver, it is useful for long-term light anesthesia by continuous infusion. Recovery is rapid following discontinuation of infusion. At antinociceptive doses, hypotension and a decrease in arterial blood pressure and heart rate are produced. Propofol decreases cerebral blood flow and cerebral oxygen consumption, and reduces intracranial pressure.[27] In rabbits, respiratory arrest may follow high doses.[28] By preadministration with an alpha-2-agonist, the propofol dose may be reduced by approximately 80%, and physiological variables can be well maintained.[29] In swine, propofol decreases myocardial contractility.

*Tribromoethanol* is a commonly used anesthetic in the production of genetically altered mice. After IP injection, surgical anesthesia is rapidly produced and maintained for 10 to 35 min.[9] Respiration and cardiovascular functions are depressed, and high mortality has been reported after repeated anesthesia with tribromoethanol. Metabolism occurs in the liver by conjugation with gluconoronic acid. Decomposition of the drug with formation of lethal products may follow improper storage.

*Chloral hydrate* is reduced to the active metabolite trichloroethanol in the liver. Sedative doses have a depressent effect on the cerebrum, with minimal effects on medullary centers and cardiorespiratory functions, whereas anesthetic doses are severely depressive.[9] Motor and sensory nerves are not affected, and the analgesic properties appear minimal. Intraperitoneal injections of high concentrations in rats may cause adynamic ileus, and large repeated doses may cause hepatic and renal damage. Intravenous administration in the dog sensitizes the heart to sudden vagal arrest or arrhythmias.

Like chloral hydrate, *alpha-chloralose* is metabolized to trichloroethanol in the liver. The hypnotic effect is of long duration, 8 to 10 h in the rat.[9] Low doses have minimal effects on reflexes and cardiorespiratory function. There are no controlled studies of the anesthetic properties, and analgesia seems to be poor. Induction is rough, and seizure-like activity can be seen in dogs and cats. After IP injection in rats and guinea pigs, a severe inflammatory reaction may occur.[30] It is only useful for nonsurvival procedures when chemical restraint with minimal cardiac and respiratory depression are needed.

*Urethane* (ethyl carbamate) produces 8 to 10 h of anesthesia after IV injection and has a wide margin of safety. Intraperitoneal injection in rats results in peritoneal effusion due to toxic effects on the mesenteric vasculature.[9] During anesthesia, cardiovascular and respiratory depression is minimal, spinal reflexes are well maintained, and effects on neurotransmission are minimal, which is why urethane is frequently used in neurophysiological studies.[31] The analgesic properties are sufficient to permit surgery in small rodents. Urethane is metabolized to carbon dioxide, ammonium, and ethanol, and has carcinogenic and mutagenic properties.

*Metomidate and etomidate* are imidazole derivates that have GABA-mimetic effects and are useful for long-term anesthesia because of their minimal cumulative effect and good preservation of cardiovascular function.[9] Metomidate has strong central muscle relaxing effects. Etomidate decreases intracranial pressure and brain oxygen consumption and is therefore beneficial for neurosurgery procedures. It is not recommended in dogs, because of the risk of serious endocrine effects, leading to Addisonian crisis.[23] For surgical anesthesia, additional analgesia is required. In mice, duration and depth of metomidate anesthesia is improved by additional SC administration of fentanyl.[32]

## Alpha-2-Agonists

Alpha-2-adrenoceptors are found in the CNS, gastrointestinal tract, uterus, kidney, and on platelets. Their activation by alpha-2-agonistic drugs results in sedation, analgesia, muscle relaxation, and anxiolysis, by inhibition of presynaptic calcium influx and neurotransmitter release. Side effects are initial hypertension due to peripheral effects, followed by slight hypotension, bradycardia, and decreased cardiac output. Intravenous fluids may be needed to counteract hypotension. If anticholinergic agents are given before the alpha-2-agonist, they may initially prevent bradycardia, but there is also the risk that the initial hypertension is potentiated.

Alpha-2-agonists depress the release of insulin, leading to elevated blood glucose levels.[23] They also decrease the release of antidiuretic hormone and have a direct renal tubular effect, leading to marked diuresis. Hypothermia is another common feature seen, and in cats, vomiting occurs, usually within 15 min of administration. To prevent aspiration of stomach contents, induction of anesthesia should be delayed until after vomiting. Requirements of intravenous and volatile anesthetics are reduced by as much as 80% after premedication with an alpha-2-agonist. The alpha-2-antagonist atipamezole reverses the effects of all alpha-2-agonists within 5 to 10 min after IM or SC injection. Intravenous administration is not recommended.

*Xylazine* is a mixed alpha-2/alpha-1-agonist, with an onset of action 3 to 5 min after IV injection and 10 to 15 min after SC or IM injection.[23] Analgesia lasts for 15 to 30 min and sedation for 1 to 2 h, whereas complete recovery may take 2 to 4 h. In most species, xylazine alone does not produce sleep or loss of the righting reflex. It should be avoided in pregnant animals due to an increase in uterus tone. Apart from atipamezole, yohimbine may be used to reverse the effects of xylazine.

*Medetomidine* is highly selective for the alpha-2-adrenoceptor, more potent than xylazine and faster eliminated.[9] It can be administered SC, IM, and IV. In the rat, peak brain levels are reached in 15 to 20 min after subcutaneous administration, and the elimination half-life is 1.6 h. Deep sedation and loss of righting reflex is produced in dogs, cats, and rats, but not in mice and rabbits (Figure 17.3).

Medetomidine has antinociceptive actions, but they are weak in rabbits, guinea pigs, and hamsters. The addition of an opioid or a benzodiazepine will enhance sedation and analgesia. The combination with ketamine for general anesthesia, offsets the bradycardia caused by medetomidine and induces a state of pronounced hypertension throughout anesthesia.[23]

## Dissociative Anesthetics

This group of lipophilic cyclohexamines has antagonistic action at excitatory *N*-methyl-D-aspartate (NMDA) receptors, and produces a "dissociative state" by depression of cortical associative areas. By stimulation of the sympathetic nervous system, blood pressure is maintained at above normal levels, despite direct myocardial depression. This is potentially useful in trauma models and in animals with hypovolemia or circulatory depression. Bronchodilation is also produced, and muscle tonus is increased. The degree of analgesia induced varies dose-dependently, and metabolism occurs in the liver.

**Figure 17.3** Cat sedated with medetomidine.

*Ketamine* is one of the most widely used anesthetics due to its wide safety margin and compatibility with other drugs. It is a racemic mixture, with L-ketamine being the likely cause of the psychomimetic side effects that occur in humans, presented as hallucinations and nightmares. Ketamine can cause tonic-clonic movements and psychomimetic emergence reactions in nonhuman primates, when used alone.[33] It can induce seizures and hallucinatory behavior in cats and dogs during emergence from anesthesia. This may be avoided by premedication with a benzodiazepine or an alpha-2-agonist.

Ketamine increases hemodynamic variables and produces good analgesia except in rodents and rabbits. Surgical anesthesia can be attained by combination with an alpha-2-agonist in rabbits and rodents, or with a benzodiazepine in rabbits. Absorption of ketamine is rapid after IV, IM, SC, and IP injection. Rectal administration has been described in cats.[4] Induction and recovery is rapid with IV administration. Ketamine solution has a pH between 3.3 and 5.5 and has been reported to cause tissue necrosis in rodents and rabbits, when administered with xylazine IM.[34,35]

Salivation and lacrimation are increased during ketamine anesthesia, with persisting laryngeal and swallowing reflexes. Metabolism by cytochrome P450 leads to formation of the hypnotic metabolite norketamine, which is responsible for the prolonged recovery and drowsiness seen after large ketamine doses.[23] Tolerance development has been demonstrated in rats.[9]

*Tiletamine* is closely related to ketamine, only longer lasting, more potent, and with more pronounced side effects. It is marketed for use in combination with the benzodiazepine zolazepam (Telazol® or Zoletil®) and may be used for surgery in cats, dogs, ferrets, and rats. Nephrotoxicity has been reported in rabbits.[9]

## Muscle Relaxants

Muscle relaxants act by blocking acetylcholine receptors on the motor end plates, in a depolarizing or nondepolarizing fashion, without affecting the CNS. Their structure is similar to acetylcholine.

The use of muscle relaxants must be limited to procedures that necessitate a fully relaxed animal during general anesthesia. The degree of blockade should be kept as low as the procedure will allow. Because muscle relaxants prevent the animal from moving or showing withdrawal responses consciously or unconsciously during noxious stimulation, great care must be taken with their use. To ensure that the anesthetic regimen is appropriate for the specific procedure, species, or strain, it should be performed without the muscle relaxant first in some animals.[36] Additionally, an adequate depth of anesthesia must be established before the muscle relaxant is added. It is not acceptable to start an infusion with a muscle relaxant and surgery, before a stable plane of surgical anesthesia has been achieved, like described in some studies.[37,38] Finally, the muscle relaxant should be allowed to wear off intermittently in order to ensure adequate depth of anesthesia throughout the procedure.

Signs of insufficient anesthesia during the use of muscular relaxants are increase of pulse rate unrelated to hemorrhage or slowing of the heart (bradycardia), increase or drop in blood pressure, pallor of the mucous membranes, pupil dilation, tear formation (lacrimation), increased salivation and twitching of the tongue and facial muscles.[39] In research protocols including hemodynamic instability or cardiopulmonary bypass, blood pressure and heart rate cannot be used to ensure adequate depth of anesthesia, and the use of muscle relaxants should therefore be avoided.

Intubation and positive pressure ventilation must be provided when using muscle relaxants, as well as a means of monitoring relaxation (response to nerve stimulus). As well as potentiating the effect of muscle relaxants, inhalation anesthesia produces muscle paralysis by a direct action on the acetylcholine receptor. In humans, this is seen increasingly with concentrations over 1 MAC and after 30 min of anesthesia (saturation effect).[40]

*Succinylcholine* is the only depolarizing paralytic in use. Duration of this hydrophilic agent is short (5 to 8 min), due to quick hydrolysis by pseudocholineesterase in the plasma and liver. Because of its depolarizing effect, initial muscle fasciculations are seen. Side effects include cardiac arrhythmia, histamine release, and muscular pain upon recovery. Succinylcholine is a potent initiator of malignant hyperthermia in pigs, dogs, and cats with a genetic predisposition, especially in combination with halothane.[36]

*Pancuronium* is a nondepolarizing muscle relaxant with a long duration of action (30 min) that blocks peripheral uptake of noradrenaline and, thus, causes hypertension and tachycardia. Part of the drug is excreted with the urine (70%), and part is metabolized in the liver (30%).[39] Its effects can be reversed with neostigmine.

*Vecuronium* is similar to pancuronium but with a shorter duration of action and no hemodynamic effects. It is eliminated with the bile.

## ANESTHETIC MANAGEMENT

### Preanesthetic Considerations

Proper preparation of the anesthetic protocol begins with the evaluation of the animal to be anesthetized. Aspects such as overall condition, underlying pathology or pre-existing instrumentation may be of influence in determining the best possible anesthetic protocol for the animal and for the experiment. Consideration with respect to the experimental protocol is relevant not only concerning immediate aspects of anesthesia and instrumentation, but also relates to late effects of anesthesia on relevant physiologic functions such as neuroendocrine balance. Proper preanesthetic assessment of the experimental animal and in-depth evaluation of the experimental protocol, with regard to possible interference from the anesthetic protocol, provide the basis for a successful experiment.

Depending on the animal species and character and the amount of handling needed before anesthesia, preanesthetic sedation may be considered. Preemptive analgesia should also be considered depending on the level of invasiveness before, during, and after anesthesia. Next to the pharmacological considerations, general supportive measures such as aspects of housing and handling, temperature control, and fluid balance, may exert an important influence on the overall well being of the animal. Together with the design and execution of the anesthetic protocol, these factors will profoundly influence the quality of recovery and thus morbidity and mortality.

### Monitoring during Anesthesia

During any anesthesia, some form of monitoring must take place. In short and relatively low-risk procedures, it can be limited to simple, nonelectronic monitoring of pulse rate, respiration rate, body temperature, and arterial oxygenation by observing the color of the skin or mucous membranes.

A simple method for obtaining relevant but only semiquantitative information on the functioning of the circulatory and the respiratory system involves the use of a pulse oxymeter (Figure 17.4). It provides information on the arterial pulse frequency as well as the level of arterial oxygenation, and thus basal information on the circulatory status and the efficiency of gas exchange. Practical applicability is limited during conditions of peripheral vasoconstriction, when the reduced signal strength prohibits proper functioning of the apparatus. Also, during oxygen supplementation, a situation of inadequate ventilation with concurrent high $CO_2$ partial pressures may exist, while arterial oxygenation is still normal or even above normal.

For more invasive procedures of long duration, involving potentially high-risk interventions, more sophisticated monitoring is indicated. Electronic monitoring of heart rate and rhythm by way of a

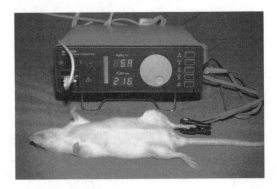

**Figure 17.4** Rat monitored with pulse oxymeter.

continuous electrocardiogram provides essential information about the electrical activity of the heart but not on the mechanical performance. For more detailed information on cardiac function, there are different hemodynamic monitoring options available. Direct arterial blood pressure measurement at different sites of the body provides such data. However, the more detailed the hemodynamic picture needs to be, the more sophisticated, expensive, and complicated the necessary apparatus will get. In principle, it can be stated that although modern hemodynamic monitoring techniques available in human medicine can be applied in animals as well, the size of the most frequently used animal species, such as rodents and rabbits, limits their application.

Respiratory monitoring in its simplest form consists of respiration rate and rhythm. More functional information is derived from parameters such as tidal volume (expiration volume) and the percentage of end-tidal (end expiratory) $CO_2$ and $O_2$. Information specifically relevant to anesthesia further includes the concentration of end-tidal $N_2O$ and the volatile agent. For all these parameters, respiratory monitors are available, but like with hemodynamic monitoring, the small size of most laboratory animals makes their application difficult. Specialized equipment for use in small animal species is increasingly becoming available on the market.

Further to monitoring the different essential organ functions, the overall quality of anesthesia must be evaluated during the procedure. Under all circumstances, assurances must be given to the adequacy of sleep (hypnosis) and pain control (analgesia). Traditionally, these aspects of anesthesia have been evaluated by observing the trends in respiration (respiration rate, breathing pattern), circulation (heart rate and rhythm, arterial and venous blood pressures, and other hemodynamic variables), and autonomic function and reflexes such as eyelid and corneal reflexes, position and movement of the eyeball (depending on species), swallowing reflex, and withdrawal reflexes (Figures 17.5 and 17.6).

More recently, newer techniques for evaluating specific elements of anesthetic quality have become available. These include several neurophysiological techniques such as the evaluation of the raw EEG, as well as methods for specifically assessing sleep (auditory-evoked responses) or analgesia (somatosen-

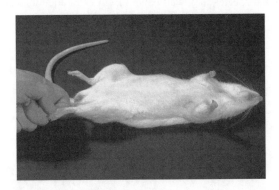

**Figure 17.5** Rat, testing pedal withdrawal response.

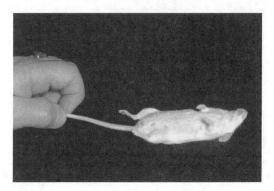

**Figure 17.6**   Mouse, assessing tail pinch reflex.

sory-evoked responses). Although a lot of work as well as progress are being made in this field of research, the clinical applicability is still limited. Further work is needed before these methods can be easily applied to produce effective and reliable information.

## Postanesthetic Care

First and most importantly when it comes to postanesthetic care, is the consideration of postintervention pain control. For this goal, opioids, NSAIDs, local analgesics, or combinations, can be employed. More detailed information is provided in relevant sections of this chapter.

Under certain circumstances, such as postanesthesia restlessness or automutilation, sedation may be indicated. This is more commonly seen in large animal species than in rodents and rabbits. Side effects of sedation, such as an increased risk of delayed recovery and hypothermia, are also potentially hazardous in small animals.

Related to the risk of hypothermia is the need for good temperature control. This starts with proper monitoring of the body temperature until the animal is fully recovered. Supporting measures, such as warm water heating blankets, infrared heating lamps, and incubators for recovery can dramatically reduce the incidence of life-threatening hypothermia. Installation of these measures does not, however, eliminate the need for regular measurement of body temperature.

For all animal species, proper housing conditions and easy access to food and water are easily realized aspects. For rodents, this may be accomplished by assuring that food is available on the floor of the cage and by using water bottles with extra long drinking nipples.

Similarly, fluid balance should be monitored during and following anesthesia, until the animal is fully recovered. Fluid therapy before and during anesthesia can greatly improve the fluid status in the postoperative period. Last, but not least, general supportive care will improve the quality and speed of the recovery period. Which measures should be part of the supportive regime depends on the circumstances and the animal species involved (Figures 17.7 and 17.8). Rodents and rabbits are often best left alone, whereas larger animal species may well benefit from extra attention and "nursing care."

**Figure 17.7**   Rats in incubator recovering after surgery.

**Figure 17.8**  Rat provided with soft bedding and soaked diet after surgery.

## PAIN MANAGEMENT

### Acute Pain

Recognition of acute pain in animals is commonly facilitated by the fact that the behavioral response is related in time to the actual tissue damage. Typical response characteristics include avoidance behavior, vocalization, or defensive behavior. Specific response patterns vary greatly among species, and for a proper evaluation of the presence and extent of the pain, an in-depth knowledge of the normal behavior of the specific species is needed.

In the acute phase of initial tissue trauma, the accompanying pain can be considered to serve a physiological function of alarming the body of the possible deleterious effects of the trauma, as well as have protective qualities, limiting the risk of further tissue damage. Related to the physiological function of acute pain, it has been suggested that its perception is proportional to the intensity of the noxious stimulus. Under more chronic circumstances, and following related neural system changes involved in recognition and processing of pain signals, the pain becomes increasingly pathological. The perception of pain no longer relates to the initial pain intensity, but is by far greater, lasts longer, and has spread over an increased area.

In the acute phase, the nociceptive stimulus is transported from the peripheral nociceptor to the spinal cord by way of afferent sensory fibers of the A delta and C class. The A delta fibers are myelinated and small in diameter and transport signals at high speed. These fibers are responsible for the immediate, sharp, and often intense pain and, through direct connection with motor neurons at the dorsal horn level, induce an instantaneous motor response, i.e., a withdrawal reflex. The activation of C fibers is held responsible for the longer lasting, dull pain that follows the initial sharp pain sensation. Relevant physiological aspects of acute pain include the nocifensive behavioral response and autonomic responses such as increase in heart rate and blood pressure, a change in respiratory rate and rhythm, and endocrine changes. The autonomic responses are traditionally used as pain indicators but are unreliable as such.

The unreliability of these parameters for detecting pain and differentiating it from other potential "negative" influences lies in the limited discriminative power of individual parameters. Only under circumstances where these parameters do not show any relevant changes can the absence of pain be postulated. The above-mentioned parameters can at best be viewed as "circumstantial evidence" of pain being present.

### General Preventative and Therapeutic Strategies in Acute Pain

Effective analgesic strategies for acute pain are primarily focused on preventing pain from occurring, rather than approaching pain once it has been established. The main advantage of preemptive analgesia is that sensitization has not yet developed. Early intervention prevents the development of the inflammatory cascade, makes pain treatment more effective, both in the short term (efficacy) and in the long term (necessary duration). Consideration should be given to the different modalities of analgesic therapy, including local analgesia, systemic use of opioid or nonsteroidal anti-inflammatory drugs, and the alpha-2-adrenergic drugs used for sedation and general anesthesia.

The choice of one (single therapy) or multiple (polymodal therapy) analgesic drugs should be determined on the basis of species- and model-dependent characteristics as well as on the underlying pathology. Each class of analgesic drugs has its own specific characteristics, and efficacy, dependent on the circumstances (see section on Analgesic Drugs).

## Chronic Pain

The pathophysiological background of chronic pain not only involves temporal aspects (i.e., long duration) of this class of pain, but also functional changes in the neurophysiological pain system. The functional changes result in an increased intensity of nociceptive stimulus perception, as well as an enlargement of the sensitive area. In the present day, the understanding of the consequences of continued nociceptive activation of the peripheral and central nervous system (i.e., pain sensitization) plays a crucial role.

Following initial nociceptor stimulation and recognition of the painful character of stimulation, the repetition of stimulation initiates a process of production and release of inflammatory mediators at the site of injury, i.e., the stimulation site.[41] Such prolonged nociceptor stimulation may be the consequence of continued tissue trauma, of long duration sympathetic stimulation, or the presence of the inflammatory response.

Vasoactive amines released from the damaged tissue, and inflammatory cells and neuropeptides released from the injured nerve endings in the traumatized area are collectively responsible for an increased sensitivity of peripheral nociceptors. The peripheral sensitization results in an increased responsiveness of these nociceptors (hyperalgesia), and also in nonnoxious stimuli to be perceived as painful (allodynia).

Next to peripheral sensitization, continued nociceptor input into the dorsal horn of the spinal cord changes the responsiveness of the dorsal horn neurons, thereby increasing the responsiveness to incoming stimuli. Consequently, low-intensity stimulation and even activation of nervous system structures normally involved in tactile sense (Aß receptors and fibers) now lead to a potentially intense, painful sensation.[41] The overall result of chronic pain is that, through peripheral and central sensitization, low-intensity stimulation of a body area previously involved in painful stimulation now produces a response that is more intense, of longer duration, and with the painful sensation spreading over a larger area.

The consequence is obvious for the clinical effectiveness of analgesic therapy. Once sensitization has set in, effective pain treatment will be harder to achieve and require a longer duration.[42]

## Analgesic Modalities for Treating Acute or Chronic Pain

For specific details (dosage, interval, and route of administration) of the different analgesic treatment modalities, the reader is referred to the relevant sections of this chapter.

The primary aspect to consider in the treatment of acute pain involves timely administration, i.e., preferably before the pain occurs. This is an option in a properly designed anesthetic protocol for performing a surgical or investigational procedure. In order to prevent pain from occurring before, during, or after surgery, it is of utmost importance to initiate treatment early and to continue treatment during and after the intervention at correct (pharmacologically determined) intervals. The aspect of the choice of drug (class) is determined by the pathophysiological background of the pain, allowing a primary choice between opioid analgesics and NSAID therapy.[43] Furthermore, specific treatment modalities such as local analgesic supplementation may be applied under specific circumstances.

For the alleviation of immediate postsurgical pain, most evaluations and comparative studies show opioids to be most effective during the first 24 to 48 h after surgery. Following this initial period, the inflammatory component of pain becomes more prominent, and NSAID therapy can be applied with good success. In circumstances where the pathophysiological background of the pain is more complex, or when single-mode therapy has proven to be inadequate, multimodal therapy with a combination of an opioid and a NSAID is indicated. Additional local analgesic supplementation might further support the reduction of pain and stress in the animal.

## OPERATOR SAFETY

Safety aspects related to anesthesia concern exposure to anesthetic gases, handling of gas cylinders, and possible misuse of drugs.

### Anesthetic Gases

Chronic exposure to trace anesthetic gases has been linked to hepatic and renal disease, abortion, infertility, birth defects, and neoplasia in humans.[25] Nitrous oxide can inhibit bone marrow function and cause abortion. Several countries have legislation related to the use of anesthetic gases. In the United Kingdom, the Control of Substances Hazardous to Health regulate the control of anesthetic gas pollution. Regulations include risk assessment, prevention and control of exposure, installation and maintenance of control measures, monitoring of exposure, and health surveillance. Information and training of staff is also included, as are exposure limits.

In the United States, safety standards are issued by the National Institute for Occupational Safety and Health.[9] Set exposure limits for anesthetic gases and recommendations on how to limit exposure are included. Low fresh gas flows and scavenging systems minimize exposure to trace gases. By using double-mask systems connected to the exhaust system and ventilated benches and hoods, exposure is further reduced. Old-type vaporizers and syringes should only be filled with volatiles in ventilated hoods. If exposure cannot be avoided, the use of safer anesthetics, like isoflurane, is recommended before less safe, like halothane. Activated charcoal systems have a limited use and will absorb halogenated volatiles but not nitrous oxide.

### Gas Cylinders

Gas cylinders are filled to high pressures, and explosions are possible if they are mechanically damaged by falling over or being dropped or being exposed to heat.[44] Cylinders must not be stored near sources of heat or combustible material. Oxygen cylinder valves and associated equipment must not be lubricated and must be free from carbon-based oils and greases. The combination of these with high-pressure oxygen may result in explosion. Smoking or naked lights must not be allowed within the vicinity of a cylinder or pipeline outlet.

### Controlled Drugs

Many of the opioid drugs are subject to control under national legislation. In the United Kingdom, for example, all pure opioid agonists must be kept in locked cupboards and records kept of purchase and usage.[3] The drugs are subject to special prescription requirements and legislation on destruction of unwanted stock. Safekeeping requirements also apply to buprenorphine. Butorphanol is subjected to controlled legislation in the United States but not in the United Kingdom.

Other drugs with a potential for misuse are ketamine, benzodiazepines, and barbiturates. Careful storage and record of usage is recommended for these drugs as well.

## ANALGESIA AND ANESTHESIA OF DIFFERENT SPECIES

### Analgesia and Anesthesia in Rodents

#### Analgesia

Like other species, rodents in painful conditions benefit from analgesic treatment. Postoperative pain will cause inappetence and prolong recovery in rats.[45] Recent research focuses on postoperative recovery and recognition of pain-related behavior.[46] Studies on the effects of perioperative opioid or NSAID treatment are increasing.[46–49]

*Morphine* is a potent analgesic of short duration (2 to 4 h) in the rat, causing sedation and respiratory depression.[8] Apart from central effects, morphine appears to exert a peripheral effect. In a mouse model of visceral pain, a low dose of morphine provided efficient analgesia after IP but not IV administration.[50] *Butorphanol* is 30 times more potent than morphine in rats. It can be used to treat mild to moderate pain and has a duration of action lasting 1 to 4 h.[51] By combination with a NSAID, more potent analgesia of longer duration is provided.

*Buprenorphine* is about 25 times more potent than morphine but has less maximum analgesic effect, which is reached 30 min after injection. Duration of action is between 3 and 12 h, depending on species and test used.[8,51] Clinical experience suggests that readministration is needed every 8 to 12 h. Low-dose buprenorphine has a beneficial effect on the postoperative recovery in rats.[47,48,52] Systemic buprenorphine has been shown to be superior to local infiltration with bupivacaine, in terms of postoperative recovery following laparatomy in rats.[53] Reported side effects from buprenorphine in rats include ingestion of bedding and gastric distension and obstruction, seemingly related to dose and strain.[54,55]

Buprenorphine can be administered SC or orally. Higher doses are needed for oral administration, due to first pass hepatic metabolism. Rats can be trained to eat fruit or meat-flavored jelly cubes (1 cm³ per 200 g rat), containing buprenorphine for postoperative oral treatment (Figure 17.9).[6,48] Preadministration of buprenorphine reduces the need of isoflurane for surgery by approximately 20%.[56] When using injection anesthesia, it is safer to administer buprenorphine at the end of the procedure, when anesthesia reaches a light plane. After extensive surgery, buprenorphine may be combined with a NSAID as well as local anesthesia. Buprenorphine may also be administered to reverse the effect of fentanyl, after the use of Hypnorm + midazolam for surgery, while still providing postoperative analgesia.

*NSAIDs* are indicated after minor to intermediate surgery. Premedication with carprofen or ketoprofen before abdominal surgery in rats will reduce pain-related behavior for 4 to 5 h, and speed recovery.[46,53] Clinical experience suggests that analgesia may last longer. Ibuprofen treatment in mice results in a more rapid return of activity and water intake after abdominal surgery.[49] There are no reports on adverse reactions with the use of the modern NSAIDs, but prolonged use should be avoided.[6] The initial dose can safely be repeated after 24 h. Carprofen is one NSAID that is proven safe to use preoperatively, while other NSAIDs can cause renal toxicity if hypotension occurs during anesthesia.

### Local Anesthetics

In most species, maximum safe doses are 4 mg/kg of lidocaine and 1 to 2 mg/kg of bupivacaine.[6] The duration of action of bupivacaine seems shorter in rats than in larger species. Reports of toxicity in rodents relate dose rates similar to those toxic in dogs and cats. Overdosing is more easily accomplished in small animals and can be avoided by accurately calculating the dose rate. Doses can be diluted to provide a more appropriate volume. General toxic dose rates are 10 to 20 mg/kg of lidocaine and 4 mg/kg of bupivacaine IV. In adult rats, the LD50 of bupivacaine is 30 mg/kg.[21] Addition of adrenaline should be avoided in ring blocks of appendages. Immersion of the mouse-tail in a dimethyl sulfoxide (DMSO) solution with lidocaine for 2 min provides effective analgesia in the heat tail-flick test.[57]

**Figure 17.9**   Rat eating buprenorphine jelly.

## Antidepressants

Tricyclic antidepressants (TCAs) or serotonin reuptake inhibitors (SSRIs) can be considered for treatment of pain when opioids and NSAIDs are contraindicated. The analgesic effect of TCAs and SRRIs is due to a blockade of serotonergic reuptake. Amitriptyline and imipramine (TCAs) have been shown to decrease signs of pain perception in neuropathic pain models for up to two weeks in mice and rats, without development of tolerance or adverse effects.[8] The SSRI fluvoxamine shows an antinociceptive effect in the hot plate test in mice.[58]

## Anesthesia

### Premedication

The use of premedication is commonly excluded in rodent anesthesia. Sedatives and tranquilizers are usually administered together with injectable anesthetics. The use of atropine is seemingly rare but should be considered with anesthetics causing bradycardia (opioids) or excessive salivation (ketamine). There are a number of circumstances where premedication with analgesics or sedatives could be advantageous in rodent anesthesia. Sedatives may be used in combination with local anesthetics for minor superficial surgery or before induction with intravenous anesthetics. Preadministration with an alpha-2-agonist (e.g., medetomidine) before induction with propofol IV or inhalation anesthesia reduces stress and decreases the anesthetic dose needed by as much as 80% in other species,[23,29] and probably also in rodents. The alpha-2-agonistic effect can be reversed if sedation is unwanted postrecovery.

Fentanyl + fluanisone (Hypnorm), administered SC, will provide analgesia and sedation sufficient for minor surgery or, if followed by injection with a benzodiazepine, surgical anesthesia.

Premedication with buprenorphine has been shown to reduce the pentobarbital dose needed for provision of surgical anesthesia, but it does not improve respiration.[59] Buprenorphine administration before isoflurane anesthesia will reduce MAC by approximately 20%.[56] Buprenorphinebefore [6]ketamine + medetomidine anesthesia has proven dangerous in rats due to severe respiratory depression.[60] Among the NSAIDs, carprofen and ketoprofen are usually safe for preanesthetic administration.[4]

**Table 17.2 Dose Rates for Analgesics in Rodents**

| Drug | Mouse | Rat | Guinea Pig |
|------|-------|-----|-----------|
| Morphine | 2–5 mg/kg SC; 4 hourly | 2–5 mg/kg SC; 4 hourly | 2–5 mg/kg SC; 4 hourly |
| Butorphanol | 1–2 mg/kg SC; 4 hourly | 1–2mg/kg SC; 4 hourly | 2 mg/kg SC |
| Buprenorphine | 0.05–0.1 mg/kg SC; 8–12 hourly | 0.01–0.05 mg/kg SC or IV; 0.1–0.25 mg/kg by mouth; 8–12 hourly | 0.05 mg/kg SC; 8–12 hourly |
| Carprofen | 5 mg/kg SC or by mouth; 24 hourly | 5 mg/kg SC or by mouth; 24 hourly | — |
| Ketoprofen | | 5 mg/kg SC or by mouth; 24 hourly | — |
| Ibuprofen | 30 mg/kg by mouth; 24 hourly | 15 mg/kg by mouth; 24 hourly | — |
| Lidocaine | 4 mg/kg or 0.4 mL/kg of a 1% solution | | |
| Bupivacaine | 1–2 mg/kg or 0.4–0.8 mL/kg of a 0.25% solution | | |
| Amitriptyline | 1.2–5 mg/kg SC or IP; 3–12 hourly | 1–10 mg/kg SC or IP; 3–12 hourly | |
| Imipramine | 2.3 mg/kg SC or IP; 12–24 hourly | 10 mg/kg SC or IP; 12–24 hourly | |

*Source:* From Dobromylskyj, P., Flecknell, P.A., Lascelles, B.D., Pascoe, P.J., Taylor, P., and Waterman-Pearson, A., Management of postoperative and other acute pain, in Flecknell, P. and Waterman-Pearson, A., Eds., *Pain Management in Animals*, Flecknell, P. and Waterman-Pearson, A., Eds., W.B. Saunders, Philadelphia, PA, 2000; and Wixson, S.K., and Smiler, K.L., Anesthesia and analgesia in rodents, in *Anesthesia and Analgesia in Laboratory Animals*, Kohn, D.F., Wixson, S.K., White, W.J., and Benson, G.J., Eds., ACLAM and Academic Press, New York, 1997.

**Table 17.3 Drug Dose Rates for Premedication in Rodents**

| Premedication | Mouse | Rat | Guinea Pig |
|---|---|---|---|
| Atropine | 40 µg/kg SC or IM | 40 µg/kg SC or IM | 50 µg/kg SC or IM |
| Acepromazine | 2.5 mg/kg IM | 2.5 mg/kg IM | 2.5 mg/kg IM |
| Diazepam | 5 mg/kg IM or IP[a] | 2.5 mg/kg IM or IP[a] | 2.5 mg/kg IM or IP[a] |
| Midazolam | 5 mg/kg IM or IP[a] | 5 mg/kg IP | 5 mg/kg IM or IP[a] |
| Xylazine | 5–10 mg/kg IP | 1–5 mg/kg IM or IP[a] | 5–40 mg/kg IP |
| Medetomidine | 30–100 µg/kg SC | 30–100 µg/kg SC | — |
| Fentanyl + fluanisone (Hypnorm®) | 0.5 mL/kg IM or IP[a] | 0.5–1.0 mL/kg IM, SC or IP[a] | 0.5 mL/kg IM |
| Buprenorphine | 0.05–0.1 mg/kg SC | 0.01–0.05 mg/kg SC | 0.05 mg/kg SC |
| Carprofen | 5 mg/kg SC or by mouth | 5 mg/kg SC or by mouth | — |

[a] IM and SC administration of a drug are likely to give a greater effect compared to i.p injection, due to possible first-pass hepatic metabolism after IP injection.

*Sources:* From Wixson, S.K., and Smiler, K.L., Anesthesia and analgesia in rodents, in *Anesthesia and Analgesia in Laboratory Animals*, Kohn, D.F., Wixson, S.K., White, W.J., and Benson, G.J., Eds., ACLAM and Academic Press, New York, 1997; and Lukasik, V.M., Premedication and sedation, in Seymour, C. and Gleed, R., Eds., *Manual of Small Animal Anaesthesia and Analgesia*, Seymour, C. and Gleed, R., Eds., BSAVA, 1999.

## Inhalation Anesthesia

Most volatile agents are safe to use in rodents. Hepatotoxicity has been reported after low concentrations of halothane in guinea pigs.[8] Hemodynamic parameters are often more stable during inhalation anesthesia compared with injectable anesthesia,[61] and due to safety and stable physiological effects, inhalation anesthesia is recommended in mouse cardiovascular, cerebral, and noninvasive imaging studies.[62,63] Premedication with buprenorphine, carprofen, or medetomidine will enhance analgesia and reduce the volatile dose needed for surgical anesthesia. By using local anesthesia as an adjunct, the concentration of the volatile agent can be kept to a minimum, with less cardiorespiratory depression as a result.

Induction is best achieved in a plexiglas anesthetic chamber, equipped with a gas inlet and an outlet for exhaled and excess gases (Figure 17.10). With halothane or isoflurane, induction is complete after 2 to 3 min. Anesthesia can be maintained by gas delivery via a face mask (Figure 17.11) or an endotracheal tube. Successful intubation has been described in rats, guinea pigs, hamsters, and mice (Figure 17.12).[8,64,65] After 30 min of isoflurane or halothane anesthesia, animals will regain their righting reflexwithin 5 o 10 min.

A low fresh gas flow is beneficial for keeping costs down and protecting the environment. In an open breathing system, a flow of three times the animal's minute volume is necessary in order to meet fresh gas requirements. The minute ventilation in rats and mice is approximately 700 mL/kg/min,[26] resulting in a required fresh gas flow of approximately 0.06 L/min for a 30 g mouse and 0.6 L/min for a 300 g rat. Some vaporizers (e.g., Mark III tecs) deliver accurate tconcentrations of volatile anesthetics

**Figure 17.10** Rat in induction chamber: inlet for anesthetic gas (bottom) and outlet for scavenging (top).

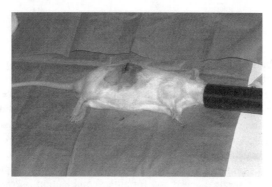

**Figure 17.11** Rat maintained on face mask.

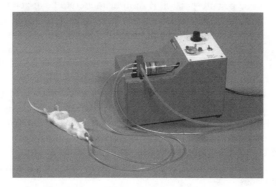

**Figure 17.12** Rat on Harvard ventilator.

at very low flows (0.2 L/min).[44] Another type of rodent anesthetic machine, managing even lower gas flows (0.05 L/min), works by injection of the volatile anesthetic into the fresh gas stream.[66]

Oxygen may be used as sole carrier gas, except in very long procedures. If near physiological blood oxygen levels are needed, air or nitrous oxide can be mixed with oxygen in a 2:1 relation. In order to prevent the development of hypoxia, the inspired gas should always consist of 33% oxygen as a minimum. Arterial $CO_2$ levels will be elevated during anesthesia, unless mechanical ventilation is used.

For brief procedures in mice, anesthesia may be induced with a volatile agent in a glass jar. Under standard temperature and pressure conditions, 1 mL of liquid isoflurane will produce 182 mL of gas.[9] In a 500 mL jar, 0.11 mL of liquid isoflurane will quickly evaporate to reach a concentration of 4%. A mouse placed in the jar will be anesthetized within a few minutes and wake up 30 to 60 sec after removal from the jar. The animal must never come in contact with the liquid anesthetic, or local irritation will occur.[8] If several mice are anesthetized after each other in the jar, care must be taken to prevent a shortage of oxygen. For protection of staff, this method should only be used in a ventilated hood.

**Table 17.4 Induction and Maintenance Concentrations for Volatile Anesthetics in Rodents**

| Volatile Agent | Induction Concentration | Maintenance Concentration |
| --- | --- | --- |
| Isoflurane | 2–4% | 5–3% |
| Sevoflurane | 4–8% | 3–4% |
| Halothane | 2–4% | 1.5–2% |

*Sources:* From Wixson, S.K. and Smiler, K.L., Anesthesia and analgesia in rodents, in *Anesthesia and Analgesia in Laboratory Animals*, Kohn, D.F., Wixson, S.K., White, W.J., and Benson, G.J., Eds., ACLAM and Academic Press, New York, 1997; and Brunson, D.B., Pharmacology of inhalation anesthetics, in *Anesthesia and Analgesia in Laboratory Animals*, Kohn, D.F., Wixson, S.K., White, W.J., and Benson, G.J., Eds., ACLAM and Academic Press, New York, 1997.

## *Injection Anesthesia*

Because most injectable anesthetics depress ventilation, oxygen supplementation is recommended during injection anesthesia. Intubation and mechanical ventilation may be necessary during long-term anesthesia. First choice combinations for surgical procedures, from the aspect of safety and efficiency, are Hypnorm + midazolam, or ketamine + medetomidine. Each combination can be mixed and administered with one injection IP or SC. The recommended IP dose should be reduced by 30 to 50% for SC administration because no immediate hepatic metabolism occurs after SC injection.

Hypnorm provides sedation and analgesia sufficient for minor superficial procedures, although muscle relaxation is poor. By combination with midazolam or diazepam, surgical anesthesia with good muscle relaxation is produced. Unlike midazolam, diazepam is not compatible with Hypnorm and must be administered separately. Duration of surgical anesthesia for either combination is 20 to 40 min. Additional SC doses of Hypnorm may be given to prolong anesthesia for several hours. The effects of fentanyl can be antagonized with buprenorphine or butorphanol, if mild or moderate postoperative pain is expected.

Ketamine produces immobility but poor analgesia. In combination with medetomidine or xylazine, surgical anesthesia is consistently attained, except in guinea pigs.[8] Duration of surgical anesthesia is 15 to 30 min in rats, and up to 80 min in mice, depending on dose and strain. Side effects are less severe than during pentobarbital anesthesia (see below). If anesthesia proves insufficient, it is best improved by adding a low concentration of inhalation anesthesia. Medetomidine and xylazine can be reversed with atipamezole, and if surgery has been undertaken, an analgesic should be provided before reversal. Long-term anesthesia (12 h) with ketamine and xylazine IV infusion has been described in rats.[67]

Pentobarbital is best used to provide light anesthesia only, due to its narrow safety margin. Dilution with saline is recommended before IP injection to prevent peritoneal irritation. Severe respiratory depression and moderate circulatory depression are produced even at light planes of anesthesia. Local anesthetics may be added to provide analgesia, or anesthesia may be improved by the addition of an inhalant agent (e.g., isoflurane).

Intravenous administration of thiopental, methohexital, and propofol can be used for short-term anesthesia. Propofol can be given repeatedly IV, or better by continuous infusion, for procedures lasting at least 3 h. Analgesia is poor with propofol but can be improved by premedication with Hypnorm in rats.[68] Alphaxalone + alphadolone (Saffan®) is another anesthetic useful for repeated IV injections or long-term infusion, except in guinea pigs, which may develop pulmonary edema.[8] Intravenous anesthesia with isoflurane has been described in mice. An average infusion rate of 1.6 µL/min of isoflurane in Intralipid was required for maintenance.[69]

Chloral hydrate can be used in rats and mice to produce surgical anesthesia but with a narrow safety margin. Severe acidosis and cardiovascular and respiratory depressions are produced. Tribromoethanol produces approximately 15 min of surgical anesthesia, after IP administration in mice. Induction and recovery is fast. Exposure to light or improper storage will cause tribromoethanol to decompose and cause inflammation if injected. Proper storage is in the dark, at 4°C.[70] Urethane will produce prolonged deep surgical anesthesia and 24 h of sleep after IP injection in rats.[8] Progressive side effects are acidosis, hypotension, bradycardia, peritoneal fluid accumulation, and decline in renal function. Urethane is a proven carcinogen and mutagen in rodents.

## Analgesia and Anesthesia in Rabbits

## *Analgesia*

There are few controlled clinical studies published on NSAIDs and opioids in rabbits, but experience shows that a single dose of the NSAID carprofen provides up to 24 h of postoperative analgesia and the opioid buprenorphine 8 to 12 h.[6] Rabbits seem to benefit from a single dose of buprenorphine, administered toward the end of a surgical procedure. After more extensive surgery, a combination of a NSAID, an opioid, and, if needed, a local anesthetic, will provide excellent analgesia.

Local anesthetics may be used in similar ways as in other species. EMLA-cream provides analgesia before venipuncture or catheter placement in the ear. After clipping the fur, the cream is applied for 1 h under an adhesive tape, providing skin-deep anesthesia for 1 h (see Figure 17.1).

**Table 17.5 Injectable Anesthetics in Rodents**

| Drug | Mouse | Rat | Guinea Pig |
|---|---|---|---|
| Fentanyl + fluanisone (Hypnorm®) + midazolam | 10–13 mL/kg IP of a premixed solution[a] | 2.7–4 mL/kg IP of a premixed solution[a] | 8 mL/kg IP of a premixed solution[a] |
| Fentanyl + fluanisone (Hypnorm®) + diazepam | 0.4 mL/kg IP + 5 mg/kg IP | 0.6 mL/kg IP + 2.5 mg/kg IP | 1 mL/kg IP + 2.5 mg/kg IP |
| Ketamine + medetomidine | 50–75 mg/kg + 1–10 mg/kg IP | 75 mg/kg + 1 mg/kg IP | 40 mg/kg + 0.5mg/kg IP |
| Atipamezole | 1–2.5 mg/kg IM or SC | 1 mg/kg IM or SC | 1 mg/kg IM or SC |
| Ketamine + xylazine | 80–100 mg/kg + 5–10 mg/kg IP | 75–100 mg/kg + 10 mg/kg IP | 40 mg/kg + 5 mg/kg IP |
| Pentobarbital | 40–50 mg/kg IP | 40–50 mg/kg IP | 37 mg/kg IP |
| Thiopental | 30–40 mg/kg IV | 10–15 mg/kg IV | — |
| Methohexital | 10 mg/kg IV | 10–15 mg/kg IV | — |
| Propofol | 26 mg/kg IV in repetitive boluses | 10 mg/kg IV + 44–55 mg/kg/h IV | — |
| Alphaxalone + alphadolone (Saffan®) | 14 mg/kg IV + 4–6 mg/kg every 15 min | 10–15 mg/kg IV + 0.25–0.45 mg/kg/min IV of a 1:10 dilution | 16–20 mg/kg IV, no repetition |
| Tribromoethanol | 250 mg/kg IP (1.2% solution) | 300 mg/kg IP (0.25% solution) | — |
| Chloral hydrate | 450 mg/kg IP + 40 mg/kg/h IP (5% solution) | 300–450 mg/kg IP + 40 mg/kg/h IP (5% solution) | 400 mg/kg IP |
| Alpha-chloralose | 114 mg/kg IP (5% solution) | 31–65 mg/kg IP (5% solution) | — |
| Urethane | — | 1200–1500 mg/kg IP (50% solution) | — |

[a] Mix one part Hypnorm® with two parts water for injection first, then add one part midazolam (5 mg/mL).

*Sources:* Wixson, S.K. and Smiler, K.L., Anesthesia and analgesia in rodents, in *Anesthesia and Analgesia in Laboratory Animals*, Kohn, D.F., Wixson, S.K., White, W.J., and Benson, G.J., Eds., ACLAM and Academic Press, New York, 1997; Flecknell, P.A., Anesthesia and analgesia for exotic species — rabbits, rodents, and ferrets, in *Manual of Small Animal Anaesthesia and Analgesia*, Seymour, C. and Gleed, R., Eds., BSAVA, 1999; Cruz, J.I., Loste, J.M., and Burzaco, O.H., *Lab. Anim.*, 32, 18, 1998; and Tanaka, N., Dalton, N., Mao, L., Rockman, H.A., Peterson, K.L., Gottshall, K.R., and Hunter, J.J., *Circulation*, 94, 1109, 1996.

**Table 17.6 Dose Rates for Analgesics in Rabbits**

| Analgesic | Dose |
|---|---|
| Buprenorphine | 0.01–0.05 mg/kg IM, SC, or IV, 6–12 hourly |
| Butorphanol | 0.1–0.5 mg/kg IM, SC, or IV, 4 hourly |
| Morphine | 2–5 mg/kg IM or SC, 4 hourly |
| Carprofen | 4 mg/kg SC daily or 1.5 mg/kg by mouth twice daily |
| Ketoprofen | 3 mg/kg SC daily |
| Lidocaine | 4 mg/kg or 0.4 mL/kg of a 1% solution |
| Bupivacaine | 1–2 mg/kg or 0.4–0.8 mL/kg of a 0.25% solution |

*Source:* From Dobromylskyj, P., Flecknell, P.A., Lascelles, B.D., Pascoe, P.J., Taylor, P., and Waterman-Pearson, A., Management of postoperative and other acute pain, in *Pain Management in Animals*, Flecknell, P. and Waterman-Pearson, A., Eds., W.B. Saunders, Philadelphia, PA, 2000.

## *Anesthesia*

### *Premedication*

Most preanesthetic drugs may be used in rabbits. Many rabbits have high plasma levels of atropinase, and glycopyrrolate may therefore be a better choice if an anticholinergic is indicated. Glycopyrrolate elevates the heart rate for 60 min in rabbits and will prevent bradycardia during ketamine + xylazine anesthesia.[71]

**Table 17.7 Drug Dose Rates for Premedication in Rabbits**

| Drug | Dose |
|------|------|
| Acepromazine | 0.25–0.75 mg/kg IM or SC |
| Atropin | 40 µg/kg SC or IM |
| Glycopyrrolate | 0.1 mg/kg SC or IM or 0.01mg/kg IV |
| Fentanyl + fluanisone (Hypnorm®) | 0.2–0.5 mL/kg IM or SC |
| Diazepam | 1–2 mg/kg IM or IV |
| Midazolam | 1–2 mg/kg IV or IM |
| Xylazine | 3–5 mg/kg IM or IV |
| Medetomidine | 0.1–0.5 mg/kg IM or SC |

*Sources:* From Flecknell, P.A., Anesthesia and analgesia for exotic species — rabbits, rodents, and ferrets, in Seymour, C. and Gleed, R., Eds., *Manual of Small Animal Anaesthesia and Analgesia,* Seymour, C. and Gleed, R., Eds., BSAVA, 1999; and Lipman, N.S., Marini, R.P., and Flecknell, P.A., Anesthesia and Analgesia in Rabbits, in *Anesthesia and Analgesia in Laboratory Animals,* Kohn, D.F., Wixson, S.K., White, W.J., and Benson, G.J., Eds., ACLAM and Academic Press, New York, 1997.

## Inhalation Anesthesia

Rabbits react with extended breath holding and severe struggling upon mask or chamber induction with most volatile agents.[10,11,12] The result of breath holding is a marked decrease in heart rate and arterial oxygen levels, as well as an elevation in blood carbon dioxide levels, and induction with inhalation anesthesia is thus best avoided in rabbits. Premedication with a sedative or tranquilizer will eliminate struggling but not prevent breath holding. Anesthesia is best induced with an injectable agent, after which the rabbit can be intubated and maintained on inhalation anesthesia (Figure 17.13)

## Injection Anesthesia

Fentanyl + fluanisone (Hypnorm) alone may be sufficient for minor superficial surgery. When used in combination with midazolam, anesthesia sufficient for invasive surgery is produced. It is best to administer Hypnorm SC first, and after 10 to 15 min, the ear vein can be cannulated and midazolam administered IV, until a plane of surgical anesthesia is achieved. After completion of surgery, fentanyl can be reversed with butorphanol or buprenorphine.

Another commonly used combination is ketamine + medetomidine. The drugs can be mixed and injected together, or medetomidine can be administered SC first for sedation, followed by ketamine IM, SC, or IV. Arterial oxygen levels are significantly reduced by this combination, and supplemental oxygen is recommended.[29,72] The heart rate is reduced, but arterial blood pressure is maintained. Medetomidine may be reversed with atipamezole. If surgery has been undertaken, an analgesic should be administered before reversal. Ketamine may also be combined with acepromazine, midazolam, or diazepam, for

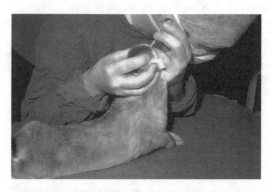

**Figure 17.13** Blind endotracheal intubation of an anesthetized rabbit.

**Table 17.8 Doses for Injectable Anesthesia in Rabbits**

| Drug | Dose |
|---|---|
| Fentanyl + fluanisone (Hypnorm®) + midazolam | 0.3 mL i.m or SC + 1–2 mg/kg IV |
| Ketamine + medetomidine | 5–15 mg/kg + 0.25–0.35 mg/kg IM or SC |
| Atipamezole (antidote) | 1 mg/kg IM |
| Ketamine + xylazine | 35–50 mg/kg + 5–10 mg/kg IM |
| Ketamine + acepromazine | 50 mg/kg IM + 1 mg/kg IM |
| Ketamine + midazolam | 25 mg/kg IM + 5 mg/kg IM |
| Ketamine + diazepam | 25 mg/kg IM + 5 mg/kg IM |
| Thiopentone | 1–2% dilution IV to effect (6–50 mg/kg) |
| Methohexital | 5–15 mg/kg of a 1% solution IV |
| Propofol | 7–10 mg/kg IV + 0.9 mg/kg/min IV |

*Source:* From Flecknell, P.A., Anesthesia and analgesia for exotic species — rabbits, rodents, and ferrets, in *Manual of Small Animal Anaesthesia and Analgesia*, Seymour, C. and Gleed, R., Eds., BSAVA, 1999.

provision of light to moderate surgical anesthesia. Each sedative may be administered prior to ketamine or mixed with ketamine.

Pentobarbital is hazardous in rabbits and should not be used. Thiopental and methohexital can be administered IV for short procedures, or to allow intubation for maintenance with inhalation anesthesia. Propofol administered IV produces light anesthesia, insufficient for surgery. Injections can be repeated or better, continuous infusion administered. In combination with local anesthesia, superficial surgery may be performed. By premedication with medetomidine, less propofol is needed, and surgical anesthesia can be provided.

## Analgesia and Anesthesia in Pigs

### Analgesia

The NSAID carprofen may be used to control postsurgical pain in pigs by IV or SC administration for up to 3 days without side effects.[6] Acetylsalicylic acid and acetaminophen may be administered orally for acute or chronic, mild to moderate pain. Ketorolac (NSAID) has been reported effective in relieving pain after major surgery.[73]

The opioid buprenorphine has been shown to be effective in thermal analgesiometry tests.[6] Buprenorphine and butorphanol are long acting, with few side effects. For minor surgery, repeated administration of analgesics may not be necessary, whereas a few days of opioid treatment alone or in combination with a NSAID is needed after major surgery.

Local anesthetics may be used like in other species. Combinations of local anesthetics infiltrated along the incision line with parenteral opioid analgesics and general anesthesia, reduce effects from noxious stimulation as well as recovery times in pigs.[18]

### Anesthesia

#### Premedication

Atropine and glycopyrrolate are used to decrease secretion, abolish the vaso-vagal reflex during intubation, and prevent bradycardia.[18] Acepromazine, diazepam, midazolam, and azaperone can be used to relieve anxiety and will decrease the amount of general anesthetic required. The duration of action is 8 to 24 h for acepromazine and 20 to 50 min for azaperone

#### Inhalation Anesthesia

Isoflurane, desflurane, and sevoflurane are all good for use in pigs.[18] They provide better control of anesthesia and shorter recovery periods than many of the injectable agents, and isoflurane provides the

**Table 17.9 Dose Rates for Analgesics in Pigs**

| Drug | Dose |
|------|------|
| Carprofen | 2–4 mg/kg SC or IV daily |
| Acetylsalicylic acid (enteric-coated aspirin) | 10 mg/kg by mouth; 4–6 hourly |
| Ketorolac | 1 mg/kg IM or IV |
| Buprenorphine | 0.005–0.05 mg/kg IV or IM; 6–12 hourly |
| Butorphanol | 0.1–0.3 mg/kg; 6 hourly |
| Lidocaine | 4 mg/kg or 0.4 mL/kg of a 1% solution |
| Bupivacaine | 1–2 mg/kg or 0.4–0.8 mL/kg of a 0.25% solution |

*Sources:* From Dobromylskyj, P., Flecknell, P.A., Lascelles, B.D., Pascoe, P.J., Taylor, P., and Water-man-Pearson, A., Management of postoperative and other acute pain, in *Pain Management in Animals*, Flecknell, P. and Waterman-Pearson, A., Eds., W.B. Saunders, Philadelphia, PA, 2000; and Swindle, M., *Surgery, Aneshthesia & Experimental Techniques in Swine*, Iowa State University Press, 1998.

**Table 17.10 Dose Rates for Premedication in Pigs**

| Drug | Dose |
|------|------|
| Atropine | 0.05 mg/kg IM or 0.02 mg/kg IV |
| Glycopyrrolate | 0.004–0.01 mg/kg IM |
| Acepromazine | 0.1–0.2 mg/kg IM, SC, or IV |
| Diazepam | 0.5–1 mg/kg IM or 0.44–2 mg/kg IV, 1 mg/kg/h IV infusion |
| Midazolam | 0.1–0.5 mg/kg IM or IV, 0.6–1.5 mg/kg/h IV infusion |
| Azaperone | 2–5 mg/kg IM |

*Source:* From Swindle, M., *Surgery, Anesthesia & Experimental Techniques in Swine*, Iowa State University Press, 1998.

**Table 17.11 Dose Rates for Injectable Anesthetics in Swine**

| Drug | Dose |
|------|------|
| Ketamine | 11–33 mg/kg IM, 3–10 mg/kg/h IV infusion |
| Ketamine + acepromacin | 25–30 mg/kg + 1.1 mg/kg IM |
| Ketamine + diazepam | 15 mg/kg + 2 mg/kg IM |
| Ketamine + azaperone | 15 mg/kg + 2 mg/kg IM |
| Ketamine + xylazine | 20 mg/kg + 2 mg/kg IM |
| Ketamine + medetomidine | 10 mg/kg + 0.05–0.1 mg/kg IM |
| Ketamine + midazolam | 33 mg/kg IV bolus, 8–33 mg/kg/h IV + 0.5–1.5 mg/kg/h IV infusion |
| Fentanyl + isoflurane | 0.03–0.1 mg/kg/h IV infusion + 0.050 mg/kg IV bolus + 0.5% isoflurane |
| Sufentanil + isoflurane | 0.015–0.030 mg/kg/h IV infusion + 0.007 mg/kg IV bolus + 0.5% isoflurane |
| Pancuronium | 0.02–0.15 mg/kg IV, 0.003–0.030 mg/kg/h IV infusion |
| Vercuronium | 1 mg/kg IV |

*Source:* From Swindle, M., *Surgery, Anesthesia & Experimental Techniques in Swine*, Iowa State University Press, 1998.

least myocardial depressant effect of all the inhalants. Mask induction followed by endotracheal intubation is possible in a well-ventilated area, for protection of staff (Figure 17.14). Nitrous oxide may be used in 1:1 or 1:2 combinations with oxygen, in order to reduce the amount of inhalant needed (20 to 25% reduction).

## Injection Anesthesia (Table 17.11)

Endotracheal intubation should always be performed during general anesthesia in pigs (Figure 17.15).[18] Ketamine and tiletamine + zolazepam (Telazol or Zoletil) are the most common injectables used in swine anesthesia. Ketamine is not sufficient for surgical procedures,[74] but in combination with acepromazine, diazepam, midazolam, azaperone, xylazine, or medetomidine, it may be used for minor procedures. Intravenous infusion with ketamine and midazolam can provide visceral analgesia suitable for major

**Figure 17.14** Face mask induction of piglet with isoflurane.

**Figure 17.15** Endotracheal intubation of piglet with a 3.0 mm endotracheal tube.

surgery,[18] but it also produces profound hypothermia. A loading dose of ketamine precedes continuous infusion. Tiletamine + zolazepam provide 20 min of anesthesia for minor surgery.

Barbiturates can be administered IV after induction of anesthesia with ketamine. Thiopental, administered by IV infusion, has the least cardiovascular depressant effect of all barbiturates. Pentobarbital infusion can be used for nonsurvival procedures in artificially ventilated animals. By adding fentanyl, profound analgesia is provided.

For cardiac surgery, a combination of IV opioid infusion and inhalation anesthesia may be used.[18,75] Midazolam or ketamine is given preoperatively to allow catheter placement. Opioid infusions have the advantages of not decreasing myocardial contractility and coronary blood flow. By starting the infusion before the bolus is given, muscle rigidity and sudden bradycardia are avoided. Propofol has a narrow margin of safety in swine and can produce severe hypotension. Paralytic agents may be indicated for some procedures, but should not be administered until it is established that surgical anesthesia has been obtained. Paralyzed animals must be monitored for changes in heart rate or blood pressure as an indication of inadequate anesthesia.

**Table 17.12 Dose Rates for Analgesics in Sheep**

| Drug | Dose |
| --- | --- |
| Acetylsalicylic acid | 50–100 mg/kg by mouth, 6–12 hourly |
| Carprofen | 1.5–2.0 mg/kg SC or IV daily |
| Buprenorphine | 0.005–0.01 mg/kg IM or IV, 4 hourly |
| Butorphanol + diazepam | 0.05–0.1 mg/kg IV + 0.05–0.20 mg/kg IV |
| Xylazine | 0.08–0.10 mg/kg IM, 0.03–0.04 mg/kg IV |

*Source:* From Dobromylskyj, P., Flecknell, P.A., Lascelles, B.D., Pascoe, P.J., Taylor, P., and Waterman-Pearson, A., Management of postoperative and other acute pain, in *Pain Management in Animals*, Flecknell, P. and Waterman-Pearson, A., Eds., W.B. Saunders, Philadelphia, PA, 2000.

**Table 17.13 Dose Rates for Premedication, Anesthetics, and Antidotes in Small Ruminants**

| Drug | Dose |
|------|------|
| Diazepam | 0.2–0.4 (sheep) or 5–10 (goats) mg/kg IV |
| Xylazine | 0.02–0.15 mg/kg IV |
| Medetomidine | 0.003–0.01 mg/kg IV |
| Atipamezole | 0.02–0.04 mg/kg IV |
| Midazolam | 0.4–1.0 mg/kg IV |
| Flumazenil | 1 mg total dose IV in sheep |
| Thiopental | 10–15 mg/kg IV |
| Propofol | 4–6 mg/kg IV for induction, then 0.4–0.6 mg/kg/min |
| Ketamine + xylazine | 2.0–7.5 + 0.1–0.2 mg/kg IV |
| Ketamine + diazepam | 2.0–5.0 + 0.5–1.0 mg/kg IV |

*Source:* From Dunlop, C.I. and Hoyt, R.F., Anesthesisa and analgesia in ruminants, in *Anesthesia and Analgesia in Laboratory Animals*, Kohn, D.F., Wixson, S.K., White, W.J., and Benson, G.J., Eds., ACLAM and Academic Press, New York, 1997.

## Analgesia and Anesthesia in Small Ruminants

### Analgesia

NSAIDs can be used safely and efficiently in ruminants.[3] Due to rapid metabolism, opioids have a short duration of action in sheep. Butorphanol in combination with diazepam is useful for sedation and pain relief of small ruminants. Continuous infusion of alpha-2-agonists can provide prolonged analgesia, but severe hypoxia may develop.[76]

### Anesthesia

Ruminants have a large capacity to redistribute anesthetic drugs in their large gastrointestinal tract.[77] Thus, thiopental has a shorter duration of action in ruminant than in nonruminant species. Unlike goats, sheep are not easily stressed by handling, show no excitement during recovery from anesthesia, and do not need routine premedication. Preadministration of xylazine increases the incidence of regurgitation and should be avoided before endotracheal intubation. Diazepam or local anesthesia may be used to prevent excitement during recovery in goats.

Xylazine and medetomidine are sedatives that can be used in combination with local anesthesia for surgical procedures and followed by reversal with atipamezole. Xylazine reduces fetal oxygenation and can cause premature parturition if used during the last trimester. Midazolam can also be used for sedation and is reversed with flumazenil. Ketamine is commonly combined with xylazine or diazepam for 15 to 30 min of general anesthesia in small ruminants. Intravenous thiopental injection produces 15 min of anesthesia and is used to allow intubation, followed by inhalation anesthesia. Induction with propofol IV followed by continuous infusion may be used for long-term anesthesia. Additional analgesia is necessary if surgery is undertaken.

Face-mask induction of halothane or isoflurane anesthesia can easily be accomplished after sedation. This method necessitates efficient gas scavenging. During maintenance with inhalation anesthesia, a cuffed tube should be used to prevent the large amounts of saliva that are produced from reaching the airways.

## Analgesia and Anesthesia in Cats, Dogs, and Ferrets

### Analgesia

Dogs are particularly sensitive to the adverse renal effects of NSAIDs during hypotension, and cats have an increased susceptibility to the toxic effects of acetylsalicylic acid and other NSAIDs.[6] Toxicity is avoided by carefully selecting a drug and keeping correct dosing intervals. Morphine is effective for the control of severe postoperative pain. Buprenorphine is less effective but has a longer duration of action (6 to 8 h) and may be used for at least 24 h in cats and 48 h in dogs without side effects. Low doses of ketamine can be used for short-lasting analgesia postoperatively and has been shown to prevent hyperalgesia in dogs, but side effects such as increased muscle activity may limit its use.

## *Anesthesia*

### *Premedication*

Anticholinergics are used to prevent bradycardia or increased salivation. Commonly used sedatives and tranquillizers are acepromazine, diazepam, and midazolam.[23] Opioids used for premedication are morphine, butorphanol, and buprenorphine, usually in combination with a tranquilizer. They increase sedation

**Table 7.14  Dose Rates for Analgesics in Cats, Dogs, and Ferrets**

| Drug | Dog | Cat | Ferret |
|------|-----|-----|--------|
| Acetylsalicylic acid | 10–25 mg/kg orally, 8–12 hourly | 10–25 mg/kg orally, 48 hourly | 20 mg/kg orally |
| Carprofen | 4 mg/kg SC or IV | 2–4 mg/kg SC or IV | — |
| Ketoprofen | 2 mg/kg SC daily for up to 3 days, up to 5 days | 1 mg/kg orally daily for | — |
| Morphine | 0.1–1 mg/kg IM, SC, or IV, 4–6 hourly | 0.1–0.2 mg/kg IM, SC or IV 6–8 hourly | 0.5–2.0 mg/kg IM, SC, 4–6 hourly |
| Buprenorphine | 0.005–0.020 mg/kg IM, SC, or IV, 6–12 hourly | | 0.01–0.03 mg/kg IM, SC, or IV, 6–12 hourly |

*Sources:* From Dobromylskyj, P., Flecknell, P.A., Lascelles, B.D., Pascoe, P.J., Taylor, P., and Waterman-Pearson, A., Management of postoperative and other acute pain, in *Pain Management in Animals*, Flecknell, P. and Waterman-Pearson, A., Eds., W.B. Saunders, Philadelphia, PA, 2000; and Harvey, R.C., Paddleford, R.R., Popilskis, S.J., and Wixson, S.K., Anesthesia and Analgesia in Dogs, Cats, and Ferrets, in *Anesthesia and Analgesia in Laboratory Animals*, Kohn, D.F., Wixson, S.K., White, W.J., and Benson, G.J., Eds., ACLAM and Academic Press, New York, 1997.

**Table 7.15  Dose Rates for Premedication in Cats, Dogs, and Ferrets**

| Drug | Dog | Cat | Ferret |
|------|-----|-----|--------|
| Atropine | 0.02–0.04 mg/kg SC or IM or 0.01–0.02 mg/kg IV | | 0.05 mg/kg SC or IM |
| Acepromazine | 0.02–0.075 mg/kg SC or IM, total maximum dose: 3 mg | | 0.2 mg/kg SC or IM |
| Diazepam | 0.1–0.6 mg/kg IV | 0.1–0.5 mg/kg IV | 1–2 mg/kg IM |
| Midazolam | 0.1–0.5 mg/kg IM or IV | | 1 mg/kg IM |
| Morphine | 0.1–2 mg/kg SC, IM, or IV | 0.05–0.1 mg/kg SC or IM | |
| Buprenorphine | 0.01–0.02 mg/kg SC, IM, or IV | 0.005–0.02 mg/kg SC or IM | |
| Butorphanol | 0.2–0.8 mg/kg SC, IM, or IV | 0.2–0.4 mg/kg SC, IM, or IV | |
| Acepromazine + buprenorphine | 0.4 + 0.05 SC or IM | | |
| Midazolam + buprenorphine | 0.1 + 0.01 mg/kg SC, IM, or IV | | |
| Xylazine | 0.2–1.1 mg/kg SC, IM, or IV | | 1 mg/kg IM or SC |
| Medetomidine | 0.01–0.04 mg/kg SC, IM, or IV | 0.04–0.08 mg/kg SC, IM, or IV | 0.1 mg/kg SC |

*Sources:* From Lukasik, V.M., Premedication and sedation, in *Manual of Small Animal Anaesthesia and Analgesia*, Seymour, C. and Gleed, R., Eds., BSAVA, 1999; and Harvey, R.C., Paddleford, R.R., Popilskis, S.J., and Wixson, S.K., Anesthesia and Analgesia in Dogs, Cats, and Ferrets, in *Anesthesia and Analgesia in Laboratory Animals*, Kohn, D.F., Wixson, S.K., White, W.J., and Benson, G.J., Eds., ACLAM and Academic Press, New York, 1997.

and provide pre- and intraoperative and sometimes also postoperative analgesia. Preoperative drug combinations used in young healthy dogs and cats are, for instance, acepromazine + butorphanol or acepromazine + buprenorphine. Pediatric, geriatric, and physiologically compromised animals are best premedicated with a benzodiazepine and butorphanol or buprenorphine.

Premedication with carprofen in dogs has no effect on the concentration of isoflurane needed. Butorphanol alone, or in combination with carprofen, reduces the isoflurane concentration needed by approximately 20%.[78] Alpha-2-agonists (e.g., medetomidine) reduce the dose of subsequent inhalation and injection anesthetics by 50%.[23] By adding butorphanol or ketamine to medetomidine in dogs, sedation is enhanced, but blood gas levels are significantly poorer than when medetomidine is given alone.[79]

### Inhalation Anesthesia

Inhalation anesthesia is the first choice in terms of safety, especially for long-duration anesthesia. All volatiles are safe to use in cats, dogs, and ferrets. Mask induction is possible, but induction by injection is preferred, followed by endotracheal intubation and inhalation anesthesia. Ferrets may be induced in a chamber and maintained on a mask. By adding 50% $N_2O$, the required isoflurane concentration is reduced by approximately 20%, which improves cardiovascular function. Nitrous oxide also speeds up induction, which is beneficial for mask induction. When using $N_2O$ as part of the anesthetic regimen, it is important to let the animal breathe pure oxygen for 5 to 10 min after discontinuation of $N_2O$ to prevent diffusion hypoxia.

### Injection Anesthesia

Thiopentone can be used as an IV induction agent, given to effect. The recommended dose should be reduced by 50 to 75% after premedication with an alpha-2-agonist such as medetomidine, and by up to 50% after other premedication. Methohexital may be used in the same way but is also suitable for maintenance of anesthesia by continuous infusion. Alphaxalone + alphadolone (Saffan®) has a wide margin of safety and is useful for IV induction and maintenance of anesthesia in the cat. Ketamine + midazolam (cat, dog), ketamine + diazepam (dog, ferret), ketamine + xylazine (ferret), and ketamine + medetomidine (cat, dog, ferret) are all useful combinations for short surgical procedures. Xylazine + butorphanol + ketamine has been described as an effective anesthetic combination in ferrets, however, hypoxemia and ventricular arrhythmias were observed. [80]

Propofol is useful for induction and maintenance by IV infusion in cats and dogs. In contrast to thiopental and volatile agents, propofol causes little reduction in renal blood flow and is especially useful

**Table 17.16 Dose Rates for Injectable Anesthesia in Dogs, Cats, and Ferrets**

| Drug | Dog | Cat | Ferret |
|---|---|---|---|
| Thiopentone | 10–15 mg/kg IV | | — |
| Methohexital | 5 mg/kg IV + 0.3 mg/kg/min IV infusion of 1–2.5% solution | | — |
| Alphaxalone + alphadolone | | 6 mg/kg IV + 0.24 mg/kg/min IV infusion | 8–12 mg/kg IV |
| Ketamine + medetomidine | 5.0–7.5 mg/kg + 0.04 mg/kg IM | 2.5–7.5 mg/kg + 0.08 mg/kg IM or 1.25 mg/kg + 0.04 mg/kg IV | 8 mg/kg + 0.1 mg/kg IM |
| Ketamine + diazepam | 5 mg/kg + 0.25 mg/kg IV | | 25 mg/kg + 2 mg/kg IM |
| Ketamine + midazolam | | 10 + 0.2 IM or 5 + 0.2 IV | — |
| Propofol | 5–6 mg/kg IV + 0.2–0.5 mg/kg/min IV infusion | 6–7 mg/kg IV + 0.2–0.5 mg/kg/min IV infusion | — |

*Sources:* From Reid, J. and Nolan, A.M., Intravenous anesthetics, in *Manual of Small Animal Anaesthesia and Analgesia*, Seymour, C. and Gleed, R., Eds., BSAVA, 1999; and Harvey, R.C., Paddleford, R.R., Popilskis, S.J., and Wixson, S.K., Anesthesia and Analgesia in Dogs, Cats, and Ferrets, in *Anesthesia and Analgesia in Laboratory Animals*, Kohn, D.F., Wixson, S.K., White, W.J., and Benson, G.J., Eds., ACLAM and Academic Press, New York, 1997.

in geriatric animals. Premedication with medetomidine reduces the propofol dose by approximately 75%. Heart rate is significantly lower, but recovery is smoother with medetomidine + propofol compared to medetomidine + ketamine in dogs.[81] Sedation with medetomidine followed by intravenous fentanyl is unsuitable for surgical anesthesia in spontaneously breathing dogs, due to severe respiratory depression.[82] The combination of medetomidine with propofol or ketamine is not as respiratory depressive and may be used without mechanical ventilation.

## Analgesia and Anesthesia in Primates

When selecting methods for anesthesia, the diversity of nonhuman primates used for research must be considered.[33] The wide range in body size plays a role when choosing anesthetics and delivery method. Considerable species differences in the reaction to certain drugs make simple extrapolation of dosages dangerous.

### Analgesia

Acetylsalicylic acid may be used to treat mild pain. Carprofen and ketoprofen are useful for postoperative pain, as are buprenorphine, oxymorphone, and butorphanol. The benefit of using a NSAID is lack of sedation and cardiopulmonary depression.

### Anesthesia

Halothane, isoflurane, and enflurane are all safe to use in primates.[33] Induction of anesthesia is accomplished with a short-acting IV injectable agent, like thiopental or propofol. Indwelling vascular catheters with implanted subcutaneous access ports allow intravenous induction of anesthesia. Some species (e.g., macaques) can be trained to present an arm or a leg for venous access. Others need to be chemically restrained for intravenous access. Ketamine IM is used for this purpose and as part of an anesthetic regime. It has a wide margin of safety but can produce psychotomimetic emergence reactions; therefore, a sedative, e.g., a benzodiazepine, should be administered as well.

Xylazine, medetomidine, and midazolam are commonly used with ketamine, providing muscular relaxation and analgesia sufficient for minor surgical procedures. Tiletamine + zolazepam (Telazol or Zoletil) have the same properties. Alphaxalone + alphadolone (Saffan®) can be used for induction by intramuscular injection, followed by intravenous infusion for maintenance of surgical anesthesia. Propofol can be used for IV induction and maintenance. During surgery, analgesia is provided or enhanced by addition of an opioid or a local anesthetic. Intravenous infusion of fentanyl is used. for this purpose.

**Table 17.17 Dose Rates for Analgesics in Nonhuman Primates**

| Drug | Dose |
|---|---|
| Acetylsalicylic acid | 20 mg/kg orally, 6–8 hourly |
| Carprofen | 3–4 mg/kg SC daily for up to 3 days |
| Ketoprofen | 2 mg/kg SC daily |
| Oxymorphone | 0.075–0.15 mg/kg IM, 4–6 hourly |
| Buprenorphine | 0.005–0.010 mg/kg IM, SC, or IV, 6–12 hourly |
| Butorphanol | 0.01 mg/kg IV, 3–4 hourly |

*Sources:* From Dobromylskyj, P., Flecknell, P.A., Lascelles, B.D., Pascoe, P.J., Taylor, P., and Waterman-Pearson, A., Management of postoperative and other acute pain, in *Pain Management in Animals*, Flecknell, P. and Waterman-Pearson, A., Eds., W.B. Saunders, Philadelphia, PA, 2000; and Popilskis, S.J., and Kohn, D.F., Anesthesia and analgesia in nonhuman primates, in *Anesthesia and Analgesia in Laboratory Animals*, Kohn, D.F., Wixson, S.K., White, W.J., and Benson, G.J., Eds., ACLAM and Academic Press, New York, 1997.

**Table 17.18 Dose Rates for Anesthetics Used in Primates**

| Drug | Dose[a] |
|------|---------|
| Ketamine | 5–20 mg/kg IM |
| Ketamine + xylazine | 7–20 mg/kg + 0.25–2 mg/kg IM |
| Ketamine + medetomidine | 2–6 mg/kg + 0.03–0.1 mg/kg IM |
| Ketamine + midazolam | 15 mg/kg IM+ 0.05–0.09 mg/kg IV |
| Tiletamine + zolazepam (Telazol) | 1.5–10 mg/kg IM |
| Propofol | 1–5 mg/kg IV + 0.13–0.6 mg/kg/min IV infusion |
| Thiopental | 10–17 mg/kg IV |
| Alphaxalone + alphadolone | 11.5–19 mg/kg IM+ 6–12 mg/kg IV |
| Fentanyl | 5–10 µg/kg IV + 10–25 µg/kg/hr IV + low dose isoflurane |

[a] Doses required vary widely between NHP species.

*Source:* From Popilskis, S.J., and Kohn, D.F., Anesthesia and analgesia in nonhuman primates, in *Anesthesia and Analgesia in Laboratory Animals*, Kohn, D.F., Wixson, S.K., White, W.J., and Benson, G.J., Eds., ACLAM and Academic Press, New York, 1997.

## Analgesia and Anesthesia — Birds, Fish, Reptiles, and Amphibia

### *Analgesia*

Birds possess nociceptors and all the opioid receptors as well as alpha-2-receptors in the brain. Analgesic treatment after surgery in birds speeds up return to normal food consumption, which is important due to their high rate of metabolism.[83] Carprofen, ketoprofen, and butorphanol are safe to use in birds. Fish, reptiles, and amphibia show avoidance reactions to noxious stimuli. Amphibians possess endogenous opioids and show an increased nociceptive threshold after opioid treatment. Review articles on pain and analgesia in birds and other nonmammalian vertebrates are recommended for further reading.[84,85]

Studies of clinical pain are scarce, but there are dose recommendations for analgesics in birds and reptiles, based on experience.

**Table 17.19 Dose Rates for Analgesics in Birds and Reptiles**

| Drug | Birds | Reptiles |
|------|-------|----------|
| Carprofen | 1–4 mg/kg IM | 2–4 mg/kg SC, IV, IM, or orally once, then 1–2 mg/kg 24–72 hourly |
| Ketoprofen | 2–4 mg/kg IM | 2 mg/kg SC or IM every 1–2 days |
| Buprenorphine | 0.01–0.05 mg/kg IM | 0.01 mg/kg IM |
| Butorphanol | 1–4 mg/kg IM, 2–4 hourly | 25 mg/kg IM (tortoises) |

*Sources:* From Dobromylskyj, P., Flecknell, P.A., Lascelles, B.D., Pascoe, P.J., Taylor, P., and Waterman-Pearson, A., Management of postoperative and other acute pain, in *Pain Management in Animals*, Flecknell, P. and Waterman-Pearson, A., Eds., W.B. Saunders, Philadelphia, PA, 2000; and Malley D., Anesthesia and analgesia for exotic species — reptiles, in *Manual of Small Animal Anaesthesia and Analgesia*, Seymour, C., and Gleed, R., Eds., BSAVA, 1999.

**Table 17.20 Dose Rates for Anesthetics in Birds**

| Drug | Dose |
|------|------|
| Ketamine | 5–30 mg/kg IV or IM |
| Diazepam | 0.5 mg/kg IM |
| Ketamine + diazepam | 5–20 mg/kg + 1.0–1.5 mg/kg IV or IM |
| Ketamine + midazolam | 5–20 mg/kg + 0.2 mg/kg IV or IM |
| Ketamine + medetomidine | 3–6 mg/kg + 0.15–0.35 mg/kg IM or IV |

*Source:* From Forbes, N.A., Anesthesia and analgesia for exotic species — birds, in *Manual of Small Animal Anaesthesia and Analgesia*, Seymour, C. and Gleed, R., Eds., BSAVA, 1999.

## Anesthesia

### Birds

Birds show great variation between species in the reaction to analgesics and anesthetics.[83] The avian respiratory system is markedly different than that of mammals, with a gas exchange ten times more efficient. Endotracheal intubation is recommended in birds over 100 g in all but the shortest procedures, to provide intermittent positive pressure ventilation (IPPV). Intubation also prevents aspiration of crop contents. Isoflurane is safe to use and preferred over halothane. Sevoflurane may also be used.

Injectable anesthetics are administered by the IM or IV route. For sedation, ketamine or diazepam may be used. Ketamine can be used in combination with diazepam, midazolam, or medetomidine to provide surgical anesthesia for 20 to 30 min.

### Fish

Administration of anesthetics to fish is usually via a bath solution.[84] For maintenance of anesthesia, the gills must be irrigated with anesthetic in water. Tricaine (MS222) was developed for fish anesthesia and is the most common anesthetic. It is available as a powder, which gives an acidic solution in water. Buffering with sodium bicarbonate (200 to 250 mg/100 mg tricaine) or Tris buffer is necessary, except when seawater is used. Tricaine is more toxic in young fish and in soft warm water, and dosage should be adjusted accordingly. Sedation is achieved with a dose of 15 to 20 mg/mL and anesthesia with 50 to 100 mg/mL.[84,85] It is important that the fish be kept moist during the whole procedure. For recovery, the fish is placed in a separate tank with water only, before it is returned to its regular tank.

### Reptiles

Premedication with a sedative or an analgesic is useful in reptiles.[86] Acepromazine, midazolam, carprofen, and ketoprofen may all be used. Analgesics are beneficial if surgery involves the skin, coelomic cavity (chest and abdominal cavity in one), or skeletal system.

Because respiratory muscles are relaxed during general anesthesia, artificial ventilation is necessary. It can be achieved by manually compressing the rebreathing bag every 20 to 30 sec, so that the cranial two-fifth of the coelomic cavity is seen to rise. Artificial ventilation must be continued until spontaneous respiration has returned.

Propofol or alphaxalone + alphadolone (Saffan®) may be used for induction, followed by intubation and administration of oxygen or air by intermittent positive pressure ventilation (IPPV). Ketamine + medetomidine IM may also be used for induction and minor surgery.[87] After completion of surgery, the effects of medetomidine can be reversed with atipamezole.

**Table 17.21 Dose Rates for Anesthetics in Reptiles**

| Drug | Dose and Route |
|---|---|
| Acepromacin | 0.1–0.5 mg/kg, IM |
| Midazolam | 2 mg/kg, IM or SC |
| Propofol | 10 (snakes)-13 (lizards)-15 (chelonians[a]) mg/kg, IV |
| Alphaxalone + alphadolone | 6 (lizards)-9 (chelonians[a], snakes) mg/kg, IV |
| Ketamine + midazolam | 20–40 mg/kg + 2 mg/kg, IM or SC (turtles) |
| Ketamine + medetomidine | 10 mg/kg + 100 (snakes)-150 (chelonians[a]) µg/kg, IM |

[a] Chelonians: tortoises, turtles, and terrapins.

*Source:* From Malley D., Anesthesia and analgesia for exotic species — reptiles, in *Manual of Small Animal Anaesthesia and Analgesia*, Seymour, C. and Gleed, R., Eds., BSAVA, 1999.

**Table 17.22 Dose Rates for MS222 in Amphibians**

| Tadpole<br>Newt<br>0.2–0.5 g/L | Frog<br>Salamander<br>0.5 g/L | Toad<br>1–3 g/L |
| --- | --- | --- |

*Source:* From Schaeffer, D.O., Anesthesia and analgesia in nontra-
ditional laboratory animal species, in *Anesthesia and Analgesia in
Laboratory Animals*, Kohn, D.F., Wixson, S.K., White, W.J., and
Benson, G.J., Eds., ACLAM and Academic Press, New York, 1997.

Isoflurane is the most commonly used volatile in reptiles. Mask induction, followed by intubation
and artificial ventilation, is useful in lizards. Snakes, chelonians (tortoises, turtles, and terrapins), and
large lizards may be intubated while conscious, and anesthesia can be induced by artificial ventilation.
Chamber induction is of limited use in reptiles. If 10% of the inspired gas mixture consists of carbon
dioxide, the speed of recovery is increased.[86]

## Amphibians

Most adult amphibians develop and breathe with lungs. Some salamanders, however, are lungless and
breathe through the skin and mouth cavity. The skin is permeable and allows amphibia to be anesthetized
by immersion methods, with tricaine methane sulfonate (MS222) in deionized water being the anesthetic
of choice.[88] When used in concentrations over 500 mg/L, buffering with sodium bicarbonate (450 to
1000 mg/L) or 0.5 M dibasic sodium phosphate ($Na_2HPO_4$) is necessary. Using high concentration
unbuffered solutions is stressful due to acidity and prolonged induction time. After removing the animal
from the induction bath, anesthesia is maintained for 5 to 20 min. Recovery can be accelerated by rinsing
the animal with clean deionized water.

Inhalation anesthesia in terrestrial amphibians may be induced in a chamber (4 to 5% halothane or
isoflurane) and maintained on lower concentrations and assisted ventilation after intubation. Isoflurane
may be bubbled through the water to anesthetize aquatic species.

Hypothermia should not be used for invasive procedures because it will not induce adequate anesthesia. Studies indicate that poikilotherm sensory input remains intact at almost freezing temperatures.[88]

## EUTHANASIA

### Goals and Definition of Terms in Euthanasia

As a consequence of the correct use of terminology, euthanasia refers to the process of inducing death
with minimal or no pain or distress in the animal involved. For such humane euthanasia, criteria include
a rapid loss of consciousness; a minimum of distress; minimal restraint; and an irreversible and reliable
CNS depression, resulting in the death of the animal. Relevant operator-related criteria are operator
safety and aesthetic aspects of euthanasia. Finally, aspects such as costs and level of complexity of the
procedure should be taken into account. As an extra requirement for animals that are part of a research
protocol, the potential consequences of the euthanasia technique for the subsequent (histo)pathological
or pharmacological analysis should be taken into consideration.

Under all circumstances, death must be confirmed on the basis of circulatory and respiratory
failure, a lowering of body temperature, and overall body stiffness (rigor mortis). The aims and
goals for euthanasia of animals are defined in documents from the veterinary[89] and the laboratory
animal communities.[90,91]

### Methods of Euthanasia

The different techniques used for euthanizing animals can be divided into two groups, i.e., euthanasia
by pharmacological–chemical methods and euthanasia by mechanical–physical methods. Whereas the

former usually involves the administration of an overdose of an anesthetic agent, is relatively fast and aesthetically acceptable, the latter provides a method that eliminates pharmacological "contamination." Based upon combined arguments such as species, age and body weight, available technique and expertise, and specific consideration of the further analyses foreseen after euthanasia, the choice between (and within) the different techniques is made.

## Pharmacological–Chemical Methods

With few exceptions, this group involves an overdose of an injectable (e.g., pentobarbital) or an inhalant anesthetic, resulting in loss of consciousness and in the cease of respiration. Consequently, the heart will stop, and death will ensue. Administration of injectable agents in larger species is preferably by the intravenous route. In smaller species, intraperitoneal injection or, after prior sedation or anesthesia, intracardiac injection is employed. Frequently used agents include the barbiturates, of which pentobarbital (100 to 150 mg/kg) is the most common. Loss of consciousness following intravenous administration is rapid, quickly followed by respiratory depression and failure. A typical phenomenon seen with the use of barbiturates is enlargement of the spleen, which may interfere with the postmortem evaluation. In larger animals, terminal gasping may be seen following loss of consciousness. This is not encountered with the use of T61®, a specific euthanasia preparation that combines a muscle relaxant with a hypnotic agent and local anesthetic. Although research supports that this preparation induces loss of consciousness before muscle relaxation sets in,[92] there are no other advantages over pentobarbital except for the aesthetic aspect.

Although any injectable anesthetic agent administered at a high dose may principally be used for euthanasia, the only agent specifically used for euthanasia is potassium chloride (KCl). Because administration is painful, it is only acceptable when performed under general anesthesia. KCl is cardiotoxic, and intravenous or intracardiac administration will result in cardiac failure. Like with barbiturates, enlargement of internal organs such as the spleen may be seen in conjunction with KCl usage.

Volatile anesthetic agents can be used for inducing euthanasia as well, when administered at sufficiently high concentrations. The use of ether is discouraged because it induces excitation and irritation of mucous membranes. Halothane, enflurane, and isoflurane can be used for euthanasia when administered under circumstances of adequate ventilation. Next to the volatiles, the use of carbon dioxide ($CO_2$) has been advocated for euthanasia. Obvious advantages such as relative safety for the user, limited costs, and availability in compressed gas cylinders are countered by the discussion on the preferred technique of administering $CO_2$. Lower concentrations (<80% $CO_2$) may result in a relatively slow induction of unconsciousness,[93] and be accompanied by histologic lesions in the pulmonary system. High concentrations of $CO_2$ (i.e., hypoxic mixtures) have been reported to cause distress, whereas establishing normoxia prevented this distress.[94] In contrast to these findings, it has been reported that signs of distress could not be documented if $CO_2$ administration takes place in the animal's home cage.[95] The relative resistance of neonates and fetuses to high concentrations of $CO_2$ renders this method less suitable for these age groups.

## Physical Methods

Physical methods are indicated when histologic or pharmacological analysis is foreseen following euthanasia and primarily employed in small laboratory animal species. Methods employed include decapitation and cervical dislocation, which can be used in mice, rats, and small rabbits. For decapitation, special apparatus (guillotine) is available, although, with certain limitations to the animal's weight, sharp scissors with adequately long blades may be used. Cervical dislocation involves the severing of the spinal cord, whereby the connection between the brain and the vital organs is disrupted. Following cervical dislocation, it is advisable to exsanguinate the animal in order to confirm death.

Although both of these methods may be aesthetically displeasing, they are acceptable when applied properly and are without the "contamination" involved with pharmacological methods. Other physical methods that may be used under specific conditions and following necessary precautions include stunning, maceration, freezing in liquid nitrogen, and microwave irradiation.[90]

## Special Considerations with Regard to Scientific Purpose

When determining the most optimal method of euthanasia, the specific method-related changes need to be taken into consideration. Barbiturate-induced splenic enlargement and hypoxemia-induced histological changes due to hyperextension of the lungs (high $CO_2$ concentrations) are potentially disrupting to proper histological analysis. Inflammatory changes in the peritoneum following injection of irritating substances or pharmacochemical interaction with postmortem analysis need to be taken into consideration, as is the extent of normal hepatic metabolism of injectable as well as volatile (halothane > enflurane > isoflurane) agents.

## Euthanasia of Different Species

### Euthanasia of Rodents and Rabbits

For the euthanasia of rodents and rabbits, most of the physical and pharmacological techniques described above are applicable. Body weight limitations apply to physical methods such as decapitation and cervical dislocation, and specific expertise is needed in certain species, such as the hamster and the guinea pig, where the short neck and heavy muscles make it more difficult to assert a proper technique.

Cervical dislocation is commonly used in small rodents (<200 g) and rabbits, resulting in a quick death due to physical damage to the brain stem. Depending on the expertise and strength of the operator, the use of this method should be limited to animals weighing less than 1 kg. For a full description of the technique, the reader is referred to the appropriate literature.

Decapitation can easily be performed in rats and mice, but should be limited to the smaller representatives of species such as rats and rabbits (<1 kg). Although special equipment is available, sharp scissors with suitably long blades may be used in smaller animals. Larger animals may need sedation to facilitate handling.

Other physical methods (concussion, captive bolt in rabbits) are only conditionally acceptable depending on the specific circumstances and expertise of the operator. Euthanasia by injection can be achieved with the use of pentobarbital or T61, whereby the use of the latter is limited to intravenous administration. Because no obvious advantages have been described for T61, and considering the ease of administration, pentobarbital is considered drug of first choice. Although intraperitoneal injection is more feasible in small rodents, intravenous administration is preferred in rabbits. The use of KCl is only acceptable when applied under full anesthesia.

Inhalant anesthetics such as halothane, enflurane, and isoflurane can be used to induce and deepen anesthesia in rodents. Rabbits are distressed from induction with inhalants. (See "Analgesia and Anesthesia in Rabbits".) The use of ether is discouraged because it induces excitation and irritation of the mucous membranes. Whereas the use of CO is not advised based upon the health hazard for the operator, $CO_2$ can be used to anesthetise and subsequently euthanize rodents but not rabbits. Despite numerous efforts and ongoing discussion, there is no clear consensus as to the preferred method for applying $CO_2$ for euthanasia. The discussion focuses on the (initially applied) concentration of $CO_2$ the animals are exposed to. Exposure to high concentrations (>80%) induces fast onset of loss of consciousness, with some authors reporting undue (hypoxemia-related) stress. Only $CO_2$ obtained from compressed gas cylinders should be used for this purpose.

### Euthanasia of Small Ruminants and Pigs

The choice of the most preferred method of euthanasia depends on the size of the animal. Whereas physical methods, such as concussion, may be applied under strict conditions to render young animals unconscious, adequate efficacy cannot be assured in large-sized animals. Electrical stunning can be used as a first step toward euthanasia, when it is executed properly. The electrodes should be placed properly, and when applicable, the skin should be shaven to ensure good electrical contact. Under all circumstances, the operator must ensure adequate voltage and current to be applied to induce immediate unconsciousness. Following stunning, the animal must be exsanguinated immediately to ensure death.

Other physical methods to be considered, such as penetrating captive bolt or even shooting should be considered only when other, more acceptable methods, have been found to be impossible to use. Euthanasia by injection can be achieved swiftly and reliably by intravenous administration of pentobarbital, but in larger animals, the volume of injection may limit its use. More concentrated formulations of the drug are available in some countries, but with their use, the aspect of operator safety should be taken into consideration. Other injectable agents used for euthanasia such as T61 also have to be administered intravenously. With the use of KCl, a state of general anesthesia is mandatory.

Euthanasia by inhalation can be performed with the aid of any of the common volatile anesthetics. Alternatively, $CO_2$ at concentrations of >70%, can be employed to induce unconsciousness as it is done in the process of slaughtering. After the animal is rendered unconscious, it must be exsanguinated immediately to ensure death.

## Euthanasia of Cats, Dogs, and Ferrets

While physical methods have been described for use in these species, their application in general is not recommended. The use of the captive bolt technique or electrocution has been described in dogs, but in general, prior sedation will be mandatory. Thereby, the potential advantage of eliminating pharmacological contamination is eliminated, making chemical euthanasia methods a better choice.

Intravenous injection of pentobarbital and T61 are viable alternatives. Intracardiac administration of pentobarbital may only be performed under full anesthesia. Acceptable methods of euthanasia by inhalation include the use of an overdose of the commonly used volatile anesthetics (halothane, enflurane, and isoflurane). Controversy exists as to the ethical acceptability of using $CO_2$ or CO for euthanasia of carnivores.

## Euthanasia of Primates

The use, and consequently, the euthanasia of nonhuman primates, will only be allowed under the strictest of regulations in most countries. Only scarce literature data is available on primate euthanasia. Most importantly, measures need to be taken to minimize stress and anxiety prior to euthanasia. This can adequately be achieved by allowing the intervention to take place in the animal's own surroundings and out of sight or hearing from other primates.

Sedation should be administered to facilitate handling and the further administration of drugs (see section on Anesthesia and Analgesia of Primates). As a general rule, an overdose of a general anesthetic or hypnotic, such as pentobarbital, is considered the only acceptable method of euthanasia in primates.[91]

## Euthanasia of Birds, Fish, Reptiles, and Amphibians

Physical methods such as concussion, decapitation, and cervical dislocation in birds of <3 kg and maceration of young chicks (<72 h) have been described and are considered to be ethically acceptable under proper conditions. Alternatively, inhalation of commonly used volatile anesthetics, $CO_2$ or CO, can be employed, whereby the last alternative involves clear operator health hazards. Intraperitoneal injection of pentobarbital results in rapid death, and alternatively, T61 may be injected intramuscularly in the pectoral muscle of small birds.

Fish and amphibians can be euthanized by a blow to the head followed by immediate destruction of the brain. Cervical dislocation of small fish may be performed but must also be followed by destruction of the brain. Chemical means of euthanizing fish or amphibians include the use of the commonly used volatile anesthetics, $CO_2$ and pH-corrected tricaine methane sulfonate (MS-222) or benzocaine dissolved in water in which the animal is placed.

When single animals are to be euthanized, injections by the intravenous or intraperitoneal routes (fish), or in the dorsal lymph sac (frog) with pentobarbital or T61 may be preferred. The major concern in euthanizing reptiles involves the resistance of their brain to hypoxic conditions. It has been shown that the severed head can respond to stimuli for some time following decapitation. Therefore, proper techniques for euthanasia other than by chemical methods should involve destruction of the brain, like with the captive bolt or shooting. Chemical methods include the administration of pentobarbital, intravenously or intraperitoneally.

# REFERENCES

1. Morton, D.B., Jennings, M., Buckwell, A., Ewbank, R., Godfrey, C., Holgate, B., Inglis, I., James, R., Page, C., Sharman, I., Verschoyle, R., Westall, L., and Wilson, A.B., Refining procedures for the administration of substances, *Lab. Anim.*, 35, 1, 2001.

2. Skarda, R.T., Local and regional anesthetic and analgesic techniques: dogs, in *Lumb & Jones' Veterinary Anesthesia*, third ed., Thurmon, J.C., Tranquilli, W.J., Benson, G.J., Eds., Williams & Wilkins, Baltimore, MD, 1996.

3. Nolan, A.M., Pharmacology of analgesic drugs, in *Pain Management in Animals*, Flecknell, P. and Waterman-Pearson, A., Eds., W.B. Saunders, Philadelphia, PA, 2000.

4. Hanna, R.M. et al., Pharmacokinetics of ketamine HCL and metabolite I in the cat: a comparison of IV, IM, and rectal administration, *J. Vet. Pharmacol. Therap.*, 11, 84, 1988.

5. Robertson, S.A., Taylor, P.M., Bloomfield, M., and Sear, J.W., Buprenorphine disposition after buccal administration in cats: preliminary observations, Proceedings, AVA, Spring 2001.

6. Dobromylskyj, P., Flecknell, P.A., Lascelles, B.D., Pascoe, P.J., Taylor, P., and Waterman-Pearson, A., Management of postoperative and other acute pain, in *Pain Management in Animals*, Flecknell, P. and Waterman-Pearson, A., Eds., W.B. Saunders, Philadelphia, PA, 2000.

7. Claassen, V., Neglected factors in pharmacology and neuroscience research, Vol. 12, in *Techniques in the Behavioural and Neural Sciences*, Huston, J.P., Ed., Elsevier, Amsterdam; New York, 1994.

8. Wixson, S.K. and Smiler, K.L., Anesthesia and analgesia in rodents, in *Anesthesia and Analgesia in Laboratory Animals*, Kohn, D.F., Wixson, S.K., White, W.J., and Benson, G.J., Eds., ACLAM and Academic Press, New York, 1997.

9. Brunson, D.B., Pharmacology of inhalation anesthetics, in *Anesthesia and Analgesia in Laboratory Animals*, Kohn, D.F., Wixson, S.K., White, W.J., and Benson, G.J., Eds., ACLAM and Academic Press, New York, 1997.

10. Hedenqvist, P., Roughan, J.V., Antunes, L., Orr, H., and Flecknell, P.A., Induction of anaesthesia with desflurane and isoflurane in the rabbit, *Lab. Anim.*, 35, 172, 2001.

11. Flecknell, P., Roughan, J.V., and Hedenqvist, P., Induction of anaesthesia with sevoflurane and isoflurane in the rabbit, *Lab. Anim.*, 33, 41, 1999.

12. Hedenqvist, P., Roughan, J.V., Antunes, L., Orr, H., Flecknell, P.A., Induction of anaesthesia with desflurane and isoflurane in the rabbit, *Lab. Anim.*, 35, 172, 2001.

13. Kolesnikov, Y., Chereshnev, I., and Pasternak, G.W., Analgesic synergy between topical lidocaine and topical opioids, *JPET*, 295, 546, 2000.

14. Baamonde, A., Alvarez-Vega, M., Hidalgo, A., and Menendez, L., Effects of intraplantar morphine in the mouse formalin test, *Jpn. J. Pharmacol.*, 83, 154, 2000.

15. Lascelles, B.D., Cripps, P.J., Jones, A., and Waterman, A.E., Post-operative central hypersensitivity and pain: the pre-emptive value of pethidine for ovariohysterectomy, *Pain*, December, 73, 3, 461–471, 1997.

16. Lascelles, B.D., Waterman, A.E., Cripps, P.J., Livingston, A., and Henderson, G., Central sensitization as a result of surgical pain: investigation of the pre-emptive value of pethidine for ovariohysterectomy in the rat, *Pain*, 62, 201, 1995.

17. Heavner, J.E., Pharmacology of analgesics, in *Anesthesia and Analgesia in Laboratory Animals*, Kohn, D.F., Wixson, S.K., White, W.J., and Benson, G.J., Eds., ACLAM and Academic Press, New York, 1997.

18. Swindle, M., Surgery, *Anesthesia & Experimental Techniques in Swine*, Iowa State University Press, 1998.

19. Hedenqvist, P., Roughan, J.V., and Flecknell, P.A., Sufentanil and medetomidine anaesthesia in the rat and its reversal with atipamezole and butorphanol, *Lab. Anim.*, 34, 244, 2000.

20. Hu, C., Flecknell, P.A., and Liles, J.H., Fentanyl and medetomidine anaesthesia in the rat and its reversal using atipamazole and either nalbuphine or butorphanol, *Lab. Anim.*, 26, 15, 1992.

21. Van Ham, L.M., Nijs, J., Mattheeuws, D.R., and Vanderstraeten, G.G., Sufentanil and nitrous oxide anaesthesia for the recording of transcranial magnetic motor evoked potentials in dogs, *Vet. Rec.*, 138, 642, 1996.

22. Bowser, P.R., Anesthetic options for fish, in *Recent Advances in Veterinary Anesthesia and Analgesia: Companion Animals*, Gleed, R.D. and Ludders, J.W., Eds., International Veterinary Information Service, 2001, www.ivis.org.

23. Lukasik, V.M., Premedication and sedation, in *Manual of Small Animal Anaesthesia and Analgesia*, Seymour, C. and Gleed, R., Eds., BSAVA, 1999.

24. Skolleborg, K.C., Gronbech, J.E., Grong, K., Abyholm, F.E., and Lekven, J., Distribution of cardiac output during pentobarbital versus midazolam/fentanyl/fluanisone anaesthesia in the rat, *Lab. Anim.*, 24, 221, 1990.

25. Ludders, J.W., Inhalant anaesthetics, in *Manual of Small Animal Anaesthesia and Analgesia*, Seymour, C. and Gleed, R., Eds., BSAVA, 1999.

26. Steffey, E.P., Inhalation anesthetics, in *Lumb & Jones' Veterinary Anesthesia*, third ed., Thurmon, J.C., Tranquilli, W.J., and Benson, G.J., Eds., Williams & Wilkins, Baltimore, MD, 1996.

27. Reid, J. and Nolan, A.M., Intravenous anesthetics, in *Manual of Small Animal Anaesthesia and Analgesia*, Seymour, C. and Gleed, R., Eds., BSAVA, 1999.

28. Glen, J.B., Animal studies of the anaesthetic activity of ICI 35 868, *Br. J. Anaesth.*, 52, 731, 1980.

29. Hellebrekers, L.J., de Boer, E.J., van Zuylen, M.A., and Vosmeer, H., A comparison between medetomidine-ketamine and medetomidine-propofol anaesthesia in rabbits, *Lab. Anim.*, 31, 58, 1997.

30. Silverman, J., Muir, III, W.W., A review of laboratory animal anesthesia with chloral hydrate and chloralose, *Lab. Anim. Sci.*, 43, 210, 1993.

31. Albrecht, D. and Davidowa, H., Action of urethane on dorsal lateral geniculate neurons, *Brain Res. Bull.*, 22, 923, 1989.

32. Green, C.J., Knight, J., Precious, S., and Simpkin, S., Metomidate, etomidate and fentanyl as injectable anaesthetic agents in mice, *Lab. Anim.*, 15, 171, 1981.

33. Popilskis, S.J. and Kohn, D.F., Anesthesia and analgesia in nonhuman primates, in *Anesthesia and Analgesia in Laboratory Animals*, Kohn, D.F., Wixson, S.K., White, W.J., and Benson, G.J., Eds., ACLAM and Academic Press, New York, 1997.

34. Smiler, K.L., Stein, S., Hrapkiewicz, K.L., and Hiben, J.R., Tissue response to intramuscular and intraperitoneal injections of ketamine and xylazine in rats, *Lab. Anim. Sci.*, 40, 60, 1990.

35. Beyers, T.M., Richardson, J.A., and Prince, M.D., Axonal degeneration and self-mutilation as a complication of the intramuscular use of ketamine and xylazine in rabbits, *Lab. Anim. Sci.*, 41, 519, 1991.

36. Hildebrand, S.V., Paralytic agents, in *Anesthesia and Analgesia in Laboratory Animals*, Kohn, D.F., Wixson, S.K., White, W.J., and Benson, G.J., Eds., ACLAM and Academic Press, New York, 1997.

37. Van Woerkens, E.C., Trouwborst, A., Duncker, D.J., Koning, M.M., Boomsma, F., and Verdouw, P.D., Catecholamines and regional hemodynamics during isovolemic hemodilution in anesthetized pigs, *J. Appl. Physiol.*, 72, 760, 1992.

38. Schou, H., Kongstad, L., Perez de Sa, V., Werner, O., and Larsson, A., Uncompensated blood loss is not tolerated during acute normovolemic hemodilution in anesthetized pigs, *Anesth. Analg.*, 87, 786, 1998.

39. Jones, R.S., Neuromuscular blockade, in *Manual of Small Animal Anaesthesia and Analgesia*, Seymour, C. and Gleed, R., Eds., BSAVA, 1999.

40. Stenqvist, O., Inhalationsanestesi, in Halldin, M.A B., Lindahl, S.G.E., Eds., *Anestesi, Liber*, 2000.

41. Woolf, C.J., Chong, M.S., Pre-emptive analgesia — treating post-operative pain by preventing the establishement of central sensitisation, *Anest. Analg.*, 77, 372, 1993.

42. Hellebrekers, L.J., Pathophysiology of pain in animals and its consequences for analgesic therapy, in *Animal Pain, A Practice-Oriented Approach to an Effective Pain Control in Animals*, Hellebrekers, L.J., Ed., Van der Wees Publishers, Utrecht, The Netherlands, 2000.

43. Lascelles, B.D.X., Clinical pharmacology of analgesic agents, in *Animal Pain, A Practice-Oriented Approach to an Effective Pain Control in Animals,* Hellebrekers, L.J., Ed., Van der Wees Publishers, Utrecht, The Netherlands 2000.

44. Clutton, R.E., Anaesthetic equipment, in *Manual of Small Animal Anaesthesia and Analgesia*, Seymour, C. and Gleed, R., Eds., BSAVA, 1999.

45. Liles, J.H. and Flecknell, P.A., The effects of buprenorphine, nalbuphine and butorphanol alone or following halothane anaesthesia on food and water consumption and locomotor movement in rats, *Lab. Anim.*, 26, 180, 1992.

46. Roughan, J.V. and Flecknell, P.A., Behavioural effects of laparotomy and analgesic effects of ketoprofen and carprofen in rats, *Pain*, 90, 65, 2001.

47. Jablonski, P., Howden, B.O., and Baxter, K., Influence of buprenorphine analgesia on postoperative recovery in two strains of rats, *Lab. Anim.*, 35, 213, 2001.
48. Flecknell, P.A., Roughan, J.V., and Stewart, R., Use of oral buprenorphine ("buprenorphine jello") for postoperative analgesia in rats — a clinical trial, *Lab. Anim.*, 33,169, 1999.
49. Hayes, K.E., Raucci, J.A., Gades, N.M., and Toth, L.A., An evaluation of analgesic regimens for abdominal surgery in mice, *Contemp. Top. Lab. Anim. Sci.*, 39, 18, 2000.
50. Reichert, J.A., Daughters, R.S., Rivard, R., and Simone, D.A., Peripheral and preemptive opioid antinociception in a mouse visceral pain model, *Pain*, 89, 221, 2001.
51. Gades, N.M., Danneman, P.J., Wixson, S.K., and Tolley, E.A., The magnitude and duration of the analgesic effect of morphine, butorphanol, and buprenorphine in rats and mice, *Contemp. Top. Lab. Anim. Sci.*, 39, 8, 2000.
52. Liles, J.H. and Flecknell, P.A., A comparison of the effects of buprenorphine, carprofen and flunixin following laparotomy in rats, *J. Vet. Pharmacol. Ther.*, 17, 284, 1994.
53. Liles, J.H. and Flecknell, P.A., The influence of buprenorphine or bupivacaine on the post-operative effects of laparotomy and bile-duct ligation in rats, *Lab. Anim.*, 27, 374, 1993.
54. Jacobson, C., Adverse effects on growth rates in rats caused by buprenorphine administration, *Lab. Anim.*, 34, 202, 2000.
55. Clark, Jr., J.A., Myers, P.H., Goelz, M.F., Thigpen, J.E., and Forsythe, D.B., Pica behavior associated with buprenorphine administration in the rat, *Lab. Anim. Sci.*, 47, 300, 1997.
56. Flecknell, P.A., Anesthesia and analgesia for exotic species — rabbits, rodents and ferrets, in *Manual of Small Animal Anaesthesia and Analgesia*, Seymour, C. and Gleed, R., Eds., BSAVA, 1999.
57. Kolesnikov, Y., Chereshnev, I., and Pasternakjpet, G.W., Analgesic synergy between topical lidocaine and topical opioids, *JPET*, 295, 546, 2000.
58. Schreiber, S., Backer, M.M., Yanai, J., and Pick, C.G., The antinociceptive effect of fluvoxamine, *Eur. Neuropsychopharmacol.*, 6, 281, 1996.
59. Roughan, J.V., Ojeda, O.B., and Flecknell, P.A., The influence of pre-anaesthetic administration of buprenorphine on the anaesthetic effects of ketamine/medetomidine and pentobarbitone in rats and the consequences of repeated anaesthesia, *Lab. Anim.*, 33, 234, 1999.
60. Hedenqvist, P., Roughan, J.V., and Flecknell, P.A., Effects of repeated anaesthesia with ketamine/medetomidine and of pre-anaesthetic administration of buprenorphine in rats, *Lab. Anim.*, 34, 207, 2000.
61. Chaves, A.A., Weinstein, D.M., and Bauer, J.A., Non-invasive echocardiographic studies in mice: influence of anesthetic regimen, *Life Sci.*, 69, 213, 2001.
62. Balaban, R.S. and Hampshire, V.A., Challenges in small animal noninvasive imaging, *ILAR J.*, 42, 248, 2001.
63. Rao, S. and Verkman, A.S., Analysis of organ physiology in transgenic mice, *Am. J. Physiol. Cell. Physiol.*, 279, C1, 2000.
64. Dalkara, T., Irikura, K., Huang, Z., Panahian, N., and Moskowitz, M.A., Cerebrovascular responses under controlled and monitored physiological conditions in the anesthetized mouse, *J. Cereb. Blood. Flow. Metab.*, 15, 631, 1995.
65. Berul, C.I., Aronovitz, M.J., Wang, P.J., and Mendelsohn, M.E., *In vivo* cardiac electrophysiology studies in the mouse, *Circulation*, 94, 2641, 1996.
66. Univentor 400 Anaesthesia Unit, AgnTho´s AB, Lidingö, Sweden, http://www.agnthos.se.
67. Simpson, D.P., Prolonged (12 hours) intravenous anesthesia in the rat, *Lab. Anim. Sci.*, 47, 519, 1997.
68. Brammer, A., West, C.D., and Allen, S.L., A comparison of propofol with other injectable anaesthetics in a rat model for measuring cardiovascular parameters, *Lab. Anim.*, 27, 250, 1993.
69. Eger, R.P. and MacLeod, B.A., Anaesthesia by intravenous emulsified isoflurane in mice, *Can. J. Anaesth.*, 42, 173, 1995.
70. Papaioannou, V.E. and Fox, J.G., Efficacy of tribromoethanol anesthesia in mice, *Lab. Anim. Sci.*, 43, 189, 1993.
71. Olson, M.E., Vizzutti, D., Morck, D.W., and Cox, A.K., The parasympatholytic effects of atropine sulfate and glycopyrrolate in rats and rabbits, *Can. J. Vet. Res.*, 58, 4, 254, 1994.
72. Hedenqvist, P., Roughan, J.V., and Orr, H.E., Antunes LM, Assessment of ketamine/medetomidine anaesthesia in the New Zealand White Rabbit, *Vet. Anaesthesia and Analgesia*, 28, 18, 2001.

73. Andersen, H.E., Fosse, R.T., Kuiper, K.K., Nordrehaug, J.E., and Pettersen, R.J., Ketorolac (Toradol) as an analgesic in swine following transluminal coronary angioplasty, *Lab. Anim.*, 32, 307, 1998.

74. Boschert, K., Flecknell, P.A., Fosse, R.T., Framstad, T., Ganter, M., Sjostrand, U., Stevens, J., and Thurman, J., Ketamine and its use in the pig. Recommendations of the Consensus meeting on Ketamine Anaesthesia in Pigs, Bergen, 1994. Ketamine Consensus Working Group, *Lab. Anim.*, 30, 209, 1996.

75. Husby, P., Heltne, J.K., Koller, M.E., Birkeland, S., Westby, J., Fosse, R., and Lund, T., Midazolam-fentanyl-isoflurane anaesthesia is suitable for haemodynamic and fluid balance studies in pigs, *Lab. Anim.*, 32, 316, 1998.

76. Grant, C., Summersides, G.E., and Kuchel, T.R., A xylazine infusion regimen to provide analgesia in sheep, *Lab. Anim.*, 35, 277, 2001.

77. Dunlop, C.I. and Hoyt, R.F., Anesthesisa and analgesia in ruminants, in *Anesthesia and Analgesia in Laboratory Animals*, Kohn, D.F., Wixson, S.K., White, W.J., and Benson, G.J., Eds., ACLAM and Academic Press, New York, 1997.

78. Ko, J.C., Lange, D.N., Mandsager, R.E., Payton, M.E., Bowen, C., Kamata, A., and Kuo, W.C., Effects of butorphanol and carprofen on the minimal alveolar concentration of isoflurane in dogs, *J. Am. Vet. Med. Assoc.*, 217, 1025, 2000.

79. Ko, J.C., Fox, S.M., and Mandsager, R.E., Sedative and cardiorespiratory effects of medetomidine, medetomidine-butorphanol, and medetomidine-ketamine in dogs, *J. Am. Vet. Med. Assoc.*, 216, 1578, 2000.

80. Ko, J.C., Smith, T.A., Kuo, W.C., and Nicklin, C.F., Comparison of anesthetic and cardiorespiratory effects of diazepam-butorphanol-ketamine, acepromazine-butorphanol-ketamine, and xylazine-butor-phanol-ketamine in ferrets, *J. Am. Anim. Hosp. Assoc.*, 34, 407, 1998.

81. Hellebrekers, L.J., van Herpen, H., Hird, J.F., Rosenhagen, C.U., Sap, R., and Vainio, O., Clinical efficacy and safety of propofol or ketamine anaesthesia in dogs premedicated with medetomidine, *Vet. Rec.*, 142, 631, 1998.

82. Hellebrekers, L.J. and Sap, R., Medetomidine as a premedicant for ketamine, propofol or fentanyl anaesthesia in dogs, *Vet. Rec.*, 140, 545, 1997.

83. Forbes, N.A., Anesthesia and analgesia for exotic species — birds, in *Manual of Small Animal Anaesthesia and Analgesia*, Seymour, C. and Gleed, R., Eds., BSAVA, 1999.

84. Rodger, H., Anesthesia and analgesia for exotic species — fish, in *Manual of Small Animal Anaesthesia and Analgesia*, Seymour, C. and Gleed, R., Eds., BSAVA, 1999.

85. Bowser, P.R., Anesthetic options for fish, in *Recent Advances in Veterinary Anesthesia and Analgesia: Companion Animals*, Gleed, R.D. and Ludders, J.W., Eds., International Veterinary Information Service, 2001, www.ivis.org.

86. Malley D, Anesthesia and analgesia for exotic species — reptiles, in *Manual of Small Animal Anaesthesia and Analgesia*, Seymour, C. and Gleed, R., Eds., BSAVA, 1999.

87. Greer, L.L., Jenne, K.J., and Diggs, H.E., Medetomidine-ketamine anesthesia in red-eared slider turtles (Trachemys scripta elegans), *Contemp. Top. Lab. Anim. Sci.*, 40, 9, 2001.

88. Schaeffer, D.O., Anesthesia and analgesia in nontraditional laboratory animal species, in *Anesthesia and Analgesia in Laboratory Animals*, Kohn, D.F., Wixson, S.K., White, W.J., and Benson, G.J., Eds., ACLAM and Academic Press, New York, 1997.

89. AVMA, Report of the AVMA Panel on euthanasia, *JAVMA*, 218, 669, 2001.

90. Working party report, Recommendations for euthanasia of experimental animals (Part 1), *Lab. Anim.*, 30, 293, 1996.

91. Working party report, Recommendations for euthanasia of experimental animals (Part 2), *Lab. Anim.*, 31, 1, 1997.

92. Hellebrekers, L.J., Baumans, V., Bertens, A.P., and Hartman, W., On the use of T61 for euthanasia of domestic and laboratory animals, an ethical evaluation, *Lab. Anim.*, 24, 200, 1990.

93. Raj, A.B.M. and Gregory, N.G., Effect of rate of induction of carbon dioxide anaesthesia on the time of onset of unconsciousness and convulsions, *Res. Vet. Sci.*, 49, 360, 1990.

94. Coenen, A.M.L., Drinkenburg, W.H.I.M., Hoenderken, R., and van Luijtelaar, E.L.J.M., Carbon dioxide euthanasia in rats: oxygen supplementation minimizes signs of agitation and asphyxia, *Lab. Anim.*, 29, 262, 1995.

95. Hackbarth, H., Küppers, N., and Bohnet, W., Euthanasia of rats with carbon dioxide — animal welfare aspects, *Lab. Anim.*, 34, 91, 2000.

# CHAPTER 18

# Welfare Assessment and Humane Endpoints

David B. Morton and Jann Hau

## CONTENTS

## INTRODUCTION

There are ethical and legal obligations to treat animals in a humane manner and to ensure high standards with respect to their husbandry and care, resulting in maximizing their well-being under the constraints of their use in science. The use of animals by humans is always associated with responsibility for the

0-8493-1086-5/03/$0.00+$1.50
© 2003 by CRC Press LLC

welfare of the animals concerned. However, when animals are used for research and subjected to procedures that may cause them pain, distress, suffering, and lasting harm, there are additional obligations to refine those techniques so as to cause only the minimum of pain and distress. It is important to separate the pain and distress that may be an unavoidable component of the research from that which is avoidable through, for example, better experimental technique or husbandry. Russell and Burch as long ago as 1959[1] referred to them as direct (necessary) and contingent (avoidable) inhumanities. They considered contingent inhumanity as being incidental or inadvertent (which may not be quite the same as avoidable), but it is generally agreed that any animal suffering not required to achieve the scientific objective can only confound the experiment being conducted. Many scientists are aware that severely stressed animals are not suitable for valid biomedical research, or as Russell and Burch wrote: "For some time it has been obvious to experimental biologists that severe pain, fear or in general distress in experimental animals, is an unmitigated nuisance in nearly all kinds of investigation…the effect of the treatment under study is utterly confused by the effects of the distress, and the animal ceases to be of the slightest use as a model of normal physiological function." While this is not always true (e.g., gaining manual skills under a badly administered anesthetic), maintaining high standards of animals' well-being is still an ethical, legal, and scientific obligation.

Welfare can be defined in different ways.[2,3,4] Some definitions focus on an animal's emotional perception of its environment, i.e., the stimuli that emanate from that environment.[5,6] Other definitions focus on how animals cope with their environment and how they have to compensate for hostile changes in their environment by behavioral and physiological adjustments. This latter definition has the advantage in that coping mechanisms may be identified and measured, allowing practical advantageous changes to an animals' environment to be made, thus improving animal well-being.[7,8] A third approach is to compare an animal in its captive environment with its counterpart in the wild or in a more natural environment.[9]

Well-being and welfare are not synonymous, although they tend to be used as if they were. Well-being may be seen as the quality of life of an animal and is about how it is faring in the short term,[10] whereas welfare is a longer-term average (whole life) and, of course, like animal well-being, can be good or bad and so has always to be qualified. For example, in order to increase the welfare of an animal, it may be necessary to temporarily reduce its well-being while it is being restrained and vaccinated, but its welfare is increased because of a reduced risk of serious disease, and so the animal has a longer life with better quality. Note also that animal health is a subset of welfare and not the other way around. Animals can be perfectly healthy but still have a poor quality of life (e.g., confinement in too small a cage), i.e., quality of life is a mixture of psychological, physical, and physiological well-being. However, it can be generally stated that an animal's welfare is compromised when its physiological or physical health or its psychological well-being, in relation to its cognitive capacity, are affected negatively (Morton, personal communication).

Some people take the view that causing animals to suffer is a breach of the rights of the animal or a failure of humans to meet their obligations. Those who do not believe that animals have rights will still require some justification for imposing poor welfare. This justification has to be balanced by the perceived benefits from that use of animals and be in proportion to the harms done. This is why ethical debate is so important, as it tries to help rationalize and justify our uses of animals (see Chapter 2). In many countries, the use of animals in research has to be seriously ethically justified before their use is allowed, and a simple or trivial gain in scientific knowledge may not be enough. Notwithstanding that, any use of animals should only cause the minimum of suffering, and ethical frameworks help in that analysis as they start to evaluate key points in the debate.

## RECOGNITION AND ASSESSMENT OF ANIMAL WELL-BEING: THEORY

Unlike humans who can tell you when they are in pain or distress, or feel uncomfortable, animals cannot. The same applies to human babies, deaf mutes, and others who are conscious but unable to communicate. However, it would be foolish to deny that they could not experience pain and distress. The observer has therefore to recognize certain signs that indicate when such humans and animals are suffering. This places an obligation on scientists and animal technical staff that they should learn to do so, as only when they can recognize the signs of pain and distress in an animal can something be done about it, such as providing pain relief, or withdrawing the animal from the study, or humanely killing it. Fortunately, we can make certain predictions that hold good for all sentient beings (i.e., all animals that are able to experience pain and pleasure, etc.) and a good start is to ask whether a human in that condition would be experiencing pain or distress. But we need to be careful and apply what has been termed a "critical anthropomorphism."[11,12,13] This means that one has to take into account the biological characteristics of the species of animal and also, if possible, to have some idea of what an individual animal has experienced in its life. For example, signs of pain and distress will vary according to the species in so far as their anatomy differs. Animals that have been raised in the wild or always been in a captive environment may respond differently. Animals that have already been subjected to or conditioned to experimental procedures, such as injections or blood removal — with or without food reward — may behave differently from those that are naïve or unconditioned. Moreover, the experience an animal has had of a particular human being, e.g., the one doing the injection, will undoubtedly influence its behavior and physiological homeostasis. Publications examining the interactions between humans and experimental animals have clearly shown major effects,[14,15] including increased weight gain in gentled rats as compared with rats unaccustomed to humans. These become important issues when handling an individual animal, but of increasing importance is how we can measure animal pain and distress in other situations, e.g., after carrying out a scientific procedure, and whether this can be done reliably. In addition to being an ethical mandate, avoiding animal suffering and discomfort during the experiment is likely to produce better science,[16,17,18,19] as so convincingly presented by the many examples listed by Chance as early as 1957.[20] These included the incidence of cancer in mice being related to numbers in a cage; blood eosinophils levels in mice being altered by sounds; crowding having an impact on the susceptibility of rats to tuberculosis; and stress influencing neoplastic events and immunocompetency.

## ANIMAL SUFFERING

Suffering can be defined as a negative emotional state that derives from adverse physical, physiological, and psychological circumstances, in accordance with the cognitive capacity of the species and of the individual being (Morton, personal communication). Suffering is an emotive word and some believe only humans can truly suffer. In older literature, the expressions "ill-at-ease" and "uncomfortable" were used to describe what we today might call suffering. An example of this is the situation for dogs kept in cages slightly too small for them, resulting in a disturbed metabolism rendering them useless for "biochemical" studies.[21] Suffering incorporates notions of self-awareness in relation to others and over time. It is still the subject of some debate about whether this is a uniquely human attribute or if animals have this capacity, and if so to what degree? Increasingly, scientific evidence is accumulating that some of the higher nonhuman primates, such as chimpanzees, gorillas, and baboons, are self-aware, and there is plenty of anecdotal evidence for other species including parrots and dolphins.[22,23] Self-awareness and self-consciousness add another dimension to the "simple" physical experience of pain and mental distress, such as anticipating adverse outcomes and frustrations of future life plans. In this chapter, we use "suffering" to cover all those feelings an animal might experience that we as humans would normally find unpleasant or aversive in some way, such as pain, mental distress, fear, and boredom. Such suffering can vary in intensity and duration and is difficult to measure and quantify. In many countries, it is intensity that is the primary concern, and that is reported in annual statistics, rather than some mathematical approximation of total suffering such as Intensity × Duration. The impact of these adverse states on an animal is probably best assessed using clinical signs and determining how far an animal has deviated from normality, rather then trying to quantify physiological responses, such as the level of endorphins or corticosteroids. The latter measurements are useful in a research setting into animal welfare, but when animals are being used in medical research and the like, then a holistic approach looking at an animals' ability to cope with the experimental conditions, is the most appropriate and most feasible.

Notwithstanding this holistic approach, it is worth understanding some of the physiological ways in which animals respond to adverse conditions. In addition to the immune system, there are at least four

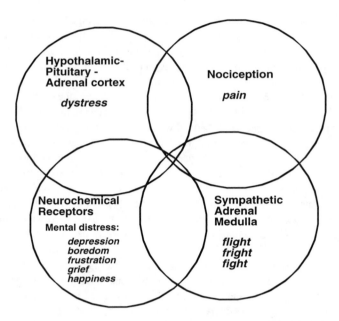

**Figure 18.1**    Venn diagram showing the various physiological response mechanisms in an animal and how they interact.

integrated body systems that come into play to varying degrees according to the stimulus and what is experienced by an animal (Figure 18.1).We ask the readers to always have at the back of their mind as they read this what they might feel themselves, how they might respond, and to acknowledge that animal research into these adverse mental states has resulted in the development of successful treatments for humans and animals such as analgesic, antidepressive, and anxiolytic drugs.

## PAIN

Pain is a highly conserved mechanism evolved to protect animals from serious injury. The pain perceiving or nociceptive system is common for mammals and probably birds and possibly throughout the vertebrates (Table 18.1). Potential tissue  damage is detected by free nerve endings called nociceptors that respond to excessive heat, pressure, and chemical stimulation. The nociceptors are linked to the dorsal horn of the spinal cord by myelinated A-delta and nonmyelinated c fibers. The nociceptive information is subsequently transferred to supraspinal nuclei in the thalamus by ascending neurons that cross over to the opposite side of the cord in the dorsal horn. These ascending neurons also synapse with motor efferent nerves in the dorsal horn, and together, they form the spinal reflex. This rapid response (milliseconds) withdraws the affected body part away from the adverse threat, thus helping to protect the animal from injury. The release of nerve transmitters from the neurons ensure that the message proceeds to the brain and up to the cerebral cortex, when the animal perceives this release as the feeling of pain. Various neuronal links with other parts of the brain collate this feeling with memory, and the animal then has choices of how to respond, so-called retroduction,[24] so the animal decides what course of action to take, e.g., to flee, to fight, to vocalize, etc. At this point in time, several seconds after the stimulus, the animal is likely to be aware of what it is doing and not responding reflexly or even instinctively.

The pain perception is accompanied, and is mitigated, by the release of endorphins in the brain and spinal cord, as they raise the threshold for pain perception. In addition to this release of endorphins in the central nervous system (CNS), there are descending inhibitory fibers that pass down the spinal cord from the brain and synapse with neurons in the dorsal horn; these also raise the threshold for upward transmission to the brain. These descending fibers do not appear to be present at birth in many species (e.g., rodents, humans), even though the ascending nerves are present and active, and so young, immature

**Table 18.1  Comparison Between Animals of Some Anatomical Structures and Biological Mechanisms of Importance to Pain Perception**

|  | Invertebrates | | | Vertebrates | | | | |
|---|---|---|---|---|---|---|---|---|
|  | Earthworm | Insects | Octopus | Fish | Amphibian | Reptiles | Birds | Mammals |
| Nociceptive system | ? | — | ? | ? | ? | ? | + | + |
| Brain structure analogous to human cerebral cortex | — | — | ? | + | + | + | + | + |
| Nociceptors connected to higher brain structure | — | — | ? | ? | ? | ? | ? | + |
| Opioid-type receptors | + | + | ? | + | + | + | + | + |
| Response modified by analgesia | ? | ? | ? | ? | ? | ? | + | + |
| Response to noxious stimuli persists | — | — | +/− | ? | ? | ? | + | + |
| Associates neural with noxious stimuli | — | + | + | + | + | + | + | + |

(+) present, (−) absent, (?) not known, (+/−) some weak evidence

*Source:* Adapted from Smith, J.A. and Boyd, K.M., Eds., Lives in the Balance: The Ethics of Using Animals in Biomedical Research, the report of a working party of the Institute of Medical Ethics, Oxford University Press, Oxford, 1991.

**Table 18.2  Stage of Development and Likely Sentience**

| Adverse effect | Zygote | Embryo | Fetus (Early) | Fetus (Late)[a] | Perinatal | Preweaning | Prepubertal | Adult |
|---|---|---|---|---|---|---|---|---|
| Pain | — | — | — | + | +++ | +++ | ++ | ++ |
| Dystress | — | — | — | ++ | ++ | +++ | +++ | +++ |
| Mental distress | — | — | — | + | + | + | ++ | +++ |
| Fear | — | — | — | + | + | + | ++ | +++ |

*Note:* Notional scale of absence [−−] or presence [+/++/+++] of sense-associated structures. NB for pain, the complete presence in a fully adult sentient form [++] is likely to mean that less pain is felt than in an immature form, when the nervous system's pain inhibitory pathways are not yet fully developed [hence, +++ cf ++]. For dystress, in some species, the system matures with age [++ to +++]. For mental distress and fear, an animal will learn from experience as it gets older [++ to +++] and as its system develops [+ to ++].

[a] Depends if it has breathed.[115]

animals may well feel more pain than older animals, in which the nervous system is fully formed and functional.[25] Table 18.2 indicates the likelihood of animals feeling as measured by the development of the various physiological response mechanisms.

There are also nociceptors in the organs of the body that link to the central nervous system by afferent neurons of the autonomic nervous system. These are sensitive to stretch (e.g., the gut) and also ischemia (heart). Visceral pain is less easily diagnosed and controlled than the somatic pain described above.

In most experimental work, pain is an unwanted side effect, and so it is necessary to relieve it. This is best achieved by exercising what is known as "preemptive analgesia" (see Chapter 17). If pain is not relieved, then neurons adjacent to the activated one in the dorsal horn of the spinal cord will also be sensitized, and the painful area will extend laterally, and so the animal will feel more pain (hyperalgesia); moreover, stimuli that are normally not painful become painful (allodynia). Pain probably cannot be measured from endorphin levels or nerve activity, but parametric observation of behavior, for instance, may be helpful. Behaviors exhibited will largely depend on the site affected (e.g., immobility, posture when abdominal muscles have been traumatized, lameness after surgery or arthritis, depth of respiration after thoracotomy, vocalizing when the site is pressed) as well as other clinical signs that can be measured, such as eating and drinking, body weight, and temperature.

## DYSTRESS*

When animals are exposed to environmental stressors such as extremes of temperatures, or when they perceive a threat by virtue of their senses (e.g., sound and sight), neurons in the CNS are stimulated, and this results in activation of the hypothalamic pituitary adrenal (HPA) axis leading to, among other things, the release of adrenocorticotrophic hormone (ACTH). This in turn will stimulate the cortex of the adrenal gland to synthesize and release corticosteroids that are predominantly either cortisol or corticosterone, depending on the species. Thus, if high levels of corticosteroids are present in the circulation and they persist for long periods of time, despite their short half-lives, it would not be unreasonable to conclude that the animal was not coping and was being subjected to a "bad stress" — hence, the word "dystress" (compare with distress below). If the animal does not adapt, then other activities of the hypothalamic axis are affected, and this impacts on nearly all the homeostatic mechanisms of the body — hence, the potential impact on the scientific objective. Such animals show poor growth rates, even body weight loss, as well as disturbances of the endocrine system to the point where the animal is no longer fit in a biological sense. It may not be able to breed and reproduce, and it may show considerable pathology, and in general, the animal becomes more susceptible to infectious and noninfectious diseases due to suppression of the immune system.[26] Dystress is, by and large, an unconscious phenomenon and can be measured by looking at the level of circulating ACTH and corticosteroids, the response of the adrenal gland to ACTH, signs of pathology, as well as the functioning of whole body systems such as the endocrine system, the immune system, and the reproductive system.

## FEAR

Fear that stimulates an animal to flee or fight is a short-term stress that again involves the adrenal gland, but this time, the medulla, where catecholamines such as adrenalin, noradrenalin, and dopamine are released. It helps prepare the animal physiologically for whatever action it decides to take, and there is a redistribution of blood to vital organs such as muscle, increased heart rate and blood pressure, increased acuity of the sense organs for hearing and sight, and in some species, piloerection (cats and dogs) or a tonic immobility (rabbits, chickens). Fear can be measured through measuring end organ functions such as heart rate, muscle tone, as well as circulating catecholamines and behavior.

---

* Dystress is coined from the Greek word "dus" having a connotation of bad, i.e., bad stress.

## MENTAL DISTRESS

Mental distress is the product of an animal sensing and responding to a hostile environment and involves a combination of memory of early experiences and diverse feelings. As such, it reflects the psychological well-being of an animal and feelings such as boredom, frustration, anxiety, and anticipating adverse events. The other side of this coin is positive well-being and feelings such as contentment (cat purring), happiness (dog wagging tail), and so on. These feelings are recognizable mainly from the behavior of an animal, although research in neuroscience is uncovering some of the underlying mechanisms such as receptor molecules for benzodiazepines, endorphins, serotonin, and noradrenaline. Animals may show stereotyped behavior when they are kept in environmental conditions that are inadequate in some way or another. It is not always possible to decide what an environment should provide, and possibly one that bears some relationship to that in which a species has evolved would be a good starting point (without the predators!). After that, one can realistically only compare types of husbandry systems e.g., caging versus pen housing. Measurements of the level of diversity of behaviors through compiling an ethogram (during the day and the night, as many laboratory animals, such as rodents, are crepuscular or nocturnal), what environments animals will choose to inhabit, how hard animals will work to reach such an environment, and the percentage of time they spend carrying out stereotypies will all provide important clues as to how well the environment is meeting an animal's needs.[27] A stereotyped behavior is defined as one that is sometimes abnormal, i.e., not part of an animal's normal behavioral repertoire, is frequently repeated, is unvarying in its form, and appears to serve no useful purpose within that animal's current environment, e.g., pacing, chewing bars, digging, and even prolonged periods of inactivity. These behaviors are thought to reflect inner feelings such as frustration (digging) and boredom (inactivity, pacing, chewing).[28,29] To be able to carry out some behaviors seems important to some animals at some times, and they will work hard to gain access to environments that enable them to carry out such activities, e.g., nest building in pregnant sows a few days before birth. These behaviors may be seen as physiological needs for keeping certain essential variables within a more or less narrow range, necessary for maintaining life,[30] as opposed to simply being part of an animals' normal repertoire. It is obviously important to be able to identify these needs in order not to inadvertently stress an animal. Interestingly, by providing a more natural or diversified environment, some basic behaviors are carried out in more natural ways. For example, rodents having access to the same food in a hopper and on the floor in the substrate will preferentially forage in the substrate rather than take it from the hopper.

Not meeting these needs may impact adversely on the development of the brain of animals and so on the science.[31,32] Recent research demonstrated that the CNS neuroreceptor density is changed in animals showing certain stereotypic behavior patterns, and some studies claim to have reversed these types of behaviors by the use of antidepressive drugs such as serotonin reuptake inhibitors.

In conclusion, it is not possible to recognize and measure all these adverse states on the basis of a single simple measure. There is no level of hormone or simple behavior, even in humans, that correlates with what human patients report as their experience. All vertebrate animals have integrated genetically programmed bodies, conditioned by experience, with adaptability for survival. One cannot look at such complicated organisms as a single system but as one that responds to adverse states according to its intrinsic biology and experience. Take for example a dog in pain. It may well be frightened of what will happen next when the scientist who operated on her comes in through the door, she may change her behavior and become aggressive as a result of what she is feeling and what she sees as potential outcomes. All of these responses are a result of how that animal feels at that time, in those circumstances, its memory and its well-being.

## RECOGNITION AND ASSESSMENT OF ANIMAL WELL-BEING: PRACTICAL APPLICATION

Recognition and assessment of adverse states in animals will vary according to the species, the system of husbandry, and, of course, the individual response of an animal [biological variation is such that animals will inevitably differ even though what is being done to them appears to be the same (e.g., lethal dose tests where at the LD50 dose, 50% of the animals die and 50% survive)]. Nevertheless, systems for recognizing and scoring animal suffering in a controlled laboratory environment, and in the context of good husbandry systems that impose a minimum of stress, and in developing reliable scientific and humane endpoints can be made, and these are described below.

Evaluation of clinical signs such as whether an animal is eating and drinking, its body weight, its appearance, posture, and behavior become crucial for underpinning the science and the well-being of that individual animal.[33-38] Some scientists see assessment of clinical signs as not being objective and as a matter of subjective interpretation; nothing could be further from the truth. What is true is that some clinical signs are metric and can be quantified (like body weight, body temperature, respiration and heat rates, behavior), whereas others are parametric and cannot. These would include signs like quality of respiration (deep, shallow, labored), posture, appearance such as closed eyes, disturbed pelage as with ruffled fur or feathers, diarrhea, coughing, and convulsions (Table 18.3). Such parametric signs tend to be present or not present, though it is often possible to grade them in some way or another. For example, pain in the foot such as a broken bone causing an animal (or human) to limp is a repeatable clinical observation, backed up in this case from verbal evidence in humans that the limping is painful because it hurts to bear weight on the foot. Moreover, an animal may be hopping lame or bear some weight and be limping. Both of these signs are objective observations in exactly the same way that one can observe the stars or that oil floats on water. The cause of the lameness may indeed vary, but the impact on the animal is what matters in this context, and if that pain affects the animal's appetite and it loses body weight, or its body temperature is raised, then we start to get an idea of the overall impact of the experimental or clinical condition on that animal, and its ability to cope.

The following is a list (based on OECD, Environmental Health and Safety Publications Series on Testing and Assessment No. 19, Guidance Document on the Recognition, Assessment, and Use of Clinical Signs as Humane Endpoints for Experimental Animals Used in Safety Evaluation, 14, 2001) of common conditions and clinical signs that may be indicative that an animal is experiencing pain and distress. By and large, the list is based on observations in rats and mice, but many of the signs also apply to other mammals used in toxicity testing. When one or more signs or conditions are observed, these should be documented in a written record with the dates of initial and subsequent observations and all treatments. If the animal is not humanely killed, a more detailed examination of the animal should be performed, the frequency of observation should be appropriately increased, and the cage or pen should be clearly marked. This list is not all-encompassing for every possibility that may occur, and animal care facilities should add other clinical signs and conditions that may be appropriate for specific studies.

**Table 18.3  Clinical Signs and Conditions Indicating the Need for Closer Observation, Treatment, or Humane Killing**

*Abortion*: May be detected by fetal remains on bedding, blood on bedding, decrease in abdominal size.

*Agalactia*: May be observed by no milk in stomachs of nursing rodents or failure to express milk from the mammary gland. Young will die, and if not cross-fostered or provided with supplemental nutrition or milk, should be humanely killed.

*Anemia*: Indicates a loss of blood (through feces, urine, reproductive tract) or poor red blood cell replenishment, to the extent that it produces clinical signs of labored or decelerated breathing (also discernible as pale membranes, pale ears and feet, dyspnea, hyperventilation).

*Analgesia*: See Reflexes.

*Anuria*: No urine flow (anuria) due to renal failure (it may be reduced) but worth checking for urine retention (see below).

*Apathy*: See Immobile/inactive.

*Ataxia/incoordination/staggering/unbalanced*: This can be due to neuromuscular coordination, weakness (check body weight), or postseizure recovery period. Observe carefully and continue to check body weight.

*Bleeding from any orifice*: See Anemia. Some internal hemorrhaging may be detectable as blood escapes from natural orifices. The seriousness will depend on the amount and frequency of the bleeds (q.v. anemia).

**Table 18.3 (continued) Clinical Signs and Conditions Indicating the Need for Closer Observation, Treatment, or Humane Killing**

*Blepharospasm*: See Eyelid closure. The cause is usually some damage to the eye, and this should be investigated further. If incidental to the study (particularly when only one eye is affected), then veterinary advice should be sought, and the animal may be treated or withdrawn from the study.

*Blood in feces or urine*: See Anemia.

*Blood around nose and eyes*: In rodents, it is necessary to differentiate between blood and porphyrin secretion. It is often a stress-related condition in rodents, the secretions not being removed by grooming. If it is blood, the presence in only one nostril may be due to physical injury.

*Boarded abdomen*: This may be detected by holding a small animal up to ear and squeezing abdomen gently. If breathing stops, then this is indicative of abdominal pain. Causes may be peritonitis due to leakage of gut contents into the abdomen or an inflamed abdominal organ. It is *extremely painful*.

*Body temperature, abnormal*: Any alteration in body temperature could be accompanied by a lowered activity level. Hypothermia of more than 10% from normal body temperature may be associated with impending death. Hyperthermia may indicate infection and/or pain.

*Body weight loss, body condition, or emaciation*: This is particularly of concern when body weight has decreased by more than 20% compared with control animals, or body weight has decreased by more than 25% over a period of 7 days or more. It is usually accompanied by reduced or absence of food intake. Body condition should be scored as well as in chronic conditions (e.g., tumor growth), as body weight may stay the same or even increase, but loss of muscle and subcutaneous fat lead to a marked loss of body condition. This is detectable through feeling the pelvis and backbone, and one may see a square tail, as muscle atrophy reveals the square shape of the underlying vertebrae.

*Breathing difficulties (dyspnea)*: This can be presented in a variety of signs such as panting, hyperventilation, labored breathing, see-saw or abdominal-thoracic breathing, or grunting with each breath (this may also be indicative of abdominal pain).

*Cachexia*: See Body weight loss.

*Chewing, persistent*: See Compulsive behavior.

*Chromodachryorrhea*: See Blood around nose and eyes.

*Circling*: See also Ataxia. This is characterized by an animal going repeatedly round and around the cage making a track; it may be accompanied by body weight loss. This may indicate damage to the brain or to the inner ear. It may be caused by a concurrent infection, but could also be caused by test substances.

*Comatose*: See also Recumbency. The animal may be unarousable due to extreme lassitude, sedation, etc., or toxic effects of a substance.

*Compulsive behavior*: Such behaviors may include gnawing, biting at the substrate, or even biting at parts of their own body (e.g., feet).

*Constipation*: This may be indicated by lack of feces in the cage, but must be differentiated from decreased feces due to anorexia. If prolonged, the animal will become lethargic and die.

*Convulsions*: See Seizures.

*Corneal ulceration*: This may be accompanied by blepharospasm, watery eyes, and ocular and nasal discharge. It *can be particularly painful* (early stages) and may be incidental to the study or caused by a test substance, such as drug-induced decreased tear production. Seek veterinary advice, and if recovery is sought, treat under veterinary supervision.

*Coughing/sneezing*: If persistent, this may be an intercurrent infection, and veterinary advice should be sought.

*Cyanosis*: This is characterized by blue or dark red extremities, such as pinna, feet, and mucous membranes of eye and mouth.

*Dehydration*: This can be assessed by lifting and twisting the skin and observing how quickly it returns to its normal "flat" position. It usually occurs as a result of reduced water intake or inadequate water intake in the case of intestinal (diarrhea), kidney, or endocrine disease (polyuria).

*Diarrhea*: Diarrhea can present in a variety of forms from frank watery or bloody feces (dysentery) to soft stools. Increased frequency of defecation can indicate greater severity. Humane criteria listed for body weight should be considered.

**Table 18.3 (continued) Clinical Signs and Conditions Indicating the Need for Closer Observation, Treatment, or Humane Killing**

*Discharge*: Normally, animals keep themselves very clean. Discharge may be from any external orifice. Veterinary advice should be sought to differentiate between infectious etiologies and effects of test substances.

*Dyspnea (difficulty in breathing)*: See Breathing difficulties. This can be a cause of severe distress.

*Edema*: Characterized by swelling in areas such as extremities, e.g., hock, below the mandible. May be indicative of insufficient heart function or low protein levels in the blood.

*Emaciation*: See Body weight loss.

*Epistaxis (nasal bleeding)*: See Anemia, and Chromodachryorrhea.

*Excitable*: See Seizures. In this case, an animal may be difficult to restrain or catch, and it may throw itself around a cage in a type of fit, causing injuries. This may be due to excessive fear or to neuronal change altering an animal's behavior.

*Eyelid closure*: See Blepharospasm and Corneal ulceration. Eyelids may be fully or part closed.

*Eyes fixed/sunken*: This is usually observed in the presence of severe body weight loss and dehydration. It indicates that an animal is close to death, and should be treated or humanely killed.

*Fractured bone*: This may be indicated by a swollen limb or lameness.

*Gasping*: See Dyspnea.

*Grooming — failure to do so*: In rodents, this may lead to porphyrin accumulations near the eyes and nose, and there may be soiling in the anogenital region. The animal is definitely ill, *and may be in severe pain and discomfort.*

*Hunched/stiff posture*: See Boarded abdomen. This is often seen in sick animals and may be due to abdominal discomfort or just a general sign of illness.

*Hyperreflexia*: See Excitable. This is an exaggerated response to a stimulus such as noise or touch.

*Immobile/inactive*: This includes inactivity, lassitude, listlessness, and reluctance to move. The animal is ill and may be close to death if accompanied by body weight loss, dehydration, or sunken or fixed eyes. Carry out the red light response test, by turning out the normal white lights and observing the animal in the dark or under a red light, when it will carry out its nocturnal patterns of behavior. (This is normally characterized by an increase in activities such as investigation, climbing, and play within 5 min.)

*Jaundice (icterus)*: This is typically observed by the presence of yellowish-colored ears, feet, and membranes. Serum clinical chemistry (bilirubin) can assist in determining the cause, such as hemolysis (prehepatic icterus), liver damage (hepatic icterus), bile tract blockage (posthepatic icterus), or infection. It may also be accompanied by inactivity when painful condition exists.

*Joints swollen*: This painful condition may be indicated when accompanied by a strong withdrawal and vocalization response, an inability to move around freely, relative inactivity compared with controls, or if animal (rodent) remains inactive during the red light response behavior test. (This red light response behavior test is carried out by turning out the normal white lights and observing the animal in the dark or under a red light, when it will carry out its nocturnal patterns of behavior. This is normally characterized by an increase in activities such as investigation, climbing, and play within 5 min.)

*Kyphosis*: This is characterized by fixed convex/outward curvature of the spine. This may be due to spasm of the flexor muscle of the vertebral column, and if so, would be painful, and the animal should be humanely killed. If intermittent, then it may be a form of seizure (see Seizures).

*Limping/lameness*: The animal may be unable to fully bear weight on that limb due to pain in the foot, leg, or one of the joints. Fractures should be considered as a possible cause.

*Locomotory behavior*: This may be reduced (see Immobile) or abnormal in some way.

*Lordosis*: Fixed concave/inward curvature of the spine is lordosis. This may be due to spasm of the extensor muscle of the vertebral column, and if so, would be painful, and the animal should be humanely killed. If intermittent, then it may be a form of seizure (see Seizures).

*Loss of condition, body muscle*: See Body weight loss.

*Mammary gland abnormalities*: A painful condition may be present if one or more mammary glands is swollen, discolored, discharging pus or blood, or the animal is extremely sensitive to touching of the gland (vocalization, withdrawal, and overreaction).

*Moribund*: See Comatose.

**Table 18.3 (continued) Clinical Signs and Conditions Indicating the Need for Closer Observation, Treatment, or Humane Killing**

*Motor excitation*: This is an exaggerated movement or limb response to a touch; see Hyperreflexia.

*Not eating/drinking*: See Body weight loss.

*Pale mucous membranes*: See also Anemia, Cyanosis, and Dyspnea. This may be indicative of anemia or circulatory insufficiency (e.g., cardiac or pulmonary insufficiency, or shock). If accompanied by labored or accelerated breathing, it may be indicative of a severe or irreversible condition. A hematocrit can be conducted to quantify the severity of suspected anemia.

*Paralysis*: This may occur because of action of substance on the CNS or spinal cord. Any animal dragging its limbs should be humanely killed.

*Paresis*: This may occur because of action of substance on the CNS or spinal cord or musculature or neuromuscular junction. Any animal showing obvious muscle weakness that may affect its ability to eat or drink or breathe should be humanely killed.

*Pinna reflex*: See Reflexes. Pinch the ear flap, and normally an animal will shake its head.

*Piloerection*: This is noticeable when hairs of an animal's fur look harsh or starey as they are partially erect. This is a sign of not grooming and general ill health.

*Prostrate*: See Recumbent. This usually characterizes an animal that has lost its righting reflex and has been in that condition for a few hours.

*Pruritis*: See Self-mutilation. An animal may scratch or bite at itself, which may lead to superficial injury, but this can progress to deeper lesions and infection.

*Pupillary constriction/dilation*: A light responsiveness test should be carried out to determine if the condition is fixed or if there is a pupillary response. Dilatation of the pupil together with inactivity may indicate that an animal is close to death, especially with a sluggish pupil response time. Otherwise, dilation or constriction may well be a substance effect.

*Rales pulmonary*: See Dyspea. This is detected by stethoscope. This may indicate pulmonary secretions as a result of intercurrent infection (pneumonia) or a test substance. Substances inducing bronchial and bronchiolar secretions may predispose to infection.

*Rectal prolapse*: See Tenesmus, Diarrhea; This occurs when part of the rectum protrudes from the anal sphincter. The animal will have to be humanely killed, as the prolapse may become infected or the animal may self-mutilate.

*Recumbency*: See Prostrate. The animal may be lateral (on its side) or abdominal, and if it has lost its righting reflex, that is more serious. It may be temporary or prolonged, though, if for more than a few hours, the animal is likely to be close to death if it is not in any form of seizure.

*Red eye(s)/nose*: See Epistaxis and Chromodachryorrhea. This is indicative of an animal failing to groom. It may also have a soiled anogenital region.

*Reflexes*: Sluggish responses or loss of reflexes such as corneal, pupillary, pedal, righting (ability to correct to normal posture when gently pushed or overbalanced), or responses to noise may be due to unconsciousness or extreme lassitude. Used to assess anesthesia and analgesia.

*Retention of feces*: See Constipation.

*Righting reflex*: See Reflexes.

*Salivation*: This is indicative of a failure to swallow or hypersalivation in response to the test substance. If unable to swallow, a clinical examination is required to determine the etiology, as it may well affect an animal's ability to eat (see Body weight).

*Seizures*: The animal may lie on its size and tremor. The muscles may be rigid or flaccid. It may last only for a few seconds or may be longer. And, it may be brought on by interaction with the observer. If the seizure lasts for more than 1 min and is repeated for more than 5 times a day without being induced, then the animal should be humanely killed, especially if due to the substance being tested. If seizures are induced and further time for study is needed, then animals should be moved to a quiet area and handled minimally.

*Self-mutilation*: See Pruritis. This is licking, scratching, or gnawing at an area, which if persistent, may result in ulcerative dermatitis. If the extent of the self-mutilation is greater than 2 cm,[2] or whole phalanges have been removed from the digits, consider humane killing or other appropriate action.

*Skin bruising/color/crepitus*: This may be due to a subcutaneous bleed or air under the skin (if over the thorax, consider lung puncture and humane killing). If due to gas-forming organisms, treat or humanely kill.

**Table 18.3 (continued) Clinical Signs and Conditions Indicating the Need for Closer Observation, Treatment, or Humane Killing**

*Spasm*: See Seizures.

*Staggering*: See Ataxia.

*Sunken flanks*: The abdominal walls of an animal may be suddenly drawn in (writhing, squirming) and can indicate abdominal pain (as in a colicky pain), or it may also be through emaciation (see Body weight and Dehydration).

*Suppuration*: This is indicative of infection. See Discharge, although suppuration may come from sources other than natural orifices.

*Swellings*: See Joint swelling. Note the position and extent of swelling. This may indicate edema (q.v.), hernias of the inguinal or femoral rings, abscess, growth of some sort, bruising, pregnancy, etc.

*Tenesmus*: This is constant straining to pass feces. It is usually associated with diarrhea (q.v.) and rectal prolapse.

*Tetany*: See Seizures.

*Tremor*: The animal may show muscular twitching or rapid skin movements. See also Seizures and Convulsions.

*Urine retention*: If suspected, palpate hardened and distended bladder through the abdominal wall. It is often painful. This can be confused with renal failure. See Anuria.

*Vaginal prolapse*: Part of the vagina protrudes from the vulva in this condition. The animal will have to be humanely killed, as the prolapse may become infected or the animal may self-mutilate.

*Vocalization:* This may be unprovoked, result from handling, or be associated with an animal being fearful of being touched. If abnormal or persistent, it may be indicative of a painful or distressful condition.

*Vomiting:* This is rare in rodents, as they lack the physiological reflex and are anatomically unable to do so because of the arrangement of the diaphragmatic musculature. In other animals, check on frequency and volume lost (see Body weight and check for fluid loss, see Dehydration). If allowed to persist, the animal will die through dehydration and electrolyte imbalance.

## HUSBANDRY AND THE FIVE FREEDOMS

Laboratory animals are kept in confinement for 100% of their life span, whereas they may be in an experiment for only a fraction of that time. That is why we have to consider the burdens we place on them for the whole of their lives and not just the impact of the scientific procedure. The five freedoms framework, first put forward in the 1960s,[39,40] concerns the welfare of all domesticated animals but is derived from their use in farming. It is generally applicable but may not always be appropriate if the scientific objective will be frustrated, e.g., keeping animals in metabolic cages. The five freedoms can be listed in the following manner and will be dealt with sequentially.

1. Freedom from thirst, hunger, and malnutrition
2. Freedom from discomfort
3. Freedom from pain, injury, and disease
4. Freedom to express normal behavior
5. Freedom from fear and distress

### Freedom from Thirst, Hunger, and Malnutrition

Access to food and water and a diet to maintain full health and vigor are important. Laboratory animals normally have free access to clean water, but one aspect of malnutrition, that of obesity, may be a common neglected problem when animals such as rats are kept for lifetime studies. Although we know more about nutrition of the rat than of any other species, including man, overfeeding is a common problem. The general practice of *ad libitum* feeding of rats leads to them being overweight, and with a shortened life span due to an accelerated senescence and an increase in the frequency of diseases such as nephropathy and various forms of cancer. It has been found that restricted feeding, administering perhaps 75% of normal ad lib food intake, results in leaner and healthier rats with fewer diseases and longer life expectancy (see also Chapter 12). However, restricted feeding is not compatible with social housing, and systems to combine restricted feeding with social housing need to be developed. On occasions, animals may be deprived of food and water as part of an experiment or prior to surgery, but in some species such as rats and mice, starving animals can be prejudicial for good health. In these small rodents, the consequences of removing food and water even for just a few hours can lead to dehydration and physiological disturbances such as a serious depletion in liver glycogen. In any event, starvation is not necessary for anesthesia, as they are unable to vomit due to their prominent cardiac sphincter of the diaphragmatic muscle.

## Freedom from Discomfort

This can be helped by providing a suitable environment including shelter and a comfortable resting area. Chronic discomfort may be inevitably associated with certain experimental protocols and disease models. However, housing animals on wire, rather than solid floors, for prolonged periods may cause some discomfort, and rodents have indicated a preference for solid-bottomed cages as well as specific substrates such as sawdust and paper bedding for making nests.[41–48] Moreover, overcrowding or mixing unfamiliar animals with each other, or mating incompatible pairs (as in the mating of juvenile female mice with heavier mature males for embryo production in transgenesis), or disturbing neonatal mice in the nest, resulting in them not being fed, can compromise animal well-being.[49]

## Freedom from Pain, Injury, and Disease

Prevention or rapid diagnosis and treatment can help provide freedom. Freedom from pain is normally one area where therapeutic intervention can easily be employed, such as after surgery (see Chapter 17), where postoperative pain is not a desired scientific outcome but an unwanted side effect, and its elimination may well reduce the variance in the data. Other sources of pain may not be possible to relieve, as the administration of pain relieving drugs might interfere with achieving the scientific objectives, such as the use of morphine in neurotransmitter research in the central nervous system. However, this has to be justified so that it is quite certain that such harms are unavoidable i.e., are really necessary to the science and cannot be minimized in any way.

Under normal circumstances, injuries are not frequent in colonies of laboratory animals apart from occasional damage to a claw or when teeth overgrow and have to be clipped. However, the increasing practice of group housing has been associated with an increase in the frequency of bite wounds, and this usually happens when groups become incompatible or when there is some environmental disturbance that disrupts an established hierarchy. Some strains of male mice for instance are difficult to maintain in groups after puberty[50-53] as are some of the more common primate species. Nonhuman primates show severely disturbed behavior when raised in isolation, and, e.g., allogrooming must be seen as an essential need for these animals. If prevented from grooming, the male and female primates have been observed to "groom" the ground, which is a redirection in which a thwarted activity is performed on a substitute object.[54] Rhesus macaques can exhibit significant aggression and can be difficult to maintain in harmonious groups,[55] although some laboratories have used group-housing strategies successfully.[56,57] Pair-housing is obviously much better than single housing and has been used successfully for Rhesus macaques,[58-60] but group housing may still be the best system. A problem with pair-housing is when one of the animals goes into experiment or is killed before the other. It can then be difficult for the remaining animal to adjust to a new partner even if one is indeed available within the project. A similar problem occurs when a group is more or less rapidly diminishing over time because of project use of animals. This will eventually disturb the established hierarchy in the group accompanied by the problems experienced when a new group hierarchy is formed.[52,53]

Infectious diseases are normal dealt with by killing the animals, sterilizing the room, and introducing a new colony of healthy animals, particularly with rodents and rabbits. It should not be forgotten that disease *per se* can be a welfare problem due to temporary ill health or even prolonged ill health for some chronic diseases such as pneumonia. Larger animals, and now to an increasing extent genetically modified mice, are often treated as individual patients when suffering from disease. Another problem is when animals suffer from an induced or spontaneous disease when they are being used as a model for a human disorder. These animals obviously need attention, and the alleviation of pain and discomfort is no different to any other patient suffering from a similar disease, unless again such an omission is well justified and based on real validated concerns.

## Freedom to Express Normal Behavior

By providing sufficient space, proper facilities, and company of an animals' own kind unless there is a veterinary or scientific reasons for single housing, this freedom can be addressed. Surprisingly, even subdominant mice that have been injured choose to be with more dominant mice rather than be on their own.[61] The benefit for animals of being housed in an environment allowing expression of important species-specific behavior (needs) has been increasingly acknowledged during the past decade. Too often, laboratory animals have been kept in barren environments that lead them to respond by trying to cope with this deficiency through carrying out stereotypic behaviors. Small birds like quail and zebra finches are often kept in barren cages in the laboratory, and modest enrichment has been demonstrated to result in increased locomotor activity, singing, and vocalization.[62,63] Rabbits particularly have been kept confined in tactilely barren conditions, and the small amount of space for them to exercise has been shown to have pathological effects, such as osteoporosis and a predisposition to fracture.[64] Animals may substitute some activities for another, thus being provided with a plastic piece of gutter may substitute for building a proper nest, even though they will show some preferences for nesting materials.[41,42] Rabbits show evidence of boredom such as overgrooming and so are prone to develop hairballs, which may be lethal to the animals.[64] Primates can do the same and may self-mutilate to the point of serious damage such as autotomy and digit removal, mice show stereotypic escape behaviors, and dogs may continually jump and so on. There is now building up a body of evidence specifically looking at the common laboratory animals such as rodents, and experiments on choice and effort for reward are becoming more commonplace; hopefully, we will learn more about animals' behavior and their needs in the future.

In line with the comments above, the revised Appendix A to the Council of Europe Convention (ETS 123) has, as a general rule, that laboratory animals should be kept in an enriched (enhanced in some way to be more than simply a barren space) environment and that they should be housed in harmonious social groups whenever possible. Being housed with another, or other, member(s) of the same species is recognized as the single most important enrichment factor to a singly housed animal. There is thus every good reason to group house animals when possible. However, it is not without its problems, especially housing intact males of certain species. As a consequence, there seems to be a trend toward avoiding the use of male rabbits and male primates. The practice of castrating male animals to allow group housing may seem convenient and it certainly seems to work, but this is a practice that is not readily accepted in all countries where castrating animals for convenience is seen as an unacceptable infringement on their integrity. However, this has to be balanced against the quality of the lives of the animals concerned and the fact that if males are not used at all, 50% of all animals born will have to be killed.

## Freedom from Fear and Distress

By ensuring conditions that avoid mental suffering, freedom from fear and distress can be addressed. This freedom, along with freedom from pain, is particularly important and certainly one of the freedoms where all staff in contact with laboratory animals have the possibility to improve conditions by actively implementing refinement initiatives. There is no doubt that severely stressed animals attempting to cope with a stressor by dramatic physiological, and sometimes also behavioral adaptation strategies, are unsuited as laboratory animals, because their coping strategies may interfere with the usefulness of the data generated. The extra variance may well even affect the validity of the data and their interpretation. There are therefore scientific as well as humanitarian reasons for eliminating significant stressors for animals. Fearfulness, as in anticipating an intervention or simply being removed from a familiar environment to a strange one, may end up causing more pain to an animal. Thus, trivial procedures such as handling, sexing, and weighing an animal, or giving it a subcutaneous injection, may be associated with fear of the unknown and even a perception of pain in a timid animal not used to human contact. In contrast, animals confident with their surroundings and staff are less affected by simple procedures and will often welcome the contact with the staff member in spite of being dosed or receiving an injection. It is thus imperative that all staff, which will eventually handle and restrain an animal, use the necessary time, often just a few minutes every day over a week or two, to handle the animal gently, and so habituate or condition it to the circumstance so it can adjust and gain in confidence. In larger animals such as primates and dogs, they may be given a reward after dosing but for some reason this is not common for rodents although it has been practiced successfully with rats in combination with saliva sampling.[65] Finally, other routine husbandry procedures such as weaning and identification methods may be seen as routine for the humans, but they are one off events for the animals. An animal may therefore come to experience (and expect) that every time it is handled, something unpleasant is going to happen and this will inevitably cause fearfulness and some mental distress. Practical examples of this are easy to see when one visits a zoo or other animal facility in which veterinary staff wear a different color coat than other animal care staff. The reaction of the animals is often more fearful toward the veterinarian, who gives injections, than it is to the care staff in "friendly" clothes, who bring food and entertainment.

Other useful "Animal Need Indices" (ANI) like, e.g., the Austrian "Tiergerechtheitsindex," which have been developed primarily at farm level as an instrument for assessing and grading livestock housing with respect to the well-being of the animals[66] may conveniently be adapted to the use in laboratory animal facilities. The approach pursued in Austria considers the following five husbandry conditions: (1) freedom of movement, (2) social contact, (3) condition of flooring, (4) stable environment, and (5) stockpersons' care levels. A scoring system leads to a sum of points, and the ANI values have been divided into different grades of good to poor welfare. In principle, this is a similar practical approach to optimization like the five freedoms but has the advantage of including an assessment of staff competence and performance, which is relevant also to assessment of laboratory animal facilities. Poole[67-70] has also written several articles on evaluating an animal's environment and makes some interesting contrasts with the wild counterparts of a species.

## ROUTINE PROCEDURES USED IN HUSBANDRY AND RESEARCH

Animals in captivity are completely dependent on human staff with regard to their well-being. To ensure a high state of animal well-being, animal facilities must be constructed and managed with the aim to provide the best possible environment for the species in question, taking into account their physiological and behavioral needs.[71] The holding rooms for animals should be constructed and ventilated in such a way that animals of different species can be separated not only physically but also with no sight, sound, or smell of other species. This is probably most important when housing predatory species near to prey species in the facility. In this context, it should be remembered that mice react with anxiety to rats as if they were predators. A good hygienic status is imperative to high-quality animal husbandry, and it is a prerequisite to minimize the introduction and spread of infectious diseases. The behavior of rats is affected by cleaning their cage,[72] and when providing animals with clean bedding, it is now considered important not to disrupt the olfactory environment through leaving a minor proportion of the dirty bedding in the cage, although in mice this has been disputed.[73-75] As with all animal husbandry, it is important that routines be adhered to quite rigidly. Changes from daily husbandry routines upset the animals, and consistency and habit are important to allow the animals to feel confident and so thrive and reproduce.

Movement and transportation of animals must be considered stressful events for laboratory animals and should be reduced to a minimum. Consequently, facilities for clinical examination and minor procedures like blood sampling and administration of substances should be available in a nearby animal holding room. When animals are transported within an institution, e.g., from holding room to laboratory, the transport should be chosen to minimize the stress associated with it.[76] When animals are transported over longer distances, a careful choice of transport caging, transport means, and route should be made in order to make the journey as short as possible with minimum stress to the animals. Guidelines exist for transportation of animals, such as the provisions of the European Convention on the Protection of Animals during International Transport (ETS 65).

Cages, pens, and other enclosures used for housing laboratory animals should allow sufficient space for them to express a wide behavioral repertoire, and the environment should provide a reasonable complexity rendering species-important behavior possible (e.g., dust bath for chickens). It would also seem reasonable to provide sufficient space on the basis of what an animal needs to use as opposed to a simple relationship with size. Thus, young animals that like to play and carry out certain space-occupying behaviors should be given more space than lethargic older and heavier animals.[64] However, this is not normally the case, and cage size is almost exclusively based on body weight and not age. Recommended minimum cage sizes and ranges for acceptable environmental parameters like ambient temperature may be found in Appendix A to the European Convention for the Protection of Vertebrate Animals Used for Experimental and other Scientific Purposes (ETS 123). Larger species seem to be more unfortunate with respect to recommended cage sizes in various guidelines. This may be due to tradition and the way these species are customarily housed, when they are used as pets (e.g., rabbits and guinea pigs) or farm animals (e.g., horses), in enclosures allowing very little locomotor behavior.

It is now generally agreed and advocated in European and North American guidelines that laboratory animals — with the exception of animals that are naturally solitary — should be pair or group housed with members of their own species. Most animals used in research are social or gregarious animals, and housing them singly restricts their natural behavior unnecessarily. Rat pups born to mothers in isolation show significantly lower weight gain and slower locomotor development than pups born in social groups.[77] Only scientific or veterinary reasons should justify a practice of housing animals singly, such as when they are placed in metabolism cages. Housing animals in groups, however, increases the need for observant and competent staff and may increase the time they have to spend with the animals. On the other hand, that time has to be set against the time and expense saved by reducing the frequency of group as opposed to cage cleaning, and cage replacement. It is time consuming to monitor newly formed groups to ensure that aggression injuries are reduced to an acceptable level, but even after a stable hierarchy has been established, fighting can break out if a group is disturbed by, e.g., removing an animal for experimental purposes. Similarly, when animals are returned to the group after having been away to the laboratory, a hierarchy has to be reestablished. Consequently, animals like rabbits and Old World primates are often housed in pairs, as this allows social contact between animals and significantly reduces many of the problems associated with housing animals in larger group sizes.[58-60]

Cages and pens should always be provided with appropriate bedding materials and shelters in which the animals can hide or escape. The bedding serves a number of various functions. It allows the animal to remain dry and clean and to create and control its own microenvironment with regard to temperature and light if the bedding is present in quantities generous enough to allow burrowing. Bedding may be used to allow foraging behavior, which is important in, e.g., primate husbandry, and many species create a nest for sleep and rest from the bedding material provided. The provision of shelters is a mixed blessing, as it reduces the ability to monitor the well-being of animals, but meets their biological needs. Research is being carried out looking at transparent shelters and suggests that the animals will use these as much as opaque ones, thus solving the problem (Leach, M., personal communication).

Acclimation of animals to a new room or cage takes some days, and it is important that this period of acclimation be permitted, as otherwise, it may affect the scientific data. For example, Damon et al.[78] found a 60-fold difference in the LD50% dose between animals that had been acclimated to a metabolic cage and those that had not. Restraint for the administration of substances and the removal of body fluids, if not done well, can also be aversive to an animal, particularly when the procedure is repeated over several days or weeks.[79-80] In fact, animals may never habituate to the procedure and show an adrenal response from both the medulla and the cortex.

## EXPERIMENTAL PROCEDURES

The ethical framework of Russell and Burch — the Three Rs: Replacement, Reduction, and Refinement — can be modified slightly from the original concepts and turned into a series of questions when the use of animals in a research project is being considered. Are there nonsentient alternatives available that could replace the use of animals? Are the numbers of animals appropriate for statistical acceptability?

And finally, can the amount of suffering be reduced (e.g., provision of postoperative analgesia) or can the well-being of the animals be improved in any way (e.g., an enriched husbandry system)? Refinement has been called the Cinderella of the 3Rs[82,83] and should be considered much broader than simply relating it to the suffering of an individual animal during an experiment. It may even be possible to improve the experimental design so that all the animals used in the project suffer less or fewer animals are used by staging the work so that key experiments are carried out first. Other approaches to improving the design might incorporate pilot studies, a grading of stimuli (e.g., in developing an novel analgesic to give a minor nociceptive stimulus before a major one; or in cancer research to evaluate treatment on small tumors before progressing to larger ones) and that *in vitro* work precedes *in vivo* work where appropriate (e.g., cell toxicity of novel chemicals).[84]

Next, we will examine how to determine when animals are suffering and to what degree, and how this information can be used to determine and implement humane endpoints as well as to gain valuable clinical information in medical research involving animal models of human disease, including transgenic animals.

## SCORE SHEETS

### Establishing a Score Sheet

A score sheet is simply a list of those clinical signs likely to be seen in a particular scientific procedure for an individual species, or even for a particular strain or breed of that species (see Tables 18.1 and 18.4).[13,37,82] It is not possible to prepare a generalized list of clinical signs that will occur in all experiments (compare the predictable side effects for a failed skin graft with a failed liver or heterotopic heart graft) and so score sheets have to be drawn up specifically for each scientific procedure. Score sheets are therefore usually specific for each of those characteristics — experiment, species, and strain. The use of pilot studies as recommended by the CCAC[85,86] helps to determine on the first few animals what the relevant and important clinical signs are likely to be, and in this regard, the opinion of the animal care staff will be invaluable. The laboratory animal veterinarian should also be skilled in identifying objective clinical signs from his or her knowledge of the biology of the species, including the range of an animal's relevant behavioral and physiological responses.[87-91] Regularly observing animals throughout a pilot study will help identify those critical periods during the experiment where animal well-being is particularly at risk (e.g., in the immediate postoperative period, or in a study on infection after the incubation period, or at the predicted time of tumor growth or organ graft failure).

The score sheet first lists those signs that can be observed from a distance in a relatively undisturbed animal, which should be showing its natural unprovoked behavior (e.g., appearance, behavior, posture, respiratory rate and pattern). The animal is then observed at closer range followed by close examination of the animal (body weight, body condition and temperature, heart rate, dehydration). The signs are scored as being present (+) or absent (-) or, if unsure, then that too can be indicated (+/-). Clinical signs are reduced to an observation that can be scored in this binary way, otherwise misinterpretation and subjective evaluation (i.e., observer error) may creep in. This is one of the strengths of the scheme, as it leaves little room for observer error, and such errors, when they have occurred, have usually been at the lower levels of severity.[92] The convention is that negative signs indicate normality, i.e., within the normal range, and that positive signs indicate the animal is outside the normal range. However, this results in some contorted descriptors, as an animal showing convulsions would be a sign of poor well-being and be scored a (+), but if an animal was eating, then this is a good sign but could be scored also as a (+). In this latter case, the descriptor is changed to "not eating" so that if an animal is not eating it is scored as a (+), and the convention is maintained.

Using this convention, it is possible to scan a score sheet to gain an overall impression of animal well-being: the more plusses, the more an animal has deviated from normality with the inference that the scientific procedure is having more impact than before. It is not unreasonable to assume, and this would be borne out by human experience, that the greater the deviation from normality, the greater has been the impact. Thus, an animal may lose body weight because it is in pain or feeling unwell because of an infection, and so it does not eat. This weight loss reflects one impact of the experiment on that

**Table 18.4 Score Sheet from an Experiment on Renal Nerve Physiology**

RAT No. SPSHR4  DATE OF OPERATION 19/11/01  ISSUE No.  PRE-OP WEIGHT: 325g

| DATE | 19/11 | 20/11 | | 21/11 | | 22/11 | | | 23/11 | | |
|---|---|---|---|---|---|---|---|---|---|---|---|
| DAY | 0 | 1 | 1 | 2 | 2 | 3 | 3 | 3 | 4 | 4 | 4 |
| TIME | 16.45 | 7.45 | 16.35 | 8.35 | 16.15 | 8.25 | 16.05 | 22.05 | 7.25 | 12.05 | 4.25 |
| **FROM A DISTANCE** | | | | | | | | | | | |
| Inactive | - | - | - | - | - | +/- | + | + | +/- | + | + |
| Hunched posture | - | - | - | - | - | -/+ | -/+ | ? | + | + | + |
| Starey coat | -/+ | - | - | - | - | + | + | +/- | +/- | + | + |
| Rate of breathing (per minute) | 80 | 80 | 90 | 80 | 80 | 80 | 84 | 80 | 90 | 120 | 120 |
| Type of breathing (*) | N | N | N | N | N | N | N | N | N | R | R |
| Not grooming | ? | - | - | - | - | +/- | + | + | +/- | + | + |
| **ON HANDLING** | | | | | | | | | | | |
| Not inquisitive & alert | ? | - | - | - | - | + | + | + | +/- | + | + |
| Not eating jelly mash | ? | - | - | - | - | -/+ | -/+ | ? | + | + | + |
| No. of pellets given to rat | 10 | 4 | 13 | 8 | | 8 | 10 | 10 | 5 | | 5 |
| No. of pellets left | - | 4 | 3 | 4 | | 5.5 | 5.5 | 10 | 10 | | 5 |
| No. of grapes given to rat | 3 | 3 | 3 | | | 2 | 2 | 2 | 3 | 3 | 5 |
| No. of grapes left | 3 | 0 | 2 | 1.5 | | 2 | 2 | 2 | 2 | 3 | 3 |
| Grain (sprinkling) | - | Y | Y | Y | -/+ | Y | Y | Y | Y | Y | Y |
| Not drinking | - | - | - | -/+ | - | +/- | + | + | + | + | + |
| No Feces passed | - | - | - | - | - | +/- | + | + | + | + | + |
| Bodyweight (g) | | 339 | | 312 | | 296 | | | 282 | | 275 |
| Change from start (%) | | 4.3%+ | | -4.30% | | -9.02% | | | -13.20% | | -15.40% |
| Body temp (degrees C) | 37.4 | 37.5 | 38.7 | 36.2 | 37.1 | 37.5 | 35.7 | 37.5 | 36.9 | 35.9 | 34.7 |
| Crusty red eyes/nose | - | - | - | - | - | - | - | +/- | + | + | + |
| Sunken eyes | - | - | - | - | - | - | - | - | + | + | + |
| Eyes half closed | +/- | - | - | - | - | - | +/- | +/- | + | + | + |
| No red light response (not tested=NT) | NT | NT | NT | NT | NT | NT | NT | NT | + | + | + |
| Coat soiling | - | - | - | - | - | - | + | + | - | - | - |
| Pale eyes and ears | - | - | - | - | - | - | - | - | - | - | - |
| Dehydration | - | - | - | +/- | - | +/- | +/- | + | +/- | + | + |
| Diarrhea | - | - | - | - | - | - | - | - | - | - | - |
| Hyperactive | - | - | - | - | - | - | - | - | - | - | - |
| Circling | - | - | - | - | - | - | - | - | - | - | - |
| *Dragging leg L R | +R | ++ | ++ | ++ | ++ | ++ | ++ | ++ | ++ | ++ | ++ |
| Cold Limb L R | - | - | BLUE | BLUE | BLUE | BLUE | BLUE | BLUE | BLUE | BLUE | BLUE |
| Toes curled | - | - | - | - | - | - | - | - | +/- | +/- | +/- |
| Lameness | + | + | + | + | + | + | + | + | + | + | + |
| Blue extremities more than one foot | + | - | - | - | - | - | - | - | - | - | - |
| **Head wound OK** | Y | Y | Y | Y | Y | Y | Y | Y | Y | Y | Y |

| **Neck wound OK** | Y | Y | Y | Y | Y | Y | Y | Y | Y | Y | Y |
| **Right side jugular wound ok** | Y | Y | Y | Y | Y | Y | Y | Y | Y | Y | Y |
| **Right side femoral wound ok** | Y | Y | Y | Y | Y | Y | Y | Y | Y | Y | Y |
| **Flank left** | Y | Y | Y | Y | Y | Y | Y | Y | Y | Y | Y |

**DRUG for Baroroflex Curve**
NAD (Nothing Abnormal Detected)
**Other**
vocalizing on handling or approach
**SIGNATURE:**

**Special husbandry requirements:**

Animals should be kept on vetbed 24 hrs post op in a warm environment.

Animals will be used in a rodent access cage.

**Scoring details:**

*Dragging leg: score as + if can move leg forward and ++ if cannot

* Breathing R=rapid; S=shallow; L=laboured; N=normal

**Humane endpoints and actions**

If bleeding occurs at the cannula exit site this will be treated by the veterinarian.

Animals will be checked at least twice daily and daily clinical monitoring of animals (bodyweight, respiration, behavior, eating and drinking, bowel habits, etc.) will be carried out to detect any adverse effects.

Blood pressure and renal nerve activity will be measured regularly and any marked deviation from normal values and patterns would be evaluated together with the clinical signs.

If there is a clear indication of progressive deterioration the animal will be killed.

Any animal showing abnormal motor or sensory effects will not be allowed to proceed into the study without further discussion with veterinarian and animal care staff.

Any animal not returning to its starting bodyweight within 7 days will be killed.

Any animal losing greater than 20% bodyweight in a two-week period will be carefully examined with respect to the other clinical signs, and their growth rate may be an important factor.

**Scientific measures**

Blood sample to be taken prior to killing but after loss of consciousness.

animal. If this can be reversed through the provision of analgesics or other treatment, then the severity can be reduced and the well-being of the animal improved, as may also the science.

At the bottom of the sheet, there are guidance notes on what should be provided for the animals in terms of husbandry and care. There are also guidelines on how to record qualitative clinical signs (such as lameness, diarrhea, respiration), as well as criteria at which to implement humane endpoints (see below). Finally, if an animal has to be killed, there are instructions about what other actions should be taken, such as a blood sample before killing, so that the maximum amount of information is always obtained from a study.

While these sheets take time to fill in, it is relatively easy for an experienced person to see if an animal is unwell, so the NAD box (Nothing Abnormal Diagnosed) is simply ticked. However, if an animal is not normal, it takes time to score it and to make judgments over what actions should be taken; that is the price for practicing humane science.

## Interpreting a Score Sheet

It can be seen from Table 18.4 that as time goes on there are more plusses to the right than to the left. Several other points should be noted. As the animal started to show clinical signs, so it was scored more frequently. On Day 0, some clinical signs were seen as the animal had just had a 7-h operation, and the following day it seemed to be recovering well, as it had gained weight and was eating although it was lame in the right leg due to the placement of a femoral cannula. Over the next 3 days, it lost body weight and its body temperature gradually decreased. Its activity appeared to vary, but eventually it did not respond to the red light test (this test is carried out by turning out the normal white daylights and observing the animal in the dark or under a red light when it will carry out its nocturnal patterns of behavior characterized by an increase in activities such as investigation, climbing, and play within 5 min). By Day 4 the coat had become starey (ruffled), the body temperature had dropped significantly, and the breathing became more rapid and labored. Furthermore, there was a significant body weight loss (15.4%) accompanied by a progressive dehydration — a strong indication that the animal had not eaten or drunk much, if anything, during the previous 3 days. In addition, the blood supply to the limb was obviously impaired, which was a surprising observation, as in the control strains of rat this lameness is normally reversed fairly quickly, so there seemed to be a strain difference with the SPSHRs (Stroke Prone Spontaneously Hypertensive Rats). The rapid weight loss and dehydration, labored breathing, inactivity, lack of a red light response, etc., confirmed that the animal was becoming severely physiologically compromised and was not going to yield valid results in relation to the scientific objective. Even more significantly, its temperature was now at 34.7°C — a very poor sign. In our experience from following such animals through to death in earlier studies, this animal would have died over the next day or so, and consequently, it was decided to kill the animal on humane as well as on scientific grounds before the end of the experiment. Even if the animal might not have died, the level of suffering was agreed to be a sufficient reason to kill the animal on humane grounds alone. In the United Kingdom, where an ethical balance is struck between the anticipated benefits of a research project and the degree of animal suffering (as indicated by the humane endpoints), the severity limit of moderate had been exceeded.

Score sheets are important tools for identifying clinical signs indicative of treatment effect. These signs may assist in determination of subsets of characteristics for which deviations may be used to develop and introduce earlier endpoints of animal experiments.

## HUMANE ENDPOINTS

As soon as an animal ceases to be scientifically useful in an experiment, or the objective of the experiment has been achieved, then it should normally be killed. Humane endpoints can also be implemented when they reflect suffering that cannot be justified or when particular clinical signs affirm a specific outcome such as pending death (Tables 18.5 and 18.6).[13,85,86] For example, decreased body temperature has been associated with pending death in mice with bacterial infections,[93,94] slow circling movements in mice

**Table 18.5 Examples of Clinical Signs and Actions that Could be Taken**

| Analogue Objective Signs | How Measured | Indicator for Intervention | Indicator for Euthanasia | Endpoint Consideration |
|---|---|---|---|---|
| Anorexia — see also Appetite poor Body weight loss Body condition poor (cachexia, emaciation) | Food and water intake as a percent of normal intake Body weight loss Body condition score | Assess cause — If diet intake is reduced, and body weight loss is more than 10% over 2 days (NB ascites and tumor growths may mask loss of body weight and so body condition scoring becomes important) | If amounts to starvation conditions for more than 2/3 days (20% body weight loss) or 25% loss over 7days, and is likely to persist | If animal likely to become irreversibly dehydrated. If body condition score is low |
| Anemia Blood in urine or feces | Clinical biochemistry Hematology Hematocrit PCV low Pale feet, tongue, membranes, mucosae | Assess cause — If unexpected, will anemia affect results? | If animal is becoming severely clinically anemic: labored or increased breathing rate, edema | If animal is becoming unexpectedly anemic |
| Breathing rate | Count breaths/min Type of breathing Rapid, slow Deep, shallow Regular, irregular Color of membranes and extremities | Assess cause, e.g., increased rates can indicate infection, pain, low oxygen, high $CO_2$, anemia If unexpected, will it affect results? | If animal is becoming distressed, in shock (very pale) or cyanotic | If animal is becoming distressed |
| Heart rate | Beats per minute (difficult in small animals) | Assess cause — If raised for prolonged periods, may lead to heart failure over several weeks May indicate animal in pain | If prolonged or very high (50% increase) | If raised by more than 25% for more than a few days |
| Body temperature Hyperthermia Hyopthermia | Clinical thermometer Thermistors (rapid) Radiotelemeter expensive) Transponder | Assess cause Remedial action, e.g., cooling, warming | Failed remedial action and if more than 4°C away from normal for more than 12 h. More critical in smaller mammals, e.g., young mice | If more than 4°C away from normal and remedial action failed |
| Joints swollen | Measure circumference Measure weight bearing Assess pain response | If unexpected, can anti-inflammatories be given? Infection? | If unexpected and in severe pain and associated with body weight loss (see Anorexia) | If unexpected and in pain |
| Mammary glands: Swollen Pus and blood in milk | Clinical biochemistry and hematology | If unexpected, is animal still scientifically useful? Is there a dependent litter – cross foster Treat under VD | If persistent despite treatment | |

Note: VD — Veterinary direction.

**Table 18.6 Questions to Determine whether Earliest Possible Endpoints have Been Sought**

What are the scientific justifications for using the proposed endpoint?

Have all existing relevant data been evaluated?

What is the expected time course for the animals from the initial treatment to first signs of pain and distress, to the death of the animal?

When are the effects to the animal expected to be the most severe?

If the course of adverse effects cannot be determined prior to the start of the study, could they be developed through the conduct of a pilot study with appropriate observations by the animal care and veterinary staff?

Have a list of observations on which the endpoint will be based been developed?

Who will monitor the animal and maintain records of observations?

Has a chain for reporting observation findings been established?

What will be the frequency of observations during the course of the study and during those times predicted to be critical for the animals?

Do the investigators, veterinary care, and animal care staffs have the training and experience necessary to perform the observations necessary to effectively and efficiently monitor the animals?

What steps have been implemented to attend to animals that demonstrate severe signs and symptoms?

Source: From: CCAC, Guidelines on Choosing an Appropriate Endpoint in Experiments Using Animals for Research, Teaching, and Testing, 1998.

with neurotropic viruses such as rabies,[95] and many other approaches can be used. The reader is referred to the report of the first meeting on humane endpoints[96] in which there are papers in various fields of research giving not only clinical signs but also chemical and hematological markers of endpoints. This approach of using surrogate signs for scientific outcome measures such as death has to be validated by allowing some animals to die and carefully recording the clinical signs throughout the animals' lives so that a retrospective analysis can be done. This approach has been used successfully, and refinements to some of the protocols required for safety testing of vaccines have been proposed.[97,98] It is less easy to apply humane endpoints to animals that are used in areas other than safety assessment, where routine procedures are used and the only variable is the batch of vaccine for example. However, many promising attempts have been made in other research disciplines.[96] If the scientific outcome is going to be frustrated by the suffering that animal is undergoing, and this may be psychological as well as physiological, then it is appropriate to humanely kill those animals, as keeping them alive will not generate valid, reproducible and reliable scientific data.

## TRANSGENIC ANIMALS

The growth in the development, generation, and use of transgenic animals (mainly mice) raises considerable animal welfare concern,[99-101] and the numbers generated per year have increased exponentially over the past 10 years or so, and looks likely to continue. Many of our physiological functions may be under the control of more than one gene, i.e., they are polygenic, and so this will involve some considerable research to identify the important and appropriate combinations. The insertion of DNA into the genome of animals or the deletion of specific genes gives rise to unpredictable outcomes in terms of animal well-being (as well as scientific results) in the first instance, but thereafter, transgenic lines can be selected and bred or cloned to avoid or select for a specific genotype. It is not an accurate science, although the methodology is constantly improving to avoid unwanted effects. The embryo manipulation procedures do not appear to affect the welfare of offspring in the mouse,[102] and the large offspring syndrome observed in farm animals has not been reported in rodents. It is important to remember though that absence of evidence is not evidence of absence, and time will tell whether this syndrome may occur also in rodents. The potential welfare risk factors associated with gene modification in animals include possible loss of gene function, harmful effect of transgene derived protein, unexpected physiological and behavioral results of manipulation, and of course, also the pain and discomfort for the animals in which the expected disease model develops satisfactorily. Gene insertion through microinjection into the pronucleus of a fertilized egg may disrupt other genes, may be inappropriately expressed in some tissues, and the overall effect may be dependent on copy number.

Increased fetal and juvenile morbidity and mortality is a frequent finding in genetically modified animals, and there is, therefore, a need for detailed monitoring of neonatal, juvenile, and adult animals for their lifetime to identify any adverse effects. Score sheets can help standardize the signs to look for, and there have been many such publications.[7,50,103-106] There are also good scientific reasons for detailed monitoring and description of phenotype and behavior in order to study the effect of the gene under study, as well as the unexpected side effects of the manipulation. It is noteworthy that in the production of transgenic farm animals, the scientist often tests the system in mice first, and welfare problems seen in the transgenic farm animals were often present in the test mice, but they were not spotted because of suboptimal monitoring of the mice.

The generation of transgenic animals raises concerns for the animals involved in the production as well as when they are being bred to maintain the line, and some of these are detailed in Tables 18.6 and 18.8. Some of these concerns also apply to natural mutations or those induced by the use of mutagens such as ethyl-nitroso-urea. There is one difference in that the generation of transgenic animals involves more technical procedures (Table 18.7), but the maintenance of a line raises similar issues for genetic mutants as well as genetically modified animals (Table 18.8). Many of the adverse effects may be seen in the background strain, and so it is important that the effect of the mutation or modification be distinguished from that. Moreover, the effects may be subtle, and this may only be picked up after extensive testing, using, e.g., the SHIRPA testing protocol.[50,108] The clinical observations by skilled animal care staff may produce invaluable results in relation to the phenotypic effect of the genetic modification. The animal care staff are far more likely to observe significant behavior and developmental differences than the scientists trying to understand the gene function, but together, they will make a strong team. Again, score sheets are particularly useful when the genetic effects are known and when animals may have to be humanely killed before their suffering is too great. Like vaccine testing protocols, it is the repeated observation of the same scientific event that enables clinical signs to be predictable and humane endpoints to be reliably established. For each transgenic line a passport should be drawn up to accompany animals to other laboratories giving details of the clinical signs, their time of onset, incidence and methods of alleviation and humane endpoints.

**Table 18.7 Assessment of Adverse Effects and Technique or Event**

| Technique Event | Examples | Estimated Severity |
|---|---|---|
| Handling | Acclimation, frequency, time constraints (1000s animals?) | Mild |
| Anesthesia | Dystress | Mild |
| Vasectomy | Scrotal laparotomy | Mild + Moderate[a] |
| Superovulation | Injection | Mild + |
| Mating | Normal disparity | Mild<br>Mild to moderate |
| Cross-breeding of genetically modified strains | Knock-outs<br>Knock-ins<br>"Natural" mutants | Mild to substantial |
| Embryo transfer | Vaginal | Mild |
| Nuclear transfer | Laparoscope<br>Laparotomy | Mild +<br>Moderate[a] |
| Premature death | Early fetal<br>Late fetal/neonatal | Mild (dam?)<br>Mild + |
| Birth | Normal<br>Dystocia/cesarian | Moderate<br>Moderate + substantial[a] |
| Euthanasia | Physical/anesthetic gas<br>Inhalation of $CO_2$<br>Peritoneal injection of sodium pentobarbitone | Mild<br>Moderate<br>Mild to moderate |

[a] When analgesics are not given.

**Table 18.8 Some Adverse Effects Associated with Husbandry Practices and Some Genetic Variables**

| Technique | Event | Estimated Severity |
|---|---|---|
| Husbandry | Social isolation | Mild/moderate |
| | Grouping aggressive strains | Mild |
| Individual identification | Ear notching | Mild |
| | Ear tagging | Mild to moderate |
| | Microchip | Mild |
| | Hair dye | Mild |
| Genetic identification | Tail tipping | Moderate |
| | Bleeding | Mild |
| | Body cells and tissues, e.g., blood, salivary swab, hair | Mild |
| Time of identification | As soon as possible, e.g., before weaning | Depends on enrichment strategies |
| Construct characters | Number of copies | Neutral to substantial |
| | Mode of expression | |
| | Promoter | |
| | Gene character | |

# HUMANE KILLING OF ANIMALS

Nearly all research animals have to be killed at the end of a scientific procedure so that tissues can be retrieved for analysis and to comply with legislation such as the EU directive 86/609/EEC (see Chapter 3). On some rare occasion, this may not be necessary, and in those cases, it may be possible to release animals back to the wild (e.g., wild-caught animals, animals bred in captivity such as early free-feeding amphibia forms, i.e., tadpoles, birds hatched from eggs taken from the wild), as pets, or back to the farm, even return them to stock for reuse, but this has to be carefully controlled to avoid misuse of animals. There are several publications dealing with humane methods of killing (see also Chapter 17), especially involving farm animals, where billions of animals will be killed each year. In research laboratories, there is sometimes some concern that the method of killing may affect the scientific data, and if this is so, and an inferior method in regard to the welfare of the animals is being used, then that method should be validated and clearly shown to have the unwanted side effect and not assumed that it does so. While an overdose of nearly all anesthetics is acceptable, the use of carbon dioxide has recently been brought into question by several authors.[109,111]

# CONCLUDING REMARKS

We realize that many other trends and developments in refinement and optimization of animal welfare deserve recognition; like, for instance, the welcome introduction of imaging techniques and telemetry technology allowing observations of physiological parameters from a distance and thus making many invasive techniques obsolete. The use of detailed behavioral analyses, search for immunochemical and neurological markers of stress and reduced welfare, as well as the use of classical stress markers and pathology in the assessment of stress, health and welfare is an active and interesting research field. However, we had to draw the line somewhere, and we hope to have given the readers an update on developments in the humane use of animals in research. The importance of the need for enrichment of animal cages and pens, and particularly, provision of a suitable social environment, is gaining general recognition in the biomedical community,[112] and it is our hope that this chapter will give food for thought and encourage scientists to routinely include welfare assessment as an integral component of their research projects.

That the highest possible welfare state of animals goes hand in hand with the generation of the sound scientific results has been known since the 1950s,[1,20] but assessment of animals' well-being is not a simple business. The welfare state of an animal is a highly complex concept, and we have chosen a holistic and practical approach to assess and improve animal welfare. We agree with Webster,[113] who stated concern "by the misuse of scientific method by those who seek to obtain so-called objective measurement of something which they preconceive to be a stress." As we see it, there are no easy measures of welfare, but behavioral and clinical measures should be components of most, if not all, proper welfare assessments. This requires that all technical, scientific, and veterinary staff engaged in laboratory animal husbandry and experimentation must have a proper knowledge of normal biology and behavior of the species they work with. Only then will it be possible to identify abnormal behavior and clinical changes associated with a scientific procedure or system of husbandry.

Knowledge and competence obtained through education and training of all staff categories are essential for continuous refinement of animal experimentation (see also Chapter 5). In addition to competence, the way forward also requires commitment and collaboration.[114] Commitment from scientific and technical staff to improve the welfare of animals in experiments and to actively search for ways and means to terminate experiments as early as possible, where animal welfare is being compromised. This includes actively searching for behavioral, clinical, and physiological parameters that may be used as indices of effect of treatment, rather than adverse welfare states, especially those that involve death or a moribund state. Collaboration through good communication between laboratory animal technicians, scientists, laboratory animal veterinarians, institutional certificate holders, local and regional ethics committees, and the regulatory and inspection authorities is vital to progress in this area.

## REFERENCES

1. Russell, W.M.S. and Burch, R.L., *The Principles of Humane Experimental Technique*, UFAW, 1959, Special Edition, 1992.
2. Broom, D.M. and Johnson K.G., *Stress and Animal Welfare*, Chapman & Hall, London; New York, 1993.
3. Appleby, M.C. and Hughes, B.O., *Animal Welfare*, CAB International, Wallingford, Oxon, United Kingdom, 1997, pp. 316.
4. Bekoff, M. and Meaney, C.A., *Encyclopaedia of Animal Rights and Animal Welfare*, Greenwood Publishing Group, CT, 1997.
5. Duncan, I.J.H., Animal welfare defined in terms of feelings, *Acta Agriculturae Scandinavica, Section A, Anim. Sci.*, Supplement 27, 29–35, 1996.
6. Fraser, D.M., Science, values and animal welfare: explaining the inextricable connection, *Anim. Welfare*, 4, 103–117, 1995.
7. Broom, D.M., Assessing the welfare of transgenic animals, in *Welfare Aspects of Transgenic Animals*, van Zutphen, L.F.M. and van der Meer, M., Eds., Springer-Verlag, Heidelberg, 1997, pp. 58–67.
8. Fraser, A.F. and Broom, D.M., *Farm Animal Behaviour and Welfare*, 3rd ed., CAB International, Wallingford, Oxon, 1997.
9. Dawkins, M.S., *Animal Suffering: The Science of Animal Welfare,* 2nd ed., Chapman & Hall, London; New York, 1992.
10. DeGrazia, D., *Taking Animals Seriously: Mental Life and Moral Status*, Cambridge University Press, London, 1996.
11. Morton, D.B., Burghardt, G., and Smith, J.A., Critical anthropomorphism, animal suffering and the ecological context, *Hasting's Center Report Spring Issue on Animals, Science and Ethics*, 20, 3, 13–19, 1990.
12. National Research Council, *Guide for the Care and Use of Laboratory Animals*, National Academy Press, Washington, DC, 1996.
13. OECD, Environmental Health and Safety Publications Series on Testing and Assessment No. 19 Guidance Document on the Recognition, Assessment, and Use of Clinical Signs as Humane Endpoints for Experimental Animals Used in Safety Evaluation, 14, 2001.
14. Davies H. and Balfour D., *The Inevitable Bond: Examining Scientist–Animal Interactions,* Cambridge University Press, London, 1992.
15. Hemsworth, P.H. and Coleman, G.J., *Human–Livestock Interactions: The Stockperson and the Productivity and Welfare of Intensively Farmed Animals*, CAB International, 1998.

16. Chance, M.R.A., Environmental factors influencing gonadotrophin assay in the rat, *Nature*, 177, 228–229, 1956.

17. Riley, V., Psychoneuroendocrine influences on immunocompetence and neoplasia, *Science*, 212, 1100–1109, 1981.

18. Rose, M.A., Environmental factors likely to impact on an animal's well-being — an overview, *Improving the Well-Being of Animals in the Research Environment, Proceedings of the ANZCCART Conference*, Baker, R.M., et al., Eds., 1994, pp. 99–116.

19. Claassen, V., Neglected factors in pharmacology and neuroscience research: biopharmaceutics, animal characteristics, maintenance, testing conditions, in *Techniques in the Behavioural and Neural Sciences*, Vol. 12, Huston, J.P., Series Ed., Elsevier, Amsterdam; New York, 1994.

20. Chance, M.R.A., The contribution of environment to uniformity, *Collected Papers — Laboratory Animals Bureau*, 6, 59–73, 1957.

21. Russell, W.M.S., Enhancing animal comfort in the laboratory, in *Humane Innovations and Alternatives*, 1994, p. 8.

22. Dawkins, M.S., *Through Our Eyes Only. The Search for Animal Consciousness*, W.H. Freeman, San Francisco, 1993.

23. Denton, D.A., *The Pinnacle of Life. Consciousness and Self-Awareness in Humans and Animals*, Allen and Unwin Pty. Ltd., NSW, Australia, 1993.

24. Morton, D.B., Self-consciousness and animal suffering, *The Biologist*, 47, 77–80, 2000.

25. Fitzgerald, M., Neurobiology of foetal and neonatal pain, in *Textbook of Pain*, 3rd ed., Wall, P. and Melzack, R., Eds., Churchill Livingstone, London; Edinburgh, 1994, pp. 153–163.

26. Moberg, G.P. and Mench, J.A., *The Biology of Animal Stress. Basic Principles and Implications for Animal Welfare*, CABI, Wallingford, Oxon, United Kingdom, 2000.

27. Lawrence, A.B. and Rushen, J., Eds., *Stereotypic Animal Behaviour, Fundamentals and Applications to Welfare*, CAB International, Wallingford, Oxon, United Kingdom, 1993, chap. 2, pp. 7–40.

28. Mason, G.J., Forms of stereotypic behaviour, *Stereotypic Animal Behaviour; Fundamentals and Applications to Welfare*, Lawrence, A.B. and Rushen, J., Eds., CAB International, Wallingford, Oxon, United Kingdom, 1993, chap. 2, pp. 7–40.

29. Wemelsfelder, F., The concept of animal boredom and its relationship to stereotyped behaviour, in *Stereotypic Animal Behaviour, Fundamentals and Applications to Welfare*, Lawrence, A.B. and Rushen, J., Eds., CAB International, Wallingford, Oxon, United Kingdom, 1993, chap. 4, pp. 65–9531.

30. FAWC (Farm Animal Welfare Council), Second report on Priorities for Research and Development in Farm Animal Welfare, Ministry of Agriculture Fisheries and Food, Tolworth, London, United Kingdom, 1993.

31. Ashby W.R., *Design for a Brain*, Chapman & Hall, London; New York, 1952.

32. van Praag H., Kempermann, G., and Gage F.H., Neural consequences of environmental enrichment, *Nature Reviews Neuroscience*, 1, December, 191–198, 2000.

33. Kempermann, G., Kuhn, H.G., and Gage, F.H., More hippocampal neurones in adult mice living in an enriched environment, *Nature*, 386, 493–495, 1997.

34. Morton, D.B. and Griffiths, P.H.M., Guidelines on the recognition of pain, distress and discomfort in experimental animals and an hypothesis for assessment, *Vet. Record*, 116, 431–436, 1985.

35. Spinelli, J.S. and Markowitz, H., Prevention of cage associated distress, *Lab. Anim.*, 14, 19–24, 1985.

36. Soma, L.R., Assessment of animal pain in experimental animals, *Lab. Anim. Sci.*, 37, 71–74, 1987.

37. Morton, D.B., A systematic approach for establishing humane endpoints, *ILAR J.*, 41, 2, 80–86, 2000.

38. Roughan, J.V. and Flecknell P.A., Behavioural effects of laparotomy and analgesic effects of keto-profen and carprofen in rats, *Pain*, 90, 1–2, 65–74, 2001.

39. FAWC (Farm Animal Welfare Council), *Second report on Priorities for Research and Development in Farm Animal Welfare*, Ministry of Agriculture Fisheries and Food, Tolworth, London, United Kingdom, 1993.

40. Webster, J., *Animal Welfare. A Cool Eye Towards Eden*, Blackwell Scientific, Oxford, 1995, p. 163.

41. Blom, H.J.M., Baumans, V., Vorstenbosch, Van C.J.A.H.V., Zutphen, Van L.F.M., and Beynen, A.C., Preference tests with rodents to assess housing conditions, *Anim. Welf.*, 2, 1–87, 1993.

42. Blom, H.J.M., Tintelen, Van G., Baumans, V., Broek, Van Den J., and Beynen, A.C., Development and application of a preference test system to evaluate housing conditions for laboratory rats, *Appl. Anim. Behav. Sci.*, 43, 279–290, 1995.

43. Blom, H.J.M., Tintelen, Van G., Vorstenbosch, Van C.J.A.H., Baumans, V., and Beynen, A.C., Preferences of mice and rats for types of bedding material, *Lab. Anim.*, 30, 234–244, 1996.

44. van de Weerd, H., Baumans, V., Koolhaus, J., and Van Zutphen, L., Strain specific behavioural response to environmental enrichment in the mouse, *J. Exp. Anim. Sci.*, 36, 117–127, 1994.

45. van de Weerd, H.A., Broek, F.A.R. van de, and Baumans, V., Preference for different types of flooring in two rat strains, *Appl. Anim. Behav. Sci.*, 46, 3–4, 251–261, 1996.

46. van de Weerd, H.A., Van Loo, P.L.P., Van Zutphen, L.F.M., Koolhaas, J.M., and Baumans, V., Nesting material as environmental enrichment has no adverse effects on behavior and physiology of laboratory mice, *Physiol. and Behav.*, 62, 5, 1019–1028, 1997a.

47. van de Weerd, H.A., Van Loo, P.L.P., Van Zutphen, L.F.M., Koolhaas, J.M., and Baumans, V., Preferences for nesting material as environmental enrichment for laboratory mice, *Lab. Anim.*, 31, 133–143, 1997b.

48. van de Weerd, H.A., Van Loo, P.L.P., Van Zutphen, L.F.M., Koolhaas, J.M., and Baumans, V., Strength of preference for nesting material as environmental enrichment in laboratory mice, *Appl. Anim. Behav. Sci.*, 55, 369–382, 1998a.

49. van de Weerd, H.A., Van Loo, P.L.P., Van Zutphen, L.F.M., Koolhaas, J.M., and Baumans, V., Preferences for nest boxes as environmental enrichment for laboratory mice, *Anim. Welfare*, 7, 11–25, 1998b.

50. JWGR (Joint Working Group on Refinement), 6th report, The welfare of genetically modified rodents: refinements in their generation, management and care, In Press, *Lab. Anim.*, 2003.

51. JWGR (Joint Working Group on Refinement), 3rd report, Refining rodent husbandry: the mouse, *Lab. Anim.*, 32, 3, 260–269, 1998.

52. van Loo, P.L.P, Kruitwagen C.L.J.J., Van Zutphen, L.F.M., Koolhaas, J.M., and Baumans, V., Modulation of aggression in male mice: influence of cage cleaning regime and scent marks, *Anim. Welfare*, 9, 281–295, 2000.

53. van Loo, P.L.P., Mol, J.A., Koolhaas, J.M., van Zutphen, L.F.M., and Baumans, V., Modulation of aggression in male mice: influence of group size and cage size, *Physiol. and Behav.*, 72, 675–683, 2001.

54. Bastock M., Morris, D., and Moynihan M., *Behaviour* 6, 66–84, 1953.

55. Augustsson, H. and Hau, J.A., Simple ethological monitoring system to assess social stress in group-housed laboratory rhesus macaques, *J. Med. Primatol.*, 28, April, 2, 84–90, 1999.

56. Bernstein, I.S., Social housing of monkeys and apes: group formations, *Lab. Anim. Sci.*, 41, 329–333, 1991.

57. Guhad, F., Augustsson, H., and Hau, J., The Torneby Primate Facility: optimisation of housing conditions for rhesus macaques in Sweden, *Scand. J. Lab. Anim. Sci.*, 25, 173–176, 1998.

58. Reinhardt, V., Advantages of housing rhesus monkeys in compatible pairs, *Scientist Center for Animal Welfare Newsletter*, 9, 3, 3–6, 1987.

59. Reinhardt, V., Behavioral responses of unrelated adult male rhesus monkeys familiarized and paired for the purpose of environmental enrichment, *Am. J. Primatology*, 17, 243–248, 1989.

60. Reinhardt V., Pair-housing rather than single-housing for laboratory Rhesus macaques, *J. Med. Primatology*, 23, 426–431, 1994.

61. van Loo, P.L.P., de Groot, A.C., Van Zutphen, L.F.M., and Baumans, V., Do male mice prefer or avoid each other's company? Influence of hierarchy, kinship and familiarity, In Press, *JAAWS*, 2002.

62. Jacobs, H., Smith, N., Smith, P., Smyth, L., Yew, P., Saibaba, P., and Hau, J., Zebra Finch behaviour and effect of modest enrichment of standard cages, *Anim. Welfare*, 4, 3–9, 1995.

63. Hawkins, P., Morton, D.B., Cameron, D. et al., Laboratory birds: refinements in husbandry and procedures, 5th report of the BVAAWF/FRAME/RSPCA/UFAW Joint Working Group on Refinement, *Lab. Anim.*, 35: Supplement 1, 2001.

64. JWGR (Joint Working Group on Refinement), 2nd report, Refinements in rabbit husbandry, *Lab. Anim.*, 27, 301–329, 1993.

65. Guhad, F.A. and Hau, J., Salivary IgA as a marker of social stress in rats, *Neurosci. Lett.*, 27, 216, 2, 137–140, 1996.

66. Bartusek, H., A review of the animal needs index (ANI) for the assessment of animals' well-being in the housing systems for Austrian proprietary products and legislation, *Livestock Production Sci.*, 61, 2–3, 179–192, 1999.

67. Poole, T., Identifying the behavioural needs of zoo mammals and providing appropriate captive environments, *RATEL*, 24, 6, 200–211, 1997.
68. Poole, T.B., The nature and evolution of behavioural needs in mammals, *Anim. Welfare*, 1, 203–220, 1992.
69. Poole, T.B., Criteria for the provision of captive environments, in *A Primate Responses to Environmental Change*, Box, H.O., Ed., Chapman & Hall, London; New York, 1991, p. 265.
70. Poole, T.B., Meeting a mammal's psychological needs: basic principles, *Second Nature — Environmental Enrichment for Captive Animals*, Shepherdson, D.J., Mellen, J.D., and Hutchins, M., Eds., Smithsonian Institution Press, Washington, DC; London, 1998, pp. 83–97.
71. Poole, T., Identifying the behavioural needs of zoo mammals and providing appropriate captive environments, *RATEL*, 24, 6, 200–211, 1997.
72. Saibaba, P., Sales, G.D., Stodulski, G., and Hau, J., Behaviour of rats in their home cage: daytime variations and effects of routine husbandry procedures analyzed by time sampling techniques, *Lab. Anim.*, 30, 13–21, 1996.
73. Bayne, K.A.L. Mench, J.A., Beaver, B.V., and Morton, D.B., Laboratory animal behavior, in *Laboratory Animal Medicine*, 2nd ed., Academic Press, 2002, Chapt. 32.
74. Ambrose, N. and Morton, D.B., The effects of environmental enrichment on cage-cleaning aggression in male laboratory mice, *BandK Sci. Now.*, 6, 1, 1–3, 1997.
75. Ambrose, N. and Morton, D.B., The use of cage enrichment to reduce male mouse aggression, *J. Appl. Anim. Welfare Sci.*, 3, 117–126, 2000.
76. Tuli, J., Smith, J.A., and Morton, D.B., Stress measurements in mice after transportation, *Lab. Anim.*, 29, 132–138, 1995.
77. Young, L.A., Pavlovska-Teglia, G., Stodulski, G., and Hau, J., Effect of oral administration of corticosterone and group housing on body weight gain and locomotor development in the neonatal rat, *Anim. Welfare*, 5, 167–176, 1996.
78. Damon, E.G., Eidson, A.F., Hobbs, C.H., and Hahn, F.F., Effect of acclimation on nephrotoxic response of rats to uranium, *Lab. Anim. Sci.*, 36, 24–27, 1986.
79. Tuli, J., *Stress and Parasitic Infection in Laboratory Mice*, Ph.D. thesis, University of Birmingham, AL, 1993.
80. Wadham, J.J.B., *Recognition and Reduction of Adverse Effects in Research on Rodents,* Ph.D. thesis, University of Birmingham, AL, 1996.
81. Hau, J. and Carver, J.F.A., Refinement in laboratory animal science: is it a Cinderella subject, and is there conflict and imbalance within the 3Rs?, *Scand. J. Lab. Anim. Sci.*, 21, 161–167, 1994.
82. Van Zutphen, L.F.M., Refinement, the Cinderella of the 3Rs, in *The Ethics of Animal Experimentation*, European Biomedical Research Association, 1998.
83. Morton, D.B., The importance of non-statistical design in refining animal experimentation, ANZCCART Facts Sheet, *ANZCCART News*, 11, 2, June, 1998.
84. Morton, D.B., Humane end points in animal experimentation for biomedical research: Ethical, legal and practical aspects, in *Humane Endpoints in Animal Experiments for Biomedical Research, Proc.Int. Conf.*, 22–25 November, 1998, Zeist, The Netherlands, Hendriksen, C.F.M. and Morton, D.B., Eds., Royal Soc. Med. London, United Kingdom, 1999, pp. 5–12.
85. Canadian Council on Animal Care (CCAC), Guideline on Choosing an Appropriate Endpoint in Experiments Using Animals for Research, 1998.
86. van den Heuvel, M.J., Clark, D.G., Fielder, R.J., Koundakjian, P.P., Oliver, G.J.A., Pelling, D., Tomlinson, N.J., and Walker, A.P., The international validation of a fixed dose procedure as an alternative to the classical LD50 test, *Food Chem. Toxicol.*, 28, 469–482, 1990.
87. Kuijpers, M.H.M. and Walvoort, H.C., Discomfort and distress in rodents during chronic studies, in *Animals in Biomedical Research*, Hendriksen, C.F.M. and Koeter, H.W.B.M., Eds., Elsevier, Amsterdam; New York, 1991, chap. 18, pp. 247–263, 281.
88. Morton, D.B. and Townsend, P., Dealing with adverse effects and suffering during animal research, in *Laboratory Animals — An Introduction for Experimenters*, revised version, Tuffery, A.A., Ed., John Wiley & Sons, New York, 1995, pp. 215–231.
89. Schlede, E., Mischke, U., Roll, R., and Kayser, D., A national validation study of the acute-toxic-class method — an alternative to the LD50 test, *Archives Toxicol.*, 66, 455–470, 1992.

90. Schlede, E., Diener, W., and Gerner, I., Humane endpoints in toxicology, in *Humane Endpoints in Animal Experimentation for Biomedical Research*, Hendriksen, C.F.M. and Morton, D.B., Eds., Royal Society of Medicine Press, London, United Kingdom, 1999, pp. 75–78.

91. Beynen A.C., Baumans, V., Bertens, A.P., et al., Assessment of discomfort in gallstone-bearing mice: a practical example of the problems encountered in an attempt to recognise discomfort in laboratory animals, *Lab. Anim.*, 21, 35–42, 1987.

92. Soothill, J.S., Morton, D.B., and Ahmad, A., The HID50 (hypothermia-inducing dose 50): an alternative to the LD50 for measurement of bacterial virulence, *Int. J. Exp. Pathol.*, 73, 95–98, 1992.

93. Hendriksen C.F.M., Steen, B., Visser, J., Cussler, K., Morton, D.B., and Streijger, F., The evaluation of humane endpoints in pertussis vaccine potency testing, *Humane Endpoints in Animal Experiments for Biomedical Research, Proc. Int. Conf.*, 22–25 November 1998, Zeist, The Netherlands, Hendriksen, C.F.M. and Morton, D.B., Eds., Royal Soc. Med., London, United Kingdom, 1999, pp. 106–113.

94. Cussler, K., Morton, D.B., and Hendriksen, C.F.M., Humane endpoints in vaccine research and quality control, *Humane Endpoints in Animal Experiments for Biomedical Research, Proc. Int. Conf.*, 22–25 November 1998, Zeist, The Netherlands, Hendriksen, C.F.M. and Morton, D.B., Eds., Royal Soc. Med., London, 1999, pp. 95–101.

95. Hendriksen, C.F.M. and Morton, D.B., Eds., *Humane Endpoints in Animal Experiments for Biomedical Research*, Proc. Int. Conf., 22–25 November 1998, Zeist, The Netherlands, Royal Soc. Med., London, 1999, p. 150.

96. Johannes, S., Rosskopf-Streicher, U., Hausleithner, D., Gyra, H., and Cussler, K., Use of clinical signs in efficacy testing of erysipelas vaccines, *Humane Endpoints in Animal Experiments for Biomedical Research, Proc. Int. Conf.*, 22–25 November 1998, Zeist, The Netherlands, Hendriksen, C.F.M., and Morton, D.B., Royal Soc. Med., London, 1999, pp. 114–117.

97. Krug, M. and Cussler, K., *Endotoxin in porcine vaccines: clinical signs and safety aspects, Humane Endpoints in Animal Experiments for Biomedical Research, Proc. Int. Conf.*, 22–25, November, 1998, Zeist, The Netherlands, Hendriksen, C.F.M. and Morton, D.B., Royal Soc. Med., London, 1999, pp. 114–117.

98. Hubrecht, R.C., Genetically modified animals, welfare and UK legislation, *Anim. Welfare*, 4, 163–170, 1995.

99. Poole, T.B., Welfare considerations with regard to transgenic animals, *Anim. Welfare*, 4, 81–85, 1995.

100. Mepham, T.B., Coombes, R.D., Balls M., et al., *The Use of Transgenic Animals in the European Union*, The report and recommendations of ECVAM Workshop 28, *ATLA*, 26, 21–43, 1998.

101. Van der Meer, M., Baumans, V., Olivier, B., et al., Behavioral and physiological effects of biotechnology procedures used for gene targeting in mice, *Physiol. Behav.*, 73, 5, 719–730, Sp. Iss., 2001.

102. van der Meer, M. and van Zutphen, L.F.M., Use of transgenic animals and welfare implications, *Welfare Aspects of Transgenic Animals*, van Zutphen, L.F.M. and van der Meer, M., Eds., Springer-Verlag, Heidelberg, 1997, pp. 78–89.

103. van der Meer, M., Baumanns, V., and van Zutphen, L.F.M., Measuring welfare aspects of transgenic animals, *Animal Alternatives, Welfare and Ethics*, van Zutphen, L.F.M. and Balls, M., Eds., Elsevier, Amsterdam; New York, 1997, pp. 229–233.

104. Costa, P., Production of transgenic animals: practical problems and welfare aspects, in *Welfare Aspects of Transgenic Animals*, van Zutphen, L.F.M. and van der Meer, M., Eds., Springer-Verlag, Heidelberg, 1997, pp. 68–77.

105. Jenkins, E.S. and Francis, R.F., New guidance notes for projects generating or maintaining genetically modified animals, *ATLA* 27, 1999, Suppl. 1, in press.

106. Dennis, M.B., Jr., Humane endpoints for genetically engineered animal models, *ILAR*, 41, 2, 94–98, 2000.

107. Rogers, D.C., Fischer, E.M., Brown, S.D., Peters, J., Hunter, A.J., and Martin, J.E., Behavioral and functional analysis of mouse phenotype: SHIRPA, a proposed protocil for comprehensive phenotype assessment, *Mamm. Genome*, 10, 711–713, 1997.

108. HSUS 2001, www.hsus.org

109. Leach, M., Bowell, V., Allan, T., and Morton, D.B., Aversion to gaseous euthanasia in rats and mice, *Comparative Med.*, 52(3), 249–257, 2002.

110. Leach, M., Bowell, V., Allan, T., and Morton, D.B., Aversion to various concentrations of different inhalational general anesthetics in rats and mice, *Vet. Record*, 150, 808-815, 2002.

111. Editorial, a predators' compassion, *The Lancet*, 343, 1311–1312, 1993.
112. Webster, A.J.F., What use is science to animal welfare, *Naturwissenschaften*, 85, 6, 262–269, 1998.
113. Hau, J., Humane endpoints and the importance of training, *Humane Endpoints in Animal Experiments for Biomedical Research, Proc. Int. Conf.*, 22–25 November 1998, Zeist, The Netherlands, Hendriksen, C.F.M. and Morton, D.B., Eds., Royal Soc. Med., London, 1999, pp. 106–113.
114. Mellor, D.J. and Gregory, N.G. Responsiveness, behavioral arousal and awareness in fetal and newborn lambs: experimental, practical and therapeutic implications, *New Zealand Vet. J.*, in press 2003.

## RELEVANT REFERENCE SITES ON THE WORLD WIDE WEB

AWIC Animal Welfare Information Center
http://www.nal.usda.gov/awic/
CCAC Guidelines on HEPs and Other Issues
http://www.ccac.ca/english/gdlines/endpts/appopen.htm
Various Papers on Refinement Including JWGR Reports
http://www.lal.org.uk/
OECD Guidance
http://www.oecd.org/ehs/test/mono19.pdf
ALTWEB
http://altweb.jhsph.edu/news/news.htm
Center for Alternatives — Good Links for Refinement
http://caat.jhsph.edu/
Interniche for Alternatives to the Use of Animals in Teaching
http://www.interniche.org/alt_loan.html
NORINA for Alternatives to the Use of Animals in Teaching and Other Refinement Papers
http://oslovet.veths.no/teaching/materials.html
Training Programs
http://dcminfo.wustl.edu/education/models.html
Refinement and Enrichment
http://www.animalwelfare.com/lab_animals/rhesus/pho1–10.htm
http://www.awionline.org/lab_animals/biblio/index.html
SHIRPA Protocol and Other Protocols
http://www.mgu.har.mrc.ac.uk/MGU-welcome.html
http://www.bzl.unizh.ch/de/database/formtransg/index2.html

# CHAPTER 19

# Surgery: Basic Principles and Procedures

H. Bryan Waynforth, M. Michael Swindle, Heather Elliott, and Alison C. Smith

## CONTENTS

0-8493-1086-5/03/$0.00+$1.50
© 2003 by CRC Press LLC

## INTRODUCTION

Surgery is the most invasive procedure that can be carried out on an experimental animal. It could possibly result in pain and distress or in an infection if not managed properly. It is an absolute requirement of the surgeon or investigator to minimize any pain or infection by the careful use of analgesics or antibiotics and to be considerate of the animal's welfare at all times. To minimize the trauma, it is important that the surgeon has access to a well-organized and well-equipped surgical facility and that he uses proper surgical technique. This can only be achieved by a thorough understanding of the animal's physiology and anatomy and by using excellent surgical technique, including the use of asepsis in large animals and rodents. Excellence in surgery is gained by studying the relevant literature, by judicious practice on cadavers or other appropriate models, and importantly, by initially seeking the help of a colleague experienced in the technique. This chapter gives an outline of the important principles of surgery. The reader who needs more detailed information on these subjects should consult more comprehensive texts.[1–20]

## PLANNING FOR SURGERY

If an animal is not to suffer unnecessarily, it will be important to ensure that when carrying out the surgery, however simple, all aspects are thoroughly planned and understood by everyone who will be involved. It is useful to prepare a checklist of requirements, which will not only be useful to the surgeon but also to any ancillary personnel who may be assisting. Assistants will usually be involved in large animal surgery, but it is not sufficiently appreciated that for rodent surgery, an assistant can also be invaluable. Such a person will be able to help in maintaining a sterile surgical field by taking on all the "dirty" tasks, such as preparing the animal beforehand, and in undertaking other useful jobs to support the surgeon. A checklist should include the following:

1. Choice and availability of the animal(s)
2. Preoperative evaluation of the animal's health
3. Provision of surgical and pre- and postoperative facilities
4. Choice, provision, and sterilization of surgical instruments, equipment, and ancillary supplies
5. Number of assistants required
6. Preparation of the animal for surgery
7. Management of the animal postoperatively and in an emergency
8. Written protocol of the surgical and anesthetic procedures
9. Experience of the surgeon, to include adequate practice in handling instruments, tissues, and in the surgical procedure
10. Record keeping

An important consideration is when to carry out the operation. Surgery done late in the day or at the end of the week may require that postoperative care continue well into the evening or weekend, and this could be inconvenient and lead to insufficient care being taken.

# PREPARING FOR SURGERY

## Surgical Facilities and Equipment

### Large Animals

At least three major components need to be supplied for facilities for aseptic survival surgery in large animals: an operating room, an animal preparation room, and a surgeon preparation room. These three areas need to be separate and have controlled ventilation. Other ancillary areas should include storage, animal recovery, and instrument preparation. The animal preparation room and the surgeon preparation room may be used for some of these functions. Floors, walls, and ceilings should be impervious to moisture and easily sanitized. Operating room lights should be suspended from the ceiling. It is preferable to have columns with gases such as oxygen and vacuum for evacuation and suction. Gas anesthesia machines, electrocautery, and ECG monitoring should be provided as a minimum. Other equipment that is useful for major procedures includes blood pressure monitors and monitors for exhaled $CO_2$. Intensive care units and cardiopulmonary emergency kits should also be available for major procedures. A floor plan of a suitable operating theater and animal facility is shown in Figure 19.1.

### Rodents

Surgery on rodents should only be carried out in a dedicated area. Ideally, this should be a facility that is separate from other activities, but for nonrecovery surgery, a clean dedicated area on a laboratory bench might suffice. For rodents, the most usual arrangement for survival surgery, where the animal is to recover from the anesthetic, is relatively simple and involves having a single large room, whereby part is used for anesthetizing and preparing the animal and the surgeon, part is used for the actual surgery, and a small part is kept to allow the animal to recover before returning it to its cage. A room showing a surgical station and areas for animal preparation and recovery is shown in Figure 192. This single room arrangement is suitable where space is at a premium, but ideally, as for large animals, each of the functions are best carried out in separate rooms, if these are available, so that microbiological contamination can be kept to a minimum. The surgical room(s) must be kept clean by the judicious use of disinfectants, and be adequately ventilated, with an appropriate system to scavenge waste anesthetic gases. For all but the simplest surgery, a good source of surgical lighting is important. Also necessary is the ability to sterilize instruments and other supplies such as surgical gowns and drapes.

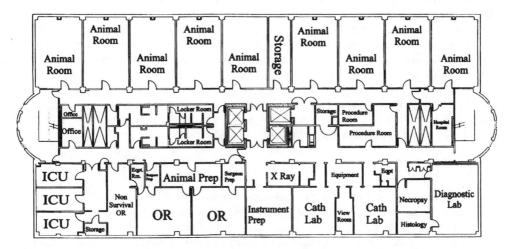

**Figure 19.1**    Floor plan of an operating theater and adjacent animal housing and support areas. (Courtesy of the Medical University of South Carolina.)

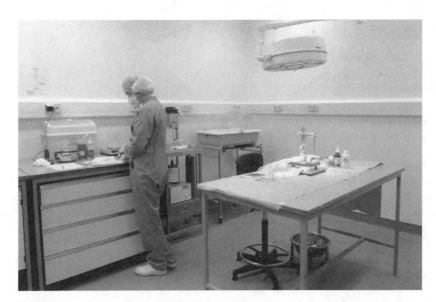

**Figure 19.2**   Rodent Operating Room. The assistant is removing the hair from the rat over the surgical site prior to the operation. In the far right-hand corner is an animal incubator for the animal to recover in.

Surgery is normally carried out on a table or bench. In many cases, it is useful to place the animal on a thermostatically controlled heat blanket and use a rectal probe to maintain and monitor its body temperature. The blanket and bench are covered with a sterile drape to maintain the sterility of the instruments subsequently placed on them. Some surgeons prefer to place a cork or rubber board on the bench first, underneath the drape, which allows pins and similar instruments to be stuck into it for securing ties to aid positioning.

## Surgical Instruments

### Large Animals

The surgical instruments required will vary with the procedure being performed. A general surgery pack should include curved and straight hemostats of various sizes, a scalpel handle, Allis tissue forceps, scissors of various sizes, forceps, needle holders, stainless steel bowls, an instrument tray, and retractors, both hand-held and self-retaining. The recommended number and size vary widely with the species and the procedure to be performed (Figure 19.3). As a general rule, more instruments should be included in the pack than the surgeon anticipates using. Consultation with a professional familiar with surgical procedures should be sought before buying the instruments.[20–22]

Instruments should be thoroughly cleaned between procedures to remove blood and tissue. Instruments should be packed in cloth or other suitable material designed to withstand autoclaving and then can be stored in a clean, dry environment, after being autoclaved in the appropriate manner. The instruments should only be reautoclaved if integrity of the pack is broken.

### Rodents

For rodent surgery, instruments are generally selected from those used in human pediatric, ophthalmic, and neurological surgery because of their small size.[8,21,22] A basic set of instruments might consist of a size 3 scalpel handle with a number 10 blade, a pair of blunt-ended Mayo scissors, a pair of pointed scissors, two pairs of straight or curved medium-fine serrated dissecting forceps, and a similar pair of rat-toothed forceps. A more satisfactory, though more expensive alternative, is to use De Bakey forceps that have fine, relatively atraumatic tips which provide a firm grip on the tissues without causing trauma. To this basic set will need to be added other instruments and accessories, which might be required for

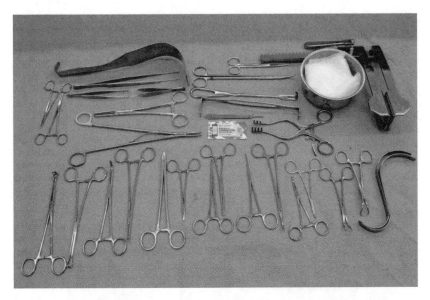

**Figure 19.3** Example of surgical instruments required for general surgery in large animals.

a specific procedure. Such instruments could include microsurgical scissors and forceps, eyelid retractors, hemostatic forceps, bulldog clips, neurological clips, stainless steel screws, etc. In addition, accessories such as cotton wool buds, drapes, gauze swabs, dental cement, plastic cannulae, steel trocar needles, etc., may be needed. Instruments to close wounds will be essential, including needle holders, needles, and suture materials of various types. For closure of the skin, stainless steel staples, automatically applied, or Michel clips together with a suitable applicator, are popular alternatives in appropriate circumstances. On occasions, other equipment may be found necessary to facilitate surgery. For example, a stereotaxic frame and drill will be required to carry out some craniotomy procedures. A particularly useful item for microsurgery, for example, is a zoom microscope operated manually or via a foot-operated control. The type of specialist apparatus required will be dictated by the particular surgical procedure being carried out.

All surgical instruments and accessories need to be sterile before use. This is achieved by sterilization using steam, heat, irradiation, or ethylene oxide gas.[23] The method chosen will generally depend on the material to be sterilized, safety considerations, and local availability. Adequate surgical sterilization cannot be obtained by boiling in water or by the use of chemical disinfectants which therefore should only be used as a last resort. Some instruments and accessories, such as needles, sutures, surgical gloves, drapes, and scalpel blades can be obtained already sterilized from commercial sources.

The instruments and accessories required for the surgical procedure should be prepackaged and, if appropriate, the complete pack(s) sterilized. Each pack is then opened, preferably by the assistant, in such a manner as to keep the contents sterile, and the contents are placed on the sterile bench drape in an orderly fashion. They are now ready for the surgeon to use (Figure 19.4).

## CARE AND PREPARATION OF THE ANIMAL BEFORE SURGERY

### Large Animals

Animals should be fasted approximately 12 h before any surgery to prevent vomiting. This period may have to be longer for some abdominal procedures. Water generally does not have to be restricted until a few hours before surgery in most cases. This can help prevent dehydration. For prolonged fasts, such as may be required for some lower gastrointestinal procedures, use of oral electrolyte/glucose solutions should be considered to maintain the animal's hydration and to prevent hypoglycemia.

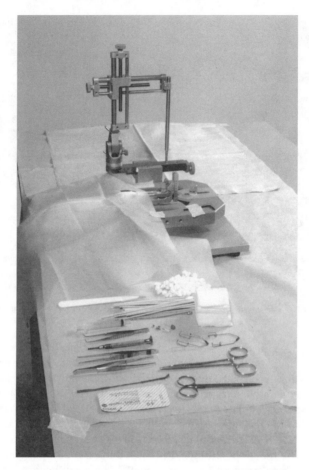

**Figure 19.4**    Arrangement of instruments placed on a sterile drape ready for surgery. Note the drape covering the whole animal, with only the surgical site over the head being exposed.

There is no agent that completely sterilizes the skin without damaging it. Thus, the agents used for skin preparation reduce the microflora to a level that minimizes danger from infection.[18,24,25] The animal should be clipped and shaved to remove all hair for a region, which is distant from the site of the surgical incision to ensure that an unprepped area cannot contaminate the incision. Antiseptic soap solutions, such as those containing iodine or chlorhexidine, are then used to scrub the area, starting along the surgical incision site and working outward in a circular fashion. This should be repeated three times. The soap can be removed with alcohol in the same manner. Some surgeons then prefer to wipe the area with iodine solution. An assistant should perform the final surgical preparation of the site with sterile instruments and gauze sponges within the operating room. Iodine impregnated adhesive drapes can also be applied to the dry scrubbed skin. The whole animal should be covered with a final sterile drape that covers all surfaces including the head and feet. Figure 19.5 shows a pig draped for aseptic surgery.

## Rodents

Rodents should be acclimated to their home environment for up to 7 days before being operated on and must be healthy. There is no need to fast the animal before surgery, as rodents are incapable of vomiting. Once the animal is anesthetized, the site of the incision and a small area around it is prepared, first, by removal of the hair using small fine clippers and then, by swabbing the area with a soap-based antiseptic cleaner followed by an antiseptic such as 0.5% alcoholic chlorhexidene or 10% alcoholic or aqueous povidone-iodine. Excess antiseptic must be removed with a sterile gauze pad, because, if left, it will inappropriately cool the surgical area by evaporation. The animal should then be draped. The sterile

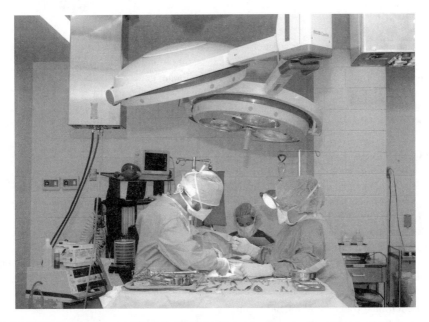

**Figure 19.5** Pig and surgeons appropriately draped and dressed for aseptic surgery.

drape, which should be used for even the smallest incision, is placed over the animal after cutting a small hole in it to correspond to the incision area. In some instances, the drape may need to be clipped to the skin to prevent it slipping. Generally, the drape will cover the whole animal, which means that with cloth drapes, it will be difficult to see that respiration is continuing normally. This should therefore be checked frequently, and semitransparent plastic drapes used if possible. Drapes should also be positioned to enable instruments to be laid on them beside the animal so that they remain sterile (Figure 19.4).

For surgical procedures that have the potential for contamination with a large microbiological load, an appropriate antibiotic should be injected prior to surgery. This will then have its peak effect while the operation is proceeding and when the contamination is likely to be at its greatest.[26]

## CARE OF THE ANIMAL DURING SURGERY

### Preparation of the Surgical Team

The surgical team should wear scrub suits, which are only used in the operating theater. Either shoes that are dedicated to the area that can be disinfected or shoe covers should be worn. A clean surgical cap and mask should be placed on the surgeon prior to performing the surgical scrub of the hands and forearms. The same precautions concerning use of disinfectant agents for the surgeons skin as discussed above for the animals skin applies. Using a sterile brush, the fingers and hands are thoroughly scrubbed and rinsed with running water. Then each forearm is scrubbed separately in the same manner. The hands are kept higher than the elbows to prevent water and soap from flowing back onto the fingers. Once the scrub is performed, a surgical gown is placed on the surgeon with the help of an assistant and tied at the back. The surgeon's fingers do not extend beyond the cuff of the sleeve until the sterile gloves are donned. A sterile assistant may hold the gloves and the surgeon then thrusts the hands into the open glove. As an alternative, the surgeon may manipulate his hands into the gloves after placing them on the forearm with the fingers pointed toward the shoulder. The outside of the gloves should never be touched by bare fingers during this procedure. Once the hands are covered and the glove is over the cuff of the sleeve, the surgeon may use the gloved hands to properly fit the gloves over the fingers. Once the surgeon is gowned and gloved, the hands and arms are never held below the waist.[20]

## Aseptic Surgery

During the surgical procedure, the surgeon should never touch a nonsterile object.[20,27] The hands and forearms should be kept over the draped area of the animal and the instrument tray, which has been opened by an assistant. For major procedures, the surgical team generally includes a surgeon, an assistant surgeon, and an instrument nurse, in addition to an anesthetist who may double as an operating room assistant to open packs, sutures, and other required supplies onto the sterile tray. The outer cover of any package opened by the circulating nurse should never touch the sterile field or the surgical team.

## Handling of Tissues and Instruments

The tissues should be handled using the principles set forth by Halsted over a century ago. The principles of surgery are asepsis, gentle handling, hemostasis, closure of dead space, careful approximation of tissues, avoidance of tension, and minimization of foreign materials. Prevention of sepsis by using aseptic technique is far more reliable than using antibiotics to try to cover breaks in asepsis. When antibiotics are indicated during a procedure, the single most important dose is one that has a blood level of the antibiotic when the skin incision is made.[20,26] Control of hemorrhage with electrocautery or ligatures is important to prevent hyponolemia as well as to prevent hematomas and seromas. Likewise, closure of dead space prevents seromas, which can be a pocket for infection. Gentle handling of tissues is essential to prevent additional trauma and inflammation.

Surgical instruments are designed to prevent undue trauma to the tissues and aid in gentle handling. The appropriate instrument should be selected when performing a particular surgical manipulation, and this requires that surgeons familiarize themselves with the use of particular instruments prior to performing the procedure. Ringed instruments should be placed to the level of the most distal knuckle of the thumb and ring finger. Instruments can then be stabilized with the two digits between the rings. The scalpel should be held firmly between the tips of the fingers and thumb, and the incision should be made with the belly of the blade, not the tip. The scalpel should be handed between personnel by the distal end of the handle, not near the blade. The tips of the instruments should be used in most cases and little surrounding tissue should be included when the instrument is closed and locked. This prevents trauma and necrosis except in the tissue or structure of interest. All tissues should be kept moist using gauze or towels wetted with warm saline. Only the edges of the incision should be exposed in the surgical field, and the drapes should be placed to prevent bare skin from being exposed to the surgeon. When using gauze on exposed tissue, avoid rubbing because it is abrasive. Rather, the wetted gauze should be used to press and hold against the bleeding edges.[20]

## Wound Closure and Suturing

The wound edges should be closed in precise anatomic correctness in layers. It is important to eliminate all dead spaces during the closure. Sutures should be placed so that the edges are in close approximation but not tied so tightly as to induce pain and inflammation. This is especially important in the skin. Inner layers of muscle and fascia, as well as vascular ligatures, should be tied tightly. Sutures should generally start at one end of the incision and proceed in a regular pattern and spacing to the opposite end. In general, sutures should be spaced twice the distance along the length as the distance of the suture bite from side to side.[8,18,20,28,29]

The suture pattern depends upon the area of the wound being closed. Simple interrupted sutures offer the advantage of security because the loss of one suture will not open the entire wound. Continuous sutures are easier and quicker to tie and offer a better seal of the edges, but if one knot comes loose, the entire layer will dehisce. Everting sutures, such as the horizontal and vertical mattress patterns, have special indications for tension relief on the wound edges. The horizontal mattress is used to relieve tension along a lengthy incision. The vertical mattress is used in the skin to reduce tension and potential scarring. The Cushing and Connell suture patterns are used to close hollow organs, such as the stomach and bladder. They are inverting patterns. The Lembert pattern is used as an inverting suture to oversew other patterns, such as the Cushing or Connell, or by itself in a single-layer closure. These inverting patterns provide a waterproof seal. The indications for the various patterns need to be appreciated by the surgeon in advance of a procedure[8,18,20,30] (Figure 19.6).

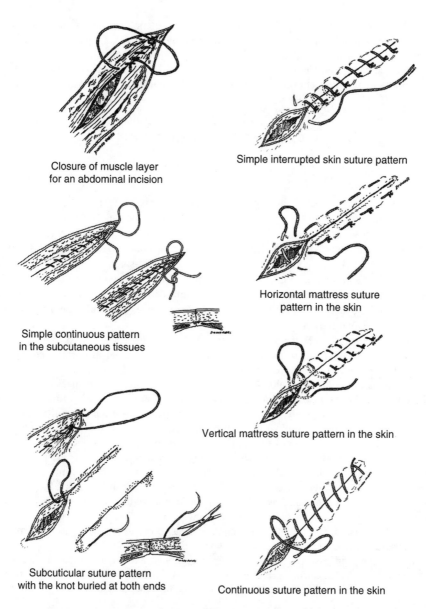

Closure of muscle layer
for an abdominal incision

Simple interrupted skin suture pattern

Simple continuous pattern
in the subcutaneous tissues

Horizontal mattress suture
pattern in the skin

Vertical mattress suture pattern in the skin

Subcuticular suture pattern
with the knot buried at both ends

Continuous suture pattern in the skin

**Figure 19.6**  Suture patterns commonly used in surgery. (Reprinted with permission from Swindle M.M., *Surgery, Anesthesia and Experimental Techniques in Swine*, Iowa State University Press, Ames, IA, 1998.)

Choice of suture material is dependent upon the surgeon's preference, the structure to be sutured, the animal species, and the size of the incision. Generally, sutures may be classified as absorbable or nonabsorbable.[8,18,20,31-34] The synthetic absorbable materials are superior to natural products such as surgical gut, because they have more tensile strength and a longer period of time before they are absorbed after implantation, as well as being less inflammatory. Synthetic nonabsorbable suture materials are also generally superior to natural products, such as silk for the same reasons. Suture material may be braided for extra strength. It may also be coated for easier passage through tissues. Monofilament sutures are synthetic sutures used widely for suturing cardiovascular tissues and skin. Suture material is sized in (0) sizes. The most commonly used suture materials for general surgery are 0–3/0 Suture diminishes in diameter as the (0) size increases with those >8/0 usually used in microsurgery. Sizes 1–3 are used in particularly large animals. Other suture materials include stainless steel, which is widely used in orthopedic procedures.

Staples have replaced suture materials in many cases and come in a variety of sizes in absorbable and nonabsorbable materials. Specialized devices are available for a wide range of procedures including bowel anastomoses, vascular transfixion, and skin closure. Using staples in the skin may lead to an increased incidence of skin infections, because foreign materials tend to get caught in them and animals can catch them on their cages. However, they can be used to particularly good effect in rodents.[6,8]

## APPROACHES TO SURGERY

### Laparotomy

The laparotomy approach selected, in large animals and rodents, depends upon the site of surgical interest. Approaches include midline, paramedian, flank, and transverse. The midline approach is the least traumatic, because the incision is made along the linea alba and thus it avoids transecting or dissecting muscles. Any other approach will involve cutting or blunt dissection through the muscles of the abdominal wall. The flank approach is usually reserved for the surgical approach to the kidneys or adrenals. Paramedian and transverse incisions are usually utilized for approaches to organs such as the spleen, biliary system, uterus, or stomach in larger animals in which adequate exposure cannot be obtained from a midline incision. Approaches for laparoscopic surgery depend upon the procedure being performed and are beyond the scope of this chapter.[18,20]

The principles of surgery are the same regardless of the approach and need to address all of the issues discussed previously. Midline incisions are made with the animal in dorsal recumbency. Large animals should be placed in a V-shaped trough or with the legs tied cranially or caudally to prevent rotation. An incision along the linea alba is relatively bloodless and in some species, such as the dog, fat may be encountered within the abdomen which requires some hemostatic procedures. The initial incision should be made completely through the skin along the entire length of the desired surgical wound. The remainder of the abdominal incision is made with scissors or an electrocautery. Next the midline is cut until the abdomen is to be entered. At the point of entering the abdominal cavity, the tissue on either side of the midline is lifted with forceps, and a stab incision is made into the abdomen. This allows air to enter the abdominal space and the organs are retracted back into the cavity, which prevents their being accidentally incised during the procedure. When using scissors, the muscles should be tented upwards so as to avoid inadvertently harming the underlying organs. At this point, the edges of the incision are covered with wetted gauze laparotomy sponges. Depending upon the site of interest and the size of the incision, self-retaining retractors, such as Balfours for large animals, are utilized to provide increased exposure.[8,18,20]

Abdominal organs and tissues should have minimal exposure to air or abrasive materials. Tissues should be frequently wetted during surgical procedures to prevent drying. The proper instruments, or fingers, should be used in handling abdominal organs, because they tend to be friable and easily damaged.[8,18,20]

### Thoracotomy

Thoracotomy is generally performed through an intercostal space or via a median sternotomy.[18,20] The choice of the incision site is the determining factor in most cases, and approaches to various thoracic organs and structures will vary among species. A ventilator is required for thoracic procedures. An experienced anesthetist is necessary in order to adequately assess oxygenation, blood gas measurements, and ventilatory rates and pressures. Cardiopulmonary emergency drugs and devices such as a defibrillator for large animals should be available.

A lateral thoracotomy is performed in an intercostal space parallel and between two ribs. Because of the difference in the angles of the ribs of the thoracic cages among species, this may be almost an oblique incision in species such as the pig. The skin and muscle are incised down to the level of the pleura with the animal in lateral recumbency. Upon reaching the pleura, a stab incision is made in the pleura as the animal exhales. This may entail turning off the ventilator momentarily. The ribs are retracted outwards to avoid damage to the lungs. After the pleura is incised, scissors are utilized to extend the incision. Self-retaining rib retractors are placed to expand the surgical exposure after applying wetted gauze pads to the edges of the incision.

Special care should be taken when handling the lungs because of their friability. They should be packed off with wetted gauze and the ventilation volume adjusted. As with other procedures, the tissues should be wetted with warm saline to prevent dessication.

The median sternotomy approach is generally utilized for exposure of the heart for procedures such as transplantation. This approach offers greater exposure of the heart without having the interference of the lungs in the surgical field. It is more readily perfomed without splitting the manubrium sterni to minimize postsurgical pain due to movement of the sternum.

After making a midline incision, the sternum is split with a surgical saw or chisel. It is best to position a metal spatula beneath the sternum when it is split to avoid damage to the heart. Self-retaining retractors are utilized as above.

Lateral thoracotomies are closed first by preplacing large sutures around the ribs on either side of the incision. A chest tube is placed through the skin and muscle layers caudal and dorsal to the incision site, placing the tip so that it does not traumatize the lungs. The rib sutures are then tied tightly. The remaining muscle, subcutaneous, and skin layers are closed routinely. The chest tube is then secured in place with a purse string suture in the skin, and the air is evacuated from the chest, either with a syringe or by using a chest tube with a one-way valve. Removal of the chest tube will depend upon the procedure performed. It may be only a few minutes if no air or blood is present in the thorax or it may have to stay in place for a number of days for traumatic procedures.

For median sternotomies, the sternum is closed with wire or nonabsorbable sutures after they are preplaced in the same fashion as described for lateral thoracotomies. The subcutaneous and skin layers are closed in a routine fashion, and a chest tube is utilized as described above.

## Catheterization

Implantation of catheters is one of the most common research surgical procedures performed.[6,8,16,18,19,33–44] Short-term catheterization of a few days is relatively easy to accomplish and maintain, however, complications increase with the amount of time required for the catheter to be maintained and are related to the site of placement. This chapter is limited in scope to the general principles of implanting and maintaining chronic intravascular catheters.

Design of the catheter is critical to the success of long-term catheterization. It is best to utilize manufactured catheters that are free of imperfections, rather than to attempt to make catheters in-house. The two most common catheter materials are silicone and polyurethane. Silicone is more flexible and less traumatic but can absorb contaminants if not handled properly prior to implantation. It is more difficult to guide into vessels because of its flexibility and is also more thrombogenic. Tapered tips are preferred over sharp or angular tips to prevent trauma to vascular walls. Catheters have to be secured tightly in place both at the entrance to the blood vessel and at the exit site on the skin to minimize movement. The length of the catheter should be premeasured to ensure that the tip placement will be correct. Use of preplaced beads and cuffs that are secured to the catheter prior to sterilization greatly facilitates surgical implantation (Figure 19.7).

In order to externalize the port of the catheter, the catheter track from the point of exit from the skin to the blood vessel should be made with a trochar (tunneling rod). The trochar is passed from the site of the blood vessel through the subcutaneous tissue to the skin, a skin incision is made over the tip of the trochar, and then the catheter is passed into the hollow tube. When the trochar is removed, the catheter should be in place with minimal trauma to the tissues. Subcutaneous vascular access ports can also be used, and the catheter track is made in the same manner. The site of exit of the catheter depends upon the blood vessel of interest, the species of animal, and the use for which the catheter is intended. The exit site should be chosen to allow ready access by personnel, while minimizing the animal's access to chew or scratch the site. The use of jackets or harnesses may be required (Figure 19.8).

Vessels that are frequently used for chronic catheterization include external and internal jugular veins, carotid artery, and femoral artery and vein. However, virtually any internal or external vessel can be catheterized if the catheter is properly designed. After the surgical approach is performed, the

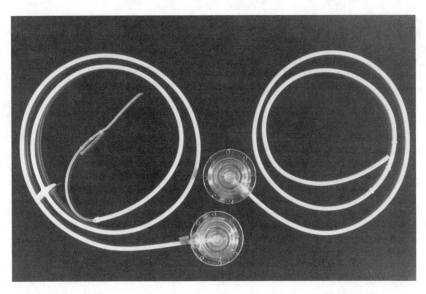

**Figure 19.7**    Examples of implantable vascular access ports with suture retention beads and cuffs.

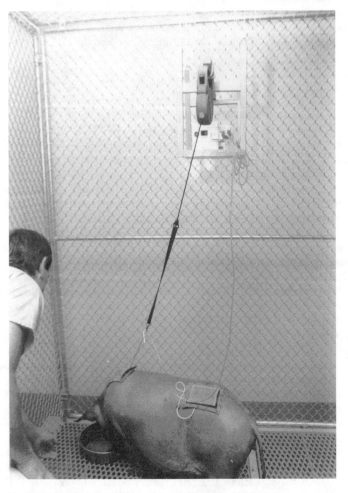

**Figure 19.8**    Pig with multiple catheterizations and a tether and harness system.

vessels are isolated and cleared of adventitious tissue. Elastic vessel loops or suture strands are placed proximally and distally to the site of entrance of the catheter into the vessel. The vessel is allowed to fill with blood, and a small nick incision is made with iris scissors or a #11 scalpel blade no more than 1/3 of the way through the vessel. A vascular pick or small probe may be used to facilitate passing the catheter into the lumen of the vessel. Catheters should be prefilled with saline. When passing the catheter toward the heart, the vascular loop that is proximal is released enough to allow the atraumatic passage. The catheters are secured in place with nonabsorbable suture, both proximally and distally. A loop of the catheter is left in the subcutaneous tissue to ensure that the catheter is not under tension during movement or growth of the conscious animal. The surgical wounds are closed in the usual manner. The catheter exit site should be tightly sealed to prevent infection and catheter migration. To prevent clotting within the catheter, a heparin-saline solution must be introduced and "locked" in place by closing the exit of the catheter with a pin or by attaching it to a device such as an access port or a swivel.

## Craniotomy

Craniotomy may be carried out in rodents, e.g., rat, mouse, guinea pig, for a variety of purposes. These are mainly associated with the study of the effects of substances on, or physical interference with, the brain, measuring or stimulating electrical activity of specific brain areas, or analyzing neurotransmitters or test compounds. This may be for the investigation of, and development of therapies for neurodegenerative and psychiatric disorders and other pathological processes, which may be centrally mediated, e.g., metabolic disease.

This surgery may be carried out to implant devices for recording electrical activity, dialysis, electrical stimulation, and administration of test substances.[47] The approach is usually via a small drill hole made in the mid-dorsal area of the skull, although other approaches may be used in certain circumstances, e.g., zygomatic approach to access the middle cerebral artery.

Where it is required to target a specific area, the location of this should be mapped and stereotaxic coordinates obtained. Atlases with this information are available,[48,49] however, it is advisable to validate these for the particular species, strain, and body weight of the animal to be used. This can be done using cadavers by positioning cannulae stereotactically, injecting dye, and examining brain sections histologically to determine the optimal position.

The procedure for a mid-dorsal craniotomy follows a standard pattern for most of the purposes mentioned above. The animal is anesthetized by injection or inhalation. Where inhalation anesthesia is used, consideration must be given to ensuring efficient delivery and scavenging of anesthetic gas, e.g., by use of a purpose-made face mask to fit on the stereotaxic frame or by endotracheal intubation.

The skin is clipped and cleaned at the surgical site, taking care not to contaminate the eyes. The animal is positioned in the stereotaxic frame in ventral recumbency with provision to maintain body temperature, e.g., on a thermostatically controlled heat blanket. A drape is placed over the whole animal with only the surgical site exposed (Figure 19.4). An incision is made in the skin from behind the eyes to between the ears. It is recommended that local anesthetic be injected into the periosteum, which is then scraped away. The position of the entry site, e.g., of implant or needle, is determined stereotaxically, and holes are carefully drilled just through the skull at this point and at two to four points around this to take steel anchoring screws. Where implants are being used, up to four anchoring screws are placed in the skull to support the placing of the head cap (Figures 19.9 to 19.11). The implant is positioned to the required depth, and then a dental cement head cap is constructed around the screws. The skin is sutured at the front and rear of the head cap, leaving the cement and implant exteriorized.

Care must be taken during surgery with gentle tissue handling, judicious use of local anesthetic, and smooth sculpting of the dental cement to an oval shape to prevent tissue irritation and swelling.

Perioperative analgesia and postoperative nursing care are given to aid recovery. Where an intracerebroventricular cannula has been implanted, its position may be confirmed before being used in studies. This can be done by administering angiotensin via the cannula and observing a drinking response.

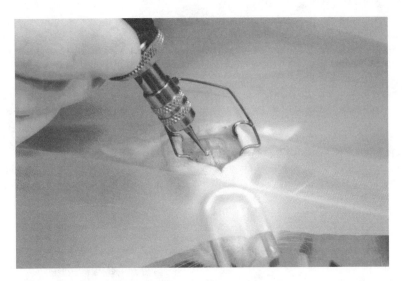

**Figure 19.9**    Drilling a hole in the skull.

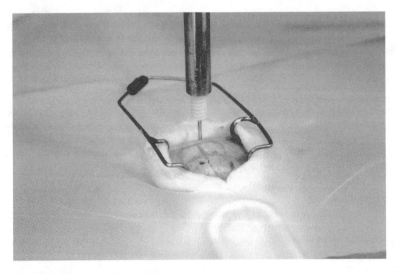

**Figure 19.10** Stereotactically determined drill hole in the skull.

## POSTOPERATIVE CARE

Provision of good postoperative care is an extension of good surgical technique and requires appropriate facilities, equipment, and trained personnel. All animals recovering from surgery and anesthesia require monitoring and specialized care, which should be predicated on species-specific needs and the type of operative procedure. This is best achieved by having a separate, dedicated recovery room. The focus of postoperative monitoring should be on returning the animal to, and then maintaining normal physiologic parameters because anesthetics disrupt homeostasis.[50] If homeostatic abnormalities are inadequately addressed postoperatively, recovery may be prolonged, resulting in further physiologic exacerbations or even death.

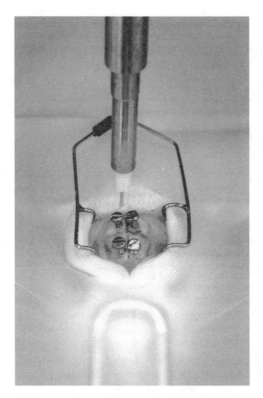

**Figure 19.11**  Stainless steel screws placed into the skull for anchoring and aiding the formation of the head cap.

## Large Animals

### The Recovery Environment

A dedicated recovery room should be warm and quiet. It should provide lighting adequate for staff to observe animals and perform nursing procedures. An ambient room temperature of 21 to 25°C is generally suitable for most mature animals, as long as supplemental heat sources such as circulating water blankets and heat lamps are available. Covering large animals with blankets or drapes also helps to warm the animal. Intensive care units may be utilized to control the ambient temperature for animals less than 25 kg, with the additional advantage of easy administration of oxygen. Frequent monitoring of rectal temperature should be performed postoperatively until it can be determined not only that the animal has returned to normothermia, but also that it can maintain a normal body temperature in the absence of supplemental heat. If an animal becomes hypothermic after removal of supplemental heat and having previously reached a normal body temperature, the animal is still unstable physiologically, and the real possibility of additional underlying problems exists.

Transport cages for dogs are suitable for recovering most large animals up to 50 kg. For larger animals, recovery in runs is more practical. In either case, the recovery area should be equipped to handle potential complications during recovery. Necessary items include intubation supplies in the event of apnea or airway obstruction, suction equipment to clear airway passages of excessive secretions, supplies for fluid administration, and equipment for supplemental oxygen administration. Additional equipment might also include an ECG monitor and pulse oximeter.

### Monitoring and Nursing Care

It is useful to divide the postoperative period into the time of recovery from anesthesia and the period of time that includes recovery from the surgical procedures, i.e., generally, healing of surgical incisions.

Vital signs and surgical incisions should be monitored and documented frequently during recovery from anesthesia. Once stabilized, observations should be performed at a minimum of twice daily. When a surgical procedure is performed with the intent of producing chronic disease, monitoring of project-specific parameters should continue for the life of the animal.

During recovery from anesthesia, greatest attention should be given to the cardiovascular and respiratory systems. Observations of depth, rate, and pattern of respiration, mucous membrane color, capillary refill time, pulse quality, and auscultation of the chest are clinical observations that can be used collectively to assess cardiopulmonary function. Clinical observations are augmented by ECG monitoring, use of pulse oximetry to monitor hemoglobin saturation with oxygen, and noninvasive blood pressure monitoring. Documentation of vital signs and pertinent clinical observations should be made in the animal's medical record. Extubation should be performed after return of a strong swallowing reflex.

During recovery, animals should be kept clean and dry. Surgical incisions should be monitored for drainage and any complications reported immediately. Administration of warmed fluids supports the cardiovascular system and also helps warm the animal. Urine and fecal output should be recorded. Reduced urinary output is cause for concern and can result from dehydration, urinary tract injury, or pain. Reduction of fecal output is not unusual postoperatively due to preoperative fasting, however, if prolonged, it may indicate the presence of more serious problems.

Once stabilized, postoperative monitoring should continue until all surgical incisions have healed. Vital signs, observations of surgical incisions, monitoring appetite, water intake, production of urine and feces, as well as evaluation for pain should be made daily. Comparing body weight postoperatively with that measured preoperatively can be used to assess adequacy of fluid intake. Although some reduction of body weight postoperatively can be anticipated due to reduced food intake, most weight loss is due to fluid deficit. Efforts should be made to encourage animals to eat postoperatively. Offering palatable foods and hand feeding are generally successful for dogs and cats, but can also be used for swine and small ruminants, provided they have been accustomed to human contact. Incisions should be cleansed if necessary and evaluated for signs of abnormal healing or infection. Postoperative medications should be given at the orders of the attending veterinarian. It is important that all information and treatments be documented in the animal's medical record.

Acclimation to handling and human contact preoperatively will help reduce an animal's anxiety postoperatively. In some cases, animals not used to human contact will require more intensive socialization prior to surgery. Preoperative acclimation should also involve specific dietary modifications, use of restraint techniques, or wearing of jackets or harnesses that will be implemented postoperatively. Ultimately, this aids the animal and the caregivers during the postoperative period when close human contact is necessary.

Monitoring of animals that have chronic induced disease as a result of surgery should be tailored to the needs of the project with the objective of minimizing animal distress. Monitoring parameters should be defined at the time of protocol design and modified as clinical experience with the model develops. Parameters may include periodic echocardiography for heart failure models or periodic serum chemistry profiles for various metabolic disease models. Surgically implanted catheters or vascular access ports should be used for projects requiring repetitive blood sampling or drug administration. Administration of analgesics or other drugs that could alleviate distress should also be utilized if not contraindicated scientifically.[18]

## Rodents

Because all rodents will require some degree of special attention in the postoperative period, it is preferable, as with large animals, to provide a separate recovery area. This not only enables more appropriate environmental conditions to be maintained, but also encourages individual attention and special nursing.[51,52]

### Warmth and Comfort

Supplemental heating to prevent hypothermia should be provided, and animal incubators are excellent for this purpose. The temperature for adult animals should be maintained between 25 to 30°C, and for neonatal animals, between 35 to 37°C. Care must be taken not to overheat the animal, and a thermometer should be placed near the animal's body.

The bedding should be comfortable and provide effective insulation. Rats should never be allowed to recover from the anesthetic on sawdust or similar bedding, which could stick to the animal's nose, eyes, and mouth. Special animal bedding material made out of cloth or synthetic fur is available commercially and is excellent for all species.

### Respiratory Depression

Respiratory depression can develop postoperatively and be dangerous for the animal. It should be monitored continuously by observation and treated appropriately if it occurs.

### Fluid Therapy

Excessive loss of fluid can occur quickly intraoperatively, particularly in rats and mice, and in addition, the animal may not drink for up to 24 h in the postoperative period. Fluid should be given orally if possible or otherwise by subcutaneous, intraperitoneal, or intravenous administration, to replace estimated losses and to provide sufficient water intake for the 12 to 24 h period. Monitoring of body weight pre- and postoperatively can provide a good indication of the adequacy of food and fluid intake. The urine and fecal output should be recorded and any abnormalities investigated. Reduced urine output may be due to dehydration, urinary tract injury, or because the animal is in pain. Failure of the animal to defecate must always be investigated, as it could be due to paralytic ileus if there has been excessive handling of the bowel or entrapment in any intraabdominal cannulae, which may be present.

### General Care

Companion animals such as dogs and cats respond well to personal contact, but rats and mice may find this stressful and prefer to be in a warm environment with subdued lighting and disturbed as little as possible. All clinical observations and drugs administered must be recorded on a patient record card kept near the animal for rapid reference. Observations should be made over the first few days postoperatively to ensure that animals are recovering normally, e.g., body weight, activity, wound site.

## POSTOPERATIVE PAIN

All surgery in which the animal is to recover from the anesthetic could cause pain and in some cases may be considerable. It is now generally considered essential to give pain relief for nearly all surgical procedures, however brief, including surgery on rodents. It is important to make an assessment of the degree of pain to enable the analgesic regime to be tailored appropriately. In many cases, this may have to be done anthropomorphically, although clinical schemes for assessment have been published. The management of pain in laboratory animals has been discussed widely, and this, together with the appropriate drug regimens to use, is well described in the scientific literature.[53-55]

## REFERENCES

1. De Boer, J., Archibald, J., and Downie, H.G., *An Introduction to Experimental Surgery*, Elsevier, Amsterdam; New York, 1975.
2. Dougherty, R.W., *Experimental Surgery in Farm Animals*, Iowa State University Press, 1981.
3. Hecker, J.F., *Experimental Surgery on Small Ruminants*, Butterworth Publishers, Stoneham, MA, 1974.
4. Lumley, J.S.P., Green, C.J., and Angell-James, J.E., *Essentials of Experimental Surgery*, Butterworth Publishers, Stoneham, MA, 1990.

5. Knecht, C.D., Allen, A.R., Williams, D.J., and Johnson, J.H. *Fundamental Techniques in Veterinary Surgery,* W.B. Saunders Company, Philadelphia, PA, 1981.

6. Swindle, M.M. and Adams, P.J., *Experimental Surgery and Physiology*, Williams and Wilkins, Baltimore, MD, 1988.

7. Slatter, D.H., *Textbook of Small Animal Surgery*, W.B. Saunders Company, Philadelphia, PA, 1985.

8. Waynforth, H.B. and Flecknell, P.A., *Experimental and Surgical Technique in the Rat*, 2nd ed., Academic Press, New York, 1992

9. Waynforth, H.B., Basics of surgery, in *Laboratory Animals. An Introduction for Experimenters,"* 2nd ed., Tuffrey, A.A., Ed., John Wiley & Sons, New York, 1995.

10. Bradfield, J.F., Schachtman, T.R., McLaughlin, R.M., and Steffan, E.K., Behavioural and physiological effects of inapparent wound infection in rats, *Lab. Anim. Sci.*, 42, 572, 1992.

11. British Laboratory Veterinary Association, Experimental Surgery Slide Programme, Slide Sets and Notes, Obtained from P.A. Flecknell, Comparative Biology Centre, Medical School, Framlington Place, Newcastle upon Tyne, United Kingdom, 1992.

12. Cunliffe-Beamer, T.L., Biomethodology and surgical technique, in *The Mouse in Biomedical Research*, Vol. 3, Foster, H.L., Small, J.D., and Fox, J.G., Eds., Academic Press, New York, 1983.

13. Cunliffe-Beamer, T.L., Surgical techniques, in *Guidelines for the Well-Being of Rodents in Research,* Guttman, H.N., Ed., Scientists Center for Animal Welfare, 1989.

14. Green, C.J. and Simpkin, S., *Basic Microsurgical Techniques. A Laboratory Manual*, Obtained from the Surgical Research Group, MRC Clinical Research Centre, Northwick Park Hospital, Harrow, Middlesex, United Kingdom, 1990.

15. Lambert, R., *Surgery of the Digestive System of the Rat*, Charles C. Thomas, Springfield, IL, 1965.

16. Petty, C., *Research Techniques in the Rat*, Charles C. Thomas, Springfield, IL, 1982.

17. Van Dongen, J.J., Remie, R., Rensema, J.W., and van Wannik, G.H.J., Eds., *Manual of Microsurgery on the Laboratory Rat. Part 1. General Information and Experimentals*, Elsevier, Amsterdam; New York, 1990.

18. Swindle, M.M., *Surgery, Anesthesia and Experimental Techniques in Swine*, Iowa State University Press, Ames, IA, 1988.

19. Swindle, M.M., Smith, A.C., and Goodrich, J.A., Chronic cannulation and fistulation procedures in swine: a review and recommendations, *J. Invest. Surg.,* 11, 7, 1998.

20. Swindle, M.M., *Basic surgical exercises using swine,* Praeger, New York, 1983.

21. College of Animal Welfare, *Veterinary Surgical Instruments, An Illustrated Guide*, Butterworth-Heinemann, Oxford, 1997.

22. Hurov, L., *Handbook of Veterinary Surgical Instruments and Glossary of Surgical Terms*, W.B. Saunders, Philadelphia, PA, 1978.

23. Eshleman, J.R., Methods used for sterilization or disinfection of instruments, *J. Dent. Educ.*, 32, 330, 1968.

24. Collins, C.H., Allwood, M.C., Bloomfield, S.F., and Fox, A., *Disinfectants: Their Use and Evaluation of Effectiveness*, Academic Press, New York, 1981.

25. Ghosh, J., Maisels, D.O., and Woodcock, A.S., Preoperative skin disinfection, *Br. J. Surg.,* 54, 551, 1967.

26. Morris, T., Antibiotic therapeutics in laboratory animals, *Lab. Anim.,* 29, 16, 1995.

27. Strachan, C.J.L. and Wise, R., Eds., *Surgical Sepsis*, Academic Press, New York, 1979.

28. Westaby, S., *Wound Care*, William Heinemann Medical Books Ltd., London, 1985.

29. Zederfeldt, B.H. and Hunt, T.K., Wound Closure, Davis & Geck, New Jersey, 1990.

30. Tera, H. and Aberg, C., The strength of suture knots after one week *in vivo*, *Acta Chir. Scand.,* 142, 301, 1976.

31. Irvin, T.T., Koffman, C.G., and Duthie, H.L., Layer closure of laparotomy wounds with absorbable and nonabsorbable suture materials, *Br. J. Surg.,* 63, 793, 1976.

32. Kjaergaard, J., Laursen, N.P., Madsen, C.M., Tilma, A., and Zimmermann-Nielsen, C., Comparison of dexon and mersilene sutures in the closure of primary laparotomy incisions, *Acta Chir. Scand.,* 142, 315, 1976.

33. Craig, P.H., Williams, J.A., Davis, K.W., Magoun, A.D., Levy, A.J., Bogdansky, S., and Jones, J.P., A biologic comparison of polyglatin 910 and polyglycolic acid synthetic absorbable sutures, *Surgery,* 141, 1, 1975.

34. Postlethwait, R.W., Willigan, D.A., and Ulin, A.W., Human tissue reaction to sutures, *Ann. Surg.*, 181, 144, 1975.

35. Bamstein, J.J., Gilfillan, R.S., Pace, N., and Rahlmann, D.F., Chronic intravascular catheterization, *J. Surg. Res.*, 6, 6, 1966.

36. Bailie, M.B., Vascular-access-port implantation for serial blood sampling in conscious swine, *Lab. Anim. Sci.*, 36, 431, 1986.

37. Girardet, R.E. and Benninghoff, D.L., Surgical techniques for long-term effects of thoracic duct lymph circulation in dogs, *J. Surg. Res.*, 15, 168, 1973.

38. Manolas, K.J., Farmer, H.M, Cussen, J., and Welbourn, An experimental model for simultaneous chronic sampling of portal and systemic blood and gastrointestinal lymph via cannulae in conscious swine, *Cornell Vet.*, 73, 333, 1983.

39. Nelson, A.W. and Swan, H., Long-term catheterization of the thoracic duct in the dog, *Arch. Surg.*, 98, 83, 1969.

40. Snow, H.D. and Tyner, J.G., Chronic arterial and venous catheterization in sheep, *Am. J. Vet. Res.*, 30, 2241, 1969.

41. Witzel, D.A., Littledike, E.T., and Cook, H.M., Implanted catheters for blood sampling in swine, *Cornell Vet.*, 63, 432, 1973.

42. Farins, L.R., Woodle, E.S., Frey, C.F., Nakayama, S.I., and Ward, R.E., A simple technique for experimental hepatic vein catheterization in swine, *Lab. Anim. Sci.*, 36, 406, 1986.

43. Faulkner, R.T., Czajkowski, W.P., Rayfield, E.J., and Hickman, R.L., Technique for portal catheterization in Rhesus monkey (Macaca mulatta), *Am. J. Vet. Res.*, 37, 473, 1976.

44. Olesen, H.P., Sjontoft, E., and Tronier, B., Simultaneous sampling of portal, hepatic, and systemic blood during intragastric loading and tracer infusion in conscious pigs, *Lab. Anim. Sci.*, 39, 429, 1989.

45. Santiesteban, R., Hutson, D., and Dombro, R.S., Chronic catheterization of the portal vein in dogs, *Lab. Anim. Sci.*, 33, 373, 1983.

46. Sirek, A. and Sirek, O.V., A new technique for hepatic portal sampling in the conscious dog, *Proc. Soc. Exp. Biol. Med.*, 172, 397, 1983.

47. Remie, R., Experimental surgery, in *The Laboratory Rat*, Krinke, G.J., Ed., Academic Press, New York, 2000.

48. Franklin, K.B.J. and Paxinos, G., *The Mouse Brain in Stereotactic Coordinates*, Academic Press, New York, 1997.

49. Paxinos, G. and Watson, C., *The Rat Brain in Stereotactic Coordinates*, 2nd Edition, Academic Press, New York, 1986.

50. Flecknell, P.A., *Laboratory Animal Anaesthesia*, Academic Press, New York, 1996.

51. Lane, D.R., *Jones' Animal Nursing*, 4th Edition, Pergammon Press, Oxford; Elmsford, NY, 1985.

52. Taylor, R. and McGehee, R., *Manual of Small Animal Postoperative Care*, Williams & Wilkins, Baltimore, MD, 1995

53. Flecknell, P.A., The relief of pain in laboratory animals, *Lab. Anim.*, 18, 147, 1984.

54. Crawford, R.L., A Reference Source for the Recognition and Alleviation of Pain and Distress in Animals, http://www.nal.usda.gov/awic.

55. Liles, J.H. and Flecknell, P.A., The influence of bupronorphine or buvivacaine on the post-operative effects of laparotomy and bile-duct ligation in rats, *Lab. Anim.*, 27, 374, 1993.

# CHAPTER 20

# Microsurgical Procedures in Experimental Research

**Daniel A. Steinbrüchel**

## CONTENTS

## INTRODUCTION

The history and development of microsurgical procedures essentially reflect recurrent attempts over several decades to solve the problem of establishment or reestablishment of vascular continuity in vessels of decreasing diameter. At the same time, this development illustrates the mutual beneficial influence of clinical experience and experimental microsurgical results, where clinical data initiated detailed investigations in microsurgical animal models, while experimental experience could directly be applied to clinical procedures.

0-8493-1086-5/03/$0.00+$1.50
© 2003 by CRC Press LLC

Carrel and Guthrie were the first who demonstrated the feasibility of vascular anastomosis.[1-3] Before this time, a major vascular lesion in an extremity usually resulted in amputation. In spite of the introduction of the microscope to clinical use in 1921[4] and a gradual development of specialized instruments and accessories (clamps, suture materials), successful vascular anastomosis of vessels in the 2 to 3 mm range was technically not feasible until 1960, when Jacobson and Suarez[5] demonstrated the successful anastomosis of blood vessels of 1 mm in external diameter.

Subsequently, the application of microsurgical and microvascular procedures progressed rapidly, both clinically and experimentally. Successful replantation of amputated extremities[6-8] and the transfer of free skin flaps as composite grafts were reported,[9-14] and a variety of microsurgical models were described from different laboratories, focusing on the transplantation of primary vascularized organs in rodents.[15-24]

Today, microsurgical techniques are widely used in ophthalmologic, otologic, and reconstructive and plastic surgeries. Experimentally, microsurgical models are applied in studies focusing on physiological aspects and processes in the microvasculature subsequent to free tissue transfer, and in transplantation research, where microvascular models (performed as routine procedures) form the basis of testing new immunosuppressing and immunomodulating treatment strategies and permit a more detailed study of the processes involved in allogenic and xenogeneic rejection of transplanted organs.

## BASIC REQUIREMENTS FOR EXPERIMENTAL MICROSURGICAL PROJECTS

Performing microsurgical procedures with success, in terms of generating valid and relevant experimental data, is dependent on several factors, including not only the necessary technical equipment, but also research assistants with motivation, a certain persistence, and a genuine scientific interest. It needs training, qualified supervision and a positive attitude to teamwork because scientifically interesting projects, which include microsurgery, will consist of an interdisciplinary approach to often complex problems, in collaboration with immunologists, pathologists, physiologists, and clinicians.

### Technical Requirements

#### *Microscopes*

Preference has to be given to the operating microscope compared to magnifying glasses. It offers the advantage of greater magnification (stepwise or zoom function), built-in illumination, and possibility of documentation (video, photograph). On the other hand, the field of view is limited and the depth of focus moderate, which can partly be compensated for, however, by foot switch control. The cost is clearly a disadvantage, but it will prove to be a good investment in the long run. Furthermore, diploscopes make the performance of complex microsurgical procedures easier and allow detailed training supervision.

#### *Microsurgical Instruments and Accessories*

A few high-quality instruments that you are familiar with and use routinely are sufficient. An extensive variety of different instruments usually have no beneficial effect on technical accuracy and efficiency. As a rule, as simple as possible is best. The choice of optimal clamps, suture material, and needles for different vessel and tissue types must be taken into account.

Figure 20.1 illustrates a simple set of essential microsurgical instruments, consisting of a needle holder, a dissecting scissors, and a pair of microsurgical forceps. A satin finish to avoid glare is an advantage, and the instruments should have a sufficient length to make a convenient pen grip possible. Two single clamps and a twin-clip approximating clamp are shown in Figure 20.2, and the size is illustrated by a match. Microsurgical instruments are rather expensive and very delicate, but appropriate use and care guarantees excellent function for many years (for further reading concerning more detailed instrument description, see References 25 and 26).

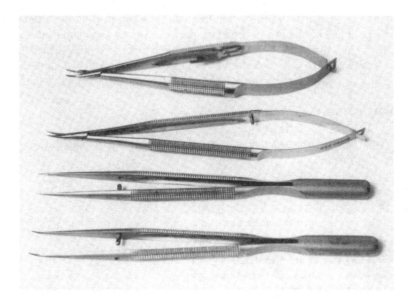

**Figure 20.1**  A basic set of microsurgical instruments with needle holder, dissecting scissor, and a pair of microsurgical forceps.

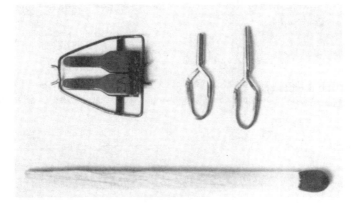

**Figure 20.2**  Two single clamps and a twin-clip approximating clamp (match for comparison of size).

## *Laboratory Facilities*

The importance of optimal facilities for observation and housing of animals (with respect to microsurgical models, most often rodents) must not be underestimated. Advanced projects often include intensive monitoring, not limited to the immediate postoperative period, but for several months. Qualified full-time technical assistance is therefore necessary to achieve optimal benefits in terms of valid, complete, and reproducible experimental results.

## BASIC TECHNIQUES OF MICROSURGICAL VASCULAR ANASTOMOSIS

Atraumatic technique is an essential prerequisite for successful microsurgical procedures. The preparation of arteries and veins includes dissection from surrounding tissue, where a sharp division of structures (with scissors) on the basis of knowledge of natural cleavages is the principle. Blunt dissection and unnecessary manipulation of the vessels must be avoided. The choice of clamps has to be adapted to the type and size of vessel; an optimal clamp exerts a minimal necessary degree of compression, diminishing the risk of endothelial damage. After division of the vessel, with one clear cut, the stumps are rinsed with saline solution. Addition of heparin is not decisive; the important factor for a successful anastomosis is the surgical technique. Subsequently, adjacent adventitia is removed and the stumps can, if necessary, be gently dilated. Any instrumental manipulation with risk of intimal damage during preparation as well as suturing increases the possibility for thrombosis of the anastomosis.

## End-to-End Anastomosis

The infrarenal aorta or the femoral artery in rats are the most suitable objects for the beginner in experimental microsurgery.

### Interrupted Suture Technique

Several methods have been described in the literature.

### Eccentric Biangulation Technique

This technique, first described by Cobbet,[27] is today very popular and can be recommended as the initial type of suture technique for beginners in the field of microsurgical vessel anastomosis. As illustrated in Figures 20.3A and 20.3B, the initial two stay sutures are applied 120 degrees apart. One or two interposed sutures will finish the anterior wall, after which the vessel is rotated 180 degrees. The posterior wall can now be sutured. The use of a twin-clip approximating clamp is a clear advantage for this type of suture technique.

### Biangulation Technique

It is not always possible to place the initial stay sutures in the exact eccentric position, especially when there is a major discrepancy between vessel diameters. The use of the biangulation technique[28]

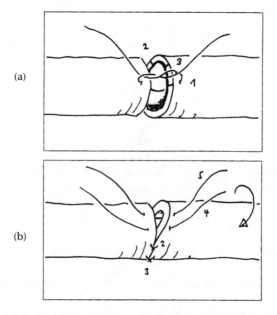

(a)

(b)

**Figure 20.3**    (A) Initial stay sutures are placed 120 degrees apart (1,2). (B) The vessel is turned 180 degrees and the anastomosis can be accomplished from the front.

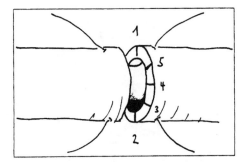

**Figure 20.4** Initial stay sutures are placed 180 degrees apart (1,2), and the remaining sutures are hereafter interposed.

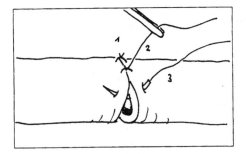

**Figure 20.5** The former stay suture is used to assist the placement of the next suture in a continuous way.

(Figure 20.4) can therefore be an advantage, where the two initial stay sutures are placed 180 degrees apart, hereby defining an appropriate adaption of the two different vessel diameters. Interpositioning of sutures in the anterior wall, and after rotation, in the posterior wall, will complete the anastomosis.

### Successive Interrupted Sutures (Ship's Wheel Type)[29]

The technique is especially suitable where interrupted sutures are preferred but anatomical circumstances do not allow a free rotation of the vessel (Figure 20.5). It is an advantage to place the first suture in the posterior wall. After the suture has been tied, one end is cut near the knot, while a gentle traction of the long end will facilitate the exact positioning of the next adjacent suture. This procedure is repeated until the anastomosis is finished.

### Running Suture Technique[30]

This technique shortens anastomosis time and improves primary hemostasis, but includes the risk of stenosis in less trained hands (Figure 20.6).

## End-to-Side Anastomosis

The principles of suture technique are basically identical to those of end-to-end anastomosis. Interrupted or running sutures can be used. However, as microsurgical organ transplantation models usually do not permit a free rotation of vessels, anastomosis technique includes suturing of the posterior wall from the luminal side, which is done more easily using a running type of suture. The technique is illustrated in Figure 20.7.

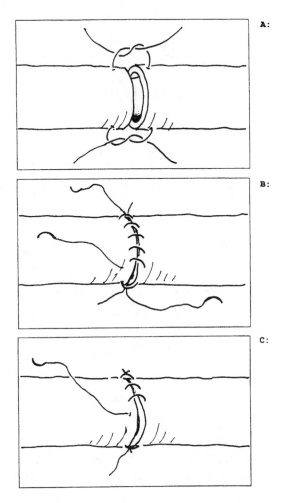

**Figure 20.6**    (A) Positioning of primary sutures, biangulated or eccentric. (B) Suturing the anterior wall. (C) Completing the posterior wall after 180 degree rotation.

## Vein Anastomosis

Veins are fragile in rodents and must be handled carefully. The technique of anastomosis is essentially the same as that applied in arteries.

## Cuff Technique[31,32]

This is an alternative, nonsuture method for vascular anastomosis (Figure 20.8). In principle, one vessel end is everted over a polyethylene cuff, the other is pulled over the endothelialized cuff, and the anastomosis is secured by a circular ligation.

## Splint Technique

This useful method has preferably been used for ureter and bile duct end-to-end anastomosis or the insertion of a stented ureter or choledochus into the recipient bladder or duodenum, respectively.[33,34]

The suture and anastomosis techniques illustrated here are the basic approach to microvascular surgery, which will naturally be the object of modifications, preferences, and improvements in accordance with personal experience and increasing surgical skill.

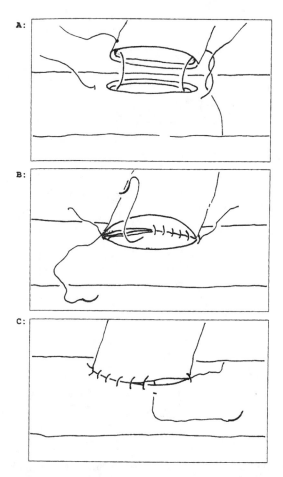

**Figure 20.7**　(A) Placing of angle sutures. (B) Suture of the posterior wall from the inside. (C) Completing the anastomosis from the front.

## SELECTION OF OFTEN USED MICROSURGICAL MODELS
## FOR TRANSPLANTATION RESEARCH

The introduction of microsurgical procedures to organ transplantation research has made the investigation of specific immunological questions and immunosuppressing or immunomodulating treatment strategies possible, to study specific problems of donor organ preservation, or to focus on more physiologic aspects of solid organ transplantation.

The use of inbred rodent strains (mainly rats and mice) combines the possibility of technically feasible whole organ transplantation in donor-recipient combinations, where the genetic disparity in respect to major histoincompatibility is identical between random individuals from the same strain or where genetically manipulated animals (transgenic or gene knockout) are available. The outcome of organ transplantation between two, for all practical reasons genetically identical, rodent strains is therefore reproducible and predictable.[35]

At the same time, rats (and to a lesser degree mice) have a size making it possible for trained microvascular surgeons to achieve patency rates (preferably >95% for more simple procedures) that satisfy the statistical and scientific demands for reproducible and valid results, where technical failure does not cover real observation data.

Transplantation research using microsurgical animal models will often include the use of polyclonal or monoclonal antibodies, genetic transfection, different sera, immunological reagents, and a variety of

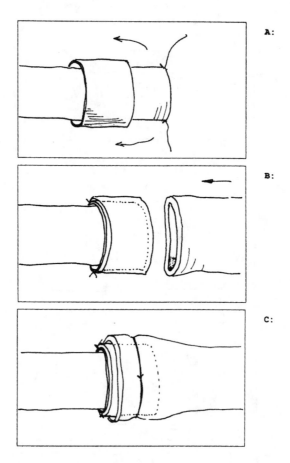

**Figure 20.8**  (A) A plastic cuff of adequate size is pushed over the vessel end. (B) Evertion of the vessel stump over the cuff. (C) Completing the anastomosis with a circular ligation.

methods and procedures, applied as organ recipient treatment, as well as for immunological and histo-logical analysis. Many of these reagents are commercially available for rats and mice, which is an enormous advantage when planning transplantation research projects in laboratories, which initially do not have the possibility of producing these reagents.

From a more practical but no less important point of view, rats and mice are cheap, housing is uncomplicated, and the animals seem to tolerate surgical and anesthesiological stress well and are highly resistant to postoperative infection.

This section will shortly deal with the two major microsurgical transplantation models (heart and kidney), and finally, a few more complex procedures will be mentioned.

## Heterotopic Heart Transplantation in the Rat

Cardiac transplantation in rats is the most often used model in transplantation research, and it consists, in principle, of a short circuit of normal heart hemodynamics. The donor heart is excised after ligation of the inferior and superior caval veins, and the pulmonary veins, and after division of the aorta and pulmonary artery. On the recipient side, the donor aorta is anastomosed end-to-side to the infrarenal aorta, and the pulmonary artery end-to-side to the inferior caval vein. As recipient vessels, the common carotid artery and external jugular vein can be used when preference is given to the heterotopic cervical heart transplantation model. Figure 20.9 illustrates the result after heterotopic heart transplantation to the recipient abdominal vessels.

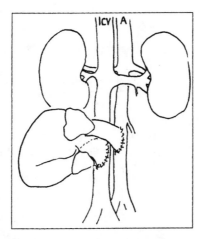

**Figure 20.9**  Heterotopic heart transplantation to recipient abdominal vessels with an aorta-to-aorta and a pulmonary artery-to-inferior caval vein anastomosis.

The transplanted heart is perfused via the coronary arteries, draining to the right atrium and through the right ventricle to the venous system of the recipient. But the left ventricle has no physiological function because there is no ventricular inlet. The model is suitable for immunological and histological studies, or investigations focusing on cardioplegic methods and organ preservation. Graft function, in terms of palpable heartbeat, is easily monitored. However, heterotopic cardiac transplantation is a less adequate model for hemodynamic or functional studies.[36,37] The creation of an atrial septal defect gives some left ventricular inlet, preventing hereby the often-observed formation of an intraventricular thrombus. But this model is not truly functional.[38]

Several modifications have been described using the recipient abdominal or cervical vessels for anastomosis, or placing the donor heart as left ventricular bypass,[17,22,23,31,39,40] however, the original techniques have shown to be highly resistant to innovation, and recent studies use the same techniques described for more than 20 years.[41,42]

## Kidney Transplantation in the Rat

In contrast to the heterotopic heart transplantation model, renal transplantation offers, besides histological and immunological monitoring, the possibility of differentiated functional assessment of a transplanted organ.[43–45] After bilateral nephrectomy of the recipient's own kidneys, renal function will exclusively depend on the graft.

Several techniques of organ harvesting and reimplantation have been described, basically differing on the site of anastomosis.

In the heterotopic renal transplantation model, the donor kidney is harvested with or without an aortic or caval vein segment or patch, and anastomosed end-to-side to the recipient infrarenal aorta and caval vein.[18,19,46] The procedure is illustrated in Figure 20.10.

The orthotopic model makes use of an end-to-end anastomosis between the donor and recipient renal vessels, hereby replacing the recipient's own kindey with a graft[47,48] (Figure 20.11).

The urinary tract can be reconstituted by ureteric implantation into the recipient bladder, bladder-to-bladder anastomosis, or a direct end-to-end ureter anastomosis.

Personally, I think that by any of these different techniques, excellent results can be achieved, and that major or minor variations are less important compared to the necessity of atraumatic microsurgical practice and perfectionism, as well as a reduction of warm ischemia time to zero.

## Other Microsurgical Transplantation Models

This section summarizes a few more complex microsurgical models used for transplantation research. For more detailed description of techniques and methods, see the references at the end of this chapter:

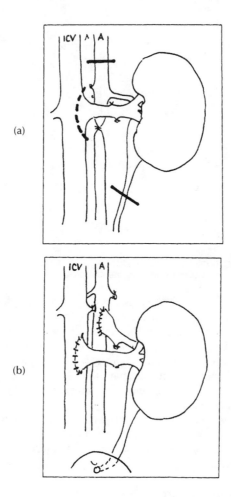

(a)

(b)

**Figure 20.10** (A) Donor nephrectomy for heterotopic left kidney transplantation with an aortic cuff and a vein patch of the inferior caval vein. (B) End-to-side anastomosis from donor to recipient aorta and from the renal vein to the inferior caval vein. Insertion of the ureter into the bladder.

1. Lung transplantation[49,50]
2. Orthotopic liver transplantation[32,51–53]
3. Small intestine transplantation[24,54]
4. Pancreas transplantation[55,56]
5. Esophagus replacement[57,58]
6. Multivisceral grafts, including liver, pancreas, stomach, omentum, small intestine, and colon[59]

## CONCLUDING REMARKS AND COMMENTS

Technical procedures and methods used in experimental microsurgery are essentially the same as applied in the clinic; however, the basic approach is principally different. In the clinic, the surgeon faces an individual problem with a subsequent optimal solution to this problem. Experimental microsurgery tries to elucidate a specific biological question where the microsurgical procedures per se are not actually interesting, and this implies a preferably 100% standardization and reproducibility of the individual experiments. The aim of these studies is the observation and determination of biological variation, and not the monitoring of intraoperative modifications. In other words, detailed repetition of procedures is necessary, and only experimental series with high success rates are scientifically acceptable.

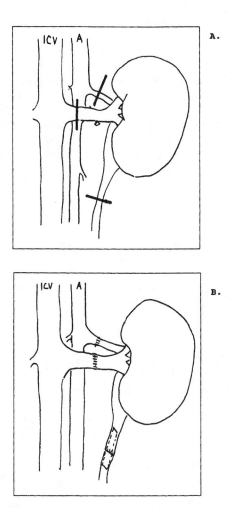

**Figure 20.11** (A) Donor nephrectomy for orthotopic left kidney transplantation with division of the renal artery, vein, and ureter. (B) End-to-end anastomosis of renal vessels after recipient nephrectomy, and end-to-end ureter anastomosis using the splint technique.

There is no mystery about successful performance of microsurgical procedures. It is the result of training, perfectionism, and persistence. The basic rule is: as simple as possible, as fast as possible.

Projects of major scientific interest and value that include microsurgical models are based on an interdisciplinary approach to often complex problems. Cooperation and teamwork are therefore essential. Care has to be taken with respect to animal observation and housing. Time-consuming and difficult microsurgical procedures can end up with surgical success but disappointing and useless results from a scientific point of view, if optimal laboratory facilities and full-time technical assistance are not available or are neglected.

## REFERENCES

1. Carrel, A., La technique opératoire des anastomoses vasculaire et la transplantation des viscères, *Lyon Med.*, 98, 859, 1902.
2. Carrel, A. and Guthrie, C.C., Complete amputation of the thigh with replantation, *Am. J. Med. Sci.*, 131, 297, 1906.
3. Guthrie, C.C., Some physiologic aspects of blood vessel surgery, *JAMA*, 51, 1658, 1908.
4. Nylén, C.O., The otomicroscope and microsurgery 1921–71, *Acta Otolaryngol.*, 73, 453, 1972.

5.  Jacobson, J.H. and Suarez, E.L., Microsurgery in anastomosis of small vessels, *Surg. Forum*, 11, 243, 1960.
6.  Kleinert, H.E. and Kasdan, M.L., Salvage of devascularized upper extremities including studies on small vessel anastomosis, *Clin. Orthop.*, 29, 29, 1963.
7.  Horn, J.S., Successful reattachment of a completely severed forearm, *Lancet*, 1, 1152, 1964.
8.  Bunke, H.J., Jr. and Schulz, W.P., Experimental digital amputation and reimplantation, *Plast. Reconstr. Surg.*, 36, 62, 1965.
9.  Krizek, T.J., Tani, T., DesPrez, J.D., and Kiehn, C.L., Experimental transplantation of composite grafts by microsurgical vascular anastomoses, *Plast. Reconstr. Surg.*, 36, 38, 1965.
10.  Strauch, B. and Murray, D.E., Transfer of composite graft with immediate suture anastomosis of its vascular pedicle measuring less than 1 mm in external diameter using microsurgical techniques, *Plast. Reconstr. Surg.*, 40, 325, 1967.
11.  McLean, D.H. and Buncke, H.J., Autotransplant of omentum to a large scalp defect with microsurgical revascularization, *Plast. Reconstr. Surg.*, 49, 268, 1972.
12.  McGregor, I.A. and Morgan, G., Axial and random pattern flaps, Br. J. Plast. Surg., 26, 202, 1973.
13.  O'Brien, B.M., MacLeod, A.M., Hayhurst, J.W., and Morrison, W.A., Successful transfer of a large island flap from the groin to the foot by microvascular anastomoses, *Plast. Reconstr. Surg.*, 52, 271, 1973.
14.  Acland, R. and Smith, P., Microvascular surgical techniques used to provide skin cover over an ununited tibial fracture, *J. Bone. Joint. Surg.*, 58, 471, 1976.
15.  Lee, S.H. and Fisher, B., Portacaval shunt in the rat, *Surgery*, 50, 668, 1961b.
16.  Miller, B.F., Gonzales, E., Wilchins, L.J., and Nathan, P., Kidney transplantation in the rat, *Nature*, 194, 310, 1962.
17.  Abbott, C.P., Lindsey, E.S., Creech, O., Jr., and DeWitt, C.W., A technique for heart transplantation in the rat, *Arch. Surg.*, 89, 645, 1964.
18.  Fisher, B. and Lee, S., Microvascular surgical techniques in research, with special reference to renal transplantation, *Surgery*, 58, 904, 1965.
19.  Lee, S., An improved technique of renal transplantation in the rat, *Surgery*, 61, 771, 1967.
20.  Reemtsma, K., Gialdo, N., Depp, D.A., and Eichwald, E.J., Islet cell transplantation, *Ann. Surg.*, 168, 438, 1968.
21.  Mikaeloff, P.P., Levrat, R., Nesmoz, P., Rassat, J.P., Philippe, M., Dubernard, L.M., and Bel, A., Heterotopic liver transplantation in the rat. Value, technique, results of about 70 cases, *Lyon Chir.*, 68, 133, 1969.
22.  Ono, K. and Lindsey, E.S, Improved technique of heart transplantation in rats, *J. Thorac. Cardiovasc. Surg.*, 57, 225, 1969.
23.  Lee, S., Willoughby, W.F., Smallwood, C.J., Dawson, A., and Orloff, M.J., Heterotopic heart and lung transplantation in the rat, *Am. J. Pathol.*, 59, 279, 1970.
24.  Monchick, G. and Russel, P.S., Transplantation of small bowel in the rat: technical and immunological considerations, *Surgery*, 70, 693, 1971.
25.  Engemann, R., Deltz, E., and Thiede, A., *Nahtmaterialien und Nahttechniken in der experimentellen Mikrochirurgie, in Nahtmaterialien und Nahttechniken*, Thiede, A. and Hamelmann, H., Eds., Springer-Verlag, Heidelberg, 1982, p. 90.
26.  Silber, S.J., Microsurgical technique, in *Microsurgery*, Silber, S.J., Ed., Williams and Wilkins, Baltimore, MD, 1979, p. 1.
27.  Cobbet, J., Small vessel anastomosis, *Br. J. Plast. Surg.*, 20, 16, 1967.
28.  Harashina, T., Use of the untied suture in microvascular anastomosis, *Plast. Reconstr. Surg.*, 59, 134, 1977.
29.  Fujino, T. and Aoyagi, F., A method of succesive interrupted suturing in microvascular anastomoses, *Plast. Reconstr. Surg.*, 55, 240, 1975.
30.  Biemer, E. and Schmidt-Tintemann, U., *Anatomische und funktio- nelle Grundlagen für die Wahl von Nahtmaterialien und Nahttechniken in der klinischen Mikrochirurgie, in Nahtmateria lien und Nahttechniken*, Thiede, A. and Hamelmann, H., Eds., Springer-Verlag, Heidelberg, 1982, p. 400.
31.  Heron, I., A technique for accessory cervical heart transplantation in rabbits and rats, *Acta. Pathol. Microbiol. Scand.*[A], 79, 366, 1971.
32.  Kamada, N. and Calne, R.Y., Orthotopic liver transplantation in the rat, *Transplantation*, 28, 47, 1979.

33. Daniller, A., Buchholz, R., and Chase, R.A., Renal transplantation in rats with use of microsurgical techniques: a new method, *Surgery*, 63, 956, 1968.

34. Zimmermann, F.A., Obermüller, K., Gokel, J.M., and Dorn-Kling, S., Die Gallengangrekonstruktion bei der Ratte durch Choledocho-Choledochostomie über einen verlorenen Drain, *Z. Exper. Chirurg.*, 14, 241, 1981.

35. Günther, E., Immunogenetic aspects of organ transplantation in the rat, in Microsurgical Models in *Rats for Transplantation Research*, Thiede, A., Deltz, E., Engemann, R., and Hamelmann, H., Eds., Springer-Verlag, Heidelberg, 1985, p. 83.

36. Konertz, W., Thiede, A., and Bernhard, A., Heterotopic heart transplantation in rats — an improved technique of functional evaluation, *Excerpta Med. Int. Congr. Ser.*, 465, 359, 1980.

37. Bernhard, A. and Konertz, W., Experimental heart transplantation, *J. Thorac. Cardiovasc. Surg.*, 86, 314, 1983.

38. Steinbrüchel, D.A., Nielsen, B., Salomon, S., and Kemp, E., Heterotopic cardiac transplantation in rodents: a new model with graft atrial septectomy, in *Recent Advance in Experimental Microsurgery*, Kajimoto, T. and Kitatani, H., Eds., PSKMU, Kanazawa, 1996, pp. 183–186.

39. Steinbrüchel, D.A., Madsen, H.H., Nielsen, B., Larsen, S., Koch, C., Jensenius, J.C., Hougesen, C., and Kemp, E., Treatment with total lymphoid irradiation, cyclosporin A and a monoclonal anti-T-cell antibody in a hamster-to-rat heart transplantation model, *Transplant. Int.*, 3, 36, 1990.

40. Konertz, W., Semik, M., and Bernhard, A., Left ventricular bypass in inbred rats — a new experimental model in microsurgery. Operative technique and hemodynamic evaluation, *Thorac. Cardiovasc. Surg.*, 28, 277, 1980.

41. Niimi, M., The technique for heterotopic cardiac transplantation in mice: experience of 3000 operations by one surgeon, *J. Heart Lung Transplant*, 20, 1123, 2001.

42. Richter, M., Richter, H., Skupin, M., Mohr, F.W., and Olbrich, H.G., Do vascular compartments differ in the development of chronic rejection? AT1 blocker Candesartan versus ACE blocker Enalapril in an experimental heart transplant model, *J. Heart Lung Transplant*, 20, 1092, 2001.

43. Salomon, S., Steinbrüchel, D.A., Nielsen, B., and Kemp, E., Hamster to rat kidney transplantation: technique, functional outcome and complications, *Urol. Res.*, 24, 211, 1996.

44. Steinbrüchel, D., Dieperink, H., Kemp, E., Starklint, H., and Larsen, S., Rat kidney allotransplantation without warm ischemia. Postoperative recovery and glomerulotubular function, *Eur. Sur. Res.*, 19-S1, 80, 1987.

45. Steinbrüchel, D.A., Larsen, S., Kristensen, T., Starklint, H., Koch, C., and Kemp, E., Survival, function, morphology and serological aspects of rat renal allografts. Effect of short-term treatment with cyclosporin A, anti-CD4 and anti-interleukin-2 receptor monoclonal antibodies, *A.P.M.I.S.*, 100, 682, 1992.

46. Jakubowski, H.D., Renal transplantation in the rat, in *Microsurgical Models in Rats for Transplantation Research*, Thiede, A., Deltz, E., Engemann, R., and Hamelmann, H., Eds., Springer-Verlag, Heidelberg, 1985, p. 47.

47. Fabre, J., Lim, S.H., and Morris, P.J., Renal transplantation in the rat: details of a technique, *Austr. NZ. J. Surg.*, 41, 69, 1971.

48. Kamada, N., A description of cuff technique for renal transplantation in the rat, *Transplantation*, 39, 93, 1985.

49. Asimacopoulos, P.J., Molokiha, F.A.S., and Peck, C.A.S., Lung transplantation in the rat, *Transplant. Proc.*, 3, 583, 1971.

50. Mark, K.W. and Wildevuur, C.R.H., Lung transplantation in the rat. I. Technique and survival, *Ann. Thorac. Surg.*, 34, 74, 1981.

51. Lee, S., Charters, A.C., Chandler, J.G., and Orloff, M.J., A technique for orthotopic liver transplantation in the rat, *Transplantation*, 16, 664, 1973.

52. Hansen, H.H., Kim, Y., and Lie, T.S., Orthotopic liver transplantation in the rat — special reference to arterialization, *Excerpta Medica Int. Congr. Ser.*, 465, 394, 1979.

53. Houssin, D., Gigou, M., Franco, D., Szekely, A.M., and Bismith, H., Spontaneous long-term survival of liver allografts in inbred rats, Transplant. Proc., 11, 567, 1979.

54. Deltz, E. and Thiede, A., Microsurgical technique for small-intestine transplantation, in *Microsurgical Models in Rats for Transplantation Research,* Thiede, A., Deltz, E., Engemann, R., Hamelmann, H., Eds., Springer-Verlag, Heidelberg, 1985, p. 51.

55. Lee, S., Tung, K.S.K., Koopmans, H., Chandler, J.G., and Orloff, M.J., Pancreaticoduodenal transplantation in the rat, *Transplantation*, 13, 421, 1972.

56. Nolan, M.S., Lindsey, N.J., Savas, C.P., Herold, A., Beck, S., Slater, D.N., and Fox, M., Pancreatic transplantation in the rat. Long-term study following different methods of management of exocrine drainage, Transplantation, 36, 26, 1983.

57. Parsa, F.D. and Spira, M., Experimental oesophagal reconstruction in rats with a free groin flap, *Plast. Reconstr. Surg.*, 62, 271, 1978.

58. Uchida, L. and Harii, K., Experimental replacement of the cervical oesophagus in rats with a jejunal free transplantation, *Laryngoscope*, 99, 837, 1989.

59. Murase, N., Demetris, A.J., Kim, D.G., Todo, S., Fung, J.J., and Starzl, T.E., Rejection of multivisceral allografts in rats: a sequential analysis with comparison to isolated orthotopic small-bowel and liver grafts, *Surgery*, 108, 880, 1990.

# CHAPTER 21

# Postmortem Procedures

**Ricardo E. Feinstein and Kimberly S. Waggie**

## CONTENTS

0-8493-1086-5/03/$0.00+$1.50
© 2003 by CRC Press LLC

## INTRODUCTION

Postmortem procedures (PMP) include necropsy, collection of samples and tissue specimens, and recording of lesions observed. The major uses of PMP are in diagnosis, health quality control, and toxicology studies. In biomedical research, PMP are also essential to many experiments, not merely as techniques for collecting samples and specimens for examination but also as means for improving experimental reliability. PMP are indispensable to evaluate and characterize mutant animals. Detailed descriptions of PMP in work with transgenic animals have been recently published.[1-4] Sampling techniques may considerably influence the results of an experiment. It is best to define what is being looked for and, in collaboration with the study pathologist, how to evaluate the change *before* starting the study. The protocols for the sacrifice and necropsy of the animals, tissue collection, time required to perform PMP, and type of fixatives and fixation times are some of the topics that must be thoroughly considered in the design of animal experiments (for details on animal euthanasia, see Chapter 20). Necropsies should follow a standard operating procedure (SOP), which is available when PMP are being performed. Obviously, any deviation from the SOP must be documented.[5,6]

In laboratory animals, infections by pathogenic organisms and the resulting lesions are uncontrolled variables that should always be investigated. The additional costs to research of diagnostic necropsies would probably be very low compared to the adverse effects of a permissive attitude. Diagnostic necropsies, however, should be entrusted to individuals with specialized training or diagnostic laboratories, because diseases cannot be investigated without knowledge of lesions, causative factors, and disease mechanisms. The availability of specialized labs such as histology, microbiology, clinical chemistry, and serology is also essential for disease diagnosis. Although most diagnostic necropsies are prompted by detection of sick or dead animals during the course of experiments, clinical signs may or may not be observed. Apparently healthy animals are not necessarily free from lesions that could hamper procedures or influence experimental results. The impact of infections will depend on the nature and aims of experiments, but researchers should be aware that a careful postmortem examination of the animals, including seemingly healthy individuals, is the most effective way to answer the question of whether complicating lesions are present. The systematic use of PMP in diagnosis and health quality control will result in better laboratory animals.[7] Defined, high-quality animals and carefully designed experiments also further reduce the number of animals required for experimentation and testing.

We are aware that some experiments will require more detailed necropsies than described in this chapter, but in most situations, a necropsy that is less thorough than presented here will suffice. A partial necropsy will be better than no necropsy at all. We want with this chapter to encourage investigators to spend more time in gross morphologic evaluation of their animals, as well as guide them in taking meaningful samples.

## NECROPSY AND LABORATORY SAFETY

Necropsies should be performed in a specially equipped room because cadavers and tissue specimens are potential sources of infection to man and animals.[8] The environment of the necropsy room requires special attention. Good light (intensity and spectrum of light and position of light sources), ventilation and air movements, and temperature are some of the factors to be considered. Due to the risk of microbial contamination, and also of exposure to allergens and high concentrations of harmful substances such as anesthetics, fixatives, and solvents, containment facilities and strict adherence to hygiene practices are necessary. Minimal containment standards for postmortem rooms, necropsy, and cleaning and disposal procedures have been summarily described.[9] A most useful biosafety manual may be found at http://www.safety.ubc.ca/Biosafety/manual2000/index.htm.

Methods of decontamination, cleansing routines, and personal hygiene should be described in the SOP of the necropsy laboratory. PMP should be performed according to accepted laboratory practices.[10,11] It is not within the scope of this chapter to provide a complete list of safety measures in necropsy work, but some such measures will be described.

Briefly, in necropsy work, closed-front protective clothing and surgical gloves, as well as protective glasses, should be worn when working close to the surface of organs. Specialized gloves may be necessary when working with toxic agents (a review of these may be found at http://phy-schem.ox.ac.uk/MSDS/glovesbychemical.html). Wristwatches, bracelets, or rings should not be worn. Instruments and all other necessary equipment should be prepared before starting the necropsy. The exterior of tubes, containers, plastic bags, etc., must be protected from contamination and spills. These items should be placed within reach but not beside the cadaver. Other objects, such as telephone, doorknobs, pencils, etc., should not be touched during the necropsy. No one performing a necropsy should ever pipette by mouth, touch unprotected body areas, or apply contact lenses or wear them without goggles. Also, the person performing a necropsy should not eat, drink, smoke, or apply cosmetics.

The necropsy of animals known or suspected to harbor zoonotic agents or hazardous organisms for other animals may require additional containment systems and safety measures, such as the use of biological safety cabinets, laminar flow hoods, or personal protection equipment like a closely fitting ventilated helmet provided with high-efficiency filters.[12] After the necropsy, hands should be washed thoroughly. Protective clothing and gloves must always be taken off when leaving the necropsy room.[10,11]

Local regulations regarding disposal procedures and labeling of containers of biological material should be consulted. Cadavers and tissues can be autoclaved or incinerated. Contaminated, disposable items, such as gloves, should be sterilized or transported in leak-proof containers to an appropriate plant for sterilization and destruction. Used needles, scalpel blades, and glass waste should be placed in appropriate "sharp" containers for decontamination and destruction according to local biosafety regulations.

## Instruments and Materials

Surgical instruments are appropriate for most necropsies, although certain procedures, such as dissection of very small organs, require microsurgical instruments. Tubes and containers for samples must be identified clearly and indelibly (not on the covers). If necessary, use a code number. The following materials and instruments are commonly used:

- Sharp knife, scalpel blades, and handle
- Dissecting scissors and small operating scissors
- Bone-cutting forceps, serrated forceps, and toothed forceps
- Syringes (1 mL, 2 mL, 5 mL, and 10 mL) and needles
- Tubes for liquid samples (3 mL, 5 mL, and 12 mL)
- Container of fixative (for routine fixation in 10% buffered neutral formaldehyde solution)
- Leak-proof containers for tissue specimens (bacteriology, mycology, parasitology, virology, PCR, chemistry)
- Squeeze bottle of 70% alcohol and squeeze bottle of saline
- Swabs, for sampling purposes (see sampling techniques)
- Plastic bags of various sizes and paper towels

A stereoscopic microscope is an invaluable aid in examining small animals, organs, lesions, and some parasites. An electric drill with a cutting disk is a practical aid to opening the cranium or cutting bony structures and teeth (protective glasses must always be worn when using such drills). During the necropsy, instruments may be placed in a stainless steel instrument holder with 70% alcohol or other suitable disinfectant.

## Description of Lesions

The lesion description should allow a reader to form a mental picture of the changes. The location, appearance, number, and severity of the lesions should be described in a precise and concise manner. The location of lesions must be described according to the organ and lobe, area of the skin, portion of the intestines, etc. Anatomical structures are used as reference points. For paired organs, it should be mentioned which of them is affected. Lesions are described in terms of size, shape, color, appearance of the surface and of the cut surface, consistency, demarcation from surrounding tissues, and severity. A diagnosis of the lesion should not be made within the description. The size should be measured in two or three dimensions in linear units (mm, cm), or in volume (mL), weight (g), or relative weight. In hollow organs and lesions, the amount, appearance, and odor of the contents should be described.[13,14]

Photographing tissue specimens is a most useful aid for description, documentation, and teaching purposes. Each photograph should include a size marker such as a ruler for reference. Good photographs of medical specimens can be obtained using rather simple equipment.[15] Modern digital photo cameras take high-quality pictures and are becoming a valuable tool for photography of biomedical specimens.[16]

## Guidelines for the Necropsy of Rodents and Rabbits

Cadavers should be refrigerated until the moment of necropsy, but specimens obtained from a recently killed animal are preferable. Freezing of cadavers should be avoided, because the freeze–thaw cycle causes marked tissue damage.

The anatomical differences between rodents and rabbits do not preclude using a similar necropsy technique. The general necropsy protocol can be modified, depending on the aims of the studies, but changes of the protocol must always be documented, preferably prior to the start of the necropsy. The order in which the necropsy is conducted depends on a number of factors, including purpose of the necropsy, samples to be collected for further study, and personal preference. For example, the prosector may wish to immediately take samples for bacteriology from the respiratory tract to minimize the chance of contamination with the normal intestinal flora. If tissues that autolyze rapidly, such as the pancreas and digestive mucosa, are to be included in the study, they should be collected and preserved as soon as possible. For a general necropsy, one convenient method is to examine the organs in the following order: external examination, skin and subcutaneous tissues; pelvis, abdomen, mouth, neck, and thorax; and head, spinal cord, muscles, and joints. After removal from the body, the organs should be laid out on the necropsy table for further examination and sampling.

Detailed guidelines for the necropsy of rodents and rabbits have been published.[13] General guidelines for the necropsy and trimming of mouse tissues, that also are applicable to other rodents, are presently available on the World Wide Web at "The virtual mouse necropsy" page at http://www.ncifcrf.gov/vet-path/necropsy.html.

Special techniques are sometimes used in conjunction with the necropsy to demonstrate specific tissues or processes. For example, vital staining, the use of colored substances injected into the vascular system, has been employed to investigate vascular permeability to different substances. The vital dye Evans blue has been found useful in the search for delicate or very small organs such as the thoracic duct or the paraganglia.[17–21] Another vital dye that has been used in myocardial infarct and stroke studies to differentiate viable from dead tissue is 2,3,5-triphenyltetrazolium chloride (TTC).[22,23]

## *External Examination, Skin and Subcutaneous Tissues*

A general necropsy is started by inspecting the cadaver. The animal species, strain, animal identity (i.e., necropsy number, ear tags, tattoos, etc.), sex, and body weight are recorded. Postmortem changes are scored: in a scale ranging from one to five, one corresponds to mild decomposition of tissues, such as observed in animals sacrificed immediately before the necropsy, and five corresponds to a pronounced disintegration of tissues. The appearance of the skin, hair coat, body openings, and visible mucous membranes is observed. Abnormalities such as loss of hair, changes in the color of the skin or mucous membranes (icterus, cyanosis, etc.), presence of discharges from natural orifices, and tumors are noted. In hamsters, the flank organs, which are a male secondary sexual characteristic, are inspected. The skin and subcutaneous tissues are palpated for lesions.

The amount of fat in the body depots and the muscular volume are observed to score the nutritional state of the animal. The body condition may be rapidly scored by observing the volume of muscular masses covering osseous protuberances: (1) emaciation, absence of fat in the body depots, (2) under-nourished, bad nutritional condition, (3) good nutritional condition, (4) over condition, and (5) obese animal. This method is also applicable for body condition scoring of live animals.[24]

The cadaver is placed on its back and pinned to a dissection board (in rabbits this is not necessary). The skin is moistened with alcohol, and a midline incision is made from the symphysis of the mandible to the anus, avoiding the penis in male animals. The skin is reflected on both sides of the incision

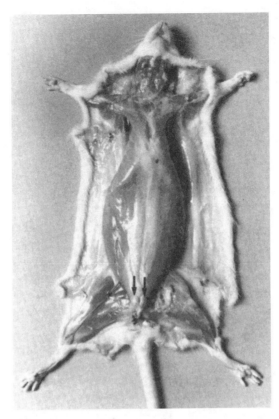

**Figure 21.1**    After doing a midline incision, the skin is reflected on both sides. The arrows point at the clitoral glands.

(Figure 21.1) The subcutaneous tissues are inspected. Skinning of the cadaver can be completed at this stage or at the end of the necropsy. In guinea pigs and autolyzed cadavers, the stomach and the intestines may rupture during the skinning procedure.

The mandibular and cervical lymph nodes, the salivary glands (mandibular, parotid, and sublingual) and the extraorbital lachrymal glands are observed. The mammary glands are inspected (mice have five pairs of mammary glands, rats six, hamsters six or seven, guinea pigs one, and rabbits four pairs). In female rodents, the clitoral glands are inspected (Figure 21.1). In males, the penis, prepuce, and the preputial glands are examined. The inguinal and axillary lymph nodes, which are usually embedded in the subcutaneous fat tissues, are observed.

### Pelvis and Abdomen

The abdomen is opened by a midline incision through the abdominal wall, from the sternum to the pelvis, and by two cuts through the muscles along the costal arcs (Figure 21.2). The floor of the pelvis is removed after making a sagittal cut on each side of the midline (Figure 21.3). The abdominal and pelvic organs are examined *in situ*. The appearance of the serous membrane and the occurrence of abnormal contents, such as serous fluid, blood, fibrin, or adhesions between organs are observed.

### Male Genital Organs and Urinary Bladder

The scrotum is cut open, and the testicles and epididymides are extracted. The fibrous ligaments anchoring the tail of the epididymis to the scrotum are cut. The vas deferens is cut, and the testicles and epididymides are removed for inspection. The ureters are cut, and the remaining genital organs are removed in block, i.e., seminal vesicles, coagulating glands, bulbourethral glands, prostate gland, urethra, and penis (male

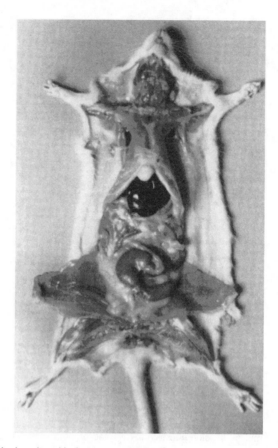

**Figure 21.2**   The abdominal cavity with the organs *in situ*. Cultural samples should be taken before handling the viscerae.

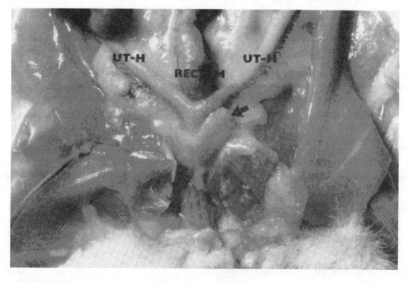

**Figure 21.3**   Pelvis with the organs *in situ*. UT-H: Uterine horn. The arrow points at the urinary bladder.

rodents have large accessory sex glands that should not be mistaken for uterine horns). The urine is collected from the urinary bladder. Rabbit and guinea pig urine is normally turbid due to its high concentration of mineral crystals. In rodents, the urethral plug, a whitish, rather hard cast, located in the lumen of the proximal urethra and sometimes extending into the bladder, is regarded as normal in males.[25]

### Female Genital Organs and Urinary Bladder

The ovaries with the oviducts are located caudal to the kidneys in the peritoneal cavity. The genital organs and supporting ligaments are partly embedded in fat, but the ovaries stand out from the fat tissues, being more reddish in color. The vulva and vagina are dissected free from the skin and rectum. The vagina is caught with a forceps, and the supporting ligaments of the vagina, uterus, oviduct, and ovary are cut. The genital organs and the urinary bladder may be opened for examination of the mucous membranes.

### Spleen and Pancreas

The spleen is removed by cutting the omentum and the ligamentum gastrolienalis along the greater stomach curvature. The pancreas is examined. It is a rather diffuse and richly lobulated organ, located in the supporting ligaments of the stomach and the small intestine. It is firmer and more grayish than fat tissues.

### Stomach and Intestines

In mice, rats, hamsters, and gerbils, the stomach is divided into two distinct regions, the forestomach, also designated as the cutaneous or proventicular region, and the glandular region. The glandular region has a thicker wall than the cutaneous region. In the guinea pig, these regions are not clearly demarcated. The rabbit stomach is not divided into distinct areas.[26] The intestines of rodents have few unique features (in guinea pigs, the cecum is a thin-walled, voluminous organ). Rabbits have long intestines, with most of the gut-associated lymphoid tissues located in the last portion of the small intestine (sacculus rotundus of the ileocecal tonsil) and in the appendix vermiformis, a cecal diverticulum. The intestinal wall is thicker at the level of the sacculus rotundus and the appendix than in the rest of the intestines.

To remove the gastrointestinal tract, the anus is dissected free from the surrounding skin. The rectum is caught with a forceps, and the mesentery and the supporting ligaments are cut. The supporting ligaments along the whole length of the large intestine, the small intestine, and the stomach are cut. It is convenient to cut the supporting ligaments as close to the intestines as possible. The cardia is caught with a forceps, and while the stomach is gently pulled away from the diaphragm, the esophagus is cut. Then the whole gastrointestinal tract is removed and placed on the necropsy table for examination and sampling (Figure 21.4). Specimens and samples for further studies may be taken at this stage or after all the other organ examinations are finished.

The stomach is opened along the greater curvature. If appropriate, the whole length of the intestines is opened, and the gastrointestinal contents are collected and examined. The digestive mucosa is inspected, including the Peyer's patches. Lesions in the mucosa, such as erosions, can be covered by adherent contents. The surface can be rinsed with saline to remove the contents. The mesentery, including the mesenteric lymph nodes, is then removed.

### Liver, Kidneys, and Adrenal Glands

The liver is removed by cutting the hepatic ligaments, and the hepatic tissues can be examined by deep cuts in different lobes. The gall bladder and its contents are examined (rats do not have this organ).

The right and left adrenals, which are located cranially and medially to the respective kidney, are removed together with the kidneys. Both kidneys are removed by cutting the ureters and the renal vessels at the level of the renal hilus. The renal capsule is removed, and the appearance of the cortical surface is observed. If possible, the kidneys are cut and the cortex, medulla, and the renal pelvis are examined. If the right kidney is cut transversally and the left kidney longitudinally, it will be possible to distinguish between them in histological sections.[2]

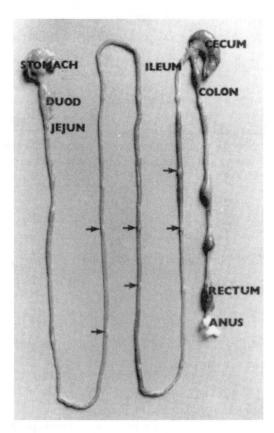

**Figure 21.4**  Gastrointestinal tract placed on the necropsy table for examination. DUOD: Duodenum; JEJUN: Jejunum. The arrows point at some of the Peyer's patches.

## *Mouth, Neck, and Thorax*

The mandibular muscles are cut on both sides, and the mandible is pulled backwards and if necessary removed. The oral cavity is inspected. The larynx, trachea, and esophagus are dissected by cutting the muscles in the ventral part of the neck. To open the thorax, the xiphoid cartilage is lifted with a forceps and the sternum removed by cutting on both sides along the costochondral junctions (Figure 21.5). The sternum is a convenient organ for histological examination of the bone marrow. The thorax and its organs are inspected *in situ*. In mice, hamsters, and rabbits, the thymus is located in the anteroventral portion of the thorax, close to the midline. Rats, in addition, have a smaller cervical portion, which lies ventrally to the trachea. In guinea pigs, the thymus is entirely located in the ventral part of the neck.

The tongue is caught with a forceps and pulled backwards, and the soft palate and pharynx are cut with a scalpel. The tongue, esophagus, larynx, trachea with thyroid and parathyroid glands, thymus, mediastinal and bronchial lymph nodes, lungs, and heart are removed in block by gently pulling backwards and after severing the thoracic aorta and the caudal vena cava at the level of the diaphragm. In small animals such as mice and hamsters, it is often best to harvest the heart and lower respiratory tract intact and with no further gross inspection if microscopic exam is also to be performed. If inspecting the heart grossly, first the pericardial sac is opened and its contents examined.

Before the heart is opened, the left and right sides should be identified. The right atrium may be opened by making an incision from the sinus venosus into the auricle. The incision is prolonged into the right ventricle, by cutting its wall parallel to the interventricular septum toward the apex, and from the apex to the pulmonary artery. The left atrium is opened, starting the incision at the entrance of the pulmonary veins. The wall of the auricle is cut, and the incision is prolonged into the left ventricle along the interventricular septum toward the apex, and from the apex into the aortic artery. The myocardial

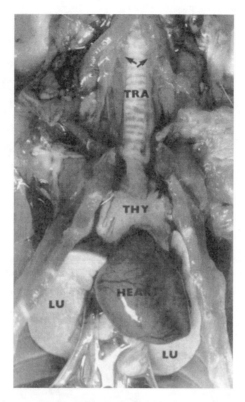

**Figure 21.5** The neck and thorax with the sternum removed and thoracic organs *in situ*. TRA: Trachea; THY: Thymus; LU: Lung. The thyroids are marked by arrows.

cut surface, the heart cavities, and the atrioventricular and semilunar valves are inspected. Clots are removed, and the endocardium is inspected.

The aortic trunk is opened, and if necessary, the whole length of the aorta. The whole length of the esophagus, larynx, trachea, and the major bronchi are then opened, and the mucosae of these organs as well as the cut surface of the lungs are examined.

### Head and Spinal Cord

The skin is cut transversally over the neck, and the cranium is skinned and severed caudally to the occipital protuberance. The cranium of young animals and of adult mice can be opened with scissors. For adult animals of other species, bone cutters are preferable. An electric drill with a cutting disk is also practical. The cranium is opened by first making two cuts from the foramen magnum to the medial part of both orbits. Then the frontal bone is cut transversally at a level just behind the orbits, which coincides with the anterior border of the cranial cavity. The calvaria is removed (Figure 21.6). To extract the brain and cerebellum from the cranium, cut the olfactory lobes with small operating scissors inserted under the anterior part of the brain. Then the optic nerves and the remaining cranial nerves on both sides are cut.

The brain and cerebellum are removed (artifacts produced by the manipulation of fresh brain tissues can be avoided by fixation of the brain *in situ*). After removal of the brain, the pituitary gland appears on the floor of the cranium, attached to the sella turcica (Figure 21.7). In rodents, there is a thin layer of dura mater covering the pituitary gland. The dura around the gland is cut with a scalpel. Using the scalpel blade as a shovel, the gland is lifted out of the sella turcica. In rabbits, the pituitary gland is covered by a bony projection (dorsum sellae), which has to be removed prior to removal of the gland.

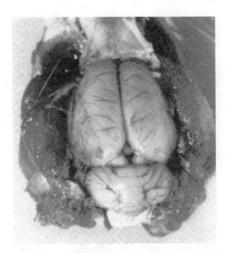

**Figure 21.6**    Cranium with the brain *in situ*. The calvaria was removed by making two lateral cuts, from the foramen magnum to the medial part of both orbits, and a transversal cut just behind the orbits.

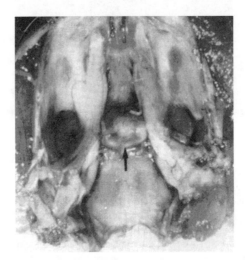

**Figure 21.7**    Dorsal endocranial view of the skull after removal of the brain and cerebellum. The pituitary gland in the sella turcica is marked by an arrow.

The tympanic cavity is inspected. The wall of the tympanic bulla is thin and can be opened ventrally with scissors (rodents) or bone cutters (rabbits), after disinfection of the surface. The middle ear can be examined histologically after decalcification of the skull.

Using small operating scissors or a scalpel inserted in the orbit, the eyeball is freed from the surrounding tissues, and the optic nerve is cut. A forceps is used to lift the eyeball, with the Harderian gland and the smaller intraorbital lachrymal gland, out of the orbit. In guinea pigs, the zygomatic salivary gland, which is located ventral to the globe of the eye, is also collected.

To inspect the nasal cavity, the cranium can be divided by a midline cut, in aboral-oral direction. However, if the nasal region is to be examined histologically, it should be cut transversally together with palatine structures for reference points, after fixation and decalcification.[27]

The spinal cord should be fixed *in situ*, because it is easily damaged if removed from the vertebral canal. In adult animals, except for mice, the spinal cord can be removed with small operating scissors or bone cutters. Starting at the first cervical vertebra, incisions are made alternately on the right and left side to remove the vertebral arches, and then the roots of the spinal nerves are cut. An alternative is to approach the spinal cord ventrally by removing the vertebral bodies (Figure 21.8).

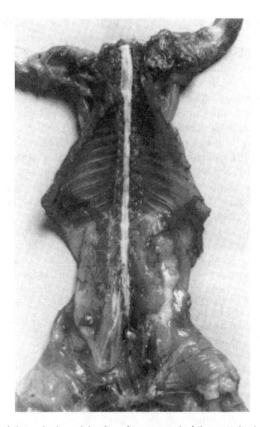

**Figure 21.8**   Ventral view of the spinal cord *in situ,* after removal of the vertebral bodies.

## *Muscles and Joints*

The sublumbar muscles and the muscles of the thigh region are inspected by longitudinal and transverse cuts. The major limb joints are examined. The periarticular muscles are removed, and the articular capsule is swabbed with alcohol, and the joints are opened with a scalpel. Samples for culturing may be obtained by scraping and swabbing the articular surfaces. If necessary, joints can be removed and fixed for histological examination.

## SAMPLING TECHNIQUES

Once the in-life phase and gross necropsies for an animal study are completed, all that remains are the accumulated records and preserved specimens.[28] Correct sampling technique is essential to postmortem studies. The results obtained from experimental studies and diagnostic work-ups depend on the care with which findings are recorded and specimens are collected.

## Sampling for Morphological Examinations

### *Histology*

Tissues for microscopic evaluation must be collected in a consistent manner according to diagnostic or experimental study protocol. Published guidelines for sampling and trimming procedures of specific organs constitute a most valuable aid for standardizing these procedures.[29] Both macroscopically normal tissues, which may contain microscopic lesions, and lesions with surrounding normal tissue should be collected. Select tissues may also be weighed prior to fixation. Prior to weighing, fat, excess blood, and other nonrelevant tissue should be removed to ensure accuracy. Tissue specimens should be obtained with a scalpel or a sharp knife and should be handled carefully to avoid artifacts caused by stretching or compression of tissues. Artifactual distortion of tissues can destroy, alter, mimic, or mask changes.

It is always advisable to check with the laboratory for their preference for fixative and fixation procedures. For routine histology, tissue specimens are fixed by immersion in 10% buffered neutral formaldehyde solution for 24 h. The volume of fixative should exceed that of the specimens at least ten times (formaldehyde solution) or 20 times (alcohol-based fixatives). For an adequate penetration of the fixative, tissue specimens should not be thicker than 5 to 6 mm (Figure 21.9). For electron microscopy, specimens not thicker than 3 mm can be fixed in glutaraldehyde or Karnovsky fluid.[30] The use of a microwave oven can considerably shorten the time required for fixation and subsequent histological procedures.[31] In addition, microwave oven heating of tissue sections often improves tissue antigen detection by immunohistochemistry in formalin-fixed, paraffin-embedded tissues.[32] *In situ* hybridization (ISH) and *in situ* polymerase chain reaction (PCR) are increasingly being used as tools for diagnostics (e.g., detection of viral pathogens) and experimental biology (e.g., gene expression). A full review of these methods is beyond the scope of this chapter. However, it should be noted that most ISH and PCR studies can employ aldehydes for tissue fixation.[33,34]

The postmortem decomposition of tissues is a hindrance to microscopic studies. It occurs rapidly in the intestines.[35,36] Intestinal specimens are best preserved by collecting them as soon as possible after the animal's death and injecting formalin into the lumen of an unopened intestinal segment. The lumen should not be left distended, because it results in altered morphology.[37] The "Swiss roll" is a practical technique that allows the examination of the whole intestine of rodents. Immediately after excision, the intestine is opened and divided in segments of a length which when rolled will easily fit into a tissue cassette. With the mucosa facing out, the intestine is rolled from end to end. The result resembles the head of a lollipop.[38] This procedure may also be done with an unopened intestine.

Optimum fixation of the lung also requires special attention. Immersion fixation can be used, but it may result in collapsed alveolar spaces that obscure detail, especially in small animals such as mice. Expansion of the lungs with fixative by perfusion through the trachea preserves the alveolar spaces and results in better sections for microscopic examination.[13,39] When using this method, care must be taken to avoid overdistension of the lungs. Small organs, such as pituitary gland, adrenals, or lymph nodes may be placed in histological cassettes directly after removal and immersed in fixative, to prevent loss and also drying.

Optimal preservation of tissue morphology and tissue antigens can be achieved by perfusion fixation. In this method, the animal is deeply anesthetized, and the fixative is injected into the vascular system. The type and amount of fixative, perfusion pressure, and injection site will depend on various factors

**Figure 21.9**    Liver specimen for histology. Tissue specimens for histology should not be thicker than 6 mm, for an adequate penetration of the fixative.

such as the aim of the study, the animal species, and body weight.[2,40,41] A different method for small tissue specimens, based on freeze substitution and low temperature plastic embedding, also results in high-quality morphology and optimum antigen preservation.[42]

## Sample Collection for Microbiology

### Parasitology

The sampling site is important in parasitological examinations, because parasites have specific predilection sites. In addition, simultaneous infection with various types of parasites is not uncommon. Thus, skin scrapings for ectoparasites should be obtained from different areas. If possible, the whole skin should be examined, including ears, eyelids, and nasal cavity. The skin should be placed in a hermetically sealed container or in a plastic bag and refrigerated until the moment of examination.

For endoparasites, samples from gastrointestinal contents should be collected during the necropsy. Samples should be placed in clean, leak-proof containers. Fresh samples are always preferable.

Microscopic examination of gastrointestinal contents or feces (wet preparation) permits the identification of parasites, larvae, and eggs. Specimens examined immediately after collection are appropriate for the demonstration of motile protozoans or trophozoite stages of protozoans. Samples for examination of protozoan oocysts, helminth eggs, and adult helminths may be refrigerated.

Most endoparasites located in organs other than the intestine are diagnosed by microscopic examination of tissue sections. For certain parasites, such as the protozoan *Encephalitozoon cuniculi*, serological methods and immunohistochemical techniques are also available.[43,44] Blood smears are useful for examining for blood parasites. Smears should be of a good quality, as no diagnostic skill can compensate for a poorly made blood film.[45] Preferably, smears should be prepared from fresh blood, recently obtained without anticoagulants. Films that are not stained immediately should be fixed and stored in a protected place. Techniques for preparing material for identification of parasites have been described.[46,47]

### Sampling for Cultivation

The results from postmortem cultures depend on the care with which specimens are collected. The time between euthanasia and necropsy is especially important in rodents because autolysis occurs rapidly. To avoid accidental contamination of tissues and the postmortem growth of contaminant bacteria, specimens for bacterial (and viral) culture and molecular techniques should be obtained before the organs are handled, i.e., as early as possible during the necropsy (Figure 21.10). The laboratory receiving the samples should be consulted regarding which tissues to examine and the best conditions for transport of the specimens.

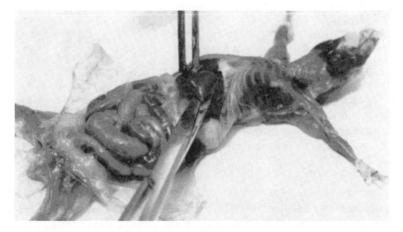

**Figure 21.10** Cutting a piece of liver for bacteriology. Samples for bacteriology should be taken with sterile instruments and preferably before the organs are handled.

## Bacteria

Most bacterial infections are diagnosed by bacterial culture. The successful isolation of bacteria depends on various factors, but a correct sampling technique is essential. Inadequate sampling techniques can result in the overgrowth of a causative agent by contaminant bacteria. Other important factors are the sampling site, type of disease, its duration, whether the animals have been treated with antibiotics, etc. The period elapsed after death should also be considered, because the viability of pathogenic bacteria and mycoplasma in tissues decreases, while bacteria of the normal flora rapidly invade the tissues and may overgrow pathogenic agents. The agonal or postmortem invasion of tissues by resident bacteria must be distinguished from an infection occurring before death. Therefore, culture results must be evaluated together with necropsy findings, such as the presence of lesions consistent with those attributed to the bacteria isolated.[11,48-50]

Tissue specimens for bacterial culture should be obtained with sterile instruments or swabs. The selected body cavity is opened aseptically, and the surface of the tissues can be seared with a red-hot spatula or a flame, or it can be washed with 70% alcohol. Bacterial specimens may consist of cut pieces of tissues (Figure 21.10). The instruments must be sterilized before collecting each specimen. In the case of hollow organs (intestine, uterus), a segment is cut after ligation at both ends. Sterile swabs inserted into the tissues are also used, but the swabs should be processed within a few hours.[11,49] Cotton swabs should be avoided, as substances present in the cotton may hinder the growth of certain bacteria. Swabbing is also convenient for sampling serosal and mucosal surfaces (e.g., pericardial sac, joints, genital tract, conjunctiva, etc.). Organs having a thick capsule, but also abscesses and pustules, can be opened with a sterile scalpel after disinfection of the surface. The contents are then sampled by thorough swabbing against the inside of the capsule. Body fluids, such as urine or blood, can be aspirated with a sterile needle and syringe.

Techniques for obtaining samples for culturing mycoplasmas and bacteria by washing the respiratory and the genital mucosa have been described in detail.[50]

Bacterial specimens from organs and serosal or mucosal surfaces can be transferred to a culture medium using a sterile loop. For print cultures, the surface of the tissues is pressed against a culture medium, or a blotting paper is pressed against the tissues and then transferred to a culture medium.

General techniques and sampling procedures for bacteriological cultures from specific organs have been described.[47,51,52] The use of transport media, and the conditions of transport of the specimens to the laboratory are of considerable importance. Commercially available swabs containing transport media for specific purposes are a convenient way to submit specimens to distant laboratories. Swabs should not be allowed to dry, because this causes the viability of many bacteria and mycoplasmas to be rapidly reduced. Transport in sterile phosphate-buffered saline (PBS) at 4°C has been found to maintain the stability of murine mycoplasmas and various pathogenic bacteria, except for *Pasteurella multocida* and *Pasteurella pneumotropica*.[53] The specimens should be refrigerated or kept on wet ice.

## Fungi

Specimens for fungal culture, except dermatological specimens, should be kept moist with sterile distilled water or saline. Specimens should be obtained aseptically, as has been described for bacterial sampling. Dermatological specimens should be free from contamination by blood. Hairs and the base of hair shafts and skin scrapings can be collected in Petri dishes after washing the skin with 70% alcohol. Skin scrapings should include the center and the periphery of lesions.[54]

## Viruses

Necropsy specimens for virus isolation should be collected aseptically (see sampling for bacteriology), placed in leak-proof sterile containers without preservative, and chilled. Specimens, except blood, should be stored frozen, preferably at -70°C. The postmortem decomposition of tissues inactivates many viruses. Thus, specimens should be obtained shortly after death, and preferably during the early stage of infection.[55]

## Sampling for Serology

Blood samples for serology should be obtained from live animals. In general, it is advisable to consult and follow guidelines from the laboratory where the serum will be submitted. Techniques to collect adequate samples have been described.[47,56-58] In recently sacrificed animals, blood can be aspirated with syringe or a Pasteur pipette from the heart, from the thorax after severing the posterior vena cava or the thoracic aorta, or from the axillary or inguinal areas after severing the arteries at these areas. Blood samples for serology should be obtained aseptically and without additives. To avoid hemolysis, the blood must be obtained and processed carefully (blood must not be forced through a small gauge needle, and the separation of serum from the rest of the blood must not be delayed). Serum samples can be maintained for a short time at 4°C, or stored frozen, preferably at -70°C. Large volumes of serum should be fractioned to avoid repeated freezing and thawing.

## Sampling for PCR in Diagnostic Microbiology

Given the high sensitivity of molecular techniques, such as the PCR, the contamination of specimens with nucleic acid is a most serious hazard during the sampling, but also during the preparation of specimens at the diagnostic laboratory. In addition to strict hygiene aimed at preventing the contamination of the necropsy room and working surfaces, the instruments used can be dipped before the sampling for at least 5 min in a 10% solution of bleach or disinfectant to destroy any remnants of nucleic acid. Specimens placed in sterile tubes or containers should be frozen rapidly at -80°C.[59]

## Sampling for Nutrient Analysis and Toxicology

Feed and bedding should be examined when nutrient excesses or deficiencies or contamination by pesticides, herbicides, heavy metals, or other substances that might influence biological processes are suspected. Serum and tissue specimens collected without chemical contamination should be placed in leak-proof, clean tubes or containers. Each organ must be sent in a separate container. Polyethylene bags are not always appropriate, because they are permeable to many organic substances, and the plasticizers used in their manufacture may contaminate the sample.[60] Serum and tissue specimens should be frozen for transportation to the laboratory. Whole blood should not be frozen. Blood, collected in tubes containing fluoride and oxalate as preservative and anticoagulant, can be transported chilled. Specimens preferred for chemical examination are liver, kidney, blood or serum, urine, and stomach with its contents. Bone should be collected when a pesticide or metal is suspected. The laboratory should be consulted about the type and amount of specimens necessary for each analysis.[61] Where possible, all toxicants to be analyzed should be defined before the necropsy in order to determine target tissues and sample volumes, appropriate containers for samples, and any other factors necessary to obtain optimal results.

## ACKNOWLEDGMENT

We thank Bengt Ekberg for the photographs presented in this chapter. The photographs shown were taken during the necropsy of a female Sprague-Dawley rat, aged approximately 3 months.

## REFERENCES

1. Wood, P.A., Phenotype assessment: Are you missing something?, *Comparative Medicine*, 50, 12, 2000.
2. Relyea M.J., Miller, J., Boggess, D., and Sundberg, J., Necropsy methods for laboratory mice: biological characterization of a new mutation, in *Systematic Approach to Evaluation of Mouse Mutations*, Sundberg, J.P., Boggess, D., Eds., CRC Press, Boca Raton, FL, 2000, chap. 5.

3. Brayton, C., Justice, M., and Montgomery, C.A., Evaluating mutant mice: anatomic pathology, *Vet. Pathol.*, 38, 1, 2001.

4. Ward, J.M., Mahler, F., Maronpot, R.R., Sundberg, J.P., and Frederickson, R.M., *Pathology of Genetically Engineered Mice*, Iowa State University Press, Ames, IA, 2000.

5. Mann, P.C., Hardisty, J.F., and Parker, M.D., Managing pitfalls in toxicologic pathology, in *Handbook of Toxicologic Pathology*, Haschek, W.M., Rousseaux, C.G., and Wallig, M.A., Eds., Academic Press, San Diego, 2001, chap. 9.

6. Bucci, T.J., Basic techniques, in *Handbook of Toxicologic Pathology*, Haschek, W.M., Rousseaux, C.G., and Wallig, M.A., Eds., Academic Press, San Diego, 2001, p. 174.

7. Homberger, F., Boot, R., Feinstein, R.E., Kornerup-Hansen, A., and van der Logt, J., FELASA guidance paper for the accreditation of laboratory animal diagnostic laboratories, Report of the Federation of European Laboratory Animal Science Associations (FELASA) Working Group on Accreditation of Diagnostic Laboratories, *Lab. Anim.*, 33, s1:19, 1999.

8. Smith, M.W., Hazards and safety aspects of animal work, in *Laboratory Animals. An Introduction for New Experimenters*, Tuffery, A.A., Ed., John Wiley & Sons, New York, 1987, chap. 7.

9. Agriculture and Agri-Food Canada Publication No. 1921/E, *Containment Standards for Veterinary Facilities*, Best, M., Ed., Nepean, 1996.

10. Grizzle, W.E. and Polt, S.S., Guidelines to avoid personnel contamination by infective agents in research laboratories that use human tissues, *J. Tissue Culture Methods*, 11, 191, 1988.

11. Du Moulin, G.C. and Love, W., The value of autopsy microbiology, *Clin. Microbiol. Newslett.*, 10, 165, 1988.

12. Biosafety in Microbiological and Biomedical Laboratories, 3rd edition, CDC- NIH, HHS Publication No. (CDC) 93–8395, U.S. Government Printing Office, Washington, 1993. This publication may also be found on the World Wide Web at: http://www.cdc.gov/od/ohs/biosfty/bmbl/bmbl3toc.htm.

13. Feldman, D.B. and Seely, J.C., *Necropsy Guide: Rodents and the Rabbit*, CRC Press, Boca Raton, FL, 1988.

14. Strafuss, A.C., *Necropsy. Procedures and Basic Diagnostic Methods for Practicing Veterinarians*, Charles C. Thomas, Springfield, IL, 1988.

15. Edwards, W.D., Photography of medical specimens: experiences from teaching cardiovascular pathology, *Mayo Clin. Proc.*, 63, 42, 1988.

16. Weinberg, D.S., Digital imaging as a teaching tool for pathologists, *Clin. Lab. Med.*, 17, 229, 1997.

17. LeVeen, H.H. and Fishman, W.H., Combination of Evans blue with plasma protein: its significance in capillary permeability studies, blood dye disappearance curves, and its use as a protein tag, *Am. J. Physiol.*, 151, 26, 1947.

18. Coleridge, H., Coleridge, J.C.G., and Howe, A., A search for pulmonary arterial choemoreceptors in the cat, with a comparison of the blood supply of the aortic bodies in the newborn and adult animal, *J. Physiol.*, 191, 353, 1967.

19. Richardson, K.C., The fine structure of autonomic nerves after vital staining with methylene blue, *Anat. Rec.*, 164, 359, 1969.

20. Clasen, R.A., Pandolfi, S., and Has, G.M., Vital staining, serum albumin and the blood-brain barrier, *J. Neuropathol. Exp. Neurol.*, 29, 266, 1990.

21. McDonald, D.M. and Blewett, R.W., Location and size of carotid body-like organs (paraganglia) revealed in rats by the permeability of blood vessels to Evans blue dye, *J. Neurocytol.*, 10, 607, 1981.

22. Bederson, J.B., Pitts, L.H., Germano, S.M., Nishimura, M.C., Davis, R.L., and Bartkowski, H.M., Evaluation of 2,3,5-triphenyltetrazolium choloride as a stain for detection and quantification of experimental cerebral infarction in rats, *Stroke*, 17, 1304, 1986.

23. Greve, G. and Saetersdal, T., Problems related to infarct size measurements in the rat heart, *Acta Anat.* (Basel), 142, 366, 1991.

24. Ullman-Culleré, M.H. and Foltz, Ch.J., Body condition scoring: A rapid and accurate method for assessing health status in mice, *Lab. Anim. Sci.*, 49, 319, 1999.

25. Kunstyr, I., Küpper, W., Weisser, H., Naumann, S., and Messow, C., Urethral plug. A new secondary male sex characteristic in rat and other rodents, *Lab. Anim.*, 16, 151, 1982.

26. Ghoshal, N.G. and Bal, H.S., Comparative morphology of the stomach of some laboratory mammals, *Lab. Anim.*, 23, 21, 1989.

27. Popp, J.A. and Monterio-Riviere, N.A., Macroscopic, microscopic, and ultrastructural anatomy of the nasal cavity, rat, in *Respiratory System*, Jones, T.C., Mohr, U., and Hunt, R.D., Eds., Springer-Verlag, Heidelberg, 1985, p. 3.

28. Noel, R.B., Toxicity testing, hazard assessment, and data quality assurance in respect to use of laboratory animals, in *Animals in Toxicological Research*, Bartosek, I., Guaitani, A., and Pacei, E., Eds., Raven Press, New York, 1982, p. 45.

29. Bahnemann, B., Maren, J., Karbe, E., Kaufmann, W., Morawietz, G., Nolte, T., and Rittinghausen, S., RITA-Registry of Industrial Toxicology Animal-data-Guide for organ sampling and trimming procedures in rats, *Exp. Toxic. Pathol.*, 47, 247, 1995.

30. Sheehan, D.C. and Hrapchak, B.B., *Theory and practice of Histotechnology*, The C.V. Mosby, St. Louis, 1980.

31. Boon, M.E. and Kok, L.P., *Microwave Cookbook of Pathology*, Columb Press Leyden, Leiden, 1987.

32. Shi, S.-R., Key, M.E., and Kalra, K.L., Antigen retrieval in formalin-fixed, paraffin-embedded tissues: An enhancement method for immunohistochemical staining based on microwave oven heating of tissue sections, *J. Histochem. Cytochem.*, 39, 741, 1991.

33. Hofler, H. and Mueller, J., In situ hybridization in pathology, *Verh. Dtsch. Ges. Pathol.*, 78, 124, 1994.

34. Teo, I.A. and Shaunak S., Polymerase chain reaction in situ: an appraisal of an emerging technique, *Histochemical Journal*, 27, 647, 1995.

35. Scheifele, D., Bjornson, G., and Dimmick, J., Rapid postmortem gut autolysis in infant rats: a potential problem for investigators, *Can. J. Vet. Res.*, 51, 404, 1987.

36. Seaman, W.J., *Postmortem Change in the Rat: A Histologic Characterization*, Iowa State University Press, Ames, IA, 1987, chaps. 1, 2, 4.

37. Fenwick, B.W. and Kruckenberg, S., Comparison of methods used to collect canine intestinal tissues for histological examination, *Am. J. Vet. Res.*, 48, 1276, 1987.

38. Moolenbeek, C. and Ruitenberg, E.J., The "Swiss roll": a simple technique for histological studies of the rodent intestine, *Lab. Anim.*, 15, 57, 1981.

39. Hayat, M.A., Method of fixation, in Principles and Techniques of Electron Microscopy, Hayat, A.M., Ed., Litton Educational Publishing, New York, 1970, p. 95.

40. Bugge, H.P. and Plöen, L., Changes in the volume of sertoli cells during the cycle of the seminiferous epithelium in the rat, *J. Reprod. Fertil.*, 76, 39, 1986.

41. Siemiatkowski, M., Plöen, L., and Björkman, N., Combined perfusion and percolation of embalmed animal bodies for removing formaldehyde, *Acta Anat.*, 133, 251, 1988.

42. Murray, G.I. and Ewen, S.W.B., A novel method for optimum biopsy specimen preservation for histochemical and immunohistochemical analysis, *Am. J. Clin. Pathol.*, 95, 131, 1991.

43. Waller, T., Lyngset, A., Elvander, M., and Morein, B., Immunological diagnosis of encephalitozoonosis from post-mortem specimens, *Vet. Immunol. Immunopathol.*, 1, 353, 1980.

44. Boot, R., Hansen, A.K., Hansen, C.K., Nozari, N., and Thuis, H.C.W., Comparison of assays for antibodies to Encephalitozoon cuniculi in rabbits, *Lab. Anim.*, 34, 281, 2000.

45. Coles, E.H., The blood film, in *Veterinary Clinical Pathology*, fourth edition, W.B. Saunders, Philadelphia, PA, 1986, p. 53.

46. Owen, G., Section 1. *Techniques and Serodiagnosis, in Parasites of Laboratory Animals, Laboratory Animal Handbooks No. 12*, Laboratory Animals Ltd., London, 1992, p. 2.

47. Kunstyr, I., *Diagnostic Microbiology for Laboratory Animals, GV-SOLAS Vol. 11*, Gustav Fischer Verlag, Stuttgart, 1992,

48. Brooks, G.F., Butel, J.S., Ornston, L.N., Jawetz, E., Melnick, J.L, and Adelberg, E.A., Principles of diagnostic medical microbiology, in *Medical Microbiology, 19th ed.*, Jawetz, E., Melnick, J.L., and Adelberg, E.A., Prentice Hall, New York, 1991, p. 587.

49. Ikram, M. and Hill, E., Laboratory procedures in bacteriology, in *Microbiology for Veterinary Technicians*, American Veterinary Publications, Inc., Goleta, 1991, chap. 5.

50. Cassell, G.H., Davidson, M.K., Davis, J.K., and Lindsey, J.R., Recovery and identification of murine mycoplasmas, in *Methods in Mycoplasmology, Vol. 2, Diagnostic Mycoplasmology*, Tully, J.G. and Razin, S., Eds., Academic Press, New York, 1983, p. 129.

51. Verstraete, A.P., Comparison of techniques for taking liver and lung samples from small rodents for bacteriological culture, *Lab. Anim.*, 7, 189, 1973.

52. Kornerup Hansen, A., *Handbook of Laboratory Animal Bacteriology*, CRC Press, Boca Raton, FL, 2000, chap. 2.

53. Shimoda, K., Maejima, K., Kuhara, T., and Nakagawa, M., Stability of pathogenic bacteria from laboratory animals in various transport media, *Lab. Anim.*, 25, 228, 1991.

54. McGinnis, M.R., *Laboratory Handbook of Medical Mycology*, Academic Press, New York, 1980, chap. 3.

55. Heuschele, W.P. and Castro, A.E., Selection and submission of diagnostic specimens, in Veterinary *Diagnostic Virology, A Practitioner's Guide*, Castro, A.E. and Heuschele, W.P., Eds., C.V. Mosby, St. Louis, 1992, p. 1.

56. Loeb, W.F. and Quimby, F.W., *The Clinical Chemistry of Laboratory Animals*, Pergamon Press, New York, 1989.

57. Flecknell, P.A., Non-surgical experimental procedures, in *Laboratory Animals, An Introduction for New Experimenters*, Tuffery, A.A., Ed., John Wiley & Sons, New York, 1987, chap. 13.

58. BVA/FRAME/RSPCA/UFAW Joint Working Group on Refinement, Removal of blood from laboratory mammals and birds, *Lab. Anim.*, 27, 1, 1993.

59. Compton, S.R. and Riley, L.K., Detection of infectious agents in laboratory rodents: traditional and molecular techniques, *Comparative Medicine*, 51, 113, 2001.

60. Clarke, C. and Clarke, M.L., Analytical evidence, in *Veterinary Toxicology*, Bailliere Tindal, London, 1975, p. 22.

61. Poklis, A., Analytic/forensic toxicology, in *Casarett and Doull's Toxicology. The Basic Science of Poisons*, fifth ed., Klaassen, C.D., Ed., McGraw-Hill, New York, 1996, chap. 31.

CHAPTER **22**

# Alternatives: Refinement, Reduction, and Replacement of Animal Use in the Life Sciences

**Alan M. Goldberg and Kimberly Drnec**

## CONTENTS

## INTRODUCTION

In 1954, University Federation for Animal Welfare (UFAW) engaged two scientists, William Russell and Rex Burch, to write a book on humane experimental technique. The book, "The Principles of Humane Experimental Technique," published in 1959[1] demonstrated with great clarity the impact of humanity on scientific experiments and comprehensively described why humane techniques are important, not only ethically, but also scientifically. They introduced essential methodology to achieve humane experimental technique. This methodology encompassed the concepts of refinement, reduction, and replacement — the Three Rs as applied to experimental protocols (see also Chapter 18).

Refinement is any method that reduces or eliminates pain in an animal during an experiment. It is more than adequate analgesia and anesthetics; it is considering every procedure carefully. Refinement also encompasses the use of species lower on the phylogenetic scale. Thus, replacing a primate with a mouse is a refinement technique.

The concepts of Reduction, decreasing the number of animals necessary in any protocol, is desirable, but the numbers should not be so low that the experiment is compromised, thus having wasted the animals that were used.

0-8493-1086-5/03/$0.00+$1.50
© 2003 by CRC Press LLC

Replacement is the use of techniques that do not use living animals and encompass *in vitro* approaches, and any other methodologies that eliminate the use of the living animal, e.g., structure–activity relationships. Thus, using primary cell culture to collect data offers the opportunity of reduction and replacement opportunities at the same time, recognizing, however, that to obtain primary cells, animals have to be used.

## CONCEPT OF HUMANENESS

That which provides for good animal welfare could be considered to be humane. Broom[2] defined animal welfare as "its state as regards its attempt to cope with its environment." The Animal Welfare Act does not define animal welfare, however, the United Federation of Animal Welfare (UFAW), among others, have provided varying aspects for the definition of welfare, and UFAW has presented that there are at least five basic needs that are necessary to provide a minimal level of welfare. These are as follows:

1. The need to have adequate food and a nutritional diet
2. The need to have access to fresh water at all times
3. The need to be pain and disease free
4. The need to be free of anxiety and fear
5. The need to be able to express normal behavior, e.g., hiding, licking, grooming

The issue of animal welfare is one of great complexity. It seems obvious that good animal welfare (as described above) leads to better scientific results. Although there is an important body of literature[3–5] on animal welfare, it is not well known in the biomedical community, and there is considerable need for refinement research and additional animal welfare investigations.

In a study carried out by Markowitz and Roberts,[6] they were able to demonstrate that animals undergoing spinal surgery recovered slower if the environment was enriched with paper for shredding and hiding places, than if a cage was empty. The reason for this is not clear, but it is possible that the animals in the enriched environment were much more comfortable and thus stayed within their hiding places, moving less and thus delaying recovery from surgery.

There are other examples where enriching the environment alters the biological outcome.[7] However, at this early stage of study, it is important to enrich the environment and maintain the best animal welfare that one can, while at the same time, making sure that whatever enrichment is provided (and it should be provided) does not interfere with the experimental protocol.

Control of pain and distress and proper animal welfare have been codified in laws pertaining to the use of laboratory animals. In fact, all Western European and American laws discuss pain and distress as the basis for the legislation.

## REGULATORY REQUIREMENTS

The Animal Welfare Act first drafted in 1966 as the Laboratory Animal Welfare Act, was the first attempt to incorporate the concept of humanness (pain and distress) into U.S. legislation and provide a degree of oversight for all animal experimentation regulated by the U.S. Department of Agriculture.

This Act has been amended many times, starting in 1973, with major amendments establishing animal care and use committees in 1985. However, the Animal Welfare Act currently does not cover the use of rats, mice, and birds in biomedical research. This is an area that is under discussion at the present time. Rats, mice, and birds and other species are covered, however, under the Public Health Service Policy that regulates all federally funded research. At the present time, all animals are covered by either or both of these laws and, in some places, with different requirements. Clearly, harmonization in the covering of all species and including rats, mice, and birds under the AWA, as originally intended,[8] is appropriate.

Table 22.1 provides a list of regulatory agencies and information resources that can be used to find information on current practices and regulatory requirements. This list goes beyond the United States

**Table 22.1 Information Resources**

| Alternative Information Resources | Web Site |
|---|---|
| AAALAC — Association for Assessment and Accreditation of Laboratory Animal Care | http://www.aaalac.org |
| APHIS — Animal and Plant Health Inspection Service | http://www.aphis.usda.gov |
| AWIC — Animal Welfare Information Center | http://www.nal.usda.gov/awic |
| CAAT — Center for Alternatives to Animal Testing | http://altweb.jhsph.edu http://caat.jhsph.edu |
| ECVAM — European Center for the Validation of Alternative Methods | http://ecvam-sis.jrc.it http://embryo.ib.amwaw.edu.pl/invittox |
| FRAME — Fund for the Replacement of Animals in Medical Experiments | http://www.frame.org.uk |
| HSUS — Humane Society of the United States | http://www.hsus.org |
| ICCVAM — Interagency Coordinating Committee on the Validation of Alternative Methods | http://www.iccvam.niehs.nih.gov |
| OECD — Organization for Economic Cooperation and Development | http://www.oecd.org |
| OLAW (PHS) — Office of Laboratory Animal Welfare | http://grants.nih.gov/grants/olaw/olaw.htm |
| RSPCA — The Royal Society for the Prevention of Cruelty to Animals | http://www.rspca.org.uk |
| UFAW — Universities Federation for Animal Welfare | http://www.ufaw.org.uk |
| USDA — United States Department of Agriculture | http://www.usda.gov |

and includes several European sites as well. These are provided in a form of a table with the name, the abbreviation, and the Web address.

## TOXIC IGNORANCE

In 1984, the National Academy of Sciences[9] published a study documenting that information on toxicity of chemicals in commerce is not truly available. Approximately 20% of chemicals in commerce have information in the public domain adequate to provide for risk assessment.

In 1997, the Environmental Defense Fund published their document "Toxic Ignorance,"[10] and in part, demonstrated this same lack of knowledge was true today. Because it was inconceivable that, in modern society, we would not have adequate information on 80% of the chemicals that we come into contact with in our daily lives, the American Chemistry Council (formerly CMA) and the Environmental Protection Agency did independent evaluations. Both confirmed that the chemicals in commerce have inadequate information available for the public to make a reasoned decision as to their hazard. As such, the Environmental Defense Fund, American Chemistry Council, and the U.S. EPA reached a voluntary agreement that a basic set of data (the OECD Screening Information and Data Set) would be used and provided on a public Web site[11] to share this information. The first chemicals chosen (approximately 3000) were those that were produced in amounts greater than 1 million pounds per year and are known as the high production volume chemicals (HPV). In all, there are approximately 80,000 chemicals on the U.S. EPA's list of priority chemicals that involve pesticides, potential endocrine disruptors, and those that might impact on children's health, among others.

This lack of information clearly is universal. In the European Union in 2000, a white paper[12] was written on chemicals they are trying to identify and classify as to their potential hazards through a system called REACH (Registration, Evaluation, and Authorization of CHemicals). The REACH list contains 30,000 priority chemicals.

Without question, there is no way that these chemicals can be adequately evaluated using current whole animal methods, two-year feeding and multigenerational studies are clearly inadequate to accomplish this task. It is an overwhelming task of time and resources (animal, space, and finance), and as such, other solutions and approaches to developing this information must be found.[13]

One approach to this problem has been the creation of a program called, "TestSmart, a humane and efficient approach to high production volume chemicals." This effort was initiated through the Johns Hopkins Center for Alternatives to Animal Testing, Environmental Defense Fund, University of Pittsburgh, and the Carnegie Mellon institutions. The outcome of this collaborative effort was in establishing an understanding that the OECD SIDS requirements could be met by techniques currently approved though the OECD, but allowing for substantial reduction in animal use. As such, the U.S. EPA at an October, 1999 meeting of TestSmart laid out a plan that would provide the same information required, but with a minimum reduction in animal use of 80%. Additionally, these assays in most cases would take less time and provide additional data for further hazard evaluation.[14,15] Similar issues are underway for endocrine disruptors and acute systemic toxicity.

An additional outcome from the HPV testing will be the development of a repository of tissues and samples to look at the genomics and proteomics counterparts to the SIDS testing. This will be one of the first, if not the first, publicly available database providing physiological, toxicological, histological, and genomics data on the same animals.[16]

## PROGRESS IN RESEARCH ALTERNATIVES

### Pregnancy Testing

One of the most obvious and early successes in the *in vitro* world is the replacement of rabbit or frog pregnancy testing with test kits that one can do in the privacy of their own home to determine whether or not an individual is pregnant. These tests, without exception, are less expensive, quicker, more convenient, and more sensitive. Test results can be obtained within minutes in an early pregnancy, rather than having to wait days for results after being pregnant for about 2 months.

### Endotoxin

The study of endotoxin, in parenteral preparations, has over the course of years undergone an evolutionary change. Early on, rabbits were used to test for the presence of pyrogenes. The test required 24 to 48 h, significant amounts of material, and required the use of live rabbits.

In order to test for short, half-live radioisotopes, rabbit testing was an impossibility, as several half-lives of the chemicals were less than the required testing period. As such, other approaches were sought.[17] In the middle 1980s, the Food and Drug Administration accepted the limulus assay (LAL test) for pyrogenes, which is still used. Most recently, Hartung and Wendel[18] have devised even more sensitive tests for pyrogenes. These are now commercially available (http://www.criver.com), and one anticipates the increase in the utilization over the next few years.

### *in vitro* Assays

Over the last 20 years, there has been a plethora of *in vitro* assays developed to study basic biology, acute cytotoxicity, and organ-specific effects.[19–22]

To date, few of these assays have found their way into routine regulatory testing. However, during the last few years, several assays on refinement or replacement have been accepted through the European Centre for the Validation of Alternative Methods (ECVAM) or the Interagency Coordinating Committee on the Validation of Alternative Methods (ICCVAM). These include the local lymph node assay (LLNA), Corrositex, and phototoxicity. During the last few years, there have been several attempts to provide reconstituted tissue equivalents for biomedical purposes (skin grafting) and for toxicological studies.

Early history of these studies was less than successful because of the conflicting desires to make a product that could be used for transplantation and toxicity testing. Most recently, reconstituted tissue equivalents have become better standardized for toxicity testing. These products derive from human cells and offer significant opportunity for hazard evaluation and toxicity testing. The two companies producing these materials are Mattek of Cambridge, Massachusetts, and SkinEthic of Nice, France.

In the case of the former, they have focused on skin and eye irritation testing and have been evaluated in numerous laboratories and publications (http://mattek.com).

In the case of SkinEthic, they hold a core technology, and that allows them to produce skin, corneal equivalents, as well as other tissues such as bucal musoca, vaginal tissue, alveoli tissue, among others (http://www.skinethic.com).

The utilization of these products is rapidly increasing, and results obtained are being used in product development, safety evaluation, and regulatory submission.[23–28] These studies clearly demonstrate the utility and acceptance of these approaches.

## NONINVASIVE IMAGING

In the refinement area, one of the great opportunities is the development of techniques that allow the study of small animals noninvasively. The use of MRI, PET scanning, x-ray techniques, etc., have become more commonplace and useful.[29]

These noninvasive approaches offer the opportunity to use each animal as its own control, thus greatly reducing the number of animals by 75 to 80%. This also reduces the study's variability because rather than using grouped data, the individual animal is being studied.

In the case of biophotonic imaging, scientists can attach luciferase, a compound obtained from fireflies or other types of glowworms, to cells, bacteria, viruses, or specific genes. Researchers can then study, for example, the effectiveness of therapeutic agents on tumor cell growth or metastases of tumors in living animals at a very early time in their development. Studies can be completed before the tumor grows to palpable size. In addition, scientists can study infectious disease in animals that have none of the clinical signs of the infection, because the disease process can be studied prior to manifestation of its clinical signs.[30]

Although there are initial start-up costs, there is little doubt that these noninvasive approaches take much less time to complete, and as a result, significantly decrease overall costs. At the same time, these more humane methods provide better, more reliable data than could have been obtained with traditional approaches.

## GENOMICS AND PROTEOMICS

The utilization of genomic and proteomic data in toxicology is just at the threshold. With the databases that are being generated at the present time, and the anticipated publicly available high production volume genomics database,[16] one can expect much. Additionally, work has begun on mapping the mouse genome for pain pathways. As this work evolves, it may be possible to construct animals with pain markers such that experiments would be able to be terminated prior to the animal experiencing pain, and in *in vitro* culture systems, it will be possible to study pain pathways and pain alleviation without creating pain. Clearly, there is huge promise for the future.

In conclusion, *in vitro* and other alternative methods have become commonplace within the field of biomedical sciences and toxicology. If one does any survey of presentations at national meetings, more than 70% utilize methodologies that are *in vitro*. Additionally, *in vitro* and other alternative approaches represent a significant contribution to these meetings.

These major advances in the *in vitro* fields cannot be interpreted to mean that whole animal studies will no longer be necessary. For the foreseeable future, the need for whole animal studies will be required to understand biology and prevent and cure diseases of man and animals. The need is for a greater awareness on the part of the scientific community to increase the use of all refinement and replacement alternatives as appropriate and whenever possible, and to make sure that all protocols and procedures are done with a focus on humanness and ways to decrease pain and distress in animal studies.[31]

As Bill Russell and Rex Burch wrote, the most humane science is the best science.[1]

## DISCLOSURE STATEMENT

Dr. Goldberg is a director and consultant to Xenogen and SkinEthic. Both companies have been mentioned in the text.

# REFERENCES

1. Russell, W.M.S. and Burch, R.L., *The Principles of Humane Experimental Technique*, Methuen, London, 1959.
2. Broom, D., Animal Welfare defined in terms of attempts to cope with the environment, *Acta Agric. Scand. Sect. A Animal Sci.* Suppl. 27, 22, 1996.
3. Duncan, I. and Petherick, J.C., Implications of cognitive processes for animal welfare, *J. Anim. Sci.*, 69, 5017, 1991.
4. Lord Soulsby of Swaffham Prior and Morton, D., Eds., Pain: its nature and management in man and animals, in *International Congress and Symposium Series,* The Royal Society of Medicine Press, London, 2000, p. 246.
5. Dol, M., et al., Eds., *Animal Consciousness and Animal Ethics*, Van Gorcum & Comp, The Netherlands, 1997.
6. Markowitz, H., New and proven enrichment techniques: research models and environmental enrichment procedures are critically interdependent, in *Proceedings of the 51ˢᵗ National Meeting of the American Association of Laboratory Animal Scientists,* San Diego, CA, 2000.
7. Guilarte, T., Environmental enrichment reverses cognitive deficits induced by lead, abstract for Society of Toxicology, *The Toxicologist*, 2002.
8. Senator Robert Dole, communication to Animal Research Defense Fund, http://caat.jhsph.edu/issues/doleletter.htm.
9. National Research Council, *Toxicity Testing: Strategies to Determine Needs and Priorities*, National Academy Press, Washington, 1984.
10. Roe, D., et al., *Toxic Ignorance,* http://www.environmentaldefense.org/documents/243_toxicignorance. htm, Environmental Defense Fund, 1997.
11. http://www.epa.gov/chemrtk/volchall.htm.
12. European Union, http://europa.eu.int/comm/enterprise/chemicals/chempol/whitepaper/report.pdf.
13. Green, S. and Goldberg, A.M., Proceedings of the 4ᵗʰ World Congress on Alternatives and Animals in the Life Sciences, In preparation.
14. Green, S., Goldberg, A.M., and Zurlo, J., The TestSmart – HPV program — development of an integrated approach for testing high production volume chemicals*, Regul. Toxicol. and Pharm.*, 33, 105, 2001.
15. Green, S., Goldberg, A.M., and Zurlo, J., TestSmart–high production volume chemicals: an approach to implementing alternatives into regulatory toxicology, *Toxicol. Sci.*, 63, 6, 2001.
16. Silbergeld, E., personal communication.
17. Wagner, H. and Cooper, J., Horseshoe crabs provide alternatives to rabbit bioassay, *CAAT Newsletter*, 2, 1, 1983, http://caat.jhsph.edu/pubs/newsletter/news_fall83/horseshoe.htm.
18. Hartung, T. and Wendel, A., Detection of pyrogens using human whole blood, *Altex*, 12, 2, 1995.
19. *Alternatives to Laboratory Animals (ATLA),* published by Fund for the Replacement of Animals in Medical Experiments, Nottingham, UK.
20. *Cell Biology and Toxicology*, Kluwer, Dordrecht, The Netherlands; *In Vitro* Animal, published by Society for *In Vitro* Biology, Largo, Maryland.
21. *In Vitro Toxicology*, Mary Ann Liebert, Inc.
22. *Toxicology In Vitro*, Pergamon Press, Oxford; Elmsford, NY.
23. Dagues, N., et al., *Correlation of in vivo studies and in vitro human epidermis model for dermal xenobiotic-induced toxicity*, Pierre Fabre Research Institute at the Controlled Release Society, San Diego, CA, 2001.
24. De Wever, B., Cappadora, M., and Rosdy, M., Prediction of acute and chronic skin irritation using human epidermis reconstituted *in vitro* in chemically defined medium, by SkinEthic Laboratories at the Eurotox Meeting, Istanbul, Turkey, *Toxicology Letters,* Elsevier, Amsterdam; New York, 2001.
25. Jones, P. et al., The effects of known eye irritants on the SkinEthic and Epiocular models of corneal epithelium, by Unilever Safety and Environmental Assurance Center (SEAC) at the European Tissue Culture Society, Granada, Spain*, Toxicology Letters*, Elsevier, Amsterdam; New York, 2001.
26. Bonowitz, A. et al., Comparative therapeutic and toxic effects of different povidone iodine (PVP-1) formulations in a aodel of oral candidosis based on *in vitro* reconstituted epithelium by the University of Munich, Germany and Mundi Pharma, *J. Drug Targeting*, 9, 75, 2001.

27. Raggen, E. et al., The use of airlifted reconstituted human lung epithelial tissues to predict *in vitro* the allergenic risk of topically applied enzyme proteins, by Novozymes Research & Development at the Eurotox Meeting, Instanbul, Turkey, *Toxicology Letters*, Elsevier, Amsterdam; New York, 2001.
28. Klausner, M., et al., The Epiocular tissue model: *in vivo* versus *in vitro* Draize scores for consumer products, *Toxicol. Sci.,* 54, 188, 2000.
29. *ILAR*, NRC, 42 (3), 187, 2001.
30. http://www.xenogen.com.
31. *ILAR*, NRC, 40, 1 1999.

# Index

## A

# J

# K

# L